Advancing Culture of Living with Landslides

Matjaž Mikoš · Vít Vilímek
Yueping Yin · Kyoji Sassa
Editors

Advancing Culture of Living with Landslides

Volume 5 Landslides in Different Environments

Springer

Editors
Matjaž Mikoš
Faculty of Civil and Geodetic Engineering
University of Ljubljana
Ljubljana
Slovenia

Vít Vilímek
Faculty of Science
Charles University
Prague
Czech Republic

Yueping Yin
China Institute of Geo-Environment Monitoring
China Geological Survey
Beijing
China

Kyoji Sassa
International Consortium on Landslides (ICL)
Kyoto
Japan

Associate editors
Mike G. Winter
Transport Research Laboratory (TRL)
Edinburgh
UK

Patrick Wassmer
Laboratory of Physical Geography (LGP), University Paris 1
Meudon
France

Jan Klimeš
Institute of Rock Structure and Mechanics
The Czech Academy of Sciences
Prague
Czech Republic

Tomáš Pánek
Faculty of Science
University of Ostrava
Ostrava
Czech Republic

ISBN 978-3-030-10414-6 ISBN 978-3-319-53483-1 (eBook)
DOI 10.1007/978-3-319-53483-1

Hiroshima landslide disasters in August 2014, Hiroshima, Japan (PASCO Corporation—Kokusai Kogyo Co., Ltd. All Rights Reserved).

Printed on acid-free paper

This Springer imprint is published by Springer Nature
The registered company is Springer International Publishing AG
The registered company address is: Gewerbestrasse 11, 6330 Cham, Switzerland

Foreword By Irina Bokova

Every year, disasters induced by natural hazards affect millions of people across the world. The loss of life is tragic, impacting on communities for the long term.

The costs are also economic, as disasters are responsible for estimated annual economic losses of around USD 300 billion. With the rising pressures of climate change, overpopulation, and urbanization, we can expect costs to increase ever more.

We cannot prevent disasters but we can prepare for them better. This is the importance of the *International Consortium on Landslides*, supported actively by UNESCO, to advance research and build capacities for mitigating the risks of landslides. Led by Prof. Kyoji Sassa, the Consortium has become a success story of international scientific cooperation at a time when this has never been so vital.

This is especially important as the world implements the *2030 Agenda for Sustainable Development* and the Paris Agreement on Climate Change, as well as the *Sendai Framework for Disaster Risk Reduction 2015–2030*—adopted in Sendai, Japan, to assess global progress on disaster risk reduction and set the priority actions.

The International Strategy for *Disaster Risk Reduction—International Consortium on Landslides Sendai Partnerships 2015–2025* is the key outcome relating to landslides from the 3rd World Conference on Disaster Risk Reduction, held in Sendai. On this basis, every member of the *International Consortium of Landslides* is redoubling efforts to understand, foresee, and reduce landslide disaster risk across the world.

Led by the Consortium, the Landslide Forum is a triennial milestone event that brings together scientists, engineers, practitioners, and policy makers from across the world—all working in the area of landslide technology, landslide disaster investigation, and landslide remediation. Meeting in Slovenia, the 4th Landslide Forum will explore the theme, "Landslide Research and Risk Reduction for Advancing Culture of Living with Natural Hazards", focusing on the multidisciplinary implementation of the Sendai Framework to build a global culture of resilient communities.

Against this backdrop, this report includes state-of-the-art research on landslides, integrating knowledge on multiple aspects of such hazards and highlighting good practices and recommendations on reducing risks. Today, more than ever, we need sharper research and stronger scientific cooperation. In this spirit, I thank all of the contributors to this publication and I pledge UNESCO's continuing support to deepening partnerships for innovation and resilience in societies across the world.

January 2017

Irina Bokova
Director General of UNESCO

Foreword By Robert Glasser

Landslides are a serious geological hazard. Among the host of natural triggers are intense rainfall, flooding, earthquakes or volcanic eruption, and coastal erosion caused by storms that are all too often tied to the El Niño phenomenon. Human triggers including deforestation, irrigation or pipe leakage, and mining spoil piles, or stream and ocean current alteration can also spark landslides.

Landslides occur worldwide but certain regions are particularly susceptible. The UN's Food and Agriculture Organization underlines that steep terrain, vulnerable soils, heavy rainfall, and earthquake activity make large parts of Asia highly susceptible to landslides. Other hotspots include Central, South, and Northwestern America.

Landslides have devastating impact. They can generate tsunamis, for example. They can bring high economic costs, although estimating losses is difficult, particularly so when it comes to indirect losses. The latter are often confused with losses due to earthquakes or flooding.

Globally, landslides cause hundreds of billions of dollars in damages and hundreds of thousands of deaths and injuries each year. In the US alone, it has been estimated that landslides cause in excess of US$1 billion in damages on average per year, though that is considered a conservative figure and the real level could be at least double.

Given this, it is important to understand the science of landslides: why they occur, what factors trigger them, the geology associated with them, and where they are likely to happen.

Geological investigations, good engineering practices, and effective enforcement of land use management regulations can reduce landslide hazards. Early warning systems can also be very effective, with the integration between ground-based and satellite data in landslide mapping essential to identify landslide-prone areas.

Given that human activities can be a contributing factor in causing landslides, there are a host of measures that can help to reduce risks, and losses if they do occur. Methods to avoid or mitigate landslides range from better building codes and standards in engineering of new construction and infrastructure, to better land use and proper planned alteration of drainage patterns, as well as tackling lingering risks on old landslide sites.

Understanding the interrelationships between earth surface processes, ecological systems, and human activities is the key to reducing landslides disaster risks.

The Sendai Framework for Disaster Risk Reduction, a 15-year international agreement adopted in March 2015, calls for more dedicated action on tackling underlying disaster risk drivers. It points to factors such as the consequences of poverty and inequality, climate change and variability, unplanned and rapid urbanization, poor land management, and compounding factors such as demographic change, weak institutional arrangements, and non-risk-informed policies. It also flags a lack of regulation and incentives for private disaster risk reduction investment, complex supply chains, limited availability of technology, and unsustainable uses of natural resources, declining ecosystems, pandemics and epidemics.

The Sendai Framework also calls for better risk-informed sectoral laws and regulations, including those addressing land use and urban planning, building codes, environmental and resource management and health and safety standards, and underlines that they should be updated, where needed, to ensure an adequate focus on disaster risk management.

The UN Office for Disaster Risk Reduction (UNISDR) has an important role in reinforcing a culture of prevention and preparedness in relevant stakeholders. This is done by supporting the development of standards by experts and technical organizations, advocacy initiatives, and the dissemination of disaster risk information, policies, and practices. UNISDR also provides education and training on disaster risk reduction through affiliated organizations, and supports countries, including through national platforms for disaster risk reduction or their equivalent, in the development of national plans and monitoring trends and patterns in disaster risk, loss, and impacts.

The International Consortium on Landslides (ICL) hosts the Sendai Partnerships 2015–2025 for the global promotion of understanding and reducing landslide disaster risk. This is part of 2015–2025, a voluntary commitment made at the Third UN World Conference on Disaster Risk Reduction, held in 2015 in Sendai, Japan, where the international community adopted the Sendai Framework.

The Sendai Partnerships will help to provide practical solutions and tools, education and capacity building, and communication and public outreach to reduce landslides risks. As such, they will contribute to the implementation of the goals and targets of the Sendai Framework, particularly on understanding disaster risks including vulnerability and exposure to integrated landslide-tsunami risk.

The work done by the Sendai Partnerships can be of value to many stakeholders including civil protection, planning, development and transportation authorities, utility managers, agricultural and forest agencies, and the scientific community.

UNISDR fully support the work of the Sendai Partnerships and the community of practice on landslides risks, and welcomes the 4th World Landslide Forum to be held in 2017 in Slovenia, which aims to strengthen intergovernmental networks and the international programme on landslides.

Robert Glasser
Special Representative of the Secretary-General
for Disaster Risk Reduction and head of UNISDR

Preface

In this Volume 5, we present four different sessions: Landslide Interactions with the Built Environment, Landslides in Natural Environment, Landslides and Water, Landslides as Environmental Change Proxies—Looking at the Past. The main objective is to draw attention to the different types of landslides with respect to communities, infrastructure and cultural heritage. Landslides in the natural environment are also covered, including all forms of aquatic environments. Recent progress in dating techniques has greatly improved the ability to determine the age of landslides allowing us to address the challenge of relating established landslide chronologies to regional paleoenvironmental changes (e.g. paleoseismic events, deglaciation, climatic changes, human-induced deforestation). The relations between climatological (and climate change) and geomorphological zones or settings are important in that they determine the dominant landslide type and the associated triggering mechanisms.

Research into landslides that are causally related to precipitation has recently been discussed in relation to climate change. Climate change may be a triggering effect for modifications to climatic parameters (weather), which are difficult to quantify and which play an important (or even key) role in the emergence of individual types of slope movement. Conditions may be different from region to region. For instance, regelation processes are of key importance for rock falls, increases in temperature for shallow slope movements in periglacial zones and increases in the sea level for coastal areas. Aridization of climate in certain areas can cause an increased frequency of fires and subsequently increase susceptibility to the formation of debris flows. Nevertheless, it is generally accepted that mountain environments are very susceptible to climatic change. The impact of changes in weather that results from climatic change, on slope stability must be studied and understood in the context of different geomorphological conditions (e.g. fluvial, glacial or periglacial types of relief) and geotechnical conditions (e.g. rock massifs, weathered mantle), as well as adaptation strategies.

However, in mountainous as well as tectonically active areas we often find a combination of different impacts generating conditions favourable to the development of slope deformations. There are steep slopes formed by intensive erosion, tectonically crushed zones, a higher degree of seismicity, and anthropogenic impacts can also be important (road construction, deforestation, agriculture, etc.) due to an increase in population. While precipitation may be the trigger, the contribution of other factors may also be important albeit difficult to quantify as their influence may be variable with time.

A clear understanding of the influence of hydraulic conditions on slope instability is necessary when water is the trigger for such movements. This may be achieved by using precipitation triggers as a proxy for the more direct factors related to the condition of the soil with respect to water. Alternatively, where a very clear, and generally straightforward, ground model exists it may be possible to use directly measured soil-moisture parameters to understand the triggering processes.

Kinematics of movements (for instance continuous movement with acceleration, seasonal movements, etc.) must also be taken into consideration. In terms of climatic factors, we cannot neglect temperature as it affects, for example, pore pressure through evapotranspiration and phase changes of precipitation.

This Volume 5 presents a wide range of papers that will make a substantial contribution to the state-of-the-art, aiding researchers in taking forward and increasing our knowledge and understanding of the effect of landslides on different environments. This is particularly the case in those environments in which humans and their infrastructure form an important part of that environment and thus comprise a significant part of the elements that are at risk from landslide hazards.

Prague, Czech Republic Vít Vilímek

Organizers

International Consortium on Landslides (ICL)

A Programme of the ICL for ISDR

International Programme on Landslides (IPL)

Univerza v Ljubljani

University of Ljubljana

GeoZS
Geološki zavod Slovenije

Geological Survey of Slovenia (GeoZS)

Co-organizers

REPUBLIC OF SLOVENIA
MINISTRY OF THE ENVIRONMENT AND SPATIAL PLANNING

Republic of Slovenia Ministry of the Environment and Spatial Planning

REPUBLIC OF SLOVENIA
MINISTRY OF INFRASTRUCTURE

Republic of Slovenia Ministry of Infrastructure

UPRAVA REPUBLIKE SLOVENIJE ZA ZAŠČITO IN REŠEVANJE

Slovenian National Platform for Disaster Risk Reduction

IZS
INŽENIRSKA ZBORNICA SLOVENIJE

Slovenian Chamber of Engineers (IZS)

- Društvo Slovenski komite mednarodnega združenja hidrogeologov (SKIAH)—International Association of Hydrogeologists Slovene Committee (SKIAH)
- Društvo vodarjev Slovenije (DVS)—Water Management Society of Slovenia (DVS)
- Geomorfološko društvo Slovenije (GDS)—Geomorphological Association of Slovenia (GDS)
- Inštitut za vode Republike Slovenije (IzVRS)—Institute of Water of the Republic of Slovenia (IzVRS)
- Slovensko geološko društvo (SGD)—Slovenian Geological Society (SGD)
- Slovensko geotehniško društvo (SloGeD)—Slovenian Geotechnical Society (SloGeD)
- Slovenski nacionalni odbor programa IHP UNESCO (SNC IHP)—Slovenian National Committee for IHP (SNC IHP)
- Slovensko združenje za geodezijo in geofiziko (SZGG)—Slovenian Association of Geodesy and Geophysics (SZGG)

Organizing Committee

Honorary Chairpersons

Borut Pahor, President of the Republic of Slovenia*
Irina Bokova, Director General of UNESCO
Robert Glasser, Special Representative of the United Nations Secretary-General for Disaster Risk Reduction*
José Graziano Da Silva, Director General of FAO*
Petteri Talaas, Secretary General of WMO
David Malone, Rector of UNU
Gordon McBean, President of ICSU
Toshimitsu Komatsu, Vice President of WFEO
Roland Oberhaensli, President of IUGS
Alik Ismail-Zadeh, Secretary General of IUGG
Hisayoshi Kato, Director General for Disaster Management, Cabinet Office, Government of Japan
Kanji Matsumuro, Director, Office for Disaster Reduction Research, Ministry of Education, Culture, Sports, Science and Technology, Government of Japan
Fabrizio Curcio, Head, National Civil Protection Department, Italian Presidency of the Council of Ministers, Government of Italy
Jadran Perinić, Director General, National Protection and Research Directorate, Republic of Croatia
Takashi Onishi, President of Science Council of Japan
Juichi Yamagiwa, President of Kyoto University
Ivan Svetlik, Rector of University of Ljubljana, Slovenia
Walter Ammann, President/CEO, Global Risk Forum Davos
*Note: Honorary chairpersons are Leaders of signatory organizations of the ISDR-ICL Sendai Partnerships. * to be confirmed.*

Chairpersons

Matjaž Mikoš, Chairman, Slovenian National Platform for Disaster Risk Reduction
Yueping Yin, President, International Consortium on Landslides
Kyoji Sassa, Executive Director, International Consortium on Landslides

International Scientific Committee

Che Hassandi Abdulah, Public Works Department of Malaysia, Malaysia
Biljana Abolmasov, University of Belgrade, Serbia
Basanta Raj Adhikari, Tribhuvan University, Nepal
Beena Ajmera, California State University, Fullerton, USA
Irasema Alcántara Ayala, Universidad Nacional Autonoma de Mexico, Mexico
Guillermo Avila Alvarez, Universidad Nacional de Colombia, Colombia
Željko Arbanas, University of Rijeka, Croatia
Behzad Ataie-Ashtiani Sharif, University of Technology, Iran
Mateja Jemec Auflič, Geological Survey of Slovenia, Slovenia
Yong Baek, Korea Institute of Civil Engineering and Building Technology, Korea
Lidia Elizabeth Torres Bernhard, Universidad Nacional Autónoma de Honduras, Honduras
Matteo Berti, University of Bologna, Italy

Netra Prakash Bhandary, Ehime University, Japan
He Bin, Chinese Academy of Sciences, China
Peter Bobrowsky, Geological Survey of Canada, Canada
Giovanna Capparelli, University of Calabria, Italy
Raul Carreno, Grudec Ayar, Peru
Nicola Casagli, University of Florence, Italy
Filippo Catani, University of Florence, Italy
Byung-Gon Chae, Korea Institute of Geoscience and Mineral Resources, Korea
Buhm-Soo Chang, Korea Infrastructure Safety and Technology Corporation, Korea
Giovanni Battista Crosta, University of Milano Bicocca, Italy
Sabatino Cuomo, University of Salerno, Italy
A.A. Virajh Dias, Central Engineering Consultancy Bureau, Sri Lanka
Tom Dijkstra, British Geological Survey, UK
Francisco Dourado, University of Rio de Janeiro State, Brasil
Erik Eberhardt, University of British Columbia, Canada
Luis Eveline, Universidad Politécnica de Ingeniería, Honduras
Teuku Faisal Fathani, University of Gadjah Mada, Indonesia
Paolo Frattini, University of Milano Bicocca, Italy
Hiroshi Fukuoka, Niigata University, Japan
Rok Gašparič, Ecetera, Slovenia
Ying Guo, Northeast Forestry University, China
Fausto Guzzetti, National Research Council, Italy
Javier Hervas, ISPRA, Italy/EU
Daisuke Higaki, Japan Landslide Society, Japan
Arne Hodalič, National Geographic Slovenija, Slovenia
Jan Hradecký, University of Ostrava, Czech Republic
Johannes Hübl, University of Natural Resources and Life Sciences, Austria
Oldrich Hungr, University of British Columbia, Canada
Sangjun Im, Korean Society of Forest Engineering, Korea
Michael Jaboyedoff, University of Lausanne, Switzerland
Jernej Jež, Geological Survey of Slovenia, Slovenia
Pavle Kalinić, City of Zagreb, Croatia
Bjørn Kalsnes, Norwegian Geotechnical Institute, Norway
Dwikorita Karnawati, University of Gadjah Mada, Indonesia
Asiri Karunawardana, National Building Research Organization, Sri Lanka
Ralf Katzenbach, Technische Universitaet Darmstadt, Germany
Nguyen Xuan Khang, Institute of Transport Science and Technology, Vietnam
Kyongha Kim, National Institute of Forest Science, Korea
Dalia Kirschbaum, NASA Goddard Space Flight Center, USA
Jan Klimeš, Academy of Sciences of the Czech Republic, Czech Republic
Marko Komac, University of Ljubljana, Slovenia
Kazuo Konagai, University of Tokyo, Japan
Hasan Kulici, Albanian Geological Survey, Albania
Santosh Kumar, National Institute of Disaster Management, India
Simon Loew, ETH Zürich, Switzerland
Jean-Philippe Malet, Université de Strasbourg, France
Claudio Margottini, ISPRA, Italy
Snježana Mihalić Arbanas, University of Zagreb, Croatia
Gabriele Scarascia Mugnozza, University of Rome "La Sapienza", Italy
Chyi-Tyi Lee, National Central University, Chinese Taipei
Liang-Jeng Leu, National Taiwan University, Chinese Taipei
Ko-Fei Liu, National Taiwan University, Chinese Taipei

Janko Logar, University of Ljubljana, Slovenia
Ping Lu, Tongji University, China
Juan Carlos Loaiza, Colombia
Mauri McSaveney, GNS Science, New Zealand
Matjaž Mikoš, University of Ljubljana, Slovenia
Ashaari Mohamad, Public Works Department of Malaysia, Malaysia
Hirotaka Ochiai, Forest and Forest Product Research Institute, Japan
Igwe Ogbonnaya, University of Nigeria, Nigeria
Tomáš Pánek, University of Ostrava, Czech Republic
Mario Parise, National Research Council, Italy
Hyuck-Jin Park, Sejong University, Korea
Cui Peng, Chinese Academy of Sciences, China
Luciano Picarelli, Second University of Naples, Italy
Tomislav Popit, University of Ljubljana, Slovenia
Saowanee Prachansri, Ministry of Agriculture and Cooperatives, Thailand
Boštjan Pulko, University of Ljubljana, Slovenia
Paulus P. Rahardjo Parahyangan Catholic University, Indonesia
Bichit Rattakul Asian Disaster Preparedness Center, Thailand
K.L.S. Sahabandu, Central Engineering Consultancy Bureau, Sri Lanka
Kyoji Sassa, International Consortium on Landslides, Japan
Wei Shan, Northeast Forestry University, China
Z. Shoaei, Soil Conservation and Watershed Management Research Institute, Iran
Mandira Shrestha, International Centre for Integrated Mountain Development, Nepal
Paolo Simonini, University of Padua, Italy
Josef Stemberk, Academy of Sciences of the Czech Republic, Czech Republic
Alexander Strom, JSC "Hydroproject Institute", Russian Federation
S.H. Tabatabaei, Building & Housing Research Center, Iran
Kaoru Takara, Kyoto University, Japan
Dangsheng Tian, Bureau of Land and Resources of Xi'an, China
Binod Tiwari, California State University, Fullerton & Tribhuvan University, USA
Veronica Tofani, University of Florence, Italy
Adrin Tohari, Indonesian Institute of Sciences, Indonesia
Oleksandr M. Trofymchuk, Institute of Telecommunication and Global Information Space, Ukraine
Emil Tsereteli, National Environmental Agency of Georgia, Georgia
Taro Uchimura, University of Tokyo, Japan
Tran Tan Van, Vietnam Institute of Geosciences and Mineral Resources, Vietnam
Timotej Verbovšek, University of Ljubljana, Slovenia
Pasquale Versace, University of Calabria, Italy
Vít Vilímek, Charles University, Czech Republic
Ján Vlčko, Comenius University, Slovak Republic
Kaixi Xue, East China University of Technology, China
Yueping Yin, China Geological Survey, China
Akihiko Wakai, Japan Landslide Society, Japan
Fawu Wang, Shimane University, Japan
Gonghui Wang, Kyoto University, Japan
Huabin Wang, Huazhong University of Science and Technology, China
Janusz Wasowski, National Research Council, Italy
Patrick Wassmer, Université Paris 1, France
Mike Winter, Transport Research Laboratory, UK
Sabid Zekan, University of Tuzla, Bosnia and Herzegovina
Oleg Zerkal, Moscow State University, Russian Federation
Ye-Ming Zhang, China Geological Survey, China

Local Organizing Committee

Biljana Abolmasov, Faculty of Mining and Geology, University of Belgrade, Serbia
Željko Arbanas, Faculty of Civil Engineering, University of Rijeka, Croatia
Miloš Bavec, Geological Survey of Slovenia
Nejc Bezak, Faculty of Civil and Geodetic Engineering, University of Ljubljana
Mitja Brilly, Slovenian National Committee for IHP
Darko But, Administration for Civil Protection and Disaster Relief, Ministry of Defence of the Republic of Slovenia
Lidija Globevnik, Water Management Society of Slovenia
Arne Hodalič, National Geographic Slovenia
Mateja Jemec Auflič, Geological Survey of Slovenia
Jernej Jež, Geological Survey of Slovenia
Vojkan Jovičić, Slovenian Geotechnical Society
Robert Klinc, Faculty of Civil and Geodetic Engineering, University of Ljubljana
Janko Logar, Faculty of Civil and Geodetic Engineering, University of Ljubljana
Matej Maček, Faculty of Civil and Geodetic Engineering, University of Ljubljana
Snježana Mihalić Arbanas, Faculty of Mining, Geology and Petroleum Engineering, University of Zagreb, Croatia
Matjaž Mikoš, Faculty of Civil and Geodetic Engineering, University of Ljubljana
Zlatko Mikulič, International Association of Hydrogeologists Slovene Committee
Gašper Mrak, Faculty of Civil and Geodetic Engineering, University of Ljubljana
Mario Panizza, University of Modena and Reggio Emilia, Italy
Alessandro Pasuto, National Research Council, Padua, Italy
Ana Petkovšek, Faculty of Civil and Geodetic Engineering, University of Ljubljana
Tomislav Popit, Faculty of Natural Sciences and Engineering, University of Ljubljana
Boštjan Pulko, Faculty of Civil and Geodetic Engineering, University of Ljubljana
Jože Rakovec, Slovenian Association of Geodesy and Geophysics
Črtomir Remec, Slovenian Chamber of Engineers
Mauro Soldati, University of Modena and Reggio Emilia, Italy
Timotej Verbovšek, Faculty of Natural Sciences and Engineering, University of Ljubljana
Sabid Zekan, Faculty of Mining, Geology and Civil Engineering, University of Tuzla, Bosnia and Herzegovina

Contents

Part I

Landslide Interactions with the Built Environment

Session Introduction—Landslide Interaction with the Built Environment

M.G. Winter, T.A. Dijkstra, and J. Wasowski

Abstract
Sub-theme 5.1 focuses on the interaction of landslides in natural and engineered slopes with the built environment. The associated socio-economic issues are particularly important given population growth and the associated demand for new and enhanced infrastructure. The papers included herein provide a wide-ranging overview of methodologies and techniques that can be used to identify the effects of landslides on buildings, infrastructure and large engineering works, as well as cultural heritage.

Keywords
Landslides • Buildings • Infrastructure • Cultural Heritage • Hazard • Risk

The main goal of Theme 5 (Volume 5) is to draw attention to the variety of landslides that have an impact on communities, infrastructure and cultural heritage. This sub-theme focuses on the interaction of landslides with the built environment, whether triggered in natural or engineered slopes. The broad relevance of this sub-theme is clear and the associated socio-economic issues are particularly important given population growth and the associated demand for new and enhanced infrastructure. It provides an overview of methodologies and techniques that can be used to identify the effects of landslides on buildings, infrastructure and large engineering works, as well as cultural heritage.

The papers in this sub-theme discuss a wide range of relevant topics with many addressing approaches to landslide hazard and risk analysis, the assessment of resilience, the evaluation of socio-economic effects of landslides, and a rational approach to the reduction of landslide risk.

Living with landslides is a recurring theme in many countries of the world and the risks that may be encountered are varied, reflecting spatial and temporal variations in vulnerability and exposure that are compounded by complexities in addressing the probability of landslide hazard occurrence. It can be argued that the most important risks are those where landslides affect life and limb. However, the risk picture is complex and landslides can compromise the functioning of the built environment in many different ways. For example, the severance of transport and utilities networks can lead to significant loss of economic, employment, educational, social and health opportunities for the populations affected. As a further example, risks need to be considered that are associated with the loss or deterioration of national monuments and cultural heritage.

In the case of severance of transport infrastructure, often the most affected populations are from communities that are dependent upon a relatively small number of transport routes —in some extreme cases there may be no alternative transport route leading to a single landslide event resulting in the isolation of remote communities. Where a sparse transport infrastructure network exists, a single, relatively small landslide event may affect a very large geographical area: i.e. the vulnerability shadow is large. As is often the case in these areas, budgets for renewal and repair may be limited leading to issues with the longer term resilience of such

M.G. Winter (✉)
Transport Research Laboratory (TRL), 10b Swanston Steading, 109 Swanston Road Edinburgh EH10 7DS, Wokingham, UK
e-mail: mwinter@trl.co.uk; mgwinter@btinternet.com

M.G. Winter
School of Earth and Environmental Sciences, University of Portsmouth, Burnaby Building Portsmouth PO1 3QL, Portsmouth, UK

T.A. Dijkstra
British Geological Survey, Nottingham NG12 5GG, Keyworth, UK

J. Wasowski
National Research Council, CNR-IRPI, Bari, Italy

M. Mikoš et al. (eds.), *Advancing Culture of Living with Landslides*,
DOI 10.1007/978-3-319-53483-1_1

infrastructure and reduced opportunities for recovery of affected communities.

In urban settings, landslides may affect and/or threaten the lives and livelihoods of a much larger exposed population. However, while the immediate (economic) consequences may be much greater, the vulnerability shadow cast by the landslide event is likely to be much smaller. Also, in such environments, there are generally better structures in place that can lead to a shorter recovery time and more secure long-term resilience.

A rational approach to the management and mitigation of future landslides hazards and risks is a basic requirement, whatever the elements at risk, the scale of the risk and associated vulnerability. The papers presented in this session contribute to this theme and cover a wide range of geographical regions. The majority is from Europe, but there are contributions from other parts of the world that range from case studies to sub-continental assessments of landslide risk. The papers cover a wide range of infrastructure and developments and include issues surrounding the interaction of landslides with roads and highways, urban and rural communities, historical buildings and complexes, and mines. Different types of landslides are evident in the context of different climatic zones, geomorphological and geotechnical settings and triggering factors. There is also a strong focus on gaining better knowledge of landslide processes in order to better mitigate and manage risk.

These contributions outline important knowledge that can be used for education of affected populations, landslide hazard/risk reduction/management and to assist in future scientific and technological progress. The key issue is to transfer this knowledge in a way that it becomes usable information not only for scientists, but also for local communities and decision-makers.

Landslide Risk Assessment for the Built Environment in Sub-Saharan Africa

Peter Redshaw, Tom Dijkstra, Matthew Free, Colm Jordan, Anna Morley, and Stuart Fraser

Abstract

This paper presents an overview of the findings from a series of country-scale landslide risk assessments conducted on behalf of the governments of five Sub-Saharan countries, the World Bank and the Global Facility for Disaster Reduction and Recovery (GFDRR). Ethiopia, Kenya, Uganda, Niger and Senegal sample a wide range of Sub-Saharan Africa's different geographies and are characterised by contrasting levels of development. Landslide hazard, exposure and vulnerability therefore differ from country to country, resulting in significant spatial variation of landslide risk. In East Africa; Ethiopia, Kenya and Uganda are characterised by mountainous and seismically active terrain which results in a relatively high landslide hazard. In conjunction with rapid urbanisation and a population which is expected to rise from around 170 million in 2010 to nearly 300 million in 2050, this means that landslides pose a significant risk to the built environment. In West Africa, a combination of low landslide hazard and lower exposure in Niger and Senegal results in comparatively low landslide risk. This paper also describes areas with perceived misconceptions with regard to the levels of landslide risk. These are areas of only low to moderate landslide hazard but where urbanisation has resulted in a concentration of exposed buildings and infrastructure that are vulnerable to landslides, resulting in higher landslide risk.

Keywords

Regional landslide risk • Built environment • Ethiopia • Kenya • Uganda • Niger • Senegal

P. Redshaw (✉) · M. Free · A. Morley
Ove Arup & Partners International Ltd, 13 Fitzroy Street, London, W1T 4BQ, UK
e-mail: peter.redshaw@arup.com

M. Free
e-mail: matthew.free@arup.com

A. Morley
e-mail: anna.morley@arup.com

T. Dijkstra · C. Jordan
British Geological Survey, Environmental Science Centre, Keyworth, Nottingham, NG12 5GG, UK
e-mail: tomdij@bgs.ac.uk

C. Jordan
e-mail: cjj@bgs.ac.uk

S. Fraser
Global Facility for Disaster Reduction and Recovery—Innovation Lab, Washington, D.C., USA
e-mail: sfraser@worldbank.org

M. Mikoš et al. (eds.), *Advancing Culture of Living with Landslides*,
DOI 10.1007/978-3-319-53483-1_2

Introduction

Sub-Saharan Africa is characterised by landscapes ranging from high mountains to low relief floodplains; by rainfall which can be desperately sparse or exceptionally intense; and by cities, towns and villages which can be both overcrowded and poorly planned or remote and isolated. The impact of landslides on these diverse environments is therefore highly variable.

In Uganda's Bududa district, near the border with Kenya, landslides have caused hundreds of fatalities and left many thousands permanently displaced due to landslide damage caused to the built environment (Kitutu 2010). In 2010 one event killed over 350 people and initiated government calls for the mass relocation of settlements away from the mountainous slopes of Mount Elgon (Terra Daily 2010). However, little subsequent action means that vulnerable communities remain at great risk. In Ethiopia, the northern highlands and many urban areas face a similar threat from landslides. The rapid expansion of Dessie Town, has resulted in unregulated or poorly planned development in areas of high landslide risk. The construction of houses, roads, bridges and utilities has in many cases been interpreted to have contributed to occurrences of landslides in population centers, indicating that landslide processes are often poorly understood (Fubelli et al. 2013).

To better understand landslide risk and to inform the provision of more detailed risk management initiatives, the World Bank and the GFDRR is supporting the development of new landslide risk information for Sub-Saharan Africa, starting with five countries: Ethiopia, Kenya, Uganda, Niger and Senegal. This study forms part of a wider initiative by the GFDRR to characterise multi-hazard risk in Sub-Saharan Africa.

Regional Landslide Risk Analysis

Regional landslide risk analysis aims to provide a better understanding of the spatial distribution of the risk posed to populations, structures, infrastructure and other assets, from damage, destruction or death as a result of a landslide. Corominas et al. (2014) summarize this process in five steps: (1) Hazard identification, (2) Hazard assessment, (3) Exposure identification, (4) Vulnerability assessment, and (5) Risk estimation.

Hazard Identification

Landsliding in Ethiopia, Kenya and Uganda is widespread and is interpreted to be influenced by topography (Ayalew and Yamagishi 2004; Kitutu 2010), geology (Kitutu 2010), anthropogenic causes (Ayelew et al. 2009; Broothaerts et al. 2012), hydrological processes (Abebe et al. 2010; Ayalew 1999; Beyene et al. 2012) and long term geomorphological evolution (Ayalew and Yamagishi 2004; Vařilová et al. 2015).

In Ethiopia, Ayalew et al. (2009) report landslides occurring preferentially in basaltic terrain and along the boundary between basalt and limestone in the Blue Nile Basin Region. Examples of major landsliding in Ethiopia include the Gembechi Village Landslide (Bechet Valley, 1960), the Wudmen Landslide (1993) and the Uba Dema Village Landslide (1994) (Ayalew 1999).

Maina-Gichaba et al. (2013) provide an overview of landslide occurrences in Kenya, identifying a number of important drivers in the generation of landslides, including anthropogenic factors related to land tenure, including unsustainable land use practices and particularly land fragmentation. Deforestation in the mountainous districts of Kenya has also been linked to increased landscape sensitivity. Ngecu and Ichang'I (1999) report on the impact of landslides in the Aberdare Mountains in Kenya, where between 1960 and 1980 around 40 major landslides occurred, mobilising approximately 1000,000 m^3 of material in an area of approximately 300 km^2. Further reports of major landslides in Kenya include the 1986 Mukurweini Landslide, the 1991 Gacharage Landslide, and the 1997 Maringa Landslide.

Kitutu et al. (2011; building on earlier work by Knapen et al. 2006) assessed farmers understanding and perception of landslides in Bududa district, Eastern Uganda. Farmers were able to provide their experiences, understanding and observations, which highlighted that steep slopes, areas with concavities and those with flowing groundwater were identified as being prone to landslides. Farmers also identified that coarse, permeable soils are prone to landsliding, responding rapidly to intense precipitation. In these areas, terraces are not popular amongst farmers because these are known to promote water infiltration and trigger landslides.

The Rwenzori Mountains form one of the regions in Uganda where landslides have made a significant impact (NEMA 2007), with 48 landslides and flash-flood events reported by Jacobs et al. (2015).

Reporting and analysis of landslides in Senegal is concentrated around the Dakar coastline. Fall et al. (2006a, b) describe six landslide locations set within a short section of cliffs which were analysed, enabling the determination of landslide zones and the geomorphological development to be interpreted. Wang et al. (2009) report a study of natural hazards in the sub-urban areas of Dakar, covering approximately 580 km^2, although the focus of the study is on coastal erosion and flooding.

To date, no studies have been identified which discuss landslide hazard or risk in Niger.

For all five countries of this study, no suitable landslide inventories were available for inclusion in the landslide hazard assessment stage. Satellite imagery was used to verify and inform interpretation of the landslide hazard.

Hazard Assessment

The hazard identification stage contributes to an interpretation of landslide susceptibility, which, when combined with approximations of landslide triggers, enables the estimation of landslide hazard.

Landslide susceptibility assessment is based on the premise that a range of parameters can be combined to obtain an approximation of the conditions in the landscape that determine the propensity of slopes to generate landslides. To obtain a score for landslide susceptibility, the components of each factor that contribute to landslide susceptibility are analysed, re-classified and then mapped across the study area. The relative contribution of an individual susceptibility factor (S_n) is regulated through multiplication with a weight (W_n). The ranked and weighted factors are then combined to derive an expression of landslide susceptibility (L_s):

$$L_s = (S_1 \times W_1) + (S_2 \times W_2) + (S_n \times W_n)$$

Due to the vast land-area and regional nature of this study, the selection of individual susceptibility factors needed to consider both the applicability across a range of different terrains and climates, and the availability of data for each country of interest. There was therefore a focus on primary geological and morphometric factors such as slope angle, bedrock lithology, and soil type. The key susceptibility factors were identified through review of relevant smaller-scale studies in Sub-Saharan Africa (e.g. Temesgen et al. 2001; Van Den Eeckhaut et al. 2009; Musinguzi and Asiimwe 2014; Meten et al. 2015).

The effect of landslide triggers was accounted for by applying a multiplier to the susceptibility score to give an expression of landslide hazard. Two landslide triggers were considered; rainfall and earthquakes. Ethiopia, Kenya, Uganda, Niger and Senegal are all subject to rainfall-triggered landslides. Ethiopia, Kenya and Uganda only are subject to earthquake-triggered landslides due to their proximity to the seismically-active East African Rift Valley. The rainfall triggering factor was based upon a weighted combination of long-term average rainfall and 100 year extreme monthly rainfall (determined from the Global Precipitation Climatology Centre monthly time series data due to the lack of available region-specific data). The earthquake triggering factor was based upon a weighting factor applied to the estimated peak ground acceleration with a 500 year return period (PGA_{500}). The use of threshold landslide triggering values (of either PGA or rainfall intensity-duration) was not possible at this scale of analysis due to the lack of available landslide inventories and high resolution region-specific data.

The population of landslide hazard scores is partitioned as a proportion of the maximum obtainable score to designate hazard classes A–E (following the rationale of Mastrandrea et al. 2010, where A: 0–10%, B: 10–33%, C: 33–66%, D: 66–90% and E: 90–100%).

Due to the lack of available complete landslide inventories it was not possible to use the probability of certain trigger events to estimate the frequency of occurrence of landslides of different sizes. In the absence of landslide inventories, an approximation of the probability of landslide occurrence was obtained using an approach similar to that used by Nadim et al. (2006, 2013). This approach uses the score of landslide hazard as an indicator/proxy for approximate annual frequency of a landslide event. This is based on expert interpretation of the likelihood of occurrence in cases where sufficient event data exists (Nadim et al. 2006, 2013).

To determine the annual frequencies of different sized landslides, published relationships between landslide size and landslide frequency are used. Van Den Eeckhaut et al. (2007) compiled and reviewed 27 landslide area/volume—frequency studies describing landslide frequency from regions around the world and found that the annual frequency of landslides versus landslide size within a region could be modelled using a negative power-law with slope β. The average value of this expression was found to give β = 2.3. It should be noted that in many areas, the annual frequency of landslides will deviate from this relationship based on local factors which cannot be captured by this analysis. However, at sub-continental scale, this methodology provides a systematic approach for estimating landslide size-frequency relationships and hence facilitating regional estimates of landslide risk.

A landslide is defined as a combination of the landslide source area and the landslide debris area. The mechanism or rate by which material moves from the source area to the debris area is not considered.

Landslide runout analysis is not typically completed for landslide hazard studies at regional or country scale because the resolution of the input data for such studies is usually too coarse to interpret flow paths (Horton et al. 2008; Corominas et al. 2014). This study estimates the probability that a given exposed asset will be affected by a landslide of a given size based on the ratio of the sum of the landslide area and the asset footprint area to the grid square area.

Figure 1 presents an example of the landslide hazard maps produced at a spatial resolution of approximately 0.25 km^2.

Exposure Identification

The inventory of elements at risk (or 'exposure') for this study was provided by a consortium comprising ImageCat Inc., CIESIN, University of Colorado and SecondMuse under a related project administered by the GFDRR. Exposure information was provided for the built environment, population and GDP. The built environment dataset comprised information on the location and structural attributes of buildings, the location of road networks and the location of rail networks. Also provided was the approximate rebuild cost associated with structures and infrastructures of different typologies.

Vulnerability Assessment

The physical vulnerability of structures and infrastructure describes the probable response to being affected by a landslide of a given size. The ability of a structure to resist damage therefore controls not just the economic losses which result from having to rebuild or repair it, but the vulnerability of the persons within it. For site specific landslide risk assessment the concept of physical vulnerability can be greatly expanded upon to consider the influence of landslide mechanism, debris type, building typology and the position of the exposed building relative to the landslide (e.g. Mavrouli et al. 2014). Region and country-scale studies such as described in this paper are better suited to

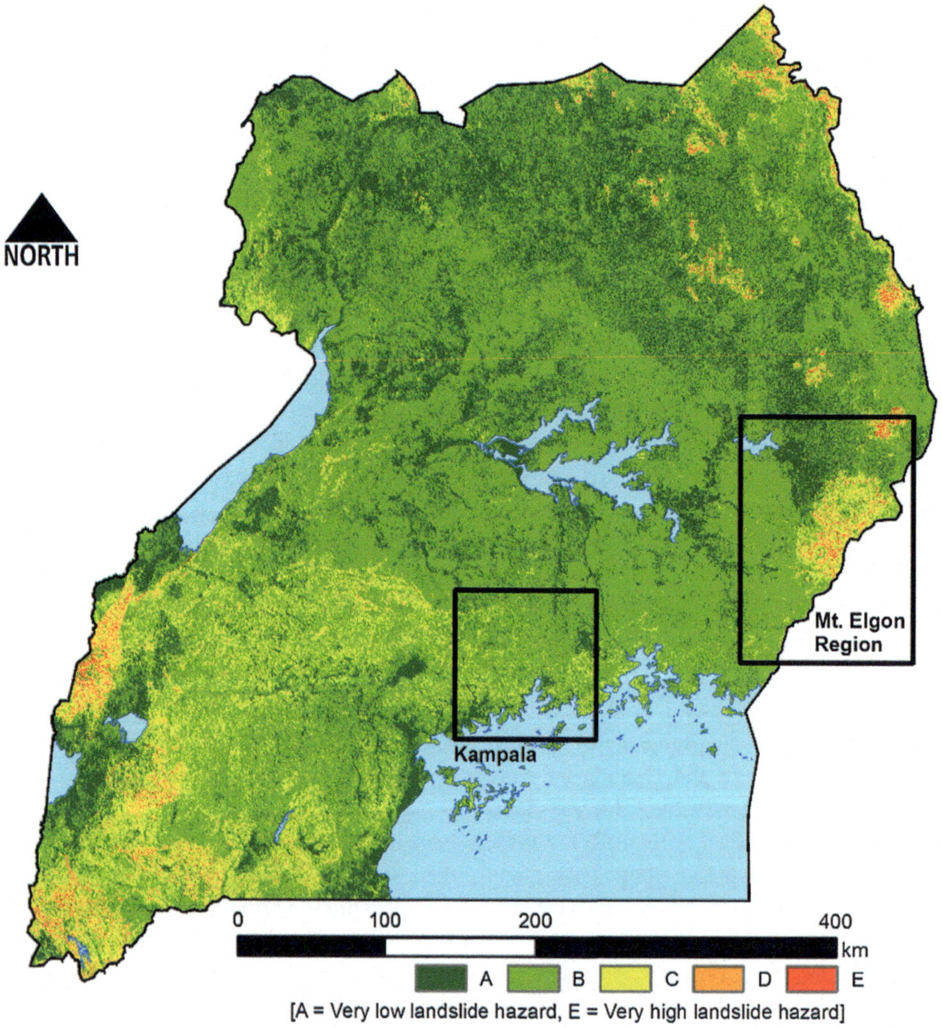

Fig. 1 Rainfall-triggered landslide hazard assessment for Uganda

methodologies which incorporate broad categories of structure typology (e.g. Du et al. 2013).

This study uses fragility functions, which estimate the probability that a certain damage state will occur as a result of a structure being affected by a landslide of a certain size. In addition, loss ratios are applied. These define what proportion of the rebuild value of a given structure is lost as a result of incurring a particular level of damage.

Risk Estimation—Results Overview

Risk is expressed as the product of the probability of hazard occurrence (e.g. a damaging landslide event) and its adverse consequences (Lee and Jones 2004).

In each of Ethiopia, Kenya and Uganda, landslides are estimated to cause approximately \$6 M–\$8 M (M = Million) worth of damage to the general building stock each year. For Niger and Senegal, average annualized losses (AAL) are estimated to be significantly less, with estimates in the region of \$1 M (Table 1).

AAL resulting from landslide damage to roads and railways is significantly less than AAL resulting from landslide damage to the general building stock. In Ethiopia, Kenya and Uganda, estimates are less than \$0.5 M, with significantly lower figures estimated for Niger and Senegal (Table 2).

Results are calculated based on a 0.25 km^2 grid and aggregated to Administrative Level 1 boundaries to allow clearer communication to key stakeholders and other intended end-users of the risk metrics. Figure 2 shows an example of the 0.25 km^2 resolution risk outputs for Uganda. Figure 3 presents the aggregated Administrative Level 1 landslide risk estimates for Ethiopia, Kenya, Uganda, Niger and Senegal (in terms of AAL in million USD). Note that the AAL estimates shown on Fig. 3 are in terms of absolute estimated annual losses to the built environment and are not shown normalized over the exposed value per administrative region. For this reason, Uganda, with its 112 Administrative Regions, shows lower AAL per region than, for example Kenya (which only has 8 official Administration Level 1 sub-divisions) despite similar nationwide estimated annual losses (Tables 1 and 2).

Discussion

The difference in landslide risk estimates between the East African countries and those in the central and western parts of Sub-Saharan Africa is interpreted to be the combined result of both increased landslide hazard in East Africa, and lower total-value of exposed assets in the West African countries of Niger and Senegal.

The relatively low AAL for transport infrastructure is possibly the result of a combination of two factors. Firstly, railway networks in Sub-Saharan Africa are generally not widely developed (and thus the total exposure is very low) and railways are constructed using low gradients, avoiding areas of steep topography (and hence typically avoiding areas of the highest landslide hazard).

Secondly, the majority of road networks in Sub-Saharan Africa are unpaved, with a typical relatively low replacement cost of \$9600/km (as estimated by ImageCat Inc., CIESIN, University of Colorado and SecondMuse). Roads are therefore comparatively cheap to repair in contrast to buildings.

Landslide risk is estimated on the basis of information on landslide hazard, exposure and vulnerability. This study indicates that exposure and vulnerability are particularly

Table 1 Estimated AAL to the general building stock from landslide-induced damage

Country	Exposure (\$M)	AAL (\$M)	AAL (% of exposure)
Ethiopia	311,834	6.115	0.00196
Kenya	537,546	8.280	0.00154
Uganda	563,621	8.915	0.00158
Niger	180,589	0.934	0.00052
Senegal	237,243	0.863	0.00036

Table 2 Estimated AAL to the roads and railways from landslide-induced damage

Country	Exposure (\$M)	AAL (\$M)	AAL (% of exposure)
Ethiopia	8044	0.302	0.00375
Kenya	7510	0.149	0.00197
Uganda	6687	0.110	0.00162
Niger	3026	0.019	0.00059
Senegal	4138	0.029	0.00065

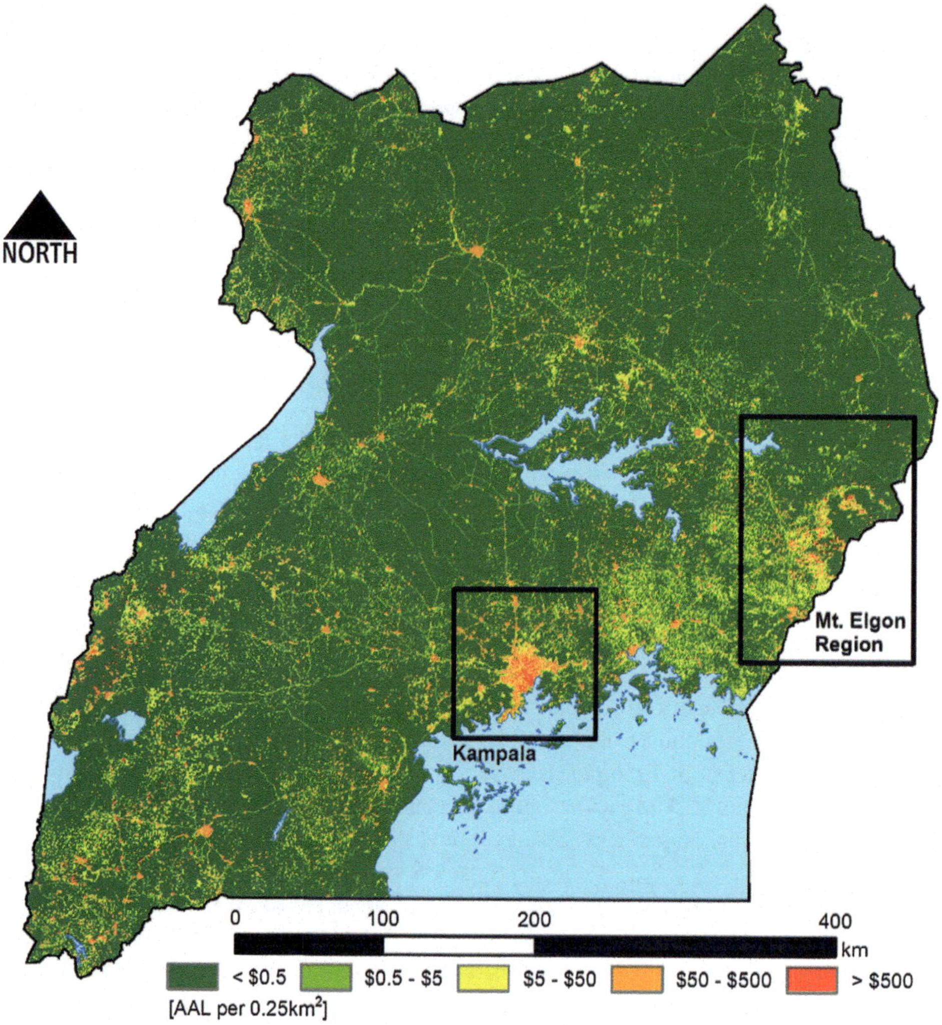

Fig. 2 Landslide risk to the built environment (general building stock, roads and railways) in Uganda

strong components of landslide risk in Sub-Saharan Africa. Landslide hazard, although high in some regions is not always coincident with locations of high exposure. As a result, landslide risk is not always high in areas of high landslide hazard. Similarly, in areas of low landslide hazard, high exposure often results in high landslide risk. For example, the area surrounding Kampala, the capital city of Uganda, is characterised by areas predominantly classified as landslide hazard class B, with some areas of hazard class C. This indicates that this urban environment is not strongly exposed to landslides (nor is it colloquially associated with landslide hazards). However, the consequences should a landslide occur are substantial due to the concentration and value of buildings and roads. By comparison, the foot-slopes of Mount Elgon carry a much greater landslide hazard, however the exposed quantity and rebuild value of the built environment in this area is significantly lower (than in Kampala), resulting in comparable levels of landslide risk and AAL. Kampala and the Mount Elgon region are highlighted on Figs. 1 and 2.

Conclusions

Widely available global datasets were used in conjunction with project-specific regional scale exposure assessments and expert elicitation, to derive estimated landslide hazard and risk assessments for the identification of regional variations in landslide interactions with the built environment in Sub-Saharan Africa.

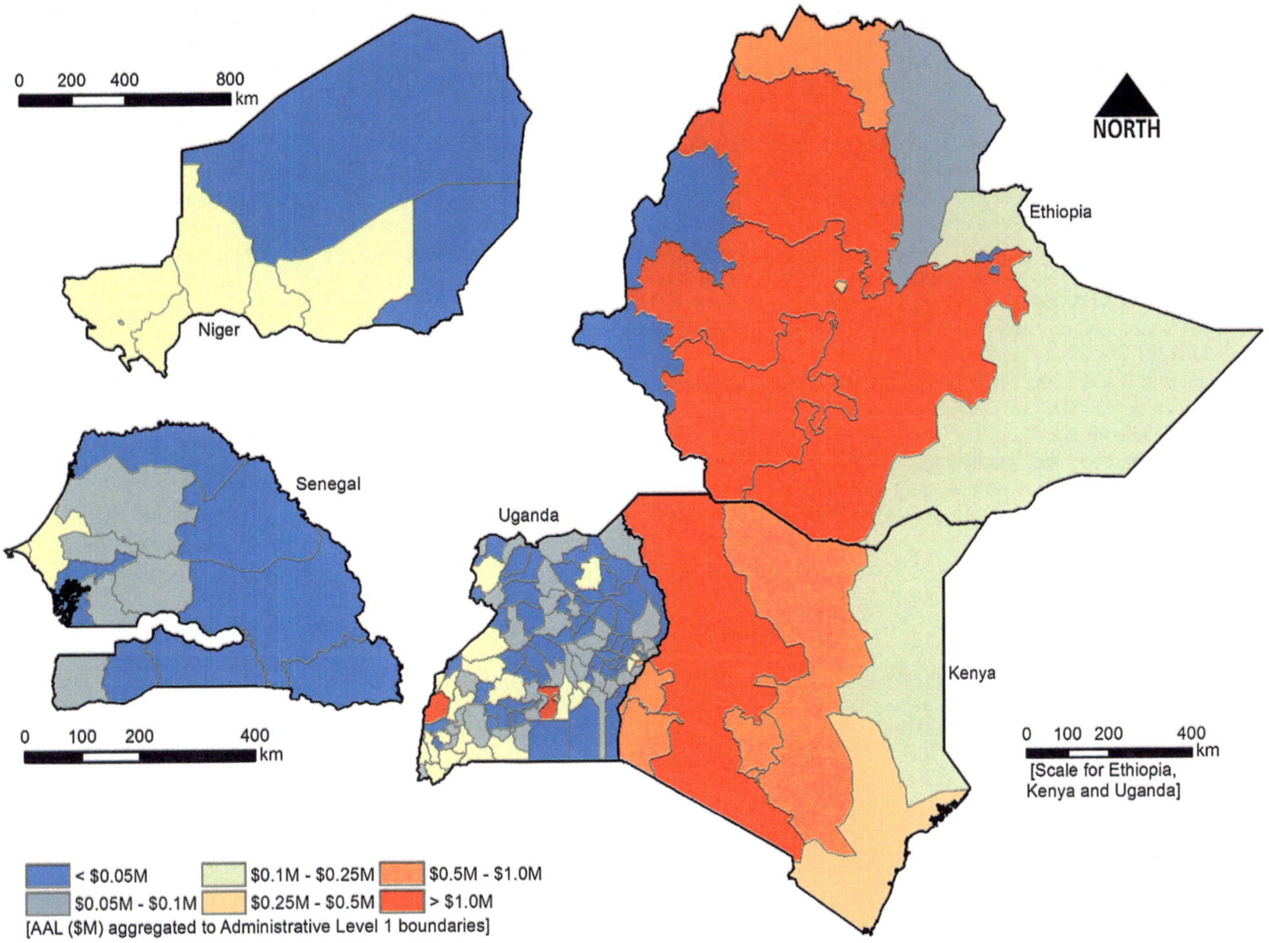

Fig. 3 Landslide risk to the built environment (general building stock, roads and railways) for Ethiopia, Kenya, Uganda, Niger and Senegal, aggregated to Administrative Level 1 boundaries. Niger and Senegal are not shown in correct location

Landslides pose a significant threat to the built environment in the eastern parts of Sub-Saharan Africa due to a combination of high landslide hazard and high vulnerability of exposed assets. Exposure and vulnerability, not hazard however are interpreted to be the key drivers of risk in the Sub-Saharan region of Africa, as illustrated by comparable estimates of expected annual losses to the built environment for low landslide hazard areas in Kampala and for high landslide hazard areas on the foot slopes of Mount Elgon. In Niger and Senegal, landslides pose a less significant threat to the built environment, with estimates of expected annual losses of <$1 M.

Acknowledgements This project was funded by the World Bank and the Global Facility for Disaster Reduction and Recovery. Exposure data was provided by a consortium comprising ImageCat Inc., CIESIN, University of Colorado and SecondMuse under a related project administered by the GFDRR. PGA_{500} data was provided by a consortium comprising Risk Engineering and Design (RED) and Evaluacion de Riesgos Naturales (ERN) under a related project administered by the GFDRR.

Dijkstra and Jordan publish with permission of the Executive Director, British Geological Survey (NERC).

References

Abebe B, Dramis F, Fubelli G, Umer M, Asrat A (2010) Landslides in the Ethiopian highlands and the Rift margins. J Afr Earth Sc 56:131–138

Ayalew L (1999) The effect of seasonal rainfall on landslides in the highlands of Ethiopia. Bull Eng Geol Env 58(1):9–19

Ayalew L, Yamagishi H (2004) Slope failures in the Blue Nile basin, as seen from landscape evolution perspective. Geomorphology 57 (1):95–116

Ayalew L, Moeller D, Reik G (2009) Geotechnical aspects and stability of road cuts in the Blue Nile Basin Ethiopia. Geotech Geol Eng 27 (6):713–728

Beyene F, Busch W, Knospe S, Ayalew L (2012). Heavy rainfall-induced landslide detection from very high resolution multi-aspect TerraSAR-X images in Dessie, Ethiopia. In: Geoscience and Remote Sensing Symposium (IGARSS), 2012 IEEE International, pp 3014–3017

Broothaerts N, Kissi E, Poesen J, Van Rompaey A, Getahun K, Van Ranst E, Diels J (2012) Spatial patterns, causes and consequences of landslides in the Gilgel Gibe catchment, SW Ethiopia. CATENA 97:127–136

Corominas J, van Westen C, Frattini P, Cascini L, Malet J-P, Fotopoulou S, Catani F, Van Den Eeckhaut M, Mavrouli O, Agliardi F, Pitilakis K, Winter MG, Pastor M, Ferlisi S, Tofani V, Hervas J, Smith JT (2014) Recommendations for the quantitative analysis of landslide risk. Bull Eng Geol Environ 73:209–263

Du J, Nadim F, Lacasse S (2013) Quantitative vulnerability estimation for individual landslides. In: Proceedings of the 18th international conference on soil mechanics and geotechnical engineering, Paris, pp 2181–2184

Fall M, Azzam R, Chicgoua N (2006a) Landslide danger assessment of large-scale natural slopes—a GIS based approach. Paper 104. IAEG 2006 Engineering Geology for Tomorrow's Cities: 10th international congress of the international association for engineering geology and the environment, 6–10 Sept 2006, Nottingham [online]. Available online: www.iaeg2006.geolsoc.org.uk. Accessed 07 Aus 2016

Fall M, Azzam R, Noubactep C (2006b) A multi-method approach to study the stability of natural slopes and landslide susceptibility mapping. Eng Geol 82(4):241–263

Fubelli G, Guida D, Cestari A, Dramis F (2013) Landslide hazard and risk in the Dessie Town area (Ethiopia). Landslide Science and Practice 6:357–362

Horton P, Jaboyedoff M, Bardou E (2008) Debris flow susceptibility mapping at a regional scale. In: Locat J, Perret D, Turmel D, Demers D, Leroueil S (eds) Proceedings of the 4th Canadian conference on Geohazards: from causes to management, Presse de l'Universite Laval, Quebec, pp 399–406

Jacobs L, Dewitte O, Poesen J, Sekajugo J, Maes J, Mertens K, Kervyn M (2015) A first landslide inventory in the Rwenzori Mountains, Uganda. In EGU General Assembly Conference Abstracts (vol 17, p 5997)

Kitutu M (2010) Landslide occurrences in the hilly areas of Bududa district in eastern Uganda and their causes. Doctoral dissertation, Makerere University, 106p

Kitutu M Goretti, Muwanga A, Poesen J, Deckers JA (2011) Farmer's perception on landslide occurrences in Bududa District, Eastern Uganda. Afr J Agric Res 6(1):7–18

Knapen A, Kitutu M Goretti, Poesen J, Breugelmans W, Deckers J, Muwanga A (2006) Landslides in a densely populated county at the footslopes of Mount Elgon (Uganda): characteristics and causal factors. Geomorphology 73(1):149–165

Lee EM, Jones DKC (2004) Landslide risk assessment. Thomas Telford Ltd, London

Maina-Gichaba C, Kipseba EK, Masibo M (2013) Overview of landslide occurrences in Kenya: causes, mitigation, and challenges

Mastrandrea MD, Field CB, Stocker TF, Edenhofer O, Ebi KL, Frame DJ, Held H, Kriegler E, Mach KJ, Matschoss PR, Plattner GK, Yohe GW, Zwiers FW (2010) Guidance note for lead authors of the IPCC fifth assessment report on consistent treatment of uncertainties. Intergovernmental Panel on Climate Change (IPCC). Available online: http://www.ipcc.ch

Mavrouli O, Fotopoulou S, Pitilakis K, Zuccaro G, Corominas J, Santo A, Cacace F, De Gregorio D, Di Crescenzo G, Foerster E, Ulrich T (2014) Vulnerability assessment for reinforced concrete buildings exposed to landslides. Bull Eng Geol Environ 73:265–289

Meten M, PrakashBhandary N, Yatabe R (2015) Effect of landslide factor combinations on the prediction accuracy of landslide susceptibility maps in the Blue Nile Gorge of Central Ethiopia. Geoenvironmental Disasters 2(1):1–17

Musinguzi M, Asiimwe I (2014) Application of geospatial tools for landslide hazard assessment for Uganda. S Afr J Geomatics 3 (3):302–314

Nadim F, Kjekstad O, Domaas U, Rafat R, Peduzzi P (2006). Global landslides risk case study. In: Arnold et al (ed) Natural disaster hotspots; case studies. Disaster risk management series. The World Bank, Washington, pp 21–64

Nadim F, Jaedicke C, Smebye, H, Kalsnes B (2013) Assessment of global landslide hazard hotspots. In: Sassa K, Rouhban B, Briceno S, McSaveney M, He B (eds) Landslides: global risk preparedness, pp 59–71

NEMA (2007). Uganda: National State of the Environment Report 2007, 357p

Ngecu WM, Ichang'i DW (1999) The environmental impact of landslides on the population living on the eastern footslopes of the Aberdare ranges in Kenya: a case study of Maringa Village landslide. Environ Geol 38(3):259–264

Temesgen B, Mohammed MU, Korme T (2001) Natural hazard assessment using GIS and remote sensing methods, with particular reference to the landslides in the Wondogenet area, Ethiopia. Phys Chem Earth Part C 26(9):665–675

Terra Daily (2010) Uganda plans to relocate 500,000 at risk of landslides: minister. Available online: http://www.terradaily.com/reports/. Hosted by Terra Daily. Accessed 20 Mar 2015

Van Den Eeckhaut M, Poesen J, Govers G, Verstraeten G, Demoulin A (2007) Characteristics of the size distribution of recent and historical landslides in a populated hilly region. Earth and Planet Sci Lett 256:588–603

Van Den Eeckhaut M, Moeyersons J, Nyssen J, Abraha A, Poesen J, Haile M, Deckers J (2009) Spatial patterns of old, deep-seated landslides: a case-study in the Northern Ethiopian highlands. Geomorphology 105(3):239–252

Vařilová Z, Kropáček J, Zvelebil J, Šťastný M, Vilímek V (2015) Reactivation of mass movements in Dessie graben, the example of an active landslide area in the Ethiopian Highlands. Landslides 12 (5):985–996

Wang HG, Montoliu-Munoz M, Gueye NFD (2009) Preparing to manage natural hazards and climate change risks in Dakar, Senegal—a spatial and institutional approach. Pilot study report, World Bank, p 101p

4thWLF2017 4th World Landslide Forum LJUBLJANA SLOVENIA EU

Rainfall-Induced Debris Flow Risk Reduction: A Strategic Approach

Mike G. Winter

Abstract

Rainfall-induced debris flows frequently cause disruption to the Scottish road network. A regional assessment of debris flow hazard and risk allows risk reduction actions to be targeted effectively. To this end a strategic approach to landslide risk reduction, which incorporates a classification scheme for landslide management and mitigation has been developed, in order to provide a common lexicon (or group of words) that can be used to describe goals, outcomes, approaches and processes related to risk reduction, and to allow a clear focus on those goals, outcomes and approaches. The focus is thus first on the desired outcome from risk reduction: whether the exposure, or vulnerability, of the at-risk infrastructure and people (and their associated socio-economic activities, which may be impacted over significant areas) is to be targeted for reduction or whether the hazard itself is to be reduced (either directly or by affecting the physical elements at risk).

Keywords

Debris flow • Rainfall • Hazard • Risk • Reduction • Management • Mitigation

Introduction and Background

The Rainfall-induced debris flow events often affect the Scottish strategic road network. After a particularly severe series of events (Winter et al. 2006, 2010; Milne et al. 2009) the Scottish Road Network Landslides Study (SRNLS) was commissioned with the overall purpose of ensuring that the hazards posed by debris flows were systematically assessed and ranked allowing sites to be effectively prioritized within available budgets (Winter et al. 2005). The hazard and risk assessment comprised three phases:

- a pan-Scotland, GIS-based, assessment of debris flow susceptibility;
- a desk-/computer-based interpretation of the susceptibility and ground-truthing to determine hazard; and
- a desk-based exposure analysis, primarily focusing on life and limb risks, but also accounting for socio-economic impacts (traffic levels, and the existence and complexity of the detour were used to estimate these impacts).

These stages determine the highest hazard ranking (risk) sites (Winter et al. 2009, 2013a).

As part of the SRNLS (Winter et al. 2009) an approach to the management and mitigation of landslides was developed. This allowed for both:

- Relatively low-cost exposure reduction (management) that allowed specific measures to be applied extensively.
- Relatively high-cost hazard reduction (mitigation) that targeted specific sites.

M.G. Winter (✉)
Transport Research Laboratory (TRL), 10B Swanston Steading, 109 Swanston Road, Edinburgh, EH10 7DS, UK
e-mail: mwinter@trl.co.uk

M.G. Winter
School of Earth and Environmental Sciences, University of Portsmouth, Portsmouth, UK

M. Mikoš et al. (eds.), *Advancing Culture of Living with Landslides*,
DOI 10.1007/978-3-319-53483-1_3

In order to facilitate a strategic approach to landslide management and mitigation a structured classification scheme has been developed (Winter 2014). This focusses on the overall goal of landslide risk reduction before homing in on the desired outcomes and the generic approach to achieving those outcomes. Only then are the processes that may be used to achieve those outcomes (i.e. the specific management and mitigation measures and remedial options) addressed. A top-down, rather than a bottom-up, approach is thus targeted. This scheme provides the main focus of this paper, drawing examples from work undertaken on debris flows. While other forms of landslide are extant in Scotland most, with the exception of rock fall, rarely impinge on infrastructure. However, it is important to note that the principles put forward in this paper, if not the detailed examples, can be equally applied to other forms of landslide and, indeed, other forms of geohazards, as well as other elements at risk.

Management and Mitigation

The primary purpose of a regional landslide hazard and risk assessment is often to enable the prioritization of sites potentially subject to risk reduction, in the light of defined budgets. However, it is important to note that it is only in cases for which the risk is deemed to be greater than that which is tolerable, or greater than the level at which the risk holder is willing to accept (Winter and Bromhead 2012), that risk reduction is required. There are many forms of landslide mitigation (e.g. VanDine 1996). However, to reduce landslide risk to acceptable levels, either the magnitude of the hazard, and/or the potential exposure (or vulnerability) or losses that are likely to arise as a result of an event must be addressed. Thus management strategies involve exposure reduction outcomes and mitigation strategies involve hazard reduction outcomes (Fig. 1). Further, it is important that those funding such works, including infrastructure owners and local governments, are able to focus clearly on goals of, the outcomes from, and the approaches to such activities rather than the details of individual processes and techniques.

To this end a strategic approach to landslide risk reduction (Fig. 1), which incorporates a classification scheme for landslide management and mitigation has been developed, in order to provide a common lexicon (or group of words) that can be used to describe goals, outcomes, approaches and processes related to risk reduction, and to allow a clear focus on those goals, outcomes and approaches.

It is designed to encourage a strategic approach to the selection of landslide management and mitigation processes (specific measures and remedial options). It is intended to aid a focus on the overall goal of landslide risk reduction, what needs to be achieved (the desired outcomes) and the generic approach to achieving that outcome rather than, initially at least, the specific measure or options (the process or processes) used to achieve that outcome. The focus is thus first on the desired outcome from risk reduction: whether the exposure, or vulnerability, of the at-risk infrastructure and people (and their associated socio-economic activities, which may be impacted over significant areas) is to be targeted for reduction or whether the hazard itself is to be

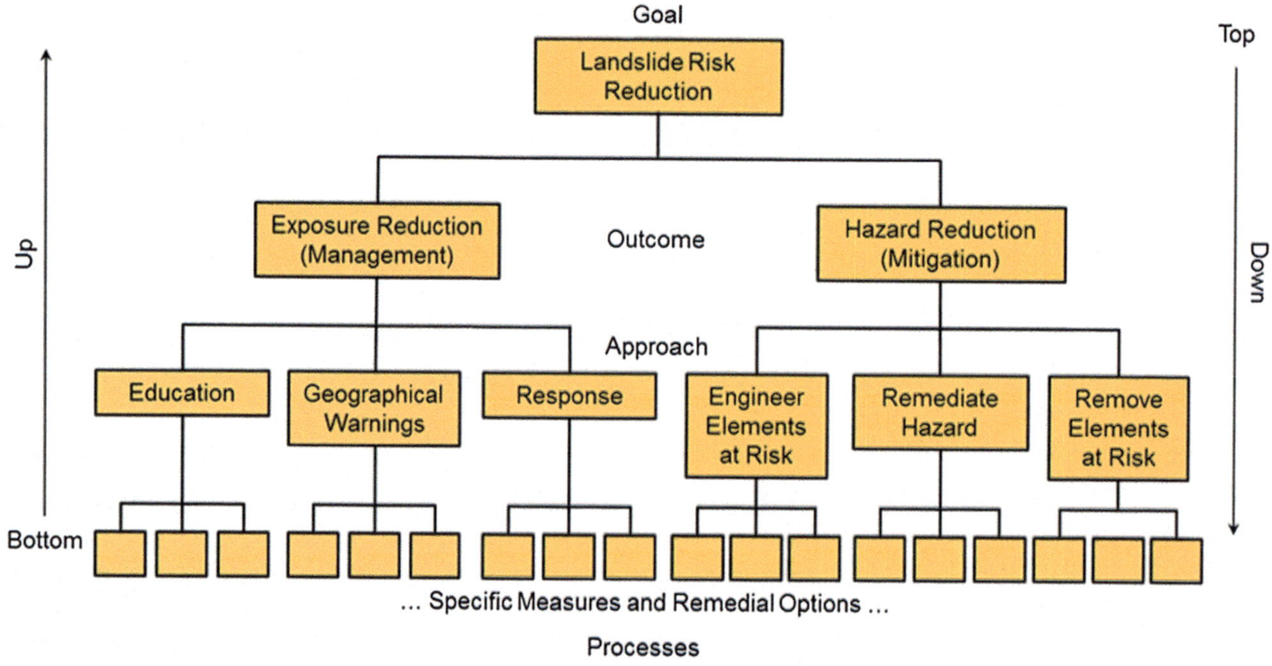

Fig. 1 Classification for landslide management and mitigation to enable a strategic approach to risk reduction

reduced (either directly and/or by affecting the physical elements at risk). In a roads environment the people at risk are road users, whereas in an urban setting they are residents and business people. The secondary focus is then on the approach(es) to be used to achieve the desired outcome before specific measures and remedial options are considered. By this means a more strategic top-down approach is encouraged rather than a bottom-up approach.

This approach also provides a common lexicon for the description and discussion of landslide risk reduction strategies, which is especially useful in a multi-agency environment. It also renders a multi-faceted (holistic) approach more viable and easier to articulate while helping to ensure appropriate responses to the hazard and risks. This approach should be especially useful for infrastructure owners and operators who must deal with multiple landslide, and other, risks, distributed across large networks. Such an approach promotes a considered decision-making process that takes account of both costs and benefits. It also encourages careful consideration of the right solution for each location and risk profile, potentially making best use of limited resources.

In the following sections approaches, specific measures and remedial options (processes) are described largely, although not exclusively, in the context of landslide hazard and risk management and mitigation on the Scottish trunk road network.

Exposure Reduction (Management)

Exposure reduction can take three basic forms:

(1) education (and information);
(2) geographical (non-temporal) warnings; and
(3) response to a period of higher risk (including temporal, or early, warnings).

Typically education in its broadest sense may form a key part of an information strategy. It may comprise leaflets, or other forms of communication, that are distributed in both electronic and hard copy. The hardcopy also may be available at rest areas for risks that relate to roads, and in retail outlets for landslide risks in urban settings. In addition, information boards may be provided in scenic rest areas, where they can be easily accessed by the public (as well as electronically). The interpretive goals embedded within the communications strategy are critical to success. These should be specific to the setting and desired outcomes, but may for example consider the development of the landscape (including geological, geomorphological and anthropogenic processes) and set the landslide consequence within that overall picture.

Considerable effort has been expended in raising the awareness of landslide issues amongst both relevant professionals, including road operators, and the public in Scotland. This has taken the form of public lectures and talks, media appearances and the development of advisory leaflets which may be accessed from the Transport Scotland website. In addition a programme to develop information signs for rest areas, lay-bys and National Park Gateways is under development. These signs are intended to set the issues surrounding landslide hazards in a balanced context. These types of activities are unlikely to affect the exposure of road users but may, of course, influence society's acceptable level of risk. However, education and information relating to desirable behaviours of, for example, drivers in areas of landslide hazard during periods of higher risk (Winter et al. 2013b) is intended to influence those desirable behaviours. These behaviours include heightening the levels of observation, moderating speeds and excluding certain stopping locations such as bridges in order to avoid likely areas of hazard, and to allow early observation and avoidance of potential hazards.

Geographical warning signs (Fig. 2) may be used in a variety of environments, to demonstrate the presence of landslide hazards. In a road environment they usually follow the standard warning sign form and include a graphic representing rock fall.

The responsive reduction of exposure lends itself to the use of a simple three-part management tool: Detection, Notification and Action (or DNA), providing a simple framework for management responses (Winter et al. 2005; Winter 2014):

- Detection of either the occurrence of an event (e.g. monitoring, observation) or by the forecast of precursor conditions (e.g. rainfall) (Winter et al. 2010).
- Notification of the likely/actual occurrence of events to the authorities (e.g. in a roads environment the Police, the Road Administration and the road operator).

Fig. 2 Wig-wag signs offer both a geographical (non-temporal) and a temporal, or early-, warning

- Action that reduces the exposure of the elements at risk to the hazard. Again, in a roads environment, this could include media announcements, the activation of geographical signs that also have a temporal aspect (e.g. flashing lights) (Winter et al. 2013b), the use of variable message signs, 'landslide patrols' in marked vehicles, road closures, and traffic diversions.

Hazard Reduction (Mitigation)

The challenge with hazard reduction in Scotland often is to identify locations of sufficiently high risk to warrant spending significant sums of money on engineering works. The costs associated with installing extensive remedial works over very long lengths of road may be both unaffordable and unjustifiable and even at discrete locations the costs can be significant. Moreover the environmental impact of such engineering work should not be underestimated. Such works often have a lasting visual impact and, potentially, impact upon the surrounding environment. Such works should be limited to locations where their worth can be clearly demonstrated.

In addition, actions such as ensuring that channels, gullies and other drainage features are clear and operating effectively are important in terms of hazard reduction. This requires that the maintenance regime is both routinely effective and also responsive to periods of high rainfall, flood and slope movement. Planned maintenance and construction should take the opportunity to limit hazards by incorporating suitable measures including higher capacity/better forms of drainage, or debris traps into the design. Critical review of the alignment of culverts (etc.) normally should be carried out as part of any planned maintenance or construction activities.

Beyond such relatively low cost/low impact options three categories of hazard reduction measures may be considered:

(1) works to engineer, or protect the elements at risk;
(2) remediation of the hazard to reduce failure probability; and
(3) removal, or evacuation, of the elements at risk.

There are many means of engineering or protecting the elements at risk and this approach accepts that debris flows will occur and makes provision to protect the road, thus limiting the amount of material reaching the elements at risk.

The potential structural forms for protection from debris flow include shelters, barriers and fences (Fig. 3). Flexible fences absorb the kinetic energy of the debris flow, thus reducing the forces that the structure must accommodate. These systems have been shown to work well, particularly

Fig. 3 Debris flow barriers

for the arrest of rock fall, but all such systems require maintenance after an impact. As part of work at the A83 Rest and be Thankful site, defined as one of the higher risk sites on the Scottish trunk road network (Winter et al. 2009), fully flexible barriers have been installed.

Debris basins are formed by large decant structures, incorporating a downstream barrier that retains debris but allows water to pass. They may be used in association with lined debris channels to move material downslope where potential storage areas on the hillside are limited; lined channels may be used in isolation if storage is limited on the hillside or available only at the foot of the slope. Rigid barriers such as check dams and baffles may slow and partially arrest flows within a defined channel, and on hillsides may protect larger areas where open hillside flows are a hazard and/or channelised flows may breach the stream course. VanDine (1996) gives design and use guidance for check dams and baffles, including low cost earth mounds.

Rigid barriers and debris basins were built as debris flow defence structures at Sarno to the east of Naples in Italy following the events of May 1998 in which 159 people were killed (Versace 2007), at a cost estimated at between €20 M and €30 M.

Debris flows are dynamic in nature, often initiated on high hillsides, and fast moving when they reach the road. Their energy has a significant impact on the engineering works that can reduce the hazard to the road and its users. Indeed, while structures can be effective in slowing and arresting flow in the debris fan area, in Scotland many roads potentially affected by debris flow are located in the high energy transport zone or the upper reaches of the debris fan. Roads on debris fans are usually close to a loch (or lake) side and the opportunity for the use of these types of measure may be limited.

The remediation of landslide hazards to reduce the probability of failure may involve alteration of the slope profile by either cut or fill, improvement of the material strength (most often by decreasing pore water pressures), or providing force systems to counteract the tendency to move (Bromhead 1997).

The engineering options applicable to prevent debris flow depend greatly upon the specific circumstances. Debris flows can be triggered from relatively small source areas, within very large areas of susceptible ground, and be initiated high on the hillside above the road. There may be particular conditions where conventional remedial works and/or a combination of techniques such as gravity retaining structures, anchoring or soil nailing may be appropriate. However, in general terms the cases where these are both practicable and economically viable are likely to be limited. The generic link between debris flows and intense rainfall is well-established and effective runoff management can reduce the potential for debris flow initiation; Winter et al. (2010) present information on the relation between rainfall and debris flow in Scotland. However, in many circumstances on-hill drainage improvement may have limited impact due to the small scale of many debris flow events. In other locations and situations positive action to improve drainage might well have a more beneficial effect. Such measures could include improving channel flow and forming drainage around the crest of certain slopes to take water away in a controlled manner.

The planting of appropriate vegetation can also contribute to the reduction of instability (e.g. Coppin and Richards 2007). Notwithstanding this, the positive effects of such measures can be difficult to quantify. The positive effects include canopy interception of rainfall and subsequent evaporation, increased root water uptake and transpiration via leaf cover, and root reinforcement. In addition, the life cycle of the vegetation planted must be considered as, depending upon the species, the climate and other conditions relevant to growth there may be a considerable period before the effects provide a meaningful positive effect on stability. In addition, future deforestation, or harvesting, must also be considered as this is widely recognised as a potential contributor to instability. Such measures do not provide instant solutions and may not always be effective in the long term, especially if commercial forestry is practised. The species planted must be appropriate to the local environment—the planting of non-native species is not allowed in most countries for example. However, the successful application of local knowledge and species can prove successful (Winter and Corby 2012).

Finally, the option of removing the elements at risk from the geographical location of the hazard remains. Typically this might involve the abandonment of a settlement (Coppola et al. 2009) or the realignment of an infrastructure route. It should, of course, be noted that decisions to adopt such extreme options are not taken in isolation. Road realignment might be undertaken as part of a road administration's route improvement activities in order to upgrade both the alignment and the layout of junctions, in particular to reduce road traffic accident risk, and to ensure compliance with current design standards. In cases where the debris flow risk is high and other factors indicate that some degree of reconstruction is required, road realignment may be a viable option. While road realignment has been undertaken in response to landslide activity in Scotland, it was also in response to a genuine need for realignment of the route to increase safety and to ensure compliance with current design standards.

Summary

Rainfall-induced debris flow is a common occurrence in Scotland with events at the most active site, the Rest and be Thankful, generally occurring more than once a year; other sites generally experience events at a much lower frequency often measured in years or decades. Major injuries and fatalities are relatively rare but the socio-economic effects can be substantial and impact over wide areas. The economic impacts typically include the severance (and/or delay) of access to and from relatively remote communities for services and markets for goods; employment, health and educational opportunities; and social activities.

The primary purpose of a regional landslide hazard and risk assessment is often to enable the prioritization of sites at which landslide risk reduction is required, in the light of defined budgets for those risks that the risk holder is not willing to accept or which are not tolerable. It is possible to consider management strategies that involve exposure reduction outcomes and mitigation strategies that involve hazard reduction outcomes. It is important that those funding such works, including infrastructure owners and local governments, are able to focus clearly on the goal of landslide risk reduction, the outcomes from that activity and the generic approach(es) to achieving the outcomes, rather than the details of individual processes and techniques.

To this end a classification scheme for landslide risk reduction has been developed to encourage a strategic approach to landslide risk reduction. It is intended to encourage a focus, in sequence, on the goal, outcome(es) and approach(es) before consideration of the specific measure(s) or remedial option(s) (the process(es)) selected for implementation. The focus is thus on landslide risk reduction whether the exposure, or vulnerability, of the infrastructure and people at risk (and their associated socio-economic activities) are to be targeted for reduction, or whether the hazard itself is to be reduced (either directly or by affecting the elements at risk). In a roads environment the people at risk are road users whereas in an urban setting they are residents and business people. The secondary focus is then on the generic approach(es) to be used to achieve the desired outcome before specific measures and remedial options are considered. By this means a more strategic top-down approach is encouraged in place of a bottom-up approach.

This approach provides a common lexicon for the description and discussion of landslide risk reduction strategies, which is especially useful in a multi-agency environment. It also renders a multi-faceted (holistic) approach more viable and easier to articulate while helping to ensure that the responses to the hazard and risks in play are appropriate. This approach should be especially useful for infrastructure owners and operators who must deal with multiple landslide risks, and other risks, that are distributed across large networks. Such an approach promotes a considered decision-making process that takes account of both costs and benefits, and encourages careful consideration of the right solution for each location and risk profile, potentially making best use of often limited resources.

References

Bromhead EN (1997) The treatment of landslides. Proc Inst Civ Eng (Geotech Eng) 125:85–96

Coppin NJ, Richards IG (2007) Use of vegetation in civil engineering. CIRIA report C708. CIRIA, London. (Reprinted from CIRIA Report B10, 1990)

Coppola L, Nardone R, Rescio P, Bromhead E (2009) The ruined town of Campomaggiore Vecchio, Basilicata, Italy. Q J Eng Geol Hydrogeol 42:383–387

Milne FD, Werritty A, Davies MCR, Browne MJ (2009) A recent debris flow event and implications for hazard management. Q J Eng GeolHydrogeol 42:51–60

VanDine DF (1996) Debris flow control structures for forest engineering. Ministry of Forests Research Program, Working paper 22/1996. Ministry of Forests, Victoria, BC

Versace P (ed) (2007) La mitigazione del rischio da collate di fango: a Sarno e negli altri comuni colpiti dagle eventi del Maggio 1998. Commissariato do Governa per l'Emergenze Idrogeologica in Campania, Naples, p 401

Winter MG (2014) A strategic approach to landslide risk reduction. Int J Landslide Environ 2:14–23

Winter MG, Bromhead EN (2012) Landslide risk—some issues that determine societal acceptance. Nat Hazards 62:169–187

Winter MG, Corby A (2012) A83 Rest and be Thankful: ecological and related landslide mitigation options. Published Project Report PPR 636. Transport Research Laboratory, Wokingham

Winter MG, Heald AP, Parsons JA, Macgregor F, Shackman L (2006) Scottish debris flow events of August 2004. Q J Eng Geol Hydrogeol 39:73–78

Winter MG, Macgregor F, Shackman L (eds) (2005) Scottish Road network landslides study. Edinburgh, Scottish Executive, p 119

Winter MG, Macgregor F, Shackman L (eds) (2009) Scottish road network landslides study: implementation. Edinburgh, Transport Scotland, p 278

Winter MG, Dent J, Macgregor F, Dempsey P, Motion A, Shackman L (2010) Debris flow, rainfall and climate change in Scotland. Q J Eng Geol Hydrogeol 43:429–446

Winter MG, Harrison M, Macgregor F, Shackman L (2013a) "Landslide hazard assessment and ranking on the Scottish road network. Proc Inst of Civ Eng (Geotech Eng) 166:522–539

Winter MG, Kinnear N, Shearer B, Lloyd L, Helman S (2013b) A technical and perceptual evaluation of wig-wag signs at the A83 Rest and be Thankful. Published Project Report PPR 664. Transport Research Laboratory, Wokingham

4th WLF 2017 4th World Landslide Forum LJUBLJANA SLOVENIA EU

RUPOK: An Online Landslide Risk Tool for Road Networks

Michal Bíl, Richard Andrášik, Jan Kubeček, Zuzana Křivánková, and Rostislav Vodák

Abstract
The landslide risk for the entire Czech road network is presented here. The risk was computed using data on landslide hazard and data on potential impacts of road blockage. Data from the official landslide database were used for landslide hazard computation combined with data from historical records on roads interrupted by landsliding. Vulnerability was computed as direct costs which are related to road construction costs and indirect costs. The latter express additional economic losses from the blocked roads. This concept was applied at II/432 road link as a case study where a landslide interrupted traffic in May 2010. Indirect losses were estimated as being 2.5 times higher than costs related to mitigation works. All data can be viewed at rupok.cz website.

Keywords
Landslide • Road • Damage • Database • Web-map application • Vulnerability • GIS

Introduction

Road blockages have serious impacts on current traffic and on the overall transportation system (e.g., Chang and Nojima 2001; Bono and Gutiérrez 2011; Bíl et al. 2015). They cause not only congestion but also a decrease in overall road network performance. The situation can be worse if the blockages are results of physical damage to road pavements or even embankments. This is often the case when landsliding along transportation corridors occurs (Fig. 1).

Heavy precipitation is the primary cause of landsliding in the Czech Republic. At least three large landslide events took place as of 1997. All of them were described in detail including the causes (Krejčí et al. 2002; Bíl and Müller 2008; Pánek et al. 2011) and their consequences for the road network were also documented (e.g., Bíl et al. 2015).

Roads Affected by Landsliding in the Czech Republic

We registered 333 landslide events which caused road blockage on the Czech road network between 1997 and 2015. Landsliding and rock falls are the cause of about 13% of the overall road blockage data due to natural hazards between 1997 and the present (Fig. 2).

Areas dominated by Tertiary flysch rocks, volcanic Tertiary rocks and Mesozoic sandstones are most prone to landsliding in the Czech Republic (Fig. 3). Roads interrupted by landslides also occur, however outside these areas. The majority of them were related to landsliding on a man-made road embankment or adjacent cut along roads.

M. Bíl (✉) · R. Andrášik · J. Kubeček · Z. Křivánková · R. Vodák
CDV—Transport Research Centre, Líšeňská 33a, 636 00 Brno, Czech Republic
e-mail: michal.bil@cdv.cz

R. Andrášik
e-mail: richard.andrasik@cdv.cz

J. Kubeček
e-mail: jan.kubecek@cdv.cz

Z. Křivánková
e-mail: zuzana.krivankova@cdv.cz

R. Vodák
e-mail: rostislav.vodak@gmail.com

M. Mikoš et al. (eds.), *Advancing Culture of Living with Landslides*,
DOI 10.1007/978-3-319-53483-1_4

Fig. 1 An example of an extremely devastating landslide which destroyed D8 motorway in June 2013. The motorway was under construction at the time of the event. The direct losses, including the damaged local railroad, were estimated at over one billion CZK. Photo: Petr Kycl, Czech Geological Survey, 2013

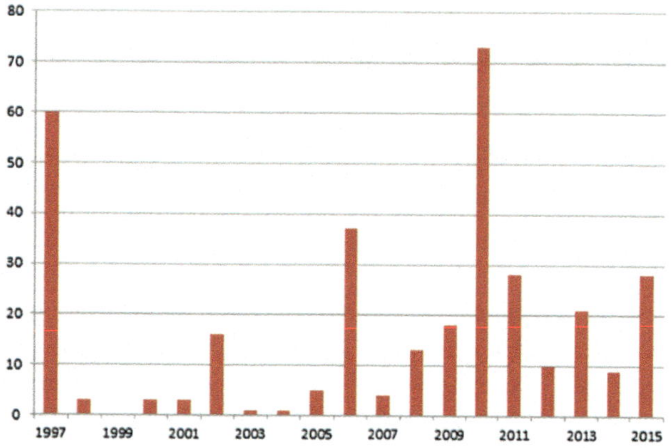

Fig. 2 Number of road link interruptions due to landsliding in respective years

Rupok.Cz

Spatial Database on Landsliding

Information on landslide occurrences is often collected in spatial databases. Many national and regional spatial databases of this kind also exist in Europe (e.g., Hilker et al. 2009; Van Den Eeckhaut and Hervás 2012; Winter et al. 2013; Taylor et al. 2015; Klose et al. 2015, 2016). The landslide database which we used is administered by the Czech Geological Survey. Data on roads damaged by landsliding come from road administrators via JSDI, i.e. the Integrated Traffic Information System of the Czech Republic (Fig. 4).

Web Map Application

We present a system where the landslide risk is computed online on the basis of new data input. Data come from a system of traffic information which is administered by the National Road Administrator (Fig. 4). We have programmed several online filters which separate incoming data, based on the causes and consequences of the incidents, into the respective databases. One of them is devoted to complete blockages due to landsliding.

Landslide Hazard

Data Available

We utilize existing information on road link exposure to landsliding. In addition, we worked with the current landslide database administered by the Czech Geological Survey (e.g., Bíl et al. 2014). These two data sources are subsequently used for landslide hazard computation. The landslide hazard is estimated using a logistic regression model and Bayesian inference which is an advantageous approach if empirical data is available. The resulting numbers determine the probabilities that an individual road link will be interrupted as a result of landsliding over the course of a year.

Methods

The hazard is the probability of occurrence of an event which causes harm or injury to people, other organisms or environmental degradation. The total hazard consists of two main classes, namely hazards induced by nature (floods,

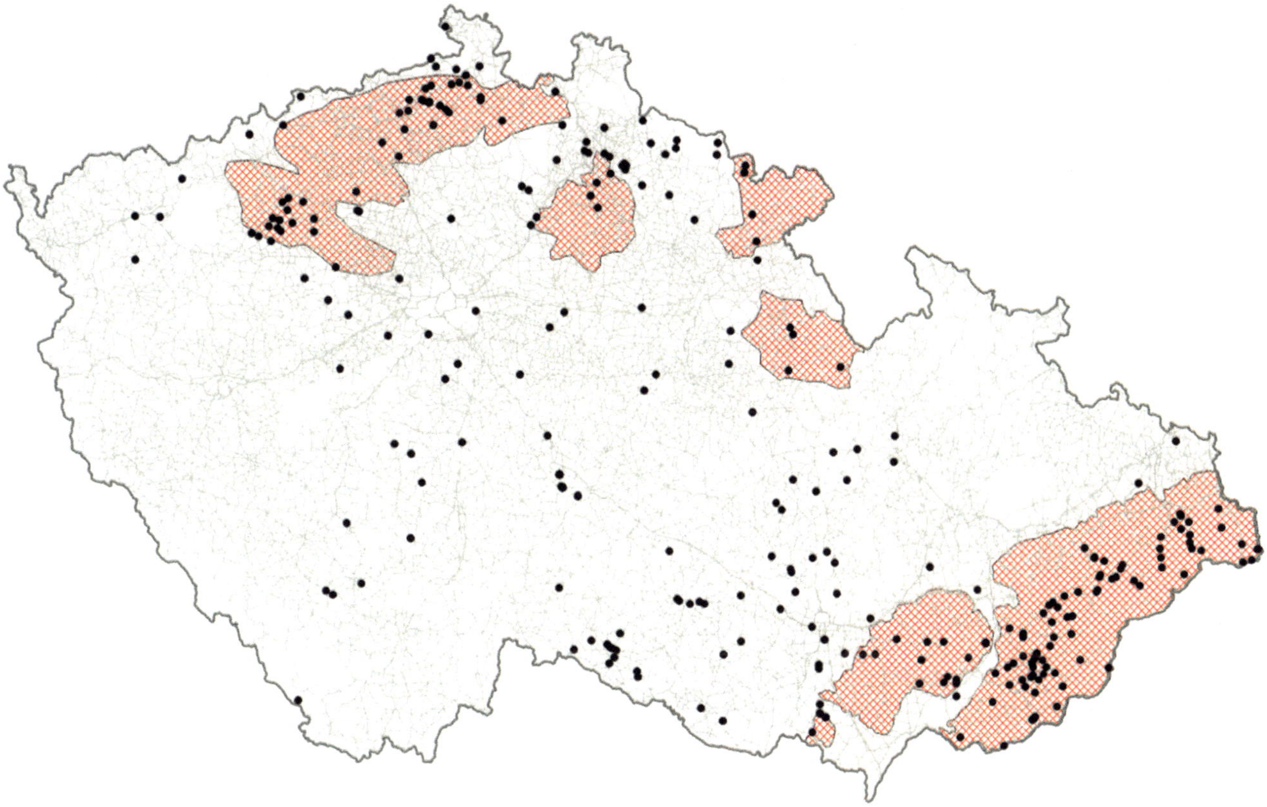

Fig. 3 Landslides (*black dots*) which caused traffic interruptions on roads. The *red areas* represent the most landslide susceptible regions in the Czech Republic. Landslides which occurred outside those regions therefore took place on embankments

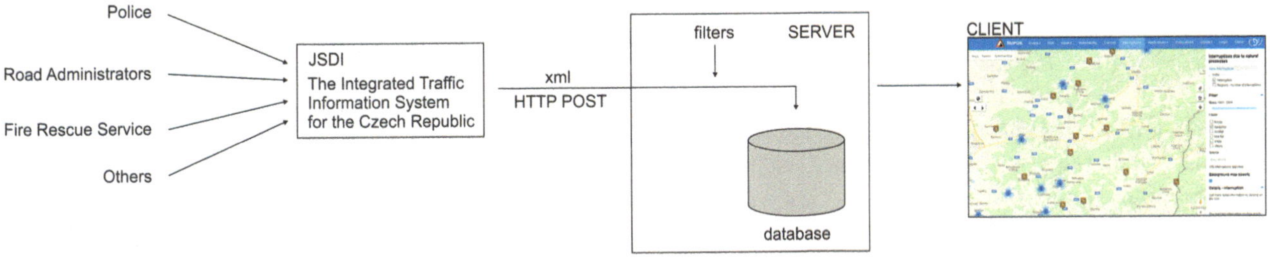

Fig. 4 A scheme of data flow into RUPOK webpage which relates to landsliding. Original data are stored within JSDI and consequently sent to a registered user via XML message. This incoming message is filtered and stored into a spatial database. The web-map application RUPOK, which is connected to this database, allows for data visualization and querying

landslides, snow, and fallen trees) and hazards produced by people. Traffic accidents are the main produced-by-people phenomena causing road disruptions.

We focused on the hazard induced by landslides. Our aim was to estimate the probability of at least one interruption of a road caused by a landslide during a year for each road within the Czech road network. The data on historical events were from the Czech Road and Motorway Directorate covering the period of 1997–2014.

Landslide hazard was estimated as follows:

1. Road segments were separated into two groups—with or without empirical information concerning the interruption caused by landslides.
2. A particular hazard for segments with a historical disruption was estimated empirically. Since an event of "at least one disruption of a road during a year" has an alternative probability distribution (yes/no), we can

estimate the probability of the road interruption in the very next year by the following ratio:

$$\frac{\text{number of years with at least one interruption} + 1}{\text{total number of years} + 2} \quad (1)$$

For instance, 3 events within 18 years led to (3 + 1)/(18 + 2) = 1/5 as a Bayesian estimate of the probability of an event. In other words, the probability of at least one interruption of a road by a landslide would be estimated as 20%. This approach is appropriate when the information concerning a possible disruption is mostly contained in the empirical data (i.e. the probability is higher in locations where a landslide already occurred). The derivation of formula (1) is shown in MacKay (2003, page 52).

3. A logistic regression model (see Hosmer et al. 2004) was constructed for all road segments. The dependent variable was "a road segment was interrupted by a landslide during the last 18 years" (yes/no). Subsequently, this model was applied only to road segments without empirical information concerning the disruption caused by landslides.

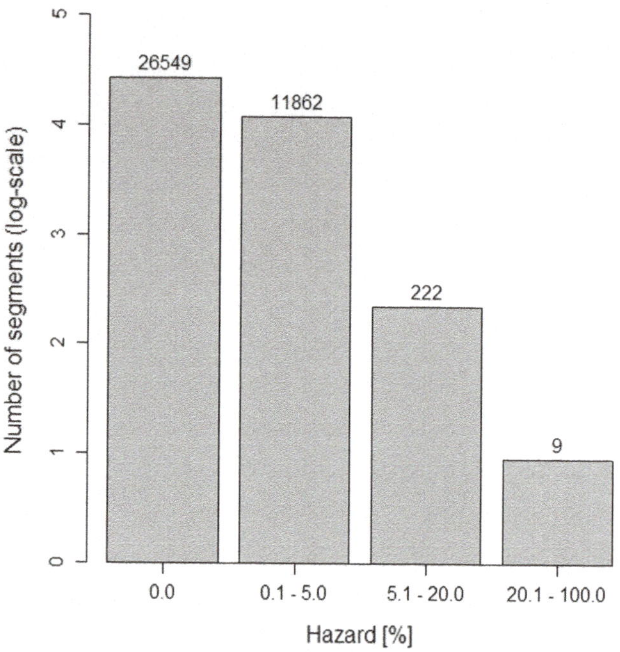

Fig. 5 Hazard frequencies for road links within the entire Czech road network. The numbers indicate probabilities that the road links will be interrupted due to landsliding in a given year

Significant explanatory variables in the logistic regression were (a) the length of a road segment in the vicinity (up to 50 m) of historical landslides and (b) the length of a road segment prone to landslides. An overview of landslide hazard along individual road links can be seen in Fig. 5.

Finally, the hazard can also be related to territorial units of any type (e.g. rectangular grid, administrative units) by

- either calculating the probability of "at least one interrupted road segment during a year" (suitable for high-resolution grids), or
- expressing the probability distribution function of "a number of disrupted road segments during a year" (appropriate for greater administrative units, see Fig. 6).

Landslide hazard along roads is visualized as grid values (Fig. 6). A detailed view indicating landslide hazards for the individual road links can be obtained after zooming in (Fig. 7—on the right).

Vulnerability

Network vulnerability can be expressed in terms of direct or indirect costs (Klose et al. 2014, 2016; Postance et al. 2015, 2016; Winter et al. 2014, 2016). Direct costs represent expenses related to mitigation works directly connected with the local source of interruption. The indirect costs are usually computed in two ways: as a ratio of the time needed for the shortest detour of a blocked road link relative to the same time for the actually blocked link and as traffic in terms of the average number of cars influenced by the road blockage.

Risk

The risk represents a product of hazard and vulnerability (Birkmann 2007). The same values of risk can therefore be achieved by two completely different situations: a place with a large landslide hazard but with only negligible potential losses and a place with a low hazard but where the potential losses are extremal. The second case is much more dangerous. A reasonable approach should therefore be to evaluate hazard and vulnerability separately. Hazard along roads, as well as vulnerabilities and risks, are visualized on the rupok.cz website. We will demonstrate the overall process of risk computation below on a concrete case of a road link interrupted by a landslide.

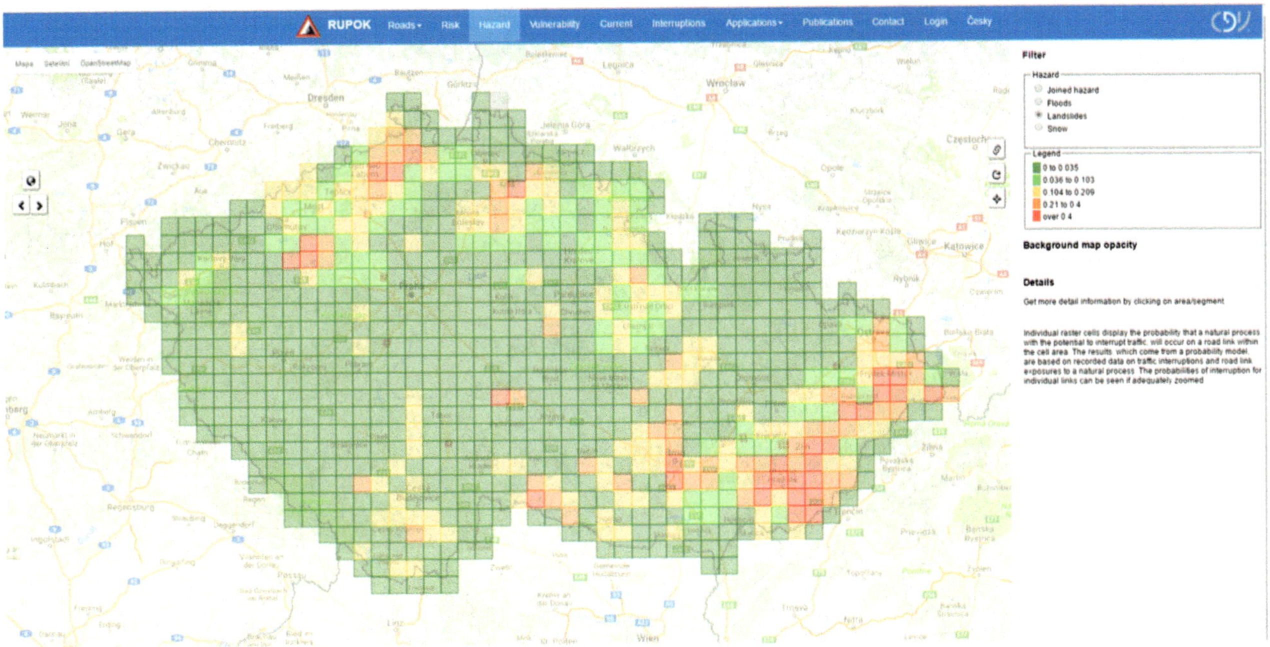

Fig. 6 A screen shot from the RUPOK website shows the grid of landslide hazard along roads in the Czech Republic. *Source* rupok.cz

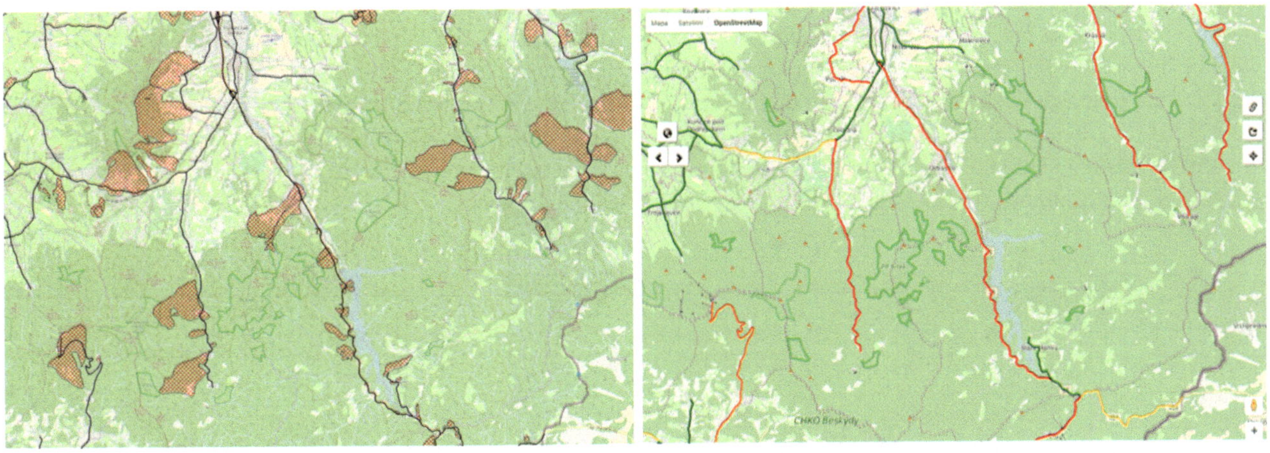

Fig. 7 Existing landslides (on the *left*) pose a threat to traffic. They also present the exposure of road links to interruption due to landsliding. Hazard can be computed on the basis of the exposure and registered interruptions (on the *right*). *Source* rupok.cz

Case Study: II/432 Road

The model of risk was applied to a II/432 road link. This secondary road is located at the eastern part of the Czech Republic (the Outer Western Carpathians). The underlying geology belongs to a Tertiary flysch which is prone to landsliding.

A landslide near II/432 road took place in May 2010 (Fig. 8). It was considered to be the most damaging result of floods in May 2010 in this region in relation to the transportation infrastructure (RSZK 2010). Road administrators originally tried to keep at least one lane of the road passable, but ongoing movement of subsoil led to complete closure for approximately five months, until the road was repaired.

Direct Costs

The direct costs were derived from the existing tables of road link construction costs. Repair costs were estimated at 21 million CZK (840,000 USD), including repairing 700 m^2 of road and construction of a retaining wall.

Fig. 8 Road II/432 between Koryčany and Jestřabice in the Zlín region destroyed by a landslide in May 2010. Cracks were up to 15 cm wide and tens of cm deep. Photo: Michal Bíl, May 2010

Indirect Costs

An official detour was immediately recommended by the road administrator (Fig. 9). The length of the detour was 38 km for passenger cars and 56 km for heavy vehicles, while the closed road link was 2 km long only. We computed the indirect cost of this detour according to (MacLeod 2005). Detour time was considered as the sum of the average time needed for passing links on the detour. Operating costs represent the average value of time, taking into account the purpose of the journey (business, travelling to work, recreation, other private), per each person multiplied by the average number of persons in the car. The respective values, including data from the National Traffic Census, were used from the Road and Motorway Directorate of the Czech Republic (RMD 2012).

Additional fuel was used as a product of average fuel consumption per 100 km in liters and the length of detour in kilometers. The price of fuel is computed as the weighted average of prices of petrol and oil fuel per one liter, with weights as the proportions of cars using petrol and cars using oil fuel in traffic.

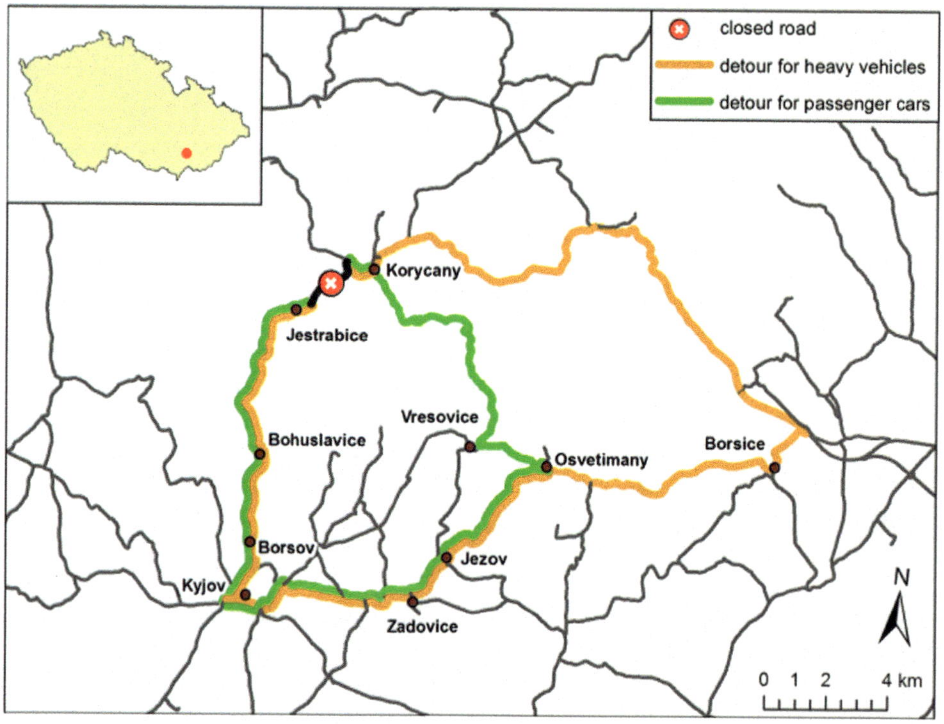

Fig. 9 An official detour of the closed road between Koryčany and Jestřabice

The number of travelers affected is represented as average daily traffic.

The road was closed for five months, until it was completely repaired, but after a month a short detour for vehicles below 3.5 tons was built only around the damaged part of the road. The total detour cost for that month could reach up to 52 million CZK (approx. 2.08 million USD). The estimated indirect losses from this road blockage were therefore 2.5 times higher than the direct losses from the road damage due to landsliding.

$$\begin{aligned} \textit{Detour cost} &= [\textit{detour time} \times \textit{operating costs} + \textit{extra fuel} \times \textit{price}] \\ &\quad \times \textit{number of travelers affected} \\ \textit{Detour cost} &= [0.85 \times (272 \times 1.9) + (6 \times 38/100) \times 35.04] \times 3334 \\ &= 1{,}730{,}916\,\text{CZK} \end{aligned}$$

The computation of the indirect costs above is based on several preconditions. The most changeable variable is the price of fuel, which depends on the price of petroleum on world markets, the currency exchange rate and taxes. Over the past ten years, the price of fuel fluctuated between 28 and 38 CZK/liter. The indirect cost then would vary from 1,677,402 CZK to 1,753,417 CZK, so the difference based on the price of fuel would be 76,015 CZK per day. The computation was only performed for passenger cars, because a number of various types of heavy vehicles exist.

II/432 Road Link Hazard, Vulnerability and Risk

The resulting risk is visualized online on the RUPOK website (www.rupok.cz). Moreover, the RUPOK application also allows for visualization of the cause, place and time of road link blockage and the entire database of blocked road links (Fig. 10).

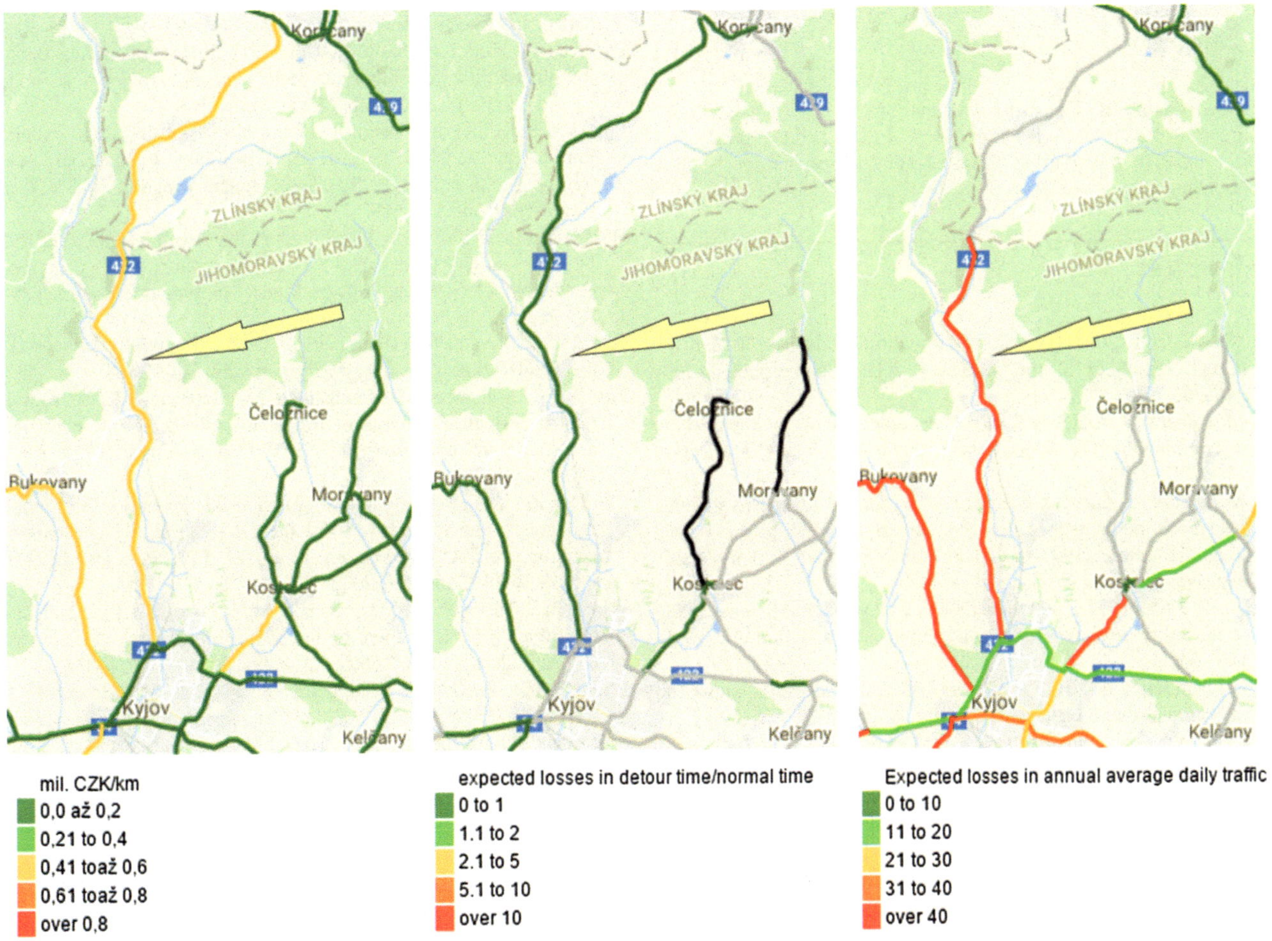

Fig. 10 Direct and indirect risk on II/432 (marked with an *arrow*). The hazard values for II/432 already include the 2010 event which caused a landslide along this road link. The respective risk values are from *left* to *right*: monetary value/unit length (106 CZK/km), additional time needed for a detour (dimensionless number) and the number of cars affected per day with the given probability of an event (number)

Conclusion

The RUPOK website is a tool which helps road administrators plan road maintenance. It visualizes online road links which are currently impassable. The landslide hazard map is part of this web-map application. The landslide hazard was computed on two levels: a spatial grid which expresses the probability that at least one road link will be interrupted due to landsliding in the current year. The second, more detailed, level relates to the individual road links. The hazard was computed using both landslide databases and data on historical road interruptions. The landslide risk is obtained by multiplying the landslide hazard with vulnerability, i.e. impacts due to road link closures. Three final risk maps are also available as three different vulnerability measures were used.

We plan, in the following work, to evaluate the precision of the landslide hazard values on the basis of the current road blockage data. This evaluation and further investigations will improve the landslide hazard modeling. The presented approaches and the entire RUPOK web map application can be easily transferred to any other region.

Acknowledgements This work was financed by the Transport R&D Centre (OP R&D for Innovation No. CZ.1.05/2.1.00/03.0064) and project LO1610. We further thank David Livingstone for English proofreading.

References

Bíl M, Müller I (2008) The origin of shallow landslides in Moravia (Czech Republic) in the spring of 2006. Geomorphology 99: 246–253

Bíl M, Kubeček J, Andrášik R (2014) An epidemiological approach to determining the risk of road damage due to landslides. Nat Hazards 73(4):1323–1335

Bíl M, Vodák R, Kubeček J, Bílová M, Sedoník J (2015) Evaluating road network damage caused by natural disasters in the Czech Republic between 1997 and 2010. Transp Res Part A: Policy Pract 80:90–103

Birkmann J (2007) Risk and vulnerability indicators at different scales: applicability, usefulness and policy implications. Environ Hazards 7:20–31

Bono F, Gutiérrez E (2011) A network-based analysis of the impact of structural damage on urban accessibility following a disaster: the case of the seismically damaged Port Au Prince and Carrefour urban road networks. J Transp Geogr 19:1443–1455

Chang SE, Nojima N (2001) Measuring post-disaster transportation system performance: the 1995 Kobe earthquake in comparative perspective. Transport Res Part A 35:475–494

Hilker N, Badoux A, Hegg C (2009) The Swiss flood and landslide damage database 1972–2007. Nat Hazards Earth Syst Sci 9:913–925

Hosmer DW, Lemeshow S (2004) Applied logistic regression (2nd edn). Wiley series in probability and statistics. Wiley, New York, 383p. ISBN 978047135632

Klose M, Highland L, Damm B, Terhorst B (2014) Estimation of direct landslide costs in industrialized countries: challenges, concepts, and case study. In Landslide Science for a Safer Geoenvironment, pp 661–667

Klose M, Damm B, Terhorst B (2015) Landslide cost modeling for transportation infrastructures: a methodological approach. Landslides 12(2):321–334

Klose M, Maurischat P, Damm B (2016) Landslide impacts in Germany: a historical and socioeconomic perspective. Landslides. 13(1):183–199

Krejčí O, Baroň I, Bíl M, Hubatka F, Jurová Z, Kirchner K (2002) Slope movements in the flysch carpathians of Eastern Czech Rep. triggered by extreme rainfalls in 1997: a case study. Phys Chem Earth 27:1567–1576

MacKay DJC (2003) Information theory, inference, and learning algorithms. Cambridge University Press, Cambridge, 628p. ISBN 9780521642989

MacLeod A, Hofmeister RJ, Wang Y, Burns S (2005) Landslide indirect losses: methods and case studies from Oregon. State of Oregon, Department of geology and mineral industries

Pánek T, Brázdil R, Klimeš J, Smolková V, Hradecký J, Zahradníček P (2011) Rainfall-induced landslide event of May 2010 in the eastern part of the Czech Republic. Landslides 8(4):507–516

Postance B, Hillier J, Dixon N, Dijkstra T (2015) Quantification of Road network vulnerability and traffic impacts to regional landslide hazards. EGU Gen Assembly Conf Abstr 17:3677

Postance B, Hillier J, Dijkstra T, Dixon N (2016) Indirect economic impact of landslide hazards by disruption to national road transportation networks; Scotland, United Kingdom. EGU Gen Assembly Conf Abstr 18:4439

RMD (Road and Motorway Directorate of the Czech Republic) (2012) Implementation guidelines for the assessment of economic efficiency of projects of road and motorway constructions, Appendix C

RSZK (Zlín Region Road Directorate) (2010) URL: http://www.rszk.cz/aktual10/tz1020.htm. Last accessed: 25th Aug 2016

Taylor FE, Malamud BD, Santangelo M, Marchesini I, Guzzetti F (2015) Statistical patterns of triggered landslide events and their application to road networks. EGU Gen Assembly Conf Abstr 17:9992

Van Den Eeckhaut M, Hervás J (2012) Landslide inventories in Europe and policy recommendations for their interoperability and harmonization. A JRC contribution to the EU-FP7 SafeLand project. Report EUR 25666 EN. 203p

Winter MG, Harrison M, Macgregor F, Shackman L (2013) Landslide hazard and risk assessment on the Scottish road network. Proc Inst Civ Eng-Geotech Eng 166(6):522–539

Winter MG, Smith JT, Fotopoulou S, Pitilakis K, Mavrouli O, Corominas J, Argyroudis S (2014) An expert judgement approach to determining the physical vulnerability of roads to debris flow. Bull Eng Geol Environ 73(2):291–305

Winter MG, Shearer B, Palmer D, Peeling D, Harmer C, Share J (2016) The economic impact of landslides and floods on the road network. advances in transportation geotechnics 3. The 3rd international conference on transportation geotechnics, vol 143, pp 1425–1434

The Impact (Blight) on House Value Caused by Urban Landslides in England and Wales

William Disberry, Andy Gibson, Rob Inkpen, Malcolm Whitworth, Claire Dashwood, and Mike Winter

Abstract

We examine how large, slow moving landslides impact urban house prices in three areas of England and Wales. 12,663 house transaction values were analysed covering all house sales 1995–2012 in Lyme Regis, Dorset; Ventnor, Isle of Wight and Merthyr Tydfil, Glamorgan. Values were analysed with respect to local landslide events and visible landslide damage. In all three study areas, individual landslide events caused little or no negative impacts on nearby property prices, though remediation is likely to have short-term positive impacts on local house prices. Localised blight and suppressed house prices to a distance of 75 m was found in areas affected by ongoing incipient movement. By comparison with other sources of property blight, the radius of influence is 25% of that expected from an abandoned property or electricity pylon and less than 5% that of a windfarm. The socio-economic environment was important in determining the degree of house price impact of landslide events and for most locations, landslides form only a minor impact compared to other factors.

Keywords

Urban landslides • House price • Lyme Regis • Ventnor • Merthyr Tydfil

W. Disberry · A. Gibson (✉) · R. Inkpen · M. Whitworth · M. Winter
University of Portsmouth, Centre for Applied Geoscience (CAG), Portsmouth, P01 3QL, UK
e-mail: andy.gibson@port.ac.uk

R. Inkpen
e-mail: rob.inkpen@port.ac.uk

M. Whitworth
e-mail: malcolm.whitworth@port.ac.uk

M. Winter
e-mail: mwinter@trl.co.uk

C. Dashwood
British Geological Survey, Nottingham, NG2 3GG, UK
e-mail: cdashwood@bgs.ac.uk

M. Winter
Transport Research Laboratory (TRL), 108 Swanston Steading, 109 Swanston Road, Edinburgh, EH10 7DS, UK

M. Mikoš et al. (eds.), *Advancing Culture of Living with Landslides*,
DOI 10.1007/978-3-319-53483-1_5

Background

It is estimated that 350,000 houses in the UK are in locations of 'significant landslide susceptibility' (Foster et al. 2012; Gibson et al. 2013). However, 'urban landslides', defined by (Petley 2009) as '*a landslide in, or directly affecting, a coherent area with a population of 2500 or more*' are concentrated in a few locations in coastal and valley locations in England and Wales for example Dorset, the Isle of Wight and the South Wales Coalfield (Jones and Lee 1994).

The evidence base for the economic impacts of urban landslides is poor. However, house price transaction data, a wide and freely available source of data have been shown to be effective at understanding the spatio-temporal impact of flooding in the UK (Lamond et al. 2010). We examine what this data might demonstrate for the pattern of landslide impact and whether it can contribute towards an evidence-base for planning and insurance purposes.

House Price as Indicators of Blight

Negative impacts on house prices, termed blight, can for instance be caused by urban decay (Han 2014), or visual impact of an object/event (Gibbons 2015). Four types are common: energy infrastructure, transport infrastructure, industrial sites and poor housing stock nearby. *Energy infrastructure*: property prices tend to decrease where wind turbines are within view, though effects can be observed as early as submission for planning consent. Impact is controlled by distance, with the greatest effect within 1 to 2 km from an installation, (Dent and Sims 2007; Gibbons 2015). Similarly, (Sims and Dent 2005) found the visual presence of electricity pylons had a negative impact on house transaction value, though impact distances were within hundreds of metres. *Transport infrastructure*: studies of the planned high speed rail link between London with Birmingham indicate losses will be up to 40% within 120 m of the line, dropping 10–20% at 500 m (Wharf 2010), though these should reduce over time. The impact of nearby *industrial sites* is dependent upon the size of the installation and its proximity (de Vor and de Groot 2010; Ready 2010). *Poor housing stock*: blight caused by neighbouring abandoned properties is also observed to decrease with distance, (Shlay and Whitman 2006) to a maximum of 100 m, though the magnitude of impact was greater if the building remained abandoned for more than three years (Han 2014). This analysis suggests a further category of blight—environmental blight as illustrated by the perception of the risk of urban landslides.

Methodology

In order to examine whether landslides can demonstrate blight, three urban landslide case study areas were selected, Ventnor (Isle of Wight), Lyme Regis (Dorset), Merthyr Tydfil (South Wales) (Fig. 1). Each area had a population qualifying it as an urban landslide setting (Petley 2009), was substantially underlain by landslides and/or landslide

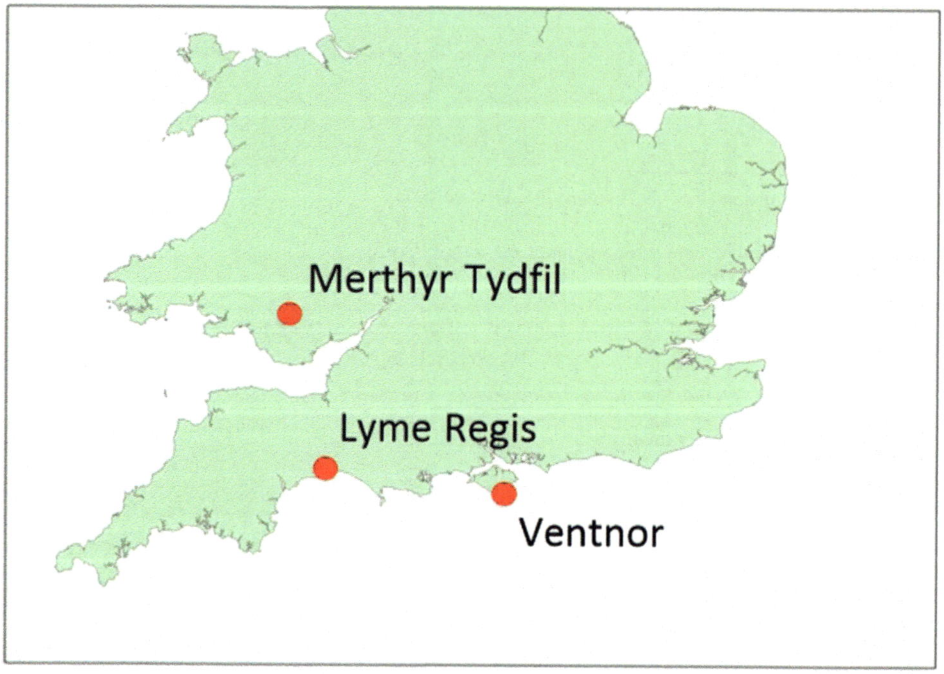

Fig. 1 Location of study sites

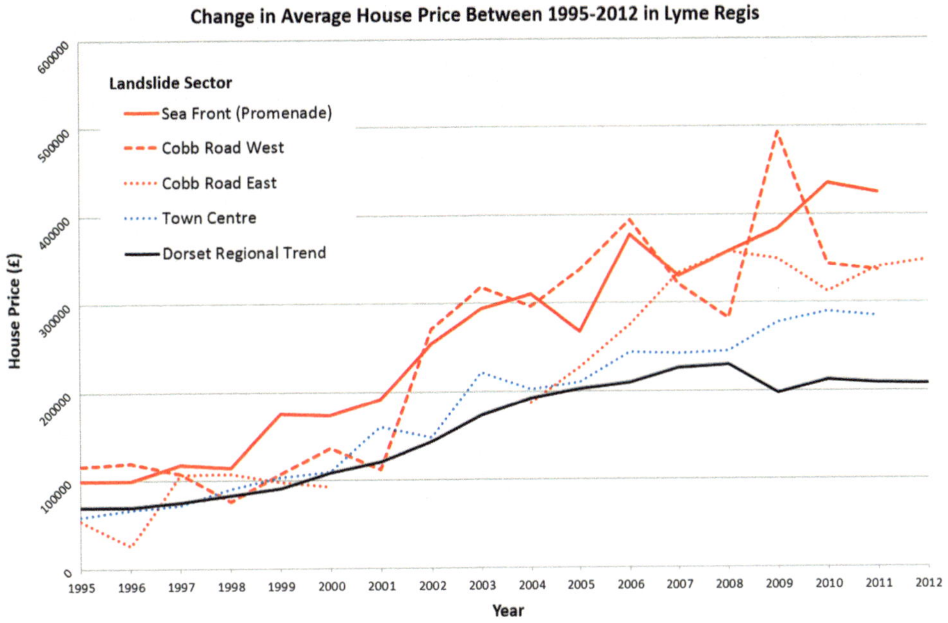

Fig. 2 Sample of transactional data presented in temporal form for sectors in Lyme Regis. By comparison to price changes in the Dorset Region and in the Town Centre, there is a clear increased trend in house prices for the three sectors adjacent to the Lyme Regis Coastal Improvements scheme which was completed in April 2007

complexes, possessed significant house price transaction data and ongoing landslide activity. Landslides in all three locations tend to be large, slow-moving rotational landslides.

The Ventnor Undercliff on the Isle of Wight has been described as the largest urban landslide complex in NW Europe. The complex of rotational landslides and their impacts are well understood, patterns of damage and the extensive programmes of remediation works have been well described (Moore et al. 2010). Landslide damage, mainly from incipient movement and maintenance is estimated to cost around £3 million annually and around 100 properties have been destroyed or demolished since 1900 (Bilbao et al. 2010). Lyme Regis is located in West Dorset on the south coast of England. The town is built upon a series of valley slopes mantled in ancient, degraded rotational landslides and flows which reactivate when affected by erosion or changes in slope profile or drainage (Brunsden and Chandler 1996). Landslide damage and remediation programmes have been reviewed extensively, (WDDC 2010 and references therein). The County Borough of Merthyr Tydfil is one of the largest urban areas in Wales and is an economically deprived area. The underlying geology and activity of the large rotational landslides and flows, (including the 1966 Aberfan disaster) that characterise the area have been reviewed in detail (Siddle et al. 2000).

Each area was sub-divided according land-use, property type and landslide type. Ventnor was divided into 8 sectors, Lyme Regis, 10 and Merthyr Tydfil 14. Transaction data between 1995 and 2012 (Land Registry 2015) was obtained from property websites www.rightmove.co.uk and www.zoopla.co.uk. For the study period, 1800 transactions values were obtained for Ventnor, 1660 in Lyme Regis and 9200 in Merthyr Tydfil. Data was collated using GIS at 1:1000 scale enabling attribution to accurate grid coordinates. Data was presented and analysed on simple timelines which classified house price change with respect to location (for example Fig. 2). Event timelines were constructed for each case study site including both macro-economic events and local events such as major landslides or engineering works. Good records of such events were found for Ventnor and Lyme Regis, a result of existing studies, however, insufficient records of events were found for Merthyr Tydfil.

A walk over survey was conducted at each study site in 2012. Every permanent property was surveyed visually from street level and exterior damage recorded using the (Cooper 2008) damage classification scheme. Light structures typically used only as holiday homes were excluded; though damage is known to have occurred in these areas, transaction data for these types of property do not fall under the remit of Land Registry records and was less reliable. An example of the coverage one of these damage surveys is shown in Fig. 3.

Temporal Impact from Landslide Events on House Value

Results show that most of the case study sectors follow regional average prices between the period of study including a period of growth during the early 2000s and a decline in values 2007–2009. These coincide with changes in the economy; the election of a new government and the onset of the 'credit crunch' and associated recession respectively.

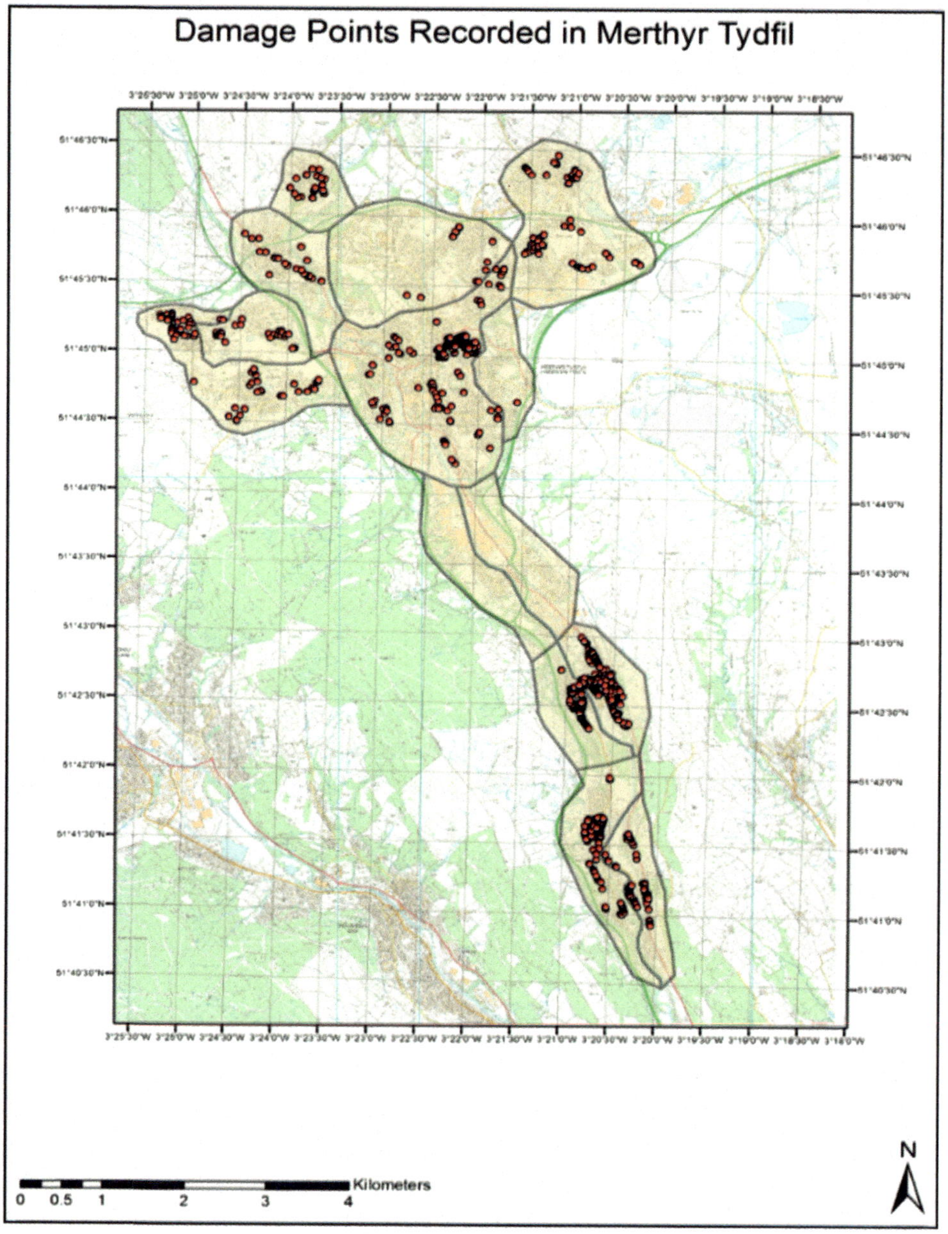

Fig. 3 Spatial variation of properties displaying evidence of damage in the town of Merthyr Tydfil as mapped in the Summer of 2012. The urbanised part of the town has been divided into different sectors according to housing type and landslide geomorphology which provided context for the recorded damage throughout the town

Few recorded discrete landslides occurred in the urban areas within the timeframe of the research. Although patterns can be determined in the datasets, it was difficult, to match discrete landslide events to discrete property transaction values. Landslides in part of Ventnor during the wet winter of 1994–95 was followed in 1995–96 by a 23% fall in house prices for this sector. House values recovered to 1995 levels in 1997. No other significant patterns were observed.

Sectors that have undergone landslide remediation show a localised positive impact on house prices. Coastal protection in Ventnor was completed in two areas in 1997. This coincided with an increase of 100% in transaction value 1997–99 in one area adjacent to the works (with a very small sample population) compared with a 30% increase 1995–97 in the same locations. A further coastal protection scheme, completed in 1999 coincides with a 47% increase in house prices in this sector 1999–2000. A third set of works completed in 2003 was followed by a 32% increase in prices in the corresponding area in 2003–04.

Lyme Regis underwent multiple phases of coastal improvement within the study period. Phase I was completed in 1995. This was followed by a 43% rise in house prices in adjacent sectors in 1995–99. Phase II of the works, in 2005, was followed by a 15% increase in house prices in adjacent areas in 2005–06 (compared to 12% for 2004–05). Prices within the improved sectors showed a 37% increase for 2005–08. This indicates that the works completed in 2007 had a significant positive impact. Emergency work was carried out in the eastern part of Lyme Regis to stabilise the slope in 2003. Despite few house transactions in this small sector, prices appear to rise as a result of the work by 45% between 2004 and 2005.

In contrast to Ventnor and Lyme Regis, Merthyr Tydfil was not subject to a large, integrated programme of landslide works during the study period, nor were any significant, discrete landslide events recorded. The most significant change in house value of any area was found in the Mount Hare District, where increased of 3 and 9% in 2005 and 2006 respectively correspond to the construction of several housing estates which make up a significant percentage of sales during this period. Landsliding in this area was not found to occur in discrete locations, though patterns of damage indicated a more dispersed nature of damage than was found in the other areas (Fig. 3).

Spatial Impact of Blight on House Value

In order to determine the presence of a spatial impact of blight, multiple ring buffer zones were created in Arcmap at equal intervals of 25 m away from each record of building damage, to capture house price transactions made at different distances away from a damaged property. These distances were chosen based on the previous blight research on the impact of abandoned or dilapidated properties by (Han 2014). House prices changes in each interval were averaged to determine spatial patterns of change with distance from damaged locations.

A clear spatial pattern of impact across all three case study locations was found. House prices increase as the distance from damage increases (Fig. 4). The increase is small, on average 11% between 25 and 75 m, followed by a greater average increase of 18% after 75 m. The index of change is almost identical across the case study areas, and

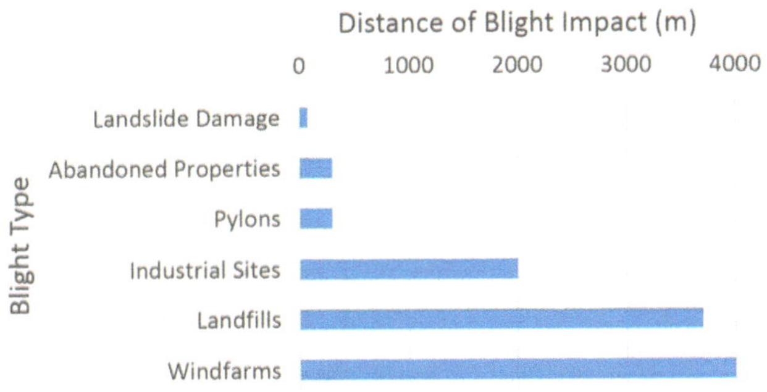

Fig. 4 Spatial impact of blight on urban house prices in comparison to other sources

that landslide damage has a localised suppression effect on house prices within 75 m of the damaged property. This is consistent with the findings of (Han 2014) who concluded that the majority of blight caused by abandoned properties was found within 75 m of said property.

Discussion and Conclusions

12,660 house transactions were analysed in three urban areas affected by slow moving landslides between 1995 and 2012. Results indicate that the impact of landslide damage is extremely localised, with a radius of influence less than 5% that of a windfarm.

Our analyses indicates that blight caused by incipient landslide damage may be considered similar to the blight caused by abandoned properties. In both cases, the source of blight likely stems from the visual condition of the properties themselves. This is complicated by socio-economic factors and broader quality of housing stock. These influence the acceptance of damage, and the speed at which repairs are completed.

Data also supports the assumption that large-scale remediation programmes to urban landslides have significant, though localised positive impacts on property value. However, whilst these increases in price are significant within the local sectors themselves, the works do not have an appreciable impact on house prices in adjusted urban zones which follow the regional trend of house price change.

This may reflect an initial perception of decrease risk of landslides in the light of specific remedial works. This perception may diminish over time as the novelty of the remedial works declines and it becomes part of the perceived general remedial activity of the area. In this case, as reflected in the data, house prices will move back towards the regional average over time. Certainly in Lyme Regis, the remedial works were integrated with a programme of general improvement to landscape and amenity which was intended to enhance the local environment.

The areas used in the analysis are known to be areas prone to landslides. It might be expected that house prices might already reflect this known hazard. The occurrence of a landslide or the visible presence of damage may merely confirm the existing knowledge of the area as a potentially hazardous one, hence the relatively small percentage impact on the house prices of an event. It could be argued that the mirroring of house prices in these landslide prone areas with regional house prices for much of the period reflects the relatively low perception and impact of the known hazard on house prices.

References

Bilbao AF, McInnes R, Mitchell L (2010) Understanding demolition costs for properties affected by coastal erosion. Scott Wilson, London

Brunsden D, Chandler JH (1996) Development of an episodic landform change model based upon the Black Ven mudslide, 1946–1995. Adv Hillslope Processes 2:869–896

Cooper AH (2008) The classification, recording, databasing and use of information about building damage caused by subsidence and landslides. Q J Eng Geol Hydrogeol 41:409–424

de Vor F, de Groot HLF (2010) The impact of industrial sites on residential property values: a hedonic pricing analysis from The Netherlands. Reg Stud 45:609–623

Dent P, Sims S (2007) What is the impact of wind farms on house prices? RICS, London

Foster C, Pennington CVL, Culshaw MG, Lawrie K (2012) The national landslide database of Great Britain: development, evolution and applications. Environ Earth Sci 66:941–953

Gibbons S (2015) Gone with the wind: valuing the visual impacts of wind turbines through house prices. J Environ Econ Management 72:177–196

Gibson AD, Culshaw MG, Dashwood C, Pennington CVL (2013) Landslide management in the UK—the problem of managing hazards in a 'low-risk' environment. Landslides 10:599–610

Han HS (2014) The impact of abandoned properties on nearby property values. Housing Policy Debate 24:311–334

Jones DKC, Lee EM (1994) Landsliding in Great Britain. HMSO, London

Lamond J, Proverbs D, Hammond F (2010) The impact of flooding on the price of residential property: a transactional analysis of the UK market. Housing Stud 25:335–356

Land Registry (2015) Land Registry House Price Index

Moore R, Carey JM, McInnes RG (2010) Landslide behaviour and climate change: predictable consequences for the Ventnor Undercliff, Isle of Wight. Q J Eng Geol Hydrogeol 43:447–460

Petley DN (2009) On the impact of urban landslides. Geol Soc London Eng Geol Spec Publ 22:83–99

Ready R (2010) Do landfills always depress nearby property values? J Real Estate Res 32:321–339

Shlay AB, Whitman G (2006) Research for democracy: linking community organizing and research to leverage blight policy. City Commun 5:153–171

Siddle HJ, Bromhead EN, Bassett MG (eds) (2000) Landslides and landslide management in South Wales. National Museum of Wales, pp 9–14

Sims S, Dent P (2005) High-voltage overhead power lines and property values: a residential study in the UK. Urban Stud 42:665–694

WDDC (2010) LREI Phase IV Project Appraisal Report. West Dorset District Council

Wharf H (2010) Property blight from HS2: pilot study, pp 24

Landslide Monitoring and Counteraction Technologies in Polish Lignite Opencast Mines

Zbigniew Bednarczyk

Abstract

The paper presents slope instabilities together with monitoring and counteraction techniques in lignite opencast mines. It includes examples of landslides in two of the largest open-pit mines and external spoil dumps in Poland. The Belchatow mine, one of the largest excavations in Europe, is located in the central part of Poland and has lignite resources of 2 bln tons and an annual production of 42 mln tons. Landslides are registered there every year. The largest with a volume of a few thousand to 3.5 mln m^3 with displacements of 2 mm-2 m per day. In the past, similar threats have occurred at the Turow mine, the second largest in Poland. It is located in the Lower Silesia District, close to the German and Czech borders. Its estimated lignite reserves are equal to 760 mln tons, with an annual production of 27.7 mln tons. In previous years, the author of this paper had the opportunity to participate in parts of landslide investigations at these mines. The research included CPTU in situ tests, laboratory tests, displacement monitoring and numerical modelling. It was difficult to come up with an interpretation of soil strength parameters. Dump soils varied in strength due to their anthropogenetic nature. The interpretation of the clayey soils parameters in the pit was complicated because of high preconsolidation and partial saturation. The new Euracoal Slopes Project conducted by an international consortium of six European countries aims for the practical implementation of new geotechnical monitoring methods. Complementary methods should allow for better prediction of landslide activity. PSI interferometry, UAV and ground-based laser scanning, in situ monitoring, shallow geophysics and laboratory triaxial and centrifuge testing should deliver new data for slope stability analysis. Although the mines have advanced monitoring systems and remediation procedures, the full elimination of hazards in mines of this size and depth is not possible.

Keywords

Mining-induced landslides • Landslide investigations • Monitoring • Counteraction works

Introduction

Lignite opencast mining has made a significant contribution to the production of electricity in a number of European countries. Germany, Greece, Poland, the Czech Republic, Bulgaria and Romania produce approximately 96% of the lignite in the European Union, a total of 433.8 million tons (Kasztelewicz 2012; Bednarczyk and Nowak 2010). However, coal mining is often associated with a number of threats related to the size and depth of the pits. This necessitates the exploitation of increasingly deeper lignite layers and the storage of large masses of overburden. Serious threats can arise from a number of geotechnical factors. The cause of these threats are complex geological structures, the

Z. Bednarczyk (✉)
"Poltegor-Institute" Institute of Opencast Mining, Parkowa 25, 51-616 Wroclaw, Poland
e-mail: zbigniew.bednarczyk@igo.wroc.pl

M. Mikoš et al. (eds.), *Advancing Culture of Living with Landslides*,
DOI 10.1007/978-3-319-53483-1_6

parameters of low-strength clayey soils, the specific geotechnical conditions of spoil dumps, precipitations, changes in groundwater levels, the use of explosives and the occurrence of karst and suffusion processes. The prevention of unfavourable phenomena is all the more important when the scale of potential failures can lead to large losses, endanger lives and adversely affect the environment. It is difficult to determine the soil strength parameters of highly overconsolidated and partly saturated soils on mine slopes. Dump soils vary in strength due to their anthropogenic nature and structure. The paper presents some examples of slope instabilities in Polish opencast mines, as well as methods of investigations and counteraction. New planned research activities inside the Euracoal Slopes research project are also described.

Localization and General Mine Characterization

The Belchatow opencast mine is located in the central part of Poland, 40 km south from the city of Lodz. It is 12.5 km long, 3 km wide and 310 m deep. The operated lignite resources are divided into two fields, Belchatow and Szczercow (Fig. 1). The PGE Company operates the mine and the nearby located power plant with a capacity of 4320 MW. The lignite exploitation requires the removal of 100–120 mln m^3 of overburden.

The second largest mine in Poland, the Turow mine, is located in south-eastern Poland in the Lower Silesia District, close to the border with Germany and the Czech Republic in area known as the "black triangle" due to its past production of heavy industrial pollution (Fig. 2). The mine is situated close to the city of Bogatynia, 55 km west of Jelenia Gora, 80 km east of Dresden, Germany and 20 km northwest of Liberec, the Czech Republic. The mine has estimated reserves of 760 mln t with an annual coal production of 27.7 mln t.

Engineering Geology Conditions

These mines are characterized by complex conditions of engineering geology. The Belchatow mine is located in the Kleszczow Tectonic Rift. This relatively young, deep structure from the Neogene and Quaternary ages is characterized by the occurrence of not entirely relaxed tectonic stresses and is built of weathered limestone and marls from the Jurassic and Calcareous ages which are involved with karst processes. The western border of the mine is located closely to the Debina salt intrusion, which, together with the faults, influences the stress state conditions. In the Turow mine, Neogene-age lignite deposits are located on both sides and under the Nysa Luzycka River on the border with Germany. These have caused geotechnical problems in the past due to the mine's proximity to the river. Lignite deposits have a complex structure. Bedrock layers built of weathered crystalline rocks from the Precambrian and Neogene/Quaternary ages create basalt intrusions in the northern part on the pit.

Examples of Reported Landslide Hazards

The mining operations in these mines are constantly accompanied by a number of mass movement problems. In the Belchatow mine, the high depth of mining operations at 200–300 m and low soil parameters have had an impact on the occurrence of landslides. Their volumes have ranged from a few thousand to 3.5 mln m^3. Most of the landslides have been activated on structural surfaces and formed on the south slope of the mine near a deep secondary rift structure with a higher thickness of lignite deposit. These slopes were

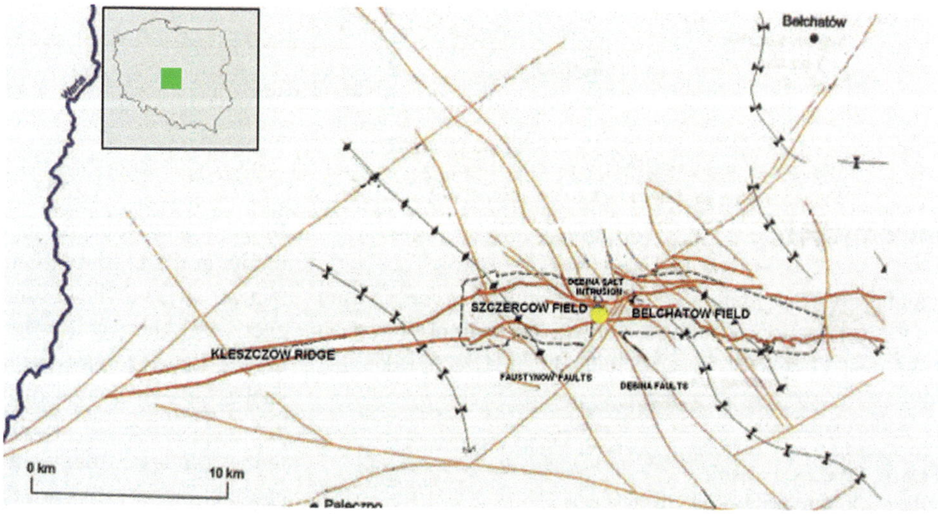

Fig. 1 Localization of Belchatow mine exploitation fields and tectonic structure

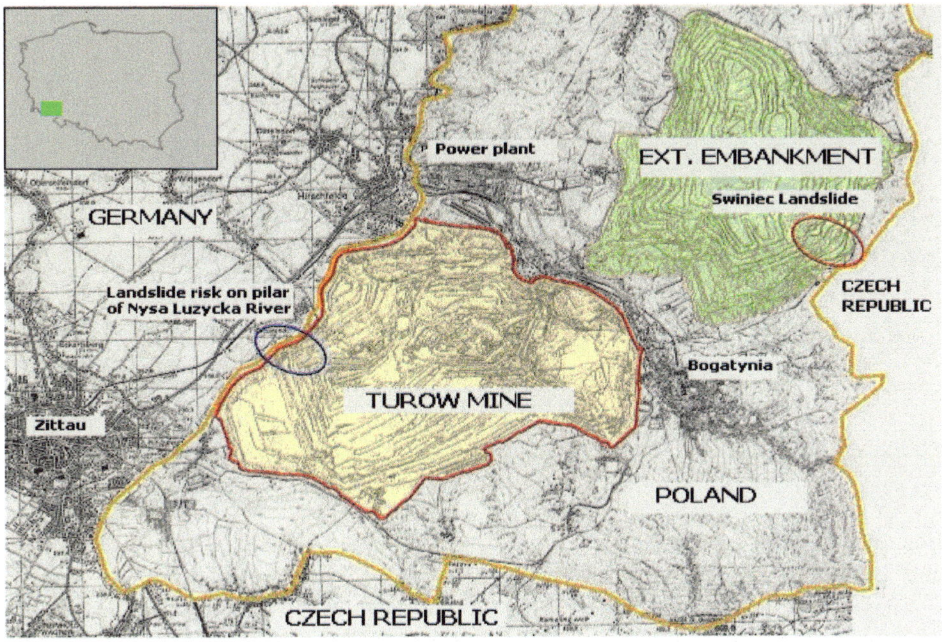

Fig. 2 Localization of the Turow mine with indications of landslide hazard zones

built of Neogene-age paleoalluvial fan deposits. Other types of landslides have been connected with Quaternary low-strength varved clays located on the north slope of the mine. In the Turow mine, a risk occurred in the 1990s on the western slopes close to the border with Germany and on the external spoil dump of lignite overburden close to the border with the Czech Republic (Fig. 2).

Landslides at the Turow Mine

The most dangerous landslide at the Turow mine occurred over 26 years ago on the pillar of Nysa Luzycka River close to the border with Germany. It was characterized by the highest thickness of coal deposits (Figs. 3 and 4). The design of the mine included a protective pillar on the western slope located between the mine and the river. In the nineties, this zone was 160–240 m wide, 100 m deep and had a general slope inclination of 19°. The mining operations in 1988 were conducted below the first coal deposit at the deepest part of the mine at a depth of 100 m. The appearance of the first cracks on the mine slopes in the protection pillar zone and the uplift of the bottom level of the mine in its western zone was detected in 1989. Cracks occurred on the Trzciniec-Sieniawka public road, located on the east side of the river. The intrusion of water

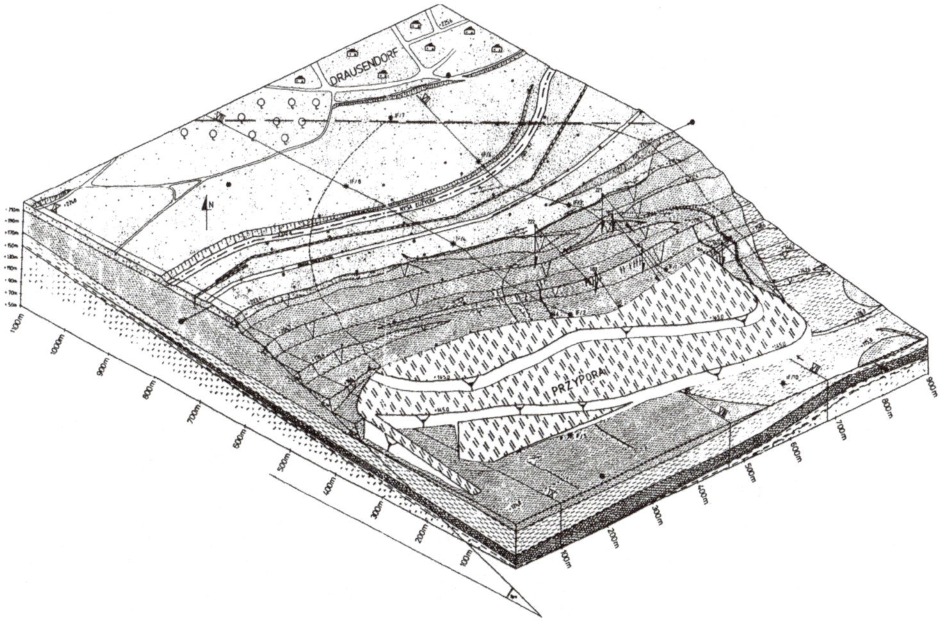

Fig. 3 3D model of pillar risk zone (Milkowski 2008)

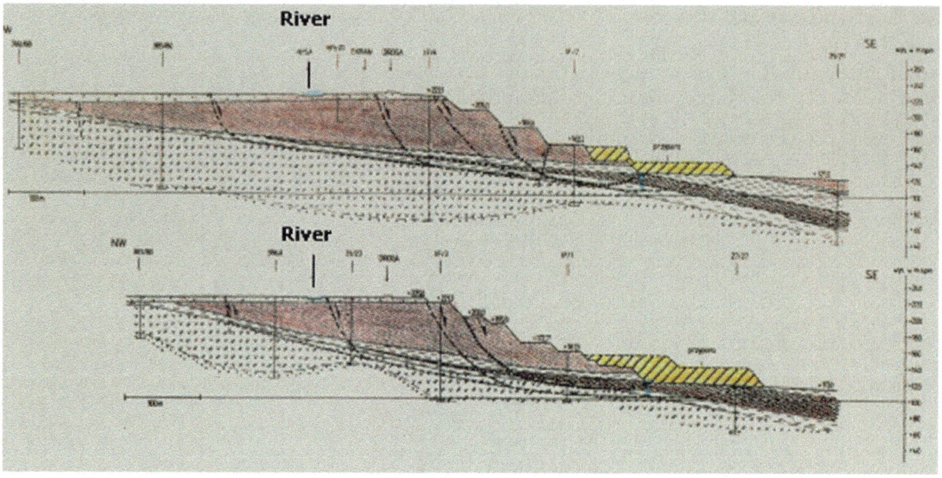

Fig. 4 Cross-section perpendicular to the pillar of Nysa Luzycka River (Milkowski 2008)

from the river to the pit could have resulted in flooding the mine excavators, conveyor belts and drainage systems.

These consequences could have led to the lack of resources in the third largest power station in Poland. In the wrong scenarios, the maximum range of the landslide area could have reached 480 ha, with an estimated volume of 12 million m^3 (Milkowski 2009). The direct causes of the risk of mass movements to the river pillar included: (1) mining operations at levels +124/140 m a.s.l., resulting in undercutting the bottom part of the second brown coal deposit; (2) a consistent slope inclination of layers; (3) the occurrence of clays and weathered rocks with low-strength parameters; (4) groundwater infiltration and seepage processes; (5) slope geometry, the long life of the slopes and the relaxation of stresses. The remediation works included measurements of the deformation of the soil masses at the pillar through measurements of: (1) surface displacement; (2) pairs of points on cracks; (3) points at the bottom of the mine; and (4) inclinometers. The monitoring network included 59 monitoring points at the slopes, the bottom of the mine and the terrain at the Polish and German side of the pillar. The two identified slip surfaces were localized at the contacts of the floor of the first and second coal deposits with clayey layers (Fig. 4). The identified strength parameters of these surfaces were c = 21 kPa, $\varphi = 8°$ (Milkowski 2009). The third slip surface was located beneath the floor of weathering crystalline rocks c ~ 0=kPa $\varphi = 8°$. Generally, the entire slope in all the variants of numerical modelling did not have sufficient stability. The horizontal displacements ranged from 100 mm a month in November 1989 to 20 mm a month in March 1990 and 9 mm a month in May 1990. The vertical displacements of 70 mm in November 1989 were limited to zero in February 1990. Inclinometer measurements detected displacements of 40–50 mm a month at the depths of 53–70 m. The rescue plan included the stabilization of the pillar zone by supporting earth buttress masses formed on a special drainage system at the length of 560 m (Dmitruk 1984, 1995). Slope stability calculations performed using the Janbu, Bishop and Fellenius methods, which took into account the support of buttresses, showed values of Fs = 1.2–1.3, guaranteeing slope stability. Buttress masses were formed in two steps. The initial step included 144,000 m^3, while the second and final one formed by the end of June 1990 included 3.5 mln m^3. After the stabilization of the risk zone, the mine implemented a special system of geohazard monitoring that included a control system of slope deformations and continuous geotechnical and hydrogeological monitoring. The angle of slope inclination was reduced to 10°–12°. The stability of the pillar zone allowed for the exploitation of lignite deposits 100 m deeper, up to 200 m below the natural terrain level in 2010.

Another example of a very dangerous landslide occurred on the external spoil dumps of this mine. In the past, the dump volume of 1.7 bln m^3 and height of 245 m caused numerous slope stability problems. The height of the dump slopes, low effectiveness of the drainage system and low strength of the clayey overburden were the main triggers. These clayey deposits were transported by a mine conveyor belt system at distances of over 15 km, which could have resulted in their softening and partial liquefaction. A landslide on the external embankments of the Turow mine overburden occurred in December 1994. The eastern part of these embankments was localized approximately 150–300 m from the border with the Czech Republic. The main cause of the geohazard risk was the storage of large masses of overburden in difficult geotechnical conditions. These were caused by inefficient drainage of bedrock layers built from low-strength parameters of saturated clayey loess loams in the early 1970s during the construction of the dump. At the end of 1992, it was decided to increase levels +348 m a.s.l and +370 m a.s.l. in the south-eastern region of the dump. These works were conducted from the east to the west from level +415 m a.s.l. In the autumn of 1993, the first limited-in-size mass movements occurred. In December

1994, a catastrophic landslide with a volume of 6 mln m^3 occurred in the region of Swiniec. The landslide was over 1300 m long, 750 m wide and occurred over an area of 68 ha (Figs. 5 and 6).

The general slope inclination was 6°. The subsidence of 0.5 m was observed at 7 December at level +415 m a.s.l. In the following days, displacements increased to up to 25 m a day between 10 and 12 December 1994, and the landslide's tongue was located 70 m from the border with the Czech Republic (Milkowski 2009). Remediation began directly after the first signs of movement. It included safeguarding the dumping machine, rescue drainage and stabilization works. However, due to the size of the moving masses, the possibilities of counteraction were limited. Remediation works were conducted by the mine continuously 24 h a day until 30 December 1994. These included the deforestation of 1 hectare of land where the landslide tongue had reached the forest. The retaining wall length of 343 m, which was built from Larsen steel elements, was hammered into the ground to a depth of 5.5–14 m (Dmitruk 1995). Special perforations were drilled to allow for the drainage of groundwater from colluviums layers at heights of 4–5 m above the ground. Three vertical boreholes pumped 24 h a day, pumping the drainage from the colluvium/bedrock contact layers. Surface displacement measurements were performed daily. The monitoring network was located on the landslide surface and in the surrounding areas. The obtained data were used for analysis on a digital map of displacements, plotted daily. The measured displacements were extremely high and varied from 5 to 22 m a day by 13 December. The head landslide part was moving towards the border with the Czech Republic. The newly built protective retaining wall was liflted due to the forces that occurred on 22 December. To strengthen the wall, ten special supporting constructions were built from steel and concrete. At the same time, other parts of the retaining wall were supported and repaired. These works were successful, and no displacements were recorded on 2 January 2014. After this time, additional remediation works included surface reclamation and drainage works. A new monitoring system that included inclinometers, pie-zometers and pore pressure transducers was built on the dump embankments. These measurements, piezocone tests and laser scanning measurements were performed for risk prediction. The external spoil dump was closed in 2006 and since that time, the overburden has been stored in the mine. This experience of performing counteraction works has shown that geohazards can seriously threaten opencast mining operations. The main triggers were a lack of effective drainage for bedrock layers and the storage of overburden in the risk area. Some other triggers were connected with the low-strength parameters of clayey overburden transported by the conveyor belt system. The high height of the dump operating levels, in some cases up to 70 m, was also important. The rescue actions reduced the scale of damages, but were very expensive. A competitive analysis of potential risks together with monitoring and numerical analysis could lower the risks.

Fig. 5 Swiniec landslide view from level +415, Dec. 1994

Fig. 6 Swiniec landslide below level +415, Dec. 1994

Examples of Landslides in the Belchatow Mine

The mining operations in this mine were constantly accompanied by landslide hazards. The Belchatow mine, with a depth of 310 m, is located in an approximately 30-km long, 2-km wide tectonic rift in a south-east direction cut by north-east to south-west faults. The rift structure between the Belchatow and the Szczercow fields is divided by the Permian-age salt dome intrusion of Debina. The coal exploitation from the east to the west is causing permanent risk for the north and south slopes tilted 1:4, and, consequently, to the slope inclinations. In the past 16 years, many landslides have been activated on structural surfaces on the south slope of the mine, which is built of Neogene-age paleolandslide deposits over upper calciferous marl detritus, north of the tectonic border of the Kleszczow Rift (Fig. 7). The landslide volumes varied from a few thousand to 3.5 mln m^3

(Kurpiewska et al. 2013). The other types of landslides were connected with Quaternary low-strength varved clays located on the north slope of the mine. Karst processes in Jurassic-age limestone have also posed a risk for the stability of the mining excavators. Landslides have posed a risk for excavators, conveyor belts and drainage power supply lines.

Another type of risk is connected with the Debina Salt Diapir. It influences the state of stress in this region (uplift), hydrogeological conditions and other related geohazard risks. All of these factors influence the engineering geology conditions and cause continuous risks for mining operations. Landslides have sometimes covered the brown coal deposits, requiring remedial works and posing problems for exploitation (Figs. 7 and 8). The author of this paper had the opportunity to investigate landslide 20S, located on the south slope of the mine. It was formed in clayey layers with the slip surface below the main coal deposits, dipping in a north-eastern direction. At that time, there was a risk that this landslide would be activated westward in the coming years.

Fig. 7 Bechatow opencast mine with landslide 20 S located on its south slope (OPGK Warsaw)

Fig. 8 Landslide 22 S, view from the terrain level, 2005

To recognize the soil conditions, 13 CPTU tests were situated west of landslide 20S (Fig. 9 and 10).

Piezocone tests were calibrated using soil samples close to the test locations. Laboratory test programs included index, oedometer CIU and CID triaxial tests (Bednarczyk and Sandven 2004). Soil in this part of the mine could be described as low plasticity (CL), clays, silty clays and clayey sands. The clay content for clays ranges from 35 to 40% (<0.002 mm). In undrained tests, high back-pressure (70–360 kPa) was used for saturating the specimens. An obtained value of friction angle and cohesion in undrained tests for clays was: $\varphi = 21.3°$, c = 8.56 kPa (in drained tests, effective values of friction angle varied from 8.1° for clays to 25.1° for sandy clays). Clay specimens taken from depths of 50 m were overconsolidated and partly saturated due to the mine water pumping system. However, testing of partly saturated specimens requires a special type of triaxial apparatus with the possibility of air pressure measurements inside the sample during the shearing stage, which was not possible during the tests. The values of the friction angle received from the CPTU tests were therefore lower compared to the laboratory test results (Fig. 11).

Interpretations of CPTU tests showed that a reasonably good comparison between field and laboratory test data could be obtained. This allowed for soil type predictions from CPTU in addition to the mechanical parameters of the soil, such as shear strength, compression moduli, preconsolidation stress, effective cohesion and friction angle. To prevent landslide risk, several monitoring methods were implemented. These included inclinometers to depths of up to 100 m (Fig. 12), piezometers, pressure cells, a geodetic monitoring network of surface displacements, CPTU tests, laboratory tests and numerical modelling (Fig. 10). The inclinometer network at the Belchatow field included 22 inclinometers, four on the north slope, twelve on the south slope and six on the western slope. Two others were located on the west slope of the Szczercow field (Jonczyk and Organisciak 2010). Ground movement inclinometer measurements had been performed there since 1999. However, few of these deep inclinometers detected the first signs of movements, some installations were damaged by large landslide movements at depths of a few to approximately 10 m (Fig. 12).

In high movement conditions, measurements of displacement were possible only through the use of standard geodesy and interferometry methods. Surface displacement monitoring networks were located on the south, north and west slopes of the mine. Additional monitoring points were located near important infrastructure, such as conveyor belt ramps, pumping stations etc. Field and laboratory tests delivered data for slope stability analysis. In this mine, remediation works included unloading portions of the slopes in unstable conditions by gathering masses off using the

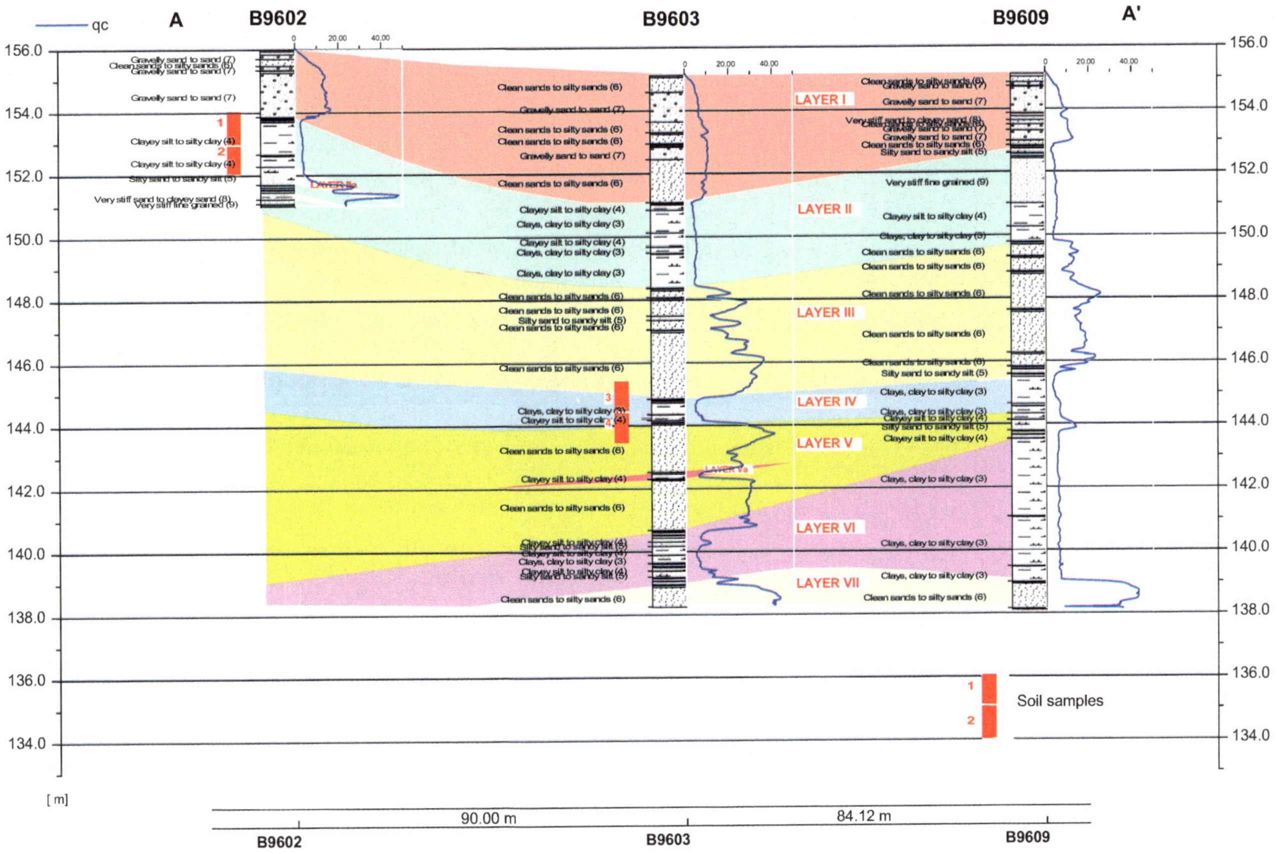

Fig. 9 Cross-section of landslide 20 S using CPTU tests

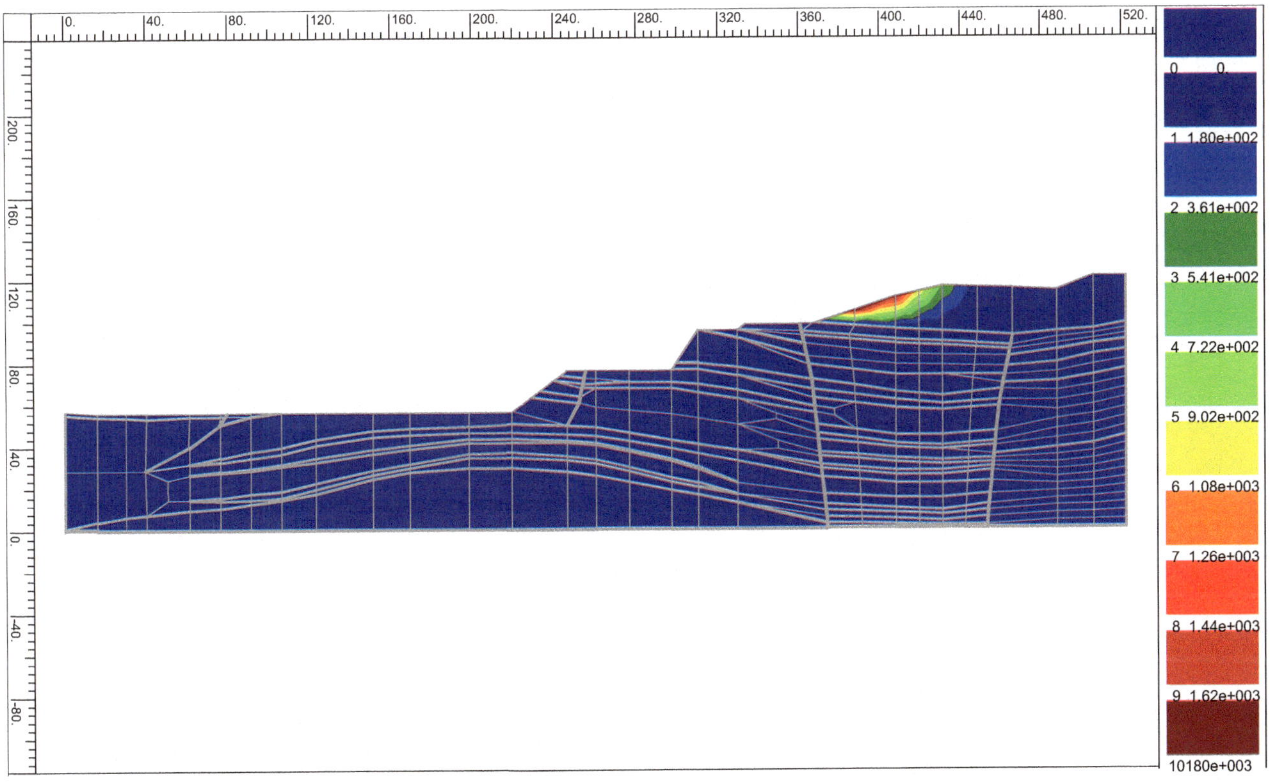

Fig. 10 Results of FEM slope stability analysis, Fs = 1.2

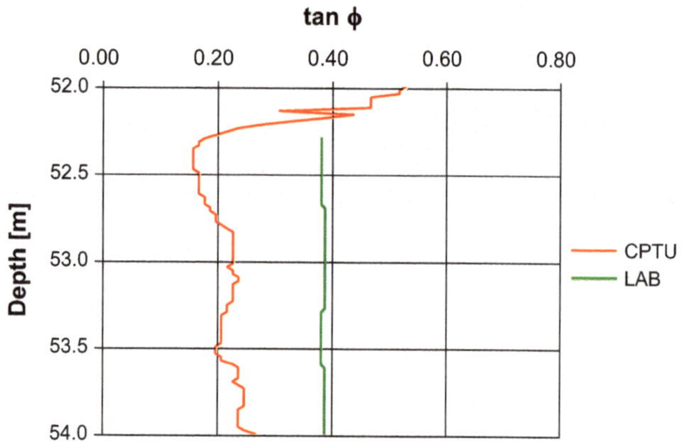

Fig. 11 Clay strength parameters using CPTU and lab results

mining excavator, resulting in improved conditions of stability. These works were followed by a detailed analysis of the geological structures in regions of potential threats. Landslide remediation works protecting stability were realized for several areas in the years 2006–2011. The remediation of the slopes at risk were conducted in the western part of the Belchatow field on the permanent south, north and western slopes. Works at the south slope included remediation of the area west of landslide 24 S. Relieving works included moving the top part of the southern slope by about 60 m to the south at a length of 1400 m by gathering and removing 4 million m^3 of overburden. At the western permanent slope, remediation works included the selective extrusion of brown coal in order to destroy the slip surface formed in the high plasticity inner coal clayey layers. Another form of remediation was connected through the gradual support of this slope by the internal spoil dump. At the north-western slope, mitigation works included unloading the portion of the western slope using mining excavators by gathering about 2 mln m^3 of overburden (Jonczyk and Organisciak 2010). These works required the implementation of complex preparatory works, construction of new conveyor lines, reconstruction of power supply lines, reconstruction of pipeline and surface drainage systems and acquisition of additional land for earth works at the terrain level. In recent years, the following regulations have been defined for safeguarding mining infrastructure and conducting counteraction works following notifications of displacement: 1) initiation velocity of 8–14 mm/day; 2) warning velocity of up to 20 mm/day; 3) critical velocity of 30 mm/day (Czarnecki and Jurczyk 2013).

Euracoal Project

The Euracoal project, Smarter Lignite Open Pit Engineering Solutions (SLOPES, 2015–2018), is being conducted by an international consortium of six European countries: the United Kingdom, Poland, the Czech Republic, France, Spain and Greece. SLOPES aims for the practical implementation of new monitoring methods of natural hazards in opencast mines in Poland (Bełchatow), the Czech Republic (the Most region) and Spain (the Aragon region). The project is coordinated by the University of Nottingham. The other members of the consortium are the University of Exeter (UK), SUBTERRA and GEOCONTROL (Spain), VUHU (the Czech Republic), CERTH (Greece), INERIS (France)

Fig. 12 Inclinometer measure, S-slope Belchatow mine

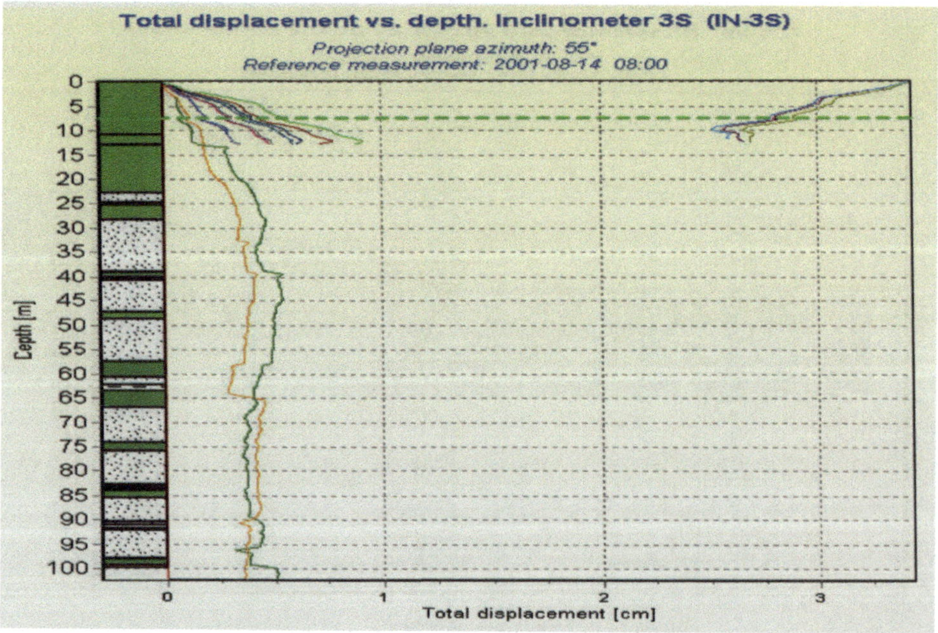

and Poltegor-Institute (Poland). The project is implementing modern technologies for the monitoring and analysis of natural hazards in the opencast mining of lignite in order to effectively predict and counteract potential failures. The project foresees the realization of 4WP-s. A careful analysis of the local natural hazard triggers will be performed on the basis of monitoring, laboratory tests and numerical modelling. The project is aiming for the implementation of a variety of monitoring systems for ground movement data collection and processing (WP-1). The main objectives of the project are geotechnical problems in many fields related to lignite mining. The results should contribute to a better understanding of hazards and risk reduction and help with counteraction (WP-2). The project should provide new information for a more secure design of the mine slopes, transport routes and spoil dumps (WP-3). In each opencast mine, there are three types of slopes: (1) constant slopes, characterized by defined boundaries for mining operations designed to obtain a license, remaining unchanged for years at a time; (2) temporary slopes, which are created in the mining process, defining the current extraction level; and (3) the spoil dump external or internal slopes. As was described in the first part of this paper, these slopes can pose a number of stability issues and threats for mining operations and in adjacent areas. In order to improve the reliability of geotechnical methods dedicated to each type of slope, the appropriate test and monitoring methods must be selected. In order to specify types of risk, the mine should have a detailed knowledge of possible triggers. This is also the important key to determining the instruments and parameters that should be subjected to monitoring. Additionally, during mining operations, observation methods should be used to verify the size of the registered movements and surface changes in comparison to those previously predicted at the design stage. Good quality core drilling, laboratory tests combined with geophysics, and numerical modelling should provide data similar to the actual conditions in the mine. Surface displacement in large areas can be identified using modern surveying methods, e.g. ground laser scanning, UAV Lidar, ground-based SAR and PSI. The SLOPES project plans to use unmanned drones to obtain Lidar data. This will be a new application of this technology in the sector of mining and RFCS projects. So far, only a few examples of the use of drones in this area can be found in the literature (Eck et al. 2011). The use of in situ online monitoring and other advanced monitoring systems together with numerical modelling should be favourable for hazard prediction. These methods will be tested for the selected mines, including the Belchatow mine. At the western slope of this mine, a continuous inclinometer system with 3D sensors every 0.5 m will be installed in the borehole to a depth of 100 m (Fig. 14). Systems installed in situ allow for more comprehensive exploration at the depth range and speed of ground movements (Bednarczyk 2012, 2013). There are, however, limitations to the maximum size of the measured displacements. Laboratory test programs will include index, IL oedometer, direct shear and triaxial tests. Recognition of landslide risk will be performed using in situ monitoring, 50 PSI satellite radar interferometry photos in high definition, geophysical surveys and numerical modelling (Figs. 13, 14 and 15). The project will also include two sessions of laser scanning and UAV Lidar performed in 2016 and 2017. These measurements are currently being performed in the Belchatow mine. The resulting measurements will provide three-dimensional images and allow for the calculation of volume using special software to process point clouds.

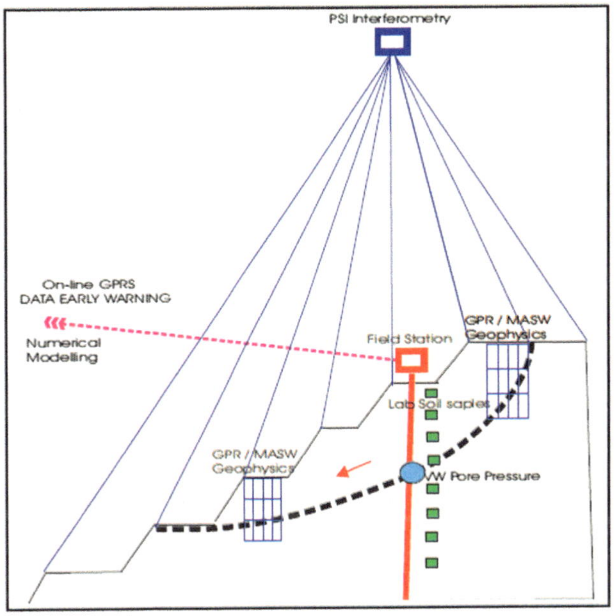

Fig. 13 Scheme of the research, Belchatow mine

Fig. 14 Automatic monitoring station (Bednarczyk 2014)

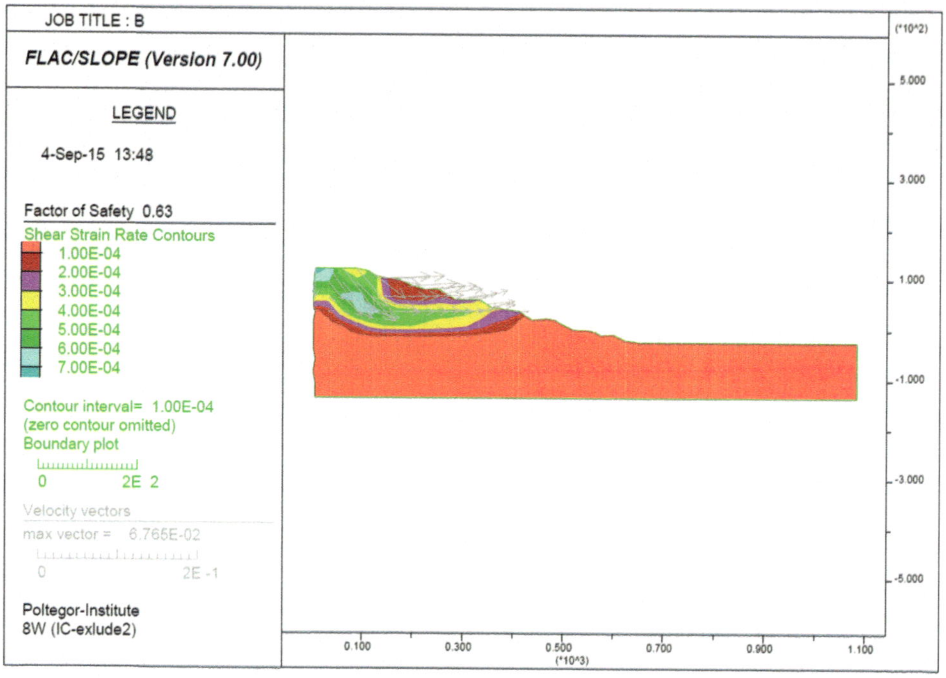

Fig. 15 Slope stability analysis, SSR method

Precipitations, fluctuations of water tables, variables of pore pressures and the partial saturation of soils caused by mine drainage systems can have a big impact on the conditions of stability. New methods of laboratory triaxial and centrifuge tests should allow for the recognition of strength parameters of the soils in conditions similar to those in the ground and on slopes that have been subjected to high overburden pressure and partial saturation. The data obtained from the monitoring will be compared with the results from numerical modelling. This will provide information for risk assessment and help develop methods for early warnings. Improved monitoring systems, data analysis and numerical modelling should allow for better management of the risks associated with the instability of open pit mine slopes and spoil dumps in order to predict the risks for exploitation and environmental risks.

Summary and Conclusions

This paper has presented examples of landslides induced by mining activity in Poland. Large opencast mines and spoil dumps are causing numerous slope stability problems. The scale of these problems is usually high enough to require the usage of different types of investigation and monitoring methods. The best practices could be obtained by monitoring, in situ tests carefully scaled by laboratory tests and numerical modelling. However, not all laboratory methods are relevant for highly overconsolidated and partly saturated specimens or spoil dump soils, and thus we should be cautious when reviewing this data. An investigation of these questions is planned through the Euracoal SLOPES project. Some methods of standard remediation and stabilization appropriate for landslides on natural slopes in opencast mines could be useless. New methods of monitoring and recognition of strength parameters for the design of safe slopes should deliver new data. However, other factors such as safe slope design and proper drainage for safe storage of overburden are also very important. Landslide remediation works in opencast mines could be made by unloading some parts of the slopes or supporting them using overburden, but these methods are usually expensive. Full elimination of geohazards is not possible. Special attention should be paid to comprehensive site investigation, monitoring and numerical modelling for safe design of exploitation.

Acknowledgements The author would like to acknowledge the European Commission Research Fund for Coal and Steel for financing the project SLOPES—RFCR-CT-2015-00,001 Smarter Lignite Open Pit Engineering Solutions. Other acknowledgments are dedicated to the Polish Energy Group (PGE) the owner of the Belchatow and Turow Mines for help in the presented research.

References

Bednarczyk Z (2012) Landslide investigation and monitoring methods. Edition of Gornictwo Odkrywkowe Wroclaw. Monograph ISBN 978-83-60905-36-4, p 213

Bednarczyk Z (2013) New real-time landslide monitoring system in Polish Carpathians. Landslide Science and Practice, vol 2. Springer, Germany. ISBN 978-3-642-31444-5, pp 3–15

Bednarczyk J, Nowak A (2010) Strategie i scenariusze perspektywicznego rozwoju produkcji energii elektrycznej z węgla brunatnego w swietle wystepujących uwarunkowan. Gornictwo i Geoinzynieria Rok 34 Zeszyt 4: 67–83 (in Polish)

Bednarczyk Z, Sandven R (2004) Comparison of CPTU and laboratory tests interpretation for Polish and Norwegian clays. International Site Char. Conference ISC-2 ISSMGE TC-16. Porto, Portugal. Millpress, pp 1791–1799

Czarnecki L, Jurczyk M (2013) Exploitation in II-order gra-ben of Belchatow brown coal mine—selection of mining works technology for securing exploitation up to coordinate -11m. Gorn Odkryw 1 (2013):62–69

Dmitruk S (1984) Problems in engineering geology processes modeling in opencast mining. Wyd Geol Warsaw 120p

Dmitruk S (1995) Analysis of geotechnical conditions at landslide on the south-east slope of dump embankments in Turow Mine, not published technical report of Wroclaw University of Science and Technology p 42 (in Polish)

Eck C, Imbach B (2011) Aerial magnetic sensing with a UAV helicopter. Spatial Inf Sci XXXVIII-1/C22:81–8

Jonczyk W, Organisciak B (2010) Natural hazard in KWB Belchatow mine identification end counteraction. Min Geoengineering 34(4):249–257

Kasztelewicz Z (2012) Wegiel brunatny na swiecie i w Polsce. Wegiel Brunatny 1/78:7–13 (in Polish)

Kurpiewska I, Wcislo A, Czarnecki L, Jurczyk M (2013) Classification of geotechnical threats area locate in open-cast mines as a tool of safety optimization of exploitation based on example of Szczercow field. Gornictwo Odkry-wkowe 1(2013):5–12

Milkowski D, Nowak J (2008) Prevention and monitoring of the landslide hazard, on the Nysa Luzycka River, and landslide “Swiniec” on the external dump, close to the Czech Republic border. XVI Mining Workshop, IGSMiE PAS, Krakow, pp 212–227

Milkowski D, Kaczerewki T (2009) Rutschungsgefahr am Neiβepfeiler und Erdrutsch “Swiniec” auf der Auβenkip-pe des Tagebaues Turow. 7 Tagung fur Ingenieurgeologie. Hohschule Zittau/Gorlitz Publication, FDGG Fachsection Ingnieurgeologie

New Perspectives on Landslide Assessment for Spatial Planning in Austria

Arben Kociu, Leonhard Schwarz, Karl Hagen, and Florian Rudolf-Miklau

Abstract

Urban issues are increasingly prominent on the national policy agenda in Austria. The biggest policy challenges consist of the development of an integrated evaluation of the threats and risks (security level, protection objectives), in the preparation of accepted standards for the consideration of landslides as well as their use in spatial planning (ÖROK 2015). The Austrian Concept on Spatial Development (ACSD), which is a strategic instrument for federal policies in regional development, was set up to create a new cooperation at expert level and to develop basic approaches for key issues in an interdisciplinary forum. The ACSD-partnership for "Risk management for gravitative natural hazards" in spatial planning concerning slope processes (landslides including rock falls) was established in 2012 under the leadership of the Federal Ministry for Agriculture, Forestry, Environment and Water Management and Geological Survey of Austria to bridge the gap between geohazard mapping, risk management and spatial planning for these relevant phenomena. The results of this partnership are based on working papers produced by the working groups Spatial Planning, Geology and Sectoral Planning. The activities of the working group "Geology", presented in this paper, consisted in the evaluation of the existing methods for the calculation of landslide susceptibility (and rock falls) and the impact area in terms of their suitability as well as in the creation of standards and guides to draw up susceptibility maps for spatial planning. Further recommendations are given in terms of the quality assurance, uncertainties, model validation and traceability.

Keywords

Geohazard • Spatial planning • Susceptibility map • Hazard index map • Risk management

A. Kociu · L. Schwarz (✉)
Department of Engineering Geology, Geological Survey of Austria, 1030 Vienna, Austria
e-mail: Leonhard.schwarz@geologie.ac.at

K. Hagen
Bundesforschungs- Und Ausbildungszentrum Für Wald, Naturgefahren Und Landschaft, 1131 Vienna, Austria

F. Rudolf-Miklau
Federal Ministry for Agriculture, Forestry, Environment and Water Management, 1030 Vienna, Austria

M. Mikoš et al. (eds.), *Advancing Culture of Living with Landslides*,
DOI 10.1007/978-3-319-53483-1_7

ACSD Partnership "Risk Management for Gravitative Natural Hazards"—Policy Objectives

Due to the geological evolution of the Alps and their topographical conditions, Austria is a country with a naturally high disposition for geohazards, such as rockfalls, landslides and debris flows. Considering the increasing of urbanization, increasing of settlement space requirements (tourism, transport, etc.) and climate change and its consequences, the need for more security and protection for the population is becoming more and more important and the economic assets are also increasing. The catastrophic events observed in recent years, (as August 2005, June 2009, June 2014) (Tilch et al. 2011b; BMLFUW 2009; Hübl et al. 2015) have led to human and large economic losses. These events are an indication that technical constructions itself are not enough to provide sufficient protection. The cooperation of all the institutions which deal with natural hazards has shown, that only with an integral solution of all experts involved, it will be possible to establish prevention measures for the protection of human lives and economic assets.

The main policy challenges of the Austrian Conference for Spatial Planning (ACSP) Partnership on risk management in spatial planning focused on the development of an integrative evaluation of the threats and risks (security level, protection objectives), a uniform planning system for the cartographic depiction of landslides as well as the use in spatial planning (see Glade and Rudolf-Miklau 2015). The key advantage of this partnerships was the involvement of a large number of experts from science, politics, administration, engineering and economy as well as national and international leading experts which have also contributed to this work. The most important objective of the partnership was the preparation of expert recommendations and their presentation to ACSP as basis for reaching policy agreement and resolutions (see Glade and Rudolf-Miklau 2015).

Another important target of the partnership was the establishment of an integrated standard procedure for assessment and mapping of hazards related to slope processes (rock fall, landslides). Within the partnership the task force "gravitational induced natural hazard processes—working group geology" focused its work on the evaluation of the existing methods for the calculation of landslide susceptibility (and rock falls) and the affected area in terms of their suitability for urban planning. In this paper we present the results of the working group geology in terms of landslide susceptibility for shallow landslides and run out.

Standardised Hazard Assessment Procedure

For susceptibility and run out assessment regarding shallow landslides and debris avalanches (see Hungr et al. 2014) there is a wide range of modelling methods available, generating different types of maps. The appropriate application and the explanatory power of these models as well as the gained results are strongly depending on

- the input data
- analysis scale
- the extend and homogeneity of study area.

In this study for every working level of spatial planning, recommendations were given for the corresponding modelling method and type of mapping, according to the requirements of spatial planning (Table 1). The proposed standards, which require the application of comparable methods, represent a prerequisite in order to obtain comparable results. Further recommendations were also given in terms of the quality assurance, uncertainties, model validation and traceability.

Because of the possible restriction of future usability of an area that is within a sensitive designated zone due to landslides, the process (methodology) for the creation of geohazard maps must be transparent and comprehensible. The comparability of the results requires a comparable methodology within the administrative unit (e.g. municipality or federal states). Only under these conditions the acceptance of the affected property owners and decision-makers can be expected. Hence there is a need of a definition of minimum requirements for the creation of susceptibility maps and hazard maps in terms of input data and methodology.

Considerations of the Quality of Landslide Susceptibility Models and Hazard Index Maps

Since the application of landslide susceptibility maps is associated with consequences for the municipality and land use planners, a detailed and transparent assessment of its quality are necessary. This very broad term of quality can be interpreted in several ways and in several stages of the process of preparing input data and landslide susceptibility maps. In general, the main inputs to modelling are the process data and the site condition parameters. Otherwise many processes and conditions leading to landslide occurrence can

Table 1 Scale levels of spatial planning and the recommended corresponding modelling method and type of mapping

Working level	Relevance	Type of map	Map scale
Regional level/Spatial development concept	Rough estimation of potentially endangered areas, area-wide	Hazard Index Map (Susceptibility map + rough run out assessment)	<1:25.000
Local level/Local development concept	Identification of endangered areas and derivation of recommendations for action, extended relevant area	Refined Hazard Index Map (susceptibility map + run out assessment)	1:25.000 1:5.000
Site specific level/Zoning plan. Proceedings according to the building law	Detailed hazard assessment (expert's report), dimensioning of protection measure planning, study area	Hazard map, proof of suitability for building land and risk assessment	>1:5000

be known and mappable, known but not collectable, or are simply unknown (see Carrara et al. 1999). The definition of minimum requirements in terms of input data is used for quality assurance of the analysis/modelling. Moreover, through well formulated minimum requirements we can guarantee a traceability, comparability and understandability of the model results by the end-user. The traceability of results is necessary for the evaluation of the results because of improvements knowledge and changes of methodologies.

Analyses of Input Data

A high quality of landslide susceptibility map premises an adequate input data set for the modelling. Besides the terrain attributes, the spatial resolution and accuracy of the geo-environmental as well as the landslide inventory data are important for the quality (see Van Westen et al. 2008). With regard to the documentation of the input data and results as well as the method description, the following requirements must be met:

- explanation of the method applied (heuristic, statistic or deterministic) and model description,
- description of the data used (such as data quality, resolution of process data etc.),
- details and justification of the used threshold values,
- cell size of the digital terrain model,
- explanation of site-parameters used,
- description of the validation method(s) used and their results,
- partitioning method of data in training and validation (in case of statistical methods),
- explanation of the selected cell size of the model used for the input parameter.

Landslide Process Data

The landslide process data (such as types of landslide, characteristics of the source area, run out) and their quality has a decisive influence on the performance of statistical modelling and is therefore more important than the model choice within the statistical methods. That is because it is necessary to provide high quality of process data, taking into account the quality criteria of the input data. The most important criteria for the quality assessment of the process data are:

Representativeness: lack of representativeness of the process data (for example, only process information based on damage areas) leads to a significant deterioration of the modelling results.

Accuracy of landslide location: a blur of positional accuracy of process points, especially data from archives, but also from field mapping, has a negative impact on the results of the modelling. At worst this data cannot be used.

Data density: an increase in the number, density and spatial homogeneity of the process data information contributes to a significant improvement of statistical modelling results. In this case a certain minimum criterion (as an example: density (D) $\geq$ 1 landslide per km^2 for a modelling cell size 50 m $\times$ 50 m; D $\geq$ 3 for modelling cell size 25 m $\times$ 25 m) has to be fulfilled for statistical modelling (see Table 2). If this condition is not met, only modelling by heuristic methods should be applied (see Tilch et al. 2011a)

Ideally, the process data is based on area-wide field mapping (B) during and immediately after the event, containing high quality process information with high density, high positioning accuracy and good information regarding land use. However, to complete the data pool for modelling, also other data sources (A-remote sensing, C-archive,) should be used (see Table 2). If A and B are available we

Table 2 Minimum requirements for the process data (Landslides) for the modelling of susceptibility maps (H = heuristic method, S = statistical method) on local and regional scale

Source of the input data	Data basis	Cell size	Regional scale (regional spatial planning)		Local scale (local spatial planning)		Comment
			H	S	H	S	
Number (N) and density (D) of landslides		25–50 m	n.s.	$N \geq 50$ $D \geq 1/km^2$	n.s.	$N \geq 50$ $D \geq 1/km^2$	cell size 50 m (A or B)
						$N \geq 150$ $D \geq 3/km^2$	Cell size 25 m (A and B)
A—remote sensing	ALS, TLS, SAT, aerial photographs	1–50 m					At least one data source (ALS, TLS, aerial photographs, SAT
B—mapping	Landslide mapping by events	Mapping scale					
C—other	GBA, WLV, ÖBB, federal states, municipalities	Mapping scale		□		□	(□) only if high data quality is available
Mandatory	Recommended						
Data source							
A—remote sensing	Across the whole study area, complete data by evaluating of ALS (Airborne Laser Scanning),TLS (Terrestrial Laser Scanning), SAT (Satellite image), areal photography						
B—field mapping	Across the whole study area, mostly complete (documentation of events)						
C—other	Archives of municipalities and other institutions [(i) incomplete, not area-wide or (ii) qualitatively different data, heterogeneous]						

recommend that the modelling can be performed in a cell size with higher resolution (25 m). To increase the process data quality it is also recommended to undertake a critical review and harmonization of the input data in terms of positioning accuracy, content of information, process type, redundancies and representativeness.

Table 2 also highlights, which data sources (A, B and C) to use in different scales (regional and local) and modelling methods [statistic (S) and heuristic (H)]. "Mandatory" means that if possible, process data have to be acquired or newly collected and in case of local scale checked for representativeness and data quality. "Recommended" means, in many cases we expect benefit by working with this data, but the use of the data is not compulsory.

Site Parameter Maps

Suitable terrain attributes (site surface parameters) such as slope angle, slope curvature, flow accumulation etc. have high relevance to landslide susceptibility and should be derived from high-resolution DTMs (MASSMOVE 2011; Tilch et al. 2011b). Airborne laser scanning digital terrain model (LiDAR-DTM) with a resolution of 1 m × 1 m is now available for all Austria and must be applied.

It is recommended that the selection of the input parameter maps should base on the following aspects:

- objective of the study
- selection of model design
- extent and heterogeneity of the study area
- resolution or scale of the input data
- terrain attribute: which parameters are useful in geological, geomorphological, hydrological, hydrogeological (geo) logical or land use-related point of view
- the parameters that are meaningful or allowed for modelling should be used (parameters with spurious correlations as well as parameters that show strong correlations to other parameters should be excluded)
- criterion of balance of forces: driving forces (for example, slope angle) and retaining forces (for example, vegetation) should be included.

The selection of the parameters for modelling run out area is carried out analogously with the criteria of susceptibility modelling.

Landslide Susceptibility Modelling

A susceptibility map provides comprehensive information about the relative likelihood of landslides (here: shallow landslides) occurring throughout a given area. The frequency or time of occurrence of the landslide process is not assessed. The estimation of the spatially variable process disposition (landslide) generally occurs based on the local terrain attributes (e.g. geology, soil, vegetation, morphology, hydrology) as well as on the process data (mandatory for statistical methods and optional on heuristic and deterministic methods, see Table 2).

Every modelling result is highly dependent on the input data. Generally, for areas with low data information density and quality the application of expert based heuristic methods to generate susceptibility maps for landslides is recommended. Statistic models should be used only when sufficient landslide inventory data of good quality and density are available. Further, statements of the landslide impact area are made by means of the Hazard Index Map (HIM). The HIM combines scarp areas, derived by the susceptibility map, and adds the run out area to it. (see Fig. 1).

For the run out assessment and for the determination of the landslide impact area comprehensive information about the start areas and the flow paths must be available. Susceptibility maps offer information on initial zones (landslide starting cells probability), by applying a critical threshold value (for example, probability of occurrence >0.5). In the case of exceedance of this value, the modelling of run out area is being started.

To guarantee a quality assurance for Hazard Index Maps, besides fulfilling the minimum requirements and reviewing the landslide inventory critically (see above), it is also important to perform several types of model validations and plausibility checks (see Tilch et al. 2011b). The validation of run out areas can only be performed based on real landslide events and expert analysis.

Results and Recommendations

The ACSD-Partnership "Risk Management for Gravitative Natural Hazards in Spatial Planning" was dedicated to the issue of how strong the territorial impact of gravitative natural hazards is, what are the possible consequences for the municipalities, what basic materials can be provided for spatial planning and how this can be implemented in spatial planning. It was possible to show that for the different working levels (regional, local, site specific) different data qualities and methodologies are required.

Depending on the input data quality and the local terrain attributes, the following recommendations can be given

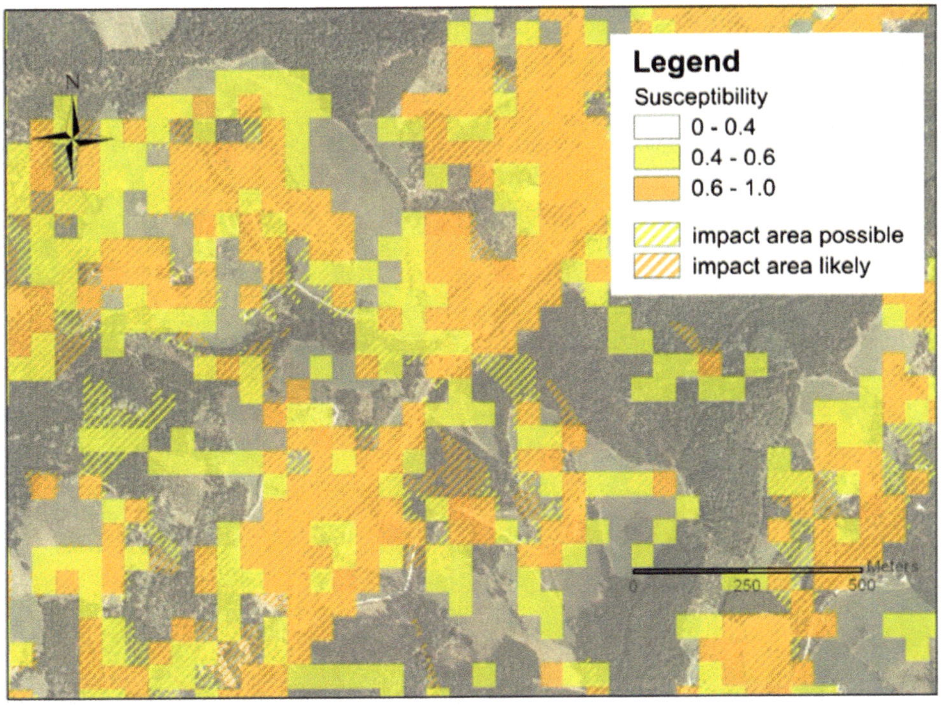

Fig. 1 Hazard index map by means of an empirical approach of Gasen, Styria for local scale (*source* BFW, ÖROK 2015)

regarding the selection of the modelling approach for the calculation of HIM. Also recommendations for the spatial visualization of gravitative natural hazard for regions and municipalities are given in the form of HIM for different planning levels.

- At the regional level the HIM offer a rough estimation of potentially endangered areas, including susceptibility map and run out assessment. Susceptibility maps should be calculated by heuristic models in case of poor data quality or by statistic models in case of high data quality. According to run out, the reach angle approach is sufficient.
- At the local level (Refined Hazard Index Maps) it is recommended to identify areas with different "needs for action" (consultation of regional planner/preliminary expert opinion/expert's report). Usually, susceptibility maps should be modelled by statistic models, unless only poor data quality is available. For these maps the estimation of the run out needs to be calculated more precisely by the application of process-orientated approaches (simplified physically based methods).
- Only at the site specific level a detailed proof of the suitability for building land by means of an expert`s report should be performed. In case of susceptibility modelling at this level, physically-based methods for the assessment of slope stability should be used. In terms of run out assessment, the estimation of frequency, magnitude and forces must be included.

The ACSD-Partnership analysed the potential consequences of natural hazards for spatial planning and formulated the recommendation and the steps required for hazard estimation and for modifications that need to be made to the existing regulations regarding the implementation of gravitative natural hazard in spatial planning. The results, including the standardized methods for hazard assessment and the technical recommendations, were agreed among partners from the federal state, the Austrian provinces and representatives of the municipalities and entered into force in April 2015.

The proposed standards and recommendations have been published in the ACSP-Publication No. 193.

References

BMLFUW (Bundesministerium für Land- und Forstwirtschaft, Umwelt und Wasserwirtschaft) (2009) Die Wildbachereignisse im Sommer 2009, Ereignisdokumentation, IAN Report 133, Institut für Alpine Naturgefahren, Universität für Bodenkultur-Wien

Carrara A, Guzzetti F, Cardinali M, Reichenbach P (1999) Use of GIS technology in the prediction and monitoring of landslide hazard. Nat Hazards 20:117–135

Glade Th, Rudolf-Miklau F (2015) ÖREK Partnership "Risk management for gravitative natural hazards in spatial planning"—In ÖROK Publication No. 193, pp 18–22

Hübl J, Beck M, Moser M, Riedl C (2015) Ereignisdokumentation 2014, IAN Report 167. Bundesministerium für Land- und Fortstwirtschaft. Umwelt und Wasserwirtschaft, Wien 80 p

Hungr O, Leroueil S, Picarelli L (2014) The varnes classification of landslide types, an update. Landslides 11(2):167–194

MASSMOVE (2011) Mindeststandards zur Erstellung von Gefahrenkarten zu Rutschungen und Steinschlägen als Werkzeug für vorbeugende Katastrophenvermeidung, Interreg IV Projekt http://www.ktn.gv.at/283664_DEThemenstartseite_Geologie_und_Bodenschutz-MassMove

ÖROK-Österreichische Raumordnungskonferenz, Hrsg (2015) Risikomanagement für gravitative Naturgefahren in der Raumplanung. Wien. (=ÖROK-Schriftenreihe 193)

Tilch N, Schwarz L, Winkler E (2011a) Einfluss der Prozessdatenqualität auf die mittels Neuronaler Netze, Logistischer Regression und heuristischer GBA-Methode erstellten Dispositionskarten hinsichtlich spontaner gravitativer Massenbewegungen im Lockergestein und die Ergebnisvalidierung. Poster im Rahmen des Geoforums Umhausen, 20, 21 Sept 2011, Niederthai

Tilch N, Schwarz L, Hagen K, Aust G, Fromm R, Herzberger E, Klebinder K, Perzl F, Proske H, Bauer C, Kornberger B, Kleb U, Pistotnik G, Haiden T (2011b) Modelling of Landslide Susceptibility and affected areas—process-specific validation of databases, methods and results for the communities of Gasen and Haslau (AdaptSlide). Endbericht des Projektes ADAPTSLIDE im Rahmen des EU-Projektes ADAPTALP, Wien, Graz, Innsbruck. http://bfw.ac.at/rz/bfwcms.web?dok=8935

Van Westen CJ, Castellanos E, Kuriakose SL (2008) Spatial data for landslide susceptibility, hazard, and vulnerability assessment: an overview. Eng Geol 102:112–131

4thWLF2017
4th World Landslide Forum
LJUBLJANA SLOVENIA EU

Characterisation of Recent Debris Flow Activity at the Rest and Be Thankful, Scotland

Bradley Sparkes, Stuart Dunning, Michael Lim, and Mike G. Winter

Abstract

The Rest and be Thankful (A83) in Scotland has been subject to frequent landslide activity in recent years and the trunk road has gained a reputation as one of the most active landslide sites in the UK. An average of two road closures per annum has been recorded over the last five years. This paper compares the site with other locations in Scotland that are prone to debris flows and explores a range of geomorphological factors using high resolution Terrestrial Laser Scanning data. The site is found to be relatively active, although normalization for mean annual rainfall makes activity at the site comparable to the likes of the Drumochter Pass. Macro-scale slope morphology is found to correspond strongly with the spatial distribution of recent activity. Channelisation is considered to be a significant factor in the overall debris flow hazard by confining flow and enabling entrainment. This was demonstrated during two recent events that mobilized at high elevations and entrained significant volumes of material along long runout paths.

Keywords

Debris flow • Geomorphology • Roads • Rest and be Thankful • Scotland • Storms

B. Sparkes (✉) · M. Lim
Northumbria University, Ellison Building, Newcastle, NE1 8ST, UK
e-mail: bradley.sparkes@northumbria.ac.uk

M. Lim
e-mail: michael.lim@northumbria.ac.uk

S. Dunning
Newcastle University, School of Geography, Politics and Sociology, Newcastle University, Newcastle upon Tyne, NE1 7RU, UK
e-mail: stuart.dunning@newcastle.ac.uk

M.G. Winter
Transport Research Laboratory (TRL), 10B Swanston Steading, 109 Swanston Road, Edinburgh, EH10 7DS, UK
e-mail: mwinter@trl.co.uk

M.G. Winter
School of Earth and Environmental Sciences, University of Portsmouth, Portsmouth, UK

M. Mikoš et al. (eds.), *Advancing Culture of Living with Landslides*,
DOI 10.1007/978-3-319-53483-1_8

Introduction

Debris flows are extremely rapid landslides that can often travel at velocities in excess of 10 m/s and may cover very large runout distances (Varnes 1978; Cruden and Varnes 1996; Hungr et al. 2014). As a result, debris flows represent a significant hazard to people and infrastructure. One estimate attributes 77,779 deaths to 213 debris flow events recorded between 1950 and 2011 (Dowling and Santi 2014).

Fortunately debris flow activity in the Scottish Highlands has not resulted in any known fatalities in recent years, however activity has been widespread and has caused damage to infrastructure and disruption to rural Scottish communities. Transport infrastructure is particularly vulnerable to debris flows, with road and rail networks having suffered disruption in recent years. In remote, rural areas road networks are essential and alternative routes may involve lengthy diversions. Winter et al. (2016) estimated that a debris flow in 2007 at the Rest and be Thankful (A83) in the west of Scotland accrued total direct costs between £1.7 m and £3 m (at 2012 prices). Even small events that block just a few tens of metres of carriageway can cast a significant vulnerability shadow often amounting to a few thousand square kilometres (Winter 2014).

Intense rainfall is considered the most common trigger of debris flows, although rapid snowmelt can also often be attributed to rapid influxes of water (Iverson et al. 2011). Antecedent conditions are however also considered an influential factor in shallow landslides, as slope material can become saturated prior to the onset of intense rainfall, allowing shorter or less intense showers to trigger failure. A relationship with rainfall goes some way towards explaining the prevalence of debris flows among some of the wettest areas of Scotland, particularly in the west. Different intensities of rainfall over varying timescales were considered by Caine (1980) who pioneered an early global Intensity Duration threshold for shallow landslides and debris flows, although it is now commonly accepted that relationships differ locally and regionally (Guzzetti et al. 2008) due to a range of diverse factors.

Debris flows in the Scottish Highlands often arise from a supply of glacigenic sediment and in many cases represent an agent of landscape denudation (Ballantyne 2008), where pre-glacial background rates of erosion are targeted. The phase in which levels of sediment reworking are elevated due to perturbation by glaciation is referred to as the paraglacial period (Ballantyne 2002). Timescales of around 10,000 years are typically ascribed to the paraglacial period (Church and Slaymaker 1989; Ballantyne 2002), however a resurgence of activity appears to have taken place at the Rest and be Thankful and at other sites around the Scottish Highlands. Reid and Thomas (2006) undertook a multidisciplinary chronostratigraphic study of Holocene slope evolution at Creagan a' Chaorainn in the Northern Highlands of Scotland and attributed protracted paraglacial response to climatic forcing. They found that a period of stabilisation c. 7.5 ka BP may have enabled the development of vegetation cover and accumulations of peat, marking a net system change from erosion to accumulation. This suggests that contemporary activity may be manifesting itself as part of a delayed paraglacial response. Further perturbations are anticipated, particularly from climate change symptoms such as increased rainfall intensities in the winter (Winter and Shearer 2014) which are heavily linked with an increased occurrence of mass movements (Stoffel and Huggel 2012).

Debris flow events of all sizes are capable of overwhelming culverts with a rapid influx of sediment, enabling overspill onto roads. Entrainment has been observed to increase debris flow volume significantly, by an order of magnitude in some cases (Milne 2008) and has also recently been shown to enhance debris flow mobility when sediment is heavily saturated (Iverson et al. 2011) as is often the case in Scotland. Such magnitude increases are reliant upon flow channelization wherein momentum can be preserved and readily erodible material resides and can be entrained. In contrast, open hillslope flows often spread out, increasing the flow surface area and the friction acting on the mass to inhibit flow. Landslide risk management efforts, particularly around transport corridors, should therefore focus on such areas where flow efficiency can be maintained and where the potential for entrainment induced magnitude and mobility increases may take place.

The Rest and Be Thankful (A83)

The Rest and be Thankful slope is above the A83 trunk (strategic) road in Glen Croe, approximately 100 km north-east of its terminus in Campbeltown in Argyll and Bute (Fig. 1). The A83 is the main route between mid-Argyll and Central Scotland and is essential for local communities, tourism and the movement of goods and services. Around 20 slope failures have been recorded at the Rest and be Thankful since 2007, resulting in around 11 road closures, equating to an average of roughly one road closure per year. Argyll & Bute is one of the wettest areas of the UK, thus contributing significantly to the rate of slope activity.

Unlike many active debris flow sites in continental Europe, such as the Illgraben torrent in Switzerland for example, the Rest and be Thankful affords a unique opportunity to holistically observe debris flow phenomena, encompassing geomorphic activity from source to sink. The Rest and be

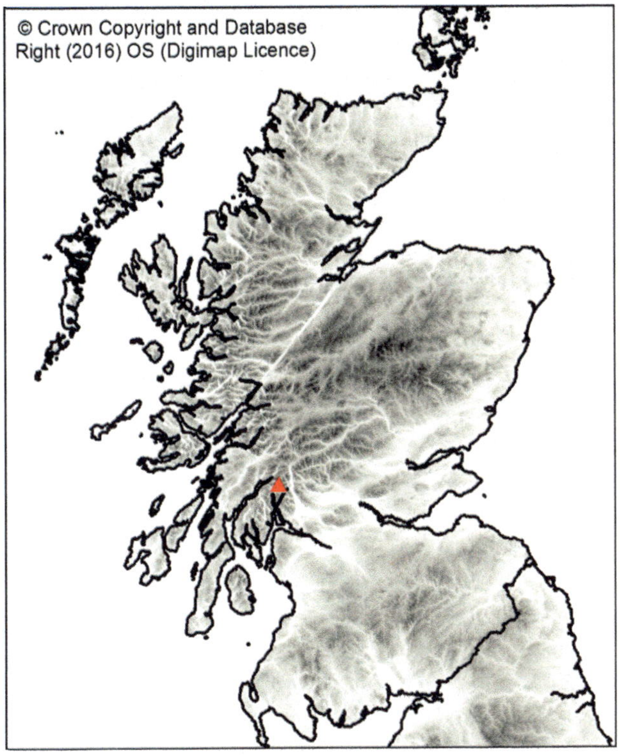

Fig. 1 The topography of Scotland and the location of the Rest and be Thankful in the west

Thankful has therefore been used as a key study site as part of an ongoing monitoring and modelling project.

The recent frequency of slope activity affecting the road at the Rest and be Thankful is more frequent compared to other UK transport routes and as a result considerable investment has been provided to fund a series of mitigation measures. A series of warning signs called wig-wags were installed along the A83 in 2011 to warn road users of an elevated landslide potential during periods of heavy rain. The most significant mitigation effort however has been the installation of debris flow nets above the A83 to prevent major flow impact with road infrastructure and its users.

Slope Morphology

Terrestrial Laser Scanning (TLS) data has been collected at the Rest and be Thankful as part of an ongoing monitoring project. Standalone data allows characterisation of slope morphology from macro features such as slope-wide morphology to the meso scale, such as relict landslide scars. Repeat surveys have enabled ongoing monitoring of slope changes to a high level of detail, helping to develop an understanding of debris flow processes at the Rest and be Thankful and similar sites.

Methodology

A Riegl LMS-Z620 has been used to collect data from three scan positions on the opposite side of the valley to the debris flow prone slope. A regular point spacing of 0.14 m at a range of 1 km has been achieved from each respective scan position. Data has been manually aligned to the central scan dataset and then registered using an iterative closest point (ICP) algorithm within the software package RiScan Pro, allowing the minimisation of occlusion. A sum of registration and device errors result in a conservative error threshold of around ±0.1 m. A relatively even distribution laterally and vertically across the slope is realised by subsampling the final point cloud to a regular point spacing of 0.2 m.

Change detection has subsequently been carried out using the Multiscale Model to Model Cloud Comparison (M3C2) algorithm within the open source software package CloudCompare (Lague et al. 2013). In this approach cylinders are generated around nodal points in the base dataset, of which the diameters are user defined. Cylinders are orientated using normal directions, which are calculated by an algorithm that considers local point cloud (PC) morphology at a user defined scale. Cylinders are extruded towards the second PC, from which the mean position of points falling within the cylinder domain are compared with corresponding points in the base dataset cylinder domain, resulting in a change value. Changes deemed significant, relative to the roughness of points within each cylinder and compounded instrument and registration errors have been retained after which a noise filter has been used to remove inherent spurious points, at vegetation margins for example. This process retains change above and below a relatively conservative error threshold of ±0.2 m. The change data presented in this paper has been detected using data collected after the winter dieback of seasonal vegetation and before its reestablishment in the proceeding spring, to minimise the change signal associated with extraneous seasonal vegetation fluxes.

Geology and Geomorphology

The main slope at the Rest and be Thankful sits at a gradient of 32°–33° with some slope regions towards the top of the slope in excess of 40°. The underlying slope is comprised principally of fine grained schists such as Pelite, with an overlying cover of glacigenic sediment and soil up to several metres in thickness. Some exposures of schist on the slope appear to be highly weathered and extremely degraded and therefore represent a likely source of progressive pedogenesis, therefore altering the hydraulic transmissivity of the soil over time and contributing towards reductions in shear strength, termed 'ripening' by Nettleton et al. (2004).

Cross profiles have been plotted using a 0.2 m Digital Elevation Model (DEM) derived from a single station TLS survey using a Riegl VZ-4000 in July 2015. The first cross profile (Fig. 2a, cross section locations shown in Fig. 3) shows the large concavity which characterises the top of the slope. This topographic depression appears to concentrate drainage towards the centre of the slope, as opposed to the convex margins where bedrock outcrops are more common. Such convergence of hydrogeological flow would explain the greater propensity of slope failures towards the centre of slope, for both inventoried events and those monitored recently (see Fig. 4).

The mid-elevation profile (Fig. 2b), approximately 300 m above the road, shows a slight decline in elevation from north-west to south-east, however the high density of gullies eroded into the slope is perhaps most noticeable. The gully density along this stretch of slope is equivalent to 18 km^{-1}, whereas other local slopes above the A83 typically contain gullies at a density of between 14 and 15 km^{-1}. Similar or greater gully densities can be observed elsewhere in the region, for example 2 km directly to the east at Coire Croe where the gully density is roughly 24 km^{-1}.

The gullies range between those eroded down to bedrock to those that are merely eroded into the superficial deposits, although such patterns vary both laterally and vertically across the slope, with some gullies scoured far more extensively than others. Given the well documented impact that entrainment can have on final debris flow magnitude, such spatial variations in gully material availability are influential to the debris flow hazard across different regions of the slope.

A number of small discontinuous superficial and sub-surface channels have also been observed on the slope, the latter in the form of soil pipes, the collapse of which has been previously linked to debris flow triggering, although may also assist in the dissipation of excessive pore water pressures in some cases (Uchida 2004).

Debris Flow Spatial Density

The high density of gullies at the Rest and be Thankful corresponds with a high rate of activity recorded at the site. At least 10 distinct debris flow gullies can be recognised within a 0.7 km slope region at the Rest and be Thankful, equating to a minimum debris flow density of around 14 km^{-1}. However taking into account observed relict evidence of debris flow prior to recorded events, an upper bound of 18 debris flow gullies/km is likely. This density is particularly high for a slope underlain by schist, with densities in similar geological units as low as 3 km^{-1} (Ballantyne 2004) although a high density of 13.3 km^{-1} (Milne and Davies 2007) to 18.2 km^{-1} (Strachan 2015) has also been observed at the Psammite schist dominated Drumochter Pass in the Scottish Cairngorms. High densities of gullies with evidence of debris flow activity are typically observed on slopes underlain by granite, such as those between 20 and 32 km^{-1} on the Isle of Skye or 14–30 km^{-1} in the Cairngorms (Ballantyne 2004; Strachan 2015).

When normalised using a mean annual rainfall value of at least 3000 mm, derived from a 1 km resolution UK Met Office map (2016) of mean annual rainfall for the period

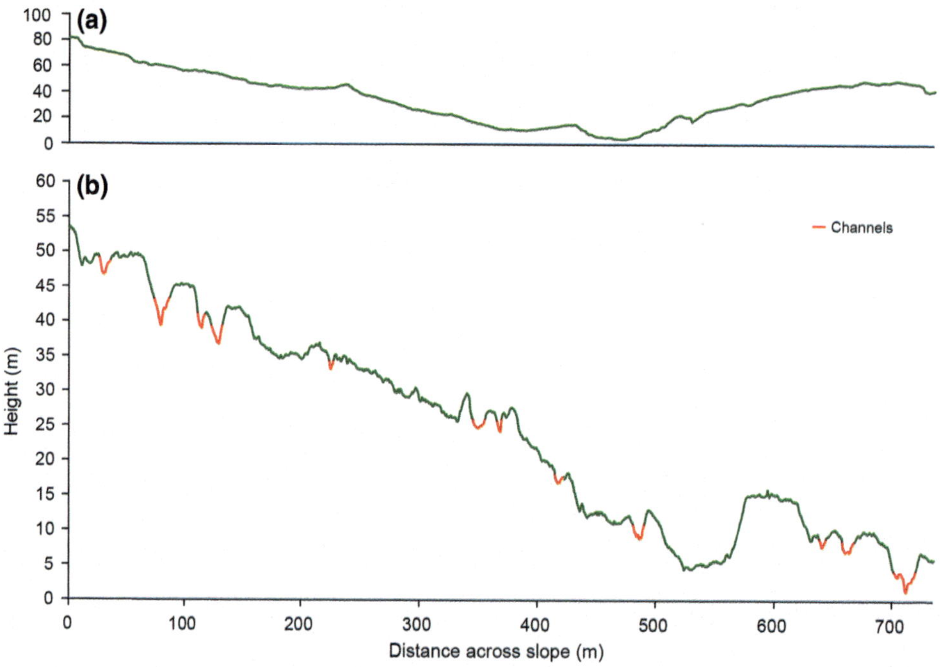

Fig. 2 Cross profiles of TLS data at the A83 Rest and be Thankful: **a** A cross profile through the top of the slope; **b** A mid-elevation profile through the slope showing the large number of major gullies

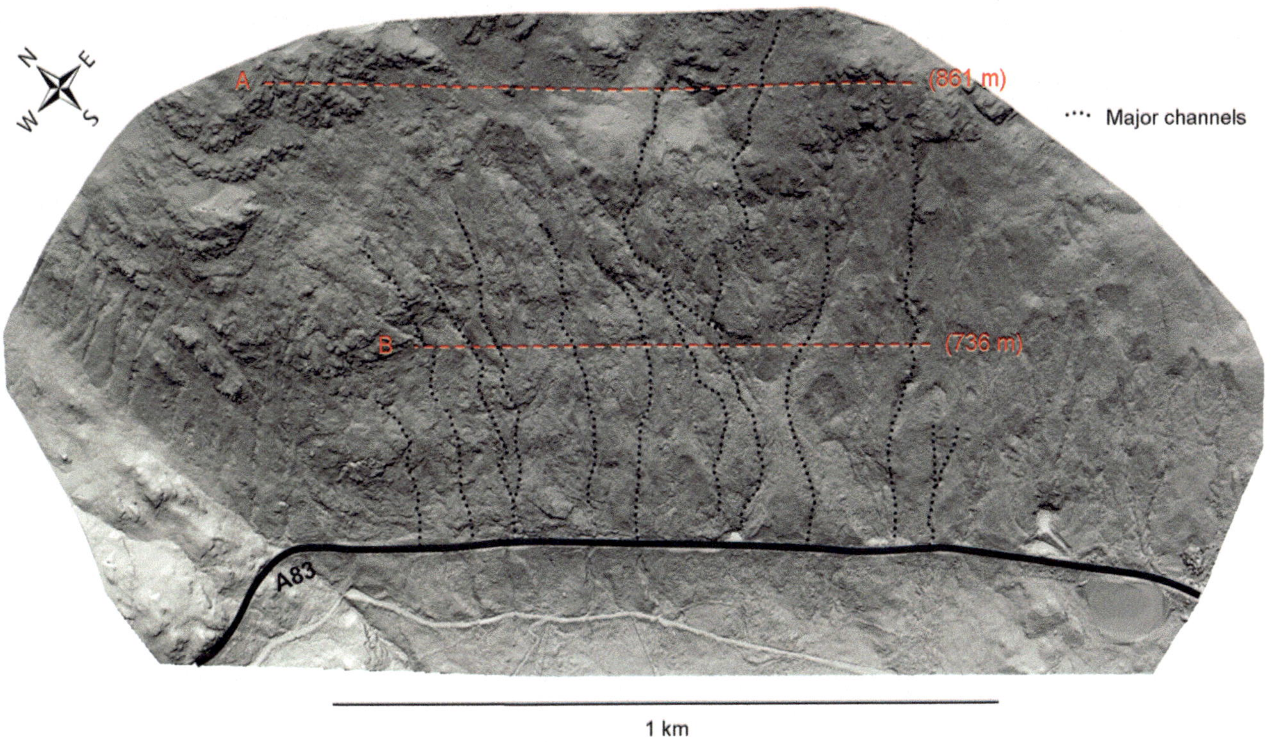

Fig. 3 A 0.2 m resolution hillshade of the Rest and be Thankful slope derived from a Riegl VZ-4000 TLS. *Red* transects A and B represent those shown in Fig. 2a and b respectively

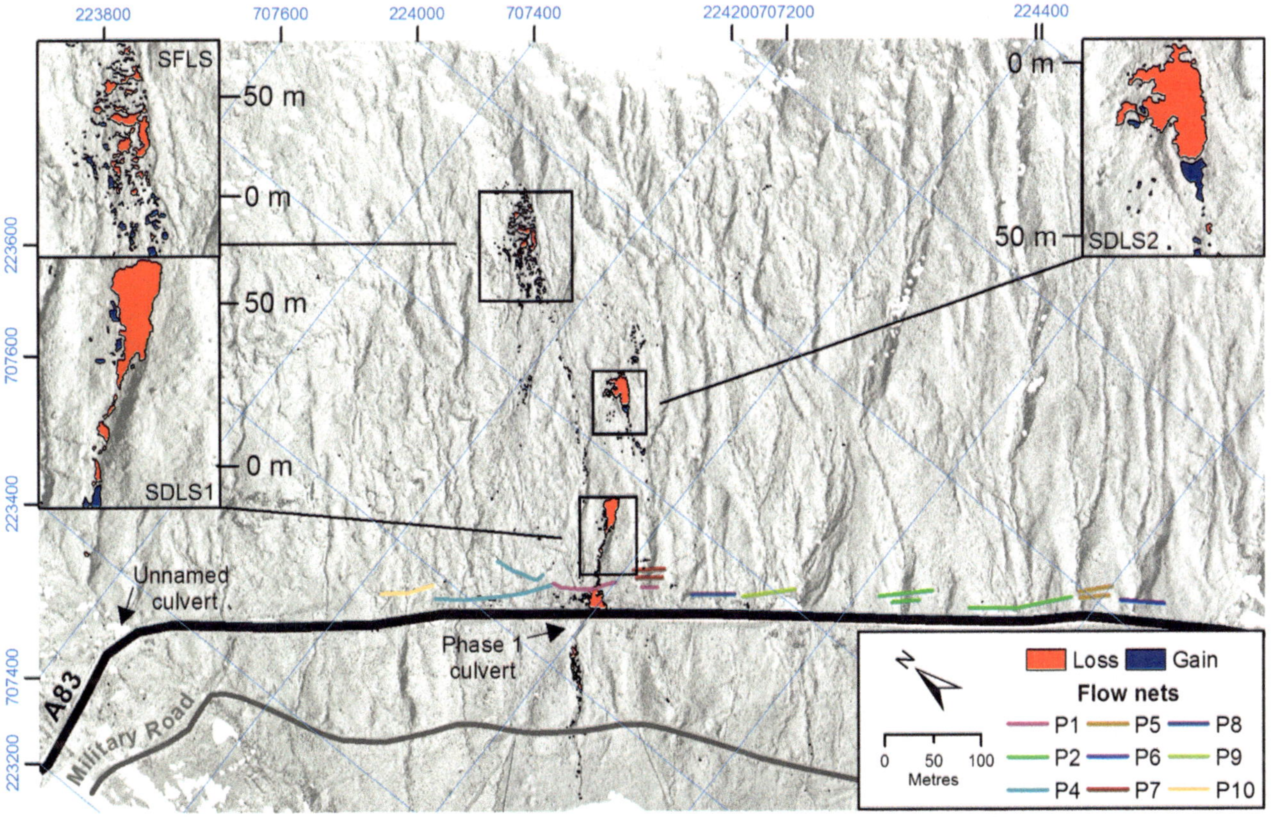

Fig. 4 Change detected between April 2015 and February 2016, after the late 2015 winter storm events. The three principle slope failures from Storm Desmond (SDLS1 and SDLS2) and Storm Frank (SFLS) are shown respectively

1981–2010, the debris flow gully density at the Rest and be Thankful equates to around 4.7 $km^{-1}/m\ a^{-1}$ (gullies per kilometre per metre of mean annual rainfall) for recorded debris flows and 6 $km^{-1}/m\ a^{-1}$ accounting for the upper bound value of 18. These values are comparable to a normalised value of 6.3 $km^{-1}/m\ a^{-1}$ at Drumochter pass, when an mid-range mean annual rainfall estimate of 2500 mm, derived from the same UK (Met Office 2016), is coupled with a mean gully density of 15.75 km^{-1}. The higher normalised value at the Drumochter pass highlights the significant role rainfall plays in stimulating slope activity at the Rest and be Thankful. Compared to normalised debris flow gully densities of 2.6 and 2 $km^{-1}/m\ a^{-1}$ at the schistose slopes of Glen Ogle and Mill Glen respectively (Milne and Davies 2007), based upon estimated mean annual rainfall values of 2000 mm and 1500 mm respectively (Met Office 2016), the Rest and be Thankful has a relatively high debris flow gully density.

Recent Events

The Rest and be Thankful has been subject to a number of debris flow events during the period of this ongoing study. The site was particularly active towards the end of 2015 during the UK winter storm events. Here available data for each event is presented and discussed.

October 2014 Debris Flow

On 28th October 2014 a large debris flow occurred at the Rest and be Thankful. An estimated 2000 tonnes of material reached the foot of the slope, the largest debris flow on record at the location. Rainfall in excess of 50 mm was recorded over the 24 h prior to the event and was followed by a peak rainfall intensity of 8.4 mm/hour at 5 AM, which appears to correspond with the approximate time of failure.

Flow nets were installed as part of an ongoing programme. The flow net at this location (part of Phase 4, P4) was installed at the foot of the gully (shown in Fig. 4). The net retained a significant volume of material however material was also deposited onto the A83 which was subsequently closed for six days. A visual comparison with other event source areas at the Rest and be Thankful suggests that the October 2014 source area is not unusually large, but it was clear that significant entrainment occurred. The high elevation of the source area, approximately 530 m above the road, appears to have granted the flow a long sediment laden path from which it could source and incorporate such further material into the flow.

3D TLS Reconstruction

Although the event took place prior to the onset of monitoring, an estimate of the initial source volume has been made by reconstructing the pre-failure source morphology. High resolution TLS data from April 2015 has been cropped around the margins of the source area and a flat pre-failure surface estimate has been produced by interpolating across the zone.

Comparison with actual post-failure morphology data yields a source volume estimate of 522 m^3, or 1044 tonnes using empirical density estimates. This is comparable with other events observed at the site. Final engineering estimates of the event deposition at road level are of over 2000 tonnes, thus representing an estimated 92% increase in volume between source and sink. Engineers estimated that the flow net (P4) at the foot of the propagation gully retained some 1200–1500 tonnes of material.

Winter 2015/2016 Storm Induced Debris Flows

The December 2015 UK winter storms Desmond, Eva and Frank caused widespread flooding across the UK, particularly in the north west of England, but also in Scotland. The storms also triggered a number of slope failures across the UK, causing damage to transport infrastructure such as the main rail route between Carlisle and Newcastle upon Tyne. The storms had a significant impact on Scottish road infrastructure, notably at the Rest and be Thankful during both Storms Desmond and Frank.

Storm Desmond (SDLS1)

Storm Desmond occurred between 3 and 8 December 2015 and triggered a low elevation debris flow approximately 120 m above the A83 Rest and be Thankful on 5 December, referred to as SDLS1. The flow subsequently propagated into the P1 flow net above the A83 (visible in Fig. 4). More than 90 mm of rain was recorded in the 20 h prior to the event. A maximum intensity of 8.8 mm/h was recorded at 4 AM, although this doesn't appear to correspond with the time of failure. A later peak of 4.4 mm/h was recorded at 1 PM, the approximate time of failure. The storm period was preceded by a 30 h lull in which very little rainfall was recorded. However 124 mm of rain was recorded at the site in the week leading up to the event, a value comparable to the preceding two winter weekly averages, therefore the lull appears unlikely to have alleviated saturated conditions prior to the onset of intense rainfall.

On 6 December a second higher elevation hillslope failure occurred, referred to as SDLS2, some of the material from which propagated into a nearby gully and some of which was retained by two small flow nets (P7) (Fig. 4). The precise timing of the event is unclear, although it appears to correspond with a peak rainfall intensity of 6.4 mm/h recorded at 5 AM. Neither event resulted in closure of the A83, although a single lane system was operated.

Storm Frank (SFLS)

Storm Frank occurred between 29 and 30 December 2015 and triggered a high elevation debris flow at the Rest and be Thankful on 30 December, referred to as SFLS. The failure was preceded by over 60 mm rainfall in 12 h, with a peak rainfall intensity of 9.2 mm/h at 1 AM, although this doesn't appear to correspond with the time of failure. Secondary peaks of over 6 mm/h were recorded between 5 and 6 AM and are considered to coincide with the approximate time of failure.

SFLS initiated higher up the slope than SDLS1 and SDLS2, approximately 370 m above the road. The flow subsequently propagated through a major gully before impacting with a flow net above the A83, which held the majority of the material, although a small volume was also deposited in the carriageway. The event also destabilised a large 150 tonne boulder within the vicinity of the source area.

Post-storm Analysis

The TLS monitoring data in Fig. 4 shows SDLS1. A relatively large mean depth of 1.3 m has been removed from the source area, although some regions such as the headscarp exhibit much larger depths. Such loss depths were likely due to the low elevation source area which coincides with a zone where the slope gradient reduces and deposits from previous events reside. SDLS1 may therefore represent a secondary failure from material deposited by a previous event. The flow path can be seen to dip to the west, aided by a subtle topographic depression. Despite not propagating into a pre-existing gully, SDLS1 has eroded and entrained material immediately below the source area, therefore demonstrating debris flow initiated channel genesis.

The higher elevation source area of SDLS2 is highlighted in Fig. 4. A mean depth of 0.72 m has been mobilised, almost half that of SDLS1. In contrast to SDLS1, the event is therefore more likely to be a primary failure of superficial slope material. A large deposit can be seen directly below the source area, demonstrating that flow arrested rapidly after failure. A small volume of material has however reached the gully approximately 20 m to the south, and some material has propagated over 100 m downslope and been deposited behind two flow nets (P7). It is likely that the volume of material that reached the gully was insufficient to mobilise any pre-existing material accumulations or to erode the gully margins. The flow deposited material in the gully, potentially providing material for subsequent mobilisation.

The source area of SFLS is relatively fragmented compared to SDLS1 and SDLS2, evidenced by the large number of soil rafts and large amounts of unconsolidated material left within the source area after initiation. A large volume of deposition can be seen directly below the source area, akin to SDLS2, although in this case a significant volume of material appears to have entered the gully approximately 10 m to the south. Figure 4 shows that the event exhibited net entrainment during propagation, although deposition has also taken place in some areas. Interestingly, the foot of the SFLS source area overlaps that of a small failure that occurred in 2009. This may therefore have contributed to destabilisation of the upslope area, perhaps demonstrating a relationship with a previous event. Furthermore a shallow landslide at the margins of the gully down which SFLS propagated occurred in 2007 and may have provided material to the gully which SFLS may have subsequently entrained. Considerable volumes of meta-stable material were found to persist within the gully during a slope walkover in September 2016. A number of deposits mount the channel margins in the form of levees, whilst other recent and potentially readily entrainable deposits reside within the gully base. This material, particularly the basal deposits, appear to be highly entrainable and could potentially be mobilised by a future mass movement.

Discussion and Conclusion

The hydrogeology and hydrology of the Rest and be Thankful (A83) appears to be influenced by a large concavity towards the centre of the slope, potentially explaining the confinement of recently recorded activity away from the margins and towards the centre. Furthermore, relatively well colonised and undisturbed vegetation towards the more convex south-eastern flank of the slope may be indicative of limited recent activity. The slope is also heavily gullied, from below the upper slope concavity to road level, exhibiting largely parallel drainage and providing several locally oversteepened regions from which material may be mobilised or into which material may be routed after which entrainment may occur and with it an increase in magnitude.

The Rest and be Thankful is found to be a relatively active site in the UK and is sensitive to intense rainfall. The predominance of rainfall to the west enhances rates of activity beyond that of similar sites such as the Drumochter Pass, although normalisation for this effect shows that the rate of activity at the site is by no means unique. The presence of a key trunk road at the bottom of the slope draws particular attention when events do occur. Analysis of recent events have shown rainfall to be the principal trigger of failure at the Rest and be Thankful. The large October 2014 debris flow occurred after lower cumulative antecedent rainfall than the 2015 storm events, but was triggered by a more intense peak in rainfall, demonstrating the balance of rainfall intensity and duration in triggering debris flows. Recent rainfall triggering values are comparable with 24 h thresholds of 60–80 mm at other studied sites in Scotland (Ballantyne 2004). The relatively brief record of debris flow activity and road closures at the site may indicate that the rate of activity may have increased in recent years, most likely as a result of increased storminess. However, the period is too short to make definitive conclusions regarding long-term trends, but climate change is forecast to increase the intensity of winter rainfall events and this trend may continue.

Channelisation of debris flows is found to magnify the overall hazard at the Rest and be Thankful, firstly by confining flow and enabling conservation of momentum but perhaps most significantly by enabling significant increases in magnitude by means of entrainment. The events of October 2014 and SFLS initiated from relatively high elevation source areas, providing a long path of entrainment, whereas lower elevation flows had limited opportunities to entrain material to the same degree. Channelisation may also increase mobility, particularly due to mixing with stream flow and saturated materials. Conversely open hillslope propagation has been observed to arrest mobility. Failures at the Rest and be Thankful may also be linked by secondary mobilisation of recent and relict deposits. Recent observations show large volumes to persist in the SFLS gully, suggesting that this could provide a source of entrainment for a future event.

Acknowledgements The work described in this paper is part of a doctoral project funded by the Scottish Road Research Board (SRRB) through the Transport Research Laboratory (TRL). This support is gratefully acknowledged. We are also grateful to Nick Rosser of Durham University for his assistance collecting and processing TLS field data.

References

Ballantyne CK (2002) Paraglacial geomorphology. Q Sci Rev 21(18–19):1935–2017

Ballantyne CK (2004) Geomorphological changes and trends in Scotland: debris-flows, Scottish Natural Heritage Commissioned Report No. 052 (ROAME No. F00AC107A)

Ballantyne C (2008) After the ice: Holocene geomorphic activity in the Scottish Highlands. Scott Geogr J 124(1):8–52

Caine N (1980) The rainfall intensity: duration control of shallow landslides and debris flows. Geografiska Ann Ser A Phys Geogr 62:23–27

Church M, Slaymaker O (1989) Disequilibrium of Holocene sediment yield in glaciated British Columbia. Nature 337:452–454

Cruden D, Varnes D (1996) Landslides: investigation and mitigation. Chapter 3: Landslide types and processes. Transportation Research Board Special Report, vol 247, pp 36–75

Dowling CA, Santi PM (2014) Debris flows and their toll on human life: a global analysis of debris-flow fatalities from 1950 to 2011. Nat Hazards 71(1):203–227

Guzzetti F, Peruccacci S, Rossi M, Stark CP (2008) The rainfall intensity–duration control of shallow landslides and debris flows: an update. Landslides 5(1):3–17

Hungr O, Leroueil S, Picarelli L (2014) The Varnes classification of landslide types, an update. Landslides 11(2):167–194

Iverson RM, Reid ME, Logan M, LaHusen RG, Godt JW, Griswold JP (2011) Positive feedback and momentum growth during debris-flow entrainment of wet bed sediment. Nat Geosci 4:116–121

Lague D, Brodu N, Leroux J (2013) Accurate 3D comparison of complex topography with terrestrial laser scanner: Application to the Rangitikei canyon (NZ). ISPRS J Photogramm Remote Sens 82:10–26

Met Office (2016) UK Climate Averages 1981–2010. Accessed: 28 Sept 2016. Available at: http://www.metoffice.gov.uk/public/weather/climate/

Milne F (2008) Topographic and material controls on the Scottish debris flow geohazard. Unpublished Ph.D. thesis, University of Dundee, Dundee

Milne FD, Davies MCR (2007) Control of soil properties on the Scottish debris flow geohazard and implications of projected climate change. In: McInnes R, Jakeways J, Fairbank H, Mathie E (eds) Landslides and climate change: challenges and solutions. Taylor & Francis Group, London, pp 249–258

Nettleton IM, Martin S, Hencher S, Moore R (2004) Debris flow types and mechanisms. In: Winter MG, Macgregor F, Shackman L (eds) Scottish road network landslide study, pp 45–119

Reid E, Thomas MF (2006) A chronostratigraphy of mid- and late-Holocene slope evolution: Creagan a' Chaorainn, Northern Highlands, Scotland. The Holocene 16(3):429–444

Stoffel M, Huggel C (2012) Effects of climate change on mass movements in mountain environments. Prog Phys Geogr 36(3):421–439

Strachan GJ (2015) Debris flow activity and gully propagation: Glen Docherty, Wester Ross. Scott J Geol 51(1):69–80

Uchida T (2004) Clarifying the role of pipe flow on shallow landslide initiation. Hydrol Process 18(2):375–378

Varnes DJ (1978) Slope movement types and processes. In: Schuster RL, Krizek RJ (eds) Special report 176: landslides: analysis and control. Transportation and Road Research Board, National Academy of Science, Washington DC, pp 11–33

Winter MG (2014) A strategic approach to landslide risk reduction. Int J Landslide Environ 2(1):14–23

Winter MG, Shearer B (2014) Climate change and landslide hazard and risk in Scotland. In: Lollino G, Manconi A, Clague J, Shan W, Chiarle M (eds) Engineering geology for society and territory—volume 1: climate change and engineering geology, pp 411–414

Winter MG, Shearer B, Palmer D, Peeling D, Harmer C, Sharpe J (2016) The economic impact of landslides affecting the Scottish road network. In: Aversa S et al. (eds) Landslides and engineered slopes. Experience, Theory and Practice, pp 2059–2064

The Use of Morpho-Structural Domains for the Characterization of Deep-Seated Gravitational Slope Deformations in Valle d'Aosta

Daniele Giordan, Martina Cignetti, and Davide Bertolo

Abstract

Deep-seated Gravitational Slope Deformations (DsGSDs) are widespread phenomena in mountain regions. In the Valle d'Aosta alpine region of northern Italy, DsGSDs occupy 13.5% of the entire regional territory. A total of 280 phenomena have been inventoried in the IFFI (Italian Landslide Inventory) project. These large slope instabilities often affect urbanized areas and strategic infrastructure and may involve entire valley flanks. The presence of settlements on DsGSDs has led the regional Geological Survey to assess the possible effects of these phenomena on human activities. This study is aimed at implementing a methodology that is based on interpreting Synthetic Aperture Radar (SAR) data for recognizing the most active sectors of these phenomena. Starting from the available RADARSAT−1 dataset, we attempt to propose a methodology for the identification of the main morpho-structural domains that characterize these huge phenomena and the definition of different sectors that make up the DsGSDs, which are characterized by different levels of activity. This subdivision is important for linking the different kinematic domains within DsGSDs with the level of attention that should be given to them in the studies that support the request for authorization of new infrastructure. We apply this method to three case studies that represent significant phenomena involving urban areas within the Valle d'Aosta region. In particular, we analyze study areas containing the Cime Bianche DsGSD, the Valtourenenche DsGSD, and the Quart DsGSD. These phenomena have different levels of evolution that are controlled by the interaction of diverse factors, and involve buildings and other infrastructure. This setting has been useful for testing the development of the methodology, which takes advantage of remote-sensing investigations, together with the local geological, geomorphological and structural setting

D. Giordan (✉) · M. Cignetti
National Research Council of Italy, Research Institute for Geo-Hydrological Protection, Strada Delle Cacce 73, 10135 Turin, Italy
e-mail: daniele.giordan@irpi.cnr.it

M. Cignetti
e-mail: martina.cignetti@irpi.cnr.it

D. Bertolo
Strutture Attività Geologiche, Regione Autonoma Valle d'Aosta, Località Amérique 33, 11020 Quart, Aosta, Italy
e-mail: davide.bertolo@regione.vda.it

M. Mikoš et al. (eds.), *Advancing Culture of Living with Landslides*,
DOI 10.1007/978-3-319-53483-1_9

of each case study that we analyzed. This method aims to produce a useful method for delineating guidelines for the building of new infrastructure, in support of the Regional Agency.

Keywords
Deep-seated gravitational slope deformation • DInSAR techniques • RADARSAT−1

Introduction

Large, slow-moving slope instabilities play an important role in the evolution of mountain landscapes and represent a significant natural hazard in urbanized areas, given their possible effects on structures and/or infrastructure. For Regional Authorities in mountainous areas, the estimation of the degree of activity of such phenomena is fundamental for natural hazard assessment and the prevention of landslides.

Deep-seated Gravitational Slope Deformations (DsGSDs) are well-known and widespread phenomena in the Alpine chain (Mortara and Sorzana 1987; Ambrosi and Crosta 2006; Martinotti et al. 2011).

In the last several decades, many authors have investigated these DsGSDs (Zischinsky 1966, 1969; Mahr 1977; Savage and Varnes 1987; Varnes et al. 1989; Crosta 1996; Crosta and Zanchi 2000; Agliardi et al. 2009; Martinotti et al. 2011), demonstrating that the process of characterizing these phenomena has been long and complex. These large slope instabilities can involve entire valley flanks, range up to several kilometers in length and hundreds of meters in depth, and present several typical morphological and structural features (e.g., scarps, trenches, double ridges, tension cracks) (Varnes et al. 1989; Agliardi et al. 2001; Tibaldi et al. 2004). These huge phenomena are the result of complex geological, geomorphological and structural settings and are often characterized by a long evolution. In particular, their evolution is controlled by interactions among the following factors: (i) lithology, (ii) geology, (iii) geomorphology, (iv) climate and weathering, (v) seismicity, and (vi) deglaciation (Crosta et al. 2013). They display evolutionary stages, and, during the advanced stages, can develop a creep mechanical behavior that leads to complete collapse. These phenomena generally display very slow or slow deformation rates that vary from a few millimeters per year to a maximum of some centimeters per year, in uncommon cases (Agliardi et al. 2012). In this context, Differential Synthetic Aperture Radar Interferometry (DInSAR) techniques represent a suitable tool for investigating slow-moving phenomena over large areas (100 × 100 km), given the long time period covered by the vast satellite system (e.g., ERS-1/2, Envisat ASAR, COSMO SkyMed, Radarsat−1).

In this paper, we propose a methodology based on Synthetic Aperture Radar (SAR) data made available by the Valle d'Aosta Regional Authority. We exploited the RADARSAT−1 dataset, which was processed using TRE S.r.l.'s SqueeSAR™ technique, covering the period from March 2003 to December 2010. Our methodology is aimed at identifying and analyzing Permanent Scatterers (PS) and Distributed Scatterers (DS), together with geomorphological and structural evidence, to identify the possible morpho-structural domains associated with three DsGSD case studies located within the Valle d'Aosta region of northern Italy. In this way, the most affected areas can be identified, providing significant information for use in natural hazard assessment and land use planning.

Valle d'Aosta Region Case Studies

Valle d'Aosta is a small alpine region (3200 km^2) located in northwestern Italy that has a complex topography ranging from 400 m a.s.l. to over 4800 m a.s.l. (Fig. 1). Due to the high topographic relief and the steep slope gradients, landslide processes are widespread and affect approximately 520 km^2 of the entire regional territory. In particular, DsGSDs occupy 13.5% (Fig. 1) of the regional territory (Trigila 2007).

The actual analysis focuses on three different case studies that represent significant DsGSDs within the Valle d'Aosta region (Fig. 1).

The first example is the Cime Bianche DsGSD, which is located in the upper part of Valtourenenche valley, above the Breuil-Cervinia settlement. This phenomenon presents a complex evolution that is characterized by the presence of recent signs of glacial activity and several active periglacial processes. The lower and marginal portions of this mass movement present the highest degrees of deformation.

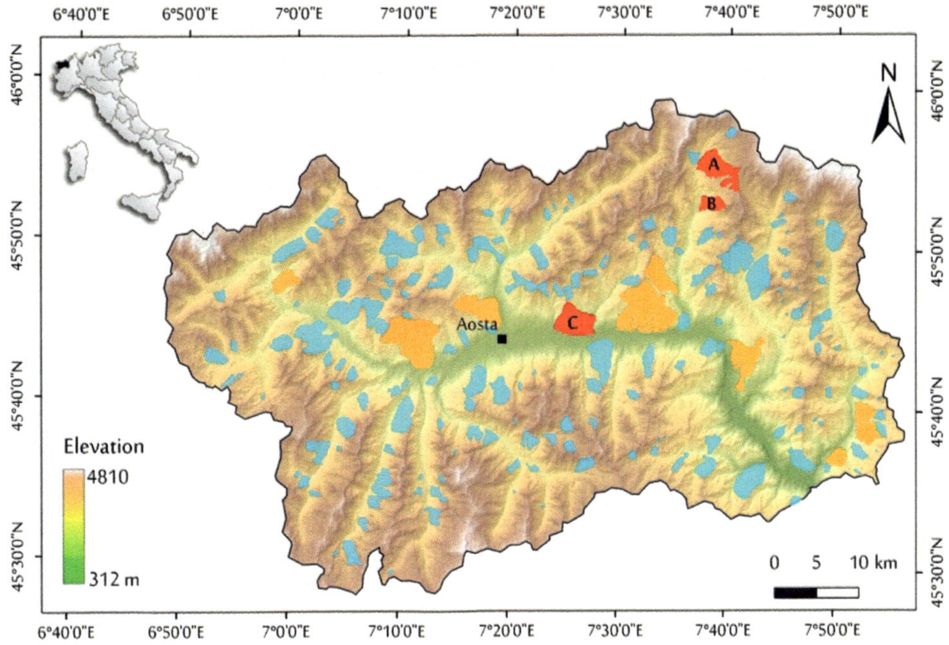

Fig. 1 Relief terrain of the Valle d'Aosta region, northwestern Italy. The map shows the distribution of DsGSDs (*blue* polygons; data from the IFFI project). The *red* polygons correspond to the three case studies: **a** Cime Bianche; **b** Valtourenenche; **c** Quart. The *orange* polygons correspond to other phenomena that involve the principal urban areas within the regional territory

The second case is located in the middle portion of the same valley and affects the town of Valtourenenche. This phenomenon is characterized by the local geomorphology, which is conditioned by the recent evolution of the stream of the same name and by the presence of several active slides superimposed on the DsGSD.

The last one is the Quart DsGSD, which is more complex than the other phenomena and is located on the left side in the middle of the main valley, not far from Aosta municipality. This DsGSD presents evidence of evolution over a long-period and is controlled by glacial activity, tectonic processes, and deep dissolution (Martinotti et al. 2011). The displacement is primarily parallel to the slope, but includes extensional and lateral components.

These three cases have been chosen because they affect several of the principal settlements of the region and present different stages of evolution.

Methods

Slow mass movements represent suitable cases for the application of DInSAR techniques (Colesanti and Wasowsky 2006; Cascini et al. 2010). The analysis uses the RADARSAT−1 dataset made available by the Regional Authority of Valle d'Aosta and processed using the SqueeSAR^{TM} technique by the TRE Srl.

As part of urban development planning, the use of SAR data [i.e., Permanent Scatterers (PS) and Distributed Scatterers (DS)] may cause problems for non-expert users. In the case of DsGSDs, the analysis of PS/DS should be a starting point in assessing the most active sectors, although the well-known intrinsic limitations of these techniques (i.e. phase decorrelation on vegetated areas, coherence loss due to large revisit time, phase decorrelation due to large and/or rapid displacement, line-of-site (LOS) measurements only) should be taken into consideration (Ferretti et al. 2001, 2011). However, to assess the complex evolution of these phenomena, the SAR data have been integrated with geological, geomorphological and structural information from the local settings. In this context, we attempt to define the morpho-structural domains within the DsGSDs. This definition requires an identification of the possible kinematic domain of the DsGSDs, while also taking into account any other active phenomena (i.e., landslides, rock glacier, talus), superimposed on the area of the DsGSDs. We perform the analysis using discretized SAR data in a GIS environment. Specifically, we interpolate the PS/DS LOS velocity values, taking into account specific barriers previously identified using knowledge of the geomorphology and the literature, as well as structural analysis. However, the discontinuous nature of the PS/DS distribution must be addressed because it can generate some limits on the applicability of interpolation functions. Appling this method, it is possible to

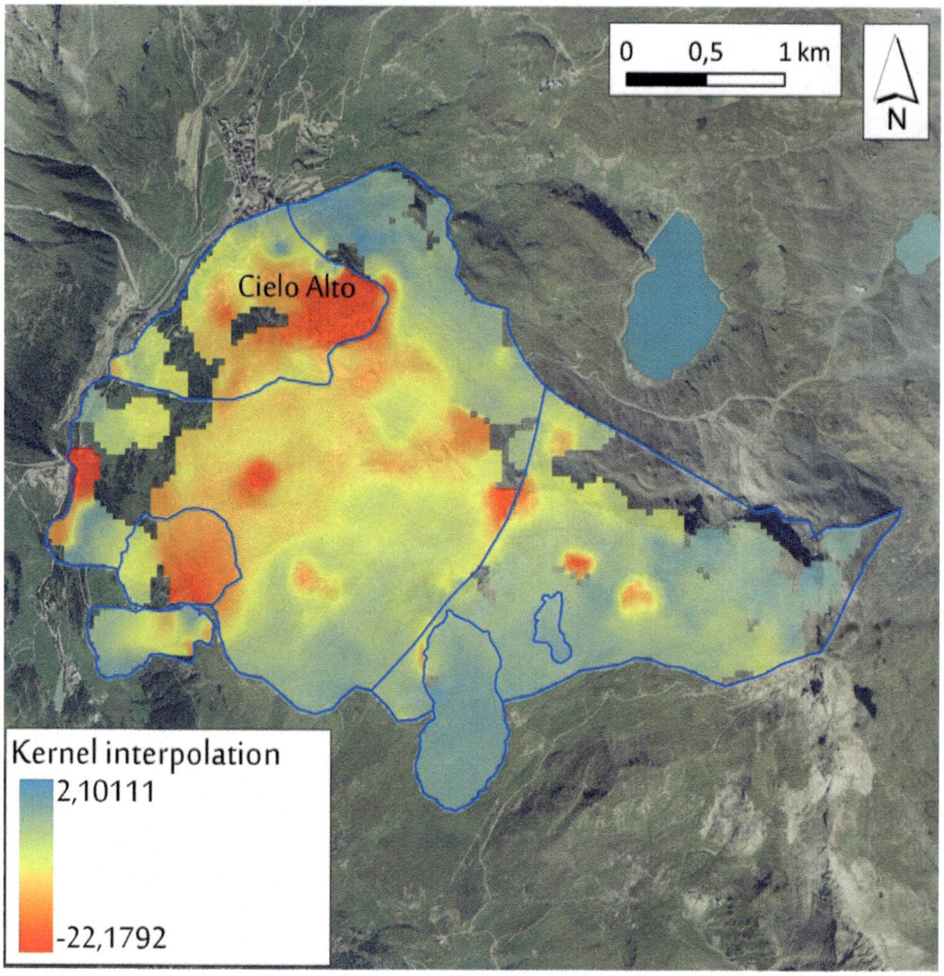

Fig. 2 Map of kernel interpolation incorporating barriers on the C.B. DsGSD, based on data from the RADARSAT−1 PS in descending orbit

subdivide the DsGSDs into diverse sectors while aggregating ground deformation data using specific limits obtained from geomorphological and structural constraints. In this way, we obtain areal deformation maps, which are more suitable than the classical PS/DS distribution maps and represent a useful tool in land-use planning and natural hazard assessment.

Results

We test our method by applying it to three specific DsGSDs of the Valle d'Aosta region (see Fig. 1) identified in the IFFI (Italian Landslide Inventory). The RADARSAT−1 images available from the Regional Authority cover the period from March 2003 to December 2010. These images processed using the SqueeSARTM technique, provide the resulting PS/DS data for this alpine region for both descending and ascending orbits.

Cime Bianche DsGSD

Considering the PS/DS data within the Cime Bianche (C.B.) DsGSD, good coverage and distribution have been observed. Specifically, the best coverage was obtained along the descending orbit. Figures 2 and 3 present the maps resulting from kernel and diffusion interpolation (cell size 40 m), respectively, applied using barriers, along the descending orbit.

We consider all the landslides superimposed on the DsGSD and subdivide this phenomenon into several distinct kinematic domains based on morpho-structural evidence and specifically on the rock mass structure and surface discontinuities.

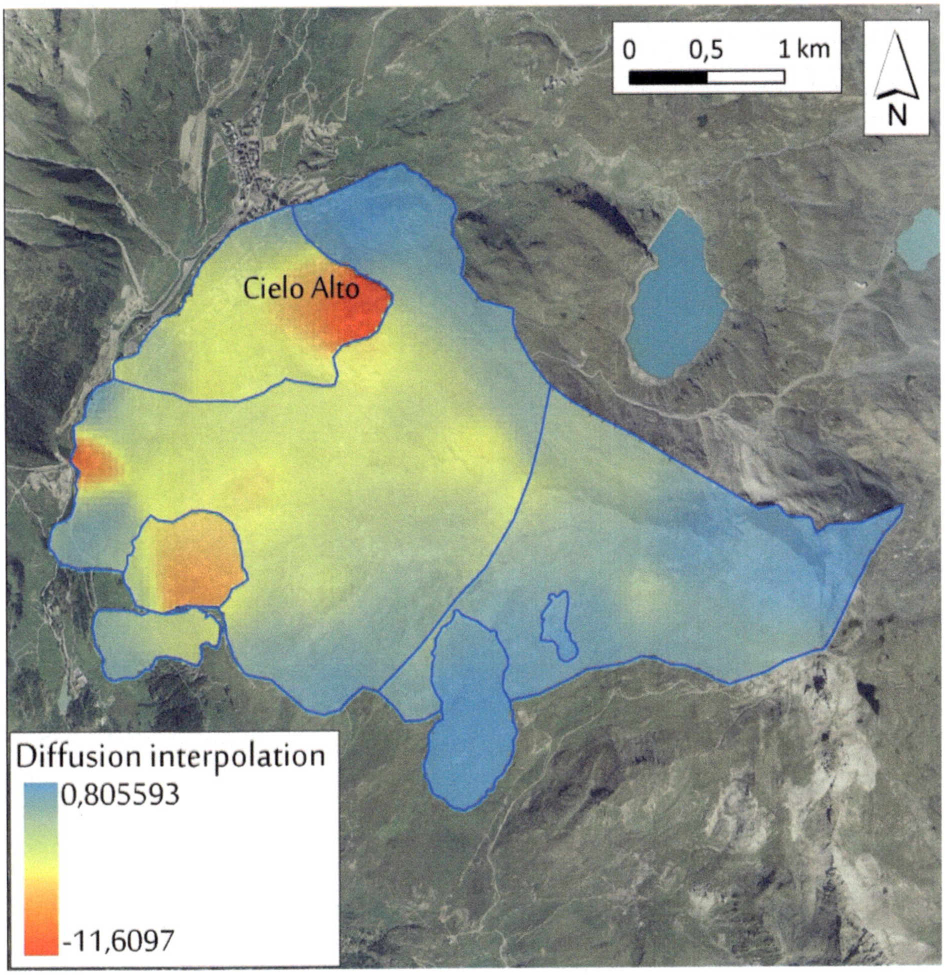

Fig. 3 Map of diffusion interpolation with barrier on the C.B. DsGSD, based on data from the RADARSAT−1 PS in descending orbit

In both cases, the most active domains correspond to the lower and marginal portions of the DsGSD, according to information from the literature. Specifically, a more active domain has been identified that corresponds to the location of Cielo Alto.

Valtourenenche DsGSD

Good PS/DS coverage and distribution have also been obtained in the case of Valtourenenche DsGSD. Figures 4 and 5 present the maps resulting from kernel and diffusion interpolation, respectively, with barrier application, using data from the descending orbit. As in the previous case, all the landslides from the IFFI have been considered, together with the identification of several other gravitational processes that are responsible for the topographic displacement measured by SAR. The identification of all of the processes that are able to generate surficial deformation is important to correctly assess the morpho-structural domains.

In the upper part, significant ground deformation that corresponds to two east-west oriented main landslide bodies has been identified, while the most active sector of the DsGSD corresponds to the lower portion, delimited by a main scarp in the upper part, close to Valtourenenche village. The presence of several landslide bodies superimposed over this sector, which further contribute to the DsGSD movement, should be noted. A middle level sector, corresponding to a transitional domain and presenting modest ground deformation, has also been identified.

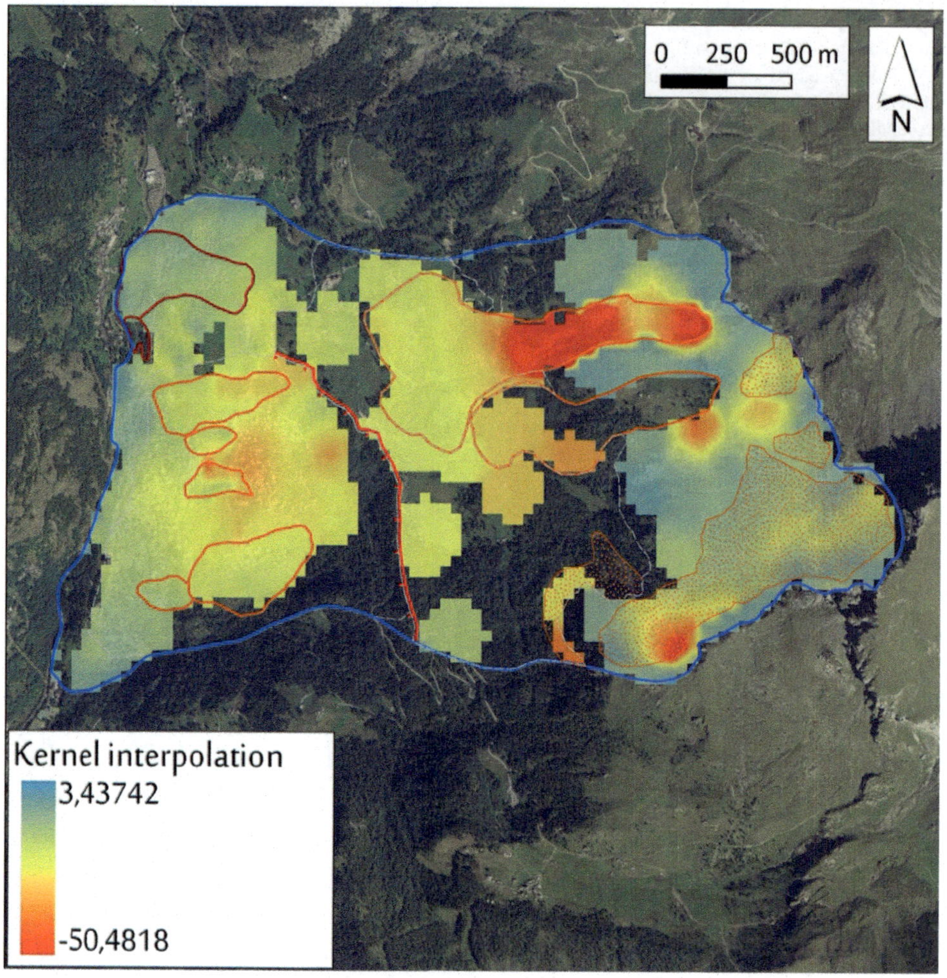

Fig. 4 Map of kernel interpolation with barriers on the Valtourenenche DsGSD, based on data from the RADARSAT−1 PS in descending orbit

In both cases, the complex evolution of this phenomenon is well-drawn by imposing as a limit all the numerous geomorphological and structural elements observed.

Quart DsGSD

In the Quart DsGSD, good PS/DS coverage and distribution have been observed for both the ascending and descending orbits.

The Quart DsGSD represents the most complex phenomenon considered in this study in terms of its long-term evolution and given the presence of a lateral valley, which causes the occurrence of two different directions of displacement. In this case, the kinematic domains proposed in the IFFI have been used for the application of kernel and diffusion interpolation (Figs. 6 and 7, respectively). A single modification has been introduced in the western portion of the DsGSD to separate the more active upper portion from the stable lower portion. In general, ground deformation decreases from the upper to the lower portions of the DsGSD. Finally, the presence of the Vollein active landslide in the eastern part of the DsGSD is outlined.

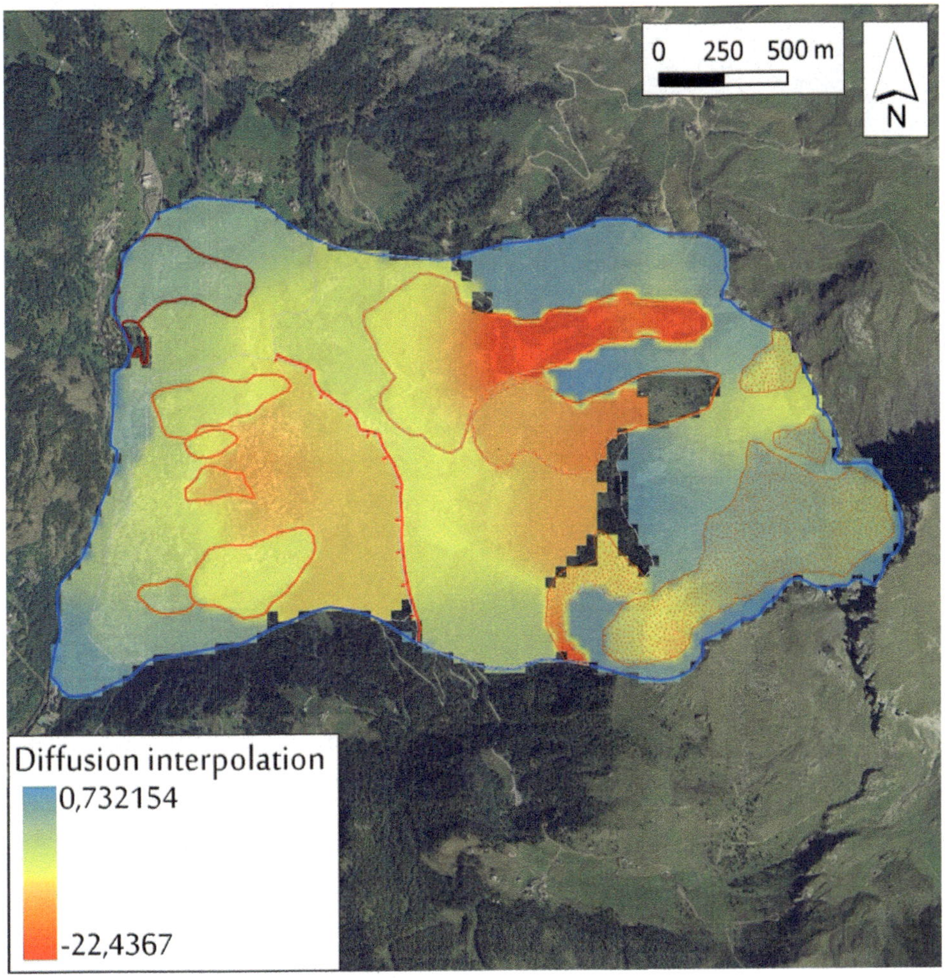

Fig. 5 Map of diffusion interpolation with barriers on the Valtourenenche DsGSD, based on data from the RADARSAT−1 PS in descending orbit

Discussion and Conclusion

The results obtained for the three case studies highlight the complexity of the DsGSDs, which are very complex phenomena that require a specific approach for their interpretation. The recognition of morphological elements is fundamental but sometimes difficult, given that DsGSDs result from long periods of composite deformation. The geological and geomorphological aspects can be used to define and characterize these phenomena, but only partially define their state of activity.

In the last decade, the introduction of the relatively new DInSAR techniques have permitted the extraction of ground deformation information over wide areas with millimeter accuracy (Ferretti et al. 2011). Nevertheless, the intrinsic limitations of these techniques (Ferretti et al. 2001) makes their application over DsGSDs suitable (Colesanti and Wasowsky 2006; Cascini et al. 2010).

The improved methodology permits assessment of the state of activity of the three DsGSD case studies. This methodology of using SAR data, in combination with geological and geomorphological knowledge, allows recognition of the morpho-structural domains of these phenomena. The rasterization of the SAR data, based on specific and reasoned limits, allows the derivation of maps of ground deformation through LOS velocities interpolation. In cases with good PS/DS coverage and distribution, the interpolation results are more reliable and best represent the real kinematic context.

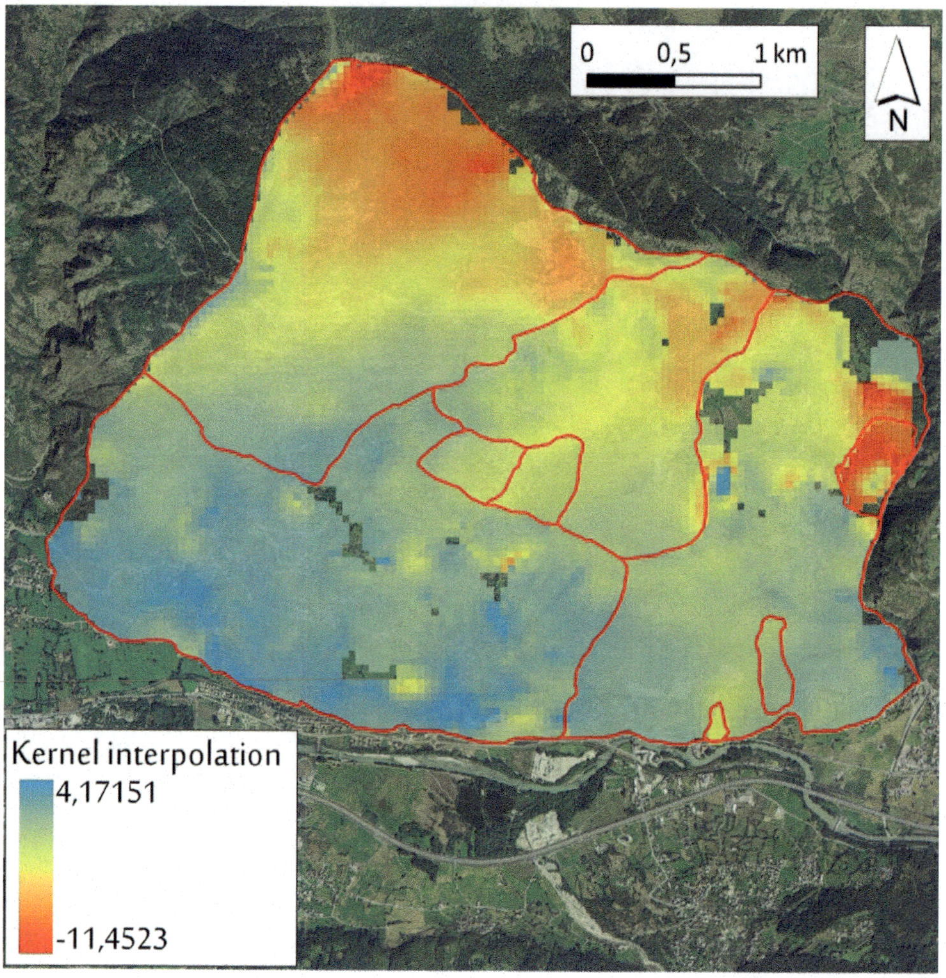

Fig. 6 Map of kernel interpolation with barriers on the Quart DsGSD, based on data from the RADARSAT−1 PS in ascending orbit

The aim of this methodology is to divide DsGSDs into sub-domains, though the definition of usage constraints that take into account the geological and geomorphological setting and the rates of mean ground deformation velocities obtained from the SAR data.

In this way, these morpho-structural domains may be used for a more efficient land management. The subdivision of DsGSDs into domains with similar characteristics can simplify the land management approach, allowing better management of hydro-geological constraints.

For the purposes of land management, the assessment of the morpho-structural domains of DsGSDs is to be viewed as a qualitative indicator, which must be integrated with field data and in situ monitoring.

A good practice will be to apply this approach over time, including constant updating of the SAR data. Integration of more recent images and data from newer satellites (i.e., Sentinel-1) should represent a good opportunity to analyze these phenomena over the coming years.

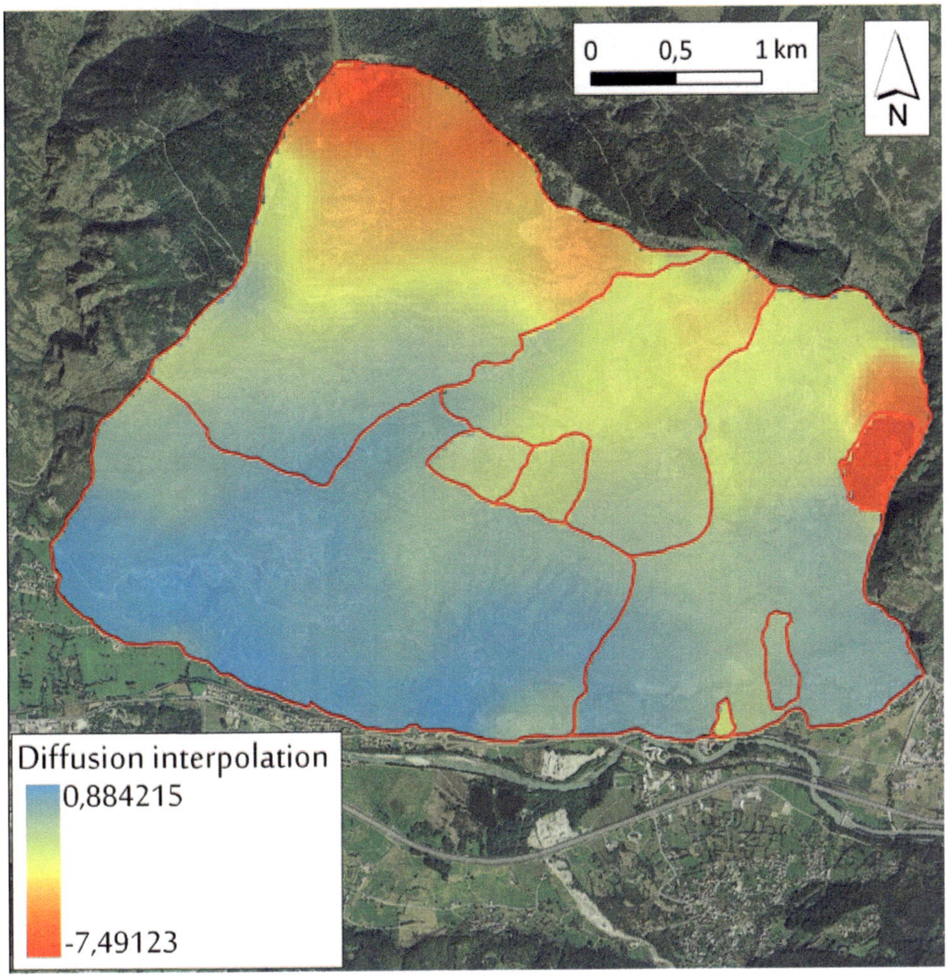

Fig. 7 Map of diffusion interpolation with barriers on the Quart DsGSD, based on data from the RADARSAT−1 PS in ascending orbit

Acknowledgements This research has been founded by Strutture Attività Geologiche of the Regione Autonoma Valle d'Aosta.

References

Agliardi F, Crosta GB, Zanchi A (2001) Structural constraints on deep-seated slope deformation kinematics. Eng Geol 59:83–102

Agliardi F, Crosta GB, Zanchi A, Ravazzi C (2009) Onset and timing of deep-seated gravitational slope deformations in the eastern Alps, Italy. Geomorphology 103:119–129

Agliardi F, Crosta G B, Frattini P (2012) Slow rock-slope deformation. In: Cague J J, Stead D (eds) Landslides types, mechanisms and modeling. Cambridge University Press, Cambridge, pp 207–221. ISBN: 978-107-00206

Ambrosi C, Crosta GB (2006) Large sacking along major tectonic features in the Central Italian Alps. Eng Geol 83:183–200

Cascini L, Fornaro G, Peduto D (2010) Advanced low- and full-resolution DInSAR map generation for slow-moving landslide analysis at different scales. Eng Geol 11:29–42

Colesanti C, Wasowsky J (2006) Investigating landslides with space-borne Synthetic Aperture Radar (SAR) interferometry. Eng Geol 88:173–199

Crosta GB (1996) Landslide, spreading, deep seated gravitational deformation: analysis, examples, problems and proposal. Geografia Fisica e Dinamica Quaternaria 19:297–313

Crosta GB, Zanchi A (2000) Deep-seated slope deformation. Huge, extraordinary, enigmatic phenomena. In: Bromhead E, Dixon N, Ibsen M (eds) Proceeding of the 8th international symposium landslides, Cardiff: Landslides in Research, Theory and Practice. Thomas Telford, London, pp 351–358

Crosta GB, Frattini P, Agliardi F (2013) Deep seated gravitational slope deformations in the European Alps. Tectonophysics 605:13–33

Ferretti A, Prati C, Rocca F (2001) Permanent scatterers in SAR interferometry. IEEE Trans Geosci Remote Sens 39:8–20

Ferretti A, Fumagalli A, Novali F, Prati C, Rocca F, Rucci A (2011) A new algorithm for processing interferometric data-stacks: squeeSAR. IEEE Trans Geosci Remote Sens 49(9):3460–3470

Mahr T (1977) Deep-reaching gravitational deformation of high mountains slopes. Bull Eng Geol Environ 16:121–127

Martinotti G, Giordan D, Giardino M, Ratto S (2011) Controlling factors for deep-seated gravitational slope deformation (DSGSD) in the Aosta Valley (NW Alps, Italy). Geol Soc London Spec Publ 351 (1):113–131

Mortara G, Sorzana PF (1987) Fenomeni di deformazione gravitativa profonda nell'arco alpino occidentale italiano. Considerazioni litostrutturali e morfologiche. Bollettino de Società Geologica Italiana 106:303–314

Savage WZ, Varnes DJ (1987) Mechanics of gravitational spreading of steep-sides ridges 8sackung). Bull Int Assoc Eng Geol 35:31–36

Tibaldi A, Rovida A, Corazzato C (2004) A giant deep-seated slope deformation in the Italian Alps studied by paleoseismological and morphometric techniques. Geomorphology 58:27–47

Trigila A (ed) (2007) Rapporto sulle frane in Italia, il Progetto IFFI—Metodologia, risultati e rapporti regionali (Rapporti 78/2007). APAT, Roma, p 681

Varnes DJ, Radbtuch-Hall D, Savage WZ (1989) Topographic and structural conditions in areas of gravitational spreading of Ridges in the Western United States. US Geological Survey, Professional Paper 1496

Zischinsky U (1966) On the deformation of high slopes. In: Proceedings of the first international conference of the international society of rock mechanics, Lisbon, vol 2. International Society of Rock Machanics, Lisbon: 179.185

Zischinsky U (1969) Uber Sackungen. Rock Mech 1:30–52

4th WLF 2017 4th World Landslide Forum LJUBLJANA SLOVENIA EU

Gediminas's Castle Hill (in Vilnius) Case: Slopes Failure Through Historical Times Until Present

Vidas Milkulėnas, Vytautas Minkevičius, and Jonas Satkūnas

Abstract

The remaining buildings of Gediminas's Castle in Vilnius stand on the top of a 40 m high hill composed of Quaternary glacial, glaciolacustrine, glaciofliuvial inter-layered deposits and technogenic (cultural layer) accumulations. The city center and castles are located in an area at the very margin at the maximum advance of the Weichselian glaciation. There was no direct erosion impact in the area of Vilnius's hills from meltwater for the formation of the upper reaches of the highest ravines. Therefore, it is proposed that the main features of the Vilnius Castles hills were formed by periglacial thermal erosion—the movement of land masses due to the thawing of permafrost at a time of climate change and the beginning of the vanishing of the Weichselian ice body. Over the course of history people reshaped the slopes of these hills for living and defense purposes. The saddle connecting Gediminas's Hill and the massif of the Hill of Three Crosses goes back to historical times when the artificial channel for the Vilnia River was dug and it became a separate hillfort. Due to the steep slopes of the hills, slope deformations and landslides have been occurring since the historical past until the present. Recently a landslide formed on the eastern slope in 2004 and reactivated in 2008. In early spring of 2016 two new landslides appeared on the northwestern slope preceded by a number of cracks on the ground surface. Causes of slopes failure and general problems of stabilization are dealt with this article, also covered is the necessity of early warning system installation, slopes surfaces permanent monitoring based on 3D laser scanning, etc.

Keywords

Gediminas's Castle Hill • Landslide • 3D scanning • Stabilization • Retaining wall • Pile

Location of Gediminas's Castle Hill

Gediminas's Castle Hill is located in the very historical centre of Vilnius, which is for its outstanding universal value included in the UNESCO World Heritage List since 1994.

V. Milkulėnas · V. Minkevičius (✉) · J. Satkūnas
Lithuanian Geological Survey Under the Ministry of Environment, S. Konarskio str. 35, Vilnius, Lithuania
e-mail: vytautas.minkevicius@lgt.lt

V. Milkulėnas
e-mail: vidas@lgt.lt

J. Satkūnas
e-mail: jonas.satkunas@lgt.lt

M. Mikoš et al. (eds.), *Advancing Culture of Living with Landslides*,
DOI 10.1007/978-3-319-53483-1_10

Vilnius historic centre began its history on the erosional hills at the confluence of the Neris and Vilnia rivers (Fig. 1) that had been intermittently occupied from the Neolithic period; a wooden castle was built around 1000 AD to fortify Gediminas's Hill. The settlement did not develop as a town until the 13th century, during the struggles of the Baltic peoples against the Teutonic Order. Later it was the centre of political and economic power in Medieval Lithuania. From the point of view of recent geological processes and phenomena, the steep slope bases were affected actively by erosion till the middle of the 20th century (Fig. 2). Later river flow was regulated by hydrotechnics.

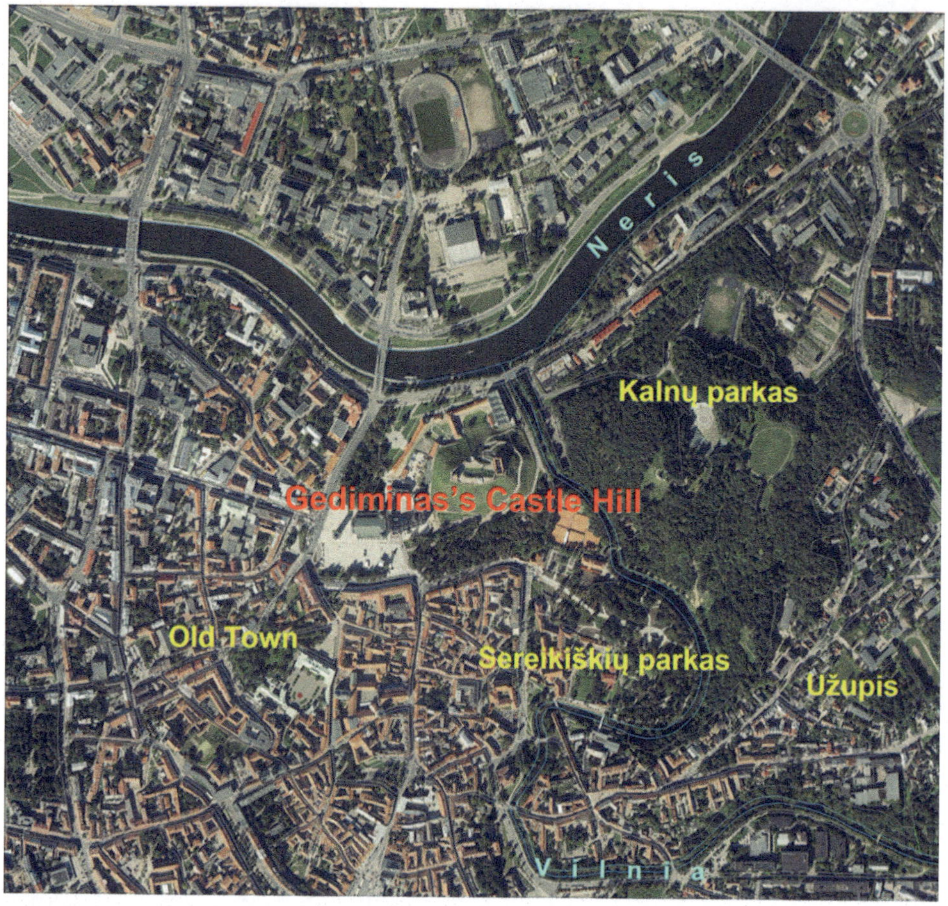

Fig. 1 Gediminas's Castle Hill location on the map of Vilnius centre

Fig. 2 Gediminas's Castle Hill in the centre of picture, view from the south east. On the right side—Bekes hill is seen, eroded by Vilnia River (photo about 1875, K. Brandel)

Origin of the Gediminas's Castle Hill

The remaining structures of the Castle stand on the top of the 40 m high hill composed of Quaternary glacial, glaciolacustrine, glaciofliuvial inter-layered deposits and technogenic (cultural layer) accumulations (Fig. 3). This hill is included in the State Cultural Reserve of Vilnius Castles that is characterised by a hilly topography with a variety of land forms, formed by glacial, fluvial erosion, suffosion and gravitational processes. The city center and castles are located in an area of the very margin of the maximum advance of the Weichselian glaciation (Satkūnas 2015). According to geological mapping data there was no direct erosion impact in the area of Vilnius's hills from meltwater for the formation of the upper reaches of the highest ravines (Guobytė 2008). Therefore, it is proposed that the main features of the Vilnius Castles hills were formed by periglacial thermal erosion—the movement of land masses due to thawing of permafrost at a time of climate change and the beginning of the vanishing of the Weichselian ice body. These masses became unstable and slid into the opening valleys of the Vilnia and Neris rivers. The saddle connecting Gediminas's Hill and the massif of the Hill of Three Crosses goes back to historical times when the artificial channel for the Vilnia River was dug and it became a separate hillfort (it was not eroded completely during formation of the first terrace of the Vilnia River because of this natural connection) (Fig. 4) (Satkūnas 2015).

Historical Landslides, Underground Constructions

Over the course of history people reshaped the slopes of suitable hills for living and defense purposes. Due to the steep slopes of the hills, slope deformations and landslides have been occurring since the historical past and until the present time. It is mentioned in the Livonian Chronicles dated back to 1396 that the southern slope of Gediminas's Hill slid down and destroyed the estate of Manvydas (Moniwid), the Voivode of Vilnius, causing the deaths of 15 people (Balinski 1836). Another landslide occurred in 1551 on the northern slope of Gediminas's Hill and the palace of Mikolaj Radziwil was damaged by this landslide (Monstvilas 1973). After that the retaining wall at the foot of slope was built. Unfortunately there are no records about the other historical landslides of the slopes of Gediminas's Hill. However, during geological investigations evidence of old landslides has found. Reinforcement remains (mainly retaining walls) on the base of slopes also indicate problems

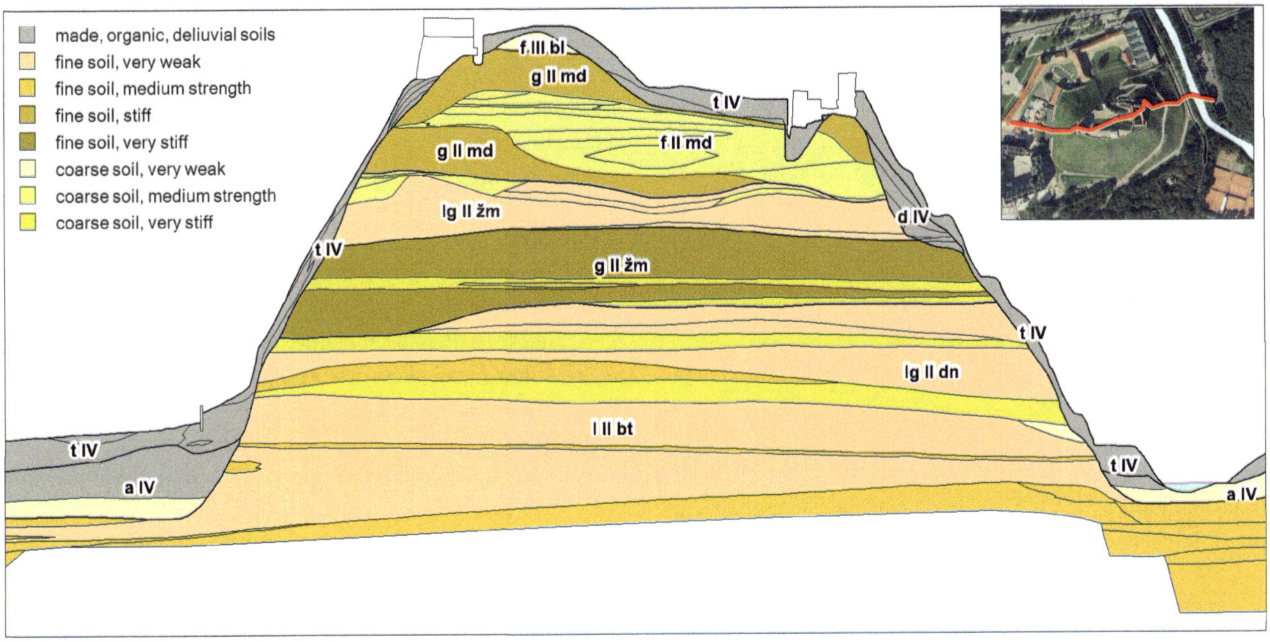

Fig. 3 Geological cross-section of Gediminas Castle Hill from W to E (acc. to Šačkus 2001; modified by V. Minkevičius, 2016). Indexes: *t IV* technogenic deposits of Holocene, *d IV* deliuvial deposits of Holocene, *a IV* aliuvial deposits of Holocene, *f III bl* graciofliuvial deposits of the Baltic stage (Weichselian), *g II md* glacial deposits of the Medininkai stage (*Upper* Saalian), *f II md* glaciofliuvial deposits of the Medininkai stage (*Upper* Saalian), *lg II md* glaciolacustrine deposits of the Medininkai stage (*Upper* Saalian), *g II žm* glacial deposits of the Žemaitija stage (*Lower* Saalian), *lg II žm* glaciolacustrine deposits of the Žemaitija stage (*Lower* Saalian), *g II dn* glacial deposits of the Dainava stage (*Upper* Elterian), *l II bt* lacustrine deposits of the Butėnai interglacial stage (Holstein)

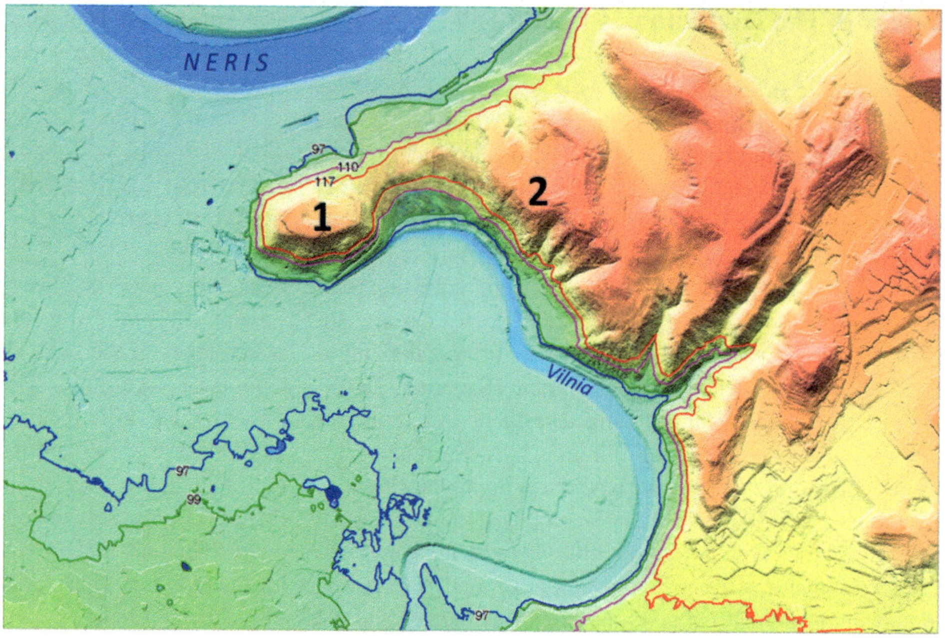

Fig. 4 The Hills of Vilnius Castles during the formation of the first Vilnia River terrace. Slopes are being modified by erosional processes. The Gediminas's Castle Hill (*1*) is still connected by a saddle with the massif of Three Crosses (*2*)

of slope deformations in 14th–17th centuries. The slopes of ancient cultural heritage objects like hillforts were and are affected by landsliding induced by both natural and anthropogenic factors (Mikšys et al. 2002).

During World War II german troops constructed the L shaped tunnel with ventilation shaft in the very body of Gediminas's Hill (Pocevičius 2016). They were under fire and collapsed in 1948. The current state and nature of the tunnel is unknown. However the tunnel could influence the overall stability of the hill and the remaining castle buildings on top. One sinkhole in the southern part of hill was observed in 2011. It was possibly due to poor technical condition of these tunnels.

Resent Constructions and Investigations, Means for Slope Stabilization

Gediminas's Hill stability is uneven because of geological, morphological peculiarities and human activities. It became clear during geological investigations in 1955–1959 that about one third of the hill's surface was lost since the Castle was established. Broader hill reinforcement works were initiated in 1985 when slopes were reinforced by a network of concrete elements anchored in the deeper layers of ground. These works were carried out until 1990. Soil slopes were reinforced with the exception of part of the eastern slope below the main path to the summit. Slopes were sown with grass and later they were overlayed by sward. 220 CFA piles (continuous flight auger) 280 mm in diameter and at depth of 2.0 to 7.0 m were installed along the main path sides in 1990. Reinforcement works were continued in 1995 and carried out until 1997 when this slope was reinforced with an additional 57 CFA piles and 55 anchors connected by pile capes. The total number of installed piles is 832 (717 CFA, 105 VCC (vibro concrete columns) and 10 driven piles) and 187 additional anchors. Thereafter the engineering geology and foundation construction company JSC "Geostatyba" installed 136 piles for the construction of a funicular in 2003 (Trumpis and Grigonienė 2010; Žoržojus and Baniulis 2010).

Despite these reinforcing works, they have remained unstable. Geodetic measurements in 1997 detected that the north-western slope started to move slowly and here a landslide occurred in April–May, 1998. The slope was again stabilized with piles.

A rather significant landslide on the eastern slope formed in 2004 and reactivated in 2008. There were notable deformations in early spring of 2003 (Fig. 5). Cracks on the walls of the upper castle and inclined trees on the slopes of Gediminas's Hill also indicated slow deformations. The slope was stabilized with geogrid in 2008 (Satkūnas et al. 2008). Works of slope stabilization were continued in 2009: all trees were removed from the slopes of the hill and, the drainage system was renewed.

A sinkhole occurred in 2010 in the northern part of square on the top of the hill. It was due to suffusion induced by ineffective removal of rainfall. Also a few subsidences appeared on the footpaths paved by boulders. Foundations of the pathway were deformed and exposed in several places

Fig. 5 Landslide in the eastern slope of Gediminas Hill formed in March, 2004 (above) and reactivated in March, 2008 (photos by Vidas Mikulėnas)

due to soil erosion. On the northern slope the pile capes and protective network of concrete elements became exposed at the surface in 2010 (Vaičiūnas 2011). The processes of deformation of these slopes reactivated in the February 2016 (Figs. 6 and 7).

The Last Slope Failure and 3D Scanning

Installation of rain drainage on Gediminas's Hill, paving of trails, cutting trees, arranging slopes cover were carried out since 2009. However in February of 2016 a few new

Fig. 6 Retaining pile caps exposed in February of 2016 by formation of shallow landslide of frontal type (photo by Vidas Mikulėnas)

Fig. 7 Slope failures of different shape and development on the representative slope of Gediminas's Hill in March 2016 (photo by Vidas Mikulėnas)

landslides of different shape appeared on the north-western slope leading by a number of cracks on the ground surface (Fig. 6). The cause of the slope failures was rapid warming of frozen weak deliuvium and technogenic soils and heavy precipitation.

The first crack on the slope surface was noticed in 1st February. Over a period of two months it developed into a set of different deformations and landslides. The most notable failure appeared next to the funicular. A circular crack on the surface appeared on 1st February, it developed into a landslide

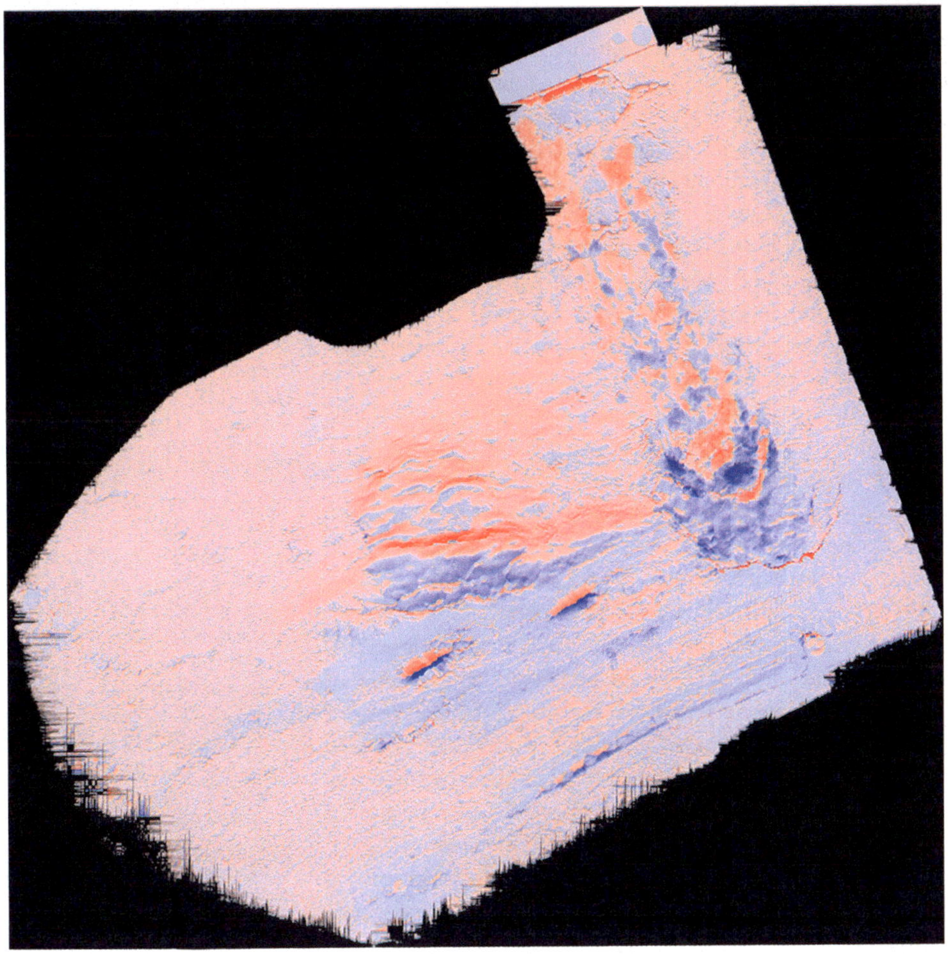

Fig. 8 An image of differentials of 3D scanning results: *red* −1.2 m, *blue* +1.4 m (after results of scanning work done by JSC "Geonovus")

in on a week and 12 cub. m. volume reached the building of the administration of the National Museum situated at a base of the slope. The direction of movement was parallel to the track of funicular 3.5 m away from it (Fig. 7).

A rotational landslide developed in 45–50 degrees steep slope. The main scarp was 12.5 m width and 1.0 m height. The landslide body was composed of cultural strata. The second landslide began to form to the west from the first one. It was very likely that the second landslide was about to develop to slumping soil massive through retaining wall (dated 16th century) at the base of the slope. Hopefully due to more favorable meteorological conditions this scenario will not came true.

The volume and morphological features of the developed landslide was estimated after by slope scanning using 3D "Trimble TX8" device. Scanning measurements were repeated after two weeks and three-dimensional models to compare difference of the scans were compiled (Fig. 8).

Relevant works have to be undertaken in order to stabilize the damaged part of slopes and prevent new deformations.

Data on these landslides were included in the State Geological Information System GEOLIS. Geological phenomena and processes subsystem recently contains records about 42 landslides occurred in territory of Vilnius City, developed both in natural and artificial slopes (Stankevičiūtė et al. 2013). When in Lithuania the most dangerous geological phenomena started to be inventoried annually in the period 1997–2003 (Bucevičiūtė and Mikulėnas 2003) and relevant subsystem developed in 2004 (Mikulėnas 2005).

Discussions

Comprehensive geotechnical solutions for slope stabilisations have be designed on the basis of detailed engineering-geological investigations of the Gediminas's Hill. Therefore, Lithuanian Geological Survey elaborated a program of coherent investigations for a period of at least two years. The program for the first year includes remote scanning and monitoring, compilation of digital terrain model from airborne or terrestrial LIDAR data; shallow geophysics (e.g. electrotomography, ground radar investigations), monitoring of slopes and building deformations, periodic remote scanning and setting of the early warning system of slope deformations. Suitable sites for direct geological research can be selected only after this phase.

During the second year monitoring has to be proceeded, geological research completed for purpose of stability assessment: engineering geological and geotechnical investigations (drilling, sampling, laboratory analysis, etc.), landslides susceptibility mapping, geotechnical modelling of slope stability. Based on the results of comprehensive investigations the relevant methods of slope stabilization could be designed thus improving the stability of Gediminas's Hill.

A problem of slope stabilization became more difficult considering status of cultural heritage. It is necessary to establish ground monitoring system and perform geotechnical investigations. Archive geotechnical data is not relevant to parameterization of soil geotechnical properties, levelling data must be renewed and provided with higher accuracy (airborne or terrestrial 3D scanning). Complex of geophysical investigations should be performed and verified by the other methods. Quantitative techniques for the evaluation of slope stability and further installation of means for stabilization of slopes are strongly recommended.

Early warning systems should be arranged and lead to permanent Gediminas's Hill monitoring. Surface methods for measuring the development of cracks, subsidence and uplift include repeated conventional surveying, installation of various instruments to measure directly, and tiltmeters to record changes of slope inclination near cracks and areas of the greatest vertical movements. Usually subsurface methods include installation of inclinometers and rock noise measurement instruments to record movements near cracks and others areas of ground deformation, etc.

Conclusions

Steep slopes are sensitive to interaction of natural processes and human activities that are witnessed by number of historical and recent deformations of the unique Gediminas's Hill as a part of the Vilnius cultural heritage.

New geotechnical investigations have to be performed because old ones lack important geotechnical parameters for numerical slope stability modelling. Modern slope modelling techniques could help to design proper slope stabilization system and could simulate various unexpected scenarios such as heavy rain, water supply or drainage system failures, impacts of additional constructions, etc.

Data of shallow geophysics has to be verified by drilling considering status and vulnerability of cultural heritage.

Acknowledgements We would like to express our gratitude to Creativity and Innovation center "LinkMenų fabrikas" and JSC "Geonovus" for 3D scanning and data processing.

References

Baliński M (1836) Historya miasta Wilna. T. 1: Zawierający dzieje Wilna, od założenia miasta aż do roku 1430: z rycinami. In: Marcinowski A (eds) Biblioteka Jagiellońska, Wilno. 285p

Bucevičiūtė S, Mikulėnas V (2003) Lietuvos karsto ir nuošliaužų informacinės duomenų bazės sukūrimas. Lietuvos geologijos tarnyba, Vilnius, 155p

Guobytė R (2008) Vilniaus pilių teritorijos egzotiškasis reljefas ir gelmių sandara = Unique topography and geology of the State Cultural Reserve of Vilnius Castles. Lietuvos pilys 2007(3):24–35

Mikšys RB, Marcinkevičius V, Mikulėnas V (2002) Human factors in landsliding processes of Lithuania. In: landslides: proceedings of the first european conference on landslides, Prague, Czech Republic, June 24–26, 2002. A. A. Balkema Publishers, Lisse, Abingdon, Exton (PA), Tokyo, pp 251–254

Mikulėnas V (2005) Lietuvos teritorijos nuošliaužų inventorizacija ir duomenų įskaitmeninimas: aiškinamasis raštas ir tekstiniai priedai. Lietuvos geologijos tarnyba, Vilnius, 35p

Monstvilas K (1973) Gedimino kalno Vilniuje sutvarkymas. Gedimino kalno ir Katedros aikštės inžineriniai geologiniai tyrinėjimai (rankraštis). ITI archyvas, Vilnius

Pocevičius D (2016) 100 istorinių Vilniaus reliktų: [nuo XIV a. iki 1944 m.]. Petrašiūnaitė L (eds) Kitos knygos, Vilnius, pp 29–32, ISBN 978-609-427-207-3

Satkūnas J (2015) Vilniaus pilių kalvyno kilmė = Origin of the Hills of the Vilnius Castles. Chronicon Palatii Magnorum Ducum Lithuaniae. Vol. III (MMXII-MMXIII) = Lietuvos Didžiųjų Kunigaikščių rūmų kronika. III tomas (2012–2013). Nacionalinis muziejus Lietuvos Didžiosios Kunigaikštystės valdovų rūmai, Vilnius, pp 282–291

Satkūnas J, Mikšys RB, Mikulėnas V (2008) Minkevičius V (2009) Geodinaminiai procesai Vilniaus pilių teritorijoje: šlaitų deformacijos = Geodynamic Process in the territory of Vilnius Castles: Deformations of the Slopes. Lietuvos pilys 4:62–68

Stankevičiūtė S, Kanopienė R, Mikulėnas V, Vaičiūnas G, Minkevičius V, Petrauskaitė L (2013) Vilniaus miesto inžinerinių geologinių duomenų bazės sudarymas = The creation of Vilnius Engineering-Geological Database // Lietuvos geologijos tarnybos 2012 metų veiklos rezultatai: [metinė ataskaita] = Lithuanian Geological Survey: Annual Report 2012, Vilnius: LGT: 18–22

Šačkus V (2001) Gedimino kalno Vilniuje tyrimo duomenų inžinerinė geologinė analizė: baigiamasis bakalauro darbas, Vilniaus universitetas. Gamtos mokslų fakultetas, Vilnius

Trumpis G, Grigonienė L (2010) Gedimino kalno ir aukštutinės pilies Vilniuje (unikalus kodas 141) atliktų geologinių ir hidrogeologinių tyrimų archyvinės medžiagos sąranka, analizė ir išvados. VĮ "Lietuvos paminklai" archyvas, Vilnius, 34p

Vaičiūnas G (2011) Gedimino kalno šlaitų stabilumo vertinimas = Evaluation of Gediminas Hill's slopes stability. Lietuvos pilys 2010(6):65–75

Žoržojus G, Baniulis D (2010) Gedimino kalno ir aukštutinės pilies Vilniuje (unikalus kodas 141) deformacijų stebėjimo ir atliktų šlaitų tvirtinimo darbų archyvinės medžiagos sąranka, analizė ir išvados. VĮ "Lietuvos paminklai," Vilnius, 35p

Design Criteria and Risk Management of New Construction in Landslide Areas: The Case of the Djendjen–El Eulma Highway (Algeria)

Mirko Vendramini, Attilio Eusebio, Fabrizio Peruzzo, Patrizia Vitale, Alessandro Fassone, and Francesca Guazzotti

Abstract

This paper presents the approach, hereafter named "dual approach", applied in reducing risks of roadworks-related landslides, based on the preliminary landslide assessment and the Observational Method (OM). This approach has been implemented to the design of a 110 km motorway project in Algeria, during the last 10 years. The North-South oriented alignment of this motorway crosses different environments starting from the plain area of the Djen Djen valley, passing through a very irregular and geomorphological composite area of the "Petit Kabylie" massif, up to the future connection with the existing East-West Highway. The geological context is characterized by the presence of several complex rock formations, including heterogeneous rock masses as Flysch. These rock formations are strongly deformed and weathered such that landslide occurrence can be often related to the poor unfavorable geological, structural and geotechnical conditions. Roadworks design and construction in such complex and irregular mountainous areas are faced with various and significant natural hazards including large extended landslides. The parallelism of and partial superimposition between the design and the construction phase had mainly two consequences; notwithstanding the fact it forced the timing of the two phases, on the other hand, it allowed designers to quickly update the base reference model and design choices in several cases, with significant improvements and advantages in terms of time and cost.

Keywords

Landslides • Observational method (OM) • Geological reference model (GRM) • Case study • Sensitivity analysis • Risk assessment • Risk management

M. Vendramini (✉) · A. Eusebio · F. Peruzzo · P. Vitale ·
A. Fassone · F. Guazzotti
Geo-Engineering Department, Geodata Engineering,
10128 Turin, Italy
e-mail: mve@geodata.it

A. Eusebio
e-mail: aeu@geodata.it

F. Peruzzo
e-mail: fpe@geodata.it

P. Vitale
e-mail: pvi@geodata.it

A. Fassone
e-mail: afa@geodata.it

F. Guazzotti
e-mail: fgz@geodata.it

M. Mikoš et al. (eds.), *Advancing Culture of Living with Landslides*,
DOI 10.1007/978-3-319-53483-1_11

Introduction

The new highway under construction develops from Djen Djen harbor up to the connection with the existing East West highway near El Eulma. The alignment develops through the Djen Djen valley, from the coastline, crossing the Atlas Mountains of the so called "*Petit Kabylie*", up to the high plateau of El Eulma. The middle part of the alignment is the most critical because of its irregular, steep and complex landscape.

The project has dealt with several critical technical aspects, mainly concerning the morphological (steep and irregular slopes) and the geological/geotechnical conditions (heterogeneous geological formations, tectonically active sectors, landslides areas). Several interferences with pre-existing roads and ongoing civil works construction (Tabellout dam) have also been faced.

Complex and heterogeneous rock formations (Flysch) are widespread in the project area and consist of shales/siltstones and sandstones, marls and limestones sequences (Tertiary and Cretaceous Flysch).

More than 60 landslides with different geometry, dimension, kinematics features and evolution degree have been detected. A dynamic and multidisciplinary approach (dual approach) has been implemented, since the earliest phases of the project (Feasibility studies). With the aim both to manage the related risks and to identify the proper design solutions, this approach focused on defining the reliable Geological and Geotechnical Reference Model (GGRM) which was progressively updated with the most efficient site investigation and monitoring methodologies (OM). The geological and geotechnical studies, among which landslide assessment and characterization, have been performed taking carefully into account technical needs and schedules to complete activities. The GGRM as well as the procedure of risks assessment represented a solid base for the assumptions and design choices, constantly updated with a continuous data flow coming from both site investigations and feedback from works construction (follow up and monitoring).

Geological Aspects

Regional Geological Framework

The project area is located in the so-called Maghrebian Cenozoic orogenic domain, which forms a 2000 km extended chain from Gibraltar to Sicily (Bouillin 1986). This domain is classically separated in two different systems: the Tell–Rif or Maghrebides and the Atlas (Fig. 1). Fringing the west Mediterranean Sea, the Tell–Rif system (Tell in Algeria and Tunisia, Rif in Morocco) is assumed to be an 'Alpine chain' (Durand-Delga and Fonboté 1980), that means a chain resulting from the closure of an oceanic domain. More precisely, Bouillin (1986) divides the Maghrebides domain in three zones:

1. the *Internal Zone* (or Kabylides), of European affinities; including the metamorphic Kabylie Massifs and the Mesozoic to Tertiary sedimentary cover of the kabylies basement (Dorsale kabyle);
2. the *Flyschs domain*, which corresponds to the sedimentary cover of the Maghrebian Tethys (Wildi 1983); The Flyschs domain refers to a deep marine depositional basin that is Middle Jurassique to Lower Tertiary in age;
3. the *External Zones* (Tell s.s.), which consist of a thick pile of allochthonous terranes, the so-called 'Tellian nappes' located between the Flysch domain and the Atlas. They represent the sedimentary cover forming thick sedimentary sequences of dominantly marly carbonate-character. According to many authors (Durand Delga 1955; Bouillin 1977; Wildi 1983; Vila 1980), this domain can be also divided, from North to South, in the following units: Ultra Tellian units, Tell units s.s., Epi Tellian units.

Geological Layout of the Project Area

The project area extends through different geological and geomorphological domains. The Miocene sedimentary deposits widely outcrop in the first 20 km of the project area and they consist in grey-blue medium plastic to plastic clayey marls with subordinate intercalations of sandy layers. A Quaternary cover (colluvial cap and stratified fine or coarse alluvial deposits) can be observed along the main rivers and drainages and above the Miocene succession as well.

Approaching the Atlas Mountains, Flyschs formations can be detected. They represent the dominant and most complex geological formations encountered in the project area. Lower Cretaceous to Oligocene in age, these formations characterizes the landscape between the Kabylies chain (Texanna) and high Plateaux (Mila). The Flyschs domain, extended for approximately 30 km, includes several kinds of Flyschs: the Cretaceous *Mauritanian and Massylian Flyschs*, which consist in deformed shales and claystones with quartz veins, limestones and sandstones interbedded (Fig. 2), and the Eocene—Lower Oligocene Numidian flysch characterized by light brown sandstones with subordinates shales and claystone layers interbedded.

Gneiss, micaschists ("*Socle Kabylie*") and evaporitic rocks (Triassic formation) outcrop between Texanna and the Tabellout Forrest, the most tectonically deformed zone crossed by the alignment. Features of geological formations are different concerning their stratigraphy (lithotypes nature, ratio between strong and weak rocks) and tectonic

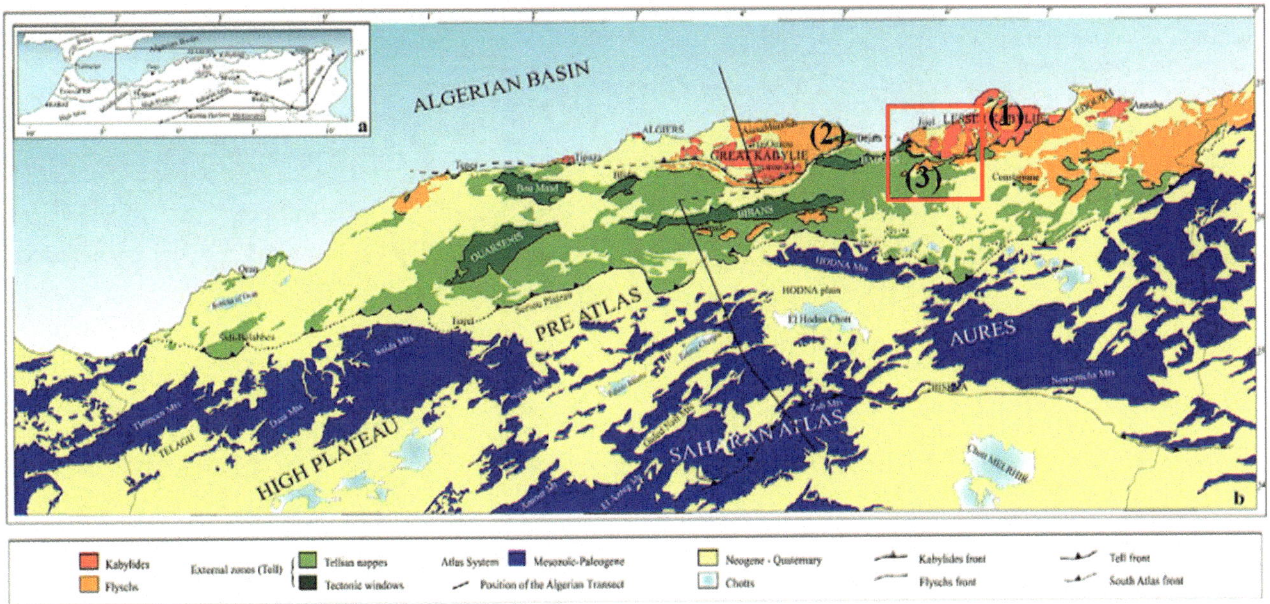

Fig. 1 Structural map of North Algeria (modified from the Geological Map of Algeria, scale: 1:500,000): Maghrebian structural domains (1, 2, 3) and location of the project area (Benaouali-Mebarek et al. 2006)

Fig. 2 *Massylian Flysch* (Aptian-Albian): tectonically deformed shales, claystones with quartzite veins, sandstones and limestone layers interbedded

disturbance. Because of their central position in the Atlas chain, Cretaceous Flyschs (*Massylian and Mauritanian flyschs*) are both strongly deformed (folded and faulted) and affected by several kinds of slope instabilities depending on the nature, the weathering and the structure of rock types (rock fall, slides, complex landslides).

On the other side, the Numidian Flysch outcrops in more external areas and it seems to be less deformed. In both cases, a well-developed weathering profile can be observed in the shallow horizons of the stratigraphic succession.

Field surveys show that landslide occurrence (slides, falls and erosion processes) increases when chemical and mechanical weathering processes are pervasive (Fig. 3), because they are responsible for the formation of a residual soils which control surface morphology and dynamics.

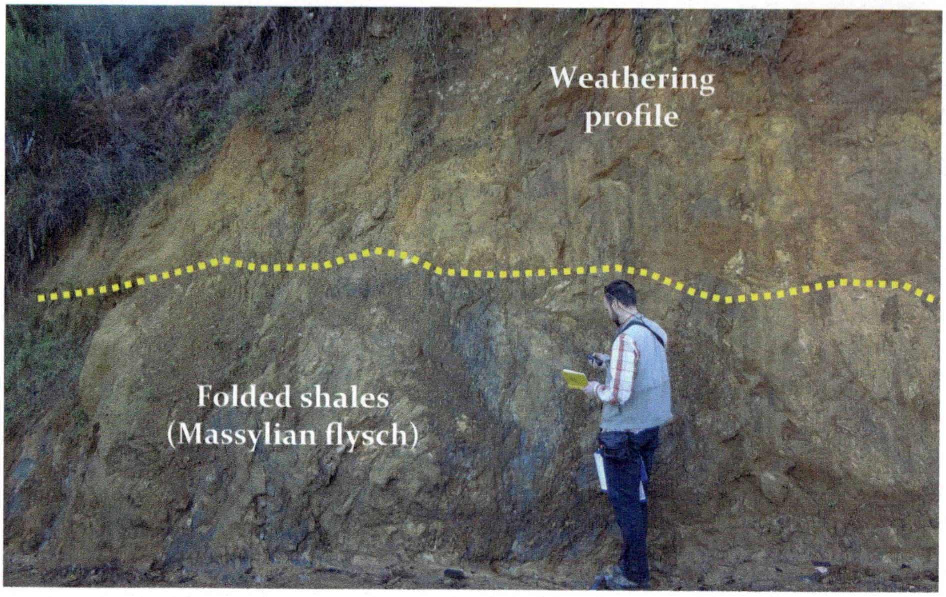

Fig. 3 Weathering profile observed in an outcrop of the Cretaceous *Massylian Flysch*

The Geological and Geotechnical Reference Model (GGRM)

According to the concept introduced by many authors in the last few years, the GGRM is considered as a conceptual framework where the collected data are comprehensively stored (factual data) and interpreted, anticipating and characterizing the ground conditions with their related risks (Riella et al. 2015).

Thus, the "reference model" can be considered a tool to help understand, define, quantify, visualize, or simulate the relevant aspects of the study area.

Ground Investigations

It is common knowledge that some geological, hydrogeological and geotechnical aspects of a project can remain partially or completely unknown prior to the actual engineering construction phase, mostly due to intrinsic and objective difficulties in performing site investigations (Riella et al. 2015).

As a consequence, planning and management of suitable site investigation campaigns is the key to obtain sufficient and reliable data for engineering project design in a timely manner while endorsing an economic design. In this case study, the method and quantity of ground investigations have been established taking into account both technical and contractual aspects, as reported below:

- design criteria and design norms of reference;
- type and importance of works (viaduct, tunnel, embankments, excavations);
- previous Site Investigation (SI) and technical data availability;
- geological and geotechnical model and residual uncertainties (i.e. landslide areas, strongly tectonically deformed zones);
- drilling and laboratory equipment availability (capability of sub-contractors);
- intrinsic difficulties of investigations, as well as access limits, time, and performance of each methods;
- overall budget.

Geological Surveys

At the beginning of the design phases, field investigation campaigns were conducted in order to elaborate a preliminary geological reference model, describing the general geological, stratigraphic and structural conditions of the project area as well as the features of natural slopes and rivers dynamics.

The field studies were focused on the following elements: soils and rocks characterization, assessment of stratigraphic succession of heterogeneous rock masses such as Flysch, their weathering degree, their intact rock strength and the characterization of discontinuity surfaces in terms of spacing, persistence and orientation. In the earliest stages of the

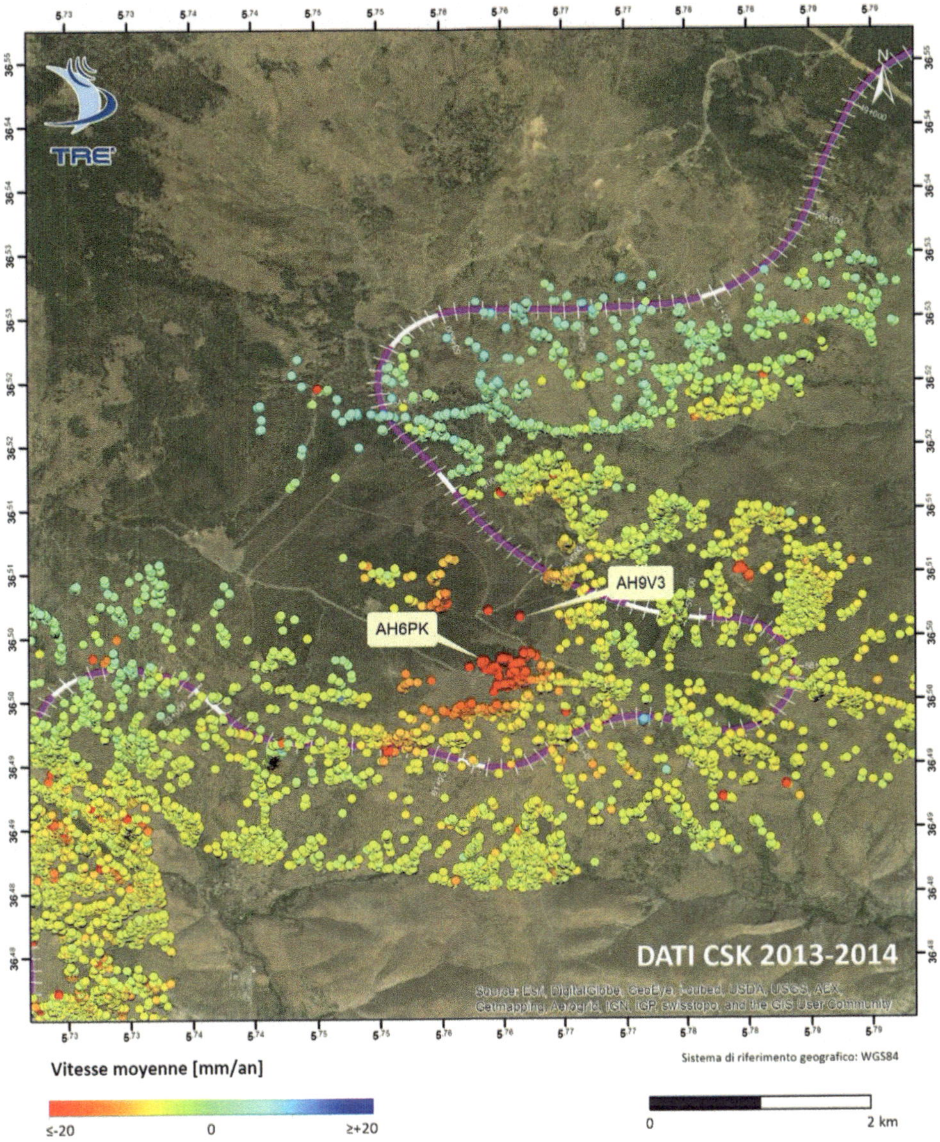

Fig. 4 Output from SAR analysis (Atlas domain)

study, landslides and potential instable sectors were identified and documented in detailed geological maps (scale 1:5000). This preliminary layout has been then used as a reference for the additional SI design.

Remote Sensing and Interferometry Analysis SAR

Analysis of remote sensing data and interferometry analysis, based on both aerial photos and satellite monitoring systems, have been used to assess soil and landslide movements and their state of evolution (active, dormant and stable areas); in particular, photointerpretation and SAR analysis allowed the investigations movements that occurred over a span of 20 years (historic series) in significant areas compared with the alignment position. Combining SAR analysis with detailed field surveys, the most critical and active areas have been detected and characterized according to their activity conditions (Figs. 4 and 5).

Geotechnical and geophysical investigations On the basis of the preliminary results of field surveys, more than 350 boreholes were drilled; Hundreds of pressiometric tests, permeability tests, SPT and thousands of laboratory tests have been also performed. As a consequence of facing with poor quality drilling and laboratory equipment, site and laboratory tests have been established in order to be able to cross check results and reduce uncertainties. Correlations between in situ and laboratory tests have then been established. Due to the difficulty in recovering intact core samples, significant geotechnical parameters (drained and undrained cohesion and friction angle) derived from Triaxial and Direct Shear tests have been compared to in situ tests outcomes. The methylene blue test helped to check reliably

Fig. 5 Landslide occurred along the national road RN77 in Miocene marine formation (weathering profile)

of the Atterberg limits and water content to correlate results of undrained shear resistance in saturated soils. This approach allows designers to adjust progressively the geotechnical characterization of soil and rock masses, yielding step by step the most reliable ranges of design parameters.

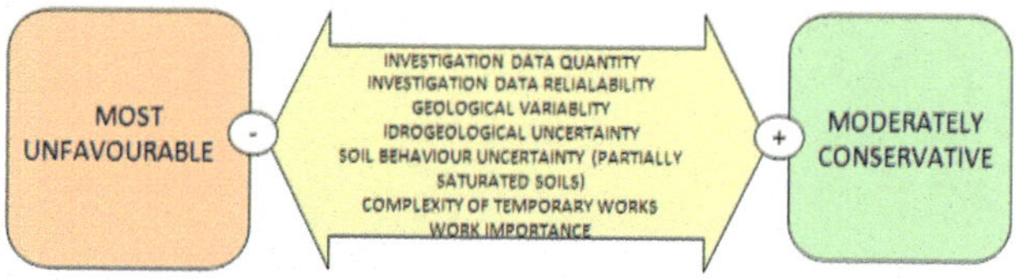

Fig. 6 Choice of design parameters values

Modeling and Design Criteria

On the basis of the geological reference model, 10 major geotechnical units (UG) have been identified. A general geotechnical characterization of UG has been carried out using both SI results and outputs of the adopted correlations. In case of a significant variability of the local geological and geotechnical conditions, design parameters have been adjusted accordingly. In order to develop a full and reliable geotechnical characterization of the detected materials, both Mohr-Coulomb and Hoek & Brown criteria were used depending on the geological, petrographic and stratigraphic nature of soils or rock masses, their weathering degree and geostructural features.

The definition of the parameters values has been done taking into account the uncertainties related to the reliability of site investigation results and the associated risks. In some cases the designer had to deal with problems linked to the very low reliability of the laboratory tests that lead to the necessity of adopting empirical correlations. Consequently, the designer was forced several times to use conservative assumptions during design stage. As a matter of fact, the design parameters values range between the most unfavorable and the moderately conservative values (Fig. 6).

In detail, the most unfavorable condition represents the 0.1% fractile of the data, while the moderately conservative one is a value that is situated between the mean value and the characteristic value defined by the Eurocode as the upper 5% fractile.

The moderately conservative parameter value is to be considered not a precisely defined value but rather a cautious estimate of a parameter (Patel et al. 2007).

Concerning the zones affected by instability or identified as barely stable, back analyses have been also performed in order to evaluate the lower bound parameter values derived from in situ tests. Hence, with the aim to avoid a too pessimistic geotechnical characterization of materials, the choice of design parameter values has been optimized both using the Observational Method (OM) and implementing sensitivity analysis as described furtherly in this paper.

Risk Assessment and Management from the Phase of Design to the Construction

Risk Assessment and Landslide Susceptibility

The effectiveness of the modern procedures of "flexible design" and "Risk Management"—today integrated inside the best practices of the geotechnical design (AFTES GT32. R2A1 2012)—is enhanced when they are based on a sound preliminary "diagnosis" phase as well as a reliable and solid reference model (Riella et al. 2015).

In the case study, the risk assessment and management procedure has been implemented as an iterative tool, from the earliest phases of design up to the construction, taking into account both design criteria and construction requirements (above all safety factors, works quality and costs). The adopted general approach follows the AFTES (2012) guidelines and recommendations in which uncertainties and risk assessment procedures have been progressively defined, detailed and updated.

More generally, the logical process of risk assessment and management has been focused on the definition of following principal aspects:

- definition of the local geological setting;
- available data quality and their reliability;
- probability of occurrence of risk sources (hazards) and their potential impacts on the project (principal risks);
- definition of countermeasures (for risk level reduction), such as complementary and secondary investigations, design parameters selection, structural measures;
- residual risks and uncertainties, such as unexpected geological conditions.

With reference to the landslides risk, it can be noted that such risk could be related to (1) old landslides thus potential reactivation due to engineering disturbance or (2) triggering of new landslides or instabilities due both to poor geological and geotechnical conditions as well as their evolution in time and to inadequate design; furthermore, such risk may manifest either at construction stage or later in the operation stage both of which require a proper risk management plan.

Field evidences show that some landslide related geological and geotechnical (GEO) factors play a key role in defining the most critical "landslide prone" areas:

- **Stratigraphic factor**: The presence of a dominant plastic clayey component both in soil and rock succession tends to reduce the effective global shear strength of the materials so that they can be involved in landslide phenomena. Additionally, different kind of landslides can be triggered also by the presence of chaotic bodies within the "ordinary" sedimentary succession (i.e. Olistostomes or tectonic breccia).
- **Weathering factor**: Weathering processes can affect deeply the geotechnical properties of soil and rock masses, reducing the physical and the strength characteristics of materials. They affect significantly the integrity and durability of a rock after excavation. According to Mulenga (2015) the original geotechnical characteristics of materials deteriorated as the degree of weathering and exposure time increases.
- **Structural factor**: a large part of detected landslides or instable areas are located in the tectonically active areas (i.e. Flyschs domain) characterized by the presence of several major faults, folds or unfavorable joints and beds orientation.
- **Hydrogeological factor**: Local water tables can be detected inside the Flysch sequences where medium to high permeable layers such as fractured sandstone and limestone layers are present. Several landslides can be triggered by water over pressures occurred in the shallowest permeable layers and in the weathering profiles of rock masses.
- **Morphologic factor**: because of the active and persistent regional tectonic activity, landscape morphology is generally composite with high and irregular steep slopes. As observed in the field, landslides susceptibility and occurrence increase in mountains areas (Atlas Chain) where the landscape shows a strong structural-control.
- **Geotechnical factor**: some landslides can be triggered by the overestimation of strength parameter especially when a fair accuracy and reliability of geotechnical data have to be faced.

Because of such strict relationship between the "GEO" relevant key factors and the landslide occurrence, the choice of proper geotechnical and geo-mechanical parameter values has been a really sensitive aspect in the project design. Especially, in case of works located in instable areas, initial risk level has been reduced to an acceptable level with a continuous process of study (additional investigations), monitoring (countermeasures) and control. In this process the Observational Method (OM) has been a suitable tool.

The Observational Method

The observational method (OM) was introduced by Peck in 1969 The OM, as it is interpreted nowadays, is slightly different but the principles remain the same. The approach proposed by Peck (1969) suggests to design with the most probable parameters values and then eventually reduce them to the moderately conservative values following the evidence of the monitoring; besides, the modern concept reverses this method, using a more conservative approach at the beginning of a project and eventually improve it to most probable conditions through field observations (Patel et al. 2007; Powderham et al. 1996; Nicholson et al. 2006). One of

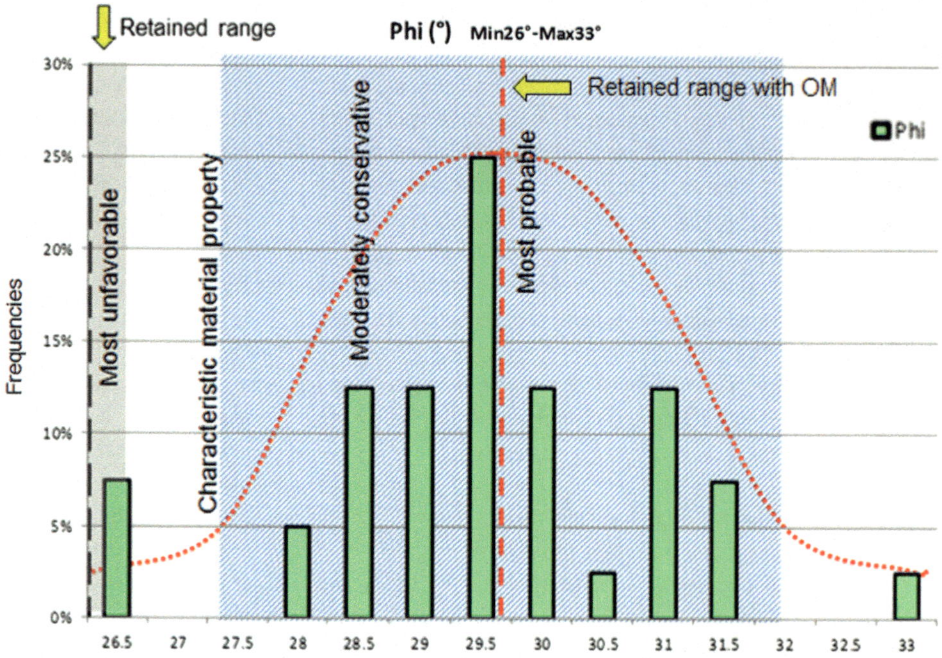

Fig. 7 Choice of parameters values: comparison between traditional and observational method

the best approaches, internationally recognized, is the one exposed in CIRIA 185; other regulations and best practices treat the subject but lack details on some of the main features of the OM process (Fig. 7).

The main benefits of using the OM are cost and time saving, control improvement and increasing scientific knowledge through feedback exchanging. It is suitable when the materials concerned in the design have sufficient ductility whilst brittle behavior does not allow sufficient warning for interventions. In some cases, when rapid deterioration of the material is an issue, a multi-stage process and incremental construction process may be applicable.

Prior to design process, an Operational Framework needs to be established. The one proposed in this project has been developed on the basis of the ones proposed by CIRIA 185 and Nicholson et al. (1999) with some adjustments, as shown in the figure below. As shown, contingency and emergency plans have to be arranged from inception of the project. The plan should define the following aspects:

- lithology, foliation and stratigraphy variability;
- soil and rock mass behavior;
- hydrogeological hazards and water pressures;
- excavations and temporary works hazards;
- risk analysis;
- trigger values definition;
- contingency works to be applied to each risk occurred;
- revision of the monitoring plan in order to assess contingency efficiency.

The use of trigger values is therefore essential to define the contingency measures to be implemented. For this purpose, the traffic light system is usually applied dividing into green, amber and red response zones, according to the degree of risk (Fig. 8).

Cases Study

Retaining Structures Sensitivity Analysis

According to the above mentioned flow chart, the Design and Planning phase constitutes the first step in the Operational Framework and it represents a stage of high sensitiveness for the project, since project success and quality critically depend on it. The final logical outcome (here intended as the finalized project) derives from a series of back-analyses on project parameters, other than from a cyclic inductive-deductive reasoning about possible variable design data and conditions. In the practical case, the design of retaining structures has been based on the process of hypothesis and interpretations. Such analysis takes into account a number of project variables:

- water table (WT) at final excavation or at higher level, for instance, at the detected geological limit;
- estimated depth of the geological limit between chaotic and landslide material (named Uf) and underlying bedrock (named Ug);

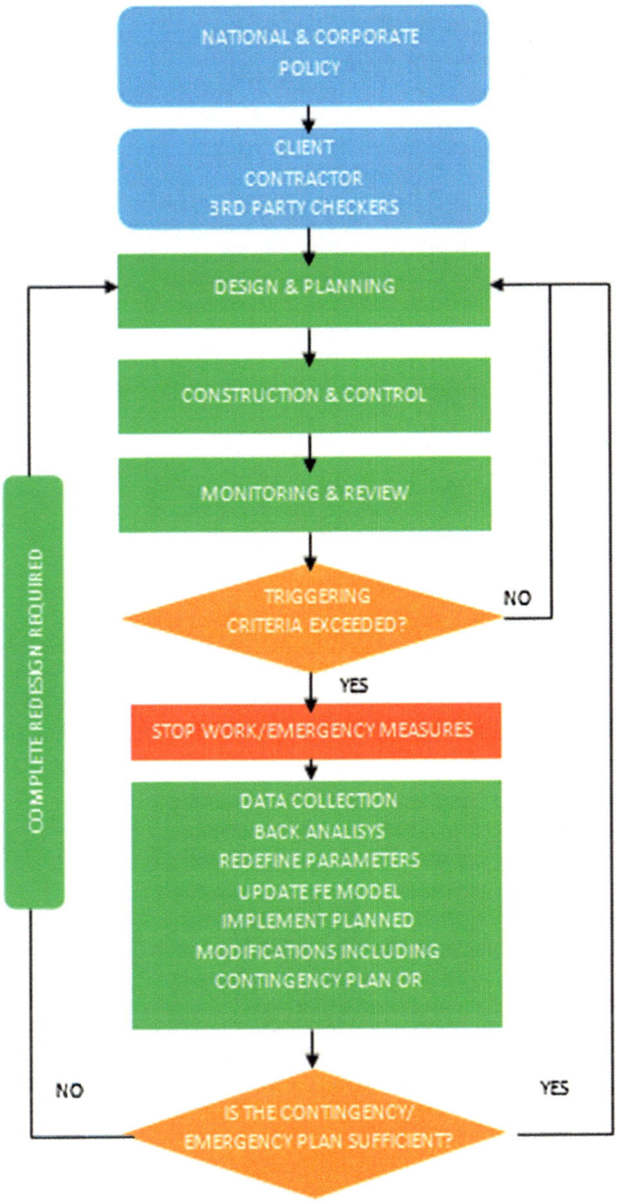

Fig. 8 The operational method framework

- inclination of ground surface over the retaining wall;
- mechanical resistance of the more competent soil unit (in terms of friction angle and cohesion).

The introduction of all these parameters in the equation makes the retaining structure design a sensitive issue. On the one hand, the designer has to take under consideration all worst possible conditions, but on the other hand, it is not acceptable to go for an oversize of the design, since this would results in anti-economic design situation.

In this way, the OM allowed designers to verify, in relative short time, the factual geological and geotechnical conditions and, if necessary, to adjust the geotechnical input data accordingly. The scenarios presented below run all afore-presented reasoning and the difference observed in lateral earth pressure on the retaining structure can be appreciated (Fig. 9).

As a consequence a reliable procedure of design control has to be implemented since it is the designer's responsibility to consider client design requests and to select the most convenient and safe design.

Observational Method Applied to Cut Slope

The Observational Method is sustainable if the amount of time spent for monitoring does not exceed the time for extra earthworks. For very large engineering projects such as a motorway, approaches based on the concept of reliability introduced by the Eurocodes, are suitable to be applied. With such approaches, cut slopes are mostly designed with characteristic values of material property parameters. However, in some cases, this would lead to extremely low slope angles, leading to large earthworks and expropriation issues. So, as an alternative, the most probable parameter approach has been used along with monitoring. Considering the large scale of the project, the monitoring has been divided into areas of hazard, the response time and trigger values have been chosen accordingly. The graphs below summarize the failures or mechanisms of failure detected in the project area; it can be noted that in most of the cases the triggering factors were mainly due to hydrogeological uncertainty and lithology variability rather than soil parameters overestimation. Since the project is mainly located in uninhabited areas, the efficiency of the method could be often optimized choosing high triggering thresholds and lower monitoring frequency. Moreover, the sizes of some instability were contained, so contingency intervention could be achieved by economical and fast solutions such as rock buttress (Figs. 10 and 11).

Conclusions

New roadworks in mountain areas often involves slope cuts, which can be either in stable soil or rock mass or in landslide prone areas. In the last case, old or new landslides could be (re) activated combining the effects of different factors such as the unfavorable geological and geotechnical conditions, the engineering disturbance and the inadequate design solutions.

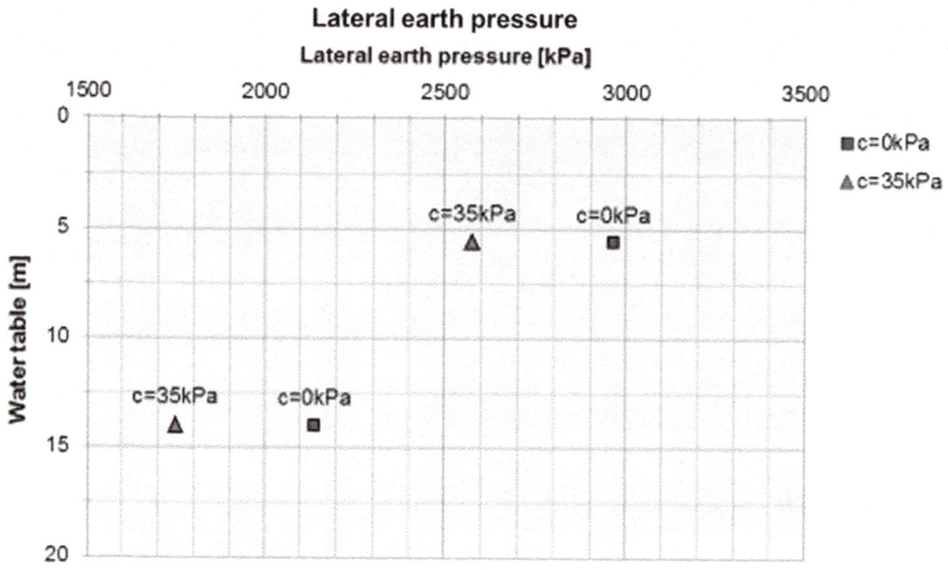

Fig. 9 Sensitivity analysis on lateral earth pressure on retaining walls, variable parameters (WT vs. cohesion)

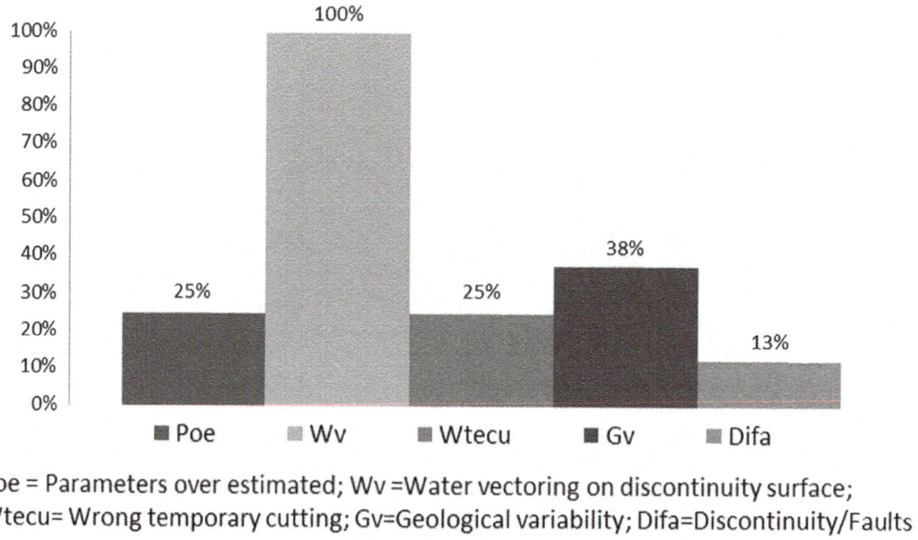

Fig. 10 Sources of instability occurrence analysis

In that scenario, the weathering process, when combined with the influence of a long exposure time, a dominant clayey fine component and the overpressure generated by water tables, has been identified as one of the main sources of landslides residual risk.

The success of a new road engineering project depends also on the ability of the designers in recognizing the factual conditions of the site and the related risks at the very design phase and in proposing effective design solutions to mitigate them. The so called "dual approach", applied in RN77 highway project, had its milestones in defining (1) a solid and reliable reference model and (2) an efficient control procedure in which the OM and the sensitivity analysis, on some key factors and parameters, had a relevant role to deal with landslide related risks. Residual risks, in fact, could be managed efficiently during construction by means of OM and sensitivity analysis if they are based on the adequate knowledge of the local geological and geotechnical conditions.

Experiences acquired in similar projects show that residual risks can be generally quantify in an increase of 5–20% of total costs, depending on several key factors among which the general context, types of works and environmental conditions. In that *scenario*, the dual approach provides a suitable example in which an efficient design can be developed, also in complex geological conditions such as landslide areas, adopting the

Fig. 11 Contingency interventions analysis

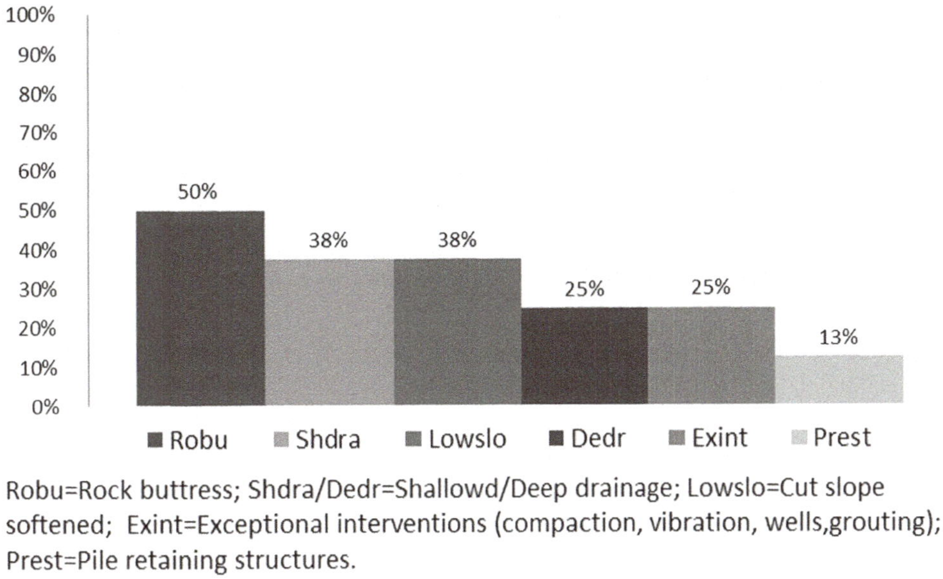

flexible tool of risk assessment and management rather than a firm too pessimistic and conservative approach.

References

AFTES Recommandation GT32, R2A1 (2012) Recommendations on the characterisation of geological, hydrogeological and geotechnical uncertainties and risks

Benaouali-Mebarek N, Frizon de Lamotte D, Roca E, Bracene R, Faure JL, Sassi W, Roure F (2006) Post-Cretaceous kinematics of the Atlas and Tell systems in central Algeria: early foreland folding and subduction-related deformation. C R Geosci 338:115–125

Bouillin JP (1977) Géologie alpine de la Petite Kabylie dans les régions de Collo et d'El Milia, Ph.D. thesis. University Pierre-et-Marie-Curie, Paris-6. p 511

Bouillin JP (1986) Le bassin maghrébin: une ancienne limite entre l'Europe et l'Afrique à l'Ouest des Alpes. Bull Soc géol France 8 (4):547–558

Durand Delga M (1955) Etude géologique de l'Ouest de la chaîne Numidique. Publ. serv. Carte géol. Algérie, Bull. n°24, P 533, Fig 143, pl. 16, pl.h.t 10

Durand-Delga M, Fonboté JM (1980) Le cadre structural de la Méditerranée occidentale. In: Aubouin J, Debelmas J, Latreille M (eds) Géologie des chaînes alpines issues de la Téthys, Colloque no 5, 26o Congrès Géologique International, Paris, in: Mem. BRGM, pp 67–85

Mulenga C, (2015) Influence of weathering and stress relief on geotechnical properties of roadcut mass and embankment fill on St. Vincent and St.Lucia. Thesis for degree of Master in Geo-Information, University of Twente, pp 108

Nicholson DP, Tse CM, Penny C (1999) The observational method in ground engineering–principles and applications. Report 185, CIRIA, London

Nicholson DP, Dew CE, Grose WJ (2006) A systematic best way out approach using back analysis and the principles of the observational method

Peck RB (1969) Advantages and limitations of the observational method in applied soil mechanics. Geotech 19(2):171–187

Patel D, Nicholson D, Huybrechts N, Maertens J (2007) The observational method in Geotechnics. Proceedings of XIV European conference on soil mechanics and geotechnical engineering: geotechnical engineering in urban environments, Madrid, vol. 2

Powderham AJ et al (1996) The observational method in geotechnical engineering ICE. Thomas Telford, London

Riella A, Vendramini M, Eusebio A, Soldo L (2015) The Design Geological and Geotechnical Model (DGGM) for long and deep tunnels. Springer International Publishing, Engineering Geology for Society and Territory, vol 6, pp 991–994

Vila JM (1980) La chaîne alpine d'Algérie orientale et des confins algéro- tunisiens. Thèse Sc Univ Paris VI, vol 3:663, Fig 199, pl. 40, pl.h.t 7

Wildi W (1983) La chaîne tello-rifaine (Algérie, Maroc, Tunisie). Structure, stratigraphie et évolution du Trias au Miocène. Rev Géol Dyn Géogr Phys 24:201–297

Numerical Analysis of a Potential Debris Flow Event on the Irazú Volcano, Costa Rica

Marina Pirulli and Rolando Mora

Abstract

The active Irazú Volcano is the highest of several composite volcanic cones which make up the Cordillera Central in Costa Rica, close to the city of Cartago. The top of the volcano is strategic for the Country, since at the height of over 3400 m sit 84 telecommunication towers used by government agencies and several TV and radio stations, which guarantee the station coverage of more than 60 percent of the national territory. Since December 2014, a series of minor tremors, or microseisms, occurred at the Irazú and some open and deep fissures formed on the upper part of the volcano associated with formation of landslides. More research is needed to determine if these fissures are directly related to recent seismic activity. However, the landslide formation has made it necessary to relocate the towers and there is evidence of the possible destabilization of a volume of about 3.5 million cubic meters of material. In particular, if the landslide triggers in conjunction with heavy rains the movement could evolve into a huge debris flow that could affect Cartago city, similar to the debris flow disaster of December 1963. The dynamics of this potential event have been analyzed using the numerical code RASH3D. The calculated flow intensities and flow paths could be used to support hazard mapping and the design of mitigation measures. The reliability of the obtained results are a function of assumptions regarding source areas, magnitudes of possible debris flows and calibration of rheological characteristics, but also digital terrain model (DTM) quality. As to this last aspect, a systematic comparison of numerical results, DTM and air photos enabled identification of various weak points of the digital terrain model and identified potentially critical zones due to the presence of man-made structures.

Keywords

Irazú • Slope stability • Debris flows • Numerical simulation

M. Pirulli (✉)
Department of Structural, Geotechnical and Building Engineering, Politecnico di Torino, Corso Duca Degli Abruzzi 24, 10129 Turin, Italy
e-mail: marina.pirulli@polito.it

R. Mora
Escuela Centroamericana de Geologia, Universidad de Costa Rica, San José, Costa Rica
e-mail: rolando.morachinchilla@ucr.ac.cr

M. Mikoš et al. (eds.), *Advancing Culture of Living with Landslides*,
DOI 10.1007/978-3-319-53483-1_12

Introduction

Gravity-driven mass flows of volcanic rock fragments mixed with water represent a very important natural hazard in volcanic areas. Such flows may occur not only prior to or during eruptions, but also during volcanically quiet periods when they can be triggered by heavy rainfall, lake outbreaks, earthquakes, or simply the progressive weakening of rocks in a volcanic edifice by hydrothermal alteration (Pierson 1998). The potential for mass movements is enhanced by both steep slopes and large quantities of non-cohesive pyroclastic materials (Valance et al. 1995).

In water-initiation cases, erosion and incorporation of sediment by flowing water on the steep upper slopes of volcanoes typically result in several-fold increases in flow volume. While clay-rich ("cohesive") debris flows can travel great distances with little or no change in rheology, debris flows containing less than approximately 5% clay-size particles in the matrix (granular or "noncohesive" debris flows) commonly become progressively more dilute as they flow and eventually evolve into flood waves of very muddy water termed hyperconcentrated flows (Pierson 1998). The term "lahar" has been recently adapted to include both debris flows and hyperconcentrated flows generated on volcanoes (Smith and Lowe 1991), because a single flow can involve both rheologic flow types and multiple flow transformations.

In densely populated areas, small- to large-scale failure events can be more destructive than any other eruptive activity (e.g. Nevado del Ruiz in Voight 1990), by destroying infrastructures, blocking rivers and burying communities located downstream from source volcanoes.

In the stratigraphic record of the Irazú Volcano, one of the most active in Costa Rica, several units of gravity-flow deposits were recognized (Alvarado 1993). Even historical eruptions, the last of which occurred in 1963–1965, were also accompanied by lahars (Waldron 1967; Alvarado and Schmincke 1993) that severely impacted the city of Cartago.

Since December 2014, a series of minor tremors, or microseisms, occurred at the Irazú and some open and deep fissures formed on the upper part of the volcano with formation of potential landslides.

The aim of this paper is to numerically analyze, with the continuum-mechanics-based RASH3D code (Pirulli 2005), the dynamics of an identified potential event with triggering volume of about 3.5 million cubic metres of material. The triggering of a landslide of this volume on the pyroclastic deposits that mantle the slopes of the volcano, for example by heavy rainfall or an eventual new volcanic eruption could have important consequences for the city of Cartago once again.

Analysis of obtained results provides an opportunity to discuss possible consequences of digital data quality and man-made structures on the shape and size of the area impacted by the simulated mass movement and to evaluate the current effectiveness of the control measures (i.e. levees) that were built in the valley bottom after the 1963–1965 events.

Description of the Study Area

Irazú Volcano is the highest (3432 m asl) of several composite volcanic cones which make up the Cordillera Central in central Costa Rica and which are located near the population and economic centers of the Country. The summit of Irazú lies only about 24 km east of San José, the largest city and capital of the Country, and only about 15 km north-northeast of Cartago, the second largest city of the Country and the capital of the Province of Cartago (Fig. 1).

The 1963–1965 Rio Reventado Debris Flows

The last major Irazú eruption started in 1963 and continued until 1965. It erupted millions of tons of ash and other

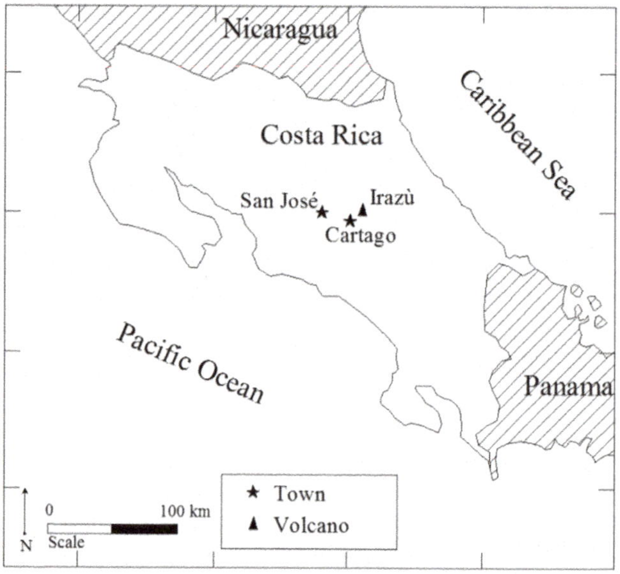

Fig. 1 Location of Irazú Volcano in Costa Rica

material which covered the slopes of the Irazú range where many streams including the Reventado river have their head waters.

The deposited material formed a hard impermeable crust on the surface destroying all vegetal cover and causing a hydrological imbalance: all the debris flow and erosion problems that have plagued the country since the eruption can be attributed to the accumulation of ash on the upper slopes of the volcano and to its effect on the hydrologic regimen of streams draining these slopes.

Some small floods and debris flows occurred in the Rio Reventado and other streams soon after the eruption began, but the most disastrous debris flow occurred in the Rio Reventado on December 9 at the end of the 1963 rainy season. An excellent description is provided by Ulate and Corrales (1966).

A violent rainstorm, which raged over much of the country for several hours, culminated in a torrent of water, mud and debris that swept down the Rio Reventado valley with high destructive force. From detailed studies of the storm and the debris flow, it was estimated that the discharge of this large flow was about 407 cubic meters per second, or approximately 29 cubic meters per second per square kilometre, and that it was composed of about 65 percent water and 35 percent sediment by volume. This large flow removed most of the accumulated debris from the valley floor, stripped the vegetation from lower parts of valley walls, and severely eroded valley walls and floors. In places the flood reached a height of more than 12 m. Near Cartago the debris flow spread out over the surface of the alluvial fan in a tongue of destruction about 1 km wide, devastating and area of nearly 3 km^2, and inundated another several square kilometres with muddy waters. More than 20 persons were killed by this disastrous debris flow and more than 300 homes were destroyed (Waldron 1967) (Fig. 2).

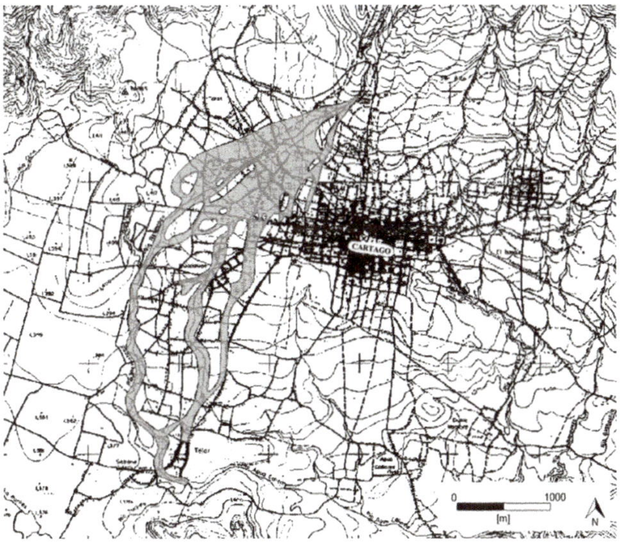

Fig. 2 The *grey* shadowed area is the main deposit of the debris flow occurred on December 9, 1963

The floods of mud and debris started again with the first storm of the 1964 rainy season and they continued to be a hazard and menace to lives and property throughout the 1965 rainy reason (Ulate and Corrales 1966). About 40 of these flows were in the Rio Reventado valley. Although none were as large as the flow in December 1963, nearly half of them were large enough to constitute a hazard to Cartago.

The physical characteristics of Rio Reventado debris flows varied greatly from one flow to another and determine the final phase of their fluid stage. Most flows became very thick and viscous during the final phase, much like a thick, flowing concrete, but some became thin and watery.

Boulders as much as 2–3 m in diameter were very common in all the flows, and several as large as 4–5 m in diameter were observed.

An average velocity of about 5 m/s was defined based on local measurements. Unfortunately, the velocities of some of the larger flows were not measured, therefore the maximum velocities attained are not known, but it seems very likely that speeds in excess of 15–20 m/s were attained by some of these larger flows (Waldron 1967).

Control Measures

Among control measures applied to the Rio Reventado after these events, were some protective engineering works constructed in the lower Rio Reventado basin to promote flood protection for Cartago and its environs. This involved the creation of two artificial levees that contain the river and extend downstream for several kilometres from where the stream debouches from the mountain front (Waldron 1967).

Five decades later, people have built houses on the levees and inside the hazardous areas. Furthermore, levees have

Fig. 3 *Dot lines* are the levees that were built after the 1963–1965 events. *Arrows* indicate the position where the roads cut the levees to cross the area. The *grey* shadowed area is the main deposit of the debris flow occurred on December 9, 1963

been locally cut to allow a new road to cross the hazardous areas (Fig. 3).

Potential Debris Flow Event

A series of minor tremors, or microseisms, occurring since December 2014 has put residents of sev-eral communities near the Irazú Volcano on alert.

Geologists from the National Emergency Commission have confirmed that some open and deep fissures have emerged on the upper part of the volcano. These fissures near the crater have led to the identification of a potential landslide (about 3.5 million cubic metres of material) near 84 telecommunication towers used by the government and emergency agencies, as well as radio and TV stations, which guarantee the station coverage of more than 60 percent of the national territory (Fig. 4).

The dynamics of this potential event (Fig. 4) has been numerically analyzed using the numerical code RASH3D (Pirulli 2005; Pirulli et al. 2007).

Numerical Analysis of the Potential Debris Flow Dynamics

Numerical analyses have been run with the numerical code RASH3D, which is based on a single-phase continuum mechanics approach and on depth-averaged St. Venant equations (Pirulli 2005; Pirulli et al. 2007).

In order to run an analysis with RASH3D, it is necessary to (1) define the topography of the study area with a Digital Elevation Model (DEM), (2) define the source volume, and (3) select the rheology.

A DEM with a 10 m grid spacing (based on 1:10,000 maps) was available for the considered study area.

The geometrical model, which is related to the estimated volume of about 3.5×10^6 m^3, was adopted in accordance with information obtained through on site surveys. A sliding surface at an average depth of 60 m was then considered (Fig. 5).

Different rheologies are implemented in RASH3D so far: Frictional, Voellmy, Quadratic and Pouliquen (Mangeney-Castelnau et al. 2003; Pirulli 2005; Pirulli and Sorbino 2008). A two-parameter Voellmy rheology was used to simulate the dynamics of this event:

$$\tau_{zi} = -\left(\rho g h \mu + \frac{\rho g \bar{v}_i^2}{\xi}\right) \frac{v_i}{\|\mathbf{v}\|} \quad (1)$$

where i = x, y, ξ is the turbulent coefficient, μ, is the apparent friction coefficient, $\bar{v}_i$ is the depth-averaged flow velocity in the x and y directions, h is the fluid depth, ρ is the mass density and g is the gravity vector.

The entrainment process is evidently an important factor during the run-out phase of debris flows on the Irazú Volcano. As a consequence this aspect has been considered in

Fig. 4 **a** Continuous line bounds the triggering area of the potential landslide with volume equal to 3.5×10^6 m^3 (Background image from Google Earth); star symbols indicate the position of the **b** and **c** details of the existing open deep fissure

Fig. 5 Numerical modelling of the topography together with the potentially unstable volume, as used in RASH3D

the analysis by defining, according to geologist interpretation, a portion of the slope where an increment in the flowing volume can occur (Fig. 6).

The following simple, yet effective, empirical formula (McDougall and Hungr 2005) for entrainment rate (Et) is implemented in RASH3D (Pirulli and Pastor 2012):

$$E_t = hvE_s \quad (2)$$

where E_s defines an average growth rate that, according to McDougall and Hungr (2005), can be obtained directly from the initial (V_0) and final (V_f) volumes of the material and the length of the erosion path (l).

Modelling Results

The modelling trials have been performed assuming a Voellmy rheology with rheological parameter values obtained from information available in technical literature, where events originating in a volcanic environment were already numerically back analysed to provide a set of calibrated cases (e.g. Sosio et al. 2012; Quan Luna 2007).

The ranges of investigated rheological parameters have been the following: 0.005–0.05 for the friction coefficient and 500–1000 m/s^2 for the turbulent coefficient (Table 1). As to the entrainment process, on the basis of the description of the December 1963 event given by Waldron (1967) and the McDougall and Hungr formula (2005), the volume average growth rate, Es, was set equal to 0.0035 m^{-1}.

Fig. 6 *Shadowed area* portion of the Reventado river channel where entrainment is considered in numerical analyses. *Continuous line*: triggering area. *Dashed line* levees in the valley bottom (Background image from Google Earth)

It is stressed that the following numerical results do not aim and do not claim to describe the exact extent of the Cartago area that could be affected but intend to identify critical aspects during the runout process that concern both the digital elevation data and the consequences of human activity on land management, and which in turn can affect a risk assessment procedure and the design of risk management protocols.

A first set of analyses was carried out ignoring entrainment along the runout path (Simulations 1–5 in Table 1). It

Table 1 Combination of rheological parameters for carried out numerical simulations

Simulation number	μ (−)	ξ (m/s^2)
1	0.05	500
2	0.02	200
3	0.01	200
4	0.01	500
5	0.05	1000
6	0.01	1000
7	0.01	500
8	0.01	200
9	0.01	500

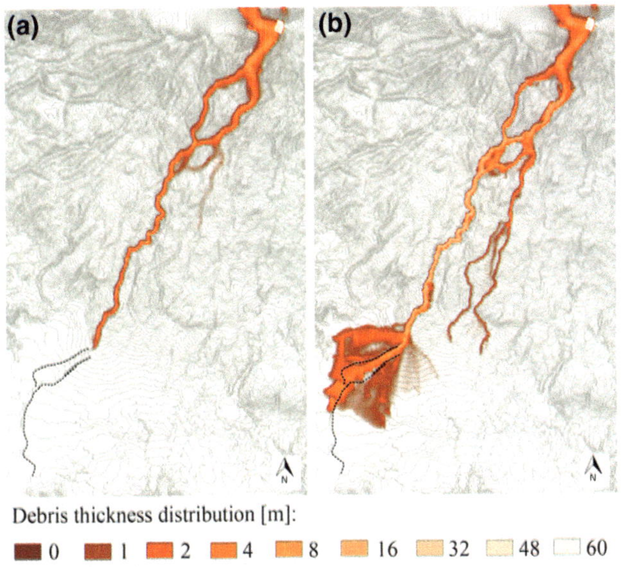

Fig. 7 RASH3D runout results. **a** no entrainment, with the rheological parameter combination of simulation n° 5 of Table 1; **b** entrainment, with the rheological parameter combination of simulation n° 8 of Table 1. *Dashed line* levees in the valley bottom

emerged that independent from the rheological parameter values the mass front stopped in the position indicated in Fig. 7a. Numerical simulations where entrainment was considered (Simulations 5–8 in Table 1), show that, as soon as the stream exits the mountain valleys, the mass rapidly spreads as indicated in Fig. 7b.

Detailed analysis of the DEM, coupled with a field investigation indicated that a bridge structure was interpreted by the model as a dam (Fig. 8). This affected both models, stopping the mass movement (no entrainment) or causing spreading of the flow (entrainment model).

The DEM was corrected and a new analysis was performed (Simulation 8 to 9 in Table 1).

The mass now flows in the channel and reaches the valley bottom where some protective engineering works (i.e. levees) were constructed to promote flood protection (Fig. 9).

The shape of the runout area in the valley bottom indicated in Fig. 9 does not represent the final deposit of the flowing mass (i.e. velocity is still not zero).

Analysis of numerical results allowed identification of the critical points where the mass spreads outside the levees; these concern both the orographic right and left of the propagation path in the levee area.

Onsite inspections support these numerical results since it was observed that:

1. the mass spreads on the orographic right exactly where the levees have been cut to allow road to cross the area (Fig. 10), and
2. the mass spreads on the orographic left due to the presence of a ramp that was built by local people to reach the river. There the flowing mass can run up the slope as indicated by the numerical simulations (Fig. 11).

Conclusions

A depth-averaged model (RASH3D) to simulate the propagation of rapid flow-like events has been applied to investigate the possible destabilization of a volume of about 3.5 million cubic metres of material that could

Fig. 8 Detail of the digital elevation model that evidences in **c** and **d** the spreading of the mass due to a dam in the channel (indicated with a circle in (**b**)) that is a digital incorrect interpretation of the bridge existing in that position (**a**)

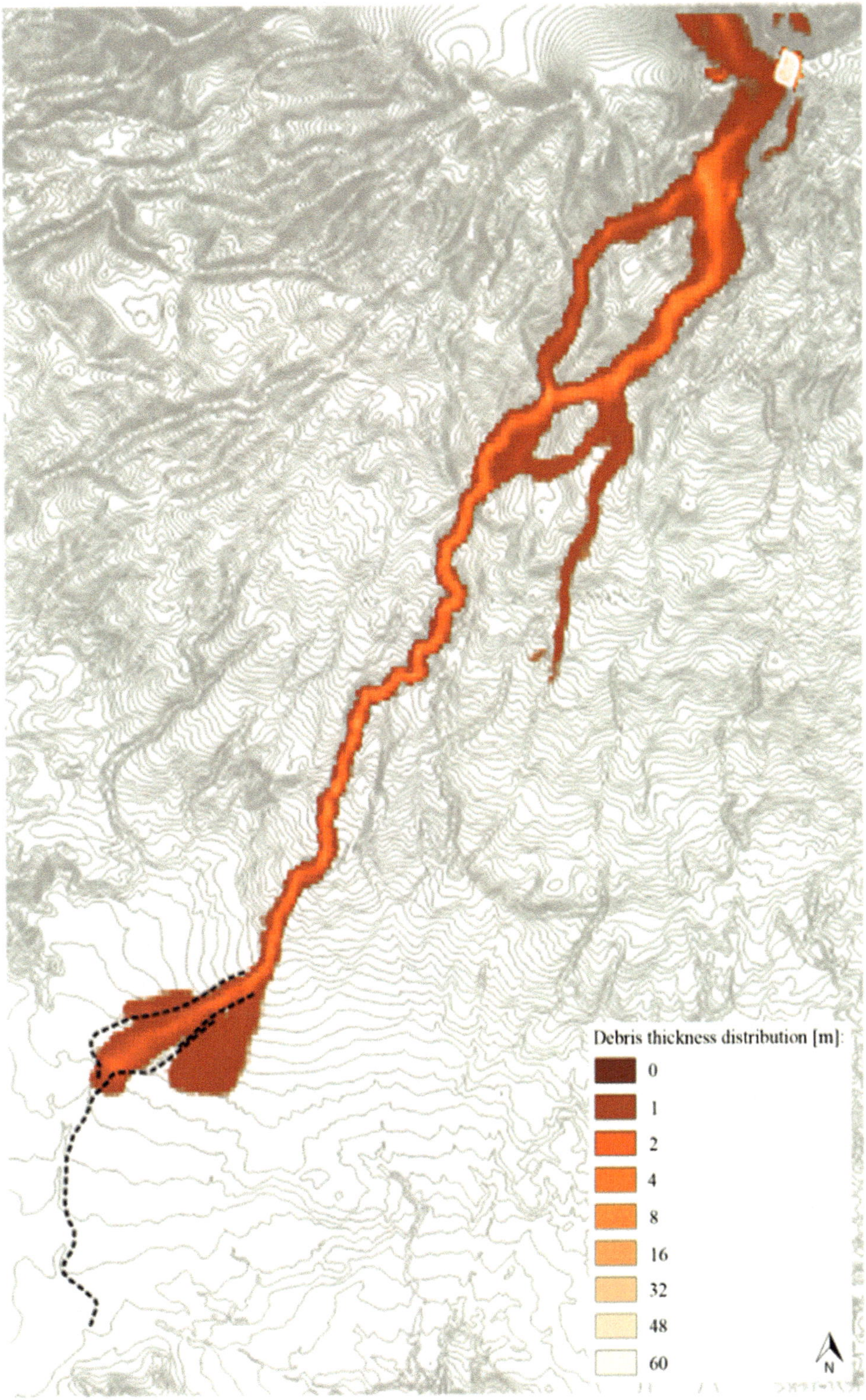

Fig. 9 RASH3D numerical results after DEM correction with the rheological parameter combination of simulation n° 8 of Table 1. *Dashed line* levees in the valley bottom

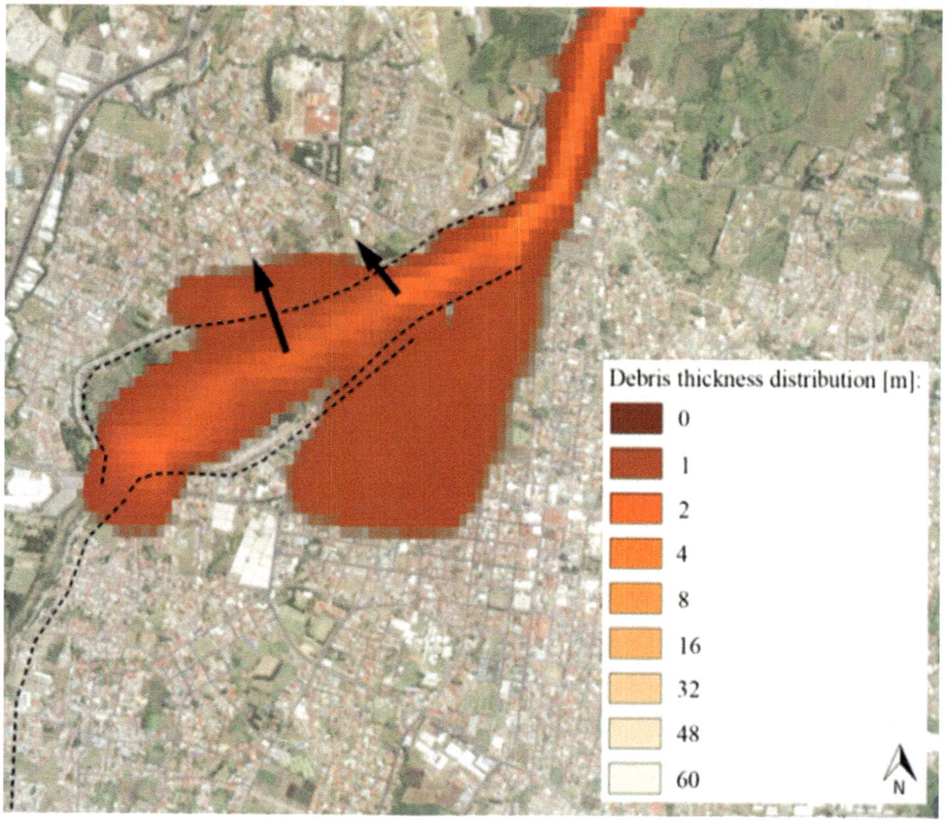

Fig. 10 RASH3D runout results in Cartago area. *Arrows* indicate where the mass flow out from the levees (*dashed lines*) where roads cross the area (Background image from Google Earth)

affect the South flank of the Irazú volcano. In particular, if the landslide triggers in conjunction with heavy rains, it could evolve to a huge debris flow and could affect Cartago city and its surroundings.

The analyses are intended to identify critical aspects during the runout process, that concern both the digital elevation data and the consequences of human activity on land management.

The flowing mass behaviour as simulated by the numerical code was confirmed to be reasonable following field inspections. In fact, the numerical results brought the attention on a mistake in the DEM: due to a wrong interpretation of a bridge, a portion of the Reventado channel resulted as obstructed in digital data. Instead, in the Cartago area, the numerical spreading of the mass with respect to the levee areas occurs due to man-made activity: the cutting of the levees to cross the area and the reduction of the river flank deeping by construction of a ramp to reach the river.

The obtained results have highlighted the capability of RASH3D to correctly simulate the dynamics of this type of phenomena on, and as a tool that can give a contribution to hazard mapping and the design of mitigation measures. It is however underlined that the reliability of calculated flow intensities and flow paths are a function of assumptions on source areas, magnitudes of possible events and calibration of rheological parameters, but also on quality of available digital data. This is why, numerical analyses always have to be combined with field surveys.

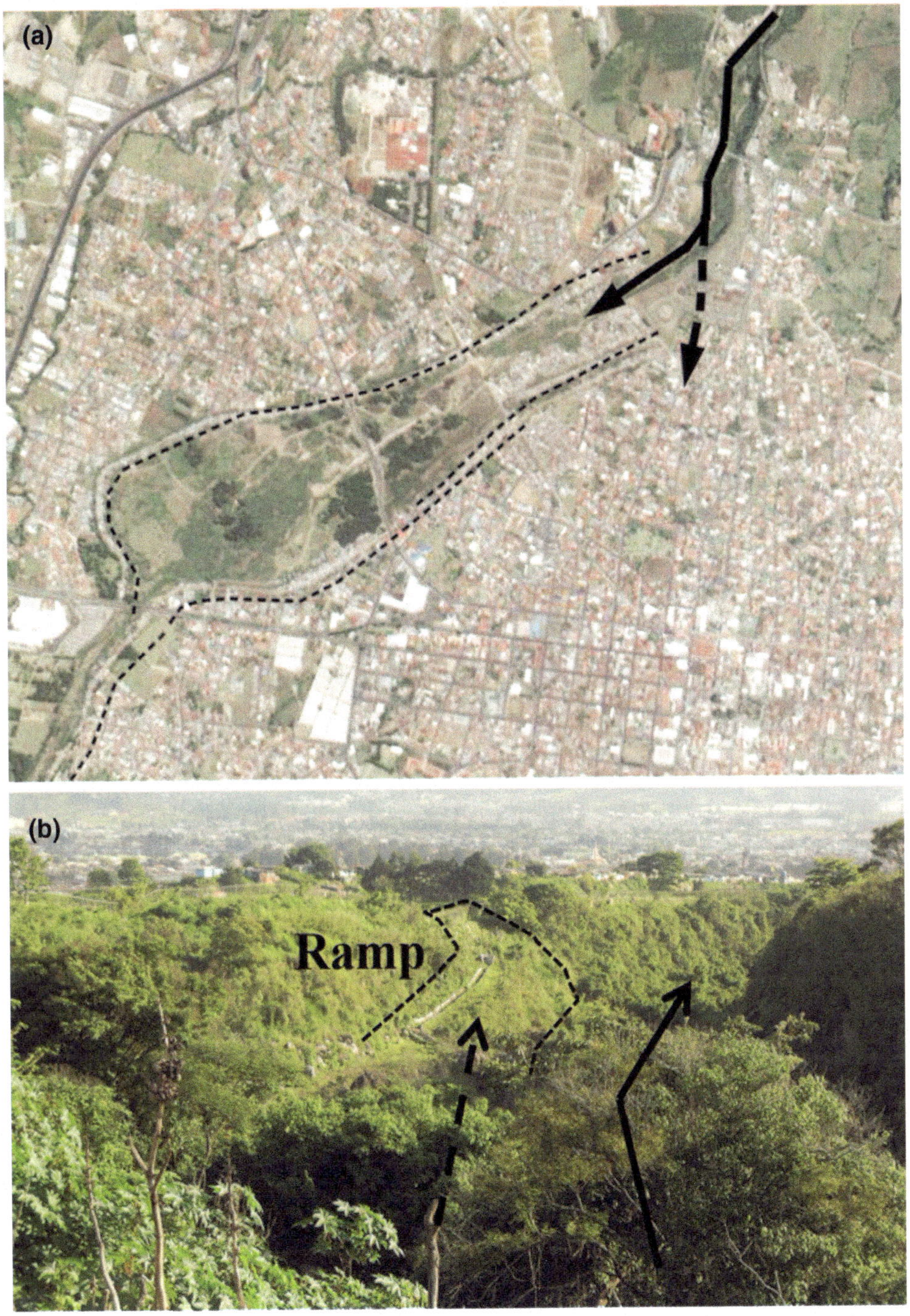

Fig. 11 RASH3D results give a spreading of the mass on bend (*dashed arrows*) with respect to the main channel path (*continuous arrows*). **a** detail of levees in the valley bottom (*dashed area*). Background image from Google Earth; **b** evidences of a ramp (*dashed area*), as seen from the bridge of Fig. 8

References

Alvarado GE (1993) Volcanology and petrology of Irazu Volcano. Thesis, Kiel Univ, Germany, Costa Rica

Alvarado GE, Schmincke HU (1993) Stratigraphic and sedimentological aspects of the rain-triggered lahars of the 1963-1965 Irazu eruption. Costa Rica Zbl Geol Paläont 1(2):513–530

Mangeney-Castelnau A, Vilotte JP, Bristeau MO, Perthame B, Bouchut F, Simeoni C, Yerneni S (2003) Numerical modelling of avalanche based on saint Venant equations using a kinetic scheme. J Geophys Res 108:B11

McDougall S, Hungr O (2005) Dynamic modelling of entrainment in rapid landslides. Can Geot J 42:1437–1448

Pierson TC (1998) An empirical method for estimating travel times for wet volcanic mass flows. Bull Volcan 60:98–109

Pirulli M (2005) Numerical modelling of landslide runout, a continuum mechanics approach. PhD Thesis, Politecnico di Torino, Torino, Italy

Pirulli M, Bristeau MO, Mangeney A, Scavia C (2007) The effect of earth pressure coefficient on the runout of granular material. Env Model & Softw 22(10):1437–1454

Pirulli M, Sorbino G (2008) Assessing potential debris flow runout: a comparison ot two simulation models. Nat Hazards Earth System Sci 8:961–971

Pirulli M, Pastor M (2012) Numerical study on the entrainment of bed material into rapid landslides. Geotechnique 62(11):959–972

Quan Luna B (2007) Assessment and modelling of two lahars caused by "Hurricane Stan" at Atitlan, Guatemala. M.S. Thesis, University of Oslo, Oslo, Norway

Smith GA, Lowe DR (1991) Lahars: volcano-hydrologic events and deposition in the debris flow–hyperconcentrated flow continuum. In: Fisher RV, Smith GA (eds), Sedimentation in volcanic settings. SEPM Spec Pub 45: 59–70

Sosio R, Crosta GB, Hungr O (2012) Numerical modeling of debris avalanche propagation from collapse of volcanic edifices. Landslides 9:315–334

Ulate CA, Corrales MF (1966) Mud floods related to the Irazu Volcano Eruptions. J Hydraulics Division 92(6):117–129

Vallance JW, Siebert L, Rose WI, Giron JR, Banks NG (1995) Edifice collapse and related hazards in Guatemala. J Volcanol Geotherm Res 66:337–355

Voight B (1990) The 1985 Nevado del Ruiz Volcano catastrophe: anatomy and retrospection. J Volcanol Geotherm Res 44:349–386

Waldron H (1967) Debris flow and erosion control problems caused by the ash eruptions of Irazu Volcano. Costa Rica USGS Bull 1241: 1–37

Landslides Impact Analysis Along the National Road 73C of Romania

Andreea Andra-Topârceanu, Mihai Mafteiu, Razvan Gheorghe, Mircea Andra-Topârceanu, and Mihaela Verga

Abstract

Landslides are the most common geomorphic hazard processes in Sub-Carpathians regions. Crossing the Getic Sub-Carpathians between Campulung Muscel city (the first royal residence of Romanian Country) and Ramnicu Valcea city, the 73C National Road is constructed for more than 90% on sloping surfaces. A wide variety of landslide types, a high landslide frequency (over 5 landslides reactivations/year and other new triggering) and a high density (between 0.5 and 3.3 landslides/km) are the main characteristics of the slope instability along the 73C National Road. Many stabilisation and repair works have been completed, but different section remain vulnerable to landslide damage. The study aims are to map the different types of landslides, to identify the landslides causes and their impact of landslides related to 73C National Road and its communities, to identify new morphological surfaces which may have a high landslide susceptibility at and to assess the impact of landslides. The main causes of landslides along 73C National Road are hydrogeologically controlled and are linked to high degree of drainage density and deforestation. A multy-disciplinary approach was taken, consisting: landslides mapping, historical maps analysis, geomorphological and geophysical methods, topographic and geotechnical surveys. The results highlight considerable vulnerability of slope stabilisation works as a consequence of an incomplete understanding of landslide processes and limited stabilisation works involving just the road embankment.

Keywords

Landslides • National road • Impact analysis • Slope stabilization works

A. Andra-Topârceanu (✉) · M. Verga
Faculty of Geography, University of Bucharest, 010041 Bucharest, Romania
e-mail: andreea.andra@geo.unibuc.ro

M. Verga
e-mail: mihaela.verga@geo.unibuc.ro

M. Mafteiu
MM Georesearch, 062152 Bucharest, Romania
e-mail: mihai_mafteiu@yahoo.com

R. Gheorghe
Faculty of Geology and Geophysics, University of Bucharest, 020956 Bucharest, Romania
e-mail: razvan_25_ageo@yahoo.com

M. Andra-Topârceanu
ISC, 030217 Bucharest, Romania
e-mail: mirceaat@gmail.com

M. Mikoš et al. (eds.), *Advancing Culture of Living with Landslides*,
DOI 10.1007/978-3-319-53483-1_13

Introduction

Landslides are gravitational geomorphological processes intensively studied because of their impact on the anthropogenic and natural environment.

The impact is defined by the relationship between the environment (geomorphologic) in its resource capacity (passive) and Man (assets) in its capacity as a modeling agent (Panizza 1996, 1999). The geomorphic processes and "landforms, both of which has become useful to man or may depending on economic, social and technological circumstances" (Panizza 1996). The landslide effects that are viewed as changes on the environment were highlighted by Highland and Schuster (2004, 2007). A comprehensive description of the types of impact of mass movements on various environment is performed by Geertsema et al. (2009). A classification method for landslide impacts was proposed by Alimohammadlou et al. (2013). Concerns over the impact of natural environmental and anthropogenic landslides (including economic impact) were conducted along roads in Curvature Carpathians (Constantin et al. 2009) and the Eastern Carpathians (Gaman 2013). The medium and long term relationship between the landslide's impact and the economic components, such as network transport, local and regional economy was highlighted by Pfurtscheller and Genovese (2013). Landslides result in different levells of impact that depend on landslide type, the area affected [natural environment, protected areas, anthropogenic environment (e.g. Geertsema et al. 2009)], the distance between triggering area and the affected surfaces, and on factors such as their frequency, duration, persistence of features and visual impact.

This paper illustrates a case study of 73C National Road, where the Rapid Impact Assessment Method (RIAM) is applied to determine the impact of landslides and assess the consequences of anthropogenic intervention (such as cutting forests, increasing stress conditions, traffic vibration, road expansion, inadequate drainage, lack of road maintenance).

Methods

The methodology consists of combining geomorphological, topographical, historical cartography, geophysical and geotechnical surveys and they are used to identify the features of landslide bodies and assess the landslide behavior. Historical analysis of the road 73C was based on georeference and analysis of historical maps. The landslides analysis included other cartographic documents, such as topographic maps at 1:25,000, 5 landslide land surveys, ortophotoplans and mapping of landslides. To compute the features of every slide were used the following software: QGIS, ArcGIS, Surfer, ACad. Geomorphological parameters as follows: elevation, slope angle and road gradient, aspect fragmentation, and curvature were assessed. Geotechnical investigations consisted of the execution of over 35 geotechnical drillings with a Beretta T41 mechanical machine, in 15 locations affected by instability phenomena. From the surveys disturbed and undisturbed sampled were collected, which were physico-mechanical analyzed, process design to determine the geotechnical parameters, which in this paper are analyzed granulometric content, humidity, density, limits of plasticity, compressibility and shear strength. Shear strength is characterized by the angle of internal friction and cohesion - basic geotechnical parameters. Standard tests (STAS 8942-2-1982) are performed using a direct shear apparatus with impose strain and measured effort. In our case, part of the samples were sheared in CU conditions (consolidated-undrained, shear speed v = 0.5 mm/minute, immersed) and other samples in UU conditions (unconsolidated-undrained, shear speed v = 0.2–0.8 mm/minute). The geotechnical data used, were part of separate commercial studies, developed punctually at different periods (2007–2015) processed by different laboratories (Fig. 1).

VES (vertical electrical sounding) using Schlumberger type array with four collinear electrodes are applicable for obtaining the geophysical data like apparent resistivity curves. Their interpretation provides physical characteristics of lithology, lithological limits, ground water, pluvial and sewage infiltrations, geomorphological processes. The electrical resistivity survey's depths varied from 6 m (into primary survey phase and to determine the shallow landslides) to 20 m (to identify multiple slide planes).

The evaluation method called RIAM (Rapid Impact Assessment Matrix) was proposed and developed by Pastakia and Jensen in 1998. Beeing a very flexible tool for EIA (Environment Impact Assessment), RIAM allow selection of analysis components or indicators related to peculiar impact analysis area, subject or activity. This are scored from −3 to 4 according with the main criteria: importance of condition (A1), magnitude of effect/change (A2), permanence of effect/change (B1), reversibility (B2) and cumulative effect (B3). Both criteria show the "measure of impact from the (chosen) component" (Pastakia 1998; Pastakia and Jensen 1998). Criterion A1 and A2 "can individually change the score obtained, and criteria B1, B2, B3 should not individually be capable of changing the score obtained". To obtain the final score (ES = environmental score) is necessary to multiply the components of criteria A (A1 × A2 = At) and the components of criteria B should summarize (B1 + B2 + B3 = Bt), and after multiply the partial result: At × Bt = ES. For comparing environmental scores it is necessary to include into ranges of different types of impact. The selected analysis components which are affected by landslides are the following: local morphology, potential landslides, phreatic levels, forest and pasture, landscape.

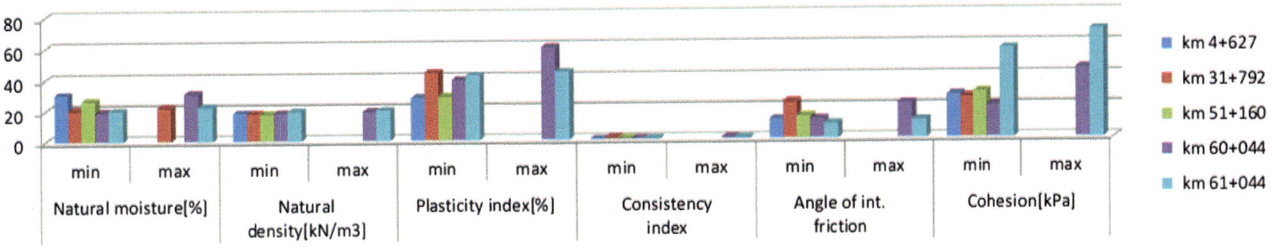

Fig. 1 Geotechnical parameters variation computed for the most frequent deposit silty clay from instability areas

Geographical and Geological Description of 73C National Road Area

The 73C is a national road running parallel to the mountainous side and it is important road for passengers and goods traffic between Arges and Valcea counties. The road passed through Getic Sub-Carpathians, with sinuous and different gradients due to morphologic substrate. The elevation vary between 250 and 1110 m (Fig. 2). Geographical unit consists of two longitudinal subunits of hills and depressions. The contact between them is given by the corridor Tigveni—Curtea de Arges—Valea Iasului—Domnesti—Slanic—Berevoiesti—Schitu Golesti, followed by 73C National Road, strongly affected by slope processes (Dinu 1999) (Fig. 3).

The slopes declivity (0–50°) increase from west to east, the intensity and frequency of landslides does not change accordingly; much more, the areas with frequent landslides that creates permanent damage and destruction of the road are in the west side. On both sides of 73C National Road

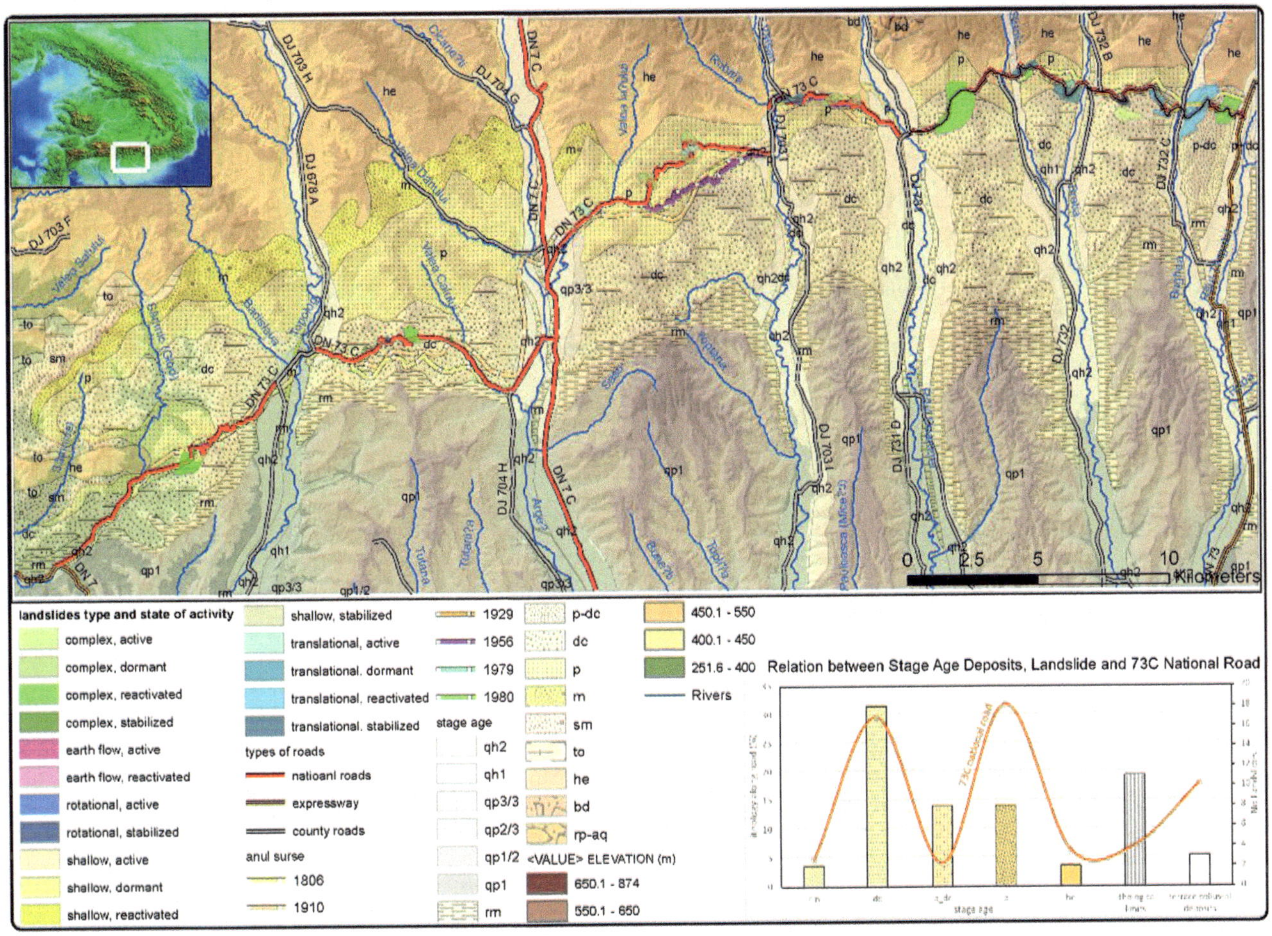

Fig. 2 Geological and geomorphological map of 73C national road area. Relation between lithological deposits, landslides and 73C road

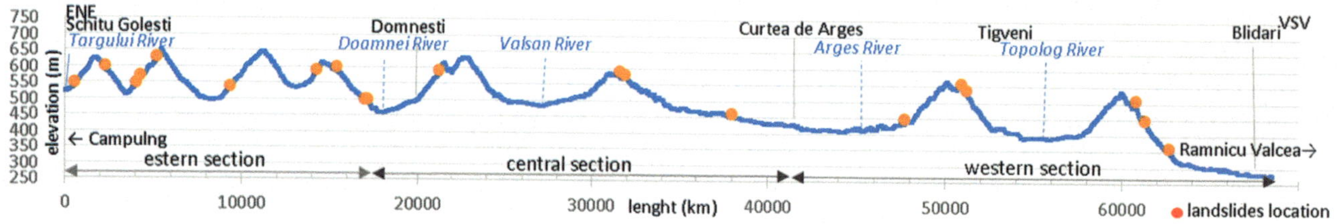

Fig. 3 Topographic profile along 73C national road and landslides location

(about 10 m distance each), the slopes are quasi-horizontal or in 7–12° range (Fig. 4). The morphology enroll in a structural type with large structural surfaces oriented E, SE and S, angular cuesta fronts, consequent main valleys and subsequent tributary valleys.

The route of 73C National Road spreads on Helvetian, Pontian, Dacian and Romanian deposits (sand, gravel, marl, marly limestones, clay) (Fig. 1). Deposits are willing monoclinal with angle about 10–15° to the S and SE (Dragos 1953; Mihăila 1971).

Historical Overview of 73C National Road

Roads, unlike highways, inherit old historic trails, connecting administrative and economic medieval cities. Such a road, important in the Middle Ages, linked the oldest city in Wallachia, Campulung Muscel (1215–1292) with Ramnicu Valcea (1389). This road was known as the medieval Voivodal/Royal Road, currently classified as 73C national road. The historic road linked three major cities: Campulung —the first capital of Wallachia Country (1330) and the

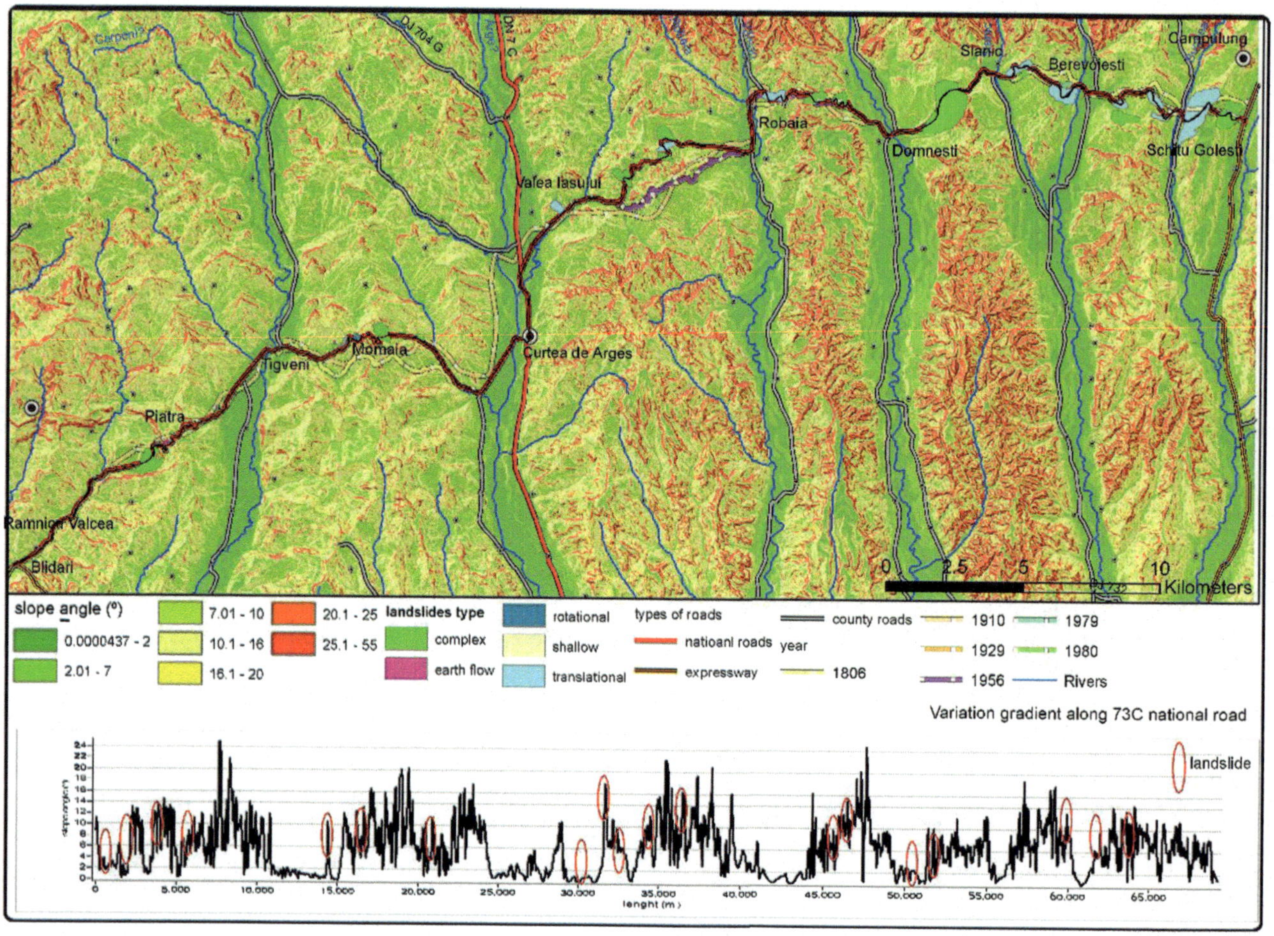

Fig. 4 Slope map and variation gradient along 73C national road

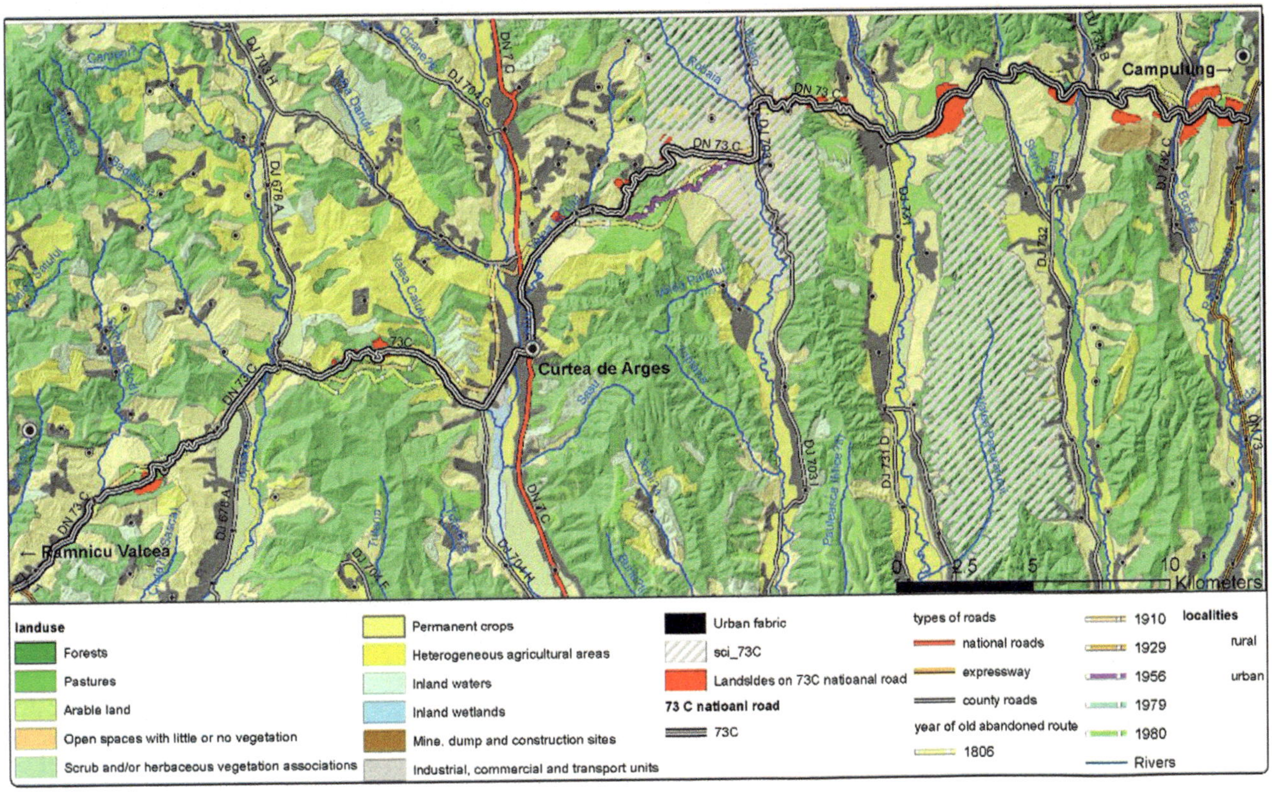

Fig. 5 Landuse map and SCI protection areas

residence of the first Romanian Mitropoly (1359), Curtea de Arges—second capital of the Wallachia Country (1369) and the residence of Mitropoly (until 1517) and Ramnicu Valcea—residence of Archdiocese of Râmnic and Arges (1505).

Among the first references to this road were recorded in old cartographic documents or charters. In the 13th century, Seneslau, Voivode of Arges (1247) "masters [...] subcarpathic way to Muscelele Campulung and Olt" (Mulţescu 2010). The road is random represented in old cartographic documents, such as *Nova et Recens Emendata Totius Regni Ungariae una cum Adiacentibus et Finitimis Regionibus Delineatio* (Davit de Meijine Map, 1596). At the end of the 18th-century Map of Specht indicate the presence of the road between Campulung, Curtea de Arges and Tigveni as the main road (Osaci 2004). The Austrian Maps (1806, 1867, 1910) capture the route of the road with some spatial differences (Table 1).

In 1910 the road was situated in a different location in most of its sections, compared to previous routes. "Directories Shooting Plans" Romanian Map under Lambert-Cholesky show new resettlement of the road in the eastern and central sectors (Fig. 6). On topographic maps 1:25,000 (1979–1980) the road overlaps the current route in western and central sector (over 35 km), but there are changes in the route with geometric improvements, in the eastern sector.

The main considerations of the historic road routes were military, strategic and economic and follow the configuration of interfluves or presence of springs and wells (Popp, 1938 quoted Osaci-Costache 2004). At the beggining it was drawn related with the shelters, crossing the forests and rivers (Târgului, Bratia, Doamnei, Vâlsan, Argeş, Topolog).

Since the forestry areas decreased from 55% (18th century) to 36% (20th century) (Fig. 5) the old landslide were reactivated and new once occurs along and across the road. During this study were mapping almost 50 slides affecting more or less the 73C road. Due to the slope mass movement some segments of the road have been modified according with increased traffic volume and capacity and also runway velocity.

Currently, running speed on the road is 80 km/h and its maximum sinuosity is 350°/km in the area of Piatra complex of landslides.

The newest constuctions and maintenance technologies (1985–2010) couldn't stop the damages caused by landslides.

Over 75% from morphological support of 73C national road are slope surfaces. Geomorphologic and geologic analysis of environmental features crossed by the road 73C made it necessary to split into 3 relative homogeneous sections route: Schitu Golesti - Domnesti in the east, Domnesti

Table 1 Historical changes of 73C road has been reported in curent route (SC-D: Schitu Golesti—Domnesti estern section, D–CA: Domnesti—Curtea de Arges central sections, CA–B: Curtea de Arges—Blidari western section)

Source		Year	Length (km)			Abandoned sections		Deviation km	Sinuosity index
			SC–D	D–CA	CA–B	Orientation			
Military survey map	I	1806	61.4			18		37.8	1.315
			17.1	18.1	26.2	9 N	9 S		
	III	1910	54.6			11		46.5	1.305
			14.2	17.3	23.3	9 N	3 S		
Lambert-Choleski plan		1929	63.9			12		–	–
		1956	16.6	23.1	24.2	4 N	8 S		
Topographic map		1979–80	65.8			13		15.0	1.37
			17.3	24.3	24.2	9 N	4 S		
Ortophotopl. topo. survey		2011 2014–16	68.1			–		–	1.428
			18	25.9	24.2				

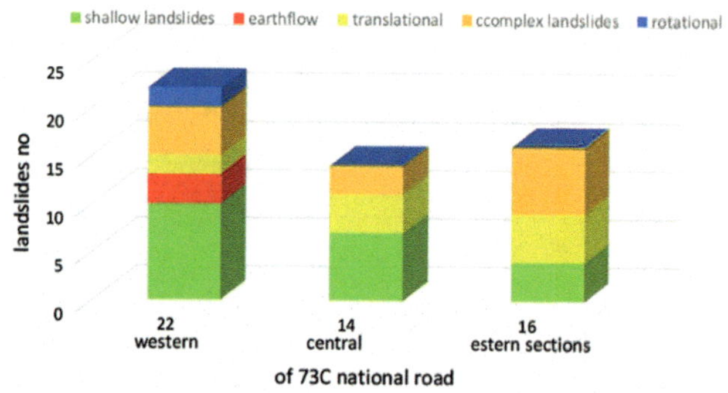

Fig. 6 Landslide types into thre sections of 73C national road

—Curtea de Arges in center, Curtea de Arges—Blidari in west (Table 1).

Five type of landslide were mapping in the western sections: shallow landslides (e.g. Curtea Arges), translational landslides (e.g. Burlusi), rotational landslides (e.g. Piatra and Pietroasa), triggered into Dacian deposits and almost every ones combine with gully and torrential erosion (Fig. 6). Some landslides are not quite large, but very active and because of that, its produce permanent damages of road. The earthflow and complex landslide (rotational and eartflow) shaped on Dacian—Romanian lithological limits are peculiar for western section. Their dynamics is very intense, mostly for subsequent and obsequent landslide. The consequent landslides have different slide mechanism and their behavior consist in reactivated, dormant and suspended dynamic (e.g. Barbalatesti, Berevoiesti, Domnesti).

The main features of complex and rotational landslides are multiple slide planes outlined by geophisical measurements (Momaia-Tigveni, Piatra) and deep-seated (slide planes, more then 20 m depth) (Fig. 6).

The interpretative cross-section of Momaia landslides shows the four landslide body thickness (high resitivity values) and their retrogresive dinamic and the failure planes. The other cross section present Berevoiesti consequent landslide with many slide segments (Andra and Mafteiu 2007) (Fig. 7).

Landslides Impact Analysis. Features, Types, Solutions and Quantification

Road rehabilitation solutions consisted in position or curvature changed of the road segments, some of them with a negative impact on the road. Technical solutions for road development include: excavation works, support works, planning topographic surface and drainage: leveling and filling, embankment cut and retaining wall, vertical drain, drainage ditch, gabion wall, piles works. The lack of interest in the stabilization works of slopes before rehabilitation road, leads repeatedly to damage of road. The construction of the new sections on suspended or dormant landslide, increase the stress condition of slope and reactivate its as a result.

Direct and Indirect Impact on the Natural Environment

The impact assessment was done with adapted version of RIAM matrix for identifying the impact range based on analyzed components scores. The direct impact of landslides prepared and triggered anthropic is new profile of slopes through the cracks, scarps, deformation. Landslide effect is shown by the curvature values of slopes (profile curves from −0.3 to 0.2, plan curvature from −0.2 up to 0.7). As a

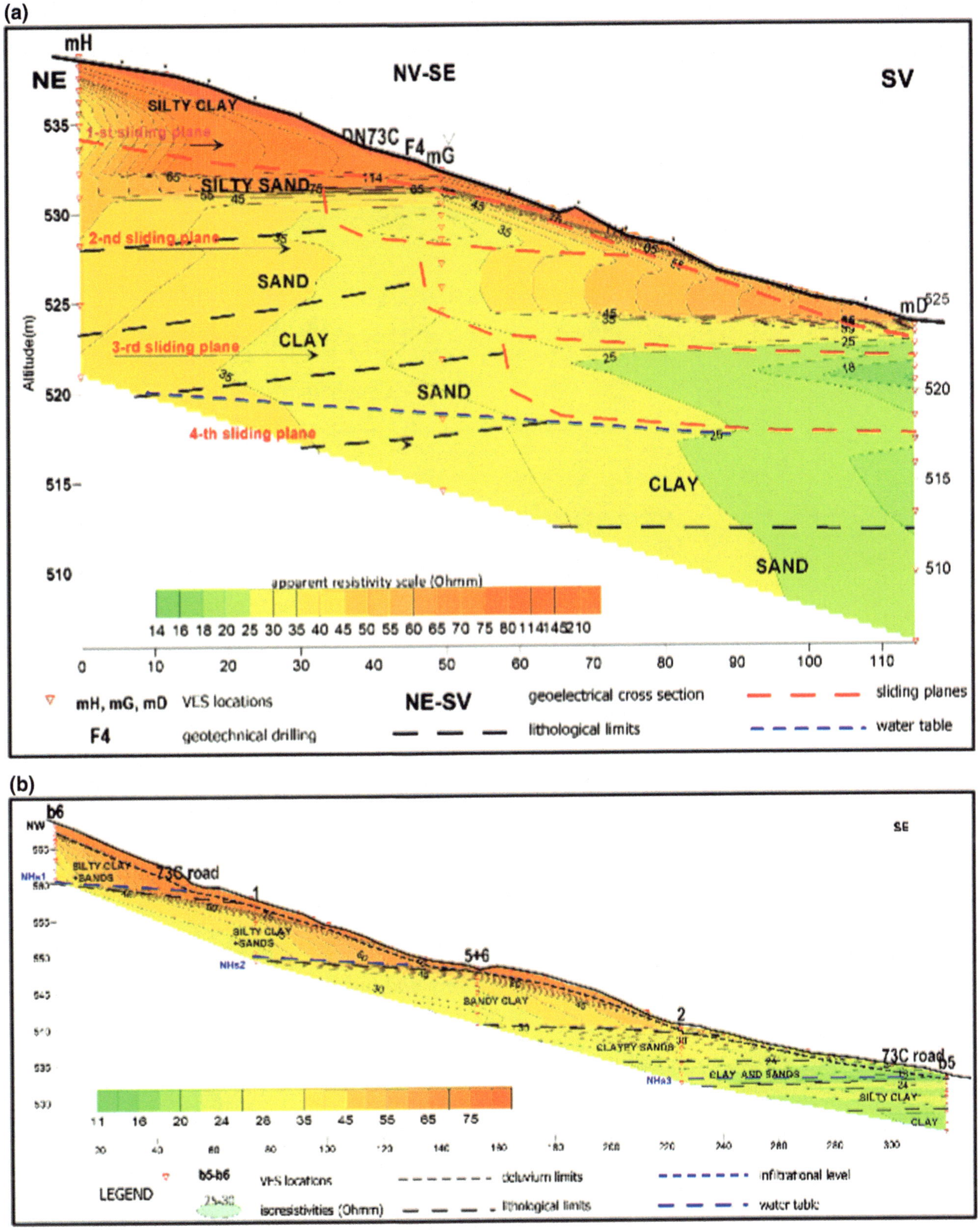

Fig. 7 Interpretative geoelectrical cross sections along 73C national road in western (**a**) Tigveni and estern (**b**) Berevoiesti section show different type of landslides

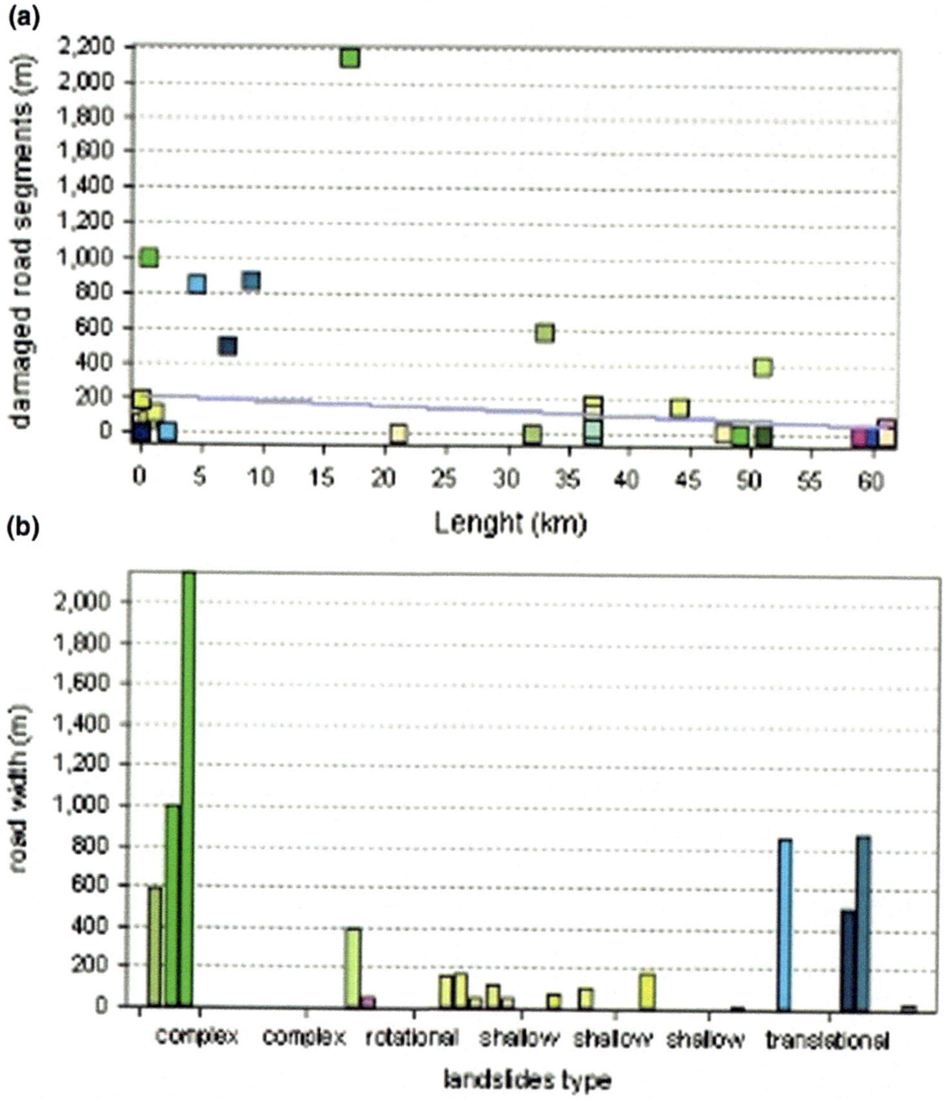

Fig. 8 Variation lengths of road segments damaged by landslides (**a**). The relationship between the lengths of road damaged and types of landslides (**b**)

consequence, the geo-mechanical conditions change and preparing for future slide processes.

These affect the original shape of the relief locally and in a irreversible way. The indirect effects occur on post-impact phase of the landslide as a resulte of the road construction works (such as repairing and consolidation or strengthening of the road, by executing backfilling and filling, either by excavation and leveling works) (Momaia, about 19,000 m^2). The phreatic level may fluctuate in narrow limits. Slope stabilization works by vertical drains will determine a long term decrease, local, of phreatic level). Complex landslides and earthflows have high impact on arborescent and herbaceous vegetation (e.g. the Momaia forest area over 15,000 m^2 was destroyed; after 10 years the vegetation has not been regenereted).

Direct and indirect socio-economic impacts

Technical features of 73C are: 80 km/h design speed, annual average daily traffic of 2400–3500 vehicles/day for passengers and merchandise (20–25%) and the capacity of 7.5 t. Landslides distributed over the entire length of 73C, with sizes between 143 and 2160 m, summarizing 7400 m, which exceeds 10%. The length of the affected segments decrease from East to West (Fig. 8). Social impact is considered over

Table 2 Landslide impact analysis (RIAM) on 73C national road

impact direct, indirect	A1	A2	B1	B2	B3	At	Bt	Es	
natural environment				*Significant negative impact*				***-63***	*-D*
local morphology as road suport	1	-1	3	3	3	-1	9	-9	-A
land stability /potential landslides	2	-1	2	2	3	-2	7	-14	-B
phreatic level effectes	2	-1	2	2	3	-2	7	-14	-B
forestry	2	-1	2	2	3	-2	7	-14	-B
pastures grassland	1	-1	2	2	2	-1	6	-12	-B
protection areas	1	0	0	0	0	0	0	0	N
Social environment				***Significant negative impact***				***-38***	***-D***
public and private services accesibility	3	-2	2	2	3	-6	7	-42	-D
temporary isolation of comunities	2	1	2	2	3	2	7	14	B
visual quality of landscape	2	-1	2	2	1	-2	5	-10	-A
Economic environment				**Moderately *negative impact***				***-28***	***-D***
landuse	1	-1	2	0	3	-1	5	-5	-A
traffic velocity	3	-1	2	2	2	-3	6	-18	-B
constructions, mentenaince cost	3	1	2	3	3	3	8	24	C
Traffic and business development	1	-1	2	3	3	-1	8	-8	-A
tourism	3	-1	2	2	3	-3	7	-21	-D
Total impacts				**Major negative impacts**				**-129**	**-E**

score	*impacts / changes*
72 to 108	*+E Major positive*
36 to 71	*+D Significant positive*
19 to 35	*+C Moderately positive*
10 to 18	*+B Positive*
1 to 9	*+A Slightly positive*
0	*N Statu quo*
-1 to -9	*-A Slightly negative*
-10 to -18	*-B Negative*
-19 to-35	*-C Moderately negative*
-36 to -71	*-D Significant negative*
-72to-108	*-E Majore negative*

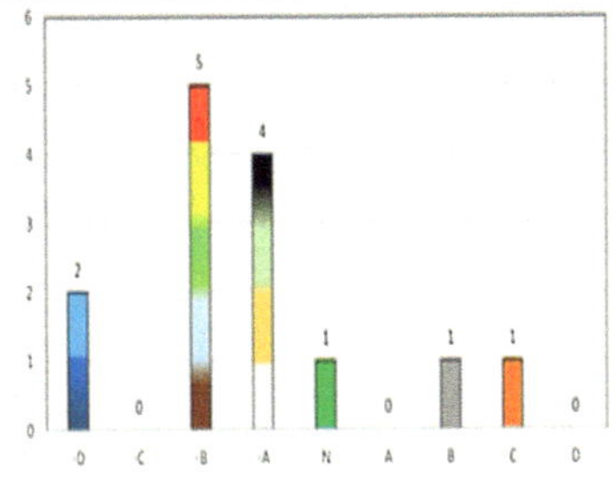

time through: displacement of the road affecting the integrity of private property (houses, expropriated land, farm land (Berevoiesti).

The traffic disruption and temporary isolation of residents, temporary deprivation of social services (road repair work sometimes takes months or even years) and use of detours routes constitutes impacts over resident life with direct connotation.

The visual impact of landslides along the road 73C is local and reflected on the slopes landscape. It's imposed by viewshed and related with complex and rotational landslides, creating rupture in landscape. Visual impact is given by the combined effect on morphology, vegetation and anthropic and it is evaluate being slightly negative. The economic impact of landslides on the road 73C is insignificant negative and was generated by additional costs necessary for crossing or bypassing damaged road segments (detours national road 7C), needed to repair infrastructure, private assets (houses from Piatra and Valea Iasului) and public assets. Transport restrictions are concerning both speed limits and carrying capacity. The frequency of these restrictions are monthly or even weekly. The density of sectors with speed restrictions at 30 km/h is 4–10 segments/km. Speed restrictions, combined with the consequence of variation in average fuel consumption, reduce the efficiency of the road. Currently, the road has an average efficiency with reduced efficiency sectors. Tourists avoid to use the road damaged by landslides. This is the main reason of poor economic development of the region. The economic impact has regional importance and has reversible character.

Conclusion

Landslides impact frequencies on the 73C road are monthly and annual. By applying adapted RIAM matrix the landslides impact on natural environment is significant negative impact (Table 2). Most intense impact has been considered being on the physical components of the natural environment.

The social components fits in the same category, but with less intense negative impact. The economic impact is moderate, even with the annual performance of the construction and maintenance and related costs. A condition for an efficient use of the road 73C in relation to the slope morphodynamics would address the whole slope stabilization works prior to construction and maintenance specific works. Otherwise, the effects of the relationship between natural and anthropogenic environment along national road 73C will be permanent. RIAM is a good tool to identify the importance and nature of each component participating to the impact condition.

Acknowledgements We want to thank GEO-SERV llc for the geotechnical data and also our thanks for all persons that choose to remain anonymous and were consulted in the time of study.

References

Alimohammadlou Y, Najafi A, Yalcin A (2013) Landslide process and impacts: A proposed classification method. Catena 104:219–232

Andra A, Mafteiu M (2007) Tigveni - Momaia Landslide, Revista de Geomorfologie, vol. 9. Editura Universităţii din Bucureşti, Bucuresti, pp 73–86

Constantin M, Trandafir AC, Jurchescu M, Ciupitu D (2009) Morphology and environmental impact of the Colti–Alunis,landslide (Curvature Carpathians). Romania, Environ Earth Sci. doi:10.1007/s12665-009-0142-1

Dinu M (1999) Subcarpatii dintre Topolog si Bistrita Valcii. Studiol proceselor actuale de modelare a versantilor. Ed Academiei Romane, Bucuresti

Dragos V (1953) Cercetări geologice asupra regiunii dintre râurile Topolog şi Olt. D.S. ale şedinţelor, vol XXXVII (1949–1950). Comitetul Geologic, Bucureşti, pp 55–76

Gaman C (2013) The impact of landslides on the DN15 national Road in the area of the Izvoru Muntelui Bicaz Reservoir. Present Environ Sustain Dev 7(1)

Geertsema M, Highland L, Vaugeoouis L (2009) Environmental Impact of Landslides. In: Sassa K, Canuti P (eds) Landslides–disaster risk reduction. Springer, Berlin 2009, pp 589–607

Mihaila N (1971) Stratigrafia depozitelor pliocene şi cuaternare dintre Valea Oltului şi Valea Vîlsanului (sectorul Rîmnicu Vîlcea – Curtea de Argeş – Vîlsăneşti), Studii tehnice şi economice, seria 3, Stratigrafie, nr. 7, IG., Bucureşti

Mulţescu Al. Mulţescu M (2010) Curtea de Argeş - Evoluţia ansamblului, ARGESIS, *Studii si comunicari*, seria Istorie, tom XIX Muzeul judetean Arges, Consiliul Judetean Arges, Edit Orgessos, Pitesti, pp 193–206. ISSN:1453 – 2182

Osaci-Costache G (2004) Muscelele dintre Dambovita si Olt in documete cartografice, Reconstituirea si dinamica peisajului geographic in secolele XVIII – XX. Editura Universitara, Bucuresti

Panizza M (1996) Environmental geomorphology, developments in earth surface processes 4. ISBN:0-444-89830-1

Panizza M (1999) Floods and landslides: Integrated risk assessment. In: Casale R, Margottini C (eds) Relationships between environment and man in terms of landslide induced risk, chapter 12. Springer, Berlin, p 375. ISBN:978-3-642-63664-6

Pastakia CM (1998) The rapid impact assessment matrix (RIAM)—A new tool for environmental impact assessment. In Jensen K (ed) Environmental impact assessment using the Rapid Impact Assessment Matrix (RIAM) Fredensborg. Olsen & Olsen, Denmark

Pastakia CM, Jensen R (1998) The rapid impact assessment matrix (RIAM) for EIA, EIA Procedure, Environ impact assess rev, Elsevier Science, Inc, NewYork, pp 461–482

Pfurtscheller C, Genovese E (2013) The Felbertauern landslide of 2013: impact on transport network, regional economy and policy decisions. Trans Policy

Schuster RL, Highland LM (2004) Impact of Landslides and innovative landslide – mitigation measures on the natural environment. In: Int Conf On Slope Eng. Hong Kong, China, Dec. 8-10, 2003. Proceedings 29

Schuster RL, Highland LM (2007) Overview of the effects of mass wasting on the natural environment. Geological Society of America, Environmental & Engineering Geoscience 8:25–44

Evaluation of Building Damages Induced by Landslides in Volterra Area (Italy) Through Remote Sensing Techniques

Silvia Bianchini, Teresa Nolesini, Matteo Del Soldato, and Nicola Casagli

Abstract

This paper aims to detecting terrain movements in landslide-affected and landslide-prone zones and their damaging effects on the urban fabric. The case study is the Volterra area in Tuscany region (Italy), covers about 20 km^2 and is extensively affected by diffuse slope instability. Firstly, the spatial distribution and types of the landslides were studied on the basis of the geological and geomorphological setting coupled with a geotechnical monitoring. Secondly, satellite SAR (Synthetic Aperture Radar) images acquired by ENVISAT and COSMO-SkyMed sensors respectively in 2003–2009 and 2010–2015 and processed with Persistent Scatterer Interferometry (PSI) techniques, were exploited. In particular, these satellite radar data combined with thematic data and *in situ* field surveys allowed the improvement of the geometric and kinematic characterization of landslides, as well as allowing a deformation and damage assessment to be undertaken on built-up zones. The classification of damage degree and building deformation velocity maps of the study area were also evaluated through PSI displacement rates. Furthermore, as a single building-scale analysis, maximum differential settlement parameters of some sample buildings were derived from radar measurements, and then cross-compared with constructive features, geomorphological conditions and with field evidences of known landslide areas. This work allowed the correlation of landslide movements and their effects on the urban fabric and provided a useful stability analysis within future risk mitigation strategies.

Keywords

Landslides • SAR interferometry • Differential settlement • Building damage • Volterra

S. Bianchini (✉) · T. Nolesini · M. Del Soldato · N. Casagli
Department of Earth Sciences, University of Firenze,
Via La Pira 4, 50121 Florence, Italy
e-mail: silvia.bianchini@unifi.it

T. Nolesini
e-mail: teresa.nolesini@unifi.it

M. Del Soldato
e-mail: matteo.delsoldato@unifi.it

N. Casagli
e-mail: nicola.casagli@unifi.it

M. Mikoš et al. (eds.), *Advancing Culture of Living with Landslides*,
DOI 10.1007/978-3-319-53482-1_14

Introduction

The analysis of slope stability in urban areas affected by landslides is an important issue for landslide risk management and mitigation planning.

A comprehensive study would include the spatial and kinematic characterization of landslides and the evaluation of their impacts and damages on structure and infrastructure (Antronico et al. 2015; Gullà et al. 2016; Goetz et al. 2011; Van Westen 2013). Such investigation can be performed by means of various methods, widely used in scientific literature, i.e. conventional geomorphological approaches and modern technologies like remote sensing techniques (Guzzetti et al. 2012).

Many different multi-temporal InSAR (Synthetic Aperture Radar Interferometry) techniques, e.g. Persistent Scatterer Interferometry (PSI) have been successfully used for measuring ground deformations in order to support landslide studies over wide areas (Farina et al. 2006; Cigna et al. 2010; Bianchini et al. 2012; Cascini et al. 2013). Recently, some PSI-based procedures have been also developed for assessing building deformation and settlement at local scale (Ciampalini et al. 2014; Sanabria et al. 2014; Bianchini et al. 2015).

This work aims at detecting terrain movements in landslide-affected and landslide–prone zones and their damaging effects on the urban fabric of Volterra area (Tuscany Region, Italy).

The study area is extensively affected by diffuse landsliding and soil erosion. Moreover, two wall collapses and some slope failures occurred in the period 2011–2014 due to intense rainfall (Pratesi et al. 2015). After these events, an accurate instability analysis was required, given the cultural and historic heritage of the site, known as one of the most important Etruscan settlements and medieval city (Sabelli et al. 2012).

Optical data, geo-thematic layers, field surveys, and existing geotechnical monitoring data were firstly exploited to characterize the geological and geomorphological setting of the site. Then, SAR images acquired by ENVISAT and COSMO-SkyMed sensors respectively in 2003–2009 and 2010–2015 and processed with Persistent Scatterer Interferometry (PSI) techniques (Ferretti et al. 2001) were analyzed for quantitatively and qualitatively assess ground motions of slow-moving landslide phenomena and their impacts on the built-up zones of the study area. The outcomes of this work permitted to correlate landslide movements and their effects on urban fabric and provided a useful stability analysis within future risk mitigation strategies.

Study Area

Geographical and Geological Setting

The Volterra study area covers about 20 km^2 and it is mainly represented by the historical town of Volterra and the surrounding semi-urban and rural landscape (Fig. 1a). We distinguished 7 sub-areas: the city center, Montebradoni, Valdera, Colombaie, Fontecorrenti, St.Lazzaro, Roncolla (Fig. 1a).

Volterra is located on a *mesa* tableland at 460–500 meters a.s.l. (above sea level) and includes Etruscan, Roman and medieval settlements, enclosed by great defensive town-wall (Fig. 1b).

From a geological point of view, the sequence of the study area consists of a Pliocene marine sedimentary succession, made of Early Pliocene thick Blue clays, upper Middle-Late Pliocene cemented sands in horizontal stratigraphic continuity and uppermost calcarenites where the city center was built (Fig. 2). Recent colluvial terrigenous deposits fill the valleys, lying upon the sandy-clay units and mainly deriving from the weathering of the upper formations.

Methodology and Data

The methodological procedure consists firstly in the background characterization of the area from the landslide hazard point of view by means of geological-geomorphological interpretation and ground truth information (i.e. geotechnical monitoring and field surveys), as well as by satellite remote sensing InSAR data. Secondly we performed the evaluation of building settlement and damages through satellite InSAR data combined with background data, finally producing building deformation velocity maps and a classification of damage degree of the study area.

For the background geomorphological analysis we used:

- Digital Terrain Model (DTM) with 10 meters cell size resolution, 1:25,000;
- 1:25,000 topographic map distributed by IGM(Military Geographic Institute) and 1:10,000 map downloaded from the Regional website of Tuscany Region;
- Digital colour orthophotos from Volo Italia 2000 (not stereo) and Digital colour orthophotos referred to years 2006 and 2013 downloaded from the Regional website of Tuscany Region;

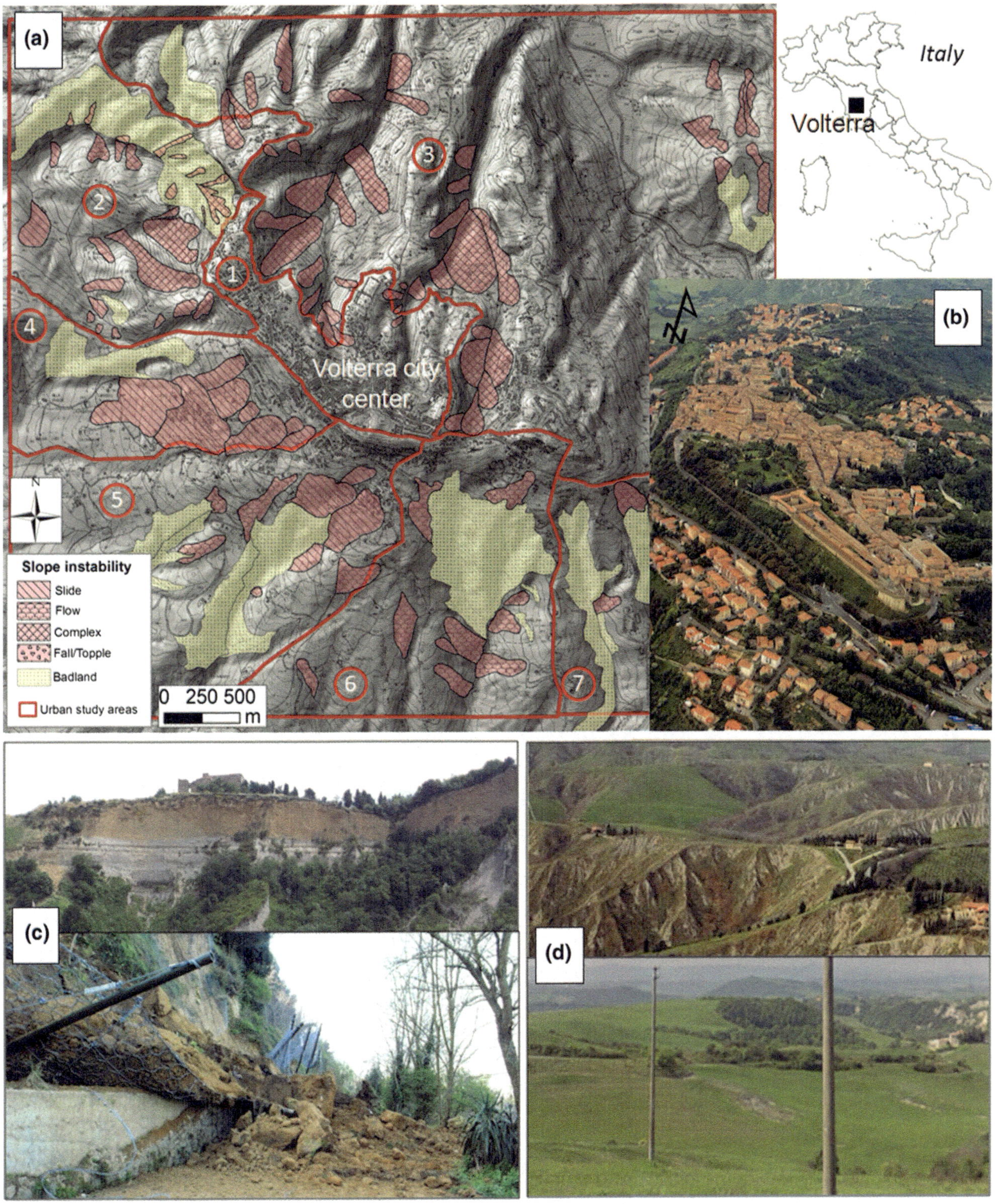

Fig. 1 **a** Location of Volterra and study areas (1: city center; 2 Montebradoni; 3 Valdera; 4 Colombaie; 5 Fontecorrenti; 6 St.Lazzaro; 7 Roncolla); **b** Overview of Volterra city center; **c** Sandy crags and fall example in the NE portion of the area; **d** Badlands and shallow soil erosion example in the SW portion of the area

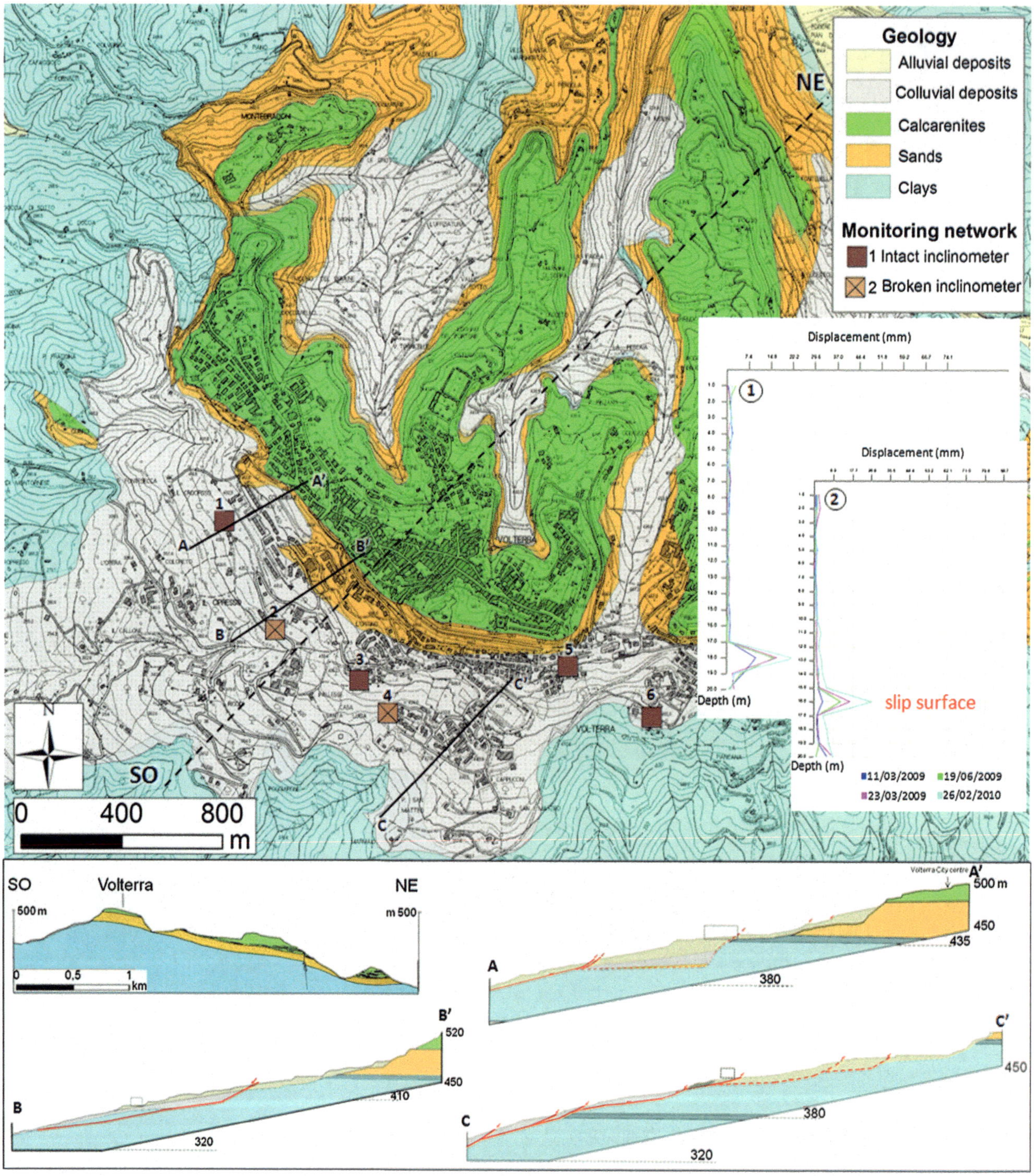

Fig. 2 Geological setting of the Volterra study area and some geological cross-sections. Inclinometer locations are displayed and labeled on the map. Two displacement graphs of inclinometers n°2 and n°4 are shown in the inserts (modified from GEOPROGETTI 2010)

- CORINE Land Cover map (Perdigao and Annoni 1997) at 1:50,000 scale (3rd classification level), published in 2000 and distributed by ISPRA (*Istituto Superiore PRotezione Ambiente*);
- Geotechnical data from inclinometers, provided by GEOPROGETTI and GEOSER companies referred to monitoring activity in 2009–2010 (GEOPROGETTI 2010);
- Recent in situ field surveys, spanning the time interval 2010–2015.

Multi-temporal remote sensing SAR data employed in this work are:

- ENVISAT ASAR (Advanced SAR) images acquired in C-band (5.6 cm wavelength) along ascending and descending orbits in 2003–2009 processed by e-GEOS (an ASI Company) by means of the PSP (Persistent Scatterer Pairs; Costantini et al. 2000) technique;
- COSMO-SkyMed SAR scenes acquired in ascending and descending geometries in X-band in the time span 2010–2015, processed with the SqueeSAR™ approach (Ferretti et al. 2011).

Analysis and Results

Slope Instability Background

The geometry of the Volterra tableland, which is slightly dipping 2–10° towards NE and the different outcropping lithotypes influence the type and the spatial distribution of landslides in the study area (Fig. 1b–d).

On one hand, on the southwestern slope of the Volterra mesa, where Blue clays mainly outcrop (areas 4–7), the morphology is gentle and affected by badlands, that are typical landforms of clayey lithotypes, surface flows and soil creep erosion (Fig. 1a, d). The shallow colluvial, made up of chaotic detritus with thickness up to 20 meters in the SW sector of the study area, contribute to determine diffuse superficial landsliding (Pratesi et al. 2015; Bianchini et al. 2015).

On the other hand, on the opposite hillslope of the Volterra mesa (areas 1–3), sands and calcarenites crop out more widely and so the morphology is steeper and sharper, characterized by sub-vertical cliffs (Sabelli et al. 2012). The main types of landslides are complex movements and falls caused by the undermining of the basal clayey of the hills and consequent retrogressive slope failures (Fig. 1a, c). The recent colluvial deposits derived from the weathering of the sandy/calcarenitic lithotypes contribute also in this area to surface creep downslope and ground instability.

The existing landslide inventory available for the Volterra area, derives from a landslide inventory, which was provided by Tuscany region at regional scale, improved by GEOPROGETTI company at local scale on the SW portion of Volterra city (GEOPROGETTI 2010). This database includes 116 phenomena, among which the most active and hazardous are located on the SW portion of the study area (Fig. 1a).

Geotechnical Monitoring

The monitoring network provided by GEOPROGETTI and GEOSER companies consist of six vertical inclinometers on the SW portion of Volterra within sub-areas 4–7 (GEOPROGETTI 2010). The inclinometer data were collected from January 2009 to February 2010 (Fig. 2). Displacement data reveal the depth of sliding surfaces in that zone affected by slope instability. In particular, inclinometers n°2 and n°4 located in Colombaie and Fontecorrenti (sub-areas 4 and 5) showed well-defined sliding surfaces at depth of 18 and 16 m respectively, allowing to identify the shear planes of the mapped landslide bodies (Fig. 2).

These two inclinometers recorded the highest movements with displacement rates of about 22 mm/yr, that are classified as "very slow/slow" according to the Cruden & Varnes classification (Cruden and Varnes 1996). Nevertheless, inclinometers n°1 and n°4 were found to be broken during the last field campaigns in 2014–2015.

Inclinometer n°1 did not seem to present sliding surfaces, even if a displacement of few millimeters per year across monitoring time was evident. Inclinometer n°3 in Fontecorrenti sub-area also did not reveal any defined shear surface; at the depth of 12 m a differential displacement of 4 mm/yr was detected, in correspondence of a change in the geotechnical properties of Blue Clays. Inclinometers n°5 and n°6 showed overall stability characterized by no distinct shear surface and almost negligible variations of 1–2 mm along the whole depth of the instrument.

Remote Sensing SAR Analysis

Advanced differential interferometric SAR techniques were exploited to derive information about slope stability conditions and building displacements of the study area.

ENVISAT PSI data acquired in C-band in both ascending and descending orbits show stability, set as ±2 mm/yr (Cigna et al. 2010), on the Volterra city center as well as in

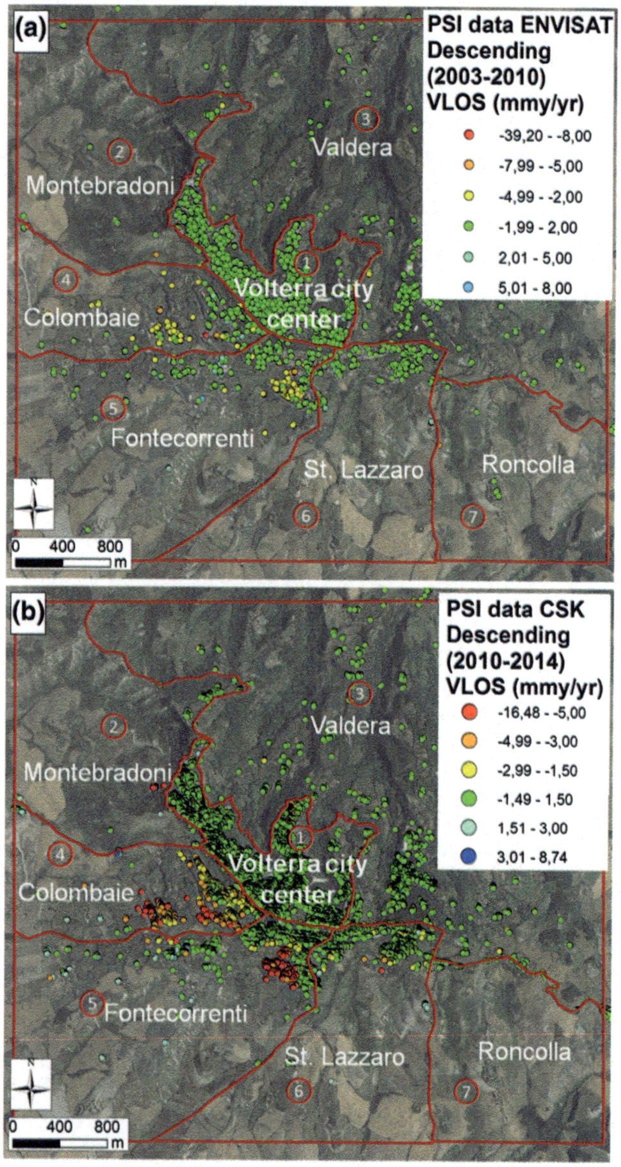

Fig. 3 **a** Spatial distribution of mean yearly velocities along the Line of Sight (VLOS) of ENVISAT data in descending geometry; **b** Spatial distribution of mean yearly velocities along the Line of Sight (VLOS) of COSMO-SkyMed (CSK) data in descending geometry

Valdera, Roncolla and St. Lazzaro sub-areas. On Colombaie and Fontecorrenti sub-areas, some ground motions are recorded, in agreement with the translational landslides mapped in these zones (Fig. 3a).

COSMO-SkyMed PSI data acquired in X-band in ascending orbit and descending orbit, respectively in the spanning time 2011–2015 and 2010–2014, reveal highest terrain movements on Colombaie and Fontecorrenti sub-areas with mean yearly velocities up to 10–15 mm/yr (Fig. 3b).

In the Colombaie sub-area, the moving radar targets are related to the translational and rotational slides mapped in this area, involving blue clays and colluvial deposits, and thus permit to derive the annual velocities of these phenomena. On Fontecorrenti sub-area, some landslide phenomena were mapped upward the wide badland area where the highest motion rates are recorded. These ground deformations are likely induced and maintained, at least partially, by gully erosion and active retrogressive movement of badlands. Some evidences of retreat of the landslide crown scarps were recognized on field and integrated with PS rates, allowing us to better trace the present boundaries of the slide movements. Also on Montebradoni and on St. Lazzaro sub-areas CSK data detected some motions, represented respectively by falls on *balze* crags or by superficial slides on badland landscape.

Given the west-facing orientation of the Volterra south-western zonation, which is the most hazardous one, movements measured in descending geometry are the most suitable to properly analyze the real motions downslope, and therefore we only show interferometric data from descending geometry in Fig. 3.

Evaluation of Building Damages

A single building-scale analysis was performed on some critical sample buildings on the most critical landslide-affected sub-areas (i.e. Colombaie and Fontecorrenti), by deriving the maximum vertical differential settlement and angular distortion parameters, cross-compared with constructive features, geo-morphological conditions and with field evidences on the known landslide areas (Fig. 4).

In particular, relying on the work by Bianchini et al. (2015), an Inverse Distance Weighted (IDW) interpolation was performed on PS cumulative displacements of time series in order to create a continuous displacement-surface from the sample set of PSI locations. For this detailed local evaluation we only consider the CSK PSI data that, due to their density and high resolution, show up on building facades and roofs, fitting well the typical scale of constructive elements. The IDW raster map is characterized by a cellsize of 3 meters, according to the 3 × 3 m cell size of COSMO-SkyMed satellite images and converted in pixel centroids (Fig. 4c).

The maximum vertical differential settlement (δ_v) of a structure is the unequal settling of a building and it is defined as the maximum difference of vertical displacement between the two IDW centroids with the maximum and the minimum cumulative displacement within the building foundation

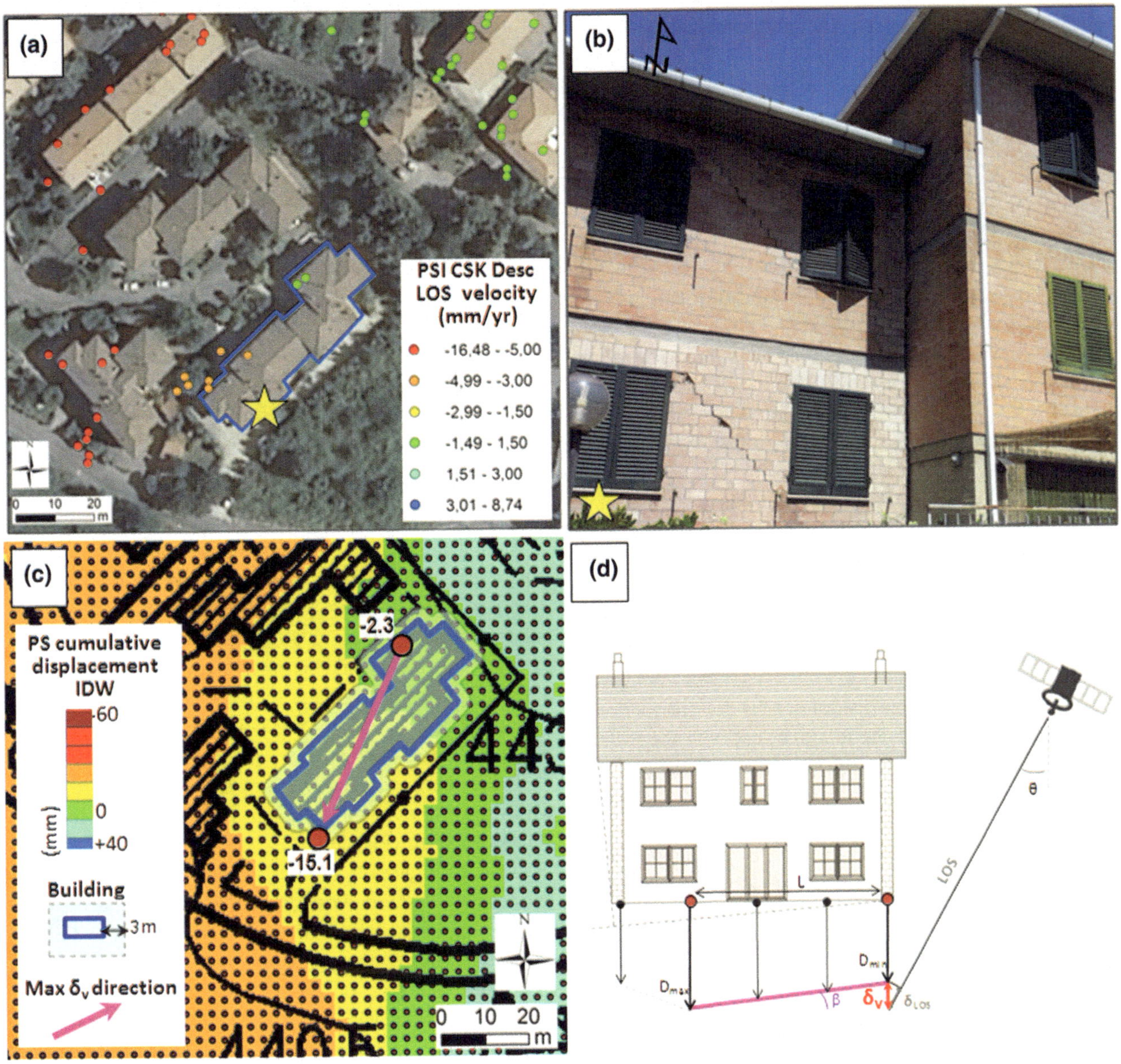

Fig. 4 Example of building deformation assessment in the Fontecorrenti sub-area: **a** VLOS velocities of CSK PSI data in descending orbit; **b** Photo of a façade; **c** IDW interpolation of cumulative displacements and δv computation; **d** Sketch of the differential settlement parameters used within the analysis

(Eurocode 1994, 2010; Bowles 1977; Bjerrum 1963). The δ_v value defines the direction along which differential settlement is dominant and it is computed as follows:

$$\delta_v = \left(\frac{|D_{\min\_LOS} - D_{\max\_LOS}|}{\cos\theta}\right) = \frac{|\delta_{LOS}|}{\cos\theta} \quad (1)$$

where D_{min_LOS} and D_{max_LOS} are the minimum and maximum cumulative displacements measured the building along the satellite LOS during the monitoring period; δ_{LOS} is the maximum differential settlement between these two measurement points along the LOS; θ is the satellite look angle (Sanabria et al. 2014; Bianchini et al. 2015) (Fig. 4d).

The angular distortion (β) is related to the measured δ_v, and it is computed as the ratio of δ_v and the distance (L) between the maximum and the minimum D_{min_LOS} and D_{max_LOS} cumulative displacement, as follows:

$$\beta = \frac{|\delta v|}{L} \qquad (2)$$

The identification of the pixel centroids as D_{min_LOS} and D_{max_LOS} is found within a buffer area drawn around the plain-edge of the building and dimensioned accordingly to the cell size resolution (3 m). The use of this tolerance area allows taking into account even PS benchmarks that do not lie within the building foundation, but that are the result of a back-scattered signal mainly influenced by the structure itself, and also permits avoiding possible shifts in the layer georeferencing procedures.

In Fig. 4 we show results of the building deformation assessment on an example building in the Fontecorrenti sub-area.

The building is a masonry structure (Fig. 4b) located in an area characterized by the terrain instability of the colluvial debris and by a translational landslide that represents the retrogressive area of influence of the badland downward. The building, initially built in the 1970s with direct foundations, were underpinned with piles, but it is still affected by settlement and suffered consequent damages (GEOPROGETTI 2010). The maximum differential settlement δv shows a considerable value with a SE-NW vector orientation (14.8 mm) and an angular distortion of 3.3×10^{-4}. These values and their direction are in agreement with the pattern of the centimetric cracks damages, normal to tension stresses, surveyed on the building during on-field campaign (Fig. 4a–c).

By the way, the measured δv and β on this structure resulted to be lower than the maximum allowable settlement set as $\delta_v = 25$ mm and $\beta = 3.0 * 10^{-4}$, for civil building on clayey/sandy soil (EUROCODE 2010). Higher values were found on Colombaie sub-area.

By the way, the measured δv and β on this structure resulted to be lower than the maximum allowable settlement set as $\delta_v = 25$ mm and $\beta = 3.0 * 10^{-4}$, for civil building on clayey/sandy soil (EUROCODE 2010). Higher values were found on Colombaie sub-area.

Classification of Building Damages

Over the whole Volterra area, at a larger scale, a classification of damage degree and building deformation velocity maps were also evaluated through PSI displacement rates, according to the procedure proposed by Ciampalini et al. (2014).

Table 1 Ranking scheme of building damage categories (modified after GSL 1991 and Cooper 2006)

Class	Name	Synthetic description
0	No damage	Intact building
1	Negligible	Hairline cracks
2	Slight	Occasional cracks
3	Moderate	Widespread cracks
4	Serious	Extensive cracks
5	Severe	Major structural damage
6	Collapse	Partial/Total collapse

Firstly, after recent field surveys in 2014–2015, a damage mapping of all the buildings of the study area was completed, following the ranking scheme of GSL (1991) and Cooper (2006) for the degree of damage. According to the crack pattern, seven classes were defined (Table 1).

A total number of 1371 buildings were digitized in the study area. The 94% of the analyzed structures (1285 structures) turned out to be in class 0 (no damage), 49 buildings fall in class 1, while 15 and 16 buildings were ascribed in class 2 and 3, respectively. £ buildings were affected by serious damage (class 4) and 3 other ones by severe damage (class 5). Any building experienced total or partial collapse (class 6).

The Volterra city center is not affected by damage at all, while most critical sub-areas are Colombaie, Fontecorrenti and St. Lazzaro (Fig. 5a).

Then, an attempt to provide a velocity value of buildings by means of PS radar data was carried out. We calculated the velocity of each single building, considering the PSI data included within its extension; in order to increase the amount of PS inside a higher number of buildings, a buffer of 2 m is defined following Ciampalini et al. (2014). This tolerance also overcomes potential problems of shift due to georeferencing process.

The velocity rates of every strcuture were computed by taking into account all the PSI targets (in ascending and descending geometry) from ENVISAT satellite and from CSK satellite (Fig. 5b, c).

A quite good correlation of the Building velocity maps computed with ENVISAT (Fig. 5c) and CSK (Fig. 5d) data with the Building damage classification map (Fig. 5a) can be assessed, the most damaged buildings are the ones with high displacement and velocity.

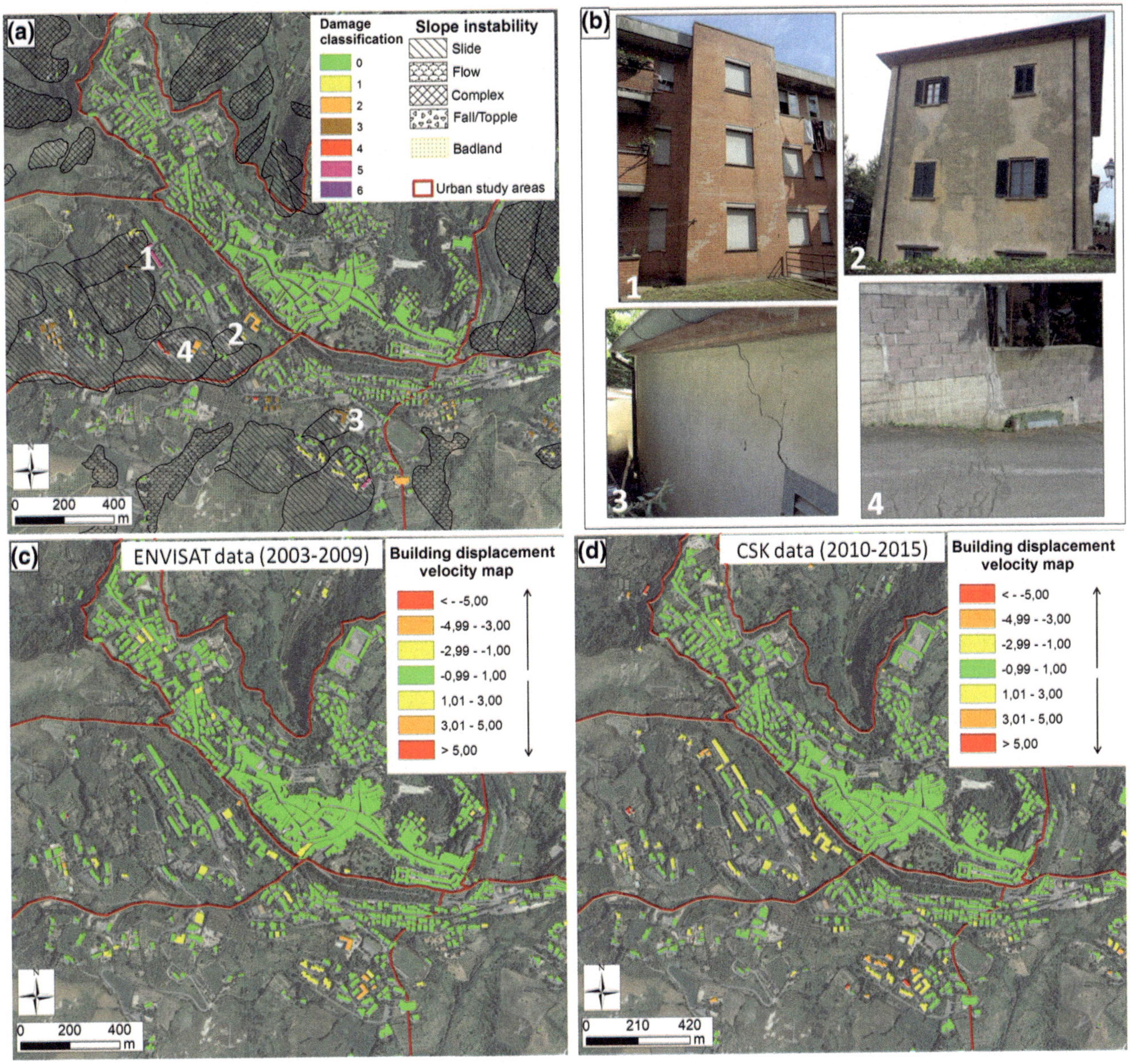

Fig. 5 **a** Building damage classification map; **b** Photos of some buildings whose locations is displayed and labeled in (**a**); **c** Building displacement map through ENVISAT data; **d** Building displacement map through COSMO-SkyMed (CSK) data

Discussion and Conclusions

The instability analysis on the Volterra study area was performed through interpretation of geological-geomorphological data and ground truth information, such as inclinometric monitoring network referred to years 2009–2010 and recent field surveys (2014–2015). PSI satellite radar data acquired in the periods 2003–2009 (ENVISAT satellite) and 2011–2015 (CSK satellite), in combination the other data, permitted to enhance the knowledge, extension, rate of displacements of the mapped landslide movements in the study area. The analysis was carried out by considering 7 sub-areas (the city center and surroundings): the most critical sub-areas characterized by the highest ground motion rates were Colombaie and Fontecorrenti located on the south-western portion of the study area, where translational/complex slides and badland gully erosion extensively occur.

In order to evaluate the impacts of mass movements on the urban fabric, we performed a building deformation and damage assessment on some structures of the study areas,by means of the PSI-based computation of differential settlement parameters. Values and directions of these settlement features were cross-compared with background data (i.e. geological data) and crack patterns detected during a recent in situ survey. This issue could be deeply investigated in the future, by considering the role and interaction with foundation types, comparing PSI results with geotechnical computations and by assuming a rating system for admissibility of differential settlements (SLS—Serviceability Limit State—criterion).

Overall, the use of data in X-band increases the level of detail of the analysis, since most of the PS targets show up on housetops and facades of buildings, enabling a very site-specific investigation and damage assessment.

This work scans active ground displacements on the area around the historic town Volterra and surroundings, and detects deformation and damage consequences on the urban fabric at local scale. The outcomes could be useful as a preliminary step for further risk management strategies and, as future improvement, studies on exposure and vulnerability of the elements at risk would be performed in the study area.

Acknowledgements The authors would like to thank GEOPROGETTI company (geologists Francesca Franchi and Emilio Pistilli) for making the geological and geotechnical data on Volterra available, and Dr. Fabio Pratesi for his efforts in field campaigns. Further data and information on the investigated area of the Volterra site are available on the municipality website: http://www.comune.volterra.pi.it.

References

Annoni A Perdigao V (1997) Technical and methodological guide for updating CORINE Land Cover Database. European Commission, EUR 17288EN, Space Application Institute of Joint Research Centre, Ispra, Italy

Antronico L, Borrelli L, Coscarelli R, Gullà G (2015) Time evolution of landslide damages to buildings: the case study of Lungro (Calabria, southern Italy). Bull Eng Geol Env 74:47–59

Bianchini S, Cigna F, Righini G, Proietti C, Casagli N (2012) Landslide hotspot mapping by means of persistent scatterer interferometry. Environ Earth Sci 67(4):1155–1172

Bianchini S, Pratesi F, Nolesini T, Casagli N (2015) Building deformation assessment by means of Persistent Scatterer Interferometry analysis on a landslide-affected area: the Volterra (Italy) case study. Remote Sensing 7:4678–4701. doi:10.3390/rs70404678

Bowles JE (1977) Foundation analysis and design. Mc Graw Hill Publications, New York

Bjerrum L (1963) Allowable settlement of structures. Proceedings of the 3rd European Conference on Soil Mechanics and Foundation Engineering. Wiesbaden, Germany, pp 15–18

Cascini L, Peduto D, Pisciotta G, Arena L, Ferlisi S, Fornaro G (2013) The combination of DInSAR and facility damage data for the updating of slow-moving landslide inventory maps at medium scale. Natural Hazards and Earth System Science 13:1527–1549

Ciampalini A, Bardi F, Bianchini S, del Ventisette C, Moretti S, Casagli N (2014) Analysis of building deformation in landslide area using multi-sensor PSInSARTM technique. Int J Appl Earth Obs Geoinf 33:166–180

Cigna F, Bianchini S, Righini G, Proietti C, Casagli N (2010) Updating landslide inventory maps in mountain areas by means of Persistent Scatterer Interferometry (PSI) and photo-interpretation: Central Calabria (Italy) case study. In: Malet JP, Glade T, Casagli N (eds) Bringing Science to Society. Florence, Italy, pp 3–9

Cooper AH (2006) The classification, recording, databasing, and use of information about building damage caused by subsidence and landslides. Q J Eng GeolHydrogeol 41:409–424

Cruden DM, Varnes DJ (1996) Landslide types and processes. In: Turner AK, Schuster RL (eds) Landslides: investigation and Mitigation, Sp. Rep. 247, Transportation Research Board, National Research Council. National Academy Press, Washington DC. pp 36–75

Eurocode, (1994) EC 7: Geotechnical design. Available online: URL: http://law.resource.org/pub/eur/ibr/en.1997.1.2004.pdf [Last accessed: 15 June 2016]

Eurocode, (2010) Basis of structural design. URL:http://www.unirc.it/documentazione/materiale_didattico [Last accessed: 15 June 2016]

Farina P, Colombo D, Fumagalli A, Marks F, Moretti S (2006) Permanent scatterers for landslide investigations: outcomes from the ESA-SLAM project. Eng Geology 88:200–217

Ferretti A, Prati C, Rocca F (2001) Permanent scatterers in SAR interferometry. IEEE Trans Geosci Remote Sens 39:8–20

Ferretti A, Fumagalli A, Novali F, Prati C, Rocca F, Rucci A (2011) A new algorithm for processing interfer-ometric datastacks: SqueeSAR™". IEEE Trans Geosci Remote Sens 49:3460–3470

GSL-Geomorphological Services Ltd (1991) Coastal Landslip Potential Assessment: Isle of Wight Undercliffe, Ventnor. Technical Report for the department of the Environment. Research Contract RECD 7/1/272

GEOPROGETTI, (2010) Indagini geognostiche e sismiche per l'analisi dell'assetto geologico e geomorfo-logico del versante Sud di Volterra. Report for the Volterra Municipality. URL: http://www.comune.volterra.pi.it. [Last accessed: 1 August 2016]

Goetz JN, Guthrie RH, Brenning A (2011) Integrating physical and empirical landslide susceptibility models using generalized additive models. Geomorphology 129:376–386

Gullà G, Peduto D, Borrelli L, Antronico L, Fornaro G (2016) Geometric and kinematic characterization of landslides affecting urban areas: the Lungro case study (Calabria, Southern Italy). Landslides. pp 1–18

Guzzetti F, Mondini AC, Cardinali M, Fiorucci F, Santangelo M, Chang KT (2012) Landslide inventory maps: new tools for an old problem. Earth Sci Rev 112(1):42–66

Van Westen CJ (2013) Remote sensing and GIS for natural hazards assessment and disaster risk management. Schroder JF and Bishop MP Treatise on Geomorphology Academic Press, Elsevier, San Diego. pp 259–298

Pratesi F, Nolesini T, Bianchini S, Leva D, Lombardi L, Fanti R, Casagli N (2015) Early warning GBInSAR-based Method for monitoring volterra (Tuscany, Italy) city walls. J Sel Top Appl Earth Obs Remote Sens

Sabelli R, Cecchi G, Esposito AM (2012) Mura etrusche di Volterra. Conservazione e Valorizzazione; Bientina, Italy. La Grafica Pisana, Italy

Sanabria MP, Guardiola-Albert C, Tomás R, Herrera G, Prieto A, Sánchez H, Tessitore S (2014) Subsidence ac-tivity maps derived from DInSAR data: orihuela case study. Nat Hazards Earth Syst Sci 14:1341–1360

The Resilience of Some Villages 36 Years After the Irpinia-Basilicata (Southern Italy) 1980 Earthquake

Sabina Porfido, Giuliana Alessio, Germana Gaudiosi, Rosa Nappi, and Efisio Spiga

Abstract

The aim of this study is to describe the modifications of the built environment that have occurred in 36 years following the Irpinia-Basilicata, 1980 earthquake. In particular, especially in the villages of the epicentral area, changes in the urban and territorial setting have been examined, as well as the consequences of ground effects that have influenced the choices of reconstruction, both in situ, and far from the original historical centers. The November 23, 1980 Irpinia–Basilicata earthquake (Mw = 6.9; Io = X MCS; Io = X ESI-07), killing 3000 people, hit 800 localities over a large area of Southern Italy; 75,000 houses totally collapsed and 275,000 were badly damaged. The earthquake induced primary and secondary environmental effects, over all slope movements. The total amount of surface faulting was 40 km in length with the maximum displacement of 100 cm; the total area affected by slope movements was estimated to be about 7400 km^2, with 200 landslides classified. One of the largest landslides damaged Calitri village, in Avellino province. We have examined, as case histories, the reconstruction of Calitri and San Mango sul Calore villages, that were affected by severe landslides and were rebuilt in situ; we have also studied Conza della Campania that was reconstructed far from the original location. In the so-called Anthropocene age, the role of technical experts both in the built environment and in the social and ethical context is extremely important, for rebuilding the villages destroyed by earthquakes, especially in respect of the people resilience.

Keywords

1980 Irpinia-Basilicata earthquake • Landslides • Resilience

S. Porfido (✉)
CNR-IAMC Calata Porta Di Massa, Interno Porto, 80133 Naples, Italy
e-mail: Sabina.porfido@iamc.cnr.it

G. Alessio · G. Gaudiosi · R. Nappi
Osservatorio Vesuviano Sezione Di Napoli Istituto Nazionale Di Geofisica E Vulcanologia, Via Diocleziano 328, 80124 Naples, Italy
e-mail: giuliana.alessio@ingv.it

G. Gaudiosi
e-mail: germana.gaudiosi@ingv.it

R. Nappi
e-mail: rosa.nappi@ingv.it

E. Spiga
Independent Researcher, Via Circumvallazione 187, 83100 Avellino, Italy
e-mail: spiga.efisio@gmail.com

M. Mikoš et al. (eds.), *Advancing Culture of Living with Landslides*,
DOI 10.1007/978-3-319-53483-1_15

Introduction

The November 23, 1980 earthquake, well known as the "Irpinia earthquake" was the strongest seismic event of the last 100 years in the Southern Apennines of Italy, with Mw = 6.9 (Rovida et al. 2016) I0 = X MCS (Postpischl et al. 1985) and I0 = X ESI-07, (Michetti et al. 2007; Serva et al. 2007). It was felt nearly everywhere in Italy, from Sicily in the South, to Emilia Romagna and Liguria in the North.

This earthquake was characterized by a complex main rupture, composed of three major sub-events, interpreted as a succession of normal faulting events (Bernard and Zollo 1989). Many localities in the Campania and Basilicata regions were nearly completely destroyed (I = IX-X MSK, Postpischl et al. 1985); among them Castelnuovo di Conza, Conza della Campania, Lioni, Santomenna, San Mango sul Calore, San Michele di Serino and Sant'Angelo dei Lombardi in the Avellino province, Fig. 1.

About 800 localities suffered serious damage (Balvano, Bisaccia, Calitri, etc.); 75,000 houses totally collapsed and 275,000 were badly damaged. The casualties were 3000 and 10,000 people were wounded.

A large amount of information on primary and secondary environmental effects, over all slope movements, was available on the basis of several geological surveys carried out in the area affected by this earthquake. The total amount of surface faulting was about 40 km in length and the maximum displacement about 100 cm (Pantosti et al. 1993; Blumetti et al. 2002; Fig. 2), while the total area interested by 200 slope movements was estimated in 7400 km^2 (Serva et al. 2007). About 47% of the landslides were rock fall/toppling, 20% rotational slides, 20% slump-earthflows, 3% rapid earthflows, 9% left undefined (Cotecchia 1986; Esposito et al. 1998; Porfido et al. 2002; Porfido et al. 2007). The largest rockfalls occurred mostly in the epicentral area, with volumes ranged by 1000–10,000 m^3, as well as slump-earth flows that affected some important historical centers. One of the largest landslides (23 million m^3) affected Calitri and its urban area of recent increase. In the rural area even larger mudflows were triggered at Buoninventre, near Caposele, (30 million m^3) and Serra d'Acquara, near Senerchia, (28 million m^3) (Wasowski et al. 2002). Many others secondary effects were observed in the near and far field such as fractures, liquefaction phenomena and hydrological changes (Porfido et al. 2007). The aim of this

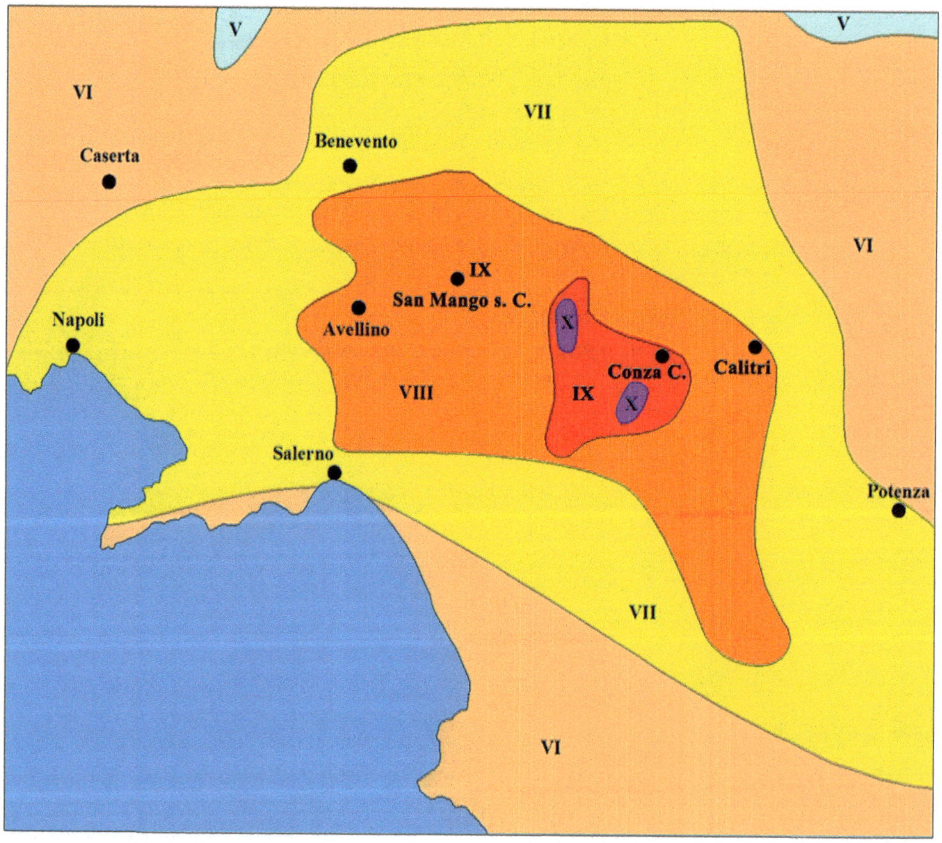

Fig. 1 Isoseismal lines of the November 23, 1980 Irpinia-Basilicata earthquake (modified after Postpischl et al. 1985)

Fig. 2 The 1980 main surface fault trace along Mt. Marzano. On the left side the original field map with the fault trace (courtesy of PFG-CNR, 1981), on the right the photo of the fault 24 years later (courtesy of Rosa Nappi, 2004)

study is to describe the modifications of the built environment occurred in 36 years following the 1980 earthquake. In particular, it has been examined the resilience of the communities (sensu CARRI Report 2013) in the villages of the epicentral area, which means the ability of local people to recover after disasters. In fact the earthquake strikes heavily the inhabitants of villages and the total destruction of the urban environment implies a significant loss of identity, resulting in a difficult choice of reconstruction.

Moreover the consequences of the earthquake ground effects and the changes in the urban and territorial setting both in situ, and far from the original historical centers, have also been considered.

Geological and Seismotectonic Setting

The Southern Apennines are a NW-SE-trending Neogene and Quaternary thrust and fold belt (Mostardini and Merlini 1986; Patacca and Scandone 1989; Doglioni et al. 1996). Since the Late Pliocene, after the opening of the Tyrrhenian Sea, extensional tectonics progressively shifting to the East has produced several deep tectonic basins, hosting mainly marine and volcanic deposits on the Tyrrhenian side. In the inner sectors of the Apennines many intermountain basins have developed, typically NW elongated graben up to tens of kilometers long; they are bounded by limestone steep slopes cut by normal faults, and host thick Quaternary continental sedimentation. The master faults are prevalently SW dipping, only in few basins NE dipping.

This portion of Apennines since historical times has been hit by several earthquakes with high energy (Rovida et al. 2016) and by numerous seismic sequences, recorded in the last thirty years, including: Irpinia-Basilicata 1980, Monti della Meta 1984, Sannio 1990, Potentino 1990 e 1991, Sannio-Matese 1997, Molise 2002, L'Aquila 2009.

The faults responsible for moderate to strong crustal earthquakes show typical hypocentral depths of 7–20 km (Amato et al. 1997), with slip-rates of the order of several mm/year.

Studies on active tectonics and paleoseismicity confirm that the present-day tectonic setting of the Southern Apennines is controlled by systems of Quaternary normal faults, which determine still immature basin-and-range morphologies.

Methodology

As a starting point, the seismic microzonation of the Progetto Finalizzato Geodinamica - Consiglio Nazionale delle Ricerche has been considered, that was performed soon after the November 23, 1980 earthquake. In this report the 39 municipalities of the epicentral area, which were the most damaged ones, were surveyed (AA VV 1983); they were chosen due to significant issues particularly of instability phenomena, and the choices for reconstruction were examined, on the basis not only of advices dictated by technicians, but also and especially on the basis of resilience of different populations and of political choices.

In this paper we have examined the case histories of Calitri, San Mango sul Calore and Conza della Campania villages; the first two were damaged by severe landslide phenomena, and in situ rebuilt, whereas the third one, on the basis of poor local soil condition and the suffered destruction, has been reconstructed far from its original position.

Successively, for these villages the geological and sociological storic and economic major studies of the literature were examined and field recognitions were carried out to ascertain the reconstruction stage (Alexander 1984; Guadagno 2005, 2010; Pignone et al. 2008; Porfido et al. 2010, 2016).

The Calitri Case History

The village of Calitri was hit by the November 23, 1980 earthquake, with Intensity VIII MSK and VIII ESI-07 (Postpischl et al. 1985; Serva et al. 2007), at a distance of about 16 km from the earthquake epicenter, which was located in Laviano. Six casualties occurred. This village was built on the top of a prominent hill characterized by the presence of sandstone and conglomerate rocks of Pliocene age. The middle-lower slopes are locally developed in older, intensely tectonized clay-rich units of pre-Pliocene age; these sectors are to great extent affected by landslides. A great landslide has taken place due to the 1980 event (Fig. 3), approximately 850 m long and up to 100 m deep (Cotecchia 1986; Del Prete and Trisorio Liuzzi 1981; Samuelli-Ferretti and Siro 1983a, b).

Such landslide had terrible consequences on the urban and road structure of the village (Fig. 4) and for that reason it was studied by many authors over the years to implement the village reconstruction safely.

Moreover, other significant environmental effects such as fracturing and liquefaction phenomena have occurred (Serva et al. 2007).

The seismic history of the Calitri village indicates that it has been hit in the past by several seismic events (the December 5, 1456; the August 19, 1561; the June 5, 1688; the September 8, 1694; the November 29, 1732; the June 7, 1910; the July 23, 1930) with I $\geq$ VIII MCS (Rovida et al. 2016).

Further landslides occurred in Calitri due to the 1694, 1805, 1910 and 1930 seismic events, although they had different epicentral locations (Samuelli-Ferretti and Siro 1983a, b; Porfido et al. 1991). In Fig. 5 the current panoramic view of the Calitri village has been shown, with the area of the new buildings in the foreground.

San Mango Sul Calore Case History

The village of San Mango sul Calore was almost totally destroyed by the November 23, 1980 earthquake, with Intensity IX MSK, VIII ESI-07 and 84 casualties (Postpischl et al. 1985; Serva et al. 2007; Fig. 6).

The seismic microzonation carried out immediately after the earthquake pointed out not only the poor building quality, but also the lack of foundations on soils with slope stability problems and unfavorable morphological position (Fig. 7). The village was provided in 1981–1982 with 310 light prefabricated buildings, of which 200 in one locality: S. Stefano, 4 km far from the old village center. The new residences were rebuilt locally, in spite of the poor subsoil conditions later partially on reclaimed and several hydrogeological problems, relative to the hill slope on which the village was based. The houses were realized by modern and anti-seismic technologies, also by means of Canadian and Como funding.

The seismic history of San Mango sul Calore shows that this village was violently hit by the earthquakes of the September 8, 1694 and the November 29, 1732, with high damage level, respectively VIII MCS and X MCS (Rovida et al. 2016).

The current panoramic view of the S. Mango sul Calore village has been pointed out in Fig. 8, where the complete reconstruction of the urban centre is visible.

Fig. 3 The original map drown by Samuelli-Ferretti and Siro (1983a) for the PFG-CNR microzonation of Calitri after the 1980 earthquake

Fig. 4 One of the main streets of the Calitri village immediately after the 1980 earhquake (proloco Calitri courtesy)

Fig. 5 Current panoramic view of Calitri (photo by E. Spiga 2016)

Fig. 6 Panoramic view of S. Mango sul Calore after devastation of the 1980 earthquake (http://www.comune.sanmangosulcalore.av.it)

Conza Della Campania Case History

The village of Conza della Campania was hit by the 1980 earthquake with Intensity X MSK, VIII ESI-07 (Postpischl et al. 1985; Serva et al. 2007), at a distance of about 9 km from the epicenter, with a casualties number of 189, and 90% of buildings collapsed or seriously damaged (Fig. 9). It was built on two small hills made by clay and sandy clay in the lower part, conglomerates with sands and sandstones in the middle part, and conglomerates of middle-low resistance in the upper part (Fig. 10).

The earthquake induced different environmental effects such as landslides, ground cracks and ground settlement (Serva et al. 2007).

The inhabitants of Conza, aware of their relevant seismic history due to seven earthquakes with I $\geq$ VII in the last six centuries (1466; 1517; 1692; 1694; 1732; 1910; 1930; Rovida et al. 2016), decided to rebuild their village in a different zone. Today the village center is completely relocated under the hill where the old village rested on, in a flat area named Piano delle Briglie (4 km far from the original nucleus).

Currently two villages of Conza della Campania coexist, the 'new town' where most of the inhabitants live, and the old Conza with its ruins, the archaeological and natural park (http://oasiwwflagodiconza.org/) that has become a tourist attraction (Fig. 11).

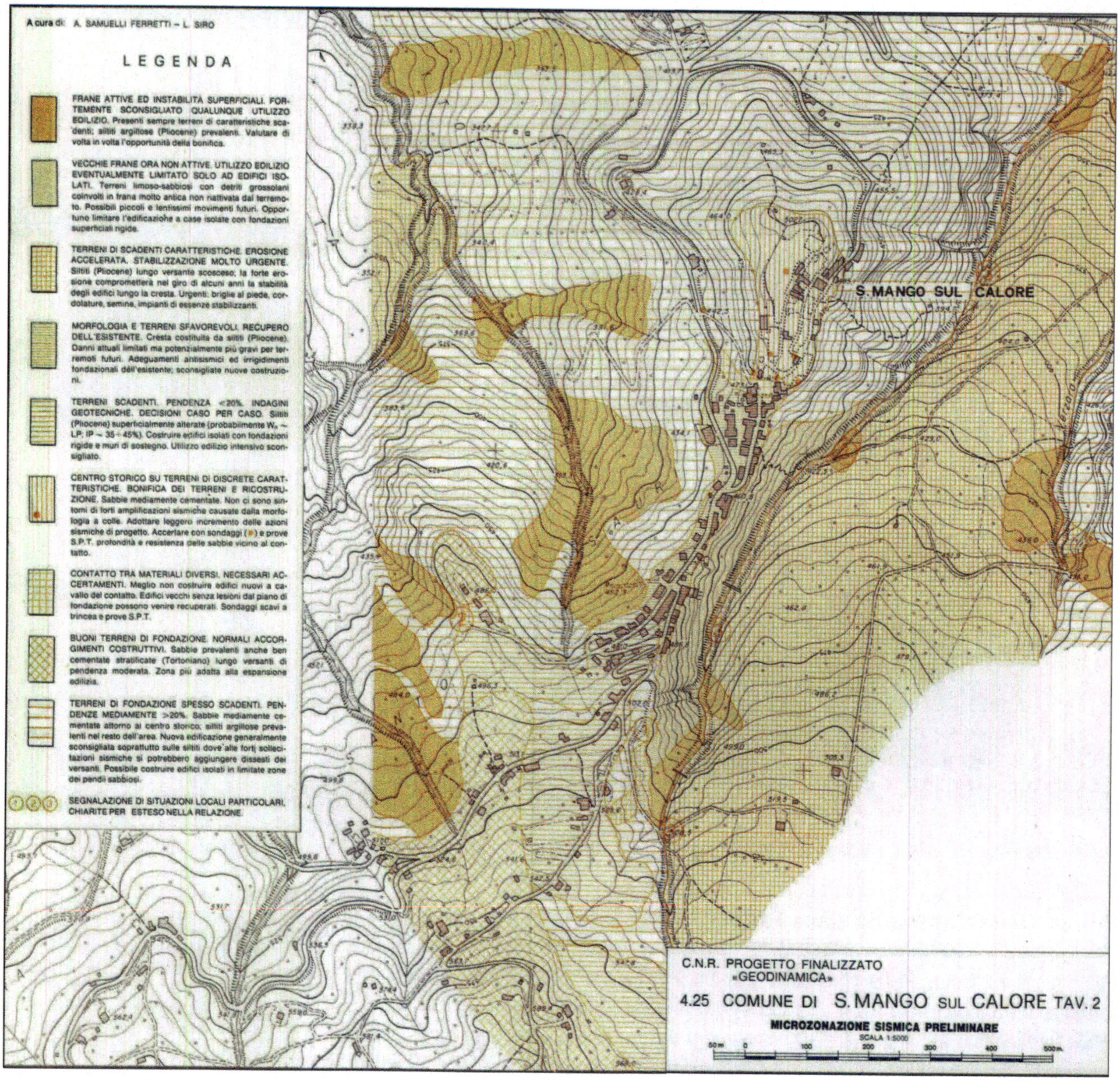

Fig. 7 The original map drawn by Samuelli-Ferretti and Siro (1983b) for the PFG-CNR microzonation of the S.Mango sul Calore village after the 1980 earthquake

Conclusions

In Italy, the restoration following the past historical seismic events has pursued different methodologies with respect to the historical heritage, mostly based on the evaluation of the Central Government and only in recent time with a more "pragmatic" intervention as in the cases of the 1976 Friuli and 1980 Irpinia earthquakes (Marchetti 2012).

The idea of resilience, in the single affected populations, is different case by case because it should take into account several environmental factors, the most relevant of them are the socio-economic characteristics of these regions and the geological conditions of these villages.

In particular, in this paper, it has been examined the resilience of the communities in the villages of Calitri,

Fig. 8 Current panoramic view of S. Mango sul Calore (photo by E. Spiga 2016)

San Mango sul Calore and Conza della Campania, that is their aptitude to recover after disasters. Calitri and San Mango sul Calore villages have been rebuilt on the same previous sites, with similar morphological and hydrological setting; only the location of Conza of Campania has been changed, being transferred in a plain area. However, the first two villages are still characterized by considerable instability problems along their slopes, such as the debris avalanche occurred in San Mango sul Calore after a rainstorm on November 2010, and the widespread phenomena of shallow landslides affecting the whole territory of Calitri (Guerriero et al. 2015).

More than the other villages, Calitri has maintained its ancient town planning, resettled on the hill with a current suggestive panoramic view, being nowadays a touristic attraction within the Apennines, especially during the summer season (Fig. 5).

On the other side, San Mango sul Calore has been completely reconstructed, with some geological intervention that have slightly modified the original elevation, and the people resilience has allowed the community to re-use the temporary village of Santo Stefano, built for people which were waiting for the reconstruction of the new village. The Santo Stefano village today is utilized as

Fig. 9 The panoramic view of Conza della Campania after the destructive earthquake of November 1980

a touristic centre, with services and recreational units, and it contributes greatly to the public budget (10–20% ordinary yearly expense) (Guadagno 2010). New industrial investments (food processing factories) have also contributed to the village revival, increasing the employment level.

For Conza della Campania, most likely, the historical memory of the destruction suffered by the community during the past earthquakes (the 1694, 1732 and 1930 seismic events) prevailed in the choice of relocation. Currently two villlages of Conza della Campania coexist: the old village that has been recovered and transformed in archaeological park (the old 'Compsa' of Roman origin) with excavation operations still continuing, through European funding, and the 'new town' where most of the population lives.

In any case, it is always difficult to say that the reconstruction of heavily damaged or destroyed villages by an earthquake is completed, even after many years as in the case of the 1980 earthquake, whose territory was based on a very poor economy and was affected also by many important hydrogeological problems.

In conclusion, although the above villages have been almost completely rebuilt, a real economic improvement is still lacking, despite the settlement of some important factories. The complex problem of reconstruction is not limited to the population resilience, but it also concerns the socio-economic context in which it is carried out.

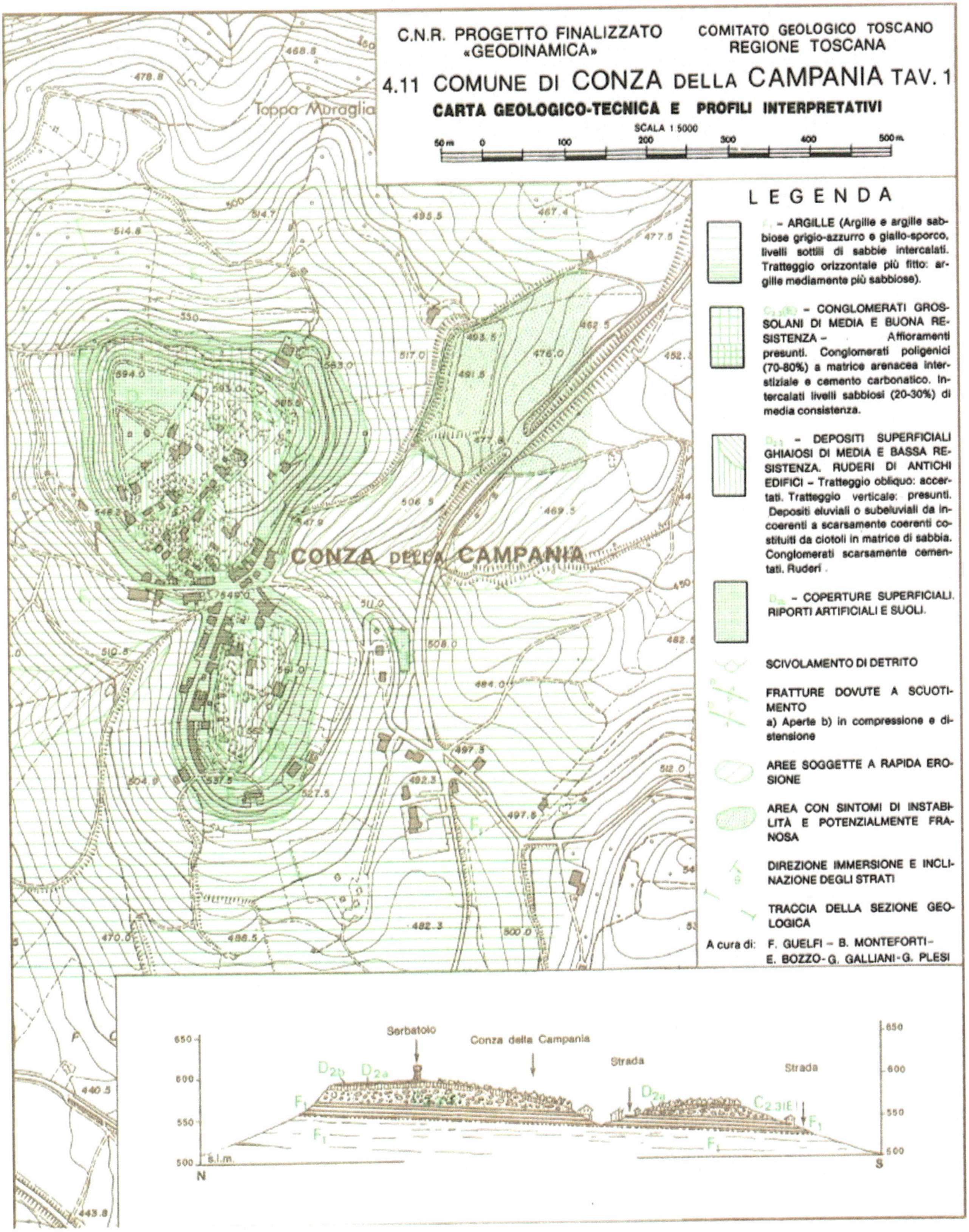

Fig. 10 The original map drawn by Guelfi et al. (1983), for the PFG-CNR microzonation of Conza della Campania, after the 1980 earthquake

Fig. 11 The new village of Conza della Campania (photo by E. Spiga 2016)

References

Alexander D (1984) Housing crisis after natural disasters: the aftermath of the November 1980 southern Italian earthquake. Geoforum, Elsevier, 15 pp 489–516

AA VV (1983) Indagini di microzonazione sismica CNR- PFG, Pubbl. n 492 CNR-PFG

Amato A, Chiarabba C, Selvaggi G (1997) Crustal and deep seismicity in Italy (30 years after). Ann Geofis 11(5):981–993

Bernard P, Zollo A (1989) The Irpinia (Italy) 1980 earthquake: detailed analysis of a complex normal faulting. J Geophys Res 94:1631–1648

Blumetti AM, Esposito E, Ferreli L, Michetti AM, Porfido S, Serva L, Vittori E (2002) Ground effects of the 1980 Irpinia earthquake revisited: evidence for surface faulting near Muro Lucano. In: F. Dramis P, Farabollini, Molin P (eds) Large-scale vertical movements and related gravitational processes, Studi Geol Camerti, pp 19–27

CARRI Report (2013) Definitions of community resilience: an analysis, 1–14

Cotecchia V (1986) Ground deformations and slope instability produced by the earthquake of 23 November 1980 in Campania and Basilicata. Geol. Appl. e Idrogeol. 21(5):31–100

Del Prete M, Trisorio Liuzzi G (1981) Risultati dello studio preliminare della frana di Calitri (AV) mobilitata dal terremoto del 23/11/1980. Geologia Applicata e Idrogeologia 16:153–165

Doglioni C, Harabaglia P, Martinelli G, Mongelli F, Zito G (1996) A geodynamic model of the Southern Apennines accretionary prism. Terra Nova 8:540–547

Esposito E, Gargiulo A, Iaccarino G, Porfido S (1998) Distribuzione dei fenomeni franosi riattivati dai terremoti dell'Appennino meridionale. Censimento delle frane del terremoto del 1980, Proc. Conv. Int. Prev. of Hydrog. Hazards: The Role of Scientific Research. CNR-IRPI Alba, 1:409–429

Guadagno L (2010) Disastri naturali e vulnerabilità sociale. Un'analisi del terremoto in Campania. Tesi di dottorato, Univ. degli Studi del Sannio, Benevento, Italia

Guelfi F, Monteforti B, Bozzo E, Galliani G, Plesi G (1983) Comune di Conza della Campania (AV). In: Indagini di microzonazione sismica CNR- PFG, n. 492

Guerriero L, Revellino P, Diodato N, Grelle G, De Vito A, Guadagno FM (2015) Morphological and climatic aspects of the initiation of the san mango sul calore debris avalanche in southern Italy. Eng. Geology Soc Territory 2:1397–1400

http://emidius.mi.ingv.it/CPTI15-DBMI15/

http://www.comune.sanmangosulcalore.av.it/

http://oasiwwflagodiconza.org/

http://www.resilientus.org

Marchetti L (2012) The 6th April, 2009 L'Aquila earthquake: restoration choices and pathways. Speciale knowledge, diagnostic and preservation of cultural heritage, II ENEA, 3–12

Michetti AM, Esposito E, Guerrieri L, Porfido S, Serva L, Tatevossian R, Vittori E, Audemard F, Azuma T, Clague J, Comerci V. Gurpinar A, Mccalpin J, Mohammadioun B, Mörner NA, Ota Y, Roghozin E (2007) Intensity scale ESI 2007. Memorie Descrittive della Carta Geologica d'Italia, Roma 74 p 53

Mostardini F, Merlini S (1986) Appennino centro meridionale: Sezioni geologiche e proposta di modello strutturale. Mem Soc Geol It 35:177–202

Pantosti D, Schwartz DP, Valensise G (1993) Paleoseismology along the 1980 surface rupture of the Irpinia fault: implications for earthquake recurrence in Southern Apennines, Italy. J Geophys Res 98(B4):6561–6577

Patacca E, Scandone P (1989) Post-Tortonian mountain building in the Apennines. The role of the passive sinking of a relic lithospheric slab. The Lithosphere in Italy. Acc Naz Lincei 80:157–176

Pignone M, Cecere G, Nostro C, Selvaggi, G, Avallone A, Zarrilli L, D'Anastasio E, Abruzzese L, Falco L, Castagnozzi A, Moschillo R, Fodarella A, Nappi R, Alessio G, Castiello A, De Rosa D;

Varriale F, Punzo M (2008) Percorso informativo sulle attività della sede irpina. Inaugurazione della sede INGV di Grottaminarda, 2 Dicembre 2008, http://www.earth-prints.org/handle/2122/4764

Porfido S, Esposito E, Luongo G, Marturano A (1991) Terremoti ed effetti superficiali: esempi nell'Appennino meridionale. Proc Conv Studi Centri Storici Instabili, CNR-Regione Marche, pp 225–229

Porfido S, Esposito E, Michetti AM, Blumetti AM, Vittori E, Tranfaglia G, Guerrieri L, Ferreli L, Serva L (2002) Areal distribution of ground effects induced by strong earthquakes in the Southern Apennines (Italy). Surv Geophys 23:529–562

Porfido S, Esposito E, Guerrieri L, Vittori E, Tranfaglia G, Pece R (2007) Seismically induced ground effects of the 1805, 1930 and 1980 earthquakes in the southern Apennines. Boll Soc Geol It 126:333–346

Porfido S (2010) Irpinia: terremoto del 1980 Cultura della prevenzione… trenta anni dopo. 6 /11/ 2010, Sorbo Serpico(AV). http://eprints.bice.rm.cnr.it/id/eprint/3066

Porfido S, Alessio G, Avallone P, Gaudiosi G, Lombardi G, Nappi R, Salvemini R, Spiga E (2016) The 1980 Irpinia-Basilicata earthquake: the environmental phenomena and the choices of reconstruction. Geoph Res Ab 18(EGU2016-9457):2016

Postpischl D, Branno A, Esposito E, Ferrari G, Marturano A, Porfido S, Rinaldis V, Stucchi M (1985) The Irpinia earthquake of November 23, 1980. Atlas of isoseismal maps of italian earthquakes, CNR-PFG N 114(2A):152–157

ProLocoConza:http://www.prolococompsa.it/Pagine_Principali/Storia.html

Rovida A, Locati M, Camassi R, Lolli B, Gasperini P (2016) CPTI15, the 2015 version of the parametric catalogue of italian earthquakes. INGV doi:http://doi.org/10.6092/INGV.IT-CPTI15

Samuelli-Ferretti A, Siro L (1983a) Comune di Calitri (AV). In: Indagini di microzonazione sismica CNR- PFG, Pubbl. n . 492

Samuelli-Ferretti A, Siro L (1983b) Comune di San Mango sul Calore (AV). In: Indagini di microzonazione sismica CNR- PFG, Pubbl. n . 492

Serva L, Esposito E, Guerrieri L, Porfido S, Vittori E, Comerci V (2007) Environmental effects from five historical earthquakes in Southern Apennines (Italy) and macroseismic intensity assessment contribution to INQUA EEE Scale Project. Quat Int 173:30–44

Wasowski J, Pierri V, Pierri P, Capolongo D (2002) Factors controlling seismic susceptibility of the sele valley slopes: The Case of the 1980 Irpinia Earthquake Re-examined. Surv Geophys 23:563

Urgent Need for Application of Integrated Landslide Risk Management Strategies for the Polog Region in R. of Macedonia

Igor Peshevski, Tina Peternel, and Milorad Jovanovski

Abstract

Results from a recent study on landslide distribution in the Republic of Macedonia have shown that the northwest part of the country is most prone to landsliding processes. In the past, there have been numerous landslides which have caused great damage to the infrastructure and endangered many villages. The last catastrophic event occurred on 3rd of August 2015 in the wider area of town Tetovo. The triggering factor for flooding and number of slope mass movements was heavy rainfall. The storms and floods were described as the worst to hit the area in over a decade. Relevant seismic events before and during the rainfall were not recorded. As a result of fast debris flow, 6 people lost their lives in Poroj village, while parts of it were covered by deep debris deposits. Roads and communal infrastructure in the entire region suffered of significant damages which caused problems in search and rescue operations. The study area Polog is located in the foothills of the Sar Mountain and divided by the river Pena. The broad area of the Municipality Tetovo itself has complex geological and tectonic conditions which also contributed to the degree of damage of the affected areas. Therefore, Tetovo and surrounding areas are vulnerable to both storm run-off and river flooding and consequently to different slope mass movements processes. In order to prevent such disasters in the future it is crucial to develop an integrated landslide risk management strategy. Efforts are made to prepare basic landslide susceptibility map of the region, which can serve as an integral part in the development of the strategy. Certain applications for risk management projects have been submitted by teams of domestic and international experts, which if approved can be of great value for the regional and local population and infrastructure.

Keywords

Polog • Catastrophic event • Rainfall • Mass movements • Landslide risk management strategy

I. Peshevski (✉) · M. Jovanovski
University Ss Cyril and Methodius, Blvd. Partizanski odredi no. 24, Skopje, 1000, Republic of Macedonia
e-mail: pesevski@gf.ukim.edu.mk

M. Jovanovski
e-mail: jovanovski@gf.ukim.edu.mk

T. Peternel
Geological Survey of Slovenia, Ljubljana, Slovenia
e-mail: tina.peternel@geo-zs.si

M. Mikoš et al. (eds.), *Advancing Culture of Living with Landslides*,
DOI 10.1007/978-3-319-53483-1_16

Introduction

Each year landslides in the Republic of Macedonia cause losses which can be measured in millions of euros. Most of these funds go on the expense of road and railway rehabilitation or cleaning due to sliding masses. This is confirmed by recent study (Peshevski et al. 2013) when it was calculated that around 60% of the registered landslides have blocked or impeded the road or railway traffic. Statistics also showed that 10% of the landslides have endangered whole settlements, while 30% have damaged individual structures both in rural and urban area. Due to intense or prolonged rainfalls, or combination of both, houses, infrastructure lifelines or even whole parts of settlements have been replaced or removed from zones of sliding (Jelovjane, Ramina). For a certain number of cases, very comprehensive and expensive geotechnical investigations had been performed, followed by design and construction of appropriate support structures.

In relation to endangered population, villages like Velebrdo, Trebishte, Rostushe, Bitushe, Skudrinje, Mogorche, Jelovjane, Bogovinje, parts of towns Veles, Prilep and Strumica with population of over 20,000 are directly affected by the negative effects of the landslide processes. According to gathered historical writings, in several cases in the period from 1995 to 2014 (Haque et al. 2016) 15 people were heavily injured or lost their life as result of landslides in these areas.

Unfortunately, on the 3rd of August 2015, after heavy rainfall that hit the villages in the Polog region, history repeated itself and 6 persons from the village Poroj were reported to have died as a result of fast debris flow.

The Study Area

Population and Infrastructure

The Polog region is situated in the northwest part of the Republic of Macedonia (Fig. 1) and it includes the Polog valley, Mavrovo upland, the mountain chain.

Bistra and the valley of river Radika. The total size of the region is 2146 km^2 or 9.7% of the territory of Republic of Macedonia.

The Polog region includes 9 municipalities: Jegunovce, Tearce, Tetovo, Bogovinje, Zelino, Brvenica, Vrapciste, Gostivar, Mavrovo-Rostusa with around 184 settlements. The population according the 2006 census is 310,178 people, and the density is 126 per km^2. Most of the settlements are placed along the foothill of the mountain ridges developed in the zone of existing road infrastructure. Interesting fact is that only 29.2% of the total population lives in the urban settlements, while the rest is in smaller villages.

The road and railway infrastructure in the Polog region consists of internal and external traffic. The total length of roads in the region is around 1036 km of local roads, 72 km main roads, 378 km regional roads and 52 km of highway as well as dense local load network connecting many villages. The railway traffic is represented with 83 km from the line Skopje-Gostivar-Kicevo.

Very important infrastructure in the region is the largest and most complex hydro system in the country "Mavrovo" consisted of the hydropower plants Vrutok, Raven and Vrben, as well as the artificial lake Mavrovo. In the total capacity of hydropower in the country this system contributes 42% and produces around 445 Gwh per year. The

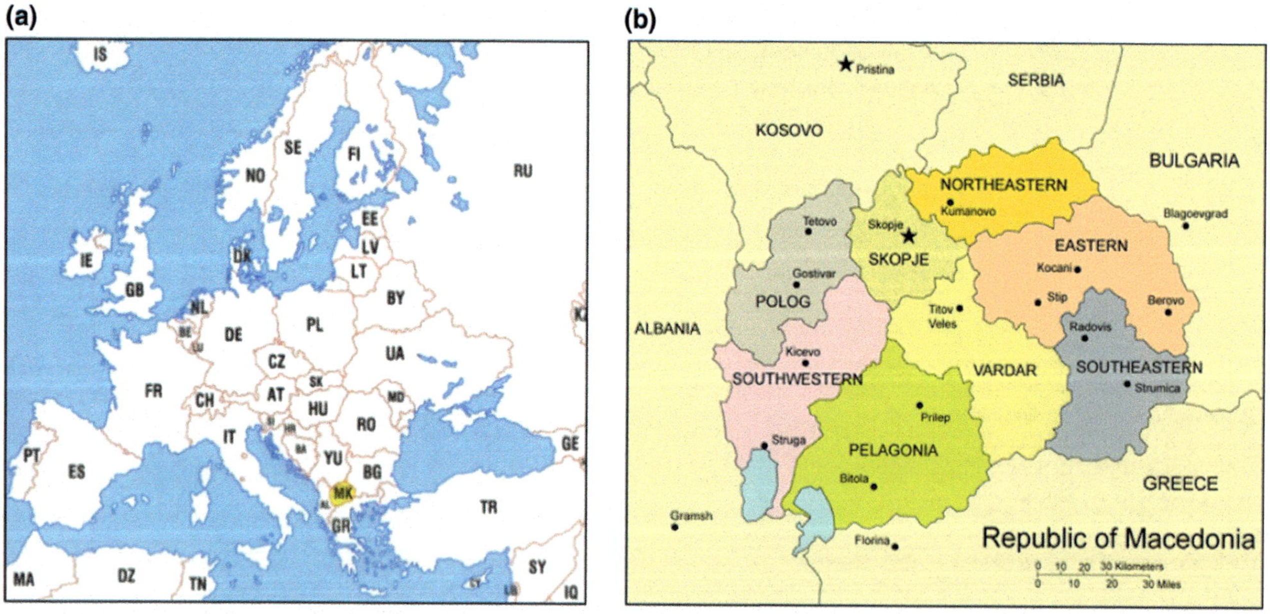

Fig. 1 **a** Position of the Republic of Macedonia in relation to Europe; **b** Polog region and position of most affected municipality Tetovo

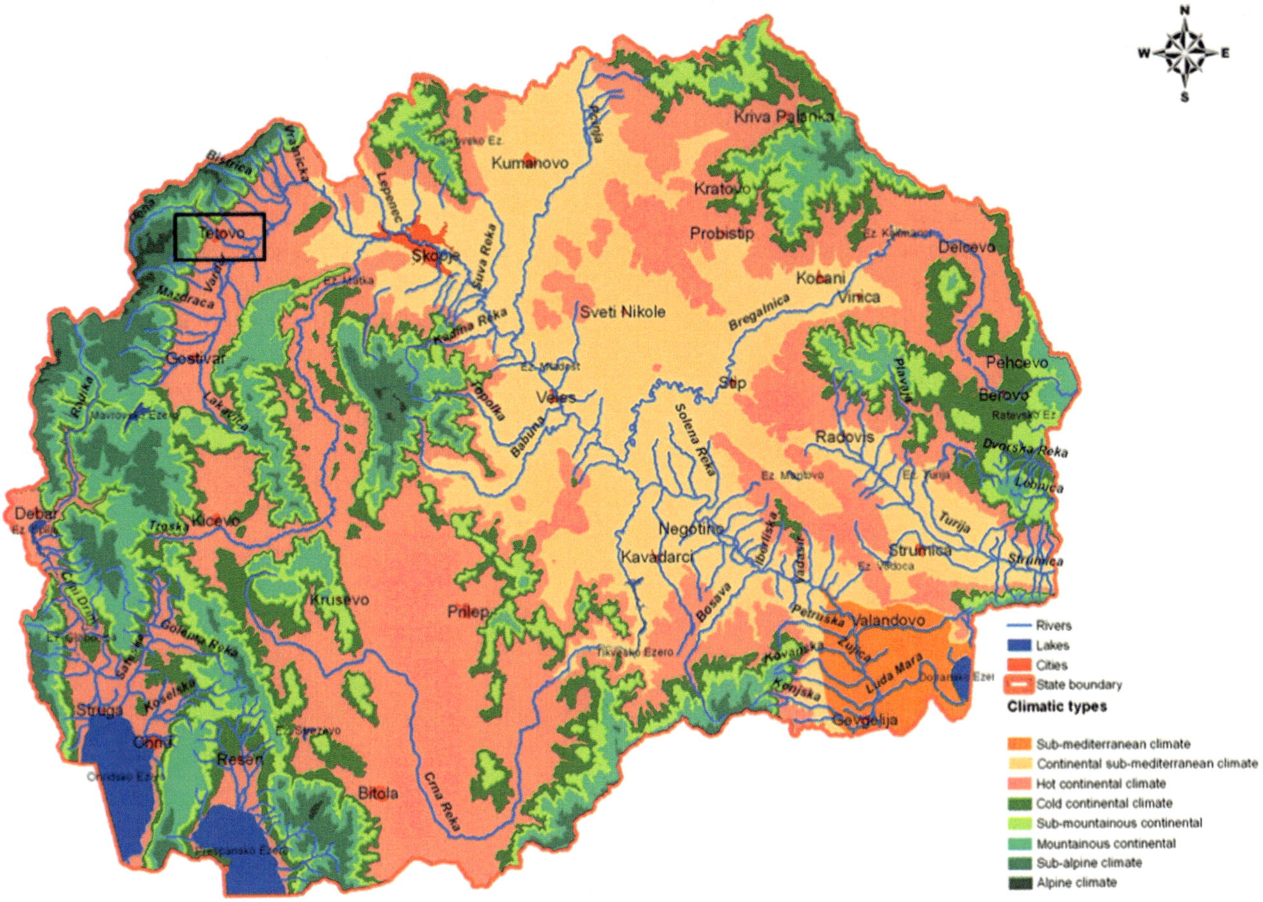

Fig. 2 Climate zones in the Polog region (*black frame*) (Reference: European Environment Agency—EEA)

waters of this hydro-system are collected from the mountains Korab and Sara (watershed area of over 500 km^2) and by systems of channels and pipelines are transported to the Mavroro lake. The length of these channels and pipelines called "Sarski vodi" is over 130 km long. The supporting service roads for this channel have a length of over 170 km. It is important to note that many sections of this hydro-system have suffered significant damage due to landsliding occurrences and costs of millions of euros. Two similar hydro-systems of smaller scale are planned to be developed in the region in the following period, which will require construction of additional 20 km of channels and over 40 km of service and access roads.

Regional Climate and Hydrographical Properties

The climate in the study area at altitude 600–900 m is hot continental, the Shar Planina Mountain has cold continental climate (900–11,000 m) and the rest has alpine mountainous climate at altitude over 2250 m. Due to the large differences in elevation, the climate throughout the region is very different (Fig. 2).

The maximum daily precipitation in Popova Shapka (large sky center on Shara Mountain) was registered in November 1979 (188 mm). On this occasion, there was huge flooding in the city of Tetovo (river Pena) and the capital Skopje (river Vardar). For comparison, the average amount of precipitation is between 530 and 900 mm per year.

Almost all rivers of the Shar Planina belong to the watershed area of Vardar River, the longest and major river in Republic of Macedonia. The river Pena which passes through town Tetovo is the biggest tributary of Vardar. It has the longest mountain watercourse with 29.7 km from its source to the city of Tetovo in the valley. The altitude difference along the flow is 1.0914, and average slope gradient 6.44%. The river Pena itself has a number of larger or smaller tributaries.

This clearly shows the potential for large quantities of water to be quickly brought to the foothills of the mountains in the case of heavy prolonged rainfall, affecting the many settlements along the river Pena.

General Geomorphology and Geology of the Polog Region

The Shar Planina mountain range is part of the Dinarides chain. Highest top is Titov vrv (Golem turchin) 2478 m. The primary mountain ridge, with length of around 80 km and width of 10–20 km extends in southwest-northeast direction. As result of the tectonics the mountain range is dissected by many faults creating its block structure. The location and orientation of gullies, torrents and rivers depend on these faults. There is presence of rock notches which are located along the slopes, the most upstream parts of the river Pena and the dominant mountain peaks (Fig. 3).

The fluvial relief is the main morphogenetic element, where fossil and recent per glacial phenomena is dominating. From the highest parts of the Sar Planina to the bottom of the Polog valley there are a numerous small and larger mountain streams with torrential character. In some instances the deep valleys have typical canyon character with deposition of big boulders in the streambeds.

The karst relief is also present on the terrains of mountain Shara, with presence of oases, stripes and smaller spots. There is also presence of sinkholes through which significant amount of water can enter into the sub-surface.

From the glacial landforms, there is presence of cirques and their deposits, also the source of river Pena is in this type of sediments.

The main present rock types in the Polog region are sedimentary and metamorphic. The torrents and the rivers are covered with quaternary deposit and most of the settlements at the toes of the mountains are raised on alluvial fans. Most of the mountainous terrain is characterized by presence of metamorphic rocks composed of weak schists and phyllites, and in some instances marbles. All these rocks are heavily deformed and weathered due to the intense tectonics from the past geological times. The rocks are covered by thinner or thicker soil debris consisted of large boulders mixed with finer particles.

The Consequences of Floods and Landslides on 3rd of August 2015

The trigger for flooding and landslides on 3 August from 17.30 to 21.00 h was heavy rainfall combined with hail.

Fig. 3 Leshnica rocky notches and cliffs

Table 1 Affected settlements in the Polog region during the rainfall event of 3 August 2015

No.	Settlement	Type of event
1	Village of Poroj	Debris flow
2	Village of Germo	Landslide
3	Village of Shipkovica	Debris flow
4	Village of Lisec	Landslide
5	Village of Bozovce	Rockfalls
6	Village of Veshala	Landslide
7	Village of Brodec	Debris flow
8	Village of Vejce	Landslide
9	Village of Golema Recica	Debris flow
10	Village of Mala Recica	Earthflow/Mudflow
11	Village of Dobroshte	Landslide

Relevant seismic events were not recorded before and during the rainfall event. Since meteorological station at Popova Sapka was out of order, there is no real measured information for the quantity of fallen rain during the event. Meteorological station in the valley of Tetovo registered 9 mm in 24 h and station at village Jazince (north of Tetovo) registered 50 mm in 24 h. People from one of the most affected villages Shipkovica witnessed that a 10 cm thick layer of hail fell during the event. Along the watercourse of river Pena there was flooding and debris flow sliding.

Many settlements were affected (Table 1) and significant degree of damage was registered (Table 2).

In Sipkovica, the water flow caused deposition with thickness of up to 5 m of material including boulders with size of 7 m in diameter. In this village due to the debris flow, fatalities were registered and more than 10 houses and supporting infrastructure was destroyed (Fig. 4).

The largest debris flow was produced on the road going from Brodec to Sipkovica and on this part it completely destroyed it. The debris flow eventually ended in the river Pena. After the debris stopped the river eroded it to a depth of 5 m mobilizing wood trunk and large boulders with maximum length of 5 m (Fig. 5).

In village Poroj approximately 200,000 m^3 of debris was deposited during the event. The main bridge in the village covered as well as the river channel (Fig. 6). Agricultural fields were also flooded in a radius of 800 m. The total area affected from the flooding and debris flow is 800 m in width and length of 500 m. The depth was estimated up to 1.5 m. The last large debris flow of river Poroj occurred in the above mentioned rainfall event in 1979. The estimated volume of the debris was then estimated on over million m^3.

The mudflow in Mala Recica covered the main road and surrounding houses. It was around 1300 m long and 100 m wide on both sides, with an estimated height up to 3 m (Fig. 7).

The estimated size of the debris flow in Golema Recica is 1200 m in length and 100 wide, with thickness of 2 m. It affected houses, road and agricultural surfaces (Fig. 8).

Rockfalls above village Bozovce had dimensions of block from 0.2 to 1.75 m in length. Fortunately on their path there were many trees which stopped the blocks above on a plateau above the houses on the outskirts of the village (Fig. 9).

The River Pena is the main stream of the area of Tetovo with a catchment of 191 km^2. During the heavy rainfall and

Table 2 Fatalities and damages due to the extreme rainfall event of 3 August 2015

No.	Type of damage
1	6 fatalities
2	11 injured
3	24 families evacuated from Sipkovica
4	169 houses directly flooded, 4 completely destroyed
5	Power cut off in Sipkovica
6	Damaged water supply systems in Tetovo, Sipkovica, Mala and Golema Recica and Poroj
7	11 damaged bridges
8	17 damaged roads
9	Additional 6 bridged damaged 4 in Tetovo and 1 in Jegunovce
10	40 ha of agricultural land flooded

Fig. 4 A debris flow path through the village Sipkovica

Fig. 5 The road between Sipkovica and Brodec was completely covered with debris flow material

Fig. 6 Debris flow deposits in the Poroj

Fig. 7 Village road in Mala Recica was covered with mud

Fig. 8 Affected area in Golema Rečica

debris flow on 3 August 2015 the tributary torrents contribute solid material to the river. The electric power plant is situated after the second protection dam around 1.5 km NW of Tetovo (Fig. 10). Behind the electric power plant is a high wall of marble, which is prone to rockfalls. The upper part of the River Pena is subjected to strong erosion. The bottom of the slope of the middle part of the River Pena was eroded and consequently destroyed the road. On 3 August 2015 The River Pena flooded in the city of Tetovo (Fig. 11). The length of the river was 1100 m, the width on the left and the right in total 200 m.

Urgent Need for Application of Integrated Landslide Risk Management Strategies

It is well known fact that the rational physical planning largely depends on the knowledge of different geohazards. In order to secure certain degree of safety of existing and new settlements in regions like the Polog in Republic of Macedonia, countries throughout the world are making efforts in preparing normative documents for procedures, methodologies and strategies for coping with floods and landslides.

At today's level of development of the science, there exist numbers of methods for analysis of hydrology and landslide data which enable definition of the flood and landslide hazard and risk level.

In relation to landslide hazard in particular, the first and most simple step is to prepare landslide inventory and susceptibility maps for particular region which will enable recognition of most critical zones that will be subjected to further investigation on higher level—landslide hazard and risk. In order to prepare such maps, appropriate landslide data base is needed.

Model for establishment and operation of a GIS databases for Republic of Macedonia is given in Pesevski et al. (2015). The model of this database is envisaged to be functional and practical, despite the limited resources of responsible institutions. In addition to the database, particular proposal on how to efficiently collect data on past and future landslides is presented, followed by possible ways of interconnection

Fig. 9 Rockfall path in Bozovce

between different parties involved in investigation, remediation, monitoring of landslides.

Since Macedonia is a developing country, the organization and coordination of these activities will be very hard process, especially from the perspective of limited funds for coping with natural hazards. In order to overcome this drawback, it is suggested that the scientific institutions need to connect with European colleagues and apply to calls that offer funding opportunities. In this relation several activities are under way, that are being realized in cooperation with colleagues from Slovenia, Bulgaria, Serbia, Albania and Croatia. Cooperation with more developed countries will be of great importance.

Once conditions are secured, the landslide inventory, susceptibility, hazard and risk maps should be included in the legal frameworks related to construction and protection of the natural environment. This will enable delineation of "risk zones" in which particular rules for construction will have to be applied.

The presented landslide events in the Polog region are clearly showing the disadvantage of non-existence of integrated landslide risk management strategies. Since catastrophes as floods and landslides from the past are repeating, there is an **urgent need for application of integrated flood risk and in particular landslide risk management strategies**.

Fig. 10 Alluvial fan above the electric power plant

Fig. 11 Flooded area in the city Tetovo

The possible sequence of steps that need to be undertaken in the Macedonia, and in particular in the Polog region are the following:

- Establishment of strong institutional relation between the Geological survey of Macedonia, the Ministry of Environment and physical Planning and the municipality of Tetovo.
- Development of project for collection and analysis of all existing data on landslides, hydrological and meteorological conditions, in particular for the Polog region.
- Realization of a regional scale project that will consist of landslide inventory preparation and susceptibility zoning according corresponding methodology relevant for the particular case study area; This will result in definition of most "landslide prone zones" throughout the region.
- Local landslide susceptibility or hazard zoning for selected areas defined in the previous stage.
- Modeling of landslide mass behavior in a extreme meteorological events: type of landslide, extent, volume, movement (including rate and path), run-out distances, etc.
- Informing the relevant institutions and local authorities for the outcomes of investigations and the modeling.
- Establishment of monitoring and early warning system.
- Education and training of local population related to the management, maintenance and use of the monitoring and early warning system.
- Expert domestic and international teams supported by state and local authorities should perform these activities in cooperation.
- Similar cascade of activities should be undertaken for flood risk management.

As mentioned earlier landslides and floods (besides earthquakes) present major natural hazard in Republic of Macedonia. In this relation, the risk management strategies related to these negative occurrences should be established as soon as possible.

Acknowledgements The study is based on the assessment report "Mission to the former Yugoslav Republic of Macedonia in the field of Civil Protection—Floods and Landslides" made by EUCP team.

References

Haque U, Blum Ph, da Silva PF, Andersen P, Pilz J, Chalov RS, Malet JP, Jemec Auflič M, Andres N, Poyiadji E, Lamas CP, Zhang W, Peshevski I, Pétursson GH, Kurt T, Dobrev N, García-Davalillo JC, Halkia M, Ferri S, Gaprindashvili G, Engström J, Keellings D (2016) Fatal landslides in Europe, Technical Note, Landslides pp 1–10 (First online: 7 May 2016)

Peshevski I (2015) Modeling of landslide susceptibility using GIS technologies. Ph.D. thesis, Faculty of Civil Engineering, Skopje

Peshevski I, Jovanovski M, Markoski B, Petruseva S, Susinov B (2013) Landslide inventory map of the Republic of Macedonia, statistics and description of main historical landslide events, Landslide and flood hazard assessment. In: Proceedings of the Regional Symposium on Landslides in the Adriatic-Balkan Region, Zagreb, Croatia, 6–9 Mar 2013, pp 207–212

4th WLF 2017
4th World Landslide Forum
LJUBLJANA SLOVENIA EU

Comprehensive Overview of Historical and Actual Slope Movements in the Medieval Inhabited Citadel of Sighisoara

Andreea Andra-Topârceanu, Mihai Mafteiu,
Mircea Andra-Topârceanu, and Mihaela Verga

Abstract

The perspective of this study is based on the relationship between geomorphological and hydrogeological environments as natural support of the Sighisoara Citadel UNESCO heritage site and the impact of human activities on both of these. Since this system of the heritage site of the Sighisoara Citadel is composed of both natural and anthropogenic elements, the slope morphodynamic are controlled by their spatial distribution. The achievement of the main objectives of our study was possible using complex methodology: historical landslides mapping, geomorphological and geophysical methods, topographic and geotechnical survey and statistical analysis. Our results show that the geomorphological environment has changed dynamic features in the last decades. Also the hydrogeological conditions at shallow depth are different in the new context of higher values of human pressure generated especially by tourism phenomenon.

Keywords

Slope movements mapping • Sighisoara citadel • Geomorphological methods • Electrometry

Introduction

Over the course of history, a series of edifices (citadels, strongholds, places of worship, palaces) has been built as a result of certain strategic, religious, social or economic requirements. These edifices are nowadays included in the national or world heritage.

A. Andra-Topârceanu (✉) · M. Verga
Department of Geomorphology, Pedology and Geomatics, Faculty of Geography, University of Bucharest, 010041 Bucharest, Romania
e-mail: andreea.andra@geo.unibuc.ro

M. Verga
e-mail: mihaela.verga@geo.unibuc.ro

M. Mafteiu
MM Georesearch, 062152 Bucharest, Romania
e-mail: mihai_mafteiu@yahoo.com

M. Andra-Topârceanu
ISC, 030217 Bucharest, Romania
e-mail: mirceaat@gmail.com

M. Mikoš et al. (eds.), *Advancing Culture of Living with Landslides*,
DOI 10.1007/978-3-319-53483-1_17

In Romania, there are over 29,500 cultural monuments in the national heritage list and among these, 32 monuments belong to the UNESCO world heritage list, including, the Inhabited Medieval Citadel Sighisoara (XIIIth century). Concerns regarding the maintenance and conservation of cultural monuments exist, but these aren't continuous nor effective for all monuments. Over the time the degradation of Sighisoara medieval citadel was caused by geomorphological processes, climate changes, historical and actual human impact such as development of urban environment, traffic, sewerage and water supply network and not at least increasing number of residents and tourists. Nowadays the historical edifices are characterized by high vulnerability because of morphodynamics and also both geological and anthropogenic factors control increase the instability of heritage site of Sighisoara Citadel.

The first stage of understanding the heritage of any territory is to know the geological and relief structure, and further to proceed with analyzing the environmental impact (how it adapted to the existing topography, material and spiritual culture), Panizza and Piacente (2000). The behavior of historical edifices at the imposed morphological and geological dynamics (natural and anthropic) was studied by Marunteanu et al. (2009), Dana (2015) (remote sensing and technologies InSAR).

The final aspects that are discussed in our paper are: the preponderance of natural processes versus anthropogenic processes of Sighisoara Citadel vulnerability variations and the focused solution for mitigation measures.

In order to assess the slope movements we consider it to be important to accomplish a comprehensive geomorphic analysis in the most important and unique UNESCO heritage from Romania. The objectives are follow: to map the natural and anthropogenic slope processes; to identify the rate of natural and anthropogenic causes of morphodynamic and their effects on the heritage site of Sighisoara.

Objectives and the Geographical Context

Sighisoara city perhaps was founded in 1191 (Kraus, G.), but only in 1280 was preserved citadel attestation document, "seniors de castro Sex" (Niedermaier 2000). Sighisoara Citadel developed spatially, creating in each stage of development another belt of fortifications, the latest being almost totally preserved. Of the 13 bastions and medieval towers only 9 are now preserved and the oldest building is Biserica din Deal (Church Hill).

Medieval Citadel Sighisoara construction was carried out on Dealul Cetatii (Citadel Hill), 432.9 m which is a fluvial confluence terrace situated in south river bank of the middle sector of Tarnava Mare Valley. In relation to the surrounding areas this hill is higher landform with less easily accessible and large visibility. Sighisoara Citadel held strategic position in the heart of major geographic unit of Transylvanian Hilly Depression. Tarnave plateau is drained ENE-WSW of Târnava Mare River and delineates the northern part of the Hârtibaciului plateau. This landform unit is formed on the Mio-Pliocene deposits with narrow interfluves and slopes affected by instability phenomena.

In Sighisoara Citadel area, Tarnava Mare River is traversing a narrow sector valley, which was deepened and formed a system of strata and fill terraces, parallel and well preserved.

Fluvial terrace treads are not covered by colluvial deposits, but they are modeled in human settlements (Rosian et al. 2011). As a result, soil and elluvial deposits were uncovered and in historical time anthropogenic deposits were formed with thickness meters, identified in drills (Marunteanu 2000). Scarps have over 20° slope angle and they are covered by delluvial deposits, and discontinuously by anthropogenic deposits.

Sighisoara Citadel morphologic support is given by fluvial terraces at the confluence of the Tarnava Mare River and Saes rivulet: T_5 (55 m relative altitude) and T_6 (70–80 m relative altitude) (Rosian et al. 2011). Slope variation is imposed by the general evolution of relief and by lithology and structure of Pannonian deposits.

Thus, the slopes surfaces and terrace scarps are indicated by values over 20°, while area of terraces tread is below 7°; value range of between 7° and 20° represent different geomorphic surfaces formed on colluvial and anthropogenic deposits.

The gravitational and hydrodynamic processes on the slope (landslides, subsidences, collapses) are generated as a result of the conditions of geodeclivity, geological, hydrogeological, bio-geographical (deforestation) and meteorological. It should be noted that the same processes can also be prepared and triggered due to anthropogenic intervention. Furthermore, the processes can also be triggered in superficial deposits and fillings.

Geomorphological processes that affect the stability of slopes and citadel buildings are gravitational (landslides, shallow slides, but also slumps, topples) and hydrodynamic (rills, gullies erosion), otherwise some of they were highlighted by Marunteanu et al. (2007). Areas affected by creep, suffusion, rills erosion, sheet wash (which affect tree roots), earth failure and earth topples were mapped (Fig. 1). Many shallow landslides and other types of landslides are active but also others are suspended or dormant if they are not affected by anthropogenic water source (sewage, leakage, water supply).

Geologically, the Citadel Hill consists in quasi-horizontal sedimentary Pannonian deposits: compact sands, silty sands, sandy clays, silty clays, alternating with fine up to coarse sands, sandstone, covered by delluvial and anthropogenic deposits. Calcareous sandstone outcrops were identified in the upper third of the southern slope, thickness layer of 1 m affected by rills erosion and dissolution processes. Pannonian deposits have a low inclination to the south.

The unconfined aquifer is said to be sporadic, first appears suspended at elevation 403.5 ISPIF Drilling 1 (altitude 431.50 m HL—28.00 m depth) near Biserica din Deal. The aquifer is dependent on rainfall and very vulnerable to pollution. Because it is not ascending, it cannot be considered as a source of seepage from the citadel and the level of collector river Tarnava Mare is approx. 342–344 m. Geotechnical data reveals a phreatic level at the base of the Northwest hill slope. The hydrostatic levels are not exploitable in Sighisoara citadel and can influence soil infiltration from construction subsoil, or the ancient walls. In the eighteenth century D. Frölich says that the hill side {Sighisoara Citadel} "has plentiful water" (Niedermaier 2000). "Aquifers" observed and reported in the citadel are actually anthropogenic aquifers as a result of sewers and hydrants

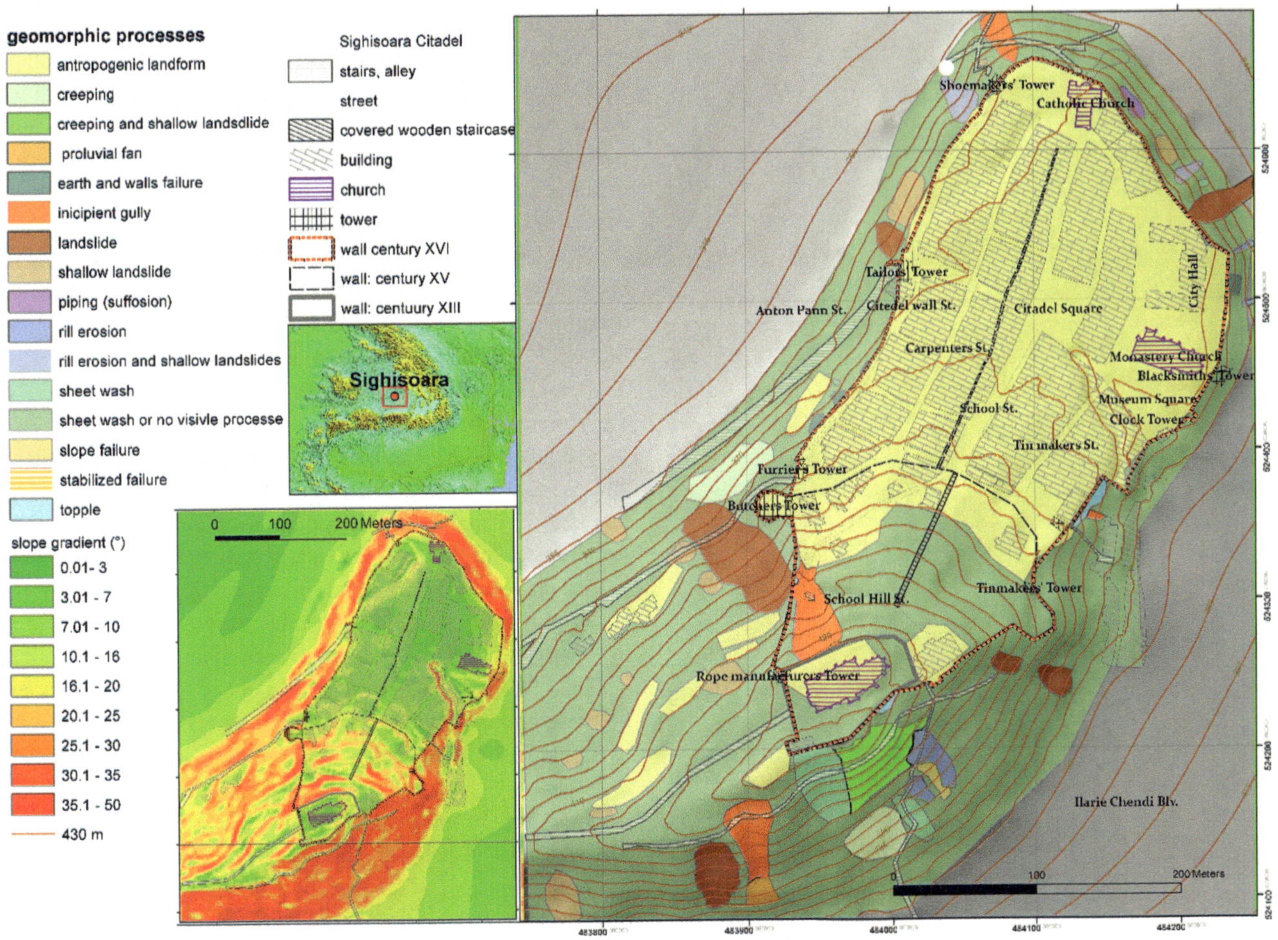

Fig. 1 Slope morphodynamic map of Sighisoara Citadel Hill

damage, loss of water and incorrect drainage of pluvial water. In addition, Şaeş rivulet is channeled from the nineteenth century from the southern side of the citadel to the western side of the hill, confluence with Târnava Mare River. Moreover to increase the efficiency of traffic a cutting was executed at western side of the castle hill through a place called "Vartejului" (Whirl) and so was interrupted continuity with the rest of the ridge, consequently the hydrostatic levels were affected.

Until the nineteenth century Dealul Cetatii (Castle Hill) has been deforested for strategic reasons (Fig. 2a–c). Starting with the second half of the nineteenth century to the present on scarps of fluvial terraces developed natural vegetation, today constituting itself as rare forest of oak, hornbeam, maple (Fig. 2d).

The Sighisoara Citadel is a touristic destination and the development of tourism infrastructure through maintenance, modernization, represents an additional factor of pressure and impact on the edifice itself. In the last decade the value of the accommodation capacity inside the citadel increased by 31% from 927 to 1347, although that the site is a protected area.

The number of tourists also increased from approximately 90,000/year in 2006 to over 131,000 in 2011 and has reached over 200,000 in 2015, and the touristic pressure increase.

Methods and Data

Available data are following: topographic maps 1:25,000 (1984), topographic plan 1:500 (1991), historical maps—First Military Survey, 1763–1787, Second Military Survey (1806–1869), Third Military Survey (1869–1887) Romanian maps under Lambert-Colesky projection system (1953), topographical survey data (2006), orthophotoplans (2011), geotechnical data 1991, slope movements GPS data acquisition, geophysical data 2006, 2016.

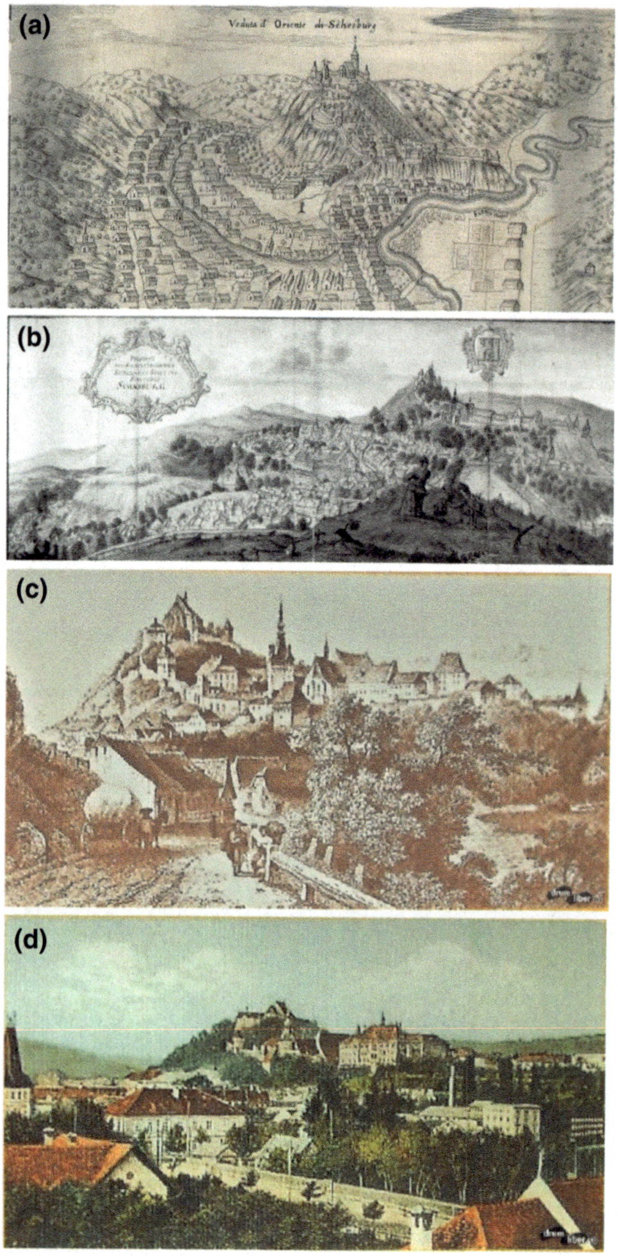

Fig. 2 **a** Perspective of Citadel Sighisoara drawn by Giovanni Morando Visconti, 1699: *Veduta di Oriente di Schesburg*. **b** Prospect der Siebenbürghischen Sächsisen Stadt und Schlosses, Schäsburg 1735; **c** Sighisoara Citadel 1860, **d** Postcard Sighisoara Citadel 1912

Geological Survey

The methods of investigating the geological environment and morphological support of monuments and heritage sites should be non-destructive. Geoelectrical method plays a non-invasive diagnostic tool useful in studies of maintenance and rehabilitation of monuments (Mafteiu et al. 2015). This is the way the increased volume of data can be provided by a network of minimal and sufficient drilling (invasive method), which otherwise are reported. Thickening geoelectrical investigation stations took into account beside the footprint of buildings to the ground and landform which they are located as well as geomorphological processes which is prone to be modeled. In addition, it allows to effectively monitoring the behavior of buildings in periods with changes in meteorological, anthropogenic and hydrological conditions.

Geoeletrical resistivity survey already is one of the most applicable, efficient and environmentally friendly methods to accomplish and validate same hidden features of geomorphic processes. Geophysical parameter investigated in our survey is apparent resistivity, an intrinsic property of how strongly the ground opposes the flow of direct electrical current. The investigated environment is not homogeneous and isotropic and its surface is not planar, resistivity value calculated using the above formula is a complex average of environment resistivities in the area measuring device. VES (vertical electrical sounding) method is using Schlumberger type array with four collinear electrodes to induce direct current in the ground to measure the variation of apparent resistivity.

The depths of investigation depend on the distance between the two current electrodes (AB) versus the other two potential electrodes (MN) and cause voltage difference (Fig. 3).

VESs are applicable for obtaining the geophysical data like apparent resistivity (roa) curves. Their interpretation highlight physical characteristics of lithology, lithological limits, ground water, failure/slide planes, pluvial and sewage infiltrations, geomorphological processes, or sewerage, foundations, old walls an fillings. We noticed as a result of analysis of 140 measure station that there is just little specific aspect of slide planes.

Mapping of anthropogenic and superficial deposits thickness consisted of geological and geophysical surveying and processing of field data. Their analysis consisted of correlating punctual lithological information from the geotechnical tests of the 16 boreholes with electrometry data of the 124 vertical electrical sounding (VES) in order to establish the limits of filling and/or superficial deposits (eluvial, delluvial, colluvial) and bedrock. The VESs network is equal distributed on over 70% of Dealul Cetatii, but 98% from Sighisoara Citadel, with a 40–50 m cell dimension. The different electrical answer of anthropogenic, surface and bedrock deposits is highlighted by each VES curve. At contact between fillings, deposits and rock, the clay

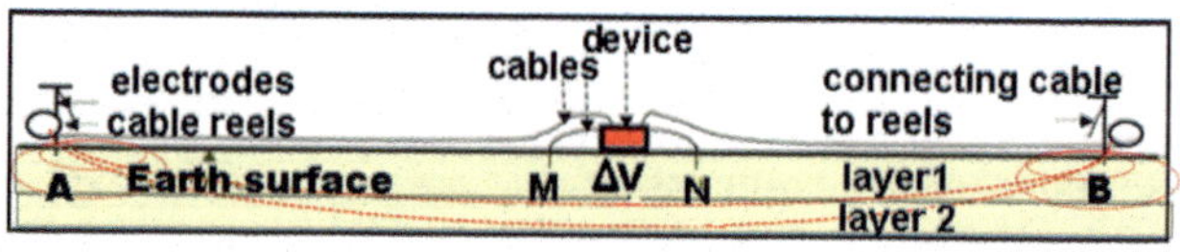

Fig. 3 VES (vertical electrical sounding) using Schlumberger type array

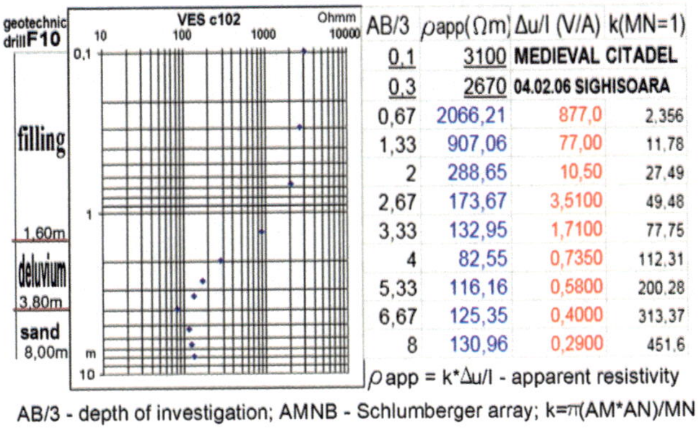

Fig. 4 Example of double inflexion points of VES curves which show three different geophysical deposits

accumulated, in presence of natural waters movement and sewage pipes, becomes very conductive and determines very small electrical resistivity values, being considered an electric landmark. First change of gradient of these curves, from the lower to higher values was verified with geotechnical data and then was drawn boundary between deposits and bedrock at a depth corresponding part of each VES. By correlating the depths of this limit with topographic surface, thickness values were obtained for natural and anthropogenic superficial deposits. Interpolating values obtained was made it in Surfer by triangulation method to create maximum accuracy around electrical measuring stations. 48.6% of VES curves have just one inflexion in which gradient of VES curves change. That means there are two different types on geological environment, one at shallow depth more resistive. The other over 50% of VES curves has two or more changes of gradient. If in the VES curves have been identified two changes of gradient, verified by drilling, then we identify three geological environments: anthropogenic deposits (fillings), delluvial and colluvial deposits and bedrock (Fig. 4).

Inventory and Geomorphological Mapping

Field survey, inventory of slope movements and geomorphological mapping are the first ways to identify the types and the nature of their cause. The areas affected by slope processes and process types were identified by geomorphological inventory and mapping and GPS survey station was applied.

Historical maps were georeferenced and through fill handling or conventional signs for maps were extracted details on the slope form, slope ruptures highlighted by inclination and direction hatch, features of vegetation. Problem encountered was on the low degree of detail of these maps. The second stage was to achieve morphometric maps and slope processes maps: nine functional slope intervals of geomorphic surfaces were based on interpretation and extrapolation values of typology slope processes and the threshold at which it starts. The resulted geomorphic processes were reclassified and summarized with land use, filings thickness, anthropic seepage areas in order to determine the slope instability categories (Table 1).

Results

Bedrock resistivity values (generally silty clay, dusty sand—the citadel) are 35–75 Ohmm, the sands of 80–200 Ohmm and diluvium and filling ranging between 300 and 5000 Ohmm sometimes higher (at the fallen walls). Areas of isolines resistivity gradient, where geophysical contrast is high, highlights amid on bedrock background, points where foundation soil, natural or anthropic is damaged.

Geoelectrical geological data interpretation in terms revealed (Figs. 5 and 6):

- filling/delluvium or colluvium limit;
- delluvium/Pannonian deposits limit;
- different thickness of the filling and the moisture degree; areas with high humidity;
- depth of placing public utility works (sewerage, water supply);
- depth foundation of buildings and citadel precinct walls;
- geoelectrical effects of cracks from foundations and walls.

Interpretative images (12 geoelectrical sections and 6 geoelectrical maps) resulting from the total of 140 boreholes and VES demonstrates presence of shallow deposits type

Table 1 Relation between actual slope morphodynamic and slope gradient, modified after Grigore, 1979

No	Slope (°)	Geomorphic processes
1	<3	Weathering, sheet wash
2	3–7	Rills erosion, soli creeping, soliflucions
3	7–10	Rills erosion, soils creeping, incipient gullies
4	10–16	Shallow landslides, mud slide, spread slide incipient gullies, soils creeping,
5	16–20	Gullies, landslides, flows
6	20–35	Slump, failure, rapid landslide
7	>35	Earth topple, fall, failure, slump

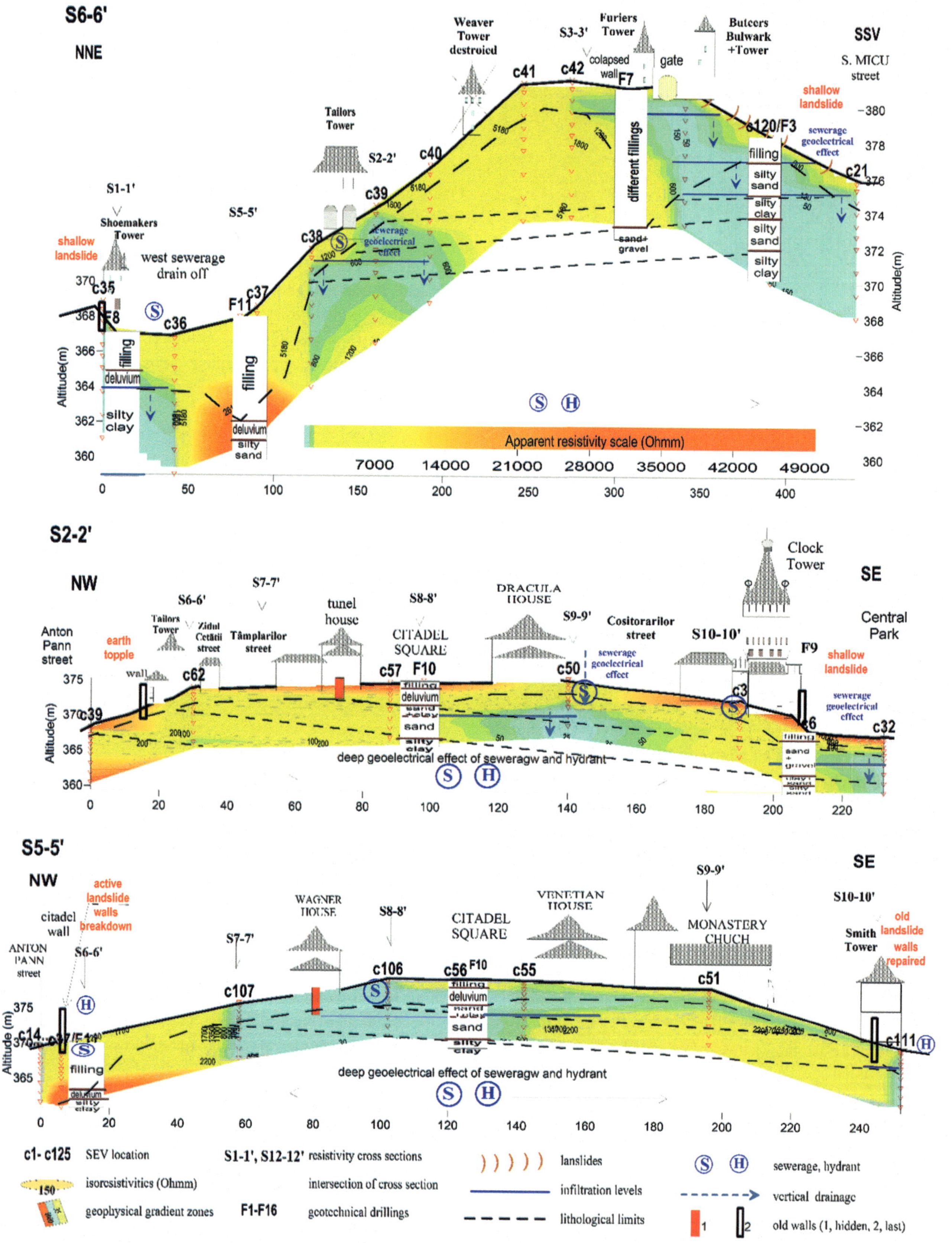

Fig. 5 Geoelectrical interpretative cross sections of Sighisoara Citadel: S2-2′ and S5-5′: transverse cross sections, S6-6′ longitudinal cross section along second ancient wall (NW of medieval Sighisoara citadel)

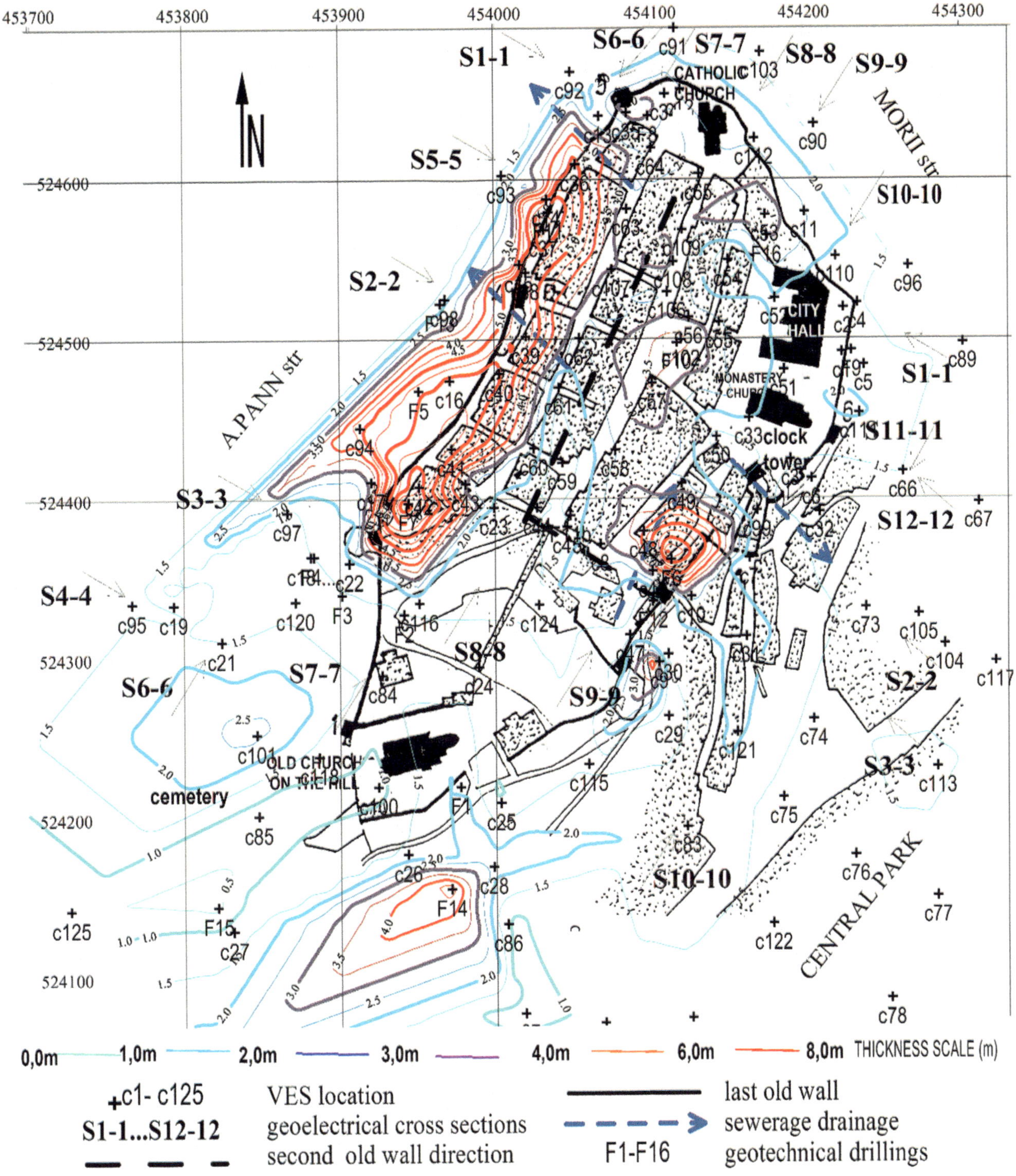

Fig. 6 Anthropogenic and superficial deposits thickness

delluvial and colluvial deposits, without anthropogenic deposits (Fig. 6) in 48.6%.

Delluvial and colluvial deposits have thicknesses of 1.5–3.5 m, even 4.7 m thick identified by Marunteanu and Coman (2000), Marunteanu et al. (2007, 2009).

We mention above because in terms of the geoelectrics have noted a number of minimum resistive anomalies specific to seepage of water having electrolyte qualities (sewage or drinking water they removed particles of clay and organic materials in fillings which lowers the resistivity of

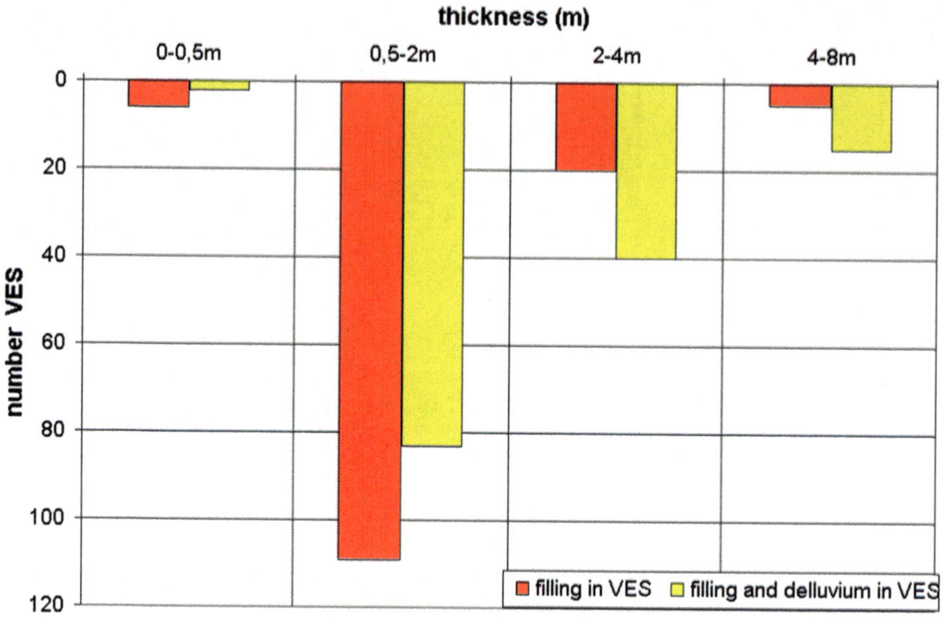

Fig. 7 Thicknes of filling and delluvium deposits highlighted by VESs and drills

the rock mass), setting the stage for the next morphodynamic episodes.

When the clay material is removed, the geoelectric effect of aerated, loose land is the maximum resistive due consistency of debris or solid elements filler and interstitial air. Geoelectric effect of deformation around the foundations and walls is often very poignant; geoelectric image is directly correlated with deformities (cracks, collapsing the walls, etc.) with differentiated subsidence effect, seepage of the mass slope and the citadel walls. In the three sections show above, doesn't identifies slip plans, as evidenced by Marunteanu et al. (2007): "the Fortres Hill is not affected by large and deep landslide." The main causes of gravitational and hydrodynamic processes at slopes level are imposed by more sandy consistency deposits that allow infiltration but in conditions of increasing demand and water consumption, waters from damaged systems sewage and water supply may generate suffusion effect. This may explain why parts of wall collapses, crumbles or sweeps.

Section S2-2′ have the following specific points with geoelectric effect:

- the effect of channeling outlet on Zidul Cetatii (Citadel Wall) St. (VES C62 depth 2.50 m), Cositorarilor (Tanners) St., corner of Dracula's House (VES C50—1.0 to 1.50 m depth) and on slope (VES C32 to −1.30 m depth);
- street fillings into the citadel (VES C573—corner House with Deer—currently already rehabilitated);

Section S5-5′ have the following points:

- rehabilitated channeling on House Wagner alley (VES C106-C107) and rehabilitated land around the Blacksmith tower (VES C111) and the presence of buried walls (Figs. 1 and 6).

Section S6-6′:

- in Western Citadel; the positive impact of the rehabilitated sewer on Zidul Cetatii St. (VES C41 and C39) and VES C36;
- major street fillings in the citadel by the western wall of the fortress whose negative impact is amplified by a sewerage system less efficient (C36-C37 and C40 VES—C41) which imposed the processes of collapse, land subsidence and destruction of the old wall [VES C40 and C42—F7 (8.9 m filling)].

Anthropogenic and delluvium deposits thickness map outlines major stuffing on western side of the fortress from 2.00 to 8.00 m and smaller on eastern side. In areas affected by landslides in natural shallow deposits, thickness of

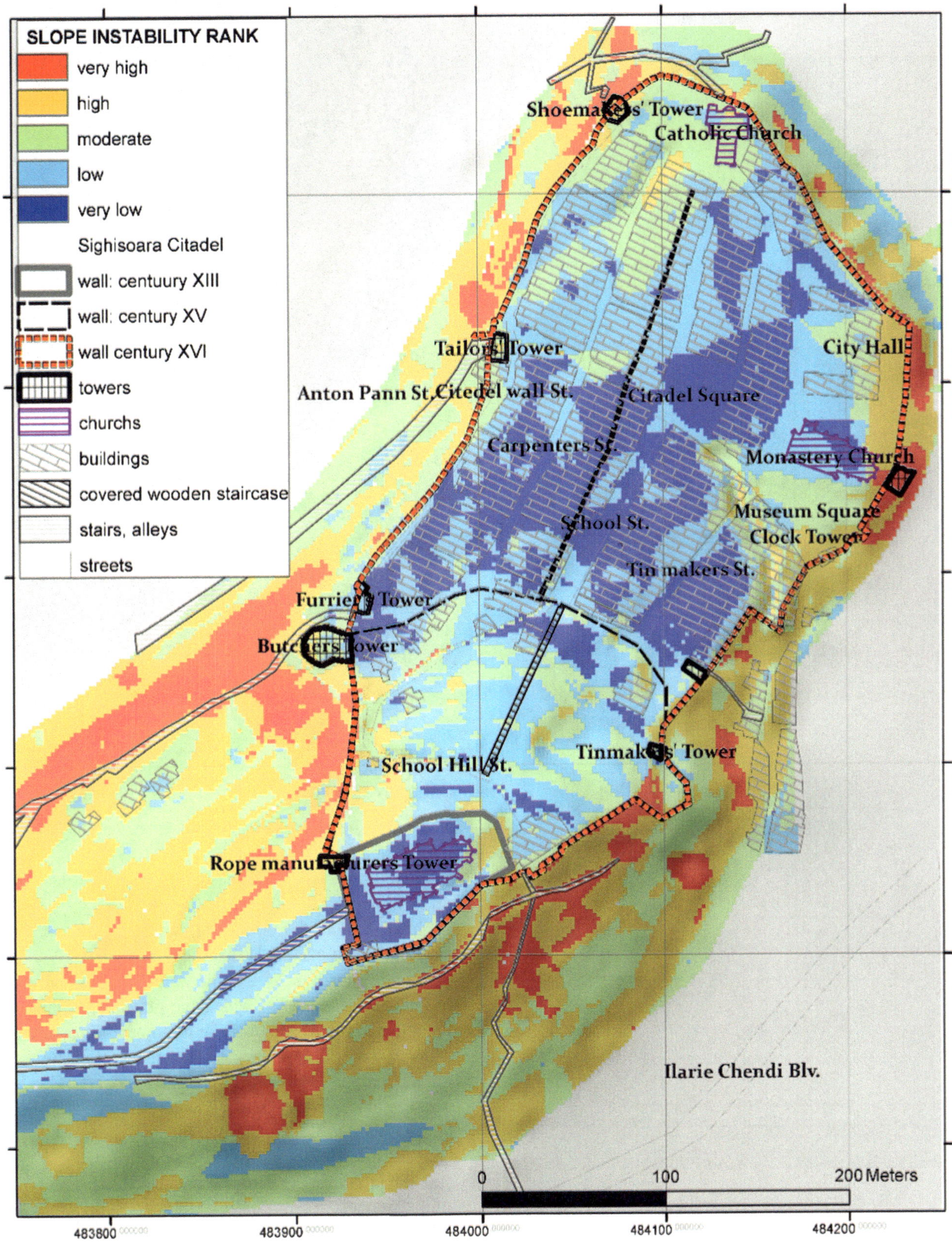

Fig. 8 Spatial distribution of slopes instability

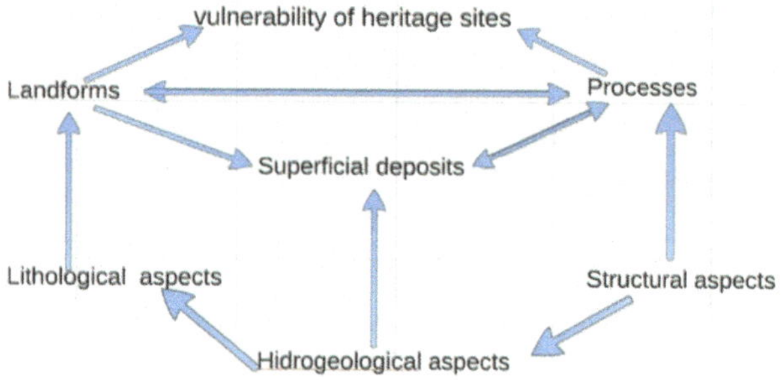

Fig. 9 Geomorphological and geological environment factors that contribute to the stability or instability of heritage sites

Photo 1 Tailor Tower and damaged wall (**a** 2006; **b** 2016)

slippery mass does not exceed 2.50–3.00 m, being less extensive, that can be stabilized by minimal improvement works.

Photo 2 Landslides affect the north-eastern wall of Sighişoara Citadel

Conclusion

Our study revealed that:

- The sandy bedrock is shaped by the deviation of 100–250 Ohmm, and the silty clay of 60–100 Ohmm;
- Rock infiltrated with sewage is highlighted by values 15–40 Ohmm, respectively 40–60 Ohmm;
- Highs of 51.000 to 1.000 Ohmm are specific to gravels and fillers with excess detrital elements;
- Maximum values are around the wall on the western side of the citadel associated with large gradients of the iso-lines, due to rapid changes from fill to bedrock or around walls breaks and differentiated settlements (Fig. 6);
- Fillings are maximized on the western slope of Citadel, where slope processes are with density and high frequency;
- On the eastern slope fillings are smaller (1.50–2.00 m) and although existing processes affects reduced surfaces (Fig. 7);
- Filler about 2.1–2.5 m thick on the north promontory (Roman Catholic Church) (VES C34, C35 and F8) determine the onset of gully erosion and ravines processes; in historical time, there was a small gully where

water leak; it was rectified with the construction of the last wall of defense (XIV–XV century); currently same process is reactivated and shape slope.

The vulnerability of Sighisoara heritage site is determined first of all by landforms (location, isolation and aspect variability) (Fig. 9). The high values of the geomorphological parameters, attributed to geological conditioning, cause the configuration of the morphological surface, and at the same time they impose a high potential of instability, favorable to the triggering of the dynamic.

In its turn, morphodynamics shapes the morphological surface by continuously altering it and by mobilizing superficial deposits (Fig. 8). As the deposits are removed towards the base of the slopes and with the participation of hydrogeological factors, new lithological surfaces are exposed to the shaping of the external agents (Fig. 9).

Where there is a pronounced morphological and hydrogeological dynamic, amplified in most cases by anthropogenic factors (inadequate sewage, water supply system, pluvial sewage), the heritage edifices are characterized by a high vulnerability (eastern part of City Hall, Shoemakers, Tailors and Butchers towers) (Photos 1 and 2.)

Inside of inhabited historical area the geomorphological and geological instability is rather low. Since the system at issue are composed by natural and anthropogenic elements, it is important to estimate the relative contribution of both factors categories to assess the instability of slopes and fillings and deluvial deposits (Fig. 8).

This results require, in the future, to assess the different type of hazard and also to estimate the vulnerability for a suitable preservation strategy of Sighişoara historical buildings.

Acknowledgements We would like to express our sincere thanks to Mr. Prof. Ph.D. Marunteanu Cristian and Mrs. eng. geol. Bugiu Sanda.

References

Dana I (2015) Monitoring of Sighisoara UNESCO World Heritage Site Using Space Technologies, World Heritage Protection: Disaster risk management and sustainable tourism planning in Romania, 29 Sept–3 Oct 2015, Sighisoara

ISPIF (1991) Informative geotechnical study for Sighisoara town especially on the instability phenomena affecting the medieval fortress hillsides, Mures County. ISPIF, Bucharest (Unpublished)

Mafteiu M, Marunteanu C, Niculescu V, Bugiu S (2015) Engineering geology for society and territory—Volume 8, Preservation of cultural heritage, Part VI. In: Lollino G, Giordan, D, Marunteanu C, Christaras B, Yoshinori I, Margottimi C (eds). Springer International Publishing, 281–285 p ISBN 978-3-319-09408-3

Marunteanu C, Coman M (2000) Natural and human induced instability phenomena in the area of medieval citadel of Sighisoara. In: Proceedings CONVEGNO GEOBEN 2000-Geological and Geotechnical Influences in the Preservation of Historical and Cultural Heritage, 7–9 June 2000

Mărunteanu C, Coman M (2005) Landslides: risk analysis and sustainable disaster management. In: Sassa K, Fukuoka H, Wang F, Wang G (eds). Springer, Berlin, 416p ISBN 978-3-540-28664-6

Marunteanu C, Mafteiu M, Niculescu V (2007) Geomorphological hazard affecting the medieval citadel of Sighisoara (Romania). Italo-Maltese Workshop on Integration of the geomorphological environment and cultural heritage for tourism promotion and hazard prevention, Malta, 24–27 Apr 2007

Mărunţeanu C, Niculescu V, Mafteiu M (2009) Engineering geology for tomorrow's cities (CD-ROM). Geological Society, London

Niedermaier P (2000) Atlasul istoric al oraselor din Romania Seria C, Transilvania, Fascicol 1 Sighisoara, Ed. Enciclopedica Bucuresti

Rosian Gh, Nita A, Mihailescu R, Baciu N, Patulea L (2011) The morphology of Tarnava Mare Corridor and it's Conditioning on the Settlements and Thoroughfares Planning (Copsa Mica—Blaj Sector), Geographia Napocensis V(1):39–47

Panizza M, Piacente S (2000) Relazioni tra scienze della terra e patrimonio storico-archelogico, Convegno GeoBen 2000. Condizionamneti Geologici e Geotecnici nella Conservazione del Patrimonio Storico Culturale, Torino, Italy, 7-9 06 200, Torino, CNR-IRPI, pp 723–730

Silvas G (2014) Managementul riscului asociat fenomenului de intabilitate a versaantilor: alunecarile de teren din zona municipiului Sighisoara. Ph.D. thesis, University of Bucharest, Bucharest, Romania

Analyze the Occurrence of Rainfall-Induced Landslides in a Participatory Way for Mid-Hills of Nepal Himalayas

Hari Prasad Pandey

Abstract

Involvement of local people in landslides disaster risk reduction planning, implementation and benefit sharing is the key to a participatory sustainable development approach at a local level. With this approach, Rolpa district which covers an area of 1879 km^2 in the foot hills of Nepal Himalayas has been chosen as the study site. Almost 103 landslides were recorded through the participatory method, analyzed and compared with rainfall data and topographic features. Linear regression model showed that occurrence of landslides is increasing significantly over time but rainfall trend is decreasing gradually. Physical infrastructures and properties such as settlement areas, arable lands, roads, forests, spring (water sources) and irrigation canals were found to be damaged. More than 80% of the landslides affected settlements whereas only 20% affected irrigation canals. The ANOVA test showed that the size of landslide has insignificant ($p > 0.05$) effect on the number of places caused damage except in settlement areas. Moreover, slope failure due to steep relief is not significant rather larger sized and higher numbers of landslides occurred in gentle slope areas (slope $\leq$ 30°). Almost 80% of landslides occurred between elevations of 1200–2400 m a.s.l. with the majority in northern aspect. This study concluded that the causative factor of occurring of landslides is rain but occurrence further accelerated by anthropogenic activities either changing the topographic reliefs or application of improper conservation measures or both reasons. Major anthropogenic activities could be construction of roads, slope farming practices, houses constructed without due consideration of conservation measures in the recent decades. These results will be helpful to guide land use related planning related to soil and its productivity conservation, and water for the government, development agencies, stakeholders of Nepal, in general, and locals of Rolpa district, in particular to get the optimum benefits from those natural resources.

Keywords

Landslide • Occurrence • Rainfall-induced • Rolpa district • Nepal Himalayas

Introduction

The landslides are considered as geomorphological hazards which occur on a terrain surface (Musat and Herban 2009) whereas the United States Geological Survey defined landslides as a wide variety of processes that result in the downward and outward movement of slope-forming materials including rock, soil, artificial fill, or a combination of these where materials may move by falling, toppling, sliding, spreading or flowing (USGS 2004). Landslides are becoming one of the major causes of environmental degradation around the world (DSCWM/PWMLGP/JICA 2010).

Nepal is regarded as one of the world's most disaster-prone countries and experiences several natural

H.P. Pandey (✉)
REDD Implementation Centre, Ministry of Forest and Soil Conservation, Babarmahal, Forestry Complex, Kathmandu, Nepal
e-mail: pandeyhp@hotmail.com; pandeyhp123@gmail.com

M. Mikoš et al. (eds.), *Advancing Culture of Living with Landslides*,
DOI 10.1007/978-3-319-53483-1_18

disasters every year and ranks 11th in the world in terms of vulnerability to earthquakes and 30th with respect to water induced disasters that cause significant losses of human lives and properties due to heavy rain and storms (UNDP 2009). Landslides are very common in hilly districts of the country locating in Siwalik, Mahabharat range, Midland, and fore and higher Himalaya regions because of steep topography and fragile ecosystem (GON/NDRRP 2016). Most of the landslides and landslides dam breaks are found in hills and mountain regions of Nepal (UNDP 2009; Dahal and Hasegawa 2008; Petley et al. 2007). Landslides are the second most hazardous disasters in the country after epidemics (UNDP 2009). About 8642 (35.68% of casualties from all disasters) people lost their lives between 1983 and 2014 due to flood and landslides in the country (DWIDP 2015) whereas about 300 people lose their lives mainly due to water induced disaster every year (GON/MoHA 2015), at least 15% of fatalities caused by landslides alone and 26% in combination to flood in Nepal between 1971 and 2007 (UNDP 2009), in 2004 monsoon, 68 of 75 districts of the country were affected by localized disasters, 192 people died and 11 were reported missing, 16,997 families were affected (Aryal 2007). In addition to these, such disasters also damage the existing infrastructures causing massive economic loss to the nation valued US$8 million per year on an average from 1983 to 2003 (Petley et al. 2007; DSCWM/PWMLGP/JICA 2010), and US$3.66 million equivalent property loss in 2014 only (DWIDP 2015).

An analysis of the 55-year record of landslides and rainfall events in the Himalayas have suggested that many landslides occurred under the influence of a wide range of rainfall durations (Dahal and Hasegawa 2008), reliefs and precipitation (Petley et al. 2007). However, unlike temperature, the precipitation data for Nepal does not reveal any significant trend (Shrestha et al. 2000; Nepal 2009). Also, small-scale, local disasters have a greater cumulative impact in terms of casualties than large-scale, national disasters (Aryal 2012). Further, in order to achieve sustainable development planning on the national and regional level, Nepal requires historical data of disaster risk vulnerability analysis (Aryal 2007). Thus, to make engineering and planning decision effectively, reliable landslides location records and susceptibility maps need to be produced (Hearn 2011). Also, to implement one of the National Strategy for Disaster Risk Management to facilitate locals to make them resilient from landslide and flood hazards (GON 2008), it is crucial to examine the historical occurrence of landslide with the involvement of local people for disaster risk preparedness, and planning and implementation of developmental projects and to save the lives.

Moreover, in the current transitional political period, the absence of district and local elected authorities over the last two decades has weakened formal community participation, empowerment and capacity building in Disaster Risk Reduction (IFRC 2011). In such situation, the role and responsibility of the local communities are further important in disaster mapping and preparedness activities at a local level. In many mountainous regions, there is a lack of fundamental data required to prepare maps of landslide hazards and risks (Hearn 2011).

There is a number of studies carried out related to landslides at local, national and international level by using the latest and advanced technologies inside and outside the country (for example, Chau et al. 2004; Devkota et al. 2013; Lee and Lee 2006). The higher level of technological methods and their availability to the local level are rare, cost ineffective and difficult to understand by the local who are implementing bodies at the local level in Nepal. These methods seemed even more difficult for the implementation in Rolpa where almost all the inhabitants are the survivals from devastating, decade-long civil war in Nepal. Therefore, it is an equally important need to cascade Geo-hazard data at the community level and use community knowledge in the development of a historical record so that communities could be aware, enabled in preparing disaster preparedness, adaptation and mitigation plan resulting in the minimalization of disaster risk exposure (Hearn 2011). Thus the participatory approach could be a tool to identify and analyze the occurrence of landslides as well as being the way forward for necessary actions for the study area. Hence, the present study aims to analyze the occurrence of rainfall-induced landslides in a participatory way for the Rolpa district in Nepal Himalayas.

Materials and Methods

Study Area

Rolpa district is among the 75 districts, the political units, in Nepal. It lies in Rapti zone of Mid-Western Development Region (MWDR). It covers an area of 1879 km^2 (DDCR 2015). The geographical location of the district is 28.08°N to 28.55°N and 82.35°E to 82.95°E (DS/GN/MLRM 2007), ranging 701–3639 m a.s.l. (DDCR 2015). It lies in foothills of the Himalaya and has a higher rate of fatalities' density and moderate type of landslides risk/exposure due climatic effect (MOE 2010).

Data Collection

The necessary data for the study were collected between February and May 2012. For the purpose of data collection, different Village Development Committees (VDCs—grass-root level government's political unit under district jurisdiction) were clustered into 20 groups consisting of 2–3 VDCs in each one. Then, as far as practicable, a center of convenient place for each cluster was identified. Participants for the meeting were purposively and inclusively requested to attend without boycotting the old aged people (>50), local inhabitants, marginalized and indigenous people (Fig. 1).

Unstructured informal group meeting and discussions of 5–10 were performed in each cluster. Meanwhile, key informants including teachers, VDC secretary, local political leaders and social workers were met separately to verify the information received from group discussion. Along with these activities, VDC secretaries and chairpersons—not restricted to age limit—were considered as respondents and also received the data from VDC records as a data source. During group discussions and the VDC visit, the participants were facilitated to recall landslides date by tallying in relation to the major rituals, events in their lives including birth, meaning, baptism, someone's death, marriage, education, festivals, occasions and of their owns or of their relatives. All together 20 group meetings and 20 key informants discussions were performed for data collection from the field. To identify the location of landslides, a topographical map of scale 1:125,000 prepared by Department of Survey (1986) was showed and facilitated respondents to locate the tentative location of landslides located in the catalog map and marked with a number and the descriptions (date, damage caused, etc.) of particular landslides were recorded in a separate notebook. In a participatory and consensus-oriented approach, the time event data with their corresponding dimension and location of landslides were ascertained during the field visit.

In the office work, tentative locations of landslides so collected during data collection were later overlaid into the geographical references and projected into coordinates in Arc GIS 10. After overlaying, the topographic data such as slope, aspect and elevation were generated for each landslide. At the same time, the information regarding physical properties damage caused by those landslides such as forests, agricultural land, water springs, roads, irrigation canals were noted down as per respondents' information and analyzed simultaneously. The dimension of the landslides was estimated using standard conversion factor from their local unit (*hat,* one hand length i.e. 1 *hat* = 45 cm) assuming of their maximum length and width caused damage.

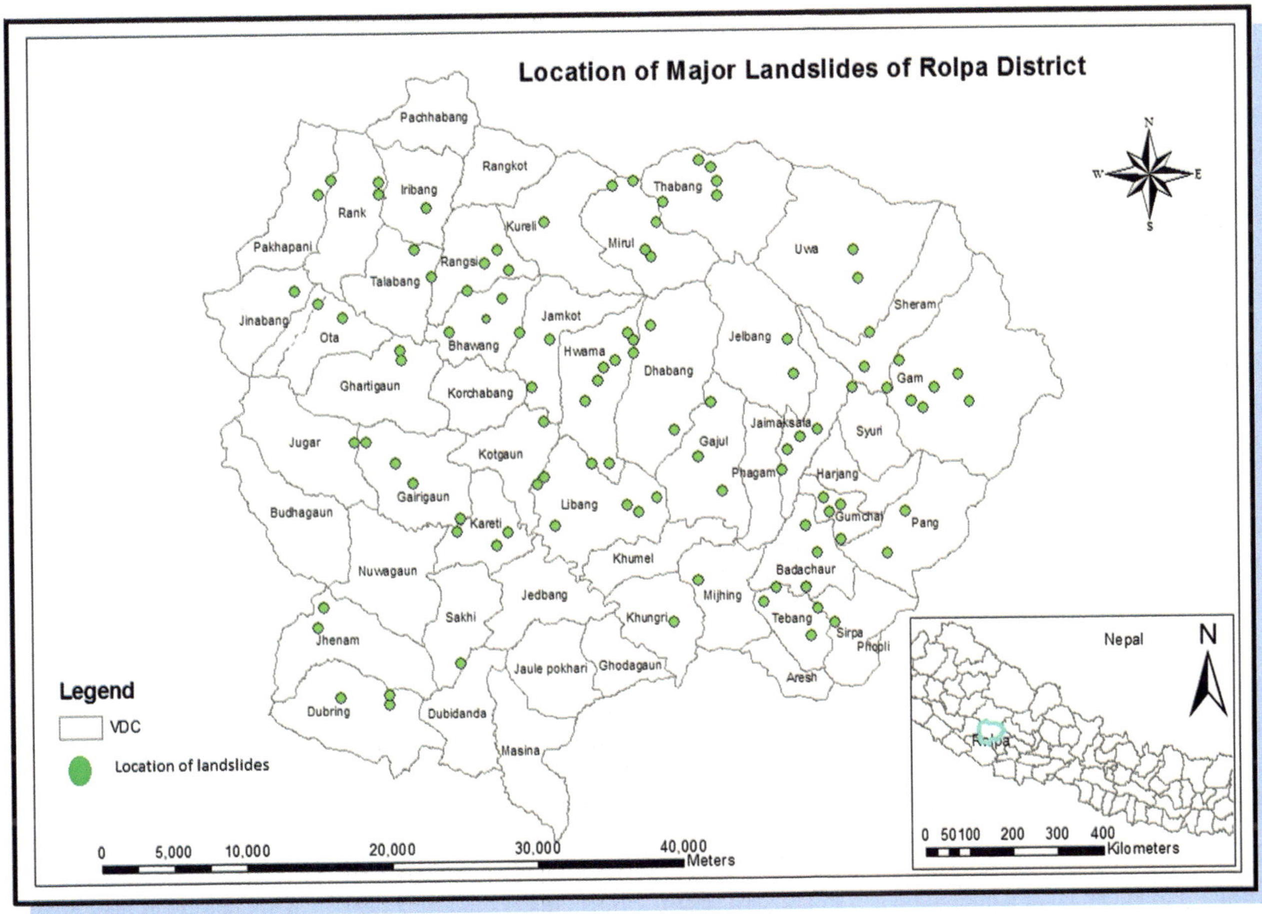

Fig. 1 Map of study area showing position of landslides

Only major types of landslides having the length of 50 m or more and the width, not less than 20 m in slope distance (not horizontal or map distance) were accounted and considered for this study. About 10% of the landslides were selected randomly for the verification and ground truthing. A meeting with 5 key informants including 3 teachers, 1 political leader and 1 local official from different fields was held to verify data from all sources of the field. From aforementioned techniques, the landslides occurrence data were ascertained from the date of 1950 to 2011. Precipitation data from 1984 to 2014 (31 years) were obtained from Department of Hydrology and Meteorology (DHM 2016). Similarly, the fatalities from landslides were received from District Police Office from 1997 to 2008 (DPO 2012) and 2009 to 2011 from Nepal Red Cross Society, Rolpa branch (RCSR 2012).

Data Analysis

Data so obtained were analyzed by using MS Excel and R-statistical packages. Linear Models (LMs) were used to analyze data and Analysis of Variance (ANOVA) was performed to see the statistical difference of the parameter tested. Tukey contrasts test was carried out to make multiple comparisons of the parameters (RCT 2015). Geographical data were presented using GIS 10.

Relationship of topographic and political (VDC) characteristics with the size of landslide was analyzed using following mathematical model as:

$$\begin{aligned} \mathit{Size\ of\ landslide} &= f(a + \mathit{VDC} + \mathit{slope\,factor} + \mathit{elevation\,factor} \\ &\quad + \mathit{aspect\,factor} + \mathit{error}) \ldots \mathit{model} \end{aligned} \quad (1)$$

And, the effect on different property due to different size of landslides was analyzed by using following mathematical model as:

$$\begin{aligned} \mathit{Size\ of\ landslides} &= f(a_i + \mathit{settlements}(\mathit{no.\ of\ household\ affected}) \\ &\quad + \mathit{forests} + \mathit{roads} + \mathit{spring}(\mathit{water\ sources}) \\ &\quad + \mathit{cultivated\ lands} \\ &\quad + \mathit{irrigation\ canals} + \mathit{error}) \ldots \mathit{model} \end{aligned} \quad (2)$$

Results

Landslide Occurrence and Annual Rainfall

The landslides occurrence in the study area with received amount of total annual rainfall is shown in Fig. 2. The result showed that number of landslides in every decade has been increasing (as indicated the graph, the highest number of landslides is accumulated in 1984 but this figure is cumulative of all previously occurred landslides rather a particular year 1984) in the later decades. In the meantime, study area received a wide annual variation of rainfall but overall trend of 31 years is decreasing (Fig. 2).

The F-statistical test showed that total annual rainfall is significantly ($p < 0.05$) decreasing from 1984 to 2014 in Rolpa district. The highest amount of total annual rainfall was received in 2000 (i.e. about 2500 mm) and the lowest in 2010 which amounted only 500 mm.

Landslide Damage upon Property

The result showed that every landslide has caused damage at a time upon different properties under study. The highest numbers of places damaged in settlements (84 places out of 103 places) followed by cultivated land (77 places) through all landslides (Fig. 3). In addition to this, forest and roads damaged were accounted in more than 40 places whereas more than 25 spring (water sources) and 18 irrigation canals were damaged by landslides under study. This figure shows the landslides are more frequent in settlements and cultivated areas in Rolpa district than other places.

Landslides and Topographic Characters

The bar graphs show the number of landslides with different topographic characters (Fig. 4). The highest number of landslides originated (occurred) in between elevations of 1600–1800 m a.s.l. Almost 80% of landslides occurred between elevations of 1200–2400 m a.s.l. in Rolpa district where few number of landslides occurred above 2500 m a.s.l. (Fig. 4). As we see the bar graphs of the slope with a number of landslides in the study area, it is clear that the relatively higher number of landslides had occurred having the slope less than 30° (Fig. 4). On the same time, the highest numbers of landslides occurred on northern aspect in the district.

Size of Landslides with Characteristics Under Study

Table 1 shows the result of variance analysis of various characteristics of topographic features and properties affected by the different size of landslides in Rolpa district. Significantly ($p < 0.05$) larger number of relatively bigger size of landslides were occurred in gentle slope (slope 30°) areas to those in steep slope (slope more than 30°) in the study area (see Fig. 4 and Table 1) whereas aspect and

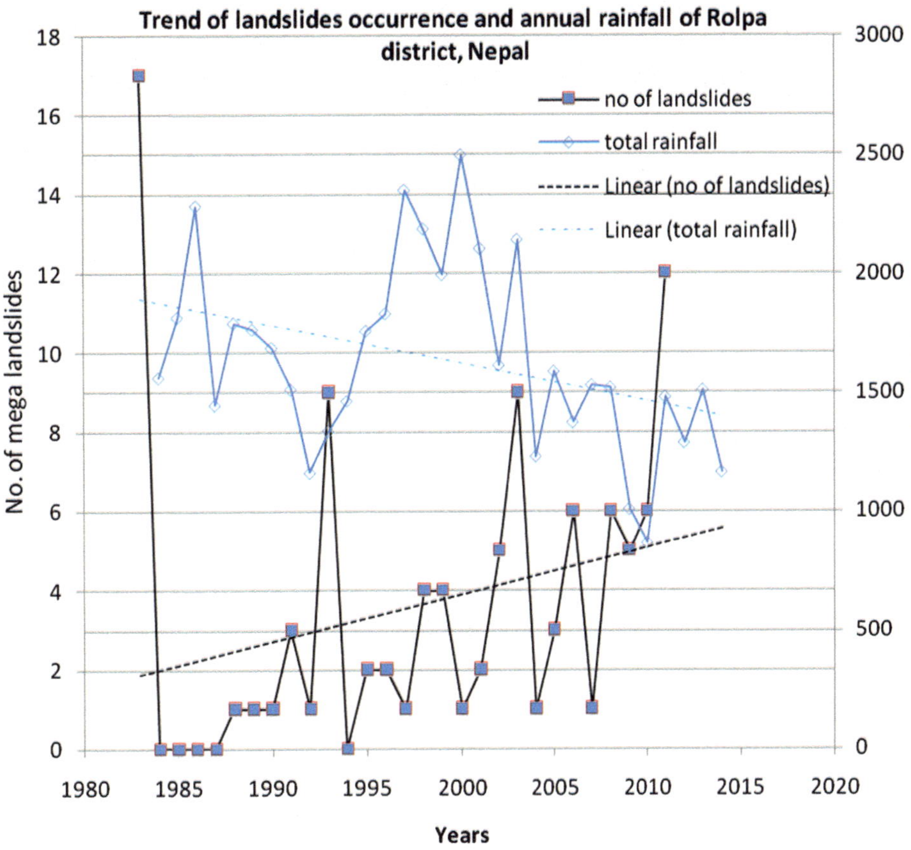

Fig. 2 Number of landslides and amount of annual rainfall (mm) of study area

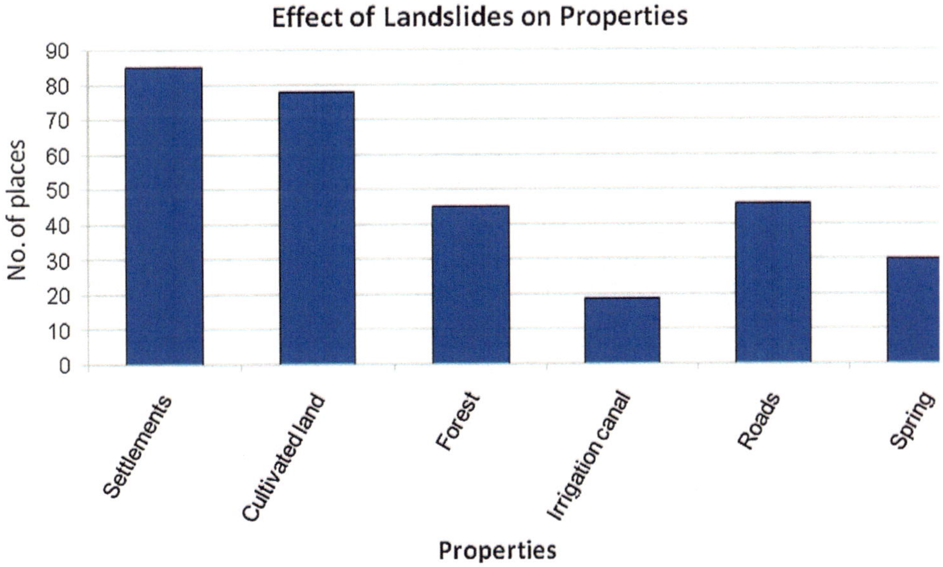

Fig. 3 Effect of landslides upon different property under study

elevation have no significant different ($p > 0.05$) regarding the size of landslides (Table 1).

Tukey contrasts test showed that the Phagam (see sample Fig. 5) and Ransi (sample Fig. 6) VDCs comprise significantly ($p < 0.05$) larger sized of landslides occurred in the history of Rolpa district.

The ANOVA result showed that there is significant ($p < 0.05$) effect of the size of landslides with a number of households affected but the study found no significant difference upon properties loss (Table 1) due to the various size of landslides. In other words, the smaller size of landslides likely to have equal chance of making an effect upon

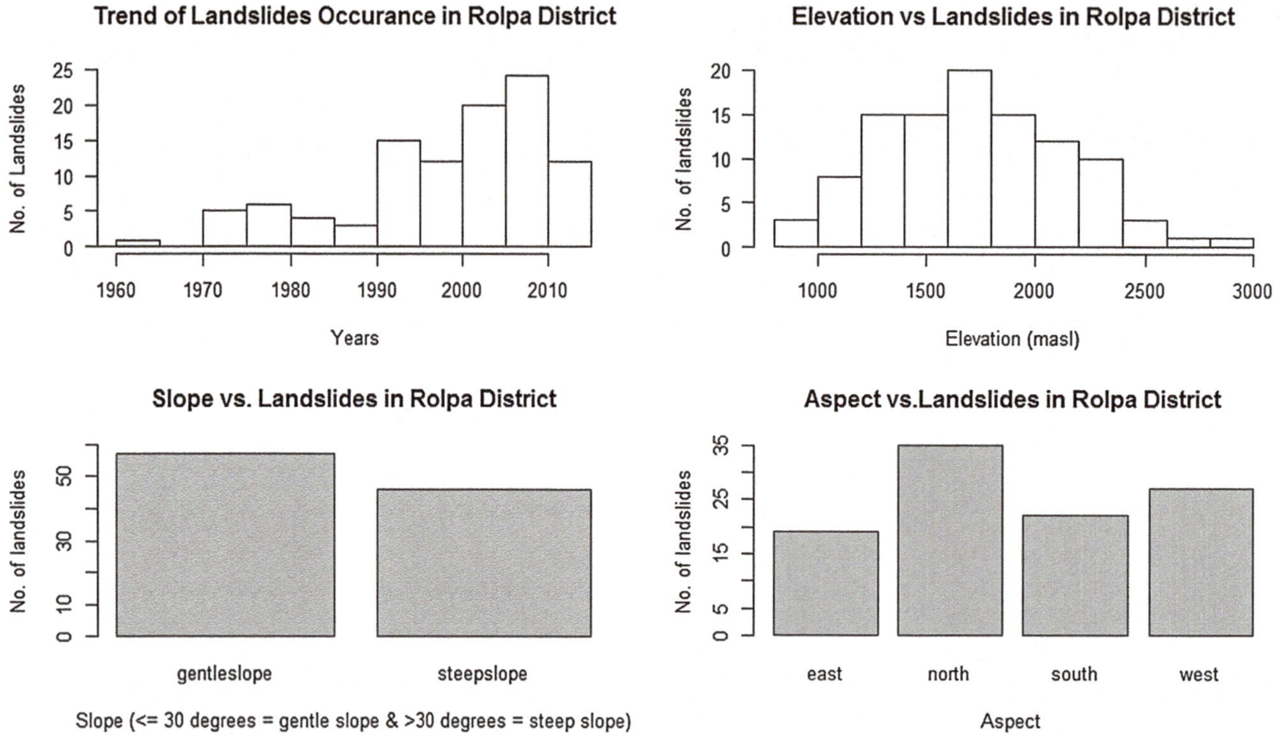

Fig. 4 Number of landslides as per topographic characteristics (elevation, slope and aspect)

Table 1 Statistical test result of the size of landslides with parameters taken under study

Characteristics	df	F-statistics	*p*-value	Remark
Aspect	3	1.9625	0.12905	
Elevation	1	0.0316	0.8594	
Slope	36	4.0307	0.04911	*
VDC	1	1.8744	0.01504	*
Cultivated land	1	1.697	0.1958	
Forest	1	0.3315	0.5661	
Settlement (No. of household)	1	17.3212	0.00007	*
Irrigation canal	1	0.147	0.7023	
Road	1	2.1597	0.1449	
Spring (water sources)	1	1.606	0.2081	

Remark *Significant code at 5% significant level

different properties as larger size of landslides had. We found that almost all landslides caused damage at least two properties at a time in the study area.

Discussion

The result showed that the occurrence of landslides occurrence was increasing in later decades (see Fig. 2). Similar result could be seen while looking for the trend of landslides occurrence of the whole country, Nepal (Petley et al. 2007; UNDP 2009). This result is not surprising that the increasing trend of landslides occurrence could be seen in developed countries as well, for example, Switzerland (Meusburger and Alewell 2008), Hong Kong (Chau et al. 2004). But the serious and alarming case is that the occurrence of fatal landslides in Nepal is increasing with time, faster than the effects of monsoonal variations (Oven 2009).

Total annual rainfall showed a wide fluctuation each year but trend is decreasing in Rolpa district (Fig. 3) but most of other studies found that there were strong positive relation between occurrence of landslides and rainfall (Petley et al. 2007; Dahal and Hasegawa 2008; Meusburger and Alewell 2008; Chau et al. 2004), landslides and topographic reliefs

Fig. 5 A picture of a landslide in Phagam VDC of Rolpa district

Fig. 6 A picture of a landslide in Ransi VDC of Rolpa district

(Petley et al. 2007), landslides and antecedent rainfall (Dahal and Hasegawa 2008). The negative relationship of occurrence of landslides and the total annual rainfall probably is the reason of anthropogenic influences such as land use change, change in reliefs, and drainage pattern without considering effective and sufficient conservation measures. As we see the data of extension of the road during the period, it seemed that there is a strong positive correlation of length of road extended during the period with the frequency of landslides occurred per decade in the study area. The length of road of Rolpa district extended from 74 km in 2002, 78 km in 2004 and continue to 207 km till 2014, seemed gradual increasing in trend (DOR 2016; RCT 2015). In addition to these figures, about 704 km village roads were constructed by 2013 in Rolpa district (DoLIDAR 2015). These attributes the possible reason for the increasing number of landslides occurrence in the study area associated even if the trend of rainfall decreasing in later decades. This reason is also supported by a study performed in Upper Bhote Koshi Valley (Oven 2009). In addition to this, associated other factors could be fragile geographical and geological formation of middle hills, inappropriate land use practice, deforestation, unplanned settlements, increasing population, economic backwardness, lack of education (DSCWM/PWMLGP/JICA 2010; GON/NDRRP 2016) may also valid for the study. Moreover, moderately vulnerable district (Rolpa) to climate change (MOE 2010) may somehow responsible for increasing number of landslides due to increasing numbers of a forest fire, sporadic rainfall, prolonged drought, soil moisture deficit to support vegetation to cover on the ground against initiating landslides.

Results show that most of the landslides occurred in gentle slope areas (slope ≤ 30°) and dominantly in northern aspect (Fig. 4). A similar result was found in Mugling-Naranghat highway of Nepal studied by Devkota et al. (2013) that 51 landslides were occurred out of 241 in the slope less than 25° and additional 86 landslides in 25–35° slope gradient, and 77 out of 241 landslides were in Northern aspect only. On contrary to these findings, most of the landslides occurred on a sloping ground of inclination from 55 to 60 in Hong Kong in between 1984 and 1996 (Chau et al. 2004). In this study, the reason of escalating the incident of landslides would be the higher density of settlements in those regions that crowdedness of settlement plays the positive and triggering role to initiate landslides. The lowest number of irrigation canals damaged (Fig. 3) is explained by the limited number of presence in upland area and most of the cultivated land probably under the water regime of rain-fed.

Moreover, soil characteristics, low angle of internal friction of fines in soil, the medium range of soil permeability, the presence of clay minerals in the soil, bedrock hydrogeology, identified by Dahal et al. (2008) were not the scope of this research but explore the way forward for new researchers. UNDP (2009) drawn a conclusion that poor people are more vulnerable to natural disasters than other socio-economic groups in Nepal, and people living in rural areas, and in the mid and far western development regions are more likely to be poorer than people from other development regions (UNDP 2009) also explains the result.

Conclusion

The result shows that occurrence of landslides was increased significantly but total annual rainfall trend was gradually decreasing in the study area. This informs the increment in landslides occurrence does not correspond directly to increase in rainfall volume received. However, anthropogenic causes such as constructional activities, roads, houses and slope agriculture practices without due consideration of conservation measures in later decades probably be the sources of triggering landslides occurrence. The properties damaged by landslides were settlements, arable land, roads, forests, spring (water sources) and irrigation canals in a descending order of magnitude of damage. Settlement areas were the most affected from landslide induced damage. At the mean time, every landslide has caused damage to at least two properties. More than 80% landslides had effects on settlements whereas only 20% affect the irrigation canal. The ANOVA test showed that the size of landslides found no significant ($p > 0.05$) effect on the number of places damage caused upon different properties except settlements (number of household affected).

Moreover, significantly ($p < 0.05$) larger number of relatively bigger size of landslides were occurred in gentle slope (slope ≤ 30°) areas to those in steep slope (slope more than 30°). Almost 80% landslides occurred in the elevations between 1200 and 2400 m a.s.l. and majority in northern aspect. This study concluded that the causative factor of occurring of landslides is rain but occurrence further accelerated by anthropogenic activities either changing the topographic reliefs or application of improper conservations measures or both reason than the amount of rainfall received only.

This study recommend that the constructional activities seemed to be confined to those areas in the days to come where the occurrence of landslides was relatively less frequent and effective. These areas could be on east, south or west facing slope at the elevation below 1200 m a.s.l. or above 2400 m a.s.l. as far as practicable, otherwise, effective and sufficient conservation measures need to be supplemented on landslides prone/risk areas to reduce the damage from landslides on properties including human lives. These results will be helpful for the government and other development agencies to guide land use related planning and applying conservation measures to protect soil and water of Nepal, in general, and for Rolpa district, in particular.

Acknowledgements The author is thankful to Bikram Manandhar (ICIMOD), Manoj Bhusal (LNPBZ SP/WWF Nepal), Kaushal Raj Gnyawali and anonymous reviewers from World Landslide Forum for their valuable comments and feedbacks on the manuscripts.

References

Aryal KR (2007) Mapping disaster vulnerability from historical data in Nepal. Disaster and Development Centre, Northumbria University (DDC, NU), Newcastle, UK

Aryal KR (2012) The history of disaster incidents and impacts in Nepal 1900–2005. Int J Disaster Risk Sci 3:147–154

Chau KT, Sze YL, Fung MK, Wong WY, Fonge L, Chan LCP (2004) Landslide hazard analysis for Hong Kong using landslide inventory and GIS. Comput Geosci 30:429–443

Dahal RK, Hasegawa S (2008) Representative rainfall thresholds for landslides in the Nepal Himalaya. Geomorphology 100:429–443

Dahal RK, Hasegawa S, Yamanaka M, Dhakal S, Bhandary NP, Yatabe R (2008) Comparative analysis of contributing parameters for rainfall-triggered landslides in the Lesser Himalaya of Nepal. Environ Geol 58:567–586

DDCR (2015) District profile of Rolpa. District Development Committee, Liwang, Rolpa, 62p

Department of Survey (1986) Topographical map of Rolpa prepared on the basis of field by Survey Division, Department of Survey, Ministry of Land Reform and Management, His Majesty's Government

Devkota KC, Regmi AD, Pourghasemi HR, Yoshida K, Pradhan B, Ryu IC, Dhital MR, Althuwaynee OF (2013) Landslide susceptibility mapping using certainty factor, index of entropy and logistic regression models in GIS and their comparison at Mugling-Narayanghat road section in Nepal Himalaya. Nat Hazards 65:135–165

DHM (2016) RE: Precipitation data of Liwang gaun. Type to PANDEY, H.P

DoLIDAR (2015) Road database. The government of Nepal, Ministry of Federal Affairs and Local Development, Department of Local Infrastructure Development and Agricultural Roads (DoLIDAR)

DOR (2016) Road statistics. The government of Nepal, Ministry of Physical Infrastructure and Transport, Department of Roads

DPO (2012) RE: Disaster victims and assistant update. Type to Dangi, K

DSCWM/PWMLGP/JICA (2010) Resource Book for Soil Conservation (Basic guide), Kathmandu, Nepal, Participatory Watershed Management and Local Governance Project (PWMLGP)

DS/GN/MLRM (2007) Land resources Map of Rolpa district. Department of Survey, Government of Nepal, Ministry of Land Reform and Management, Kathmandu

DWIDP (2015) Disaster review 2014. Government of Nepal, Ministry of Irrigation, Department of Water Induced Disaster Prevention, Kathmandu, 36p

GON (2008) National Strategy for Disaster Risk Management of Nepal. In: Nepal GO (ed). Government of Nepal, Kathmandu, Nepal

GON/MOHA (2015) Casualties reports from landslides and flood. Government of Nepal, Ministry of Home Affairs, Kathmandu

GON/NDRRP (2016) Risk profile of Nepal. Government of Nepal, Nepal Disaster Risk Reduction Portal, Kathmandu. http://drrportal.gov.np/risk-profile-of-nepal. Accessed 8 Feb 2016

Hearn GJ (2011) Slope engineering for mountains roads. https://books.google.com.np/. Accessed 1 Sept 2016

IFRC (2011) Analysis of legislation related to disaster risk reduction in Nepal. International Federation of Red Cross and Red Crescent Societies, Geneva

Lee S, Lee MJ (2006) Detecting landslide location using KOMPSAT 1 and its application to landslide susceptibility mapping at the Gangneung area, Korea. Adv Space Res 11

Meusburger K, Alewell C (2008) Impacts of anthropogenic and environmental factors on the occurrence of shallow landslides in an alpine catchment (Urseren Valley, Switzerland). Nat Hazards Earth Syst Sci 8:509–520

MOE (2010) Climate Change Vulnerability Mapping for Nepal. Ministry of Environment, Kathmandu, Nepal

Musat CC, Herban SI (2009) Geoinformation system for interdisciplinary planning of landslides areas. In: Eleventh WSEAS international conference on sustainability in science engineering, 2009, Romania, pp 257–261

Nepal PA (2009) Temporal and spatial variability of climate change over Nepal (1976–2005). Practical Action Nepal, Kathmandu

Oven KJ (2009) Landscape, livelihoods and risk: community vulnerability to landslides in Nepal. Doctors of Philosophy, Durham University

Petley DN, Hearn GJ, Hart A, Rosser NJ, Dunning SA, Oven K, Mitchell WA (2007) Trends in landslide occurrence in Nepal. Nat Hazards 43:23–44

RCSR (2012) RE: Landslides and Fire data of Rolpa. Red Cross Society, Rolpa, Type to PANDEY, H.P

RCT (2015) R: R Core Team, R Foundation for Statistical Computing, A language and environment for statistical computing, Vienna, Austria

Shrestha AB, Wake CP, Dibb JE, Mayewski PA (2000) Precipitation fluctuation in the Nepal Himalaya and its vicinity and relationship with some large scale climatological parameters. Int J Climatol 20:317–327

USGS (2004) Landslide Types and Processes: Fact Sheet 2004-3072. 4p. http://pubs.usgs.gov/fs/2004/3072/. Accessed 29 Dec 2012

UNDP (2009) Nepal Country Report: Global Assessment of Risk. United Nations Development Programme, Pulchock, Kathmandu

Part II
Landslides in Natural Environment

Session Introduction—Landslides in Natural Environment

J. Klimeš

Abstract

This session presents contributions describing landslides from highly contrasting climatic conditions ranging from boreal zone to mountains in temperate and tropical climate. Majority of the presented research is motivated by solving landslide problems with aim to prevent negative effects of landsliding on societies through improving knowledge about conditions and processes responsible for landslide occurrence and size. Large attention is paid to climate driven environmental processes (e.g. permafrost degradation) which increase landslide occurrence frequencies and magnitude.

Keywords

Landslides • Site investigations • Permafrost degradation • Soil properties

Landslides in Natural Environment

Landslides are inherent part of landscape development in different environments including even extraterrestrial conditions. Therefore understanding processes leading to their formation and affecting their size and type is crucial for any hazard or risk assessment. Special attention needs to be paid to lithological and structural characteristics which determine water distribution and strength of rock mass. Water availability is mainly affected by conditions independent of landsliding itself including extreme precipitations, change of precipitation patterns or enhanced melting of snow or permafrost and are often related to global environmental change. At the same time, landslide processes change properties of affected rocks which often create better conditions for water accumulation above possible sliding surfaces. Mutual action of these two processes may cause transition between different landslide types or affect their size and occurrence frequencies. At the same time, it is important to consider and study how landslide processes affect surrounding environmental conditions including local climate, soil properties, river network and sediment transport. In turn, these conditions determine character and type of vegetation and other biotic processes.

This session presents contributions describing landslides from boreal zone and mountains in temperate or tropical climate. They represent highly contrasting climatic and rock conditions, but majority of the presented research is motivated by solving landslide problems with aim to prevent negative effects of landsliding on societies.

Three of the studies identified hydrologically contrasting geological conditions where upper, permeable layers (e.g. sandstones, gneisses) allow water infiltration until impermeable and less strong rocks (e.g. claystones, weathered mica schists), where the sliding surfaces tend to occur. In arctic tundra, above average warm summer temperatures in last years, caused transition from translational landslides to debris flows. The latter significantly contributes to forming and enlarging thermocirques which form on slopes around lakes (Khomutov et al. 2017). Also landslide research in South American Andes describes possible effects of environmental change on number, frequency and types of landslides through processes strongly related to high mountain permafrost degradation (Moreiras and Dal Pont 2017). Other research in regions with isolated permafrost focused on landslide identification and description using complex data

J. Klimeš (✉)
Institute of Rock Structure and Mechanics, The Czech Academy of Sciences, V Holešovičkách 41, Prague, 18209, Czech Republic
e-mail: klimes@irsm.cas.cz

M. Mikoš et al. (eds.), *Advancing Culture of Living with Landslides*,
DOI 10.1007/978-3-319-53483-1_19

from field mapping, geophysical investigations and drilling with integrated monitoring of environmental conditions. Results describe character and development of landslides in this fragile environment as well as other hazardous processes (e.g. icing in road cut slopes) possibly affecting important local transportation corridors (Hu et al. 2017; Guo et al. 2017). Effects of landslides on soils conditions in boreal forests clearly shows that landslides act as natural disturbance process significantly altering natural environments across very small distances (Masyagina et al. 2017). These effects may contribute to diversity of natural environments.

Other presented research illustrates an importance of combination of reliable landslide inventories with in-depth geotechnical and geomorphological investigations of selected landslide cases. This allows describing specific stability conditions (Vasudevan et al. 2017) which may be applicable to broader regions if the stability governing geological conditions are correctly identified (Wilde et al. 2017).

Multi-methodological Studies on the Large El Capulín Landslide in the State of Veracruz (Mexico)

Martina Wilde, Wendy V. Morales Barrera, Daniel Schwindt, Matthias Bücker, Berenice Solis Castillo, Birgit Terhorst, and Sergio R. Rodríguez Elizarrarás

Abstract

During the last decade, the State of Veracruz (Mexico) experienced a series of intense rainfall seasons with more than 1000 registered landslides. As a consequence, more than 45,000 people had to be evacuated and resettled. Even though the mountainous areas of Veracruz are highly prone to landslides, neither susceptibility maps nor any other relevant information (distribution of landslides, geology, etc.) with high spatial resolution is available. The high social impact of the most recent landslide hazards points out the necessity of detailed investigations in the affected areas. The aim of this study is to improve the understanding of process dynamics for the landslides and to provide the base for future susceptibility mapping. As an example, a young landslide with a high complexity of nested processes from the year 2013 is selected for detailed investigations in the east Trans Mexican Volcanic Belt in the State of Veracruz, related to the complexity of the studied landslide a multi-methodological approach is applied, which includes geomorphological mapping, sediment characterization as well as geophysical methods (electrical resistivity tomography, seismic refraction tomography). Field results indicate that the studied landslide must be regarded as a reactivated older landslide body, with a variety of intricate

M. Wilde (✉) · B. Terhorst
Institute of Geography and Geology, University of Würzburg, Am Hubland, 97074 Würzburg, Germany
e-mail: martina.wilde@uni-wuerzburg.de

B. Terhorst
e-mail: birgit.terhorst@uni-wuerzburg.de

W.V. Morales Barrera · S.R. Rodríguez Elizarrarás
Institute of Geology, National Autonomous University of Mexico, 2376 Mexico City, Mexico
e-mail: moralesw@geologia.unam.mx

S.R. Rodríguez Elizarrarás
e-mail: srre@unam.mx

D. Schwindt
Geomorphology and Soil Science, Science Centre Weihenstephan, Technical University of Munich, 85354 Freising, Germany
e-mail: daniel.schwindt@tum.de

M. Bücker
Department of Geophysics, Steinmann Institute, University of Bonn, 53115 Bonn, Germany
e-mail: buecker.matthias@yahoo.de

B.S. Castillo
Centre for Research in Environmental Geography, National Autonomous University of Mexico, 2376 Mexico City, Mexico
e-mail: bsolis@ciga.unam.mx

M. Mikoš et al. (eds.), *Advancing Culture of Living with Landslides*,
DOI 10.1007/978-3-319-53483-1_20

processes and numerous secondary slides. Detailed investigations provide deep insights in the dynamics and interactions of landslide processes related to their natural and anthropogenic settings.

Keywords
Landslide • Electrical resistivity tomography • Seismic refraction tomography • Mexico

Introduction

Landslides in mountainous areas of Mexico annually threaten inhabitants, settlements, infrastructure, and cultivated landscapes. Reports on the loss of human lives are numerous and massive damages occur during extreme precipitation events (Alcántara-Ayala 2008).

Hazard assessment still remains difficult, especially in areas where there is a deficiency of detailed information. Lack of general data in combination with recurring events demonstrate the necessity of acquiring basic information.

With an almost annually occurrence of various landslides and only little detailed basic data (e.g. detailed geological information), the State of Veracruz, Mexico, is an optimal base to establish an exemplary study. A general overview of the geological hazards in the State of Veracruz exists in form of the 'Atlas of Natural Hazards' (scale 1:250,000), however investigations in great detail have not been carried out, so far. Between the years of 2005 and 2012 at least 1127 landslide events were identified in this state, according to the data of the State Communication Center of the Civil Protection of Veracruz (CECOM). In the year 2013, Veracruz suffered from the consequences of more than ten severe storms, resulting in over 650 landslides (Fig. 1) (Morales-Barrera and Rodríguez-Elizarrarás 2014).

The focus of this study lies on the El Capulín landslide, one of the largest events, which occurred in 2013 and caused severe damages to infrastructures and settlements. The frequent occurrence of mass movements in the State of Veracruz and particularly in the study area is not only related to climatic drivers and societal development, but is influenced by geomorphological and geological characteristics of this region. Flexible and adaptive procedures, to manage and stabilize active slopes, are required in order to prevent or reduce mass movements (Fell and Hartford 1997; Margottini et al. 2013). In order to optimize the knowledge on causative factors and processes, interdisciplinary approaches need to be applied and adapted to future requirements (Guzzetti 2000). In general multi-methodological and interdisciplinary concepts enable a closer adaption and more precise reaction to specific local and regional problems. Inevitably, the basis for any characterization of mass movements and the subsequent reconstruction of process structures depends on surface analysis by detailed geomorphological and geological analysis (a.o. Guzzetti et al. 1999; Heimsath and Korup 2012; Ferrari et al. 2005). The combination with sub-surficial studies, as electrical resistivity tomography (ERT) and seismic refraction tomography (SRT), enables deep insights into complex processes, composition, and structure of a slide mass.

The aim of the study is to identify causative factors, which are characteristic for the study area. The obtained results are the base for further investigations, such as susceptibility modeling, and as a reference for regions with comparable natural and cultural environments.

Study Area

The El Capulín landslide occurred on October 1st, 2013, as one of the largest landslides that happened in this time period in the State of Veracruz. People of the nearby village El Capulín had noticed movements in the lower part of the affected area, three days prior to the actual event. The landslide area is located on the road between the villages El Capulín and Las Sombras, only about 100 m from the western exit of El Capulín. The slide mass destroyed a section of the connecting road, with a length of about 300 m, blocking the western access to the village. Long-lasting precipitation preceded the event, however precise data is not available due to the absence of a station in this specific region.

Regional Setting

The El Capulín landslide is located in the Palma Sola-Chiconquiaco Mountain Range (SPSC) in the central part of the state of Veracruz, Mexico. The study area is situated in the municipality of Chiconquiaco, close to the villages of El Escalanar and El Capulín (c. 250 km east of Mexico City and c. 30 km west of the Golf coast).

Considering the distribution of the deposits and lava flows, the altitudes of this massif goes from 0 to 2000 m. The general geomorphology is characterized by moderately steep to very steep slopes and deeply incised valleys. The elevation differences between ridges and valleys is up to 500 m and thus, show intense morphodynamics with a high

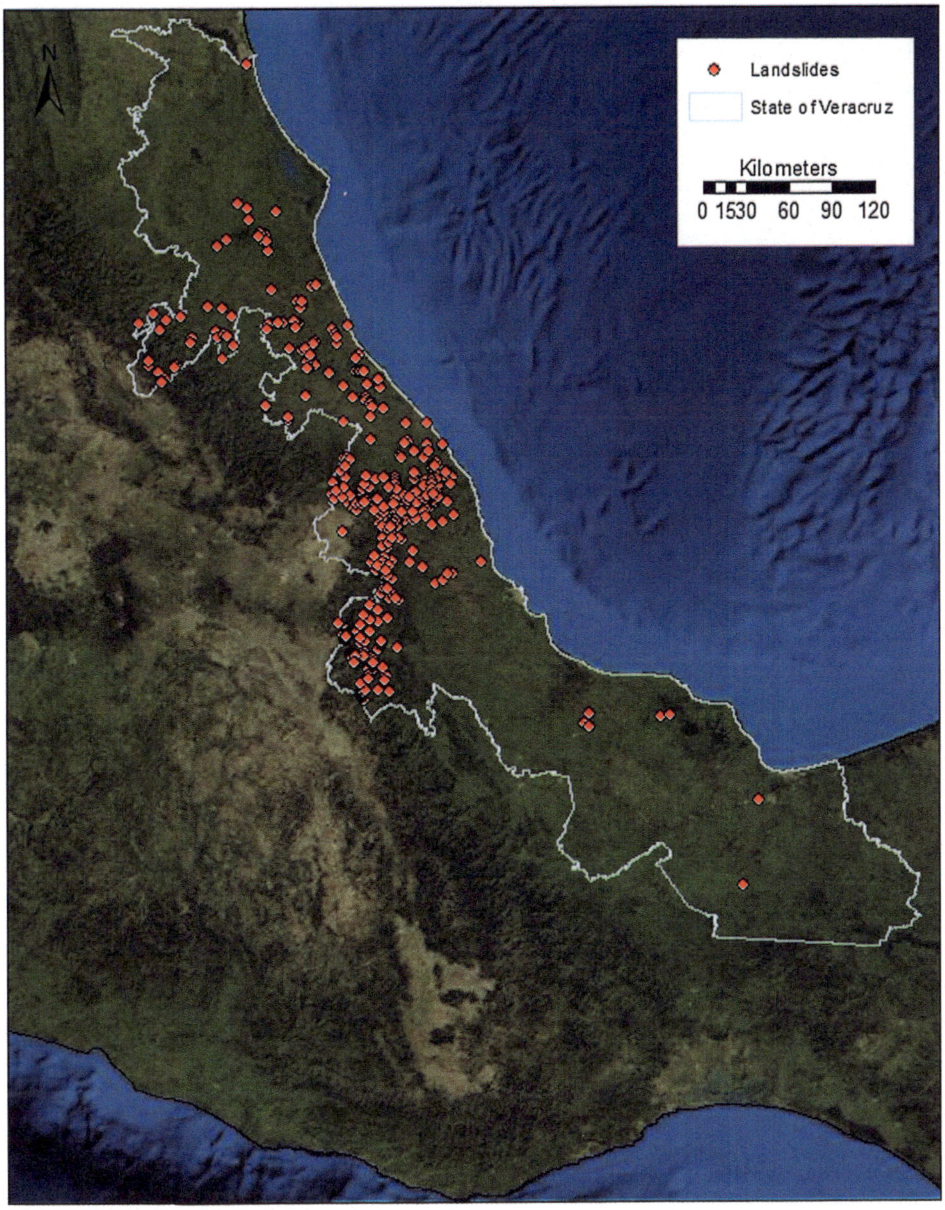

Fig. 1 Overview map of the landslides from 2013 in the state of Veracruz. Image source Blue Marble MODIS (http://www.visibleearth.nasa.gov/). Source of landslide data Morales-Barrera and Rodríguez-Elizarrarás (2014)

density of mass movements, and also, according to factors such as the climate, human influence, geology and geomorphology, there is also a very high susceptibility for landslide occurrence.

Multiple small and intermediate streams that together constitute the Nautla, Actopan, Misantla, and Juchique river basins, which discharge the water to the Gulf of Mexico, drain the area.

Geology

The Palma Sola-Chiconquiaco mountain range (SPSC) is a huge volcanic massif located near the Gulf of Mexico coastline in the eastern sector of the Trans Mexican Volcanic Belt (TMVB) (Fig. 2). The TMVB is a continental arc with an approximately distance of 1000 km and an irregular wide between 20 and 200 km (Gómez-Tuena 2002). The TMVB is the result of the subduction of two oceanic plates (Cocos and Rivera) under a continental plate (North America) (Demant 1978; Nixon 1982; Pardo and Suárez 1995; Gómez-Tuena et al. 2005; Ferrari et al. 2012), most of the active volcanoes and Quaternary monogenetic volcanic fields in Mexico are located within this E-W oriented volcanic province.

The pre-volcanic basement of the SPSC is formed by schists, volcano-sedimentary, and intrusive rocks that together constitute a structural high known as the Teziutlán massif of Paleozoic age; limestones and shales that belong to the Mesozoic Tampico-Misantla platform and finally the

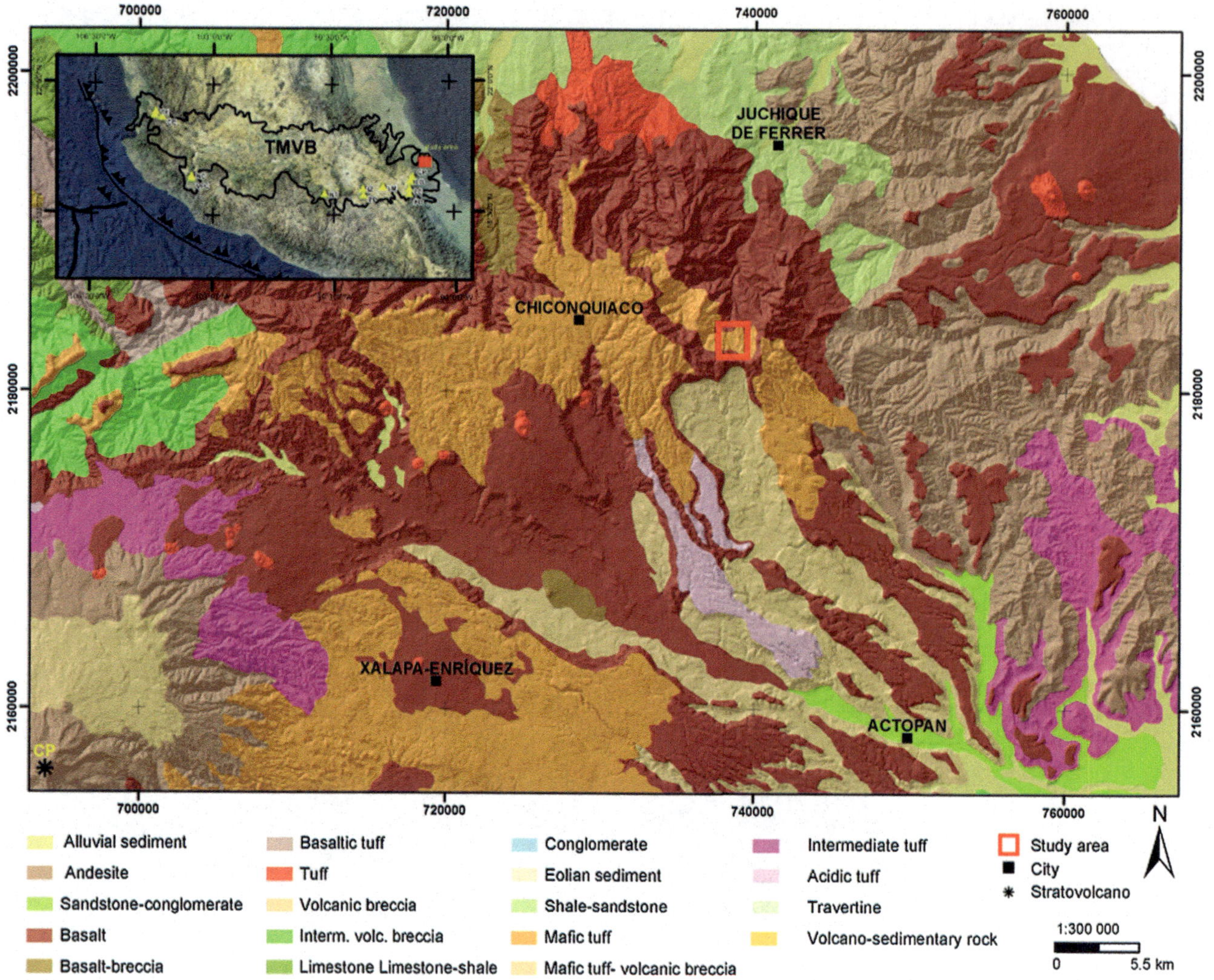

Fig. 2 Geology of the Palma Sola-Chiconquiaco mountain range. Trans Mexican Volcanic Belt in the background image. Map source INEGI 1998 (http://www.inegi.org.mx/)

thick terrigenous sequence deposited in the Veracruz basin during the Tertiary (López-Infanzón 1991), which essentially is formed by alternated beds of sandstone and shale deposits. Within the study area it is possible to observe scattered outcrops of shales and sandstones that belong to the Tertiary terrigenous sequence. According to Gómez–Tuena et al. (2003), in a time span of about 17 Ma occurred three different phases of magmatic rocks in the SPSC: Miocene calc-alkaline plutons, latest Miocene-Pleistocene alkaline plateau basalts, and Quaternary calc-alkaline cinder cones. The study area of El Capulín is characterized by a sequence of late Miocene to Pleistocene plateau basalts emitted from fissures and central vents, whose individual layers may vary between 10 and 50 m in thickness. The different lava flows are commonly separated by horizons of red deposits, thin layers of alluvial deposits or thick layers of breccia talus deposits (López-Infanzón 1991). The lavas are basaltic and andesitic in composition and are highly altered by the effects of tropical weathering. K-Ar isotopic dating have yielded ages between seven and three Ma for the Chiconquiaco plateau basalts (Ferrari et al. 2005; López-Infanzón 1991).

Massive, heterogeneous talus deposits cover the slopes of the study area and provide material for different types of mass movements.

Geomorphological Survey

The elevation range of the affected slope is between 1900 m above sea level (m a.s.l.) at the top and 1100 m a.s.l. at the toe. The steepest part here ranges between 1900 m a.s.l. and 1650 m a.s.l., followed by intermediate sections, in which the landslide itself is located.

The El Capulín landslide has a maximum length of 1000 m and is 300 m wide (max.), with a volume of

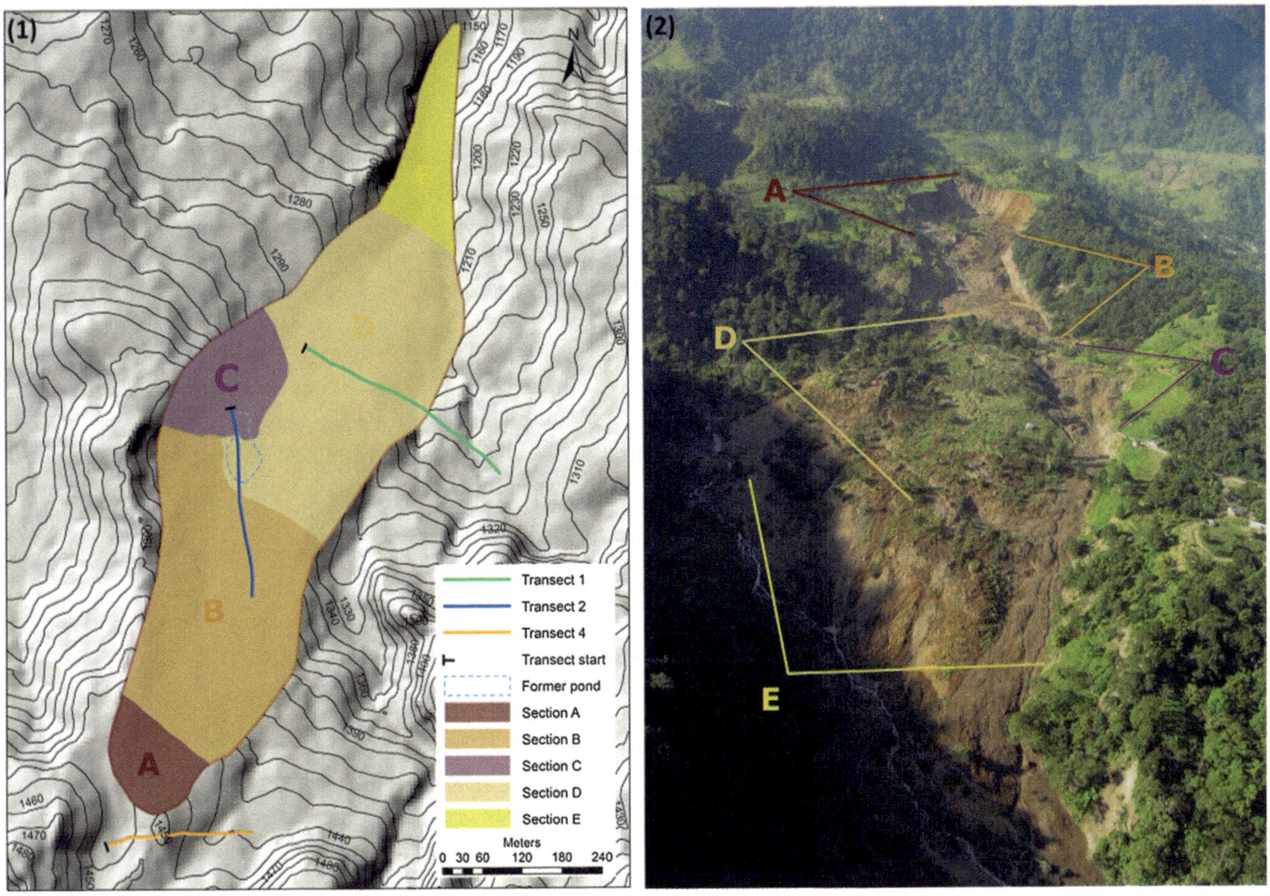

Fig. 3 Geomorphological map. *1* Hillshade (orientated North), generated before the landslide (5 m resolution); outline of the landslide divided in sections; geophysical displays 1–3 [start of the transects (0 m) marked in *black*]; outline of the location of the former pond. Picture of the El Capulín landslide. *2* Frontal view of the landslide; marked here—sections from the map. *Map source* INEGI 2011 (http://www.inegi.org.mx/), photo by S.R. Rodríguez-Elizarrarás (taken from a helicopter)

approximately 4 million m^3 (Fig. 3). In general, the slide body can be divided into five sections. The scarp area, marked here as Section A, followed by the upper part of the slide mass (Section B), the main slide mass (Section D) and the lower part (Section E) with flow characteristics. Section C marks a secondary landslide on the west flank. The landslide scarp is at 1440 m a.s.l. with an approximate length of 200 m and up to 50 m at its highest point (Fig. 3, Section A). Satellite images of the area before the event show about 200 m long cracks, which demonstrate that this part of the slope was already affected by movements, prior to the main event. A relatively plane area (c. 50 m wide) at the toe of the scarp, completes this area and is followed by Section B (Fig. 3).

With a width of approximately 100 m, Section B represents the narrowest part of the landslide. The western flank is steep and clearly defined, whereby the eastern flank shows a more rugged topography. The flanks clearly trace the upper part of the slide mass. On the western flank there is a drainage structure, which was formed after the event. Boulders mixed with fine material form a heterogeneous mass. On the eastern part the upper slide mass merges into the main slide body (Section D), whereas the movement was blocked on the western side. In general the main slide body forms a homogenous and remarkable flat area inside the landslide complex. A secondary mass movement (Section C), which is located in the central part of the left flank, reaches the main slide mass of Section D. Of great interest is the fact, that in this specific region, a former pond with a length of 100 m was present in the described contact zone, before the landslide event occurred.

To learn more on the supposed integration of the pond of the slide complex, geophysical investigations were carried out in this section.

Section D (Fig. 3) can be described as the main element of the landslide, with a maximum width of about 300 m. A newly constructed road is incised in the slide mass here. Indicators for an en-bloc movement of this section are undisturbed agricultural areas (coffee plantation) and soil profiles.

A small creek, which originally was affected by the slide mass, is located at the external part of the landslide toe (Section E).

Geophysical Survey

ERT and SRT

Direct exploration methods, such as drilling or trenching, provide true information on the subsurface configuration of slide masses. However, these methods are comparatively time consuming, expensive, and the retrieved information is limited to specific sampling points. On the other hand, the cost-efficient investigation of the subsurface by geophysical methods has been recognized as a suitable methodology for the indirect and spatially continuous characterization of the sedimentological-hydrological properties in landslide areas (e.g. Jongmans et al. 2009; Hibert et al. 2012). For the geophysical characterizing of the El Capulín landslide, electrical resistivity tomography (ERT) and seismic refraction tomography (SRT) have been used in joint application. A combination that can contribute reliable information on the subsurface configuration of slide masses as well as the detection of subterranean hydrological situation (e.g. Perrone et al. 2014; Giocoli et al. 2015).

For geoelectrical measurements two electrodes firmly attached to the ground are used to inject an electric current into the subsoil, while a second pair of electrodes is used to measure the resulting voltage. From the voltage-to-current ratio and taking the geometric configuration of the electrodes into consideration, the geometric distribution of electric resistivities in the subsoil can be reconstructed. ERT measurements were carried out using a Syscal Pro Switch 48 device (IRIS Instruments). 2D electrical resistivity sections were reconstructed using the software RES2DINV (Loke and Barker 1995), which is based on a smoothness—constrained least-squares inversion algorithm, using robust inversion (L1-norm). Outliers in the data, here defined as those measurements with anomalously low (<15 Ωm) or high (>300 Ωm for profiles 1–3 and >1000 Ωm for profile 4) apparent resistivities, were removed from the data sets prior to inversion.

Seismic refraction measurements are based on the propagation velocity of refracted seismic waves (P-waves) between a seismic source and a receiver (geophone). To achieve a 2D model of the subsurface, geophones are installed along a line and seismic signals are created at a number of shot points along the array. Measurements were conducted using a 24-channel seismograph (Geode, Geometrics). To generate the seismic signal a 5 kg sledgehammer hitting a steel strike plate was used. For data processing (picking of first arrivals) and tomographic inversion the software package SeisImager/2D (Geometrics) was used.

Aim of the geophysical approach was to get an insight into the thickness of the slide mass (Transect T1), to detect potential slip surfaces (Transect T2), and to characterize slope conditions before the mass movement occurred (Transect T4) (Fig. 3). In combination with geomorphological investigations these data contribute to an enhanced understanding of the process dynamics.

The setup of geophysical arrays has been chosen site-specific. Information on location and setup of geophysical arrays is given in Fig. 3 and Table 1.

Geophysical Results

Figures 4 and 5 illustrate the electrical resistivity and P-wave velocity sections obtained from the combined ERT and SRT survey as well as subsurface models derived from these data. A good agreement of ERT and SRT can be observed along all four profiles. High resistivities (>50 Ωm) and low p-wave velocities (<1000 m/s) generally coincide and can be attributed to the slide mass, which is underlain by a more conductive (resistivities <20 Ωm) and more consolidated (p-wave velocities >1500 m/s) unit. While the probable slip plane or the old land surface is best traced by the large p-wave velocity gradient between the two units, the electrical resistivity sections contribute additional information on smaller-scale heterogeneities within the slide mass.

Transect T1 (Fig. 4) crosses the landslide from WNW to ESE. While the SRT profile only covered the first 120 m, the much longer ERT profile reached the ESE end of the main slide body and even covered a part of the right flank. The pronounced p-wave velocity gradient encountered between

Table 1 Information on the setup of ERT and SRT arrays of transect T1, T2 and T4

	T1		T2		T4	
	ERT	SRT	ERT	SRT	ERT	SRT
Spread length	355	220	235	115	235	215
Electrode/geophone spacing	5	5	5	5	5	5
Roll along	+	+	–	–	–	+
# of electrodes/geophones	96	48	48	24	48	48
Array type/# of shot points	We/DD	40	We/DD	15	We/DD	26
# of stacks	3	10	3	10	3	10

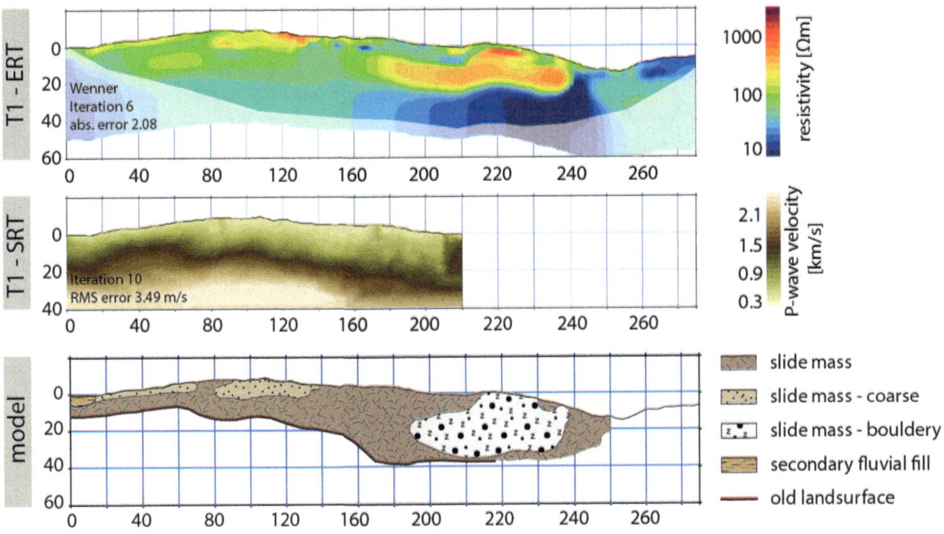

Fig. 4 Transect 1. Combined ERT and SRT survey and subsurface model

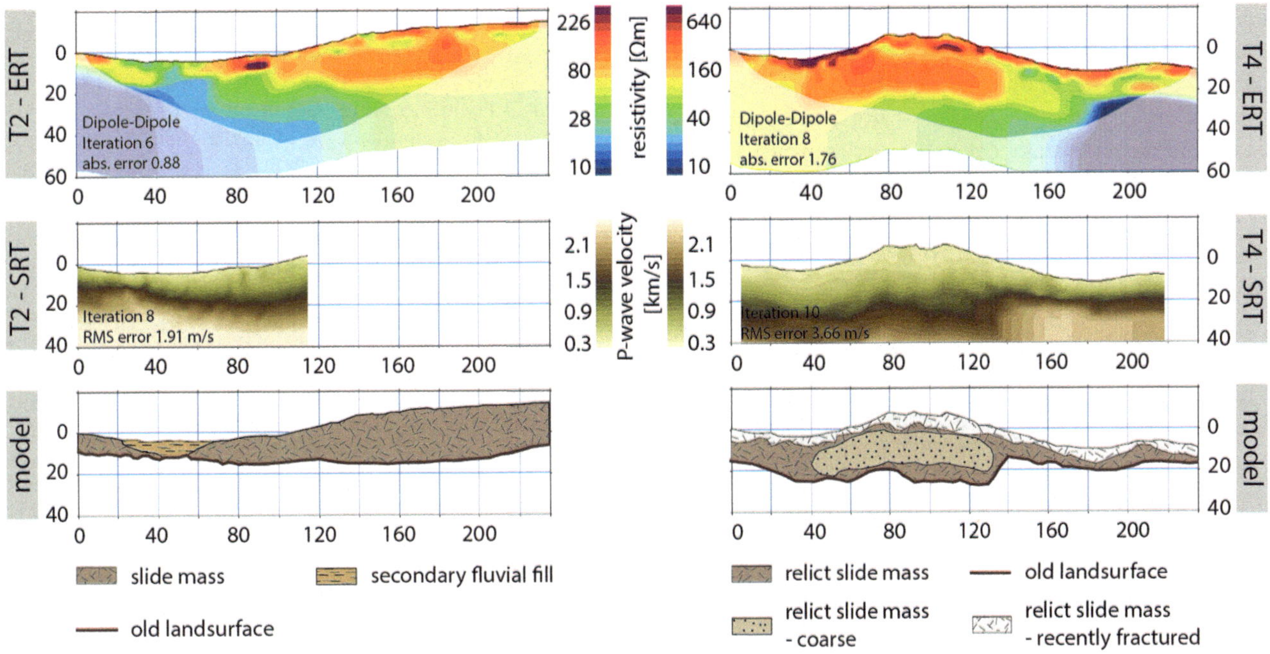

Fig. 5 Transect 2 and 4. Combined ERT and SRT survey and subsurface model

ca. 10 and 30 m depth can be interpreted as an old land surface. Due to low data coverage towards the depth >30 m, information on the substrate below the old land surface is sparse. However, it is likely that consolidated slope material, or strongly weathered calcareous shale, which can be found in the vicinity of the transects' end, is overlain by the landslide deposit. While the slide mass appears quite homogeneous in the seismic section, the electrical resistivity section reveals heterogeneities within the landslide. High resistivities point to coarse or bouldery material and the presence of rock fragments within the landslide. Between horizontal distances from 190 to 240 m the resistivity tomogram shows a larger lens of high resistivities (300–2000 Ωm), which is characterized by high p-wave velocities of up to 1500 m/s in the fringe of the seismic tomogram. This can be interpreted as a relatively massive part of the slide mass that has been moved en-bloc.

Transect T2 (Fig. 5) is located in the contact area, between the main landslide deposit (Section D) and the secondary movement (Section C). Transect T2 crosses the plain, starting from the secondary mass, reaching up to mass of the main process. Both, p-wave velocity and resistivity section indicate a reduced thickness of the slide mass in the vicinity of the former pond, i.e. between 20 and 60 m along the profile. Low resistivities in this area indicate an accumulation of moist and fine-grained material in this topographic position. The steep p-wave velocity gradient points to a shallow refractor at about 4 m depth. While the inferred slip plane is nearly horizontal along the entire profile; from 60 to 240 m, the thickness of the supposed slide mass increases significantly with the topography to a maximum of about 30 m. The much narrower range of electrical resistivities observed within the slide mass indicates a rather homogeneous composition.

Transect T4 (Fig. 5) is located above the scarp and crosses various extension cracks with a 90° angle. Although the overall resistivities are higher than those observed along the transects located on the accumulation area, both, p-wave velocity and resistivity sections again indicate a rough separation into the same two main units observed before. Indeed, the main scarp consists of relatively loose, shattered (mixed fine-grained and coarse) sediments that correspond to the geophysical observations. Furthermore the subsurface is expected to be drier above the scarp, compared to the material of the slide mass, due to a better drainage. This fact can explain the overall increase of electrical resistivity.

The geophysical results indicate the presence of an earlier landslide located on top of the semi-consolidated mudstone that serves as slip surface.

Conclusion

In general terms, the comparison of SRT and ERT results shows that both methods consistently identify the lithological contact between the slide mass and the underlying unit. While this contact is traced a bit clearer by the SRT results, the ERT sections provide interesting insights into the internal configuration of the landslide body. It is worth mentioning that the usually weathered and partly disintegrated material of the landslide body generally stands out with a lower resistivity than the underlying geological units (Perrone et al. 2014). In this respect, the El Capulín landslide represents an atypical case with an increased resistivity in the landslide body.

To support the results of the geophysical investigations, a drilling campaign is planned. In a further step the results will be used for susceptibility modeling.

Acknowledgements The authors would like to thank the German Research Foundation (DFG) (Te295/19-1) and the German Academic Exchange Service (DAAD) for funding. Furthermore, we thank the Civil Protection of Veracruz, for their cooperation. We also would like to express our appreciation to the landowners and the community of El Capulín for their help. Special thanks goes to Dr. Elizabeth Solleiro Rebolledo (UNAM) for scientific and personal support.

References

Alcántara-Ayala I (2008) On the historical account of disastrous lands in Mexico: the challenge of risk management and disaster prevention. Adv Geosci 14:159–164

Demant A (1978) Características del Eje Neovolcánico Transmexicano y sus problemas de interpretación. Revista Instituto de Geología. 2:172–187

Fell R, Hartford D (1997) Landslide risk assessment. In: Cruden D, Fell R (eds) Proceedings of the international workshop on landslide risk assessment, Honolulu, Hawaii, USA, 19–21 Feb 1997. A.A. Balkema Rotterdam, 371p. ISBN 978-9-054-10914-3

Ferrari L, Tagami T, Eguchi M, Orozco-Esquivel MT, Petron CM, Jacobo-Albarrán J, López-Martínez M (2005) Geology, geochronology and tectonic setting of late Cenozoic volcanism along the southwestern Gulf of Mexico: The Eastern Alkaline Province revisited. J Volcanol Geoth Res 146(4):284–306

Ferrari L, Orozco-Esquivel T, Manea V, Manea M (2012) The dynamic history of the Trans-Mexican Volcanic Belt and the Mexico subduction zone. Tectonophysics 522–523:122–149

Giocoli A, Stabile TA, Adurno I, Perrone A, Gallipoli MR, Gueguen E, Norelli E, Piscitelli S (2015) Geological and geophysical characterization of the southeastern side of the High Agri Valley (southern Apennines, Italy). Nat Hazards Earth Syst Sci 15(2):315–323

Gómez-Tuena A (2002) Control Temporal del Magmatismo de Subducción en la Porción Oriental de la Faja Volcánica Transmexicana: Caracterización del Manto, Componentes en Subducción y Contaminación Cortical. Ph.D. thesis, National Autonomous University of Mexico, Mexico City, Mexico

Gómez-Tuena A, LaGatta AB, Langmuir CH, Goldstein SL, Doherty L, Ortega-Gutiérrez F, Carrasco-Núñez G (2003) Temporal control of subduction magmatism in the eastern Trans-Mexican Volcanic Belt: Mantle sources, slab contributions, and crustal contamination. Geochem Geophys Geosyst 4–8:1–33

Gómez-Tuena A, Orozco-Esquivel MT, Ferrari L (2005) Petrogénesis ígnea de la Faja Volcánica Transmexicana. Boletín de la Sociedad Geológica Mexicana. Volumen Conmemorativo del Centenario, Tomo LVII. 3:227–283

Guzzetti F (2000) Landslide fatalities and the evaluation of landslide risk in Italy. Eng Geol 58:89–107

Guzzetti F, Carrara A, Cardinali M, Reichenabch P (1999) Landslide hazard evaluation: a review of current techniques and their application in a multi-scale-study, Central Italy. Geomorphology 31:181–216

Heimsath AM, Korup O (2012) Quantifying rates and processes of landscape evolution. Earth Surf Proc Land 37:249–251

Hibert C, Grandjean G, Bitri A, Travelletti J, Malet JP (2012) Characterizing lands through geophysical data fusion: example of the La Valette landslide (France). Eng Geol 128:23–29

Jongmans D, Bièvre G, Renalier F, Schwartz S, Beaurez N, Orengo Y (2009) Geophysical investigation of a large landslide in glaciolacustrine clays in the Trièves area (French Alps). Eng Geol 109:45–56

Loke MH, Barker RD (1995) Least-squares deconvolution of apparent resistivity pseudosections. Geophysics 60(6):1682–1690

López-Infanzón M (1991) Petrologic study of the volcanic rocks in the Chiconquiaco-Palma Sola area, central Veracuz, Mexico. MS thesis, Tulane University of Louisiana, New Orleans, USA
Margottini C, Canuti P, Sassa K (eds) (2013) Landslide science and practice. Volume 7: Social and economic impact and policies. Springer, Berlin, 333p. ISBN 978-3-642-31312-7
Morales-Barrera W, Rodríguez-Elizarrarás SR (2014) La Gestión del Riesgo por deslizamientos de laderas en el estado de Veracruz durante 2013. Gobierno del Estado de Veracruz. Secretaría de Protección Civil, 113p. ISBN 978-607-7527-90-9
Nixon GT (1982) The relationship between Quaternary volcanism in central Mexico and the seismicity and structure of subducted ocean lithosphere. Geol Soc Am Bull 93:514–523
Pardo M, Suárez G (1995) Shape of the subducted Rivera and Cocos plates in southern Mexico: Seismic and tectonic implications. J Geophys Res Solid Earth 100(B7):357–373
Perrone A, Lapenna V, Piscitelli S (2014) Electrical resistivity tomography technique for landslide investigation: a review. Earth Sci Rev 135:65–82

4thWLF 2017
4th World Landslide Forum
LJUBLJANA SLOVENIA EU

Cut Slope Icing Formation Mechanism and Its Influence on Slope Stability in Periglacial Area

Ying Guo, Wei Shan, Zhaoguang Hu, and Hua Jiang

Abstract

Understanding the formation and distinctive conditions that contribute to icing in cut slopes are needed to mitigate it for highway engineering. Using the K162 cut slope of the Bei'an-Heihe Expressway as a study site, we conducted field surveys, geological exploration, field monitoring, laboratory tests and numerical simulations to carry out an integrated study on the icing formation mechanisms and its influence on the slope stability. Research results show that: the surface unconsolidated Quaternary sediment and Tertiary sandstone provide passage for atmospheric precipitation infiltration; but underlying mudstone forms an aquiclude. Phreatic water forms in the loose overburden after infiltration. As the freezing front thickens, the phreatic aquifer thins and becomes pressurized. Slope cutting has exposed the phreatic aquifer. When the excess pore water pressure exceeds the strength of surface material, the pressurized water flows out of the slope, and freezes, forms icing. In the spring melt period, surface icing and shallow seasonal frozen soil melt completely, water infiltrates into the slope; but meltwater is blocked by the unfrozen soil in infiltrating process, accumulates on the interface between melted and frozen layers, increasing the water content at the mudstone interface. The mudstone reaches a saturated state, and its shear strength decreases, and forms a potential rupture surface.

Keywords

Landslide • Icing • Permafrost • Cutting slope • Stability

Y. Guo · W. Shan (✉) · Z. Hu · H. Jiang
Institute of Cold Regions Science and Engineering, Northeast Forestry University, No. 26 Hexing Road, Harbin, 150040, China
e-mail: shanwei456@163.com

Y. Guo
e-mail: samesongs@163.com

Z. Hu
e-mail: huzhaoguang008@163.com

H. Jiang
e-mail: jianghua3433@163.com

M. Mikoš et al. (eds.), *Advancing Culture of Living with Landslides*,
DOI 10.1007/978-3-319-53483-1_21

Introduction

The Bei'an-Heihe Expressway was built by widening the original second-class highway in northeastern China. K159–K184 Sections of it traverses the Lesser Khingan Mountains. Special geological, topographical, hydrological, and climatic conditions, contribute to geological hazards distributed along the route (Sun et al. 2008; Meng et al. 2001), including island-like permafrost, landslides and icing, posing serious challenges to road widening. This paper will talk about icing hazard.

Icing [also named naled (Russian) or aufeis (German)] is a distinctive cryogenic phenomenon in high latitude and high altitude deep freezing or permafrost area (Zhang et al. 2000; Li 2010; Wu et al. 2001; Iwahana et al. 2012). Depending on the source, icing can be divided into three types: river icing, spring icing and phreatic water icing (Hu and Pollard 1997). Icing that occurred on the highway, originated from groundwater flowing to the surface in the cold winter, accumulating, and covering the slope and pavement/road surface. Sun reflection from bright accumulated ice can lead to traffic jams and serious accidents. Icing melt water tends to infiltrate the road subgrade sometimes causing argilation and slope failures during spring (Wu et al. 2003, 2007; Huang et al. 2010; Seppälä 1999). Researchers have made some studies on icing. In the UK, Worsley and Gurney (1996) reported the influence of geomorphology and hydrogeology condition on the formation of pin go and icing in Karup Valley area. Hodgkins et al. (2004) described the morphology and formation process of a proglacial icing in Svalbard. In Canada, Hu and Pollard (1997) analyzed the spreading and thickening of icing layers and their relationship with hydrologic, topographic and climatic variables. French and Guglielmin (2000) discussed the relationship between icing and perennially frozen lakes in Northern Victoria land. Pollard (2005) utilized freezing-point depression experiments to explain icing hydrology and the spatial pattern of icing formation in Axel Heiberg Island. In the USA, Kane (1981) examined the role of streambank pore water pressure on icing thickness in central Alaska. Streitz and Ettema (2002) analyzed the influence of gravity and wind drag on aufeis morphologies through indoor wind tunnel tests. Vinson and Lofgren (2003), using climatologically records, proposed manufacturing snow in late fall or early winter to insulate water sources of icing on the Denali Park access road. Yoshikawa et al. (2007) employed remote sensing studies and field hydro-meteorological and geophysical investigations to characterize several aufeis fields in the Brooks Range in USA. Callegary et al. (2013) documented that aufeis is controlled by a number of factors including geology and elevation, and can cause overbank flooding and backwater effects in Alaska. In Finland, Saarelainen and Vaskelainen (1988) outlined a method to prevent icing where the groundwater should be collected in drains at a frost-free depth before reaching the road and then it should be drained into the culvert. Seppälä (1999) recommended several water drainage techniques along the road embankment to minimize the icing difficulties. In China, Zhang et al. (2000) carried out a study on the formation law of icing under the action of freezing pressure. Wu et al. (2003) proposed adopting a blind ditch, ice containing pit, and ice retaining wall to mitigate the influence of icing on a highway. Yu et al. (2005) analyzed the different ways in which flow ice could form; they also suggested new composite measures to control the icing problem on a road in the Great Hinggan Mountains forest region of China.

Icing on cut slopes belongs to an artificial geological hazard, it has unique laws of formation, development and disappearance. A full understanding of the formation and behavior of icing in cut slopes is necessary to solve the problems it brought to the highway (Wang and Liu 2006). This article took the K162 section cut slope of Bei'an-Heihe Expressway as the main research object, employed field survey, geological exploration, field monitoring, laboratory test and numerical simulation to carry an integrated study on the cut slope icing formation mechanism and its influence on the slope stability.

Study Area

The study area is located in the Northeast China, in a zone with isolated patches of permafrost (Fig. 1). The area has a continental monsoon climate, with a short, rapidly warming spring, tepid and rainy summers, the short and rapidly cooling autumns, and long, cold winters. The average annual precipitation is 530–552 mm, rainfall concentrates in July to September, accounts for about 61–67% of total annual precipitation (Local records of Heilongjiang Province). The annual average temperature is −0.6 °C, the lowest temperature is −48.1 °C, the highest temperature is 35.2 °C. The ground begins to freeze in October and melt in April. The annual average freezing index is 3000–3300 Cxd, the frost-free period is about 96 d. The ground surface of the region can form seasonal frozen soil, and the time of ground reaching its maximum frozen depth is late May, where the maximum seasonal freezing depth is 2.26–2.67 m. There are many isolated permafrost patches (Shan et al. 2012; Guo

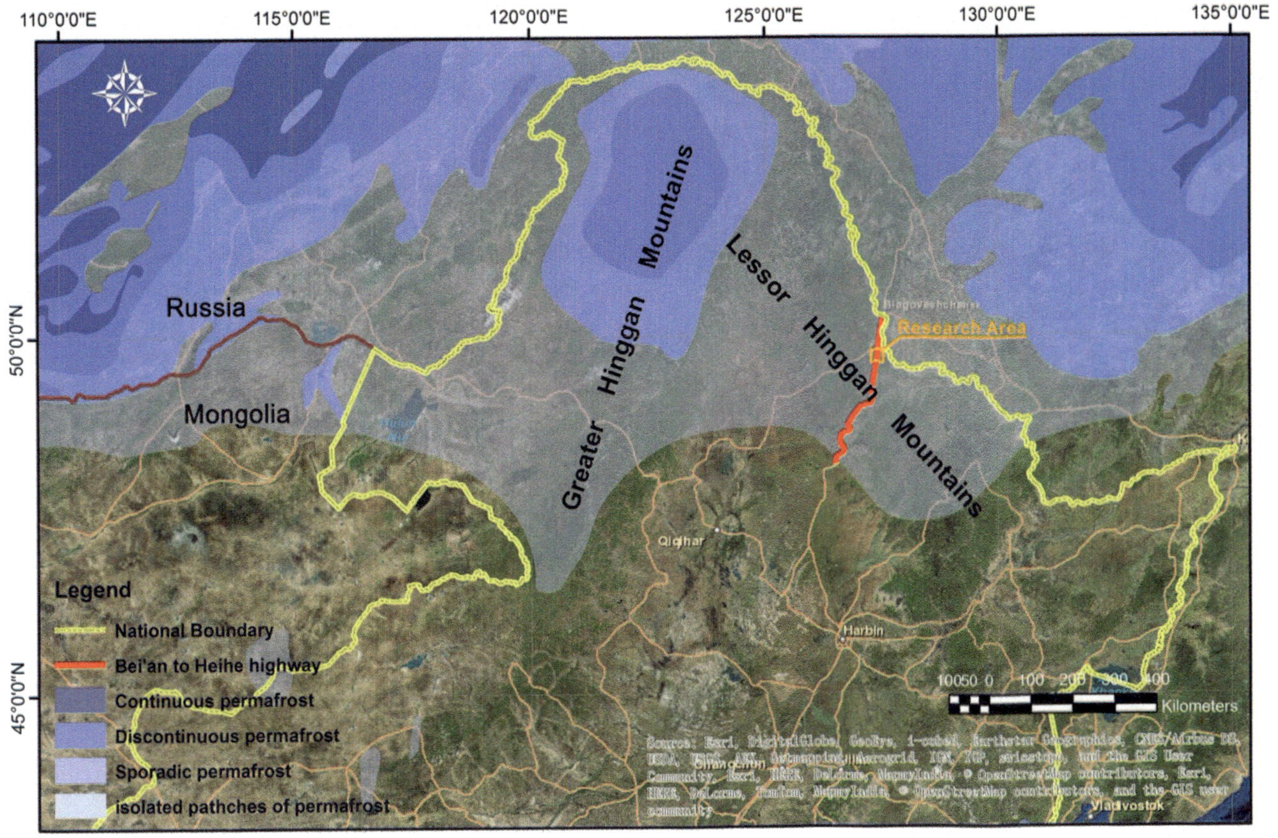

Fig. 1 Geographical location of the study area

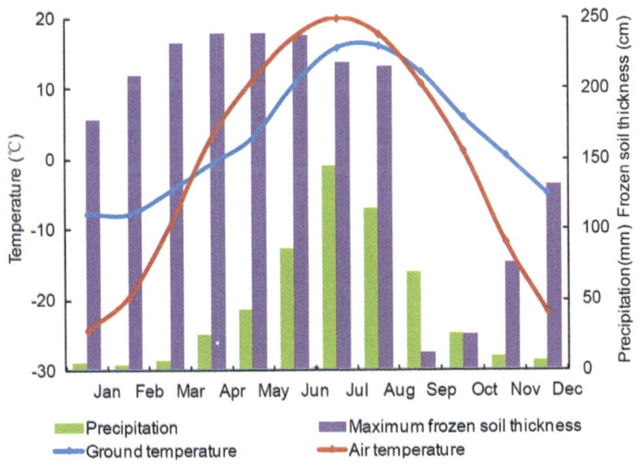

Fig. 2 Monthly average air temperature/ground temperature precipitation/maximum frozen depth (1971–2000, Sunwu County meteorological station)

et al. 2014). Figure 2 shows the 30 year monthly average air temperature, ground temperature (40 cm depth), precipitation and maximum frozen soil thickness curve from 1971–2000 at the Sunwu County meteorological station (14 km away from study area).

The geology of the study area is influenced by the "Wuyun-Jieya new rift zone". Surface exposed stratums contain the Upper Cretaceous Nenjiang formation (mainly shale), Tertiary Pliocene series of Sunwu formation (sedimentary rock having basalt), Quaternary Holocene series of modern river alluvium layers (mainly is limestone and sandstone). Upper rock-bed weakly cemented and weathering resistant. Individual strata form aquicludes such that groundwater is divided into Cretaceous, Tertiary and Quaternary pore water (Jiang and Shan 2012).

Topographical and Geological Condition of the K162 Cut Slope

K162 section cut slope of The Bei'an-Heihe Expressway is located in the central region of low hills, its upper slope has a larger catchment area (Fig. 3); the upper slope surface gradient is 3–5°, and is covered by farmland. These

Fig. 3 Topography of K162 cutting slope of the Highway (40 m wide)

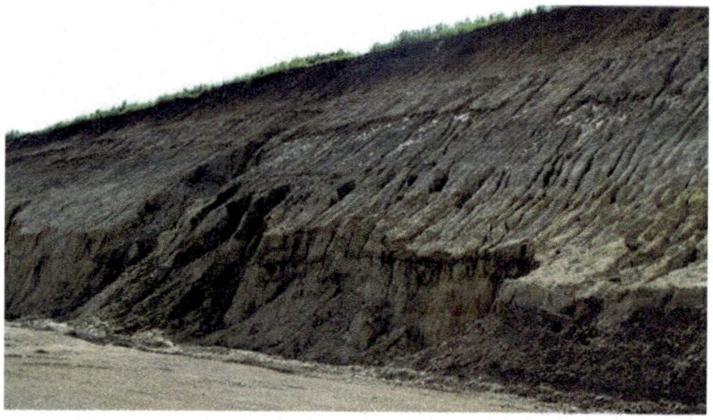

Fig. 4 Groundwater out flowed and eroded the cutting slope

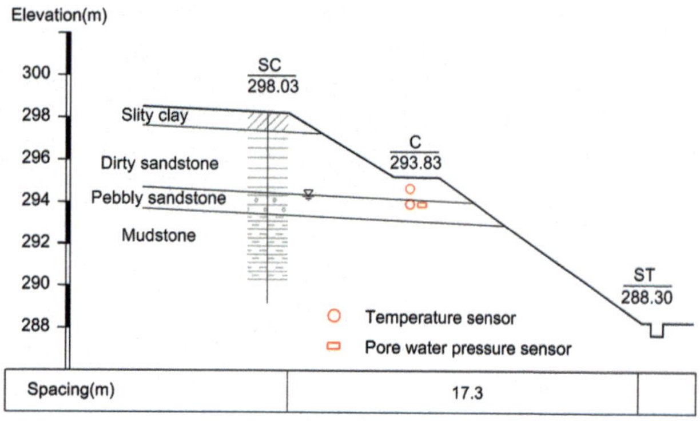

Fig. 5 Geological profile of K162 cutting slope (M1 monitoring point)

factors combine to retain atmospheric precipitation on the slope for a long time. In 2009, there is the expansion of the highway widening and on the left slope of the mechanical excavation, the slope after excavation was shown in Fig. 4. In early February 2010, the aufeis occur largely in the cutting slope excavated, which is shown in Fig. 9.

In October 2010, at the second floor of the left slope and near the toe of the slope, two drilling borehole was set (M1, M2), the distance between M1 and M2 is about 40 m, and the surface elevation of M1 is lower than M2, so the groundwater burial depth at M1 is shallower than M2, the position is shown in Fig. 5. The sensors of soil temperature, water content, pore water pressure was installed in the borehole, the position is shown in Fig. 3. From surface to depth, the layer is slity clay, dirty sandstone (muddy sandstone), pebbly sandstone and mudstone seperately (Fig. 5 and Table 1). The particle size distribution curve of soil are shown in Fig. 6.

It could be seen that the surface rock and soil has stronger infiltration ability and is more permeable than the deeper mudstone. When the atmospheric precipitation penetrates downward through surface cracks and soil, it accumulates in the sandstones and forms phreatic water. Abundant rainfall feeds the relatively shallow groundwater, which is exposed in the cut slope. The distance between the ground and phreatic surfaces is no more than 3.4 m, while the regional largest seasonal freezing depth is up to 2.7 m. This is the importent factor for icing occurence.

Monitoring Data Analysis

In order to investigate icing or landslides occurence in the K162 cut slope, we examined temperature, pore water pressure, and soil moisture at two monitoring points (M1 and M2) in the cut slope (Fig. 3). We began to collect monitoring data in the middle of November (Figs. 7 and 8).

When the monitoring began, the region had entered the freezing period. As the air temperature decreased, the freezing front in the slope gradually deepened. At M1, on 25th February of 2011, the freezing front reached the

Table 1 Soil physical indicators in monitoring section of the slope

Layer	Category	Rock types	Thickness (m)	Permeability coefficient (cm/s)
1.	Quaternary loose layer	Silty clay	0.8–1.0	3.84×10^{-8}
2.	Quaternary loose layer	Muddy sandstone	2.5–3.0	6.49×10^{-6}
3.	Tertiary sandstones	Pebbly sandstone	1.0–1.2	2.11×10^{-5}
4.	Cretaceous mudstone	Mudstone	>5	2.09×10^{-8}

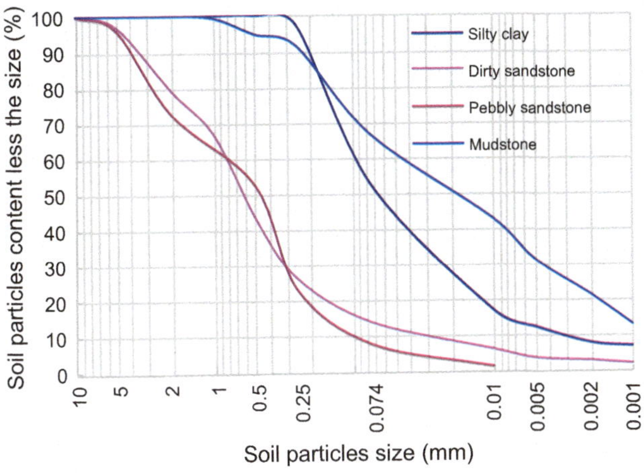

Fig. 6 Particle size distribution curve of rocks and soils

groundwater level (elevation: 295 m), because the layer beneath the ground water level has weak infiltration ability, so the water was blocked there, when the water began to freeze from above, the unfrozen water began to be pressurised and the pressure increased rapidly (Fig. 7b When the pressurized water breached the frozen ground and flowed out of the slope (through pebbly sandstone), the pore preassure dropped gradually. The outflow water then was froze on the slope resulting in the development of icing (Fig. 9). At the site M2, the groundwater level was deeper (elevation: 294 m) than at the M1. Due to this, the freezing front did not reach down to the groundwater level, and thus the pore water pressure remained more or less constant during the freezing period (Fig. 8b), and hence, no icing on the surface occured. In years with larger rainfall or colder temperature, the groundwater level could be shallower or freezing depth deeper, causing the icing occurence at M2 site or other places of the cut slope.

About on 2011.3.30, the region had entered the spring melt period, surface ice and seasonal frozen soil began to melt. Melt water infiltrated into the slope. On the 150th day (2011.4.18), icing and shallow seasonal frozen soil had melted completely, and a lot of water infiltrated into the slope. However, the deep seasonal frozen soil remained frozen, so that meltwater infiltration was blocked (Figs. 7c and 8a). The trapped water, accumulated on the interface between the melted strata and frozen layer, caused the water content to increase rapidly in this surface layer and reach a saturated water content, causing landslide occurence on the cut slope (Fig. 10).

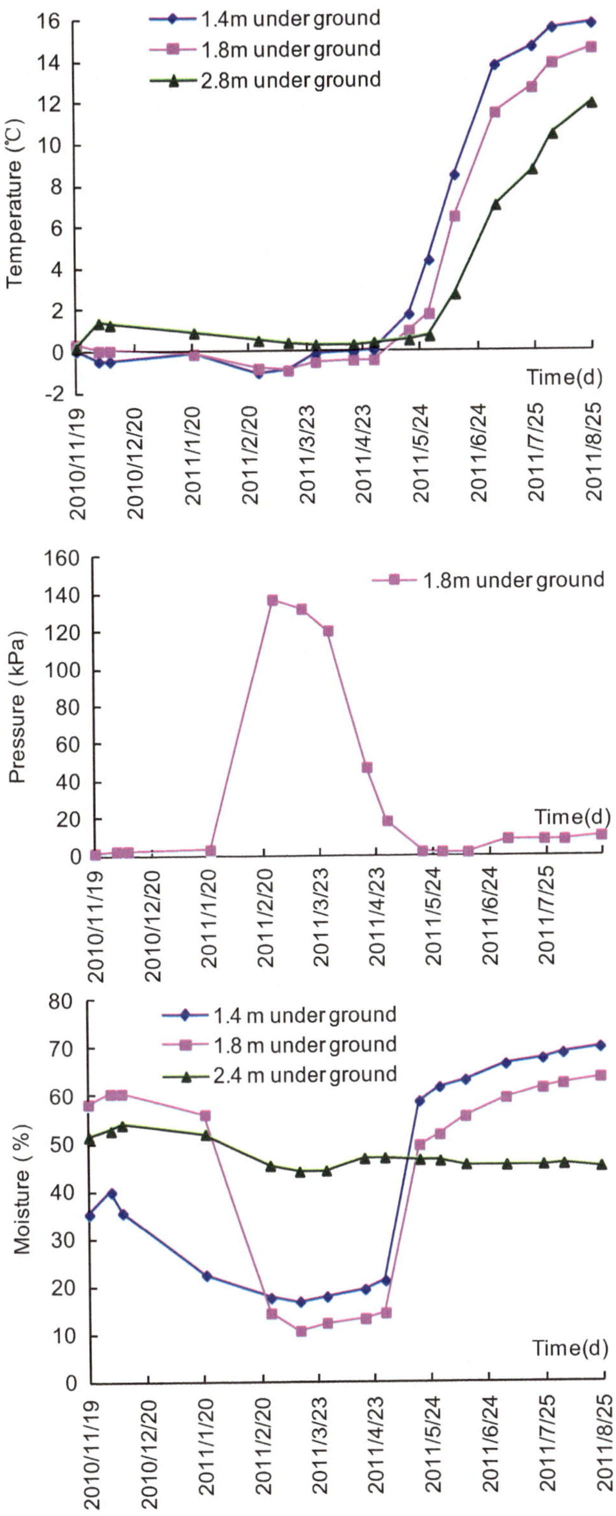

Fig. 7 M1 (1.8 m depth) monitoring point **a** underground temperature **b** pore water pressure **c** moisture changes with time

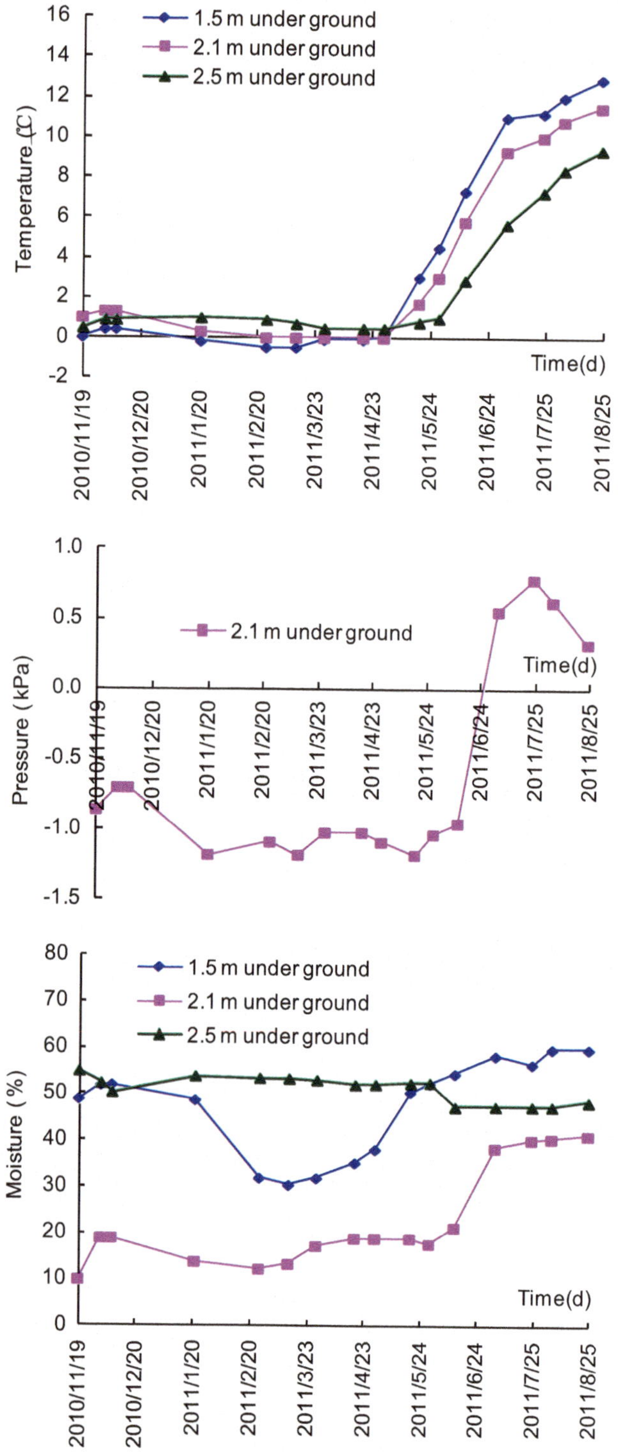

Fig. 8 M2 (2.1 m depth) monitoring point **a** underground temperature **b** pore water pressure **c** moisture changes with time

Fig. 9 Icing on the cut slope during the freezing period (February 2010) as well as location of this icing point

Fig. 10 Landslide on the cut slope (2012.4.23)

Conclusions

Through an integrated study on the cut slope, icing formation mechanism and its influence on slope stability in periglacial area, some conclusions can be drawn.

1. The surface covered by clay with cracks provides passage for infiltration of atmospheric precipitation, but underlying mudstone forms an aquiclude. Atmospheric precipitation forms phreatic water in the loose overburden after infiltration, but expressway cut slopes causes phreatic aquifer exposure and seepage. During the freezing period, the freezing front in the slope deepens gradually, thinning the aquifer and pressurizing the phreatic water. When the excess pore water pressure exceeds the strength the slope materials, pressurized water flows out of the slope through the weak point, and freezes, forming icing on the slope surface.

2. In the spring melt period, as slope surface icing and shallow seasonal frozen soil melt completely, a lot of water infiltrates into the slope. The melting water is blocked by the unfrozen soil in the infiltrating process, accumulates on the interface between melted (near surface) and frozen (deeper) layers, increasing the water content of the mudstone interface. The mudstone reaches a saturated state, and its shear strength sharply decreases, forming a potential rupture surface.

Acknowledgements We thank the science and technology research project of Heilongjiang Provincial Education Department (12533017) and Heilongjiang Postdoctoral Fund (LBH-Z14024) for funding support.

References

Callegary JB, Kikuchi CP, Koch JC, Lilly MR, Leake SAm (2013) Review: groundwater in Alaska (USA). Hydrogeol J 21(1):25–39

French HM, Guglielmin M (2000) Frozen ground phenomena in the vicinity of Terra Nova Bay, Northern Victoria Land, Antarctica: a preliminary report. Geografiska Annaler. Series A. Phys Geogr 82 (4):513–526

Guo Y, Shan W, Jiang H, Sun YY, Zhang C C (2014) The impact of freeze-thaw on the stability of soil cut slope in high-latitude frozen regions. In: Shan W (ed) Landslides in cold regions in context of climate change, Springer International Publishing (ISBN 1431-6250), p 85

Hodgkins R, Tranter M, Dowdeswell JA (2004) The characteristics and formation of a high-arctic proglacial icing Geografiska Annaler: Series A. Phys Geogr 86(3):265–275

Huang MK (2010) Study on chain-style mechanism and chain-breaking method of road disaster in cold regions. Disaster Adv 3(4):166–169

Hu XG, Pollard WH (1997) Ground icing formation: experimental and statistical analyses of the overflow process. Permafrost Periglac Process 8(2):217–235

Iwahana G, Fukui K, Mikhailov N, Ostanin O, Fujii Y (2012) Internal structure of a lithalsa in the Akkol Valley, Russian Altai Mountains. Permafrost Periglac Process 23(2):107–118

Jiang H, Shan W (2012) Formation mechanism and stability analysis of Bei'an-Heihe Expressway expansion project K178 landslide. Adv Mater Res 368–373:953–958

Kane DL (1981) Physical mechanics of aufeis growth. Can J Civ Eng 8 (2):186–195

Li MY (2010) Salivary flow ice formation mechanism in highway engineering. Commun Sci Technol Heilongjiang 33(10):12

Meng FS, Liu JP, Liu YZ (2001) Design principles and frost damage characteristics of frozen soil roadbed along the Heihe-Bei'an highway. J Glaciol Geocryol 23(3):307–311

Pollard WH (2005) Icing processes associated with high arctic perennial springs, Axel Heiberg Island, Nunavut, Canada. Permafrost and Periglacial Process 16(1):51–68

Saarelainen S, Vaskelainen J (1988) Problems of arctic road construction and maintenance in Finland. In: Proceedings of the 5th international conference on Permafrost, 2–5 August 1988. Norway, pp 1466–1491

Seppälä M (1999) Geomorphological aspects of road construction in a cold environment, Finland. Geomorphol 31(4):65–91

Shan W, Jiang H, Hu ZG (2012) Island permafrost degrading process and deformation characteristics of expressway widen subgrade foundation. Disaster Adv 5(4):827–832

Streitz JT, Ettema R (2002) Observations from an aufeis windtunnel. Cold Reg Sci Technol 34(2):85–96

Sun QH, Xu HW, Zhang YQ, Wang JM (2008) Application of high density resistivity method in highway road landslide prediction. J Guizhou University of Technol 37(6):101–105

Vinson TS, Lofgren D (2003) Denali Park access road icing problems and mitigation options. In: Proceedings of the 8th international conference on Permafrost, 21–25 July 2003. Zürich, Switzerland, pp 1189–1194

Wang Y, Liu XP (2006) Formation characteristics, condition and classification of highway salivary flow ice. Commun Sci Technol Heilongjiang 29(6):47–48

Worsley P, Gurney SD (1996) Geomorphology and hydrogeological significance of the Holocene pingos in the Karup Valley area, Traill Island, northern east Greenland. J Quat Sci 11(3):249–262

Wu H, Wang L, Zhao K (2003) Formation and control of highway salivary flow ice in cold mountain region. Commun Sci Technol Heilongjiang 26(4):32–35

Wu QB, Zhu YL, Shi B (2001) Study of frozen soil environment relating to engineering activities. J Glaciol Geocryol 23(2):200–207

Wu ZJ, Zhang LX, Ma W (2007) Influence of soil's cryogenic course on deformation of roadbed in permafrost region of Qinghai-Tibet Railway. Rock and Soil Mech 28(7):1477–1483

Yoshikawa K, Hinzman LD, Kane DL (2007) Spring and aufeis (icing) hydrology in Brooks Range, Alaska. J Geophys Res: Biogeosci (2005–2012) 112(4):1–14

Yu WB, Lai YM, Bai WL, Zhang XF, Sh Zhuang D, Li QH, Wang JW (2005) Icing problems on road in Da Hinggangling forest region and prevention measures. Cold Reg Sci Technol 42(1):79–88

Zhang BL, Liu YY, Yang M (2000) Control measures of salivary flow ice in highway engineering. For Eng 16(4):46–47

Climate Change Driving Greater Slope Instability in the Central Andes

Stella Maris Moreiras and Ivan Pablo Vergara Dal Pont

Abstract

Global climate change linked to meso-scale environment modifications such as regional above average precipitations, stronger El Niño-ENSO warm phase, global warming, permafrost degradation, and glacier retreatment could promote slope instability. However, which of these mechanisms is leading landslide activity in the high mountain landscape of Central Andes is still uncertain. Otherwise, changes of landslide features as consequence of these climate drivers is rare approached so proposal of effective preventive measures is not viable. The main concern of this research is to elucidate whether climate change is driving more frequent slope instability in the Central Andes. We focus on two key questions of our research: (1) Which landslide features are changing due to climate change? and (2) Which climate change mechanisms are certainly forcing landslides generation in the Central Andes? Our findings explain how the global environment change is shifting slope behavior in the Central Andes increasing landslide frequency and intensity, modifying landslide spatial distribution, shifting initial points of slope instability to higher topography, and generating more complex landslides. Main explanation for this shifting on slope instability behavior is intensified summer rainfall and global warming.

Keywords

Global warming • 0º Isotherm • Permafrost degradation • Above mean rainfall

Introduction

The influence of climate change on slope instability in the main mountain ranges as the Alps and the Himalayas has been referred worldwide (Evan and Clague 1994; Crozier 1997; Beniston 2003; Gruber and Haeberli 2007; Clague et al. 2012); still it has not been properly evaluated in the Central Andes. Global warming, climbing of the zero degree isotherm (Haeberli and Gruber 2009; Huggel et al. 2012), degradation of permafrost (Gruber and Haeberli 2007; Harris et al. 2009; Huggel 2009; Stoffel and Huggel 2012) and above average precipitations (Moreiras 2006; Moreiras et al. 2012a) are associated with current climate variability promoting the occurrence of landslides in the broad sense. Hence, climate change may promote landslides in many ways intensifying precipitations, shifting temperatures, increasing adverse effect of ENSO phenomenon among others. Which of these mechanisms are certainly forcing slope instability in the Central Andes is uncertain as well as feeding mechanisms among them. The main concern of this research is to elucidate how climate change is increasing landslide activity in this high mountain topography involving the highest peak of America (Aconcagua peak 6958 m). We focus on two key questions: (1) Which landslide features are changing due to climate change? and (2) Which mechanisms linked to climate change are certainly forcing

S.M. Moreiras (✉) · I.P.V.D. Pont
IANIGLA–CONICET. Geomorphology Group. Av Ruiz Leal S/N Parque Gral. San Martin S/N. Ciudad, Mendoza, 5500, Argentina
e-mail: moreiras@mendoza-conicet.gob.ar

I.P.V.D. Pont
e-mail: ivergara@mendoza-conicet.gob.ar

S.M. Moreiras
Cuyo University. Fac. Ciencias Agrarias, Mendoza, Argentina

M. Mikoš et al. (eds.), *Advancing Culture of Living with Landslides*,
DOI 10.1007/978-3-319-53483-1_22

landslides in Central Andes? Unknown this fact coupled with poorly planned developments and population growth can drastically increase landslide-associated casualties, especially in Andean communities, where pressure on land resources exists.

Study Area

At the latitude of study area, Central Andes are comprised by three different geological provinces in argentine territory: Main Cordillera, Frontal Cordillera and Precordillera, from west to east. The Main Cordillera, comprising highest peaks involves Cretaceous-Jurassic marine and volcanic rocks. The Permo-Triassic volcanic Choiyoi Group outcrops covering a Paleozoic basement in the Frontal Cordillera. Paleozoic sedimentary rocks appear in Precordillera range intruded by Permian batholiths (Fig. 1). Even though, lithology and slope are main conditioning factors for landslide distribution (Moreiras 2009), this link is not analyzed in this work.

An arid climate and high topography characterized this portion of Central Andes. However both parameters vary gradually longitudinal wise from west to east (Fig. 1). Topography of this mountain landscape decreases toward the eastern piedmont (700 m a.s.l.) forcing precipitation behavior. While solid winter precipitation predominates in highest mountain areas associated with Pacific Ocean westerlies; summer rainfall does in lower areas linked to Atlantic anticyclone behavior. Likewise, an average annual precipitation of 500 mm is measured in highest areas of the Andes diminishing until 200 mm in the Andes foothill where Mendoza city is established (see diagram on Fig. 1).

Annual to multi-annual climatic variations in the Central Andes are influenced by large-scale atmospheric circulation variations associated with the El Niño—Southern Oscillation (ENSO) phenomena (Aceituno 1990; Compagnucci and Vargas 1998). Related change in the dominant wind direction causes variations in the movement of synoptic weather systems across the region. Enhanced south-westerly air flow in El Niño years leads to higher than normal precipitation

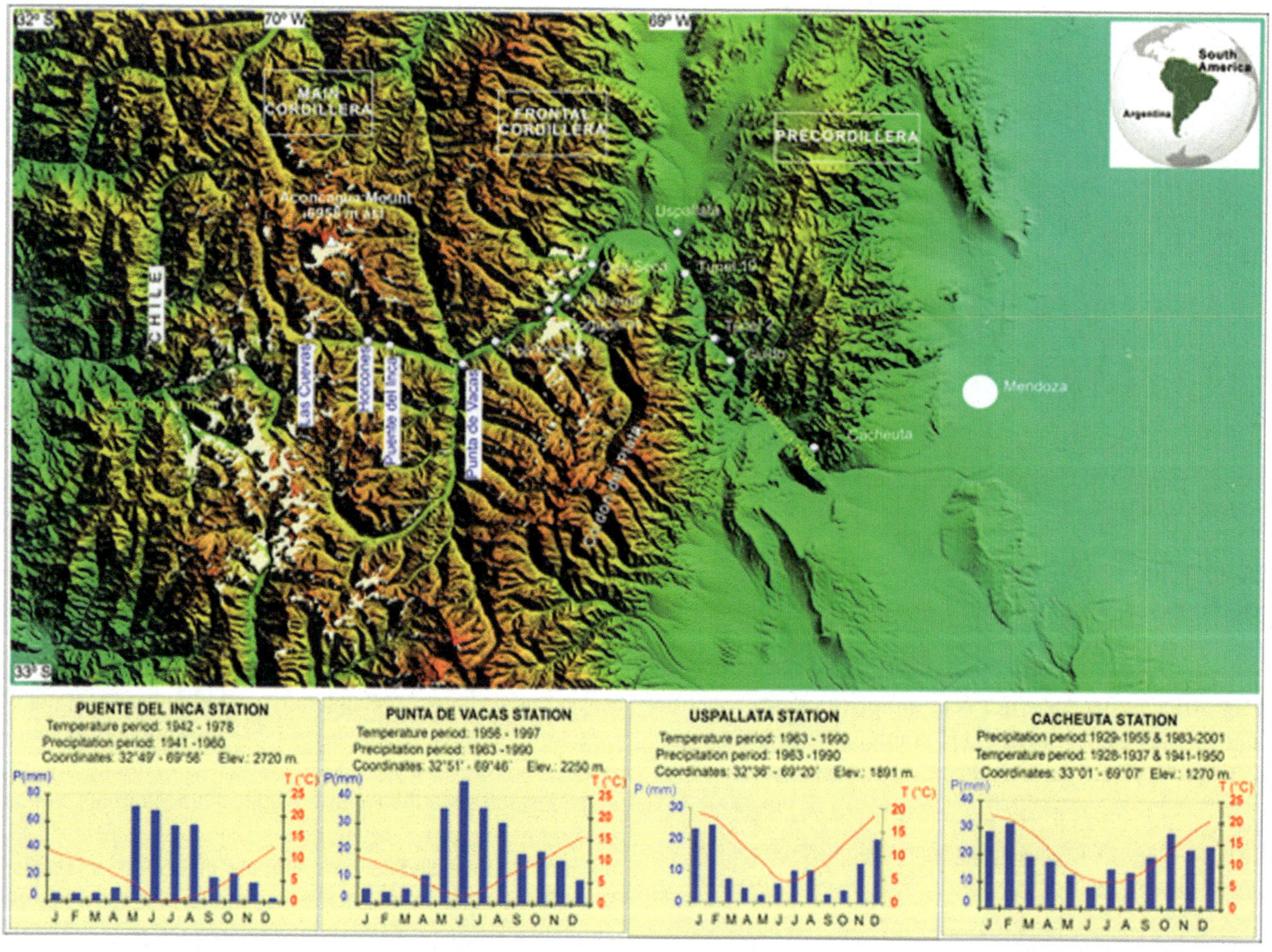

Fig. 1 Shutter Radar Topography of the study area showing main geological provinces and villages located along the Mendoza River valley. Charts show mean annual precipitation and mean annual temperatures recorded in local meteorological stations in the measuring period indicated in each diagram

and warm air temperatures in the Central Andes. The Interdecadal Pacific Oscillation (IPO) is thought to modulate ENSO-related climate variability on an interdecadal scale, leading to more frequent and more prolonged El Niño events during its positive phase.

Methods

For evaluating the coupled system climate change-landslides (CCL) we considered those events occurred during last 50 years along the Mendoza River valley. (32 S). We analysed temporal and spatial distribution of historical events to elucidate any change in frequency. Historical data were collected from newspapers and reports of official institutions updating inventories done previously (Moreiras 2004, 2005, 2006). In our inventory we contemplated rockfalls, debris flows and slides; including as well historical paths of snow avalanches. Available meteorological data (precipitation and temperatures) of local gauges were analysed to determine threshold precipitation values (Table 1). For correlation we used those records of meteorological station located not farer than 10 km. Radiosonde records taken in Santiago, Chile were used to detect changes in the altitude of the zero isotherm. We compared the elevation of zero isotherm based on geopotential data and temperatures recorded by the radiosonde during South Hemisphere summer (DJF) and winter (JJA).

Triggering Mechanisms

Even seismic triggering mechanism has been established for this high seismic area (Moreiras 2004; Moreiras and Páez 2015), main triggering mechanism is linked to slope/debris water saturation during intense summer rainstorms. A daily precipitation range of 6.5–12.9 mm has been determined for landsliding in middle elevations (1.500–2.700 m a.s.l.) during South Hemisphere summer (December–February). This low threshold could be partially explained by the reduced amount of annual precipitation (200 mm) and the abundant generation of debris in these mountain areas. Rainfall intensity is not available. Mountain meteorological stations measure 24 h accumulated precipitation. Likewise, meteorological records are scarce in the region limiting a precise determination of the threshold values. Meteorological stations are located along the Mendoza River valley, but no data exists for remote highest areas where events have been reported as well.

Antecedent precipitation plays an important role. Mean values of accumulated rainfall reach to 28 mm when a 5-day precipitation window previous to the landslide events is taken in account as they are normally debrisflows or rockfalls. In fact, 50 debris flows induced by rainfall during 2013 rainfall were associated to 29 days in the Central Andes (Moreiras and Sepúlveda 2013).

Landslides are also associated with snow/ice thawing during spring season. Snow precipitation takes place in higher topography during winter (see Fig. 1), and then greater slope instability is recorded in the following warmer seasons (Moreiras et al. 2012a). Herein, landslide triggering factors (rain/snow thawing) and temporal distribution (summer/spring) varies in the different ranges of the Argentinean Central Andes. Whereas debris flows and rockfall induced by rainfall are clustered in summer periods (December–February) along the valleys and lowest areas; debris flows and debris avalanches are consequence of snow/ice melting during spring (September–November) in highest areas of Main and Frontal cordilleras.

Above-mean precipitation recorded during the warm Pacific Ocean episodes easily saturate unstable debris material of steep mountain slopes causing severe debris flows in the Arid Central Andes affecting cities located downstream (Sepúlveda et al. 2006; Moreiras 2005; Moreiras et al. 2012b). There is a significant interannual variation in the number of landslides, with the El Niño/La Niña cycle emerging as a key control in Pacific coast of Latin America (Sepúlveda and Petley 2015).

Table 1 Local gauges: *G* Guido, *U* Uspallata, *P* Polvaredas, *PV* Punta de Vacas, *PI* Puente del Inca; *H* Horcones and *CR* Cristo Redentor. *Data source* (1) Agua y Energía. (2) EVARSA. (3) Servicio Meteorológico Nacional. (4) Secretaría Recursos Hídricos, and (5) Dirección Nacional de Irrigación

Gauge	Elev. (m)	Recording period	Data source
G	1550	1957–2007	1–2
U	1880	1963–2007	3
P	2350	1992–1997	2
PV	2400	1983–2000	4
PV	2400	1955–1997	1–2
PI	2720	1955–2005	4
PI	2720	1941–1976	3
H	3200	2001–2006	5
G	1550	1957–2007	3

Shifting Slope Instability Type

Landslides as many other natural hazards are affected by climate/weather conditions. A total of 631 fatalities resulted from the occurrence of 611 landslides in Latin America and the Caribbean between 2004 and 2013 (Sepúlveda and Petley 2015). Majority of these fatal landslides were triggered by heavy rains (73%) and 15% were triggered by extreme weather conditions like hurricane tropical storms. As rainfall pattern could be modified both temporally and spatially due to climate change, as well as frequency and intensity of extreme weather conditions, climate change influence on slope instability could not be ignored. Landslide features mainly affected by climate change is not completely understand yet.

Temporal distribution of landslides occurred along the Mendoza River valley during the last 5 decades shows a markedly arise (Fig. 2). Landslides triggered by summer rainstorms in lower areas have been triplicated during this period. Moreover, those higher areas associated with presence of discontinue permafrost above 3.200 m a.s.l. have become to be more prone to slope instability.

Otherwise, intensity of landslides resulted higher during the last decades. In this sense, material mobilized by landslides has been increasing as well as their danger and associated social impacts. Extraordinary occurrence of at least 55 landslide events triggered by convective rainstorms during a short period of 29 days in Mendoza River valley caused repeated trouble for traffic along the international road connecting Mendoza with Santiago, Chile. A first set of events happened on 13th January, 2013 spatially gathered in Guido locality (1.550 m a.s.l.) located 12 km downstream Uspallata. Fifteen debris flows were generated in Guido where a granitic intrusive crops out, known for historical rockfall processes. On 7th February 2013 another cluster of debris flows was triggered by rainstorms in a western sector located 5–15 km from Uspallata village. During this month, mean monthly precipitation exceeded in 20 mm the historical mean value (40 mm) of February for this region. Likewise, 23 danger debris flows impacted on the study area during last summer (December 2015–February 2016) generating important economic losses to the region. A bridge on the International Road to Chile was washed away by a violent debris flow generated in the headwaters of a tributary basin after an intensifier rainfall.

Topography limits the type of geomorphological processes. Summer rainstorms use to trigger rockfall and debris flows in lower areas of the Andes; while heavy snow accumulation in higher steep mountain slope used to promote snow avalanches. However, this topographic control is changing at present because of imminent current warming. The climbing of the 0 °C isotherm is migrating up this border barrier. Those slopes covered by a dense cap of snow in the past, are saturated by rainfall at present. Hence localities historically dammed by violent snow avalanches are being impacted by debrisflows (e.g., Cortaderas, Polvaredas). As a response to this warmer scenario, ancient snow avalanche tracks are currently affected by water saturated debris flows. As a consequence propagation of these channels due to repeated debris flow is driving a denser spatial distribution of this type of phenomenon.

Climate change is impacting seriously on slope instability increasing frequency-intensity and also changing distribution of landslide pattern. But also this phenomenon has contributed to multifaceted causes of landslides. The Godzilla El Niño events (December 2015–February 2016) drives 23 debris flows in the Mendoza River valley (Moreiras and Vergara Dal Pont 2016). However, triggering mechanisms were completely different for the three separated clusters of events. While recurrent daily debris flows were generated during December 2015 by the thawing of basal ice of two rock glaciers located in the headwaters of tributaries basins;

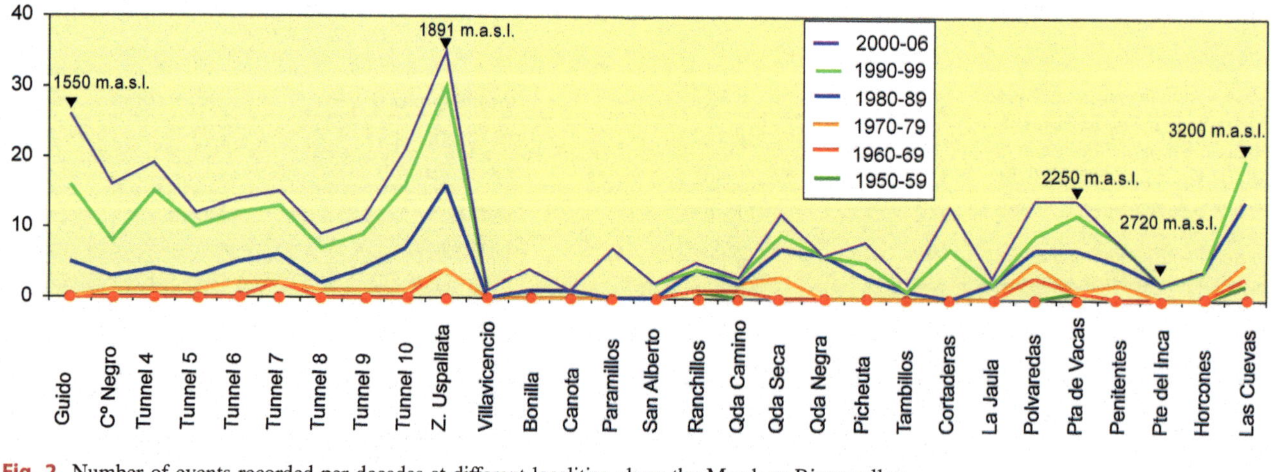

Fig. 2 Number of events recorded per decades at different localities along the Mendoza River valley

20 events were lately generated by a daily intensive precipitation on 23th January 2016. However, the most damaging events occurred when a damming landslide lake collapsed as consequence of 30 mm-rainfall being the fifth most intensive precipitation recorded in Uspallata since 1983, during warmer months (December, January, February and March).

Climate Change Mechanisms Inducing Landslides

Understanding of climate change mechanisms promoting slope instability in the Central Andes needs a proper local study. Still global warming is widely accepted by the international research community, we analyzed the regional evolution of the zero degree isotherm during last decades. According to our findings, the altitude of the 0 °C isotherm is roughly climbing since 1940. Still this rise is not very pronounced, from 4.380 to 4.420 m a.s.l., any change at this altitude compromises seriously the permafrost preservation. Discontinues permafrost line is inferred by lower toe of debris rock glaciers. These ice bodies are extremely sensitive to temperature changes.

Geopotential elevation and temperature values correlate well during South Hemisphere summer; however a disengaged of these variables exists since 1976 for winter seasons (Fig. 3). That means that during winter precipitation, the 0 ° C isotherm could be uplifted generating rainfall instead of snowfall. Main consequence of that is the rapid saturation of slope materials and their eminent collapse. Maybe climbing of the 0 °C isotherm could explain predomination of landslides processes along the Mendoza River valley during last decades when compared with snow avalanche events (Fig. 4).

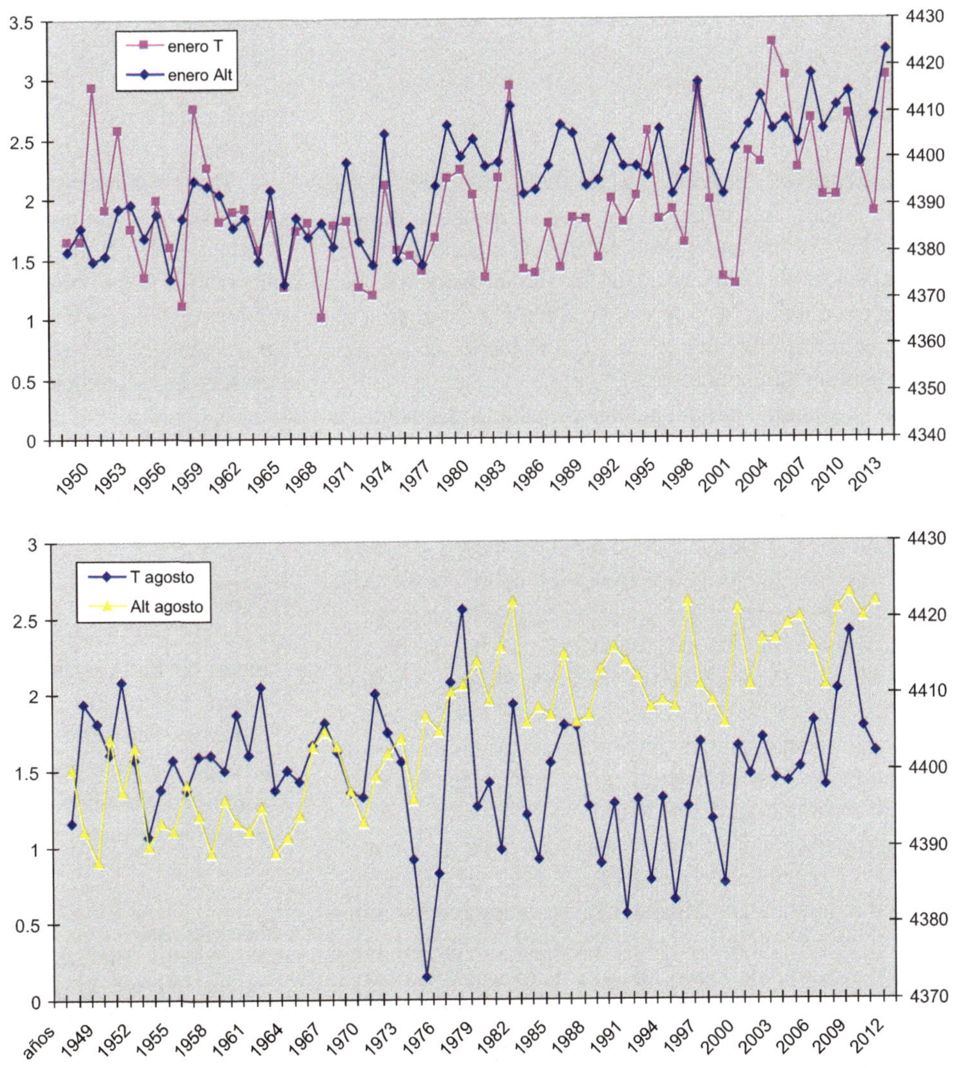

Fig. 3 Geopotential elevation and temperature of the zero degree isotherm for the study area since 1942. Above graphic shows data of January. Below graphic represents records measured in August

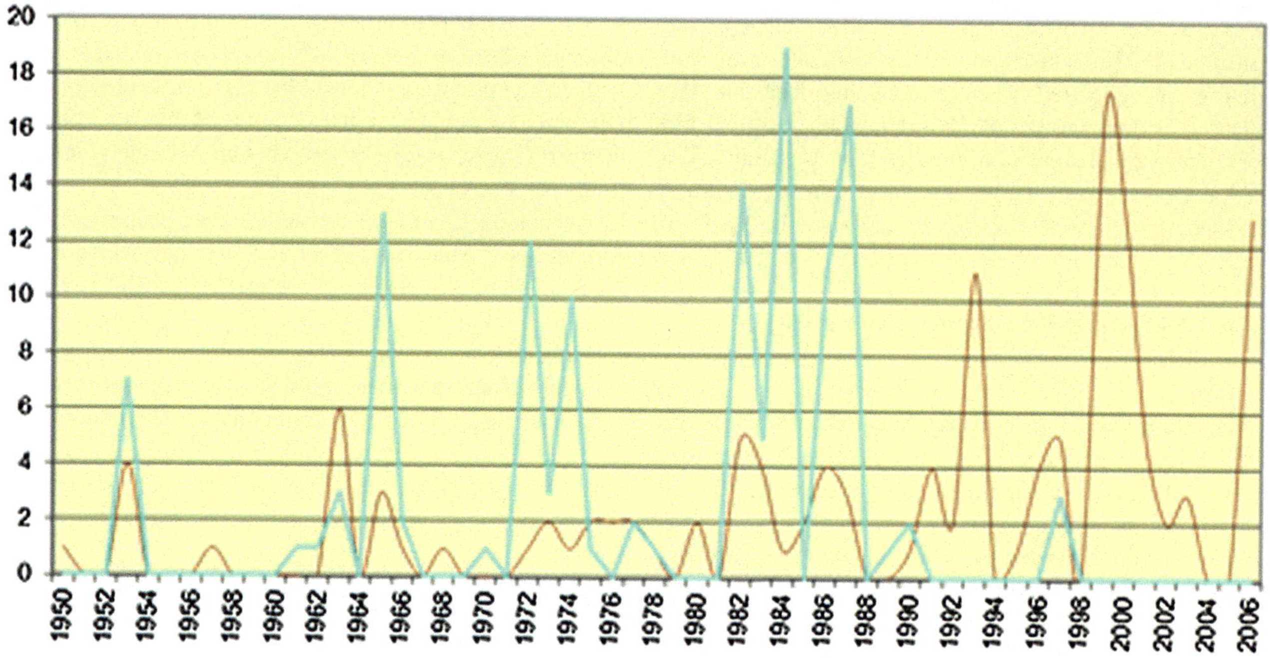

Fig. 4 Annual number of snow avalanches (*light blue line*) and landslides (*brown line*) recorded from 1950–2006

Discussion

Climate change is driving greater slope instability in the Central Andes. According to our results, landslide activity has been raising during last decades. Landslides in the Central Andes show higher frequency and-intensity, a denser spatial distribution, a progressive altitudinal migration as consequence of climbing of the 0 °C isotherm, and consequent permafrost degradation.

The key factor connecting climate change to landslides is water. Global warming provoking higher temperatures on air/soils and heating on mountain areas promotes rainfall precipitations minimizing snow precipitation in higher mountain areas. That generates a rapid soil/debris saturation promoting and favoring occurrence of debris flows. This instability prone scenario is exacerbated by the climbing of the 0 °C isotherm favoring snow/ice thawing process in headwaters of tributary basins. Besides, shifting on slope behavior (snow avalanche vs. landslides) have increased mountain Andean vulnerability as structural measures constructed in the past for mitigating snow avalanches are not efficient for retaining debris flows.

Main explanation of this landslide increase is found in changes on precipitation records associated with intensified summer rainfall. In the Mendoza River valley precipitations above average values have being recorded during last years causing catastrophic debris flow in 2013 and 2015–2016. Besides, global temperature rising is reflected in roughly climbing of the zero degree isotherm in the Andean hillslopes with a severe impact on ice degradation of debris rock glaciers. This phenomenon could be extrapolated to a general permafrost degradation in mountain areas. Based on this perspective, the climate change is both rising mean local temperatures (Rosenblüth et al. 1997), and increasing temperatures of Niño 3.4 Pacific region. The main consequence of that are changes on weather conditions. Future greater landslide activity is predicted for stronger warm ENSO phases associated with above mean precipitation in mountain region and intense summer rainstorms associated with east wetlines. Forecasting on climate change predicted a strengthen of the El Niño phenomenon (IPCC 2014) and its consequences.

Conclusions

Central Andes are among the most sensitive environments on Earth to the changes in climate that have happened over the past decades. Geomorphological processes sculpting these high mountains are strongly affected by changes in temperature and precipitation. Amplified warming at high elevations is destabilizing slopes by permafrost degradation and thawing of glacial ice.

Acknowledgements This research was founded by PIP 484, PIP 11220150100191 (2015–2017) and ANLAC (Natural Hazards in the Central Andes) Program founded by UNCU University leader by Stella Moreiras. We are gratefully to D. Araneo for discussion of results.

References

Aceituno P (1990) Variabilidad interanual en el caudal de ríos andinos en Chile Central en relación con la temperatura de la superficie del mar en el Pacífico central. Rev. de la Sociedad Chilena de Ingeniería Hidráulica 5(1):7–19

Bardou E (2011) Influence of the connectivity with permafrost on the debris-flow triggering in high-alpine environment. Italian J Eng Geol 11(3):13–21

Beniston M (2003) Climatic change in mountain regions: a review of possible impacts. Clim Change 59(1–2):5–31

Beniston M, Stephenson DB, Christensen OB, Ferro CAT, Frei C, Goyette S, Halsnaes K, Holt T, Jylhae K, Koffi B, Palutikov J, Schoell R, Semmier T, Woth K (2007) Future extreme events in European climate: an exploration of regional climate model projections. Clim Change 81(suppl. 1):71–95

Clague JJ, Huggel C, Korup O, Mc Guire B (2012) Climate change and hazardous processes in high mountains. Rev Asociación Geológica Argentina 69(3):328–338

Compagnucci RH, Vargas WM (1998) Interannual variability of Cuyo rivers streamflow in Argentinean Andean mountains and ENSO events. Int J Climatol 18:1593–1609

Crozier MJ (1997) The climate-landslide couple: a Southern Hemisphere perspective. Paläoklimaforschung Paleoclimate Res 19:333–354

Evans SG, Clague JJ (1994) Recent climatic change and catastrophic geomorphic processes in mountain environments. Geomorphology 10:107–128

Gruber S, Haeberli W (2007) Permafrost in steep bedrock slopes and its temperature related destabilization following climate change. J Geophys Res 112:F02S18

Haeberli W, Gruber S (2009) Global warming and mountain Permafrost. In Permafrost soils. Soil Biology 16, Springer Berlin Heidelberg, pp 205–218

Harris C, Arenson LU, Christiansen HH, Etzemüller B, Frauenfelder R, Gruber S, Haeberli W, Hauck C, Hoelzle M, Humlum O, Isaksen K, Kääb A, Kern-Lütschg MA, Lehning M, Matsuoka N, Murton JB, Nozli J, Phillips M, Ross N, Seppälä M, Springman SM, and Mühll DV (2009) Permafrost and climate in Europe: monitoring and modelling thermal, geomorphological and geotechnical responses. Earth-Sci Rev 92(3–4):117–171. doi: 10.1016/j.earscirev.2008.12.002

Huggel C (2009) Recent extreme slope failures in glacial environments: effects of thermal perturbation. Quatern Sci Rev 28:1119–1130

Huggel C, Clague JJ, Korup O (2012) Is climate change responsible for changing landslide activity in high mountains? Earth Surf Proc and Landf 37(1):77–91

IPCC (2014) Climate change 2014: synthesis report. In: Core Writing Team, Pachauri RK, Meyer LA (eds) Contribution of Working Groups I, II and III to the Fifth assessment report of the intergovernmental panel on climate change. IPCC, Geneva, Switzerland, p 151

Moreiras SM (2004). Landslide hazard zonification along the Mendoza River valley. Ph.D. thesis. UNSJ. p 342

Moreiras SM (2005) Climatic effect of ENSO associated with landslide occurrence in the Central Andes, Mendoza province, Argentina. Landslides 2(1):53–59

Moreiras SM (2006) Frequency of debris flows and rockfall along the Mendoza river valley (Central Andes), Argentina. Special issue Holocene environmental Catastrophes in South America. Quat Int 158:110–121

Moreiras SM (2009) Analysis of variables conditioning slope instability in valleys of Las Cuevas and Mendoza rivers. Rev Asociación Geológica Argentina 65(4):780–790

Moreiras SM, Lauro C, Mastrantonio L (2012a) Stability analysis and morphometric characterization of palaeo-lakes of the Benjamin Matienzo Basin- Las Cuevas River, Argentina. Nat Hazards 62 (2):593–611

Moreiras SM, Lisboa S, Mastrantonio L (2012b) The role of snow melting upon landslides in the central Argentinean Andes.In: Guest editors: Wohl E, Rathburn S (eds) Earth surface and processes landforms. Special issue on Historical Range of Variability

Moreiras SM, Sepúlveda SA (2013) The high social and economic impact 2013 summer debris flow events in central Chile and Argentina. Bollettino di Geofisica Teorica ed Applicata 54 (Supp. 2):181–184

Moreiras SM, Páez MS (2015) Historical damages and secondary effects related to intraplate shallow seismicity of Central Western Argentina. Geodynamic Processes in the Andes of Central Chile and Argentina. Geological Society of London, vol 399, p 369–382. doi: 10.1144/SP399.6

Moreiras SM, Vergara Dal Pont I (2016) The role of climate change on slope instability of the Central Andes during the last decades. In: Proccedings of second Central American and Caribbean Landslide Congress, Tegucigalpa, Honduras, p 4

Páez MS, Moreiras SM, Brenning A, Giambiagi LB (2013) Flujos de detritos-aluviones históricos en la cuenca del Río Blanco (32°55'–33°10' Y 69°10'–69°25'), Mendoza. Rev. Asociación Geológica Argentina 70(4):488–498

Rosenblüth B, Fuenzalida HA, Aceituno P (1997) Recent temperature variations in southern South America. Int J Climatol 17(1):67–85

Sepúlveda SA, Rebolledo S, Vargas G (2006) Recent catastrophic debris flows in Chile: geological hazard. Climatic relationships and human response. Quatern Int 158:83–95

Sepúlveda SA, Petley DN (2015) Regional trends and controlling factors of fatal landslides in Latin America and the Caribbean. Nat Hazards Earth Syst Sci 15:1821–1833

Sepúlveda SA, Moreiras SM, Lara M, Alfaro A (2015) Debris flows in the Andean ranges of central Chile and Argentina triggered by 2013 summer storms: characteristics and consequences. Landslides 12(1):115–133

Stoffel M, Huggel C (2012) Effects of climate change on mass movements in mountain environments. Prog Phys Geogr 36(3):421–439

Understanding the Chandmari Landslides

Nirmala Vasudevan, Kaushik Ramanathan, and Aadityan Sridharan

Abstract

Chandmari Hill lies in Gangtok City in the Himalayan Mountain Ranges of Northeast India. The Himalayas are particularly prone to landslides due to complex geology combined with high tectonic activity, steep slopes, and heavy rainfall. Chandmari Hill has experienced a significant number of landslides, both rainfall and earthquake triggered, during the past several decades. Recently, the Government of India commissioned Amrita University to develop and deploy a landslide early warning system at Chandmari Hill. During the initial phase of the deployment, we conducted walkover surveys at Chandmari Locality, which comprises a large portion of Chandmari Hill. We also extracted and tested soil samples, drilled a 33.5 m borehole, and analyzed rock cores from the borehole. We present the results of laboratory soil tests and use these results in mathematical models. We examine all landslides (rainfall-triggered and earthquake-induced) recorded at Chandmari Locality during the past five decades. Simple calculations demonstrate that when the input parameters of the models mimic the field conditions precursory to an actual landslide, the factor of safety of the slope is less than unity. Gangtok City lies close to the Main Central Thrust, MCT2, which separates the gneissic rocks of the Paro/Lingtse Formation from the mica schists of the Daling Formation. Our field investigations revealed that at Chandmari Locality, gneissic rock overlies highly weathered mica schist. We postulate that surface runoff infiltrates through fractures in the overlying gneiss and results in an extrusion of the finer micaceous material, leading to subsidence which is routinely observed during the monsoon season. During torrential rains, rainwater infiltration causes the sliding of the soft micaceous bands underlying the gneissic rock, leading to rockslides at the hill. We suggest that similar processes are responsible for the frequent and widespread occurrences of landslides and subsidence observed throughout the region.

N. Vasudevan (✉)
Amrita Center for Wireless Networks and Applications, Amrita School of Engineering, Department of Physics, Amrita School of Arts and Sciences, Amrita Vishwa Vidyapeetham, Amrita University, Amritapuri, 690 525, India
e-mail: nirmalav@am.amrita.edu

K. Ramanathan
Department of Civil Engineering, Amrita School of Engineering, Amrita Vishwa Vidyapeetham, Amrita University, Coimbatore-641 112, India

A. Sridharan
Deparment of Physics, Amrita School of Arts and Sciences, Amrita Vishwa Vidyapeetham, Amrita University, Amritapuri-690 525, India

M. Mikoš et al. (eds.), *Advancing Culture of Living with Landslides*,
DOI 10.1007/978-3-319-53483-1_23

Keywords
Landslide • Subsidence • Chandmari • Himalayas • Sikkim • Slope stability

Introduction

Chandmari Hill (27°20' N, 88°37' E, elevation 2015 m, Fig. 1) is located in Gangtok, the capital city of Sikkim State in Northeast India (Fig. 2), and is part of the Inner Himalayan Mountain Ranges. The Himalayan Mountain Ranges are particularly prone to landslides due to complex geology combined with high tectonic activity, steep slopes, heavy rainfall, and poorly planned and executed human activities. Chandmari Hill has experienced a significant number of landslides, both rainfall and earthquake induced, during the past several decades. The most disastrous landslide occurred on 8th/9th June 1997, and was triggered by 211 mm of rainfall in 4 h (Dubey et al. 2005). Water infiltrated through fractured rock and caused the sliding of soft micaceous bands underlying hard gneissic rock. The rock slide led to the removal of the rocky toe support and resulted in a debris slide of the supersaturated soil overburden. Eight people were killed in the landslide, and a school complex, a temple, a government vehicle workshop, and a large number of parked vehicles were damaged (Ghoshal et al. 1998; Basu and De 2003).

We had earlier developed and deployed a wireless sensor network-based early warning system to monitor the Anthoniar Colony Landslide in South India (Ramesh and Vasudevan 2012). The system has been in successful operation since 2009. Recently, the Government of India sanctioned the deployment of a similar system at Chandmari Hill aimed at providing early warnings of potential landslides to the local inhabitants.

We have phased the deployment at Chandmari Hill in two stages: a pilot deployment that has been completed, and a full-scale deployment that shall commence shortly. During the pilot deployment, walkover surveys were conducted, and a 1:1000 scale topographic map and a longitudinal profile of the landslide area were prepared. Soil samples were collected and geotechnical tests were later performed at our university laboratories. Geophones and a weather station were installed in different regions of the hill. A 33.5 m borehole was drilled into rock, rock samples were extracted and analyzed, and piezometers and inclinometers were installed at different depths in the borehole. In the full-scale deployment, we will similarly instrument several other areas of Chandmari Hill so that the region can be adequately monitored.

In the interim, we are using information obtained during the pilot deployment to deepen our understanding of the landslide processes at Chandmari Hill (Vasudevan et al. 2016). In this paper, we examine all of the landslides and occurrences of subsidence recorded during the past fifty years at Chandmari Locality, which forms a large portion of Chandmari Hill. Rainwater infiltration and subsequent build-up of pore pressure are modelled after the method proposed by Iverson (2000). The factors of safety are estimated using infinite slope stability analyses. For the earthquake-induced landslide, attenuation relations specific to the Himalayan Region provide the peak ground acceleration, while Newmark's (1965) sliding block procedure is used to compute the seismic factor of safety. Additionally,

Fig. 1 The Chandmari Hill with the scarp of the 1997 landslide, photographed on 2 June 2015

Fig. 2 India (in *light yellow*), Sikkim State (in *coral orange*), Gangtok City (in *blue*), Chandmari Hill (location indicated by the *arrow*)

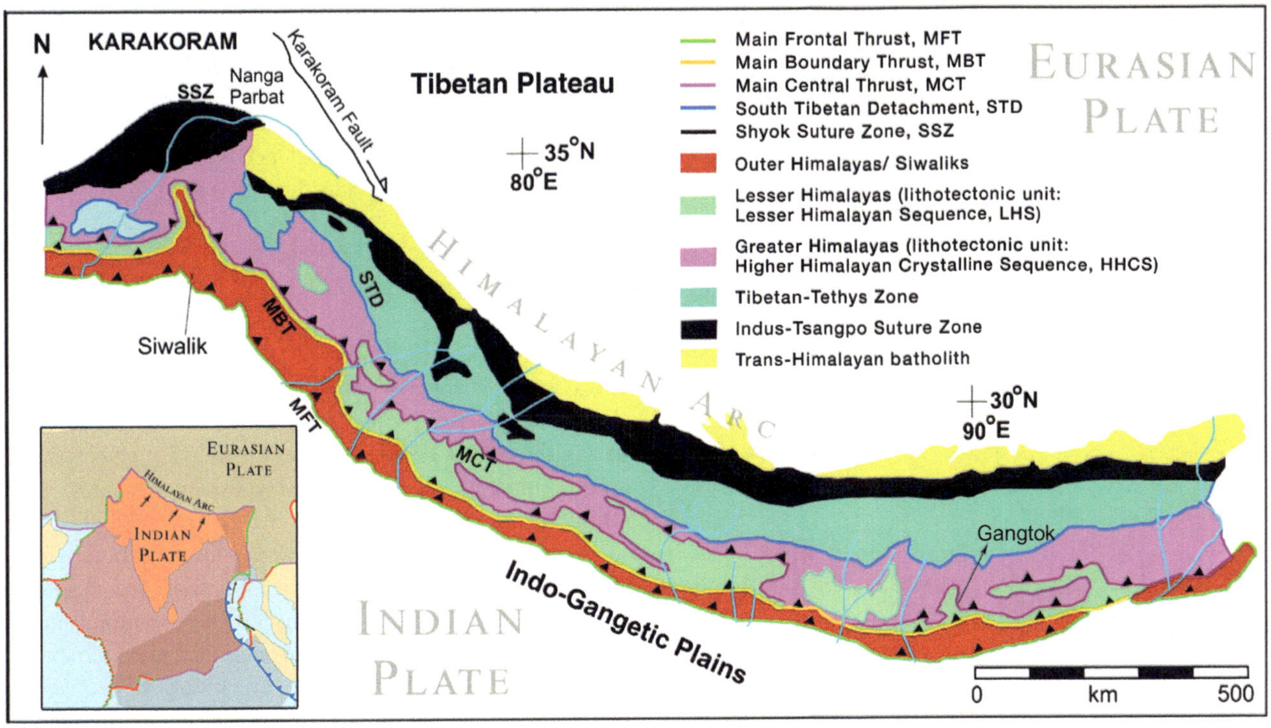

Fig. 3 General geology and tectonics of the Indian arc, with the various thrust zones, MFT, MBT, and MCT, and the mountain chains (Gupta and Gahalaut 2014) (The *black triangles* indicate thrust faults; the rocks on the side with the triangles have been pushed up and over the rocks on the other side). Chandmari Hill lies in Gangtok City, and Chandmari Locality (the focus of this paper) constitutes a large portion of Chandmari Hill. Inset: The subduction of the Indian Plate under the Eurasian Plate

the rock slides and occurrences of subsidence are explained based on our field studies and borehole investigations.

The Chandmari Landslides

Geologic Setting

Chandmari Hill lies in the Inner Himalayan Mountain Ranges. The Himalayas arise from the subduction of the Indian Tectonic Plate under the Eurasian Plate (Fig. 3, inset), and comprise several parallel mountain chains stretching along 2400 km-long arcs.

Moving north from the Indo-Gangetic plains (Fig. 3), we first encounter the Main Frontal Thrust, MFT, which separates the plains from the Siwaliks, or Outer Himalayas. Further north, the Main Boundary Thrust, MBT, separates the Siwaliks from the Inner Himalayas which are also known as the Lesser, Lower, Middle, or Mahabharata Himalayas. Next in sequence is the Main Central Thrust (MCT) zone comprising the MCT3, or Ramgarh Thrust (RT), the MCT2, and the MCT1, followed by the Greater, Higher, or Himadri Himalayas. These thrust zones and mountain chains are depicted in Fig. 3.

Chandmari Hill, in northeast Gangtok City, lies close to the MCT2 (Fig. 4) which separates the lower (greenschist)

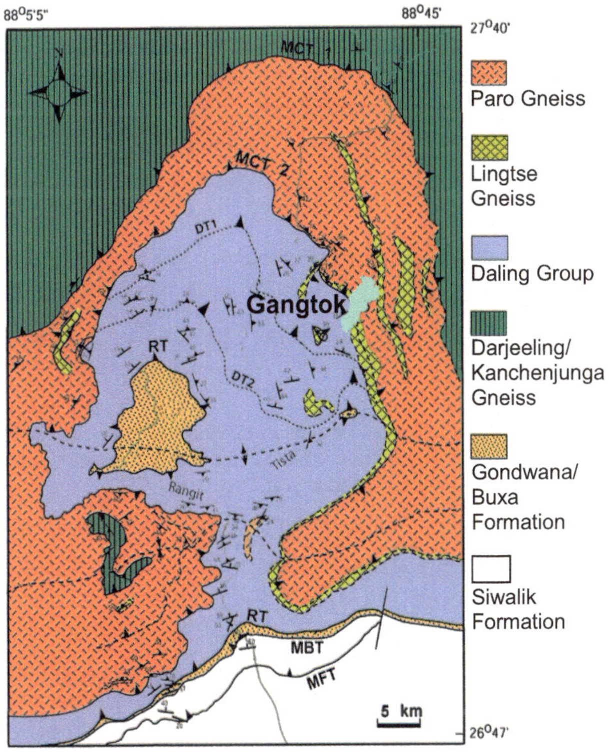

Fig. 4 Sikkim Himalayas (Bhattacharyya 2010); Gangtok City is colored in *light blue*

Fig. 5 Gangtok City, Chandmari and Tathangchen localities, Ranikhola and Roro Chu streams

grade Daling Formation from the higher (amphibolite) grade Lingtse and Paro gneisses. The Daling Group consists of highly foliated metamorphic rocks such as phyllites and schists intercalated with quartzites (Roy et al. 2015). The Lingtse gneisses are biotite-bearing orthogneisses with prominent feldspar augen, while the Paro gneisses are garnetiferous biotite-staurolite-muscovite-bearing paragneisses (Bhattacharyya 2010; Anbarasu et al. 2010).

Gangtok City is flanked by two streams—the Roro Chu to its east bordering Chandmari Hill, and the Ranikhola to its west (Fig. 5). The two streams meet at Ranipul, in Gangtok City, and flow south as the main Ranikhola River. Gangtok receives an annual rainfall of 3500 mm (Table 1) with torrential rains during the monsoon season; the city has witnessed several rainfall-triggered landslides. The mean temperature ranges from 4 °C in winter to around 22 °C in summers; snowfall is extremely rare. The region is seismically active with several thrust and strike-slip faults, and earthquake-induced landslides have occurred in the past.

Landslide History

Subsidence was first observed at Chandmari Locality in 1966 and recurred during subsequent monsoon seasons (Rawat 2005). The phenomenon was particularly noticeable during the period 1975–1976, but the hill then stabilized for a while until its reactivation as a subsidence zone in 1984 (Basu and De 2003).

Some years later, on the night/early morning hours of 8/9 June 1997, rainfall of 211 mm in 4 h triggered at least 9 landslides in and around Gangtok City resulting in 38 fatalities and over 50 injuries. This rainfall caused a rock and debris slide at Chandmari Locality, killing 8 people and damaging several buildings and vehicles (Dubey et al. 2005; Reuters 1997; Basu and De 2003).

A few years after the 1997 landslide, Bhasin et al. (2002) reported creep movement at Chandmari Locality. Subsequently, on 12 July 2007, incessant rain triggered yet another rock and debris slide that destroyed a house in its wake (Sharma 2008). Next, on 7 June 2011, heavy rains triggered two small debris slides at different locations on the hill (The Sikkim Times 2011).

A 6.9 M_W strike-slip earthquake, with a focal depth of 10 km, occured on 18 September 2011, and triggered over 1,196 landslides in the region. The earthquake occured at

Table 1 Gangtok rainfall (mm), averaged over a 50-year period (1966–2015), Indian Meteorological Department (IMD)

	Jan	Feb	Mar	Apr	May	Jun	Jul	Aug	Sep	Oct	Nov	Dec	Annual
1966–2015	25.1	63.4	125.3	310.4	494.3	645.6	649.2	570.3	441.2	167.5	39.4	17.3	3549.1

Table 2 Landslide history of Chandmari Locality. Columns 3, 4, and 5 specify the rain on the day of the landslide and 1 and 2 days before, while Columns 6, 7, and 8 specify the cumulative rain during the week (wk), month (mo), and 3 month period preceding the landslide. Rainfall data were obtained from the Indian Meteorological Department (IMD)

Date (dd-mm-yyyy)	Landslide type	Total rainfall (mm) before the landslide event						Trigger/Primary cause
		0 day	1 day	2 day	1 wk	1 mo	3 mo	
1966–1975	Subsidence							Rainfall
1975–1976	Increased subsidence							Rainfall
1984	Subsidence							Rainfall
09/06/1997	Rock & debris slide	211, 4 h						Rainfall
12/07/2007	Rock & debris slide	72.8	21.2	10.7	155.9	1021.6	1613.7	Rainfall
06/06/2011	Debris slides (2)	118.0	34.3	11.2	186.4	490.6	797.8	Rainfall
18/09/2011	Debris slide	39.3	4.1	52.1	124.8	432.3	2555.3	Earthquake-induced

(27°42' N, 88°12' E), 68 km away from Gangtok City. Gangtok received moderate rain before, during, and after the earthquake (Table 2), and around 13 landslides occured in and around the city (Chakraborty et al. 2011). A small, shallow debris slide was recorded at Chandmari Hill (SikkimNow 2011). Pertinent information is summarized in Table 2.

Field Investigations and Laboratory Experiments—Methods and Results

Walkover Surveys

We conducted walkover surveys at Chandmari Locality (Vasudevan and Ramanathan 2016). The rock formation comprised Paro gneisses (discussed in greater detail later in this paper) interbedded with micaceous material. The scarp of the 1997 landslide was visible (Fig. 1, also marked on the topographic map in Fig. 6), with debris consisting of pebbles, cobbles, and gneissic boulders (maximum observed boulder size: 9 m) embedded in a biotite-muscovite mica-rich silty-sand matrix.

The area to the west of the 1997 landslide scarp was covered with forests of Uttis trees (scientific name: Alnus nepalensis). The trees leaned downslope, indicating ongoing landslide activity. Further visual evidence of the threat of landslides was the tilting of a television tower located farther away from the forest at approximately the same elevation. That part of the hill was quite populated and housed some important government offices.

As we walked down the hill, we saw more further indications of landslide activity, such as large cracks on the ground and rock fragments that had broken away from the bedrock. Gabion walls and wire-mesh fences had been constructed but were slightly damaged by landslide activity. At one location, we observed strata of permeable coarse-grained soil confined between layers of low-permeability fine-grained soil; this contrast in permeability might be one of the causes contributing to land subsidence and landslides.

During surveys conducted in May 2015 just before the onset of the monsoons, we encountered a water spring near the crest of the hill (elevation: 1975 m). We also observed and photographed thin water channels near the toe of the 1997 scarp. The soil was moist even some distance away from the channels. At a slightly lower elevation of 1840 m in the forest of Uttis trees, we observed a water stream during an earlier visit in January 2015. Drainage canals have been constructed in the lower regions of the hill; these canals collect surface water and divert it to a nearby stream, the Lokchu Khola. The canals contained a sizable volume of water both during the dry season in January 2015, and just before the rains in May 2015.

Soil Tests

We extracted undisturbed soil samples from depths of approximately 0.8 m from two locations at Chandmari Locality (elevations: 1749 and 1851 m, Fig. 6). The locations were some distance away from the Chandmari Locality landslides (Table 3).

The Casagrande apparatus was used to determine the liquid limit. Since neither the liquid limit nor the plastic limit could be determined, we concluded that the soil was non-plastic. Undisturbed samples were used to determine the saturated hydraulic conductivity at a load of 0.1 kg/cm^2. Later, reconstituted samples with bulk density and moisture content matching field conditions were used to determine the consolidation properties, the saturated hydraulic conductivities under different loads, and the shear strength parameters. The rate of shear in the consolidated drained (CD) tests was 0.125 mm/min.

Rock Samples

During the pilot deployment, we drilled a 33.5 m borehole for the installation of the first batch of piezometers and inclinometers. The borehole yielded negligible amount of topsoil (less than 10 cm) and several meters of staurolite-garnet-biotite-muscovite gneiss occasionally intercalated with quartzite. The quartzite intercalations varied in thickness from 5 to 20 cm. Rock samples extracted from depths of 33.0–33.5 m were darker in color and displayed marked schistosity, possibly due to increased biotite content. Both intact rock samples and fractured rock were obtained with high core recovery.

An Interesting Observation

We extracted largely intact samples; some of these rock pieces were longer than 1.0 m (barrel length = 1.0 m). However, samples extracted from depths of 3.5–5.0, 6.0–9.0, and 14.4–15.0 m were weathered, with uniform fine sand/silt grain size. In Table 4, we present the results of dry sieve analysis (Vasudevan et al. 2016).

We identified the rock pieces as gneisses of the Paro Formation of the Higher Himalayan Crystalline Sequence, HHCS, and the parent rocks of the highly weathered strata as mica schists belonging to the Daling Formation of the Lesser Himalayan Sequence, LHS (Figs. 3, 4) (Bhattacharyya 2010; Anbarasu et al. 2010), which is consistent with the fact that gneiss is relatively more resistant to weathering than schist. Thus, our borehole observations indicated that gneissic rock overlies highly weathered mica schist at various depths.

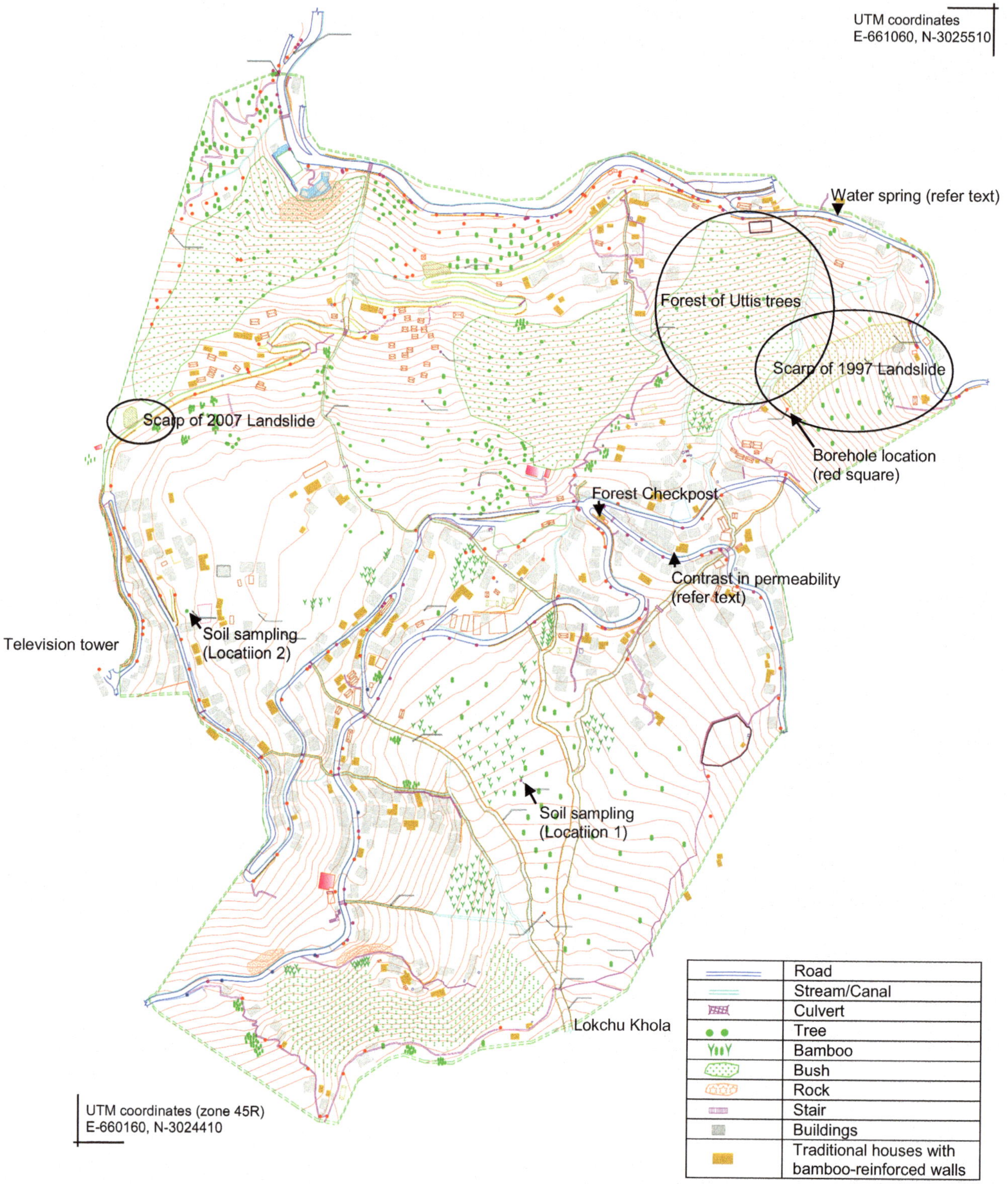

Fig. 6 Topographic map of the study area. Contour interval: 5 m, surveyed area: 0.53 sq. km

As we elaborate in the Discussion Section, this finding may have implications not only for the Chandmari landslides, but also for many of the rockslides and occurrences of subsidence common to the region. We measured the shear strength parameters of the weathered sample from 3.5 to 5.0 m using the CD direct shear test and obtained $c = 0\,\text{Pa}$ and $\varphi = 35°$.

Table 3 Results of geotechnical tests performed in accordance with the Indian Standard (Arora 2003)

	Elevation 1749 m	Elevation 1851 m
Bulk density (kg/m^3) measured at the time of extraction	1310	1450
Natural moisture content (%) using the oven-drying method	21.3	12.7
Grain size (gravel, sand, silt, clay %) wet sieve and hydrometer analysis	2, 58, 26, 14	
Plasticity index	Non-plastic	Non-plastic
Specific gravity	2.46	
Hydraulic conductivity (m/s) under a load of 1.5 kg/cm^2, using a fixed ring consolidometer	1.3×10^{-5}	1.4×10^{-5}
Coefficient of compressibility (m^2/N) under a load of 1.5 kg/cm^2, using a fixed ring consolidometer	6.4×10^{-5}	1.7×10^{-4}
Consolidated drained (CD) direct shear test		
Cohesion (Pa)	0	2800
Angle of internal friction (°)	39	36

Table 4 Grain size distribution from dry sieve analysis of weathered rock material obtained from 3.5 to 5.0 m

Constituent	Diameter limits, Indian Standard classification IS 2720 (Part 4)	Percentage
Gravel	>4.75 mm	0
Coarse sand	4.75–2 mm	1
Medium sand	2 mm–425 μm	24
Fine sand	425–75 μm	68
Silt and clay	<75 μm	7

Preliminary Analyses of the Chandmari Landslides

We modelled rainwater infiltration and the subsequent changes in soil pore-water pressure using the method proposed by Iverson (2000). The method assesses both the pore-water pressures that develop in response to individual rainstorms, and the quasi-steady background water pressures that develop in response to rainfall averaged over a much longer time period, such as the entire monsoon season.

In our analysis of the earthquake-induced landslide at Chandmari, we computed the peak horizontal acceleration at the ground surface due to the September 2011 earthquake and then applied Newmark's (1965) sliding block procedure to evaluate seismic slope stability. Finally, for each of the Chandmari Locality landslides, we used infinite slope stability analyses to evaluate the factor of safety (FS) at the time and location of the landslide.

Soil properties were presented in Table 3, landslide details were provided in the subsection, Landslide history, and analysis details are provided in the following subsections.

A Few Notes

1. We have adopted the terminology and notation of Iverson (2000).
2. The pore pressure, $\psi(Z,t)$, that develops at a depth Z and time t in response to a rainfall of intensity I_Z and duration T, can be expressed as

$$\psi(Z,t) = (Z - d_Z)\beta + Z \cdot \frac{I_Z}{K_Z} \cdot [R(t^*)] \text{ for times } t \leq T \quad (1)$$

d_Z : Steady state water table depth

$$\beta = \cos^2\alpha - (I_Z/K_Z)_{\text{steady}}$$

α: Slope angle

K_Z : Hydraulic conductivity in the Z direction

$$R(t^*) = \sqrt{t^*/\pi}\exp(-1/t^*) - \text{erfc}(1/\sqrt{t^*})$$

$$t^* = \frac{4D_0\cos^2\alpha}{Z^2}t$$

where D_0 : Hydraulic diffusivity

3. We assumed that $d_Z = 2\,\text{m}$.
4. From Tables 1 and 3, $(I_Z/K_Z)_{\text{steady}} \sim 0.01$.
5. Computed from Table 3, $D_0 = 1.4 \times 10^{-5}\,\text{m}^2/\text{s}$.
6. An expression for the factor of safety (Iverson 2000)

$$FS = \frac{\tan\varphi}{\tan\alpha} + \frac{-\psi(Z,t)\,\gamma_w \tan\varphi}{\gamma_s Z \sin\alpha \cos\alpha} + \frac{c}{\gamma_s Z \sin\alpha \cos\alpha} \quad (2)$$

c : Cohesion
φ : Angle of internal friction
γ_s : Unit weight of soil
γ_w : Unit weight of water, $\gamma_w = 9800\,\text{N/m}^3$.

The 6 June 2011 Rainfall-Triggered Debris Slides

From the description of the slides (The Sikkim Times 2011) and our 2015 topographic map of the study area, we estimated:

For Slide 1, at the final turning to Chandmari Locality,

$$\alpha = 45^\circ, Z \approx 2\,\text{m}$$

For Slide 2, near the Forest Checkpost,

$$\alpha = 44.2^\circ, Z \approx 2\,\text{m}$$

Computing FS from Eqs. (1) and (2), we find that at the time of the slides, $FS = 0.93$ for Slide 1 and $FS = 0.95$ for Slide 2.

The 18 September 2011 Earthquake-Induced Debris Slide

For the 18 September 2011 landslide,

$$\alpha \approx 55^\circ, 2\,\text{m} \leq Z \leq 5\,\text{m}$$

(Chakraborty et al. 2011; SikkimNow 2011).

The static factor of safety (James and Sitharam 2015) was calculated from Eqs. (1) and (2). As per Indian Standard (Arora 2003), we performed the CD direct shear test (Table 3) using fully saturated soil. However, since this slide occurred towards the end of the monsoons, we used $c = 11000\,\text{Pa}$ and $\varphi = 33^\circ$, values obtained using partially saturated soil. $FS(Z = 2\,\text{m}) = 1.3$.

Seismic Factor of Safety

The attenuation relation proposed by Sharma et al. (2009) was used to evaluate the peak horizontal acceleration at bedrock level, Y_{br}.

$$\log(Y_{br}) = b_1 + b_2 M_W - b_3 \log\left(\sqrt{R_{JB}^2 + b_4^2}\right) + b_5 S + b_6 H$$

The regression coefficients corresponding to the shortest period $T = 0.04\,\text{s}$ were used. As recorded earlier in this paper, $M_W = 6.9$, Joyner-Boore distance (shortest distance from the site to the surface projection of the rupture surface) $R_{JB} = 68$ km, $H = 1$ since it is a strike-slip fault, and $S = 1$ since it is a rock site.

The peak horizontal acceleration at ground surface

$$Y_{gs} = F \cdot Y_{br}, \text{where } \log F = a_1 \cdot Y_{br} + a_2 + \log(\delta_s)$$

$a_1 = 0, a_2 = 0.49, \delta_s = 0.08$ (Kanth and Iyengar 2007). Computing the critical factor of safety (Newmark 1965; James and Sitharam 2015),

$$FS_{critical} = \left(Y_{gs}/g \sin\alpha\right) + 1$$

We obtained $FS_{critical} = 5.5$.

The static factor of safety ($FS = 1.3$) was lesser than the critical factor of safety and the slope was unstable.

The 1997 and 2007 Rainfall-Triggered Rock and Debris Slides

As we noted in an earlier section on the rock samples and discuss further in the Discussion Section, our borehole observations indicated that gneissic rock overlies highly weathered mica schist. Moreover, fractures were observed in some of the gneissic rock samples. We suggest that during the 1997 and 2007 rock and debris slides (Table 2), rainwater infiltrated through fractures in the overlying gneiss and caused the sliding of the mica schist.

We applied an infinite slope stability equation (Duncan et al. 1987) to the mica schist layers:

$$FS = A\frac{\tan\varphi}{\tan\alpha} + B\frac{c}{\gamma_s Z}$$

where A and B are dimensionless coefficients.

We assumed that $Z = 5$ m, the depth of the first mica schist layer. Since there was high core loss, it was not possible to determine the unit weight of the mica schist; we assumed it to be the same as that of Chandmari undisturbed soil.

For the 1997 slide, $\alpha = 31^\circ$ (Bhasin et al. 2002), $FS = 0.3$.

For the 2007 slide, $\alpha = 49.4^\circ$ (our topographic map), $FS = 0.1(c = 0\,\text{Pa}, \varphi = 35^\circ$, Section: Rock samples).

We realize that the factors of safety are unrealistically low, but they lend credence to our hypothesis that in the rare circumstances that water infiltrates through several meters of the overlying gneissic rock and saturates the underlying mica schist, a landslide is highly likely.

Increased Subsidence During the 1975 and 1984 Monsoons

Basu and De (2003) recorded two landslides at Tathangchen (a locality on the lower regions of Chandmari Hill, Fig. 5); the landslides occurred during the 1975 and 1984 monsoons respectively. We suggest that these landslides in the lower regions of the hill caused increased subsidence at Chandmari Locality.

Discussion

Subsidence During the Annual Monsoons

An examination of the rock samples from the borehole revealed strata of fine sand/silt-sized material underlying gneissic rock. We postulate that during the monsoons, surface runoff infiltrates through fractures in the overlying rock, causes removal of the fines, and leads to subsidence.

Further, the sand/silt-sized material was identified as highly weathered mica schist. Since the Paro/Lingtse gneisses thrust southward over the mica schists of the Daling Formation along the Main Central Thrust, MCT2, it is reasonable to suppose that in many hills and mountains of this region (near the MCT2), gneissic rock overlies highly weathered mica schist, as it does on Chandmari Hill. We propose that leaching of the highly weathered mica schist by surface runoff accounts for subsidence in these hills.

Rockslides in the Region

We further submit that many of the rockslides in this region are initiated when rainwater infiltrates through fractures and causes sliding of the soft micaceous bands underlying the hard gneissic rock.

Rainfall and Landslides

An examination of each rainfall entry in Tables 1 and 2 leads us to believe that the daily rainfall in Gangtok City rarely exceeds 35 mm, but on the days when a slide occurred, the rainfall greatly exceeded this amount. On the day of the earthquake-induced landslide, there was 39.3 mm of rain; on the day of the 2007 rock and debris slide, there was 72.8 mm of rain; and on the day of the 2011 debris slides, there was 118 mm of rain. However, there are days when the total rainfall exceeded 100 mm (not shown in Tables 1 and 2) but no slide occurred.

Modelling Landslides

This paper presents work in progress; we continue our efforts to (1) update our information on the Chandmari landslides, rainfall, and soil and rock properties, and (2) refine the mathematical models used to study Chandmari Hill. During the full-scale deployment, we will drill several boreholes for the installation of sub-surface instruments. We hope to obtain undisturbed soil and rock samples from different depths and locations on the hill, further our knowledge of the sub-surface conditions, and find answers to questions that arise during interim studies.

Summary and Conclusion

We examined all landslides and occurrences of subsidence recorded at Chandmari Locality. We conclude that

1. Surface runoff infiltrates through fractures in overlying gneissic rock and causes extrusion of finer micaceous material, leading to subsidence.
2. During torrential rains, rainwater infiltration causes sliding of the soft micaceous bands underlying gneissic rock, leading to debris and rock slides.
3. Processes similar to those described in (1) and (2) are responsible, at least in part, for the frequent and widespread occurrences of rockslides and subsidence observed throughout the region.

Acknowledgements We thank Mr. Ranjith N. Sasidharan for helping us prepare the figures for this paper, Dr. H. M. Iyer, Dr. R. Dhandapani, Dr. Sreevalsa Kolathayar, Dr. P. Thambidurai, Dr. Ganesh Khanal, and Mr. Keshar Kumar Luitel for technical discussions, Mr. Kevin Degnan for help with technical writing and proof-reading the entire manuscript, and Shri. P. P. Shrivastav for his support of this work. We thank the session editor, Dr. Jan Klimeš, for reading our original submission with great care and providing insightful comments. The Amrita University landslide projects, aimed at developing wireless sensor network-based landslide early warning systems, are vast, multi-disciplinary efforts, and the team is too large for us to name every member; we gratefully acknowledge their help and support. Above all, we express our heartfelt gratitude for the immense motivation and guidance provided by the Chancellor of our University, Sri Mata Amritanandamayi Devi (Amma). This work was funded by the Ministry of Earth Sciences (MoES), Government of India.

References

Anbarasu K, Sengupta A, Gupta S, Sharma SP (2010) Mechanism of activation of the Lanta Khola landslide in Sikkim Himalayas. Landslides 7(2):135–147

Arora KR (2003) Soil mechanics and foundation engineering. Standard Publishers Distributors, Delhi

Basu SR, De SK (2003) Causes and consequences of landslides in the Darjeeling-Sikkim Himalayas, India. Geographia Polonica 76 (2):37–52

Bhasin R, Grimstad E, Larsen JO, Dhawan AK, Singh R, Verma SK, Venkatachalam K (2002) Landslide hazards and mitigation measures at Gangtok, Sikkim Himalaya. Eng Geol 64(4):351–368

Bhattacharyya K (2010) Geometry and kinematics of the fold-thrust belt and structural evolution of the major Himalayan fault zones in the Darjeeling–Sikkim Himalaya, India. Ph.D. thesis, University of Rochester, Rochester, USA

Chakraborty I, Ghosh S, Bhattacharya D, Bora A (2011) Earthquake induced landslides in the Sikkim-Darjeeling Himalayas—an aftermath of the 18th September 2011 Sikkim earthquake. Geological Survey of India, Kolkata

Dubey CS, Chaudhry M, Sharma BK, Pandey AC, Singh B (2005) Visualization of 3-D digital elevation model for landslide assessment and prediction in mountainous terrain: a case study of Chandmari landslide, Sikkim, Eastern Himalayas. Geosci J 9 (4):363–373

Duncan JM, Buchignani AL, De Wet M (1987) An engineering manual for slope stability studies. Virginia Polytechnic Institute and State University, Blacksburg

Ghoshal TB, Sringanengam S, Sengupta CK (1998) Detailed investigation of Chandmari Landslide near Gangtok. Geological Survey of India, Kolkata

Gupta H, Gahalaut VK (2014) Seismotectonics and large earthquake generation in the Himalayan region. Gondwana Res 25(1):204–213

Iverson RM (2000) Landslide triggering by rain infiltration. Water Resour Res 36(7):1897–1910

James N, Sitharam TG (2015) Macro-level assessment of seismically induced landslide hazard for the State of Sikkim, India based on GIS technique. In: IOP Conference series: Earth and Environmental Science 26(1):012027. IOP Publishing

Kanth S R, Iyengar R N (2007) Estimation of seismic spectral acceleration in peninsular India. Journal of Earth System Science. 116(3)

Newmark NM (1965) Effects of earthquakes on dams and embankments. Geotechnique 15(2):139–160

Ramesh MV, Vasudevan N (2012) The deployment of deep-earth sensor probes for landslide detection. Landslides 9(4):457–474

Rawat RK (2005) Geotechnical investigations of Chandmari landslide located on Gangtok-Nathula road, Sikkim Himalaya, India. J Himalayan Geol 26(2):309–322

Reuters (1997) Landslides claim at least 28 in India. CNN. 9 June 1997. http://edition.cnn.com/WORLD/9706/09/india.mudslides/. Last accessed 7 Sept 2016

Roy S, Baruah A, Misra S, Mandal N (2015) Effects of bedrock anisotropy on hillslope failure in the Darjeeling-Sikkim Himalaya: an insight from physical and numerical models. Landslides 12 (5):927–941

Sharma AK (2008) Landslide and its mitigation for disaster management using remote sensing and GIS technique-a case study of Gangtok area, East Sikkim. MSc thesis Sikkim Manipal University of Health, Medical and Technological sciences, Gangtok, India

Sharma ML, Douglas J, Bungum H, Kotadia J (2009) Ground-motion prediction equations based on data from the Himalayan and Zagros regions. J Earthquake Eng 13(8):1191–1210

SikkimNow (2011) Around Gangtok, 19 Sept 2011. http://sikkimnow.blogspot.in/2011/09/around-gangtok-19-sept-2011a-four.html. Last accessed 9 Sept 2016

The Sikkim Times (2011) Heavy rains trigger landslides across Gangtok. http://sikkimnews.blogspot.in/2011/06/heavy-rains-trigger-landslides-across.html. Last accessed 14 Sept 2016

Vasudevan N, Ramanathan K (2016) Geological factors contributing to landslides: case studies of a few landslides in different regions of India. In: Institute of Physics Conference Series: Earth and Environmental Science. 30(1):012011-012016. IOP Publishing

Vasudevan N, Kolathayar S, Sridharan A, Ramanathan K (2016) An investigative study of seismic landslide hazards. In: Proceedings of the international conference on recent advances in rock engineering (RARE-2016), 16–18 November 2016, Bengaluru, India, pp 195–204

Activation of Cryogenic Earth Flows and Formation of Thermocirques on Central Yamal as a Result of Climate Fluctuations

Artem Khomutov, Marina Leibman, Yury Dvornikov, Anatoly Gubarkov, Damir Mullanurov, and Rustam Khairullin

Abstract

Study area in continuous permafrost zone, characterized by tabular ground ice distribution, is known for active slope processes. In 90-s main attention was paid to translational landslides (active layer detachments). Due to climate trends summer temperature became warmer, active layer depth increased. As a result, active-layer base ice thawed and stopped development of translational landslides. At the same time, tabular ground ice table got involved into seasonal thaw and triggered earth flows at the lake shores, the second known type of cryogenic landslides found previously mainly at the sea coasts. Earth flows are the main process in thermal denudation: a complex of processes responsible for formation of thermocirques. Thermocirques are semi-circle shaped depressions resulting from massive ground ice thaw and removal of detached material downslope. Monitoring of thermocirque activation and development allows analyzing climatic controls of thermal denudation, and rates of thermocirque enlargement. At present in the Yamal Peninsula tundra predominance of processes associated with tabular ground ice thaw (cryogenic earth flows) over the processes associated with the ice formation at the bottom of the active layer (cryogenic translational landslides) is observed. This is caused by deepening of the active layer and exposure of the massive ground ice (tabular ground ice or ice-wedges) within permafrost to first seasonal and then perennial thaw. Activation of thermal denudation which started on Yamal Peninsula in summer 2012, is associated with extremely warm spring and summer

A. Khomutov (✉) · M. Leibman · Y. Dvornikov · D. Mullanurov · R. Khairullin
Siberian Branch of Russian Academy of Sciences, Earth Cryosphere Institute, Malygina Str. 86, Tyumen, 625026, Russia
e-mail: akhomutov@gmail.com

M. Leibman
e-mail: moleibman@mail.ru

Y. Dvornikov
e-mail: ydvornikow@gmail.com

D. Mullanurov
e-mail: damir.swat@mail.ru

R. Khairullin
e-mail: rustam93-93@bk.ru

M. Leibman
Tyumen State University, Volodarskogo Str. 6, Tyumen, 625003, Russia

A. Gubarkov
Tyumen Industrial University, Volodarskogo Str. 38, Tyumen, 625000, Russia
e-mail: agubarkov@gmail.com

M. Mikoš et al. (eds.), *Advancing Culture of Living with Landslides*,
DOI 10.1007/978-3-319-53483-1_24

of this year, and the warmest July of 2013. By the end of the warm season thawing of the top of icy permafrost and tabular ground ice on some slopes resulted in cryogenic landsliding in the form of earth flows and further thermocirque development. Thermocirques may form on slopes of various aspects but develop faster on south-facing slopes.

Keywords

Cryogenic earth flows • Thermocirques • Yamal • Cryogenic processes activation • Climate fluctuations • Field monitoring • Remote sensing

Introduction

North of West Siberia is known for continuous permafrost distribution. Bodies of tabular ground ice are found rather close to the surface of slopes, sometimes directly beneath the active layer. These ice bodies in Yamal Peninsula characterized by deeply dissected Quaternary plains control relief formation through cryogenic slope processes. Less widely distributed and relatively small in size (first meters high and no more than 1 m wide) are ice wedges. They penetrate into deeper sitting tabular ice bodies. Central Yamal is characterized by widely distributed cryogenic translational landslides (active layer detachments) with the last peak of activation in 1989. This process is rather well understood (Khomutov and Leibman 2014; Kokelj and Lewkowicz 1998; Leibman and Egorov 1996; Leibman et al. 2003, 2014; Lewkowicz 1990) and results from the consecutive climate cooling and respective reduction of the active layer depth and formation of ice lenses at the active-layer bed, followed by a warm summer and deeper seasonal thaw which melts this few years' ice at the active-layer base.

Climatic fluctuations over the past few years, specifically, extremely warm spring of 2012, triggered cryogenic earth flows (retrogressive thaw slumps) activity in the tundra zone of the Yamal Peninsula.

The paper focuses on the features resulting from thermal denudation: thawing of icy permafrost or pure ice and removal of the thawed material by gravitation. Thermal denudation leads to the formation of specific landforms: thermocirques: semi-circle shaped depressions remaining after massive ground ice thaw and removal of detached material downslope. Thermocirques form by a combined effect of several destructive relief-forming processes in massive (tabular and ice-wedge) ground ice-bearing areas. Leading and starting process is cryogenic earth flow.

Research Area

The Vaskiny Dachi research station (Fig. 1) is located in continuous permafrost zone. Within 28 years of monitoring, a unique database is established as well as bench marking research in periglacial geomorphology is undertaken (Leibman et al. 2015).

Research area is located at the watershed of Se-Yakha and Mordy-Yakha rivers covering a system of highly-dissected alluvial-lacustrine-marine plains and terraces where a number of hazardous processes operate, such as thermal erosion and landsliding. Deposits are sandy to clayey, most are saline within permafrost, and some are saline in the active layer. The periphery of hilltops is embraced by windblown sand hollows, covering large areas. Saddles between the hilltops are often covered by polygonal peatlands bearing ice wedges, while some convex hill tops are occupied by wellshaped rectangle sandy polygons with sand-and-ice wedges. Slopes comprise a mosaic of concave and convex surfaces, first being ancient and modern landslide shear surfaces, and second being stable slopes (Leibman et al. 2015).

Methods

Study of thermocirques involved direct observations from helicopter, land-based trips, GPS surveys, and interpretation of very-high resolution images of several years. Relation of thermocirque activation to climate fluctuations was estimated using: mean annual temperature (MAAT), sum of summer temperatures (thaw index), sum of winter temperatures (freeze index), both in degree-days, summer and winter precipitation. Summer parameters are presented in two types: by the date of measurement and for the full season.

Distribution and thermocirques activation date were obtained by direct observations from helicopter and reconnaissance trips since 1987.

Initial size of thermocirques was measured on a very-high resolution image (GeoEye-1) of July 5, 2013 (*Digital Globe Foundation©*). The only exception was thermocirque discovered and GPS-surveyed during the field trip of 2012. Area increase was determined using repeated GPS-surveys of thermocirque edge position close to the end of the warm period. Area increase in per cent and square meters was calculated in relation to both initial size and previous year's size.

Fig. 1 Research area location on Central Yamal

Old thermocirques activated by new forms were located based on previous experience and direct observations according to degree of re-vegetation of the surface (Leibman et al. 2014).

Lateral distribution of thermocirques was interpreted on GeoEye-1 (*Digital Globe Foundation©*). Part of the image was covered by clouds and coverage was improved by using WorldView-2 (*Digital Globe Foundation©*) for the covered area.

Climatic Factors of Cryogenic Landsliding Activation

Climate fluctuations over the past few years significantly affected increase of cryogenic processes activity in the tundra zone of the Yamal Peninsula. On Central Yamal a large-scale cryogenic landsliding was observed in 1989, while cryogenic earth flows were actively developing since 2012 through tabular ground ice thawing.

In general, period 2004–2011 is characterized by thaw index of 767 degree days (range 501–896 degree days). Compare, in 2012–2015 thaw index in average was 801 degree days (range 492–1057, minimum is a little bit lower and maximum is higher) (Table 1).

Thus, change of the mechanism of slope processes from translational landsliding to earth flows is explained by the following. The warm period of 2012 started on May 25, maximal average daily temperature was +18,0 °C on June 29 (Fig. 2), and thaw index calculated for the period from May 25 to September 2 (date of active layer depth measurement) was 854 degree-days with amount of precipitation 257 mm (Table 1). Extremely warm season of 2012 resulted in increased active layer. According to measurements on CALM grid (data included into GTN-P database: http://www.gtnpdatabase.org/activelayers/view/114#.U8Upj7HBaSk) active layer depth was 102 cm, 15% deeper than the average for the 1993–2011 by the end of the warm period. Early measurement on July 3, 2012 showed active layer depth averaged at 66 cm (already 76% of thaw depth

Table 1 Climatic controls of the active-layer dynamics since 2004, weather station Marre-Sale (www.rp5.ru)

Year/Summer season	MAAT	Thaw index		Summer precipitation		Winter season	Freeze index	Winter precipitation
		Till measure date	Full	Till measure date	Full			
2004	−7.7	720	796	nd	nd	2004–05	−2710	nd
2005	−5.4	661	878	146	279	2005–06	−3181	213
2006	−8.0	645	705	112	201	2006–07	−3049	136
2007	−5.0	654	896	68	146	2007–08	−2748	218
2008	−5.0	679	806	307	384	2008–09	−2974	253
2009	−8.3	nd	804	nd	243	2009–10	−3785	140
2010	−8.1	465	501	151	204	2010–11	−2788	247
2011	−4.4	496	752	108	188	Mean (2004–05)–(2010–11)	−3034	201
Mean 2004–2011	−6.5	617	767	149	235	2011–12	−2771	190
2012	−4.1	854	1057	257	305	2012–13	−3105	144
2013	−7.3	656	743	114	243	2013–14	−3262	203
2014	−7.4	393	492	115	151	2014–15	−3024	180
2015	−5.3	764	912	149	179	–		–
Mean 2012–2015	−6.0	667	801	159	220	Mean (2011–12)–(2014–15)	−3041	179

reached by late August of 2010 and 2011). Thus deeper thaw melts ground ice which accumulates at the active layer base and triggers translational landslides (Leibman 1998; Leibman and Egorov 1996). In 2012, on some slopes thawing has reached the top of icy permafrost or mono-mineral tabular ground ice. As a result, cryogenic earth flows form on lake shores and river banks, further on merging to form new or re-activate stabilized thermocirques. While translational landslide events are separated by several centuries and form landslide cirques, earth flows form thermocirques which once being triggered, develop until either ice is exhausted, or insulated by landslide bodies from further thaw.

In 2013, in spite of close to average climatic parameters for a period from 2004 to 2011, active layer was also abnormally deep (average 103 cm). It could be due to a combination of two factors: delayed refreezing in the fall 2012, and thawing of saline marine clays overlain by sandy and loamy deposits. Saline clay retains plasticity to a temperature of −1 °C and reacts to the mechanical probing as thawed. Warm season of 2013 shows maximal daily air temperature (+22.3 °C on July 19) for entire period from 2004 to 2015. Daily air temperature was higher than +14 °C twelve days running from July 16 to 27 (thaw index was 210 degree-days or 32% of thaw index till active layer measure date) (Fig. 2). Summer 2014 was much cooler, and 2015 warmer than average (Fig. 2). The cooler years after thermal denudation started would not stop it because once exposed, ice continues melting each summer. The rate of its melting and removal of thawed material out of thermocirque also depends on sum of summer precipitation which was among maximums in 2012.

Analysis of Thermocirque Distribution

By 2013, according to the field and remote sensing data (GeoEye-1, WorldView-2), there were more than 90 active thermal denudation landforms from 66 to 25,700 m^2 in size on the territory of 345 km^2 (301 km^2 of land without lakes). Area affected by thermocirques occupies 159,000 m^2, which is about 0.0005% of entire area without lakes, compare to a translational landslide-affected surface of about 1% of the land area.

Comparison of satellite images of 2010 and 2013 for the same area of 315 km^2 showed that the number of active thermal denudation forms in technogenically undisturbed environments has increased in number from 11 to 65. In 2010 there was only non-significant activity in the upper part of over-vegetated thermocirques, while in 2013 there were mostly new and re-activated thermocirques with considerable backwall retreat.

No image of 2012, when a new thermocirque was discovered is available to find indications for the date of other thermocirques' formation. Direct observations establish that there were no thermocirques in 2011. We assume that activation or re-activation of all thermocirques could have

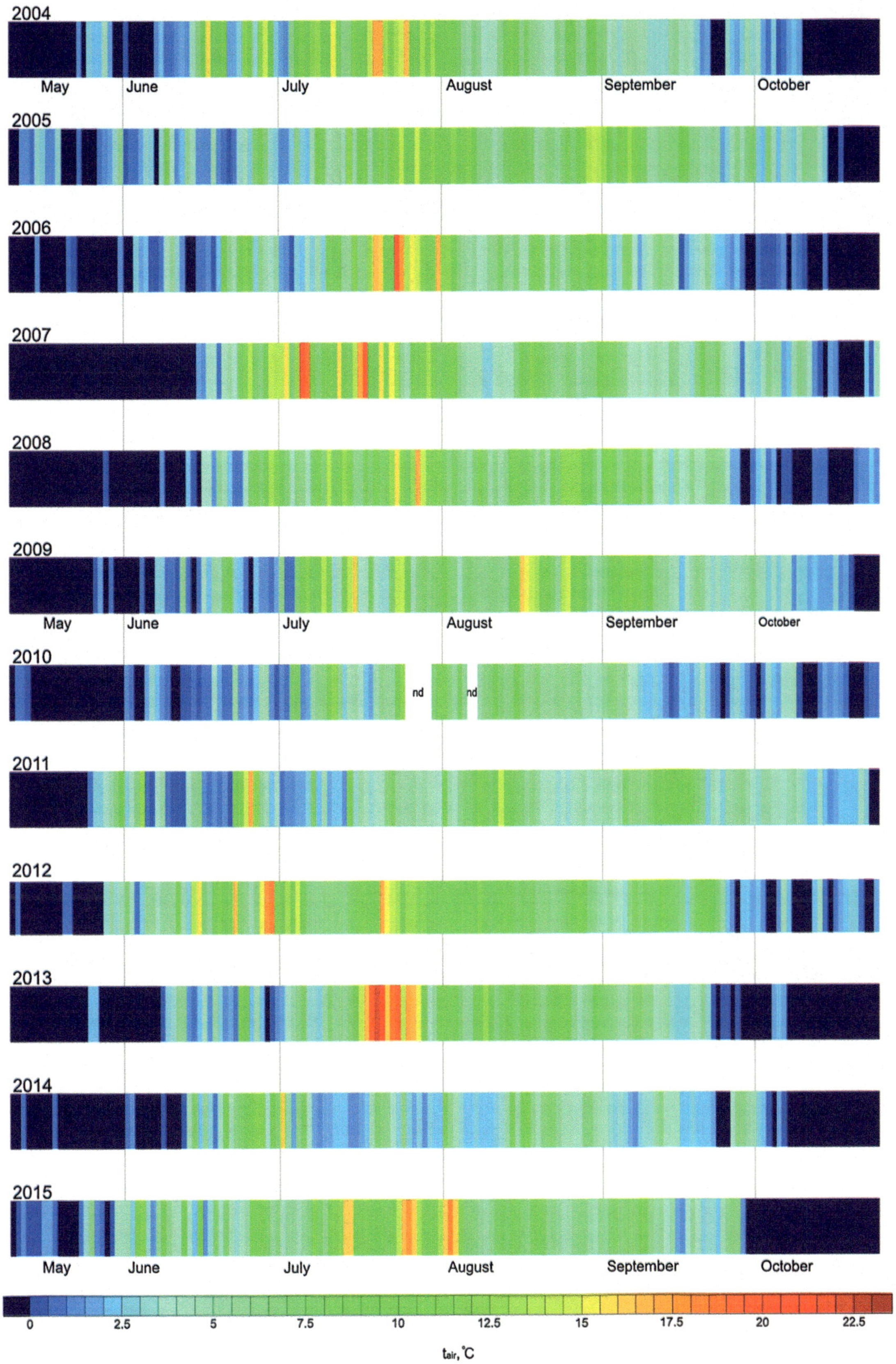

Fig. 2 Warm period daily air temperature since 2004, according to Marre-Sale weather station records (www.rp5.ru)

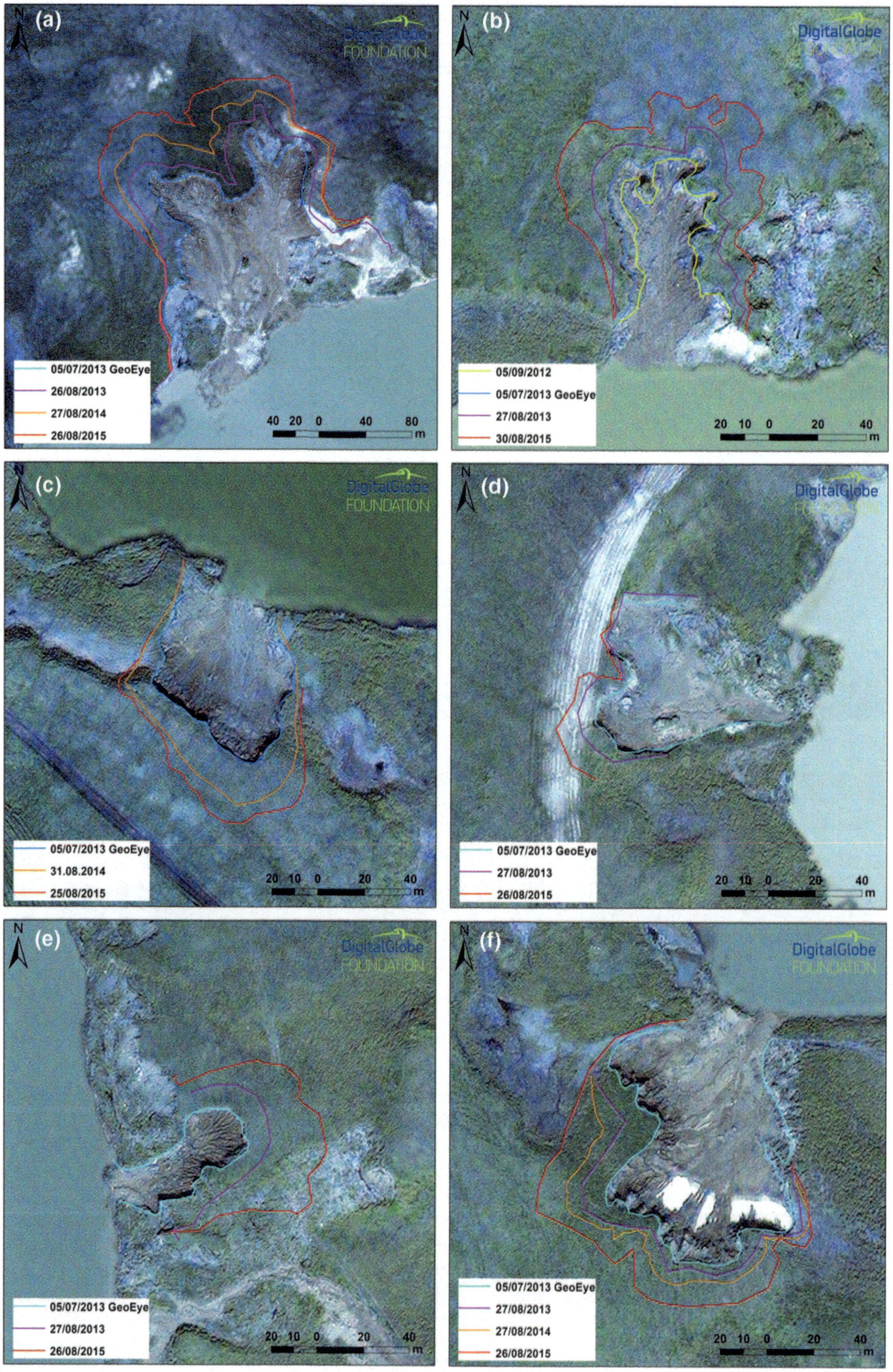

Fig. 3 Thermocirques #1–6 (a–f, respectively) area enlargement from 2012–2013 to 2015 according to field survey and interpretation of GeoEye-1 image (*Digital Globe Foundation©*)

Table 2 Thermocirque area enlargement on Vaskiny Dachi research station

Thermocirque ID	Ground ice type	Date of retreat measurement	Area of thermocirque, m^2	Area increase, in %	
				To initial size	To previous year's size
Thermocirque 1	Tabular ice	05/07/2013 26/08/2013 27/08/2014 26/08/2015	25,700 30,800 35,900 40,800	19.8 19.8 19.1	16.5 13.6
Thermocirque 2	Tabular ice and ice wedges	05/09/2012 05/07/2013 27/08/2013 30/08/2015	2300 2900 4200 6400	26.1 56.5 95.7	82.6 50.2
Thermocirque 3	Tabular ice	05/07/2013 31/08/2014 25/08/2015	3300 5000 5700	51.5 21.2	51.5 14.0
Thermocirque 4	Tabular ice and ice wedges	05/07/2013 27/08/2013 26/08/2015	2600 4100 4700	57.7 23.1	14.6
Thermocirque 5	Tabular ice	05/07/2013 27/08/2013 26/08/2015	1300 2900 5100	123.1 169.2	75.9
Thermocirque 6	Tabular ice	05/07/2013 27/08/2013 27/08/2014 26/08/2015	5800 7600 8500 10,200	31.0 15.5 29.3	11.8 20.0

started in 2012 because of the extreme summer warmth. They were discovered in field and appeared on the image of 2013. As far as the image date was early summer (05.07.2013), and thermocirque depressions still keep snow patches, our best judgment was that earth flows forming thermocirques started during the warm period of 2012 when active layer got deep enough to start massive ground ice thaw.

Thermocirque Monitoring

Six thermocirques originating from the thaw of tabular ground ice (Fig. 3a, c, e, f) or both tabular and ice-wedge ice (Fig. 3b, d), having high retreat rates, are annually monitored by GPS survey (Table 2). One thermocirque is monitored since 2012. Thermocirque #1 (Fig. 3a) is the largest, old but re-activated, while 5 others (#2–6, Fig. 3b–f) are relatively small, newly formed features. All thermocirques monitored are linked to slopes of lake depressions. Results of monitoring are as follows: Thermocirque #1 having area of 25,700 m^2 (Table 2, Fig. 3a) between the first measurement (on an image of 05/07/2013) and next measurement (in field) increased by almost 20%, up to 30,800 m^2. Taken that the average area of other thermocirques is about 1560 m^2, enlargement of this specific thermocirque is really significant. Next years the rate of area enlargement did not reduce much even though year 2014 was the coolest in the period 2004–2015.

Thermocirque #2 (Table 2, Fig. 3b) which exposed both tabular and ice-wedge ice, was first measured in 2012, the probable year of its appearance. Yet by that time its area already was 2300 m^2. Since August 2012 and till August 2013 its area increased by more than 80%. At the same time, between July 05 and August 27, 2013 thermocirque aggradation reached 1300 m^2, which more than twice exceeded its growth during the period September 5, 2012 to July 5, 2013. The other thermocirque with thawing of both tabular and ice-wedge ice (#4, Table 2, Fig. 3d) during the first stage of observation in 2013 between the satellite image date and field monitoring date also increased considerably in area by 1500 m^2 or more than by 50% of its initial area. Yet according to field monitoring in 2015, the following year's rate of its backwall retreat reduced, probably, due to the fact that its scar reached the highest point of the hilltop. The largest thermocirque out of newly formed (#6, Table 2, Fig. 3f) since June 5 till August 27, 2013 increased by more than 1800 m^2 (30%). Then by the end of summer 2014 thermocirque area increased by 900 m^2 more (15.5% of initial area), and in 2014–2015, by 1700 m^2 (29.3% of initial area). Reduced rate of growth in 2014 and its consequent increase in 2015 is explained, first, by very cool summer of 2014 preceded by a rather severe winter, and second, with rather hot and more humid summer of 2015. Two remaining thermocirques (#3 and #5, Table 2, Fig. 3c, e) with only tabular ground ice thaw have different initial area (3300 and 1300 m^2, respectively). Thermocirque #3 formed on a steep flat short slope of the lake depression and initially was relatively isometric in shape, while thermocirque #5 occurred on a longer slope, and initially had an elongated downslope shape. By the end of summer 2015 areas of these thermocirques became practically equal (5700 and 5100 m^2), Thermocirque #3 developed only upslope reaching hilltop, while thermocirque #5 grew in 3 directions within the slope at a high rate compared to its initial size.

Thermocirques #1 and #2 (Table 2, Fig. 3a, b) are facing south, while other 4 thermocirques have different aspects. This indicates that slope aspect does not control thermocirque occurrence. At the same time, the rate of thermocirque enlargement is higher for the south-facing thermocirque #2 compared to other small thermocirques. As to the largest thermocirque #1, also facing south, its retreat rate is constantly low in per cent, due to its huge initial area. In absolute values the rate of enlargement is high. Effect of southern aspect is explained by a higher rate of exposed in the backwall ice melt under direct solar radiation.

All new thermocirques increased their area at a rate from 20 to 50% in relation to their first measured size. The range is explained by differences not only in slope aspect, but also gradient and length of slope, distance of the initial area to the hilltop, and type of ice exposed in the backwall and side walls. Extremely cold summer of 2014 slowed down the rate of thermocirques' enlargement, with the higher rates observed again in warm summer of 2015.

The rate of the thermocirque area enlargement slows down with gradual increase of their total area. However, field measurements of 2015 show that backwall retreat rate is still high.

Conclusion

Thus, at present in the Yamal Peninsula tundra predominance of processes associated with tabular ground ice thaw (cryogenic earth flows playing the leading role in thermal denudation) over the processes associated with the ice formation at the bottom of the active layer (cryogenic translational landslides) is observed. It is caused by deepening of the active layer and exposure of the massive ground ice within permafrost to first seasonal and then perennial thaw.

Activation of thermal denudation which started on Yamal Peninsula in summer 2012, is associated with extremely warm spring and summer of this year, and the warmest July of 2013. By the end of the warm season thawing of the top of icy permafrost and tabular ground ice on some slopes resulted in cryogenic landsliding in the form of earth flows and further thermocirque development.

Thermocirques may start to form on slopes of various aspects but develop faster on south-facing slopes.

Acknowledgements The study is partially supported by RFBR, research project No. 13-05-91001-АНФ_а; ASF No I 1401-N29; and The Presidential Council for grants, Science School Grant No. 9880.2016.5; Russian Science Foundation Grant 16-17-10203; International projects CALM and TSP.

References

Khomutov A, Leibman M (2014) Landslides in cold regions in the context of climate change, envi-ronmental science and engineering. In: Shan W, Guo Y, Wang F, Marui H, Strom A (eds). Springer International Publishing, Switzerland. (ISBN 978-3-319-00866-0), pp 271−290

Kokelj SV, Lewkowicz AG (1998) Long-term influence of active-layer detachment sliding on permafrost slope hydrology. Hot Weather Creek, Ellesmere Island, Canada, pp 583–589

Leibman MO (1998) Active layer depth measurements in marine saline clayey deposits of Yamal Peninsula, Russia: procedure and interpretation of results. Proceedings of seventh international conference on Permafrost, 23–27 June 1998, Yellowknife, Canada. pp 635–639

Leibman MO, Egorov IP (1996) Climatic and environmental controls of cryogenic landslides, Yamal, Russia. Landslides. Balkema Publishers, Rotterdam, pp 1941–1946

Leibman MO, Kizyakov AI, Sulerzhitsky LD, Zaretskaya NE (2003) Dynamics of the landslide slopes and mechanism of their development on Yamal peninsula, Russia. Proceedings of the 8th international conference on Permafrost, 21–25 July 2003, Zurich, pp 651–656

Leibman M, Khomutov A, Kizyakov A (2014) Landslides in cold regions in the context of climate change, envi-ronmental science and engineering. In: Shan W, Guo Y, Wang F, Marui H, Strom A (eds). Springer International Publishing, Switzerland. (ISBN 978-3-319-00866-0), pp 143−162

Leibman MO, Khomutov AV, Gubarkov AA, Mullanurov DR, Dvornikov YuA (2015) The research station "Vaskiny Dachi", Central Yamal, West Siberia, Russia—a review of 25 years of permafrost studies. Fennia 193(1):3–30

Lewkowicz AG, (1990) Morphology, frequency and magnitude of active-layer detachment slides, Fosheim Peninsula, Ellesmere Island, N.W.T. Proceedings of the 5th Canadian Permafrost Conference, June 1990. Quebec, Canada. pp 111–118

4th WLF 2017 4th World Landslide Forum LJUBLJANA SLOVENIA EU

Landslide Investigations in the Northwest Section of the Lesser Khingan Range in China Using Combined HDR and GPR Methods

Zhaoguang Hu, Ying Guo, and Wei Shan

Abstract

In the northwest section of the Lesser Khingan Range located in the high-latitude permafrost region of northeast China, landslides occur frequently due to permafrost melting and atmospheric precipitation. High-density resistivity (HDR) and ground penetrating radar (GPR) methods are based on soil resistivity values and characteristics of radar-wave reflection, respectively. The combination of these methods together with geological drilling can be used to determine the stratigraphic distribution in this region, which will allow precise determination of the exact location of the sliding surface of the landslide. Field measurements show that the resistivity values and radar reflectivity characteristics of the soil in the landslide mass are largely different from the soil outside the landslide mass. The apparent resistivity values exhibit abrupt change at the position of the sliding surface in the landslide mass, and the apparent resistivity value decreased suddenly. In addition, the radar wave shows strong reflection at the position of the sliding surface where the amplitude of the radar wave exhibits a sudden increase. Drilling results indicate that at the location of the sliding surface of the landslide mass in the study area, the soil has high water content, which is entirely consistent with the GPR and HDR results. Thus, in practice, sudden changes in the apparent resistivity values and abnormal radar-wave reflection can be used as a basis for determining the locations of sliding surfaces of landslide masses in this region.

Keywords

Landslide • Drilling • High-density resistivity (HDR) • Ground penetrating radar (GPR) • Sliding surface

Introduction

In recent decades, geophysical investigations for assessing stratigraphic distribution have become a common tool in geological research. In-situ geophysical techniques are able to measure physical parameters directly or indirectly linked with the lithological, hydrological and geotechnical characteristics of the terrains related to the movement (McCann and Foster 1990; Hack 2000; Benedetto et al. 2013). These techniques, less invasive than direct ground-based techniques (i.e., drilling, inclinometer, laboratory tests, etc.), provide information integrated on a greater volume of the soil thus overcoming the point-scale feature of classic geotechnical measurements. Among the in-situ geophysical techniques, the High Density Resistivity method (HDR) and Ground Penetrating Radar method (GPR) have been increasingly applied for landslide investigation (McCann and Foster 1990; Malehmir et al. 2013; Timothy et al. 2014).

Z. Hu · Y. Guo · W. Shan (✉)
Institute of Cold Regions Science and Engineering, Northeast Forestry University, Harbin, 150040, China
e-mail: shanwei456@163.com

Z. Hu
e-mail: huzhaoguang008@163.com

Y. Guo
e-mail: samesongs@163.com

M. Mikoš et al. (eds.), *Advancing Culture of Living with Landslides*,
DOI 10.1007/978-3-319-53483-1_25

HDR is based on the measure of the electrical resistivity and can provide 2D and 3D images of its distribution in therocks. It is one of the standard methods of the geophysical prospecting for solution of shallow geological problems (Cardarelli et al. 2003; Yamakawa et al. 2010; Donohue et al. 2011; Sauvin et al. 2014). Current applications of HDR focus on landslide recognition and permafrost detection while investigations on debris thickness in arctic and alpine environments are comparatively sparse (Carpentier et al. 2012; Donohue et al. 2011; Perrone et al. 2014). GPR is based on the measure of reflection of radar waves in the underground, which mainly focus on the fields of exploration of natural resources, hydrogeology, engineering purposes and archaeological investigation (Sass 2007; Sass et al. 2008; Schrott and Sass 2008; Zajc et al. 2014).

HDR and GPR are useful to determine some characteristics of landslides such as main body, geometry, surface of rupture and they have been used in landslide investigations since late 1970s (Hack 2000; Havenith et al. 2000; Bichler et al. 2004; Drahor et al. 2006; Rhim 2011). However, the applicability of the various geophysical methods, including regional limitations and reliability, for stratum thickness estimation and sliding surface location of landslide in high latitude permafrost regions in northeast China has not been addressed in detail. Applications of geophysical prospecting on landslide in cold regions were rare in Lesser Khingan Range of China.

Using Landslide K178 + 530 (the landslide is located at K178 + 530 km stone of Bei'an-Heihe Highway) in the landslide area as an example, in this paper we present a combination of traditional methods (drilling and mapping) and geophysical techniques (HDR and GPR) on landslide in Lesser Khingan Range of northeast China, being applied to gain knowledge about its thickness and internal structure. the landslide that occur frequently due to permafrost melting and atmospheric precipitation, endeavouring to ascertain the applicability of HDR and GPR on the regional type of landslide.

Study Area

The Bei'an-Heihe Highway is located in the northwest section of the Lesser Khingan Range of northeast China, between east longitude 127°17′31″–127°21′24″ and north latitude 49°30′57″–49°41′50″ (Fig. 1). This area is situated on the southern fringe of China's high-latitude permafrost region. The discontinuous permafrost in this region belongs to residual paleoglacial deposition and is currently under degradation process (Shan et al. 2015). Hence, the geological conditions are extremely unstable (Shan et al. 2015). In 2010, the survey conducted in the project of widening the Bei'an-Heihe road and constructing the Bei'an-Heihe Highway also showed that, in the K177 + 400–K179 + 200 section, within 10 m in the northern of the roadbed there were four landslides with the surface area of over 2000 m^2, as illustrated in Fig. 1.

The geological structure of the study area belongs to the Khingan-Haixi fold belt. From the bottom up, the stratigraphy is composed by Cretaceous mudstone, Tertiary pebbly sandstone, silty mudstone, and powdery sandstone. From the late Tertiary to the early Quaternary, the Lesser Khingan Range experienced block uplift. Due to long-term erosion and surface leveling, colluvial sediments have gradually thinned, and the thickness of the current residual layer is generally only 1–2 m. The deposits accumulate mainly in the basin and valley areas with a thickness of about 10 m. The soil is mainly composed of clayey silt, mild clay, and gravelly sand, and the surface is covered with a relatively thick layer of grass peat and turf. The vegetation is represented by grassland and woodland, and there are many tilted trees in the woodlands.

The Landslide K178 + 530 on the Bei'an-Heihe Highway is located at the widened embankment on the north side of the road, as shown in Fig. 2. The roadbed soil and the surface soil slide together to the valley, with run-out of 200 m away from the main scarp. The landslide mass is 20–30 m wide, and covers an area of approximately 6000 m^2. The elevation of the landslide toe is 254 m, and the elevation of the main scarp is 285 m. The trailing edge of the landslide has an arc-shaped, which is located inside the range of the widened embankment.

Four engineering geological boreholes were drilled into K178 + 530 section for prospecting purposes. The depths of the boreholes range from 14 to 26 m (Fig. 3). Following lithologies were identified: Quaternary soils, Tertiary pebbly sandstones, siltstones, and Cretaceous mudstones (Fig. 5).

Embankment: yellow, mainly composed of loosely mixed Tertiary pebbly sandstones, Cretaceous mudstones and sandy mudstones; the soil is incompact when dry, plastic when saturated with water. Clay: yellow, soft and plastic, of high strength and toughness when dry. The upstream region of the landslide mass has 1.5–3.8 m depth distribution ofclay stratum. The downstream region of the landslide mass has 0–6.7 m depth distribution of clay stratum, and there is more than one sandwiched grit layer; the thickness of a single grit layer is approximately 1–10 cm, which greatly enhances the water seepage capacity of the soil.

Tertiary pebbly sandstones: distributed in the embankment at a depth of 2.0–3.4 m and the rear region of the landslide mass at a depth of 3.8–4.5 m, all weathered, composed mainly of feldspar stone and mineral sands,

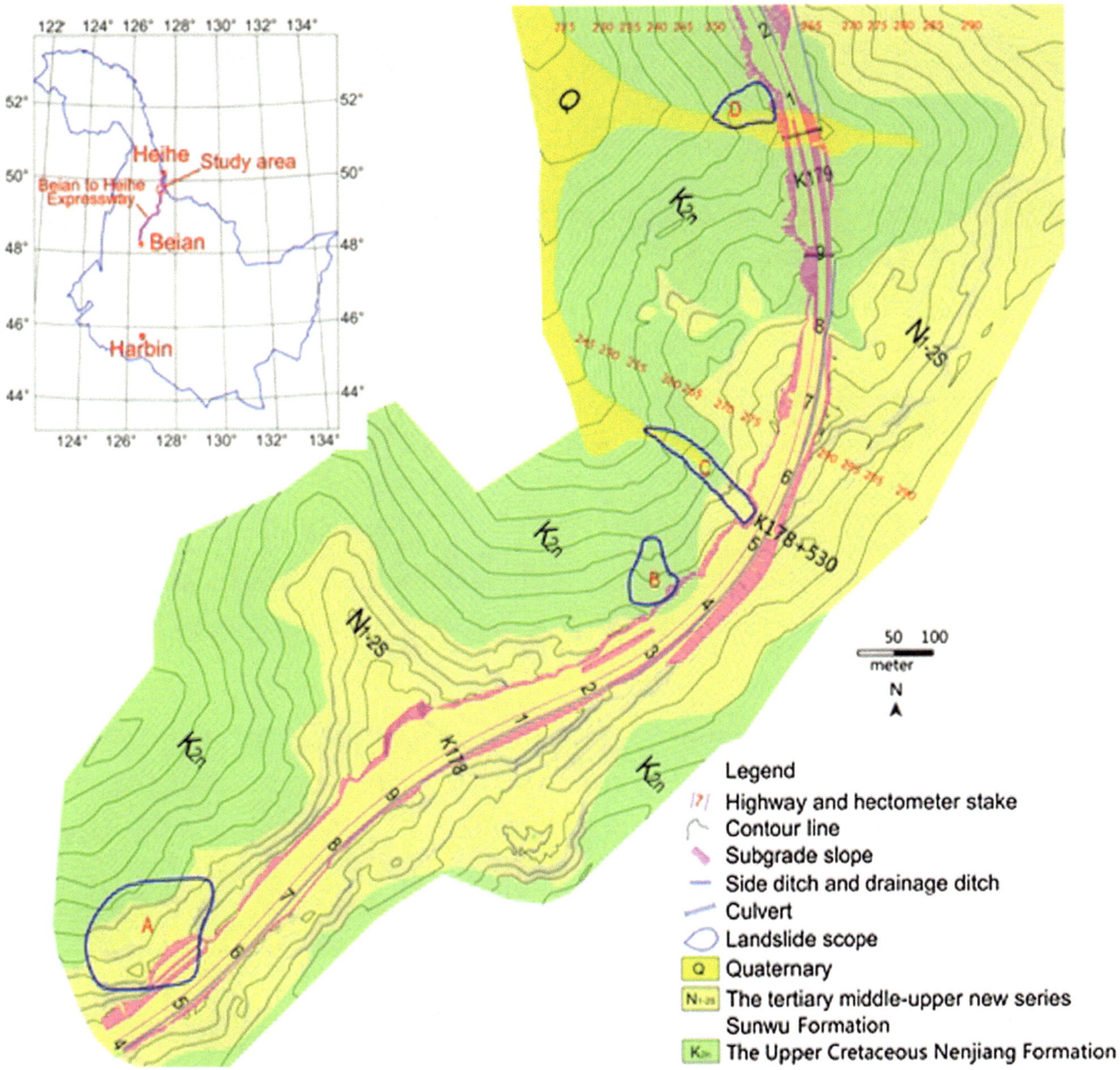

Fig. 1 Geologic and geomorphic map of the study area, four landslides are located in A, B, C and D position, respectively. C shows Landslide K178 + 530, Fig. 1 shows the geologic and geomorphic map of the landside road area of the study area plotted from the field survey conducted in June 2010

well-graded, high permeability, the moisture content is very high, which is greater than 25%. Fully weathered siltstones: yellow, distributed in the upstream region of the landslide mass at a depth of 4.5–9.7 m, sandy, of bedding structure and poor water seepage capacity.

Fully weathered mudstone: yellow or gray-green, pelite, of layered structure, easy to soften with water, of poor water seepage capacity. Strongly weathered mudstones: dark gray, pelite, of layered structure, weakly cemented rock. Moderately weathered mudstones: brown, black and gray, pelite, layered structure.

Methods

In this paper we present combination of geophysical techniques of HDR and GPR. The survey lines were established, as shown in Fig. 4.

HDR Method

The instrument used in this study was the WGMD-9 Super HDR system produced by the Chongqing Benteng

Fig. 2 Studied landslide: **a** Panoramic view of landslide (November 2014); **b** main scarp of the landslide (June 2010); **c** The landslide leading edge (October 2013)

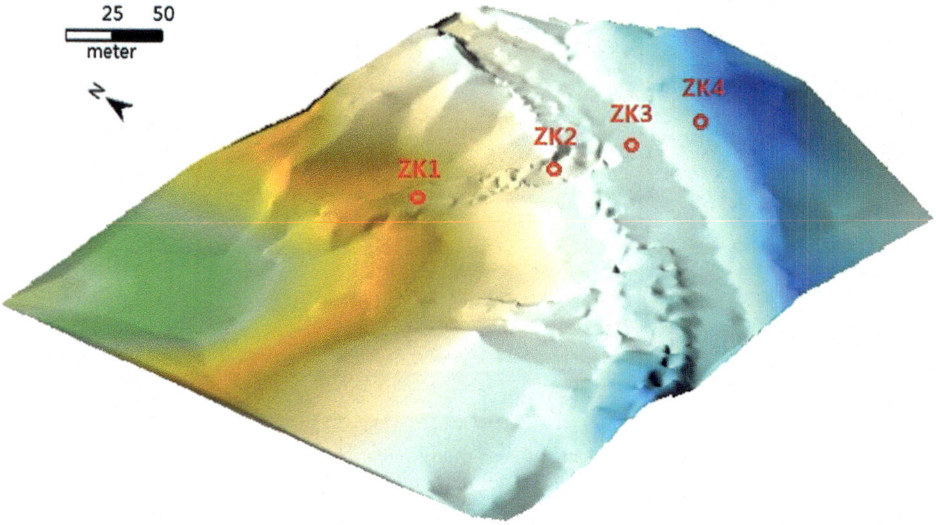

Fig. 3 Location of boreholes on Landslide K178 + 530 and the digital elevation model prepared by SURFER software using data from GPS terrain survey from (June 2010)

Digital Control Technical Institute (Chongqing, China). In this system, the WDA-1 super digital direct current electric device is used as the measurement and control host, and with the optional WDZJ-4 multi-channel electrode converter, centralized high-density cables and electrodes, centralized two-dimensional HDR measurements can be achieved. The inversion of the obtained apparent resistivity data sets was performed using the software RES_2DINV. This software package produces a two-dimensional subsurface model from the apparent resistivity pseudosection (Loke and Barker 1996). Data were acquired by a Wenner configuration. The method is based on the smoothness-constrained least squares inversion of pseudo-section data (Tripp et al. 1984; Sasaki 1992; Loke and Barker 1996). In this algorithm, the subsurface is divided into rectangular blocks of constant resistivity. Then the resistivity of each block is evaluated by minimizing the difference between observed and calculated pseudo-sections using an iterative scheme. The smoothness-constraint leads the algorithm to yield a solution with smooth resistivity changes. The calculated pseudo-sections can be obtained by either finite-difference or finite-element methods (Coggon 1971; Dey and Morrison 1979). In this

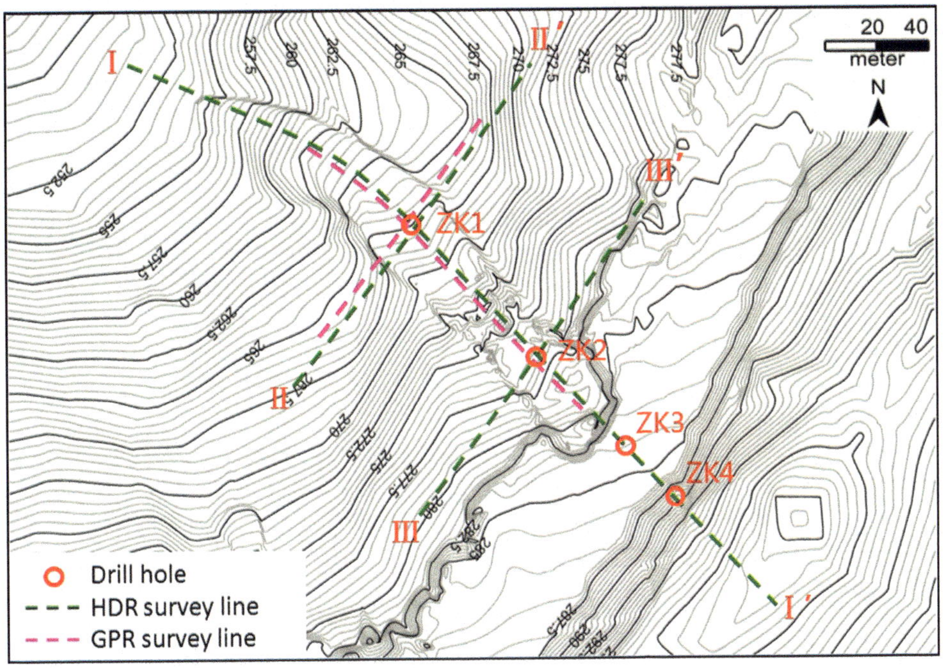

Fig. 4 The layout of the geophysical prospecting survey lines on Landslide K178 + 530. (1) HDR survey lines: on the road section of Landslide K178 + 530, a total of three HDR survey lines, i.e., I-I', II-II' and III-III', were established. Line I-I' ran along the sliding direction of the landslide mass and passed through its *center*. The starting point of the survey line was located 40 m outside landslide toe. The total length of this survey line was 357 m (horizontal distance of 300 m). Line II-II' was perpendicular to the sliding direction of the landslide mass, with point ZK1 serving as the midpoint, electrodes numbered 1–60 were arranged in order from west to east, the total length of this survey line was 177 m. Line III-III' was perpendicular to the sliding direction of the landslide mass, with point ZK2 serving as the midpoint, electrodes numbered 1–60 were arranged in order from west to east, the total length of this survey line was 177 m. The measuring date is September 3, 2012. (2) GPR survey lines: two survey lines were set up, of which the positions coincided with those of the HDR survey lines, but the start and end points were different, the GPR survey lines lengths were shorter than HDR survey lines. The lengths of the two survey lines (I-I' is GPR1, and II-II' is GPR2) were 150 and 118 m, respectively. The measuring date is October 1, 2013

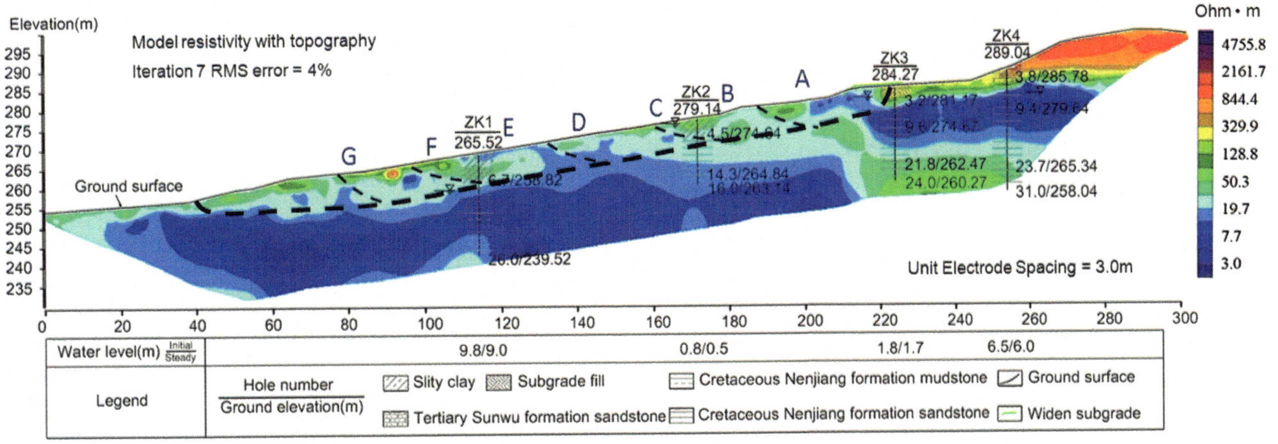

Fig. 5 The Drilling results (June 2010) compared with electrical resistivity profile (September 3, 2012) of the survey line I-I', we can see the start and end points of the profile in Fig. 4, Line I-I'. The *thick black dashed line* show the position of the major sliding surface, the *thin black dashed lines* show the secondary sliding surfaces

case, the finite element scheme was employed due to the topographical changes in the field.

The smoothness-constraint least squares method was used in the inversion model. Essentially, this method is used to constantly adjust the model resistivity through model correction to reduce the difference between the calculated apparent resistivity and the measured resistivity, and to describe the degree of fit between the two using the mean square error. The smoothness-constraint least squares method, which has been widely applied, has a number of advantages, such as adaptability to different types of data and models, relatively small influence of noise on the inversion data, high sensitivity to deep units, rapid inversion, and a small number of iterations. In tests using the HDR method, the spacing between unit electrodes was 3.0 m, and the maximum exploration depth of the survey lines was 30 m.

GPR Method

The GPR instrument used was the RIS-K2 FastWave Ground Penetrating Radar produced by IDS Corporation (Italy). The radar antenna is unshielded dual antenna with low-frequency of 40-MHz. The detection time window was set to 600 ns, the sampling rate to 1024, the data acquisition track pitch was 0.05 m. Two GPR survey lines were set up, as shown in Fig. 4, of which the positions coincided with those of the HDR survey lines, but the start and end points were different. The lengths of the two survey lines (I-I' and II-II') were 150 and 118 m, respectively. The measuring date was October 1, 2013. GPR raw data were processed with the REFLEXW software (Sandmeier Scientific Software, Germany). Coordinates for each trace were calculated at equal distances. The surface signal reflection was set to time zero. Low frequency parts and noise in the spectrum were filtered applying a dewow and bandpass filter. In a next step temporally consistent signals were eliminated using a background removal. Topographical corrections were applied. The data were exported with the attribute of the two-way travel time, the velocity of propagation of the wave in this case appears to be about 0.10 m/ns.

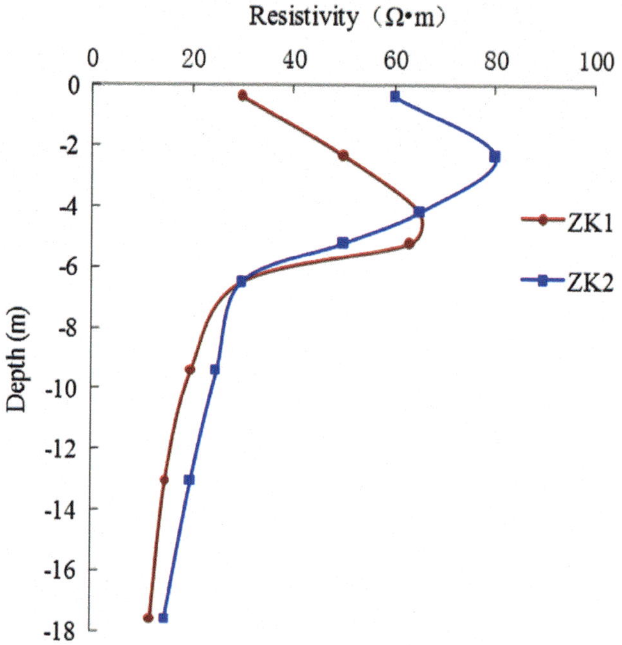

Fig. 6 The electrical resistivity curves at positions ZK1 and ZK2

Results, Analysis, and Discussion

Hdr

Survey line I-I'

As can be seen from the image, the rock resistivity values of the landslide mass exhibited distinct layering. To better analyze how soil resistivity changes with depth, the software RES2DINV can be used to extract the resistivity curve value of any point on the survey line. Figure 6 illustrates the soil resistivity versus depth curves at boreholes ZK1 and ZK2. Taken together with the drilling result, changes in the characteristics of soil resistivity are analyzed as follows.

1. The borehole ZK1. At a depth of 0–2.1 m, the soil is silty clay and rather incompact, in the surface layer containing approximately 15% grass roots and other organic matters. The resistivity value ranges from 25 to 45 Ohm m. At a depth of 2.1–6.7 m, the soil is silty clay, and there are local weathered sand layers. At a depth of 6.5–6.7 m, the moisture content is 25–40%. The resistivity value ranges from 45 to 65 Ohm m. At a depth of 6.7–8.0 m, the soil is yellow mudstone. The permeability coefficient is small, and it is difficult for the water to infiltrate downward, forming a watertight layer, water easily gathered here. The resistivity value is relatively low, i.e., 20–30 Ohm m. At a depth of 8.0–26 m, the soil is gray mudstone, close to or below the water table. The resistivity value is relatively low, i.e., 10–25 Ohm m. As shown in the curve, silty clay contacts mudstone at a depth of 6.7 m, and resistivity exhibits apparent layering, the resistivity value decreased suddenly.
2. The borehole ZK2. At a depth of 0–4.5 m, the soil is rather incompact. At a depth of 4.2–4.5 m, the moisture content is 30–40%. The resistivity values range from 45 to 80 Ohm m. The resistivity of the surface embankment soil dominated by silty clay (depth 0–3.8 m) ranges from 60 to 80 Ohm m, and that of gravelly sand (depth 3.8–4.5 m) ranges from 45 to 60 Ohm m. At a depth of 4.5–9.7 m, the soil is siltstone and composed of rather small particles. The permeability is poor, forming a watertight layer, water easily gathered here. The resistivity value ranges from 25 to 35 Ohm m. At a depth of 9.7–14.6 m, the soil is sandstone, and the resistivity value ranges from 15 to 25 Ohm m. As evident in the curve, gravelly sand contacts the siltstone at a depth of 4.5 m, and there is apparent resistivity layering, the resistivity value decreased suddenly.

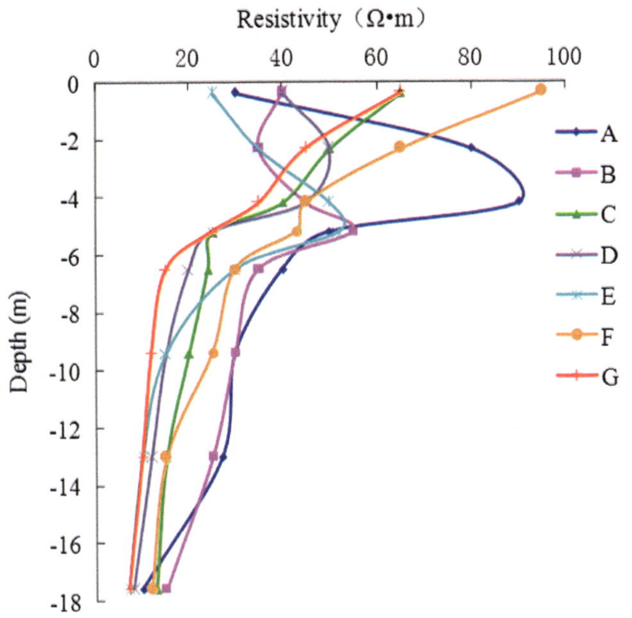

Fig. 7 The electrical resistivity curves at different points on the survey line I-I'

To better understand the changes in the soil resistivity at different positions of line I-I', the software RES2DINV was used to extract the soil resistivity curves at points A, B, C, D, E, F and G on the survey line (Fig. 7). At position A, the soil resistivity value decreased abruptly at a depth of 4.5 m. In other words, the soil resistivity values exhibited abrupt stratification at this depth. It can thus be determined that the sliding surface is located at a depth of 4.5 m. Similarly, the depths of the sliding surface at points B, C, D, E, F and G on line I-I' were determined to be 5, 4.7, 5.5, 6.5, 6, and 5.5 m, respectively.

The above HDR profiles and resistivity curves show that the soil resistivity values above and below the sliding surface of the landslide mass are clearly different and exhibit an abrupt stratification. According to this typical characteristic of the sliding surface, the positions of the major sliding surfaces along line I-I' were deduced, as shown by the thick black dashed line in Fig. 5. The sliding power originated from the trailing edge of the landslide. The slip rate of the trailing edge was the greatest, followed by the middle part of the landslide; the minimum slip rate occurred at the leading edge (Shan et al. 2015). As a result, secondary sliding occurred in the landslide mass. Combining the changes in the soil resistivity values of different positions in the landslide mass and drilling exploration, the secondary sliding surface was obtained, The soil resistivity values above and below the secondary sliding surface of the landslide mass are clearly different and exhibit an abrupt stratification, as shown by the thin black dashed line in Fig. 5.

Survey line II-II'

Based on Figs. 8 and 9a, we can know the changes in the soil resistivity values, there are apparent resistivity layering at the depths of the sliding surfaces, the resistivity value decreased suddenly. According to this characteristic of the sliding surface, the positions of the sliding surfaces along line II-II' were deduced, as shown by the black dotted line in Fig. 8. Based on the changes in the soil resistivity values, it can be inferred that the depths of the sliding surfaces at positions C, D, ZK1, E and F on line II-II' were 4, 6, 6.5, 5.5 and 3.5 m, respectively. Figure 9b shows the soil resistivity curves of positions A, B, G and H (all outside the landslide mass) on the survey line II-II'. As can be seen in Fig. 9b, the soil resistivity values of the stable soil body outside the landslide only showed stratification in the surface loose layer. As depth increased, the resistivity basically exhibited a monotonic decline, and there was no abrupt stratification.

This landslide belongs to a recurring old landslide that slipped again (Shan et al. 2015). The black dashed line in

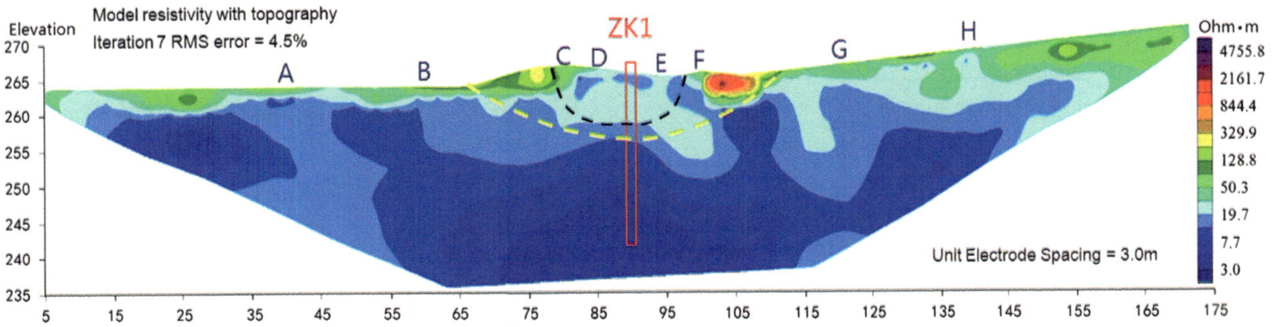

Fig. 8 The electrical resistivity profile of the survey line II-II'. we can see the start and end points of the profile in Fig. 4, Line II-II'. The *black dashed line* shows the current sliding surface. The *yellow dashed line* is likely to show the position of the sliding surface for the paleo-landslide

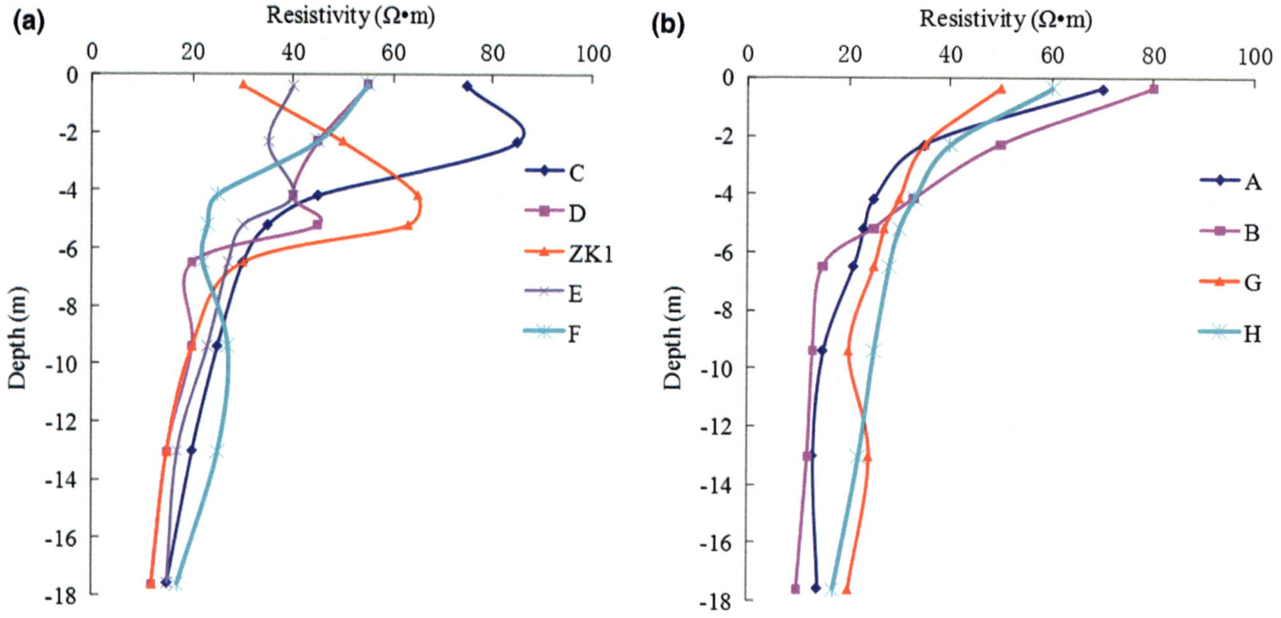

Fig. 9 The electrical resistivity curves at different points on the survey line II-II' (points C, D, ZK1, E and F are all on the landslide mass, points A, B, G and H are all outside the landslide mass)

Fig. 8 shows the current sliding surface. According to the site geological survey, combined with characteristics of resistivity changes, the position of the sliding surface for the paleo-landslide can be inferred, as shown by the yellow dashed line in Fig. 8.

GPR

Survey line I-I'

The profile determined by the GPR survey line I-I' is illustrated in Fig. 10 (due to the constraints under field conditions, because of the site of ups and downs, unable to complete the same length like HDR I in GPR survey line I, this GPR survey line can only be used to measure this long section). Using layer picking option (phase follower) in the REFLEXW software can search continuous reflector, as shown in Fig. 10 (the thick red dashed line). The intensity of the radar-wave reflection was significantly different from that of the surrounding medium. The signal of the reflected-wave is strong and shows distinctive horizon characteristics, presenting a low-frequency high-amplitude sync-phase axis, which can be inferred as the sliding surface. The continuity of the sliding surfaces is good, basically reflecting the depth range of the landslide mass development (Telford et al. 1990; Daniels 2004; Jol 2009).

Using REFLEXW radar data processing software, the radar waves amplitude values of all the data acquisition track points in the profile at different depths can be extracted. In order to better understand the changes in the intensity of the reflected radar waves, the radar-wave amplitude curves at positions A, B, ZK2, C, D, E, ZK1, F and G (as shown in Fig. 10) on the survey line I-I' were plotted (Fig. 11). A higher radar-wave amplitude value indicates a greater intensity of the reflected radar wave (Telford et al. 1990; Benedetto et al. 2013). As can be seen from Fig. 11, because the surface soil body is rather incompact in this area, most curves showed relatively large amplitudes in the depth range of 0–2.5 m. In position A, at a depth of 4.5 m, the radar-wave amplitude increased substantially, exhibiting an abrupt change. According to characteristic differences in the radar-wave reflection in different types of soil bodies (Telford et al. 1990; Benedetto et al. 2013), it can be inferred that the soil moisture content was relatively high in this position, and thus the position of the sliding surface of the landslide mass can be deduced. Similarly, in positions B, ZK2, C, D, E, ZK1, F and G on line I-I', an abrupt increase in radar-wave amplitudes occurred at depths of 4.7, 4.7, 5.5, 5.5, 6.7, 6.5, 5.6 and 5 m, respectively, and the position of the sliding surface can be inferred. This is very close to the position of the sliding surface as denoted by thick the red line in Fig. 10.

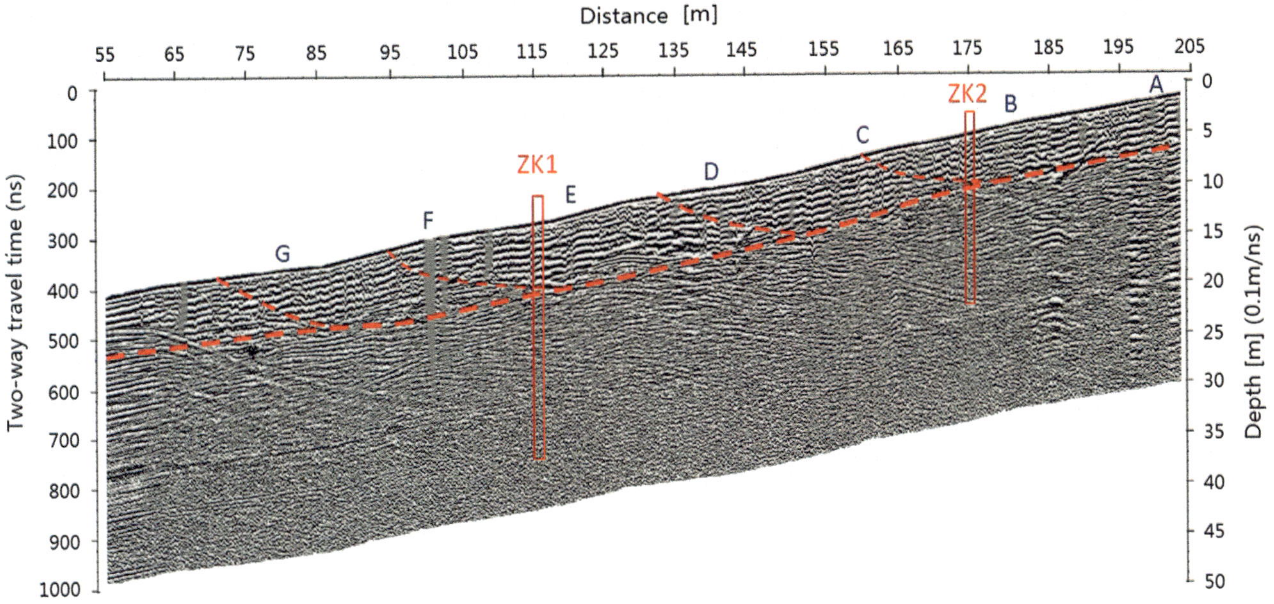

Fig. 10 The GPR profile of the survey line I-I'. The *thick red dashed line* shows the position of the major sliding surface, the *thin red dashed lines* show the secondary sliding surfaces

Survey line II-II'

The measured GPR profile of the survey line II-II' is illustrated in Fig. 12. The red dotted line on the profile is the low-frequency high-amplitude sync-phase axis, and is deduced to be the sliding surface of this profile. This reflects the depth range of the landslide mass development.

Using REFLEXW radar data processing software, we can obtain the radar-wave amplitude curves at positions A, B, C, D, ZK1, E, F, G and H (as shown in Fig. 12) on the survey line II-II', as shown in Fig. 13. As can be seen from Fig. 13, because the surface soil body is rather incompact in this area, most curves showed relatively large amplitudes in the depth range of 0–2.5 m. D, ZK1 and E are all on the landslide mass, in positions D, ZK1 and E, a sudden and substantial increase in the radar-wave amplitude occurred at depths of 3.5, 6.5 and 3.2 m, respectively, exhibiting abrupt changes. According to characteristic differences in radar-wave reflection in different types of soil bodies (Telford et al. 1990; Benedetto et al. 2013), it can be inferred that the soil moisture content in this position is relatively high, and thus the position of the sliding surface of the landslide mass can be inferred. The deduced position of the sliding surface is about the same as that denoted by the red dashed line in Fig. 12. Positions A, B, G and H were all located outside of the landslide mass. Except in the surface layer, i.e., in the depth range of 0–2.5 m, there were abrupt changes in the radar-wave amplitudes; at deeper depths, no abrupt changes were observed in the radar-wave amplitude curves.

As shown in Fig. 13, in positions C, D, ZK1, E and F, the amplitude values showed substantial increases at depths of 6, 9.5, 9, 7 and 3.5 m, respectively, exhibiting abrupt changes. These positions can be used to deduce the position of the sliding surface of the paleo-landslide, as denoted by the yellow dashed lines in Fig. 12. Meanwhile, the magnitude of the abrupt change in the radar-wave amplitudes at the position of the sliding surface of the paleo-landslide was smaller than that at the position of the current sliding surface.

The GPR results show that the moisture content of soils at the sliding surface of the landslide mass is relatively high. The drilling data also show that the moisture content of the sliding surface of the landslide mass in the study area is very high, the moisture content is greater than 25%, the maximum reach 40%, which is completely consistent with the results obtained from the GPR profile and the HDR profile.

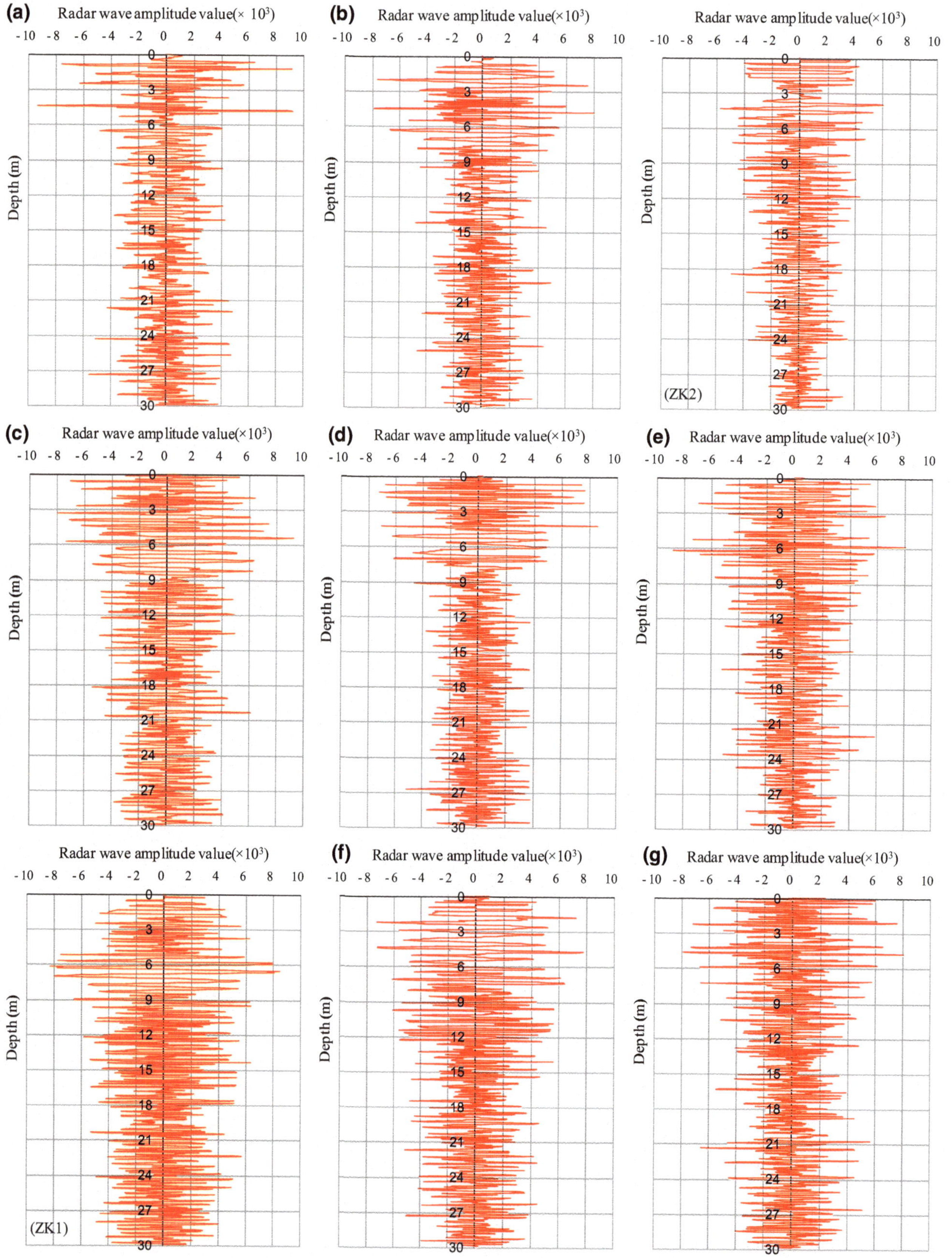

Fig. 11 The radar-wave amplitude curves on the survey line I-I'. Positions A, B, ZK2, C, D, E, ZK1, F and G are all on the survey line I-I' (as shown in Fig. 10)

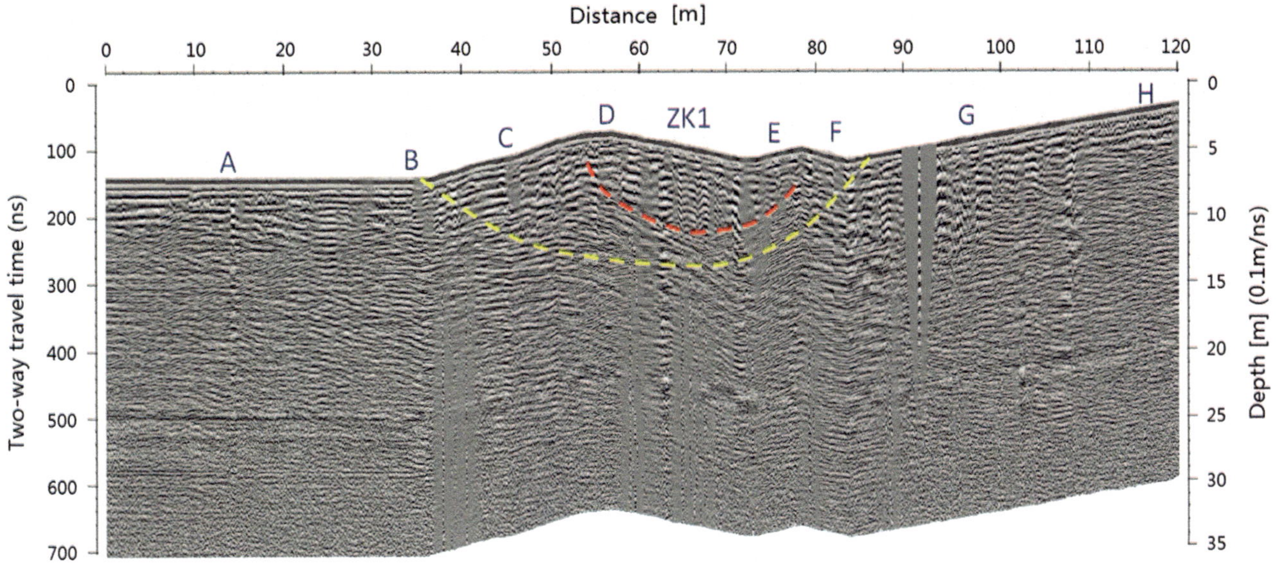

Fig. 12 The GPR profile of the survey line II-II'. The *red dashed line* shows the current sliding surface. The *yellow dashed line* shows the position of the sliding surface of the paleo-landslide

Analysis of the Mechanism Underlying Landslide Development

In May 2010, a geological survey of the study area revealed that there is permafrost in the shady slopes on the two sides of the landslides in this area (Shan et al. 2015). Permafrost was found by drilling on the profile of the survey line III-III', as shown on Fig. 14 of permafrost soil sample photos. In addition, the HDR method was used on June 2, 2010 for prospecting on the profile of the survey line III-III', based on soil resistivity characteristics (Telford et al. 1990), we can infer the permafrost layer range on the profile of the survey line III-III', the measured apparent electrical resistivity profile of survey line III-III' obtained in June 2, 2010 is shown on Fig. 15 (the permafrost layer range is shown by the black dotted line).

Due to permafrost melting and concentrated summer rainfall, the landslide mass started to slip at the end of July 2010, and the landslide was formed (Shan et al. 2015). Site surveys and borehole drilling data demonstrate that Landslide K178 + 530 is a shallow creeping consequent landslide in the permafrost region. Water seepage generated from permafrost melting together with water infiltration generated from concentrated summer precipitation increases the local moisture content of the hillside soil. This is the main reason for landslide formation in the northwest section of the Lesser Khingan Range in China (Shan et al. 2015). Instability may easily occur in the rainy season and the spring melting season. The highly permeable surface soil, the gravel and sand layer, and the silty clay containing a weathered sand interlayer provide a convenient channel for water infiltration. The mudstone and siltstone layers below have small permeability, forming watertight layers. Water generated from precipitation, melting snow and permafrost melting is blocked by the impermeable layer when it infiltrates downward, and the local moisture content increases sharply. Water thus infiltrates along the interface between the permeable layer and the impermeable layer, forming a slip zone. Combining the geophysical and drilling data, the position of the sliding surface can be determined.

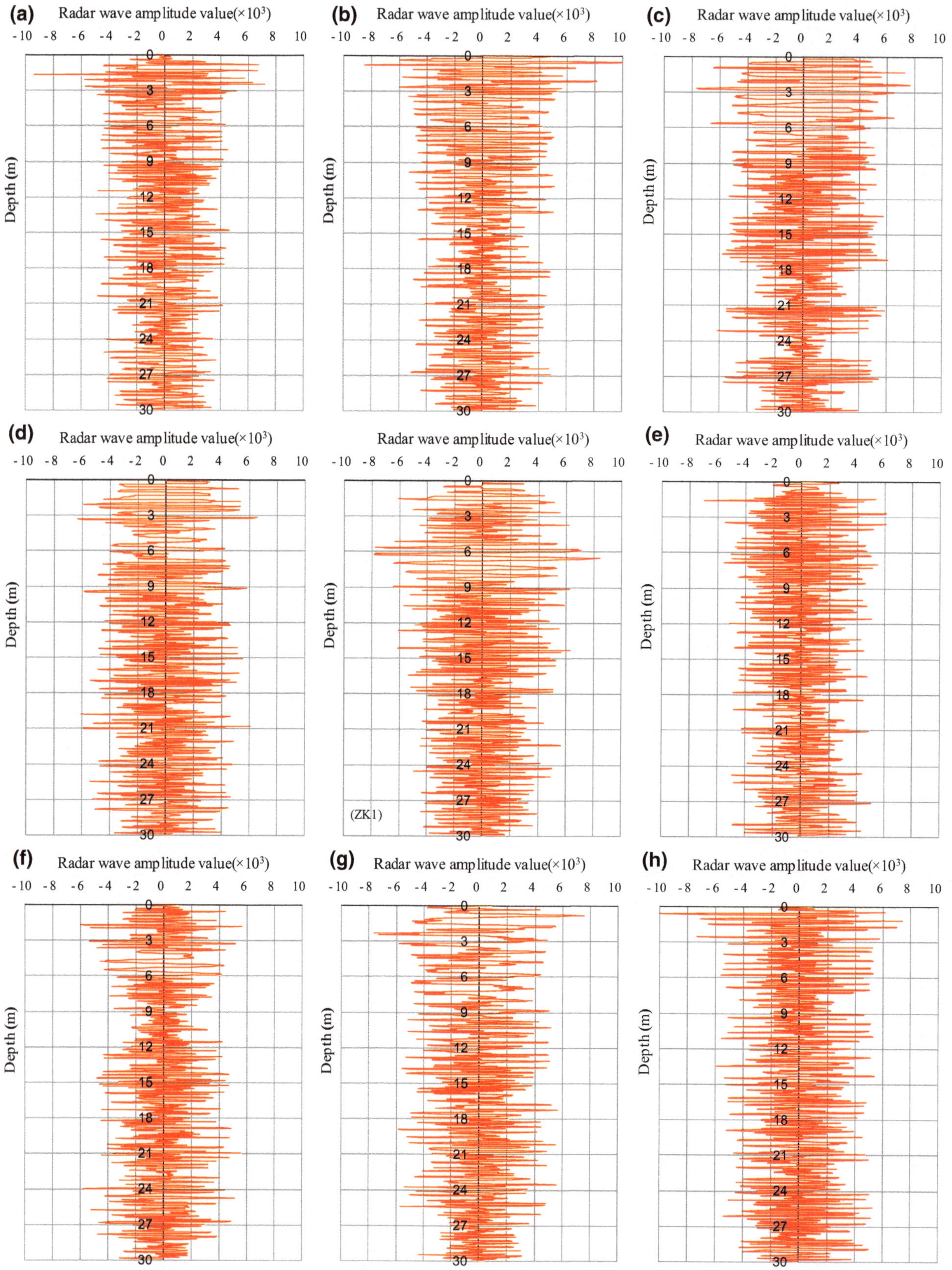

Fig. 13 The radar-wave amplitude curves on the survey line II-II'. Positions A, B, C, D, ZK1, E, F, G and H are all on the survey line II-II' (as shown in Fig. 12)

Fig. 14 The photo of high temperature permafrost. Sampling location is in the survey line III-III', at 40 m position on the X axis, the drilling time is in May 2010

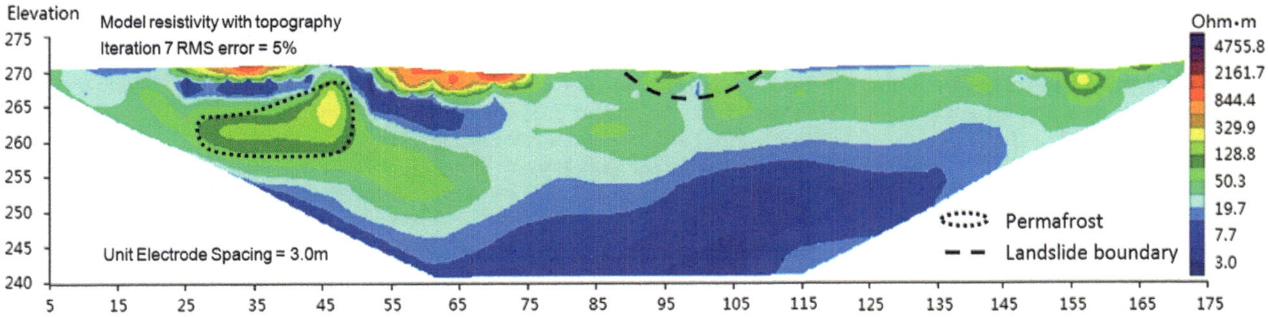

Fig. 15 The electrical resistivity profile of the survey line III-III', we can see the start and end points of the profile in Fig. 4, Line III-III', the detection data is on June 2, 2010

Conclusions

The soil resistivity values above and below the sliding surface of the landslide mass are clearly different and exhibit an abrupt stratification. There is apparent resistivity layering at the position of the sliding surface in the landslide mass, the resistivity value decreased suddenly. On the GPR profile, the sliding surface is manifested as a low-frequency high-amplitude sync-phase axis, and there is a sudden increase in the amplitude of the radar wave. In practice, the abrupt abnormal changes revealed in the HDR and GPR results can be used as characteristic markers for identifying the sliding surface position of shallow landslides in this region.

Prospecting of landslides in the study area using two methods, namely HDR, and GPR reveal basically the same sliding surface position. This suggests that HDR and GPR methods are reliable methods for site prospecting of landslides. They can be applied to shallow landslides in the high-latitude permafrost region for fast and accurate determination of the sliding surface position, providing an accurate reference for engineering projects and to ensure appropriate measures are taken.

Acknowledgements This work was supported in part by Science and Technology project of Chinese Ministry of Transport (Nos. 2011318223630) and the Fundamental Research Funds for the Central Universities (Nos. 2572014AB07).

References

Benedetto A, Benedetto F, Tosti F (2013) GPR applications for geotechnical stability of transportation infrastructures. Nondestr Test Eval 27(3):253–262

Bichler A, Bobrowsky P, Best M, Douma M, Hunter J, Calvert T, Burns R (2004) Three-dimensional mapping of a landslide using a multi-geophysical approach: the Quesnel Forks landslide. Landslides 1(1):29–40

Cardarelli E, Marrone C, Orlando L (2003) Evaluation of tunnel stability using integrated geophysical methods. J Appl Geophys 52 (2–3):93–102

Carpentier S, Konz M, Fischer R, Anagnostopoulos G, Meusburger K, Schoeck K (2012) Geophysical imaging of shallow subsurface topography and its implication for shallow landslide susceptibility in the Urseren Valley, Switzerland. J Appl Geophys 83:46–56

Coggon JH (1971) Electromagnetic and electrical modelling by the finite element method. Geophysics 36:132–155

Daniels DJ (2004) Ground penetrating radar, 2nd edn. The Institution of Electrical Engineers, London, United Kingdom

Dey A, Morrison HF (1979) Resistivity modelling for arbitrarily shaped two-dimensional structures. Geophys Prospect 27:106–136

Drahor MG, Gokturkler G, Berge MA, Kurtulmus TO (2006) Application of electrical resistivity tomography technique for investigation of landslides: a case from Turkey. Environ Geol 50 (2):147–155

Donohue S, Gavin K, Tolooiyan A (2011) Geophysical and geotechnical assessment of a railway embankment failure. Near Surf Geophy 9(1):33–44

Hack R (2000) Geophysics for slope stability. Surv Geophys 21:423–448

Havenith HB, Jongmans D, Abdrakhmatov K, Trefois P, Delvaux D, Torgoev A (2000) Geophysical investigations of seismically induced surface effects: case study of a landslide in the Suusamyr valley, Kyrgyzstan. Surv Geophys 21:349–369

Jol HM (2009) Ground penetrating radar theory and applications. Elsevier, Amsterdam The Netherlands

Loke MH, Barker RD (1996) Rapid least-squares inversion of apparent resistivity pseudosections using a quasi-Newton method. Geophys Prospect 44:131–152

McCann DM, Forster A (1990) Reconnaissance geophysical methods in landslide investigations. Eng Geol 29:59–78

Malehmir A, Bastani M, Krawczyk CM, Gurk M, Ismail N, Polom U, Persson L (2013) Geophysical assessment and geotechnical investigation of quick-clay landslides—a Swedish case study. Near Surf Geophys 11(3):341–350

Perrone A, Lapenna V, Piscitelli S (2014) Electrical resistivity tomography technique for landslide investigation: a review. Earth Sci Rev 135:65–82

Rhim HC (2011) Measurements of dielectric constants of soil to develop a landslide prediction system. Smart Struct Sys 7(4): 319–328

Sauvin G, Lecomte I, Bazin S, Hansen L, Vanneste M, LHeureux JS (2014) On the integrated use of geophysics for quick-clay mapping: The hvittingfoss case study, Norway. J Appl Geophys 106:1–13

Sass O (2007) Bedrock detection and talus thickness assessment in the European Alps using geophysical methods. J Appl Geophys 62:254–269

Sass O, Bell R, Glade T (2008) Comparison of GPR, 2D-resistivity and traditional techniques for the subsurface exploration of the Oschingen landslide, Swabian Alb (Germany). Geomorphology 93(1): 89–103

Sasaki Y (1992) Resolution of resistivity tomography inferred from numerical simulation. Geophys Prospect 40:453–463

Schrott L, Sass O (2008) Application of field geophysics in geomorphology: advances and limitations exemplified by case studies. Geomorphology 93:55–73

Shan W, Hu Z, Guo Y, Zhang C, Wang C, Jiang H, Liu Y, Xiao J (2015) The impact of climate change on landslides in southeastern of high-latitude permafrost regions of china. Front Earth Sci 3 (7):1–11

Telford WM, Geldart LP, Sheriff RE (1990) Applied geophysics, 2nd edn. Cambridge University Press, Cambridge, United Kingdom

Timothy S, Bilderback EL, Quigley MC, Nobes DC, Massey CI (2014) Coseismic landsliding during the Mw 7.1 Darfield (Canterbury) earthquake: implications for paleoseismic studies of landslides. Geomorphology 214:114–127

Tripp AC, Hohmann GW, Swift CM Jr (1984) Two-dimensional resistivity inversion. Geophysics 49:1708–1717

Yamakawa Y, Kosugi K, Masaoka N, Tada Y, Mizuyama T (2010) Use of a combined penetrometer-moisture probe together with geophysical methods to survey hydrological properties of a natural slope. Vadose Zone J 9(3):768–779

Zajc M, Pogacnik Z, Gosar A (2014) Ground penetrating radar and structural geological mapping investigation of karst and tectonic features in flyschoid rocks as geological hazard for exploitation. Int J Rock Mech Min Sci 67:78–87

Soil CO_2 Emission, Microbial Activity, C and N After Landsliding Disturbance in Permafrost Area of Siberia

Oxana V. Masyagina, Svetlana Yu. Evgrafova, and Valentina V. Kholodilova

Abstract

In boreal forests developed on permafrost, landslide processes are widespread, occur in years of above average summer-autumn precipitation and can cover up to 20% of total area of slopes adjacent to rivers. Permafrost landslides will escalate with climate change. These processes are the most destructive natural disturbance events resulting in complete disappearance of initial ecosystems (vegetation cover and soil). We have studied sites of landslides of different ages (occurred at 2009, 2001 and 1972) along with Nizhnyaya (Lower) Tunguska River and Kochechum River to analyze postsliding ecosystem changes. Just after the event (as at 1-year-old site in 2010), we registered drop in soil respiration, 3 times decreasing of microbial respiration contribution, 4 times lower mineral soil C and N content at bare soil (melkozem) middle location of a site. Results show that regeneration of soil respiration and eco-physiological status of microbial communities in soil during post disturbance succession starts with vegetation re-establishment and organic soil layer accumulation. As long as ecosystems regenerate (as at 35-year-old site), accumulated litter contains similar to control C and N content as well as the main pool of microorganisms, though microbial biomass and soil C and N content of old landslide area does not reach the value of these parameters in control plots. Therefore, forested ecosystems in permafrost area disturbed after landsliding requires decades for final successful restoration.

Keywords

Landslides • Soil respiration • Microbial respiration • Soil C and N content • Permafrost • Boreal ecosystems • Siberia

O.V. Masyagina (✉) · S.Yu. Evgrafova
Federal Research Center "Krasnoyarsk Science Center SB RAS", Sukachev Institute of Forest SB RAS, Akademgorodok 50/28, Krasnoyarsk, 660036, Russia
e-mail: oxanamas@ksc.krasn.ru

S.Yu. Evgrafova
e-mail: esj@ksc.krasn.ru

V.V. Kholodilova
Siberian Federal University, Krasnoyarsk, 660041, Russia
e-mail: besenoknaokne@mail.ru

M. Mikoš et al. (eds.), *Advancing Culture of Living with Landslides*,
DOI 10.1007/978-3-319-53483-1_26

Introduction

In boreal forests, landsliding process (e.g., solifluction) is one of the most destructive natural disturbance factor resulting in complete disappearance of initial ecosystems (vegetation cover and soil). Its impact on forest vegetation in permafrost area of Siberia can be regarded as the other two important influences—wildfires and cuttings. Unlike to latter ones, landslides result in the full destruction of initial ecosystems (Prokushkin et al. 2010). During landslide, an active layer of soil (together with vegetation) slides down on a rupture surface of permafrost. Thus, ecosystem development starts on parent material. In first years after landslide, the local temporal streams form due to melting of outcropping permafrost layer. After landslide and started erosion, large amount of soil and organic material enters streams as sediment, and results in reducing the water quality (Geertsema et al. 2009). New ecological conditions at areas disturbed by landslide lead to changes in environment, species composition, soil carbon and nitrogen content as well as to soil respiration fluctuations and soil microbial associations change (Bugaenko et al. 2005; Masyagina et al. 2013).

The most widespread landslide type in the Siberian permafrost region near Yenisei River is fast landslide at slopes with slope dip of 8–15° characterized by slow, fast and sometimes catastrophic land sliding events with groove formation (Gigarev 1967). Soil landslides usually occur on the thickest active layer (e.g., south and west slopes). Main causes of landsliding events can be wildfires or overwetting of seasonally thawed layer due to climate change (as a result of combination of high air and soil temperatures and excessive precipitation in summer, or intensive nival thawing) (Kharuk et al. 2016). In river valleys, landsliding is an important landscape-forming factor (Abaimov 1997; Kaplina 1965; Pozdnyakov 1986). At present, the formation and consequences of landslides, as well as soil-vegetation regeneration patterns in high latitudes are getting little attention though currently increasing frequency of landslide processes in permafrost area is one of the prospective consequences of global change (Bugaenko et al. 2005; Prokushkin et al. 2010; Kharuk et al. 2016).

Various heterogeneous conditions (microclimatic and edaphic) are forming at a place of a landslide, especially on permafrost soils. Biogenic elements (N, P, K, Ca), pH and carbon contents decrease, sharp changes of C/N occur especially on alluvial bare ground. All these factors initialize the first phase of regenerative succession for newly forming biogeocoenosis. Therefore soil C and N content, soil microbial activity, vegetation regeneration and soil respiration were highly variable at different microsites (e.g., depending on microtopography).

In this research, we studied how landslide disturbance influences ecological conditions (soil temperature at 5 cm depth, soil water content), soil C and N content, soil respiration and soil microbial activity.

Study Area and Methods

The study area is located within the northern part of central Siberia (Figs. 1 and 2). We have selected three sites disturbed by landslides of different age on south-east-facing slopes in a valley of Nizhnyaya (Lower) Tunguska River and Kochechum River (64°N, 100°E) as described in Bugaenko et al. (2005). One site was destroyed by landsliding in 2009 (L2009), second—in 2001 (L2001), and third—in 1972 (L1972).

The total areas affected (zone of depletion and accumulation) is about 1165 m^2 for L2009, 11200 m^2 for L2001 and 5700 m^2 for L1972 (Table 1). Our surveys have been conducted in the middle slope position (Fig. 2) on the landslide sites in 2007 (L2001 and L1972) and in 2010 (L2009). Different forest types, amounts of mineral and organic matter, and ecological conditions (Prokushkin et al. 2010) characterize the chosen zones. Successful regeneration of Gmelin larch [*Larix gmelinii* (Rupr.) Rupr.] has been noticed at L2001 and L1972 sites.

According to microtopography in the middle part of slope of landslide sites (Fig. 1), we chose following microsites:

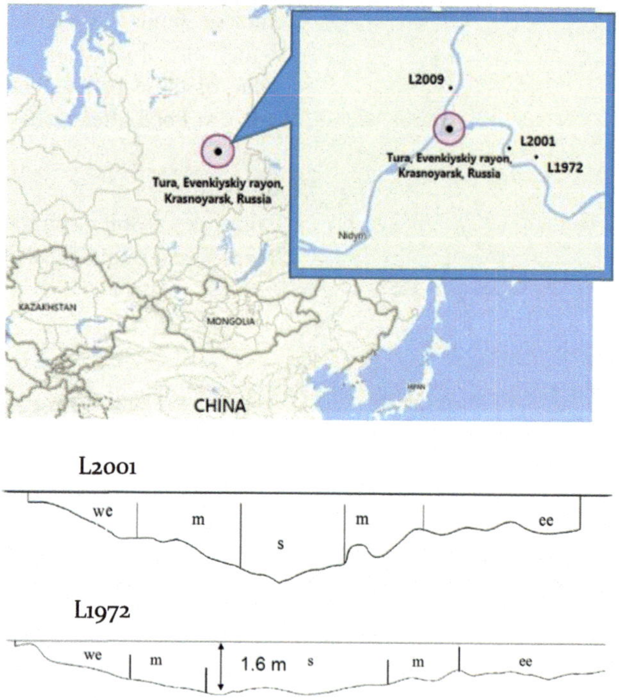

Fig. 1 Map of study sites and cross profiles of L2001 site and of L1972 site in the *middle* part of slope: we and ee—western edge and eastern edge; m—alluvial bare ground; s—temporal stream bed. Cross profile of L2009 site is not presented

Fig. 2 Studied landslides L2001 and L1972 were taken in 2007, and L2009 site was taken in 2010 (photo by Masyagina O.V.). Middle part of slope is marked by *red line*

western (four plots) and eastern edges (four plots), representing soil-vegetation complexes between the edge of landslide site and undamaged (control) forest, and central part (bare ground (melkozem)—three plots) have been selected along with the direction of landslide progressive movement. The size of plots was 10 × 10 m. Under landslide impact, edges and central zones (bare ground) have been drastically changed by concentration of nutrients and ecological conditions (Prokushkin et al. 2010). Undamaged forests neighboring landslide sites were used as a control sites (four plots of 10 × 10 m size) to estimate the effect of soil sliding on ecosystem processes.

To assess how soil sliding affects ecosystem re-establishment, soil surface CO_2 emission, soil heterotrophic respiration (basal respiration, BR) and soil microbial biomass (C_{mic}) in the mineral soil layer (0–10 cm) have been measured.

Soil respiration rates were measured at subplots (10 subplots of 1 × 1 m size) located at the middle part of slope along with the red line showed at Fig. 2 on the base of 3–5 replications with Li-Cor 6200 (LI-COR, USA) at landslide and control sites in July of 2007 and 2010. In parallel, soil temperature and water content were measured at 5 cm depth, and height of vegetation cover and litter was determined (Table 1).

Soil heterotrophic microbial biomass (C_{mic}, μg C g soil^{-1}) was assessed with kinetic method implying soil microbiota substrate-induced respiration (SIR, mg CO_2-C g^{-1} of soil hour^{-1}) determination with subsequent recalculation on carbon of microbial biomass basis, C-CO_2 according to Eq. (1) (Sparling 1995).

$$C_{mic} = 50.4 \times SIR \tag{1}$$

Basal respiration (BR, μgCO_2-C g soil^{-1} day^{-1}) of soil microorganisms as a CO_2 emission rate in 24 h of soil incubation at 22 °C and 60% of soil water capacity has been studied. Coefficient of microbial activity qCO_2 (μgCO_2-C mg^{-1}C_{mic} hour^{-1}) was calculated with the Eqn. [2].

$$\frac{BR}{C_{mic}} = qCO_2 \tag{2}$$

Mineral soil C, N contents and C/N ratio was determined on CN-analyzer after soil samples drying at 60 °C during 48 h.

Studied parameters (soil respiration, BR, C_{mic}, qCO_2, C and N content) were tested for normality prior to statistical analysis. The nonnormally distributed variables were logarithmically transformed prior to analysis to stabilize variance. Two-way factorial analysis of variances (ANOVAs) was used to test the main effects of age of landslide and microsite types (edges and bare soil): soil respiration, BR, Cmic, qCO_2, C and N content. TukeyHDS (p adjusted) was

Table 1 Sites characteristic

Parameter	L2009	L2001	L1972
Year of landslide event	2009	2001	1972
Time passed since disturbance, (years)	1 (as for 2010)	6 (as for 2007)	35 (as for 2007)
Slope dip, (°)	5–7	19–27	11–20
Slope length, (m)	95	390	290
Maximum width of a site, (m)	15	33	37
Minimum width of a site, (m)	14	12	7
Site area, (m^2)	1165	11200	5700
The volume of washout of soil, (m^3)	252	5365	3725
Height of vegetation cover and litter (as of 2007), (cm)	ND	1.1 ± 0.6	5.7 ± 0.9

ND—no data

used for Tukey multiple comparisons of means. The analyses of the obtained data were performed using RStudio version 0.99.893—© 2009–2016 RStudio, Inc.

Results and Discussion

Temperature regime (at 5 cm depth) measurements revealed increased mineral soil temperature at landslide sites compare to control sites (Table 2). Moreover, at L2009, site of initial successional stage, soil temperature was 3–4 times higher (about 15–23 °C, T-test = 7.22, p = 0.001 as for 2010) and at L2001-2 times higher (about 16–19.5 °C, T-test = 5.22, p = 0.001 as for 2007) than that of control sites. Found differences are obviously caused by the lack of tree cover especially at central part with bare soil. Soil temperature differences between the L1972 site and respective control site were even less—about 1–2 °C, but still significant (T-test = 3.45, p = 0.015) in 2007, which points possibly to the final successional stage.

Soil water content is one of the most important ecological factor influencing regeneration of disturbed ecosystem. Our previous research (Masyagina et al. 2013) revealed lower values of water content at landslide sites compared to control sites. Maximal differences in soil water content were found between L2001 and its control site, and minimal between the oldest landslide site (L1972) and its control site.

Soil respiration measurements showed significantly higher values at L1972 site than that of other landslide sites as well as control sites (Table 3, T-test, $p < 0.0001$). Thus, soil respiration at "old" landslide and control site was about 9–14 µmol CO_2 $m^{-2}s^{-1}$, whereas at "young" landslides and its control sites its values was much lower (1–9 µmol CO_2 $m^{-2}s^{-1}$). Concerning changes of height of vegetation cover and litter (Table 1) after landsliding event, its values at L1972 and control sites are comparable, whereas in L2001 and control site there is a 5–8 times difference between values, that indicates the disturbance of ecosystem. Respiration rates at L2009 and L2001 sites were significantly lower (as for 2007, T-test = 2.85, p = 0.001) than that of

Table 2 Monthly averaged soil temperature at 5 cm depth (°C) at the middle part of landslide area in July of the year of monitoring (2010—for L2009; 2007—for L2001 and L1972 sites)

Microsite	L2009	L2001	L1972
Bare ground	23.0 ± 0.5	19.5 ± 1.6	12.1 ± 0.8
Eastern edge	15.0 ± 3.1	16.2 ± 0.7	12.9 ± 1.4
Western edge	16.3 ± 5.6	16.1 ± 0.8	10.0 ± 0.7
Control site	5.4 ± 1.4	8.0 ± 0.9	10.0 ± 0.5

Table 3 Soil respiration rate and microbial respiration contribution in total CO_2 emission from soil surface in the middle part of landslide site

Microsite type	Respiration rate, µmol CO_2 m^{-2} s^{-1}			Microbial respiration contribution (%)		
	L2009	L2001	L1972	L2009	L2001	L1972
Control site	1.5 ± 0.1	6.3 ± 0.2	11.8 ± 0.2	15	11	11
Western edge	3.0 ± 0.2	6.0 ± 1.5	13.5 ± 0.2	16	24	2
Bare ground	0.9 ± 0.6	2.8 ± 0.3	12.1 ± 2.0	5	12	1
Eastern edge	1.7 ± 0.6	8.6 ± 2.6	15.8 ± 1.6	53	17	9

control sites at bare soil microsites due to vegetation cover disturbance and low root activity.

Microtopography is a factor that intensifies variation of the studied parameters, especially soil temperature and water content. Soil temperature and water content at L2001 varied depending on microsite type (Table 2) which is in contrast with L1972 site pattern of these factors. Moreover, at L1972 site values of soil respiration between microsite types did not differ and were very close to control site (Table 3). Whereas at young landslide sites (L2009 and L2001) there was high variation of CO_2 emission between microsites: maximal soil respiration was at edges (3.0 μmol CO_2 $m^{-2}s^{-1}$ at L2009 and 8.6 μmol CO_2 $m^{-2}s^{-1}$ at L2001) and minimal—at bare ground (0.9–2.8 μmol CO_2 $m^{-2}s^{-1}$).

Carbon and nitrogen content in mineral soil of all landslide sites remain significantly lower (3–5 times) than that of control sites (Fig. 3). According to Leibman (2009), it can be because geochemical processes on slopes exposed by landsliding can cause a redistribution of elements within the active layer and upper permafrost. C/N values in mineral soil (0–10 cm) at landslide sites varies wider (from 17 to 23) than that of control sites (from 21 to 23).

At control sites, contribution of microbial respiration in total soil CO_2 emission occurred to be almost the same—about 11–15% (Table 3). At young landslide sites (L2009 and L2001), this value is higher than that of control site, especially at edges (about two times higher). At the older landslide site (L1972), on the contrary, microbial respiration contribution was lower than that of control site. Low contributions of microbial respiration at the older site was due to low soil microbial biomass. The lowest contribution of microbial respiration was revealed at bare soil microsites.

Measurements of microbial biomass in mineral soil layers of landslide and control sites revealed reverse results. At control sites of young landslides C_{mic} located mainly (up to 80%) in 0–5 cm layer. Whereas at corresponding landslide sites its values at 0–5 cm layer were much lower (about 30–60%) and widely varied depending on disturbance degree: maximum at edges and minimum at bare ground. Total C_{mic} at L2009 and L2001 was higher, than that of control sites due to changed ecological conditions and enhanced substrate availability after event. At L1972, total C_{mic} was lower than that of control. Moreover, its value at bare ground at L1972 site was comparable to values at edges, which indicates gradual regeneration of microbiota in mineral soil.

One of the main parameter of eco-physiological status of microbial population is coefficient of microbiological activity qCO_2. High values of qCO_2 points to the disorder of

Fig. 3 Total carbon, nitrogen content and C/N values in organic (litter) and mineral soil at L2009 (A, D, J), L2001 (B, E, H) and L1972 (C, F, I) sites. Control—control site, edge—edge of landslide, m—bare ground (melkozem)

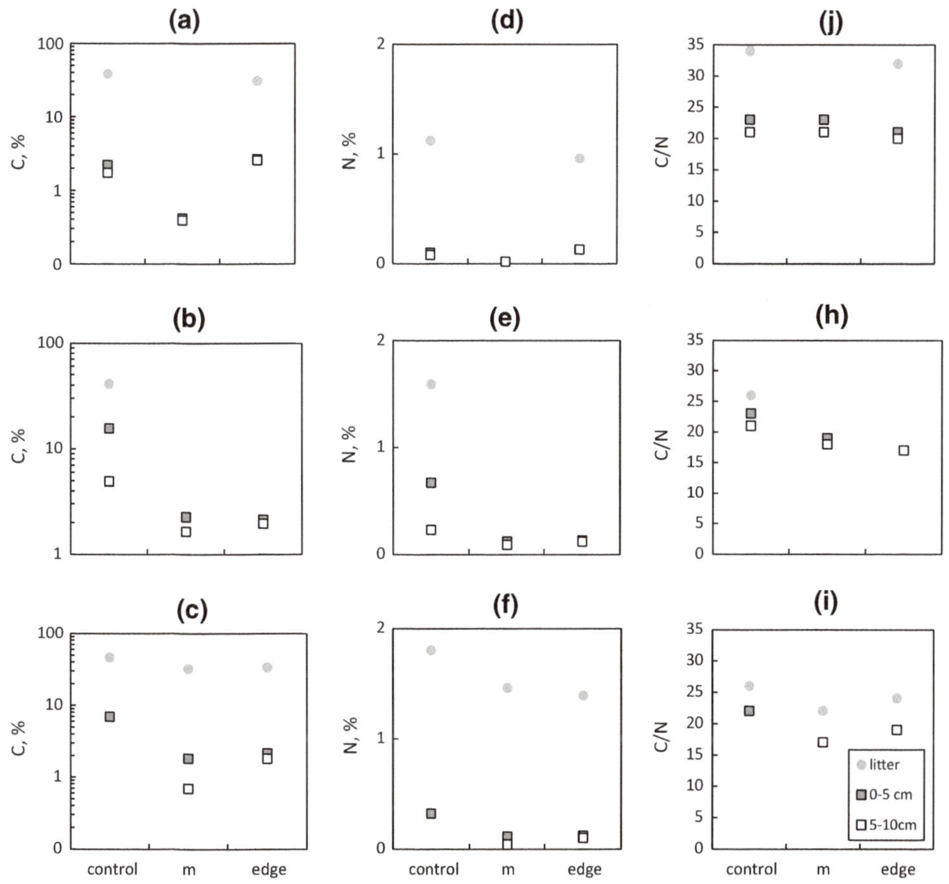

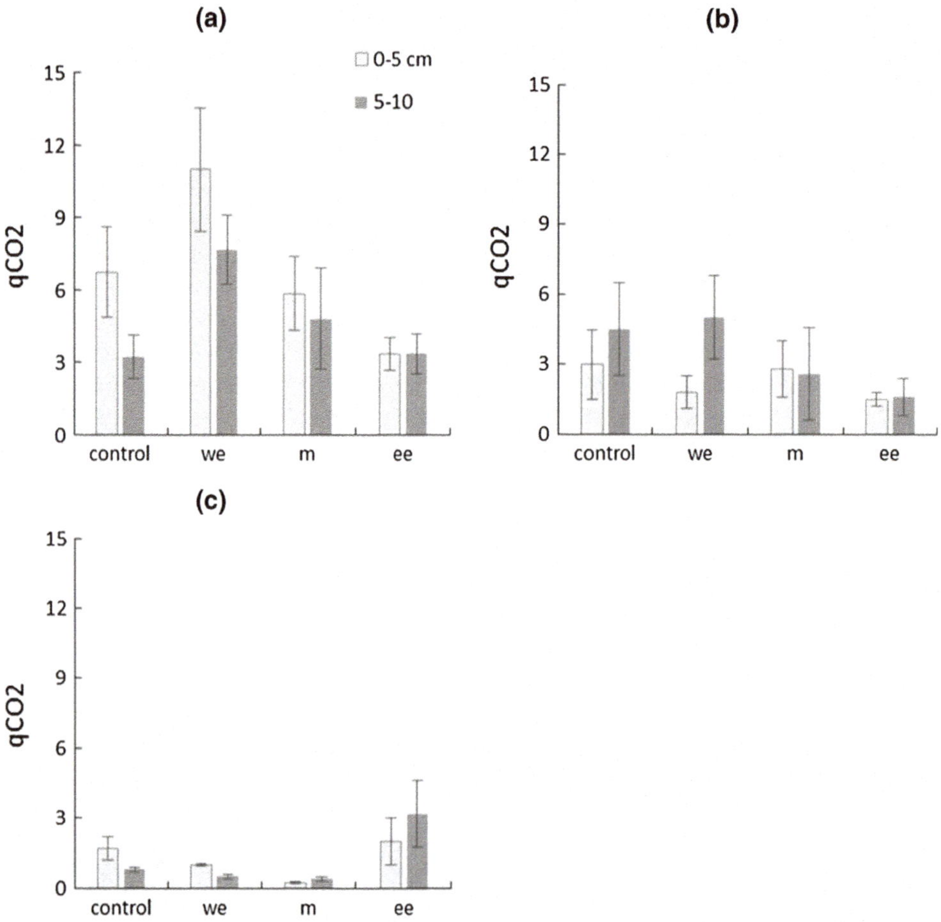

Fig. 4 Coefficients of microbiological activity (qCO_2, μgCO_2-C mg^{-1}Cmic $hour^{-1}$) in mineral soil at L2009 (A), L2001 (B) and L1972 (C) sites. Control—control site, ee and we—eastern and western edge, m—bare ground

normal functioning of soil microbiota due to some stress (Anderson and Domsch 1990). Analysis of coefficient of microbiological activity values revealed intensive disturbance of microflora functioning at control and L2009 sites, and at particular microsites of L2001 and L1972 sites (Fig. 4). The degree of disturbance of soil microbiota is much lower at L1972 site than that of the L2009 and L2001 sites. Therefore, the restoration of eco-physiological status of microflora at sites damaged by landslides is very slow.

Thus, ecological conditions of landslide sites were sharply different among types of microsites as well as control sites and were dependent on the age of landsliding disturbance event. Ecological conditions formed at landslide microsites affected regeneration processes of newly establishing forest in different ways. On the one hand, wood species (larch and *Dushekia*) regeneration occurred intensively during successional period. On the other hand, live vegetation cover, which is unrepresentative (or did not exist before) for these particular native larch stands, has been appeared. As we found in our previous research (Bugaenko et al. 2005, Masyagina et al. 2013) at young landslides (L2001), vascular plant biodiversity is 1.5 times higher (27 vs. 18) than that of control site, whereas mosses species amount at landslide site was lower compare to control (3 vs. 6). Similar tendencies on species biodiversity were found at the old L1972 site, but differences between landslide and control were not so pronounced especially regarding to species composition of mosses and lichens. This is goes well with considerations about landslides as a disturbance process, which increases site biological diversity through changing topography, soil and run-off conditions (Geertsema et al. 2009).

Conclusion

Microecological conditions formed at the initial phase of regenerating succession after landslide disturbance are far different from those in control sites. Low values of soil C and N at all landslide sites compare to control sites confirms that even "old" L1972 site is very far from the state of control site. "Young" landslide sites are also very heterogeneous spatially by ecological conditions among studied landsliding sites. High variation of hydrothermal conditions at the "young" landslide sites resulted in increasing of vascular plant biodiversity, which resulted in high soil respiration variation. Similarly, at "young" landslides there was bigger microbial biomass, qCO_2 values as well as microbial respiration contribution compare to control or the "old" L1972 site.

At the "old" landslide site, hydrothermal conditions were very close to control sites. Soil respiration and microbial respiration contribution here are comparable to control site, however microbial biomass was higher at L1972 site relate to control one. The coefficient qCO_2 points out a stabilization of microbial community at L1972 site and approaching control site state, except for eastern edge microsites. So, our results showed that in permafrost conditions of Siberia, after decades of a disturbance occurred, soil changes (like C and N content, microbial activity) may persist much longer and have greater ecological effects even if plant diversity and CO_2 fluxes are similar to control site.

Acknowledgements The reported study was funded by RFBR according to the research project № 16-04-01677 and was partly supported by Russian Government Megagrant Project No. 14. B25.31.0031.

References

Abaimov AP (1997) Larch forests and open woodlands of Siberian North (Diversity, ecological and forest development traits). Dr theses, CSBG, Novosibirsk, Russian Federation

Anderson T-H, Domsch KH (1990) Application of eco-physiological quotients (qCO_2, and qD) on microbial biomasses from soils of different cropping histories. Soil Biol Biochem 22(2):251–255

Bugaenko TN, Oreshenko DA, Shkikunov VG, (2005) Regeneration of forest vegetation after solifluction events in permafrost region. Proceedings of young scientist's conference on studies on components of Siberia forest ecosystems, 21–22 March 2005. Krasnoyarsk, Russia, pp 12–14

Geertsema M, Highland L, Vaugeouis L (2009) Landslides—disaster risk reduction. In: Margottini C, Canuti P, Sassa K (eds). Springer-Verlag Berlin Heidelberg. (ISBN 978-3-540-69970-5), p 649

Gigarev LA (1967) Reasons and mechanisms of solifluction development. Nauka, Moscow, p 197p

Kaplina TN (1965) Cryogenic slope processes. Nauka, Moscow, p 295p

Kharuk VI, Shuspanov AS, Im ST, Ranson KJ (2016) Climate-induced landsliding within the larch dominant permafrost zone of central Siberia. Environ Res Lett 11:045004. doi:10.1088/1748-9326/11/4/045004

Leibman M (2009) Mechanisms and landslides—disaster risk reduction. In: Margottini C, Canuti P, Sassa K (eds). Springer-Verlag Berlin Heidelberg. (ISBN 978-3-540-69970-5), p 649

Masyagina O, Evgrafova S, Prokushkin S, Prokushkin A (2013) Landslide science and practice vol 4: global environmental change. In: Margottini C, Canuti P, Sassa K (eds). Springer-Verlag Berlin Heidelberg. (ISBN 978-3-642-31337-0), p 431

Pozdnyakov LK (1986) Permafrost forest science. Nauka, Novosibirsk, p 192p

Prokushkin SG, Bugaenko TN, Prokushkin AS, Shkikunov VG (2010) Succession-driven transformation of plant and soil cover on solifluction sites in the permafrost zone of Central Evenkia. Biol Bull 1:80–88

Sparling GP, (1995) Methods in applied soil microbiology and biochemistry. In: Alef K, Nannipieri P (eds). London Academic Press. (ISBN 0125138407), p 404

Part III

Landslides and Water

Session Introduction—Landslides and Water

Patrick Wassmer

The present volume is devoted to landslides in various environments. Needless to remember here that landslide process is an extensive phenomenon able to affect the entire topographical context, from mountainous steep slopes to sub-horizontal area and that it is one of the major causes of disasters on earth. The role of water is ubiquitous during the triggering phase and the displacement of these movements except for the large scale movements like rockslide-avalanches that are not taken in account in this volume.

Among all the environments prone to landslides some of them are particularly sensitive. It is often the case for man-made modifications of the natural environment like the creation of a dam on a river for instance. One of the aims of an artificial dam is to reduce the flooding effect on the flat plains downstream. The water level fluctuations this aim imply induce water pressure variations within soils and weathered rocks able to destabilize the foot of the slope inducing landslides and, by hazard filiation, related tsunamis in the reservoirs. One still have in mind the Vajont disaster in northern Italy where the filling of the reservoir destabilized the foot of a slope dominating the lake. At 22:39 on October 9th 1963, a mass of 260×10^6 m^3 of rocks and debris runs into the water. The huge wave generated overpassed the dam and runs through a narrow gorge to the village of Langarone causing about 2000 fatalities. This issue is addressed by Kun Song et al. in their contribution that deals about "Landslide deformation prediction by numerical simulation in the Three Gorges, China". Along the Three Gorges Reservoir, a few ancient landslides have been reported to be reactivated by the water level fluctuation. A 3-D numerical simulation model was employed to study the spatial deformation of a landslide which was established according to its geological condition, and verified by comparisons of actual monitoring information and simulation recording data during a filling-drawdown cycle of the reservoir. The authors show that the reservoir water variation controls the deformation of forepart of the landslide.

One of the main issues related to landslides is that in mountainous area, the mass displaced on the slopes accumulates in the valleys creating more or less large landslide dams. The nature of the material accumulated and its volume will determine the resistance against the pressure of the water accumulated upstream the dam. This hazard association represent a minor risk for the population living upstream. The water accumulation is slow enough to allow population relocation. On the other hand it represents an increasing-in-time threat for the population living downstream. The more time passes, the more the water volume increases. The dam rupture is able to provoke a dramatic high energy flash flood in the valley. The hazard is then shifted spatially and postponed from the initial event. In Tajikistan, the Usoi Dam on the Murghab River was created on February 18th 1911 by an earthquake-triggered landslide. Behind the dam, with more than 60 km in length, the Lake Sarez constitutes a high threat for tens or possibly hundreds of thousands of people in the Murgab, Bartang, Panj, and Amu Darya valleys downstream. As a consequence, each landslide-dam formed, in reason of its destructive potential, must be closely monitored. Adam Emmer and Anna Juřicovà in their contribution "Inventory and typology of landslide-dammed lakes of the Cordillera Blanca (Peru)" considers that these entities require appropriate scientific attention, because: (i) they significantly influence geomorphological processes (erosion-accumulation interactions) at the catchment spatial scale; (ii) they act as a natural water reservoirs and balance stream fluctuation on different temporal scales (daily to seasonal); (iii) they may represent threat for society (lake outburst flood; LOF). The main objective of this study is to provide inventory of landslide-dammed lakes in the Cordillera Blanca, overview on their typology and discuss their geomorphological significance.

Another contribution to the topic of landslide dams is proposed by Carlo Tacconi Stefanelli et al.: "Assessing

P. Wassmer (✉)
Université de Strasbourg, Strasbourg, France
e-mail: Wassmerpat@aol.com

M. Mikoš et al. (eds.), *Advancing Culture of Living with Landslides*,
DOI 10.1007/978-3-319-53483-1_27

landslide dams evolution: a methodology review". The authors applied and reviewed a procedure to evaluate landslide dam evolution. Not less than 300 obstruction cases that occurred in Italy were analyzed with two recently proposed indexes: (i) the Morphological Obstruction Index (MOI) which combines the landslide volume and the river width, is used to identify the conditions that lead to the formation of a landslide dam or not; and (ii) the Hydromorphological Dam Stability Index (HDSI) which combines the landslide volume and a simplified formulation of the stream power, allowing a near real time evaluation of the stability of a dam after its formation. It is shown that the two indexes show a good forecasting effectiveness (61% for MOI and 34% for HDSI) and employ easily and quickly available input parameters that can be assessed on a distributed way even over large areas.

The case study of a Tailings dam failure is investigated by the contribution of Vanessa A Cuervo et al.: "Downstream geomorphological response of the August 4th Mount Polley Tailings Dam Failure" linked to a copper and gold mine in British Columbia, Canada. The study utilizes an integrated approach of geomorphological mapping, topographic analysis, historical aerial photograph analysis, and field surveys to identify and quantify the impact of the event. A first stage aims at reconstructing the pre-event hydrologic and geomorphic conditions of the Hazeltine Creek channel and floodplain. The second stage describes and reconstructs the event. Eventually, the last stage involved the evaluation and quantification of flow impacts to Hazeltine Creek. Volumes of deposition are estimated through geomorphological and surficial material mapping. Overall results indicate that the observed impacts of the breach and subsequent flow are comparable to impacts resulting from extreme flood events. Finally, the authors discuss how the analysis of the geomorphological response of dam failure events can contributes to the design of better hazard and risk assessment guidelines.

The potential destructive effect of landslides on highway infrastructures is tackled by Giuseppe Formetta et al. in their contribution: "Quantifying the performances of simplified physically based landslide susceptibility models: an application along the Salerno-Reggio Calabria highway" in Italy. Great effort has been devoted since a long time by the scientific community to develop landslide susceptibility models. Based on the fact that only few studies have been focused on defining accurate procedures for model selection, assessment, and inter-comparison, the authors applied a methodology for objectively calibrate and compare different landslide susceptibility models in a framework based on three steps: (i) calibration of the automatic model parameter based on different objective functions and compare the models results in the ROC plane; (ii) inter-comparison of a set of model performance indicators in order to exclude objective functions that provide the same information; and (iii) analysis of a model parameter sensitivity to understand how model parameter variations affect the model performances. This three-step procedure is applied in this study to compare two different simplified physically based landslide susceptibility models along the highway Salerno-Reggio Calabria.

Another important aspect of landslides is the mobilization and transportation of a sometimes huge stock of loose material (weathered rocks, soils and debris) downslopes, feeding river streams with loads of sediments easily erodible. This strong sediment flux poses significant issues downstream, increasing the erosion power of the flow and filling quickly man made water supply reservoirs. Chih Ming Tseng et al. propose here a contribution on this thematic: "The Sediment Production and Transportation in a Mountainous Reservoir Watershed, Southern Taiwan", exploring, on the base of a LiDAR-derived DTM taken in 2005 and 2010, the differences in landslide sediment production and sediment transportation abilities of reservoir watersheds in different geological environments after extreme rainfall.

The river bank failure and their consequences downstream are addressed by Archana Sarkar in his contribution on "Brahmaputra River Bank Failures—Causes and Impact on River Dynamics". The author outlines the types of river bank failure mechanism and erosion process in the Brahmaputra River, pointing out the "domino effect" induced downslope by these common processes: Drastic increasing of sediment laden and correlative erosion of the river banks leading to aggradations and widening of the river channels. Banks oscillations generated are responsible of considerable loss of good fertile land each year and causing shifting of outfalls of its tributaries, it brings newer areas under waters. Frequent changes of channel courses and bank erosion with high rate of siltation are identified as major threats to the riverine biota as they have a great bearing on the faunal composition of the river, some of which belong to the most rare and endangered species. This in turn has a negative impact on the sustainability of the wetlands and on the deterioration of flood hazard scenario in the state.

Other aspects of Landslides are explored by:

Udeni Priyantha Nawagamuwa and Lasith Pathum Perera who broach the topic of rainfall threshold for landslide triggering in Sri Lanka. In this country where the main cause of landslides is heavy and prolonged rainfall, the increasing frequency and intensity of landslides, is responsible for extensive damage to lives and properties. Therefore, it becomes a necessity to predict and warn landslide hazards before it occurs. In this research an hourly rainfall trend was proposed in order to predict occurrences of landslides in Sri Lanka considering hourly rainfall data from twelve hours before to any disastrous event.

Olena Dragomyretska et al. who explores, thanks to the help of correlation-regression methods and spectral analysis, patterns of spatial and temporal development of landslide processes in the north-west coast of the Black Sea. On the south-west of the Odessa Bay, abrasion process is described by dependence model abrasion indicators from wind wave energy, beach width and precipitation while on the east of the Bay the dependence model is stronger between abrasion indicators, sea level and precipitation. It is shown that the landslide processes in relation to climate and hydrodynamic impacts have a delay of 1–2 years, due to the inertia of the coastal geological system.

Quantifying the Performances of Simplified Physically Based Landslide Susceptibility Models: An Application Along the Salerno-Reggio Calabria Highway

Giuseppe Formetta, Giovanna Capparelli, and Pasquale Versace

Abstract

Landslides are one of the most dangerous natural hazards in the world causing fatalities, destructive effects on properties, infrastructures, and environment. A correct evaluation of landslide risk is based on an accurate landslide susceptibility mapping that will affect urban planning, landuse planning, and infrastructure designs. Great effort has been devoted by the scientific community to develop landslide susceptibility models. Only few studies have been focused on defining accurate procedures for model selection, assessment, and inter-comparison. In this study we applied a methodology for objectively calibrate and compare different landslide susceptibility models in a framework based on three steps. The first step involves the automatic model parameter calibration based on different objective functions and the comparison of the models results in the ROC plane. The second step involves the intercomparison of a set of model performance indicators in order to exclude objective functions that provide the same information. Finally the third step involves a model parameter sensitivity analysis to understand how model parameter variations affect the model performances. In this study the three-step procedure was applied to compare two different simplified physically based landslide susceptibility models along the highway Salerno-Reggio Calabria in Italy. The model M2, able to consider the spatial variability of the soil depth respect to the model M1, coupled the distance to perfect classification index provided the most accurate result for the study area.

Keywords

Landslide susceptibility analysis • Model performances evaluation • ROC

G. Formetta (✉)
Centre for Ecology & Hydrology, Crowmarsh Gifford, Wallingford, UK
e-mail: giufor@nerc.ac.uk

G. Capparelli · P. Versace
Dipartimento di Ingegneria Informatica, Modellistica, Elettronica e Sistemistica Ponte Pietro Bucci, University of Calabria, Cubo 41/b, 87036 Rende, Italy
e-mail: giovanna.capparelli@unical.it

P. Versace
e-mail: pasquale.versace@unical.it

M. Mikoš et al. (eds.), *Advancing Culture of Living with Landslides*,
DOI 10.1007/978-3-319-53483-1_28

Introduction

Landslides are one of the most significant hazards in mountain areas involving loss of life and properties (Park 2011). Brabb (1984) defined the landslide susceptibility as "the likelihood of a landslide occurring in a certain area on the basis of local terrain conditions a given area". Landslide susceptibility is a fundamental component of the landslide hazard estimation i.e. the probability of landslide occurrence within a given area and in a given period of time (Varnes 1984). A correct landslide susceptibility mapping is essential because it will affect landslide hazard quantification, wide area urban planning, and environment preservation (Cascini et al. 2005).

During the last decades many landslide susceptibility models have been developed on the base of different approaches (Corominas et al. 2014) such as qualitative methods, based on field campaigns and on expert knowledge, statistical methods (e.g. Naranjo et al. 1994; Guzzetti et al. 1999), physically based models (Montgomery and Dietrich 1994; Lu and Godt 2008; Simoni et al. 2008). In this paper we limited our analysis to simplified physically based models for landslide susceptibility.

Regardless the model used, landslide susceptibility mapping involves the knowledge of many environmental factors such as geology, hydrology, climate, land-use, that in most of the case are uncertain and present a high spatial variability. Moreover, the identification of the most suitable model parameters for the studied area makes the landslide susceptibility mapping even more uncertain.

Based on these motivations, the choice of the most appropriate landslide susceptibility model parameters and an effective model performances evaluation are still scientifically open questions (e.g. Dietrich et al. 2001; Frattini et al. 2010; Guzzetti et al. 2006). In the past landslide susceptibility model performances were based on subjective, heuristic criteria depending on the expert judgment. In most recent papers they are quantified by using a variety of indices that compare the model results with a map of occurred landslides providing a measure of the fitness (goodness of fit indices, GOF). Those indices are discussed in many papers (e.g. Bennett et al. 2013; Jolliffe and Stephenson 2012) and some of them are presented in Table 1.

In many applications one of these indices is selected and used as objective function (OF) in combination with a calibration algorithm in order to obtain the optimal model parameter set. The procedure aims to optimize the OF that is a measure of the similarity between observed and modeled landslide map. However, in most of the applications the selection of the OF arbitrary and is not based on any appropriate criteria.

Differently from previous applications, the methodology we used in this paper aims to objectively: (i) calibrate the landslide susceptibility model using a set of OFs and select the most appropriate to use for parameters estimation; (ii) execute the model using the parameter sets selected in the previous step and quantify its performances in order to identify the OFs that provide peculiar and not redundant information; (iii) perform a model parameters sensitivity analysis in order to understand the relative importance of each parameter and its influence on the model performance. The methodology allow the user to: (i) identify the most appropirate OFs for estimating the model parameters and (ii) compare different models in order to select the most performing in estimating the landslide susceptibility of the study area.

The paper is organized in four sections: the first section presents the landslide susceptibility models and the three step verification procedue, the second describes the study area, the third presents the results, and the last provides discussion and conclusions.

Table 1 List of the most used goodness of fitness indices to compare predicted and occurred landslides

Parameters	AI
Critical success index (CSI)	$CSI = \frac{tp}{tp+fp+fn}$
Equitable success index (ESI)	$ESI = \frac{tp-R}{tp+fp+fn-R}$ $R = \frac{(tp+fn)\cdot(tp+fp)}{tp+fn+fp+tn}$
Success Index (SI)	$SI = \frac{1}{2}\cdot\left(\frac{tp}{tp+fn}+\frac{tn}{fp+tn}\right)$
Distance to perfect classification (D2PC)	$D2PC=\sqrt{(1-TPR)^2+FPR^2}$
Average Index (AI)	$AI=\frac{1}{4}\left(\frac{tp}{tp+fn}+\frac{tp}{tp+fp}+\frac{tn}{fp+tn}+\frac{tn}{fn+tn}\right)$
True skill statistic (TSS)	$TSS=\frac{(tp\cdot tn)-(fp\cdot fn)}{(tp+fn)\cdot(fp+tn)}$
Heidke skill score (HSS)	$HSS=\frac{2\cdot(tp\cdot tn)-(fp\cdot fn)}{(tp+fn)\cdot(fn+tn)+(tp+fp)\cdot(fp+tn)}$
Accuracy (ACC)	$ACC = \frac{(tp+tn)}{(tp+fn+fp+tn)}$

tp, *fn*, *fp*, and *tn* indicate true positives, false negatives, false positives, and true negatives, respectively. *TPR* and *FPR* indicate the true positive and the false positive rates, respectively

Methodology

The Landslide Susceptibility Model

The landslide susceptibility model implemented and used for this study is based on the infinite slope model which assumes that: (i) the depth of the failure plane is small compared to the length of the slope, and (ii) the failure plane is a straight line parallel to the ground surface.

The model compute the FS for each pixel of the computational domain following the derivation presented in Rosso et al. (2006):

$$FS = \frac{C \cdot (1+e)}{[G_s + e \cdot S_r + w \cdot e \cdot (1-S_r)] \cdot \gamma_w \cdot H \cdot \sin\alpha \cdot \cos\alpha} + \frac{[G_s + e \cdot S_r - w \cdot (1+e \cdot S_r)]}{[G_s + e \cdot S_r + w \cdot e \cdot (1-S_r)]} \cdot \frac{\tan\varphi'}{\tan\alpha} \tag{1}$$

where FS [-] is the factor of safety, C = C′ + Croot is the sum of Croot, the root strength [kN/m^2] and C′ the effective soil cohesion [kN/m^2], φ' [-] is the internal soil friction angle, H is the soil depth [m], α [-] is the slope angle, γ_w [kN/m^3] is the specific weight of water, and w = h/H [-] where h [m] is the water table height above the failure surface [m] and H [m] is the depth of the potential failure plane, Gs [-] is the specific gravity of soil, e [-] is the average void ratio and Sr [-] is the average degree of saturation.

The formulation used in this paper for w is based on the assumptions of steady state, flow parallel to the surface and distribution of pore pressure described by the Darcy law. As proposed by O'Loughlin (1986) and Beven and Kirkby (1979) of the subsurface runoff that flows through a cell based on Darcy's law is equal its total contributing area (TCA [L2]) times the rainfall intensity (Q [L/T]):

$$\begin{aligned} Q \cdot TCA &= (b \cdot h \cdot \cos\alpha) \cdot K_{sat} \cdot \tan\alpha \\ &= b \cdot h \cdot \sin\alpha \cdot K_{sat} \end{aligned} \tag{2}$$

where K_{sat} [L/T] is the saturated hydraulic conductivity, and b [L] is the length of the contour line. The subsurface runoff that flows through a cell will be equals to its maximum value when the water table reaches the surface (h = H). Under the previous steady-state hypothesis the degree of w (h/H) is defined as:

$$w = \frac{h}{H} = \min\left(\frac{Q}{T} \cdot \frac{TCA}{b \cdot \sin\alpha}, 1.0\right) \tag{3}$$

where T [L2/T] is the soil transmissivity defined as the product of the soil thickness and the sas ration between the tp and the sum ofaturated hydraulic conductivity. In this paper we used two different models: (i) M1 implements the Eqs. (1) and (3) considering H as a constant value overall the analyzed domain; model M2, differently from M1, takes into account that soil depth can vary in space pixel by pixel. The soil depth was modeled using the parameterization presented in Eq. (4) that was used in Lanni et al. (2012):

$$H_i = p_1 - p_2 \cdot \tan\alpha_i \tag{4}$$

where α [-] is the slope angle and p1 [L] and p2 [L] are two parameters that can be derived by fitting modeled and measurements soil depth data or can be considered as calibration parameters. The use of Eq. 4 allows the user to consider the spatial variability of the soil depth pixel by pixel depending on the slope value. This was not accounted in previous applications (i.e. Formetta et al. 2014; Formetta et al. 2015) where the soil depth was considered constant over the whole computational domain. The approach proposed in Eq. (4) is one among the simplest method for modeling spatially variable soil depths. Other options involve procedures that merge geological and morphological observations with geotechnical analyses (Cascini et al. 2016), models that relates slope angle and slope convexity with soil production (Heimsath et al. 2012), and geostatistical methods such as kriging to predict the soil thickness (Kuriakose et al. 2009).

The Three Steps Verification Procedure (3SVP)

In order to objectively assess the model performance we implemented a package of indices for assessing the quality of a landslide susceptibility map (Table 1). Accurate description of those indices is presented in Bennett et al. (2013), Jolliffe and Stephenson (2012), Beguería (2006).

These indices are based on pixel-by-pixel comparison between observed landslide map (OL) and predicted landslides (PL). They are binary maps with positive pixels corresponding to "unstable" ones, and negative pixels that correspond to "stable" ones. Therefore, four types of outcomes are possible for each cell: if the pixel is mapped unstable both in OL and in PL is a true-positive (tp); if a pixel is mapped as stable in both OL and PL is a false positive (tn); if a pixel is mapped as "unstable" in PL but is "stable" in OL is a false-positive (fp); if a pixel is mapped as "stable" in PL, butt is "unstable" in OL, is a false-negative (fn).

The true positive rate TPR = tp/(tp + fn) is defined as the ratio between true positive and the sum of true positive and false negative and the false positive rate FPR = fp/(fp + tn). A model that perfectly reproduce the observed landslide has TPR = 1 and FPR = 0.

The three steps verification procedure (3SVP) presented in Formetta et al. (2015) is based on the automatic model parameter calibration and on the assessment of the model performances using three objective steps.

In this application we calibrated the landslide susceptibility model parameters using the Particle Swarm Optimization (PSO) method, a genetic model presented in Kennedy and Eberhart (1995). The PSO is inspired by social behavior and movement dynamics of insects, birds and fish. A group of random "particles" (values of parameters) is randomly initialized and then, in order to find the global optimum of the OF, each particle in the population adjusts its change according to the beast known position visited by itself and by the group.

The calibrated parameters for the model M1 and M2 are specified in the first column of Tables 1 and 2, respectively. They are hydrological (T, Q), mechanical (c, φ', and Gs), and related to the soil depth distribution (p1 and p2).

Three indices have been selected as objective function (OF) of the optimization algorithm: the success index, the distance to perfect classification, and the average index. The Success index (SI), Eq. (5), equally weight the true positive rate and specificity, defined as 1 minus false positive rate (FPR):

$$SI=\frac{1}{2}\cdot\left(\frac{tp}{tp+fn}+\frac{tn}{fp+tn}\right) \tag{5}$$

SI varies between 0 and 1 and its best value is 1 and is also named modified success rate. The Distance to perfect classification (D2PC) is defined in Eq. (6):

$$D2PC=\sqrt{(1-TPR)^2+FPR^2} \tag{6}$$

It measures the distance, in the plane FPR-TPR between an ideal perfect point of coordinates (0, 1) and the point of the tested model (FPR, TPR). D2PC ranges in 0–1 and its best value are 0. The Average Index (AI), Eq. (7), is the average value between four different indices: (i) TPR, (ii) Precision defined as ration between the tp and the sum of tp and fp, (iii) the ratio between successfully predicted stable pixels (tn) and the total number of actual stable pixels (fp + tn) and (iv) the ratio between successfully predicted stable pixels (tn) and the number of simulated stable cells (fn + tn):

$$AI=\frac{1}{4}\left(\frac{tp}{tp+fn}+\frac{tp}{tp+fp}+\frac{tn}{fp+tn}+\frac{tn}{fn+tn}\right) \tag{7}$$

In turn, the three indices presented in Eqs. (5, 6 and 7) were optimized as objective functions (OF) providing three different set of optimal model parameters and three different model results.

To assess their performances, the 3SVP involves the following three steps.

The first step aims to evaluate the performance of each single objective function on the Receiver Operator Characteristic (ROC, Goodenough et al. 1974) graph. It is a Cartesian plane with the FPR on the x-axis and TPR on the y-axis. The optimizations of the three objective functions provide three different parameter sets and, consequently, three different models results. Each of them is represented in the ROC plane by its own TPR and FPR. The performance of a perfect model corresponds to the point P(0, 1) on the ROC plane; points that fall on the bisector are associated with models considered random because they predict stable or unstable cells with the same rate.

The second step aims to verify the information content of each optimized OF, checking if it is analogous to other metrics or it is peculiar of the optimized OF. Let's assume that the model is executed with the optimal parameter set provided by optimizing SI. The performance of this parameter set is evaluated using all the remaining indices presented in Table 1. The result of this procedure is a vector of the 8 model performance values presented in Table 1 and related to optimization of SI (MP_{SI}). MP_{SI} has 16 elements: 8 for the calibration and 8 for the verification dataset. The two datasets of occurred landslides are presented in Fig. 1b. Repeating the same procedure for AI and D2PC provides MP_{AI} and MP_{D2PC}, respectively. Assessing the correlation between the vectors MP can be verified if each OF metric has its own information content or if it provides information analogous to other metrics (and unessential).

Finally, the third step of the procedure involves a sensitivity analysis of each optimal parameter set by perturbing optimal parameters and by evaluating their effects on the indices presented in Table 1.

Table 2 Optimal parameter set for model M1 and for the three OFs: average index (AI), distance from perfect classification (D2PC), and success index (SI)

Parameters	AI	D2PC	SI
T [m^2/d]	65.43	38.22	84.54
c [kPa]	25.17	16.94	30.01
ϕ [°]	29.51	32.30	24.57
Q [mm/d]	236.14	153.61	294.70
Gs [-]	2.11	2.44	2.77
H [m]	2.35	2.44	2.74

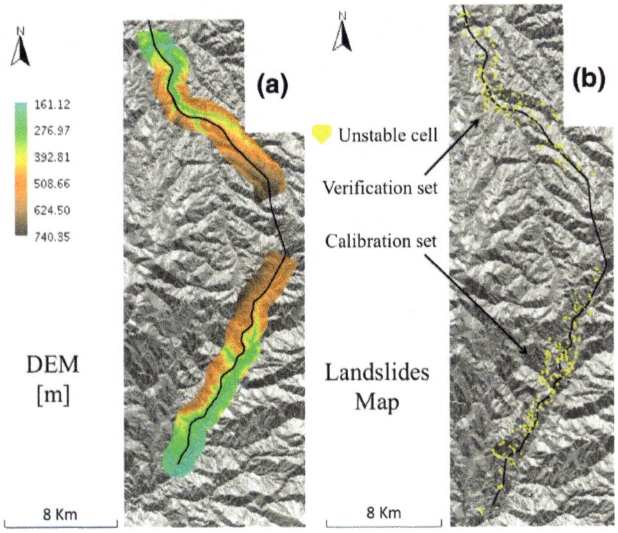

Fig. 1 Study area: digital elevation model (**a**) and map of occurred landslides (**b**)

Case Study

The landslide susceptibility analysis was carried out along the highway Salerno-Reggio Calabria in Calabria region (Italy), between Cosenza and Altilia municipalities. In the study area, Fig. 1a the average elevation is around 450 m a. s.l. and slopes ranges between, range from 0° to 55°. The mean annual precipitation is about of 1200 mm and mean annual temperature of 16 °C. We used a 30-m digital elevation model from which we extracted the map of slopes and of the total contributing areas for the study area.

A map of occurred landslides, Fig. 1b, that refers only to their initiation area was prepared on the based of aerial photography interpretation and extensive field survey (Conforti et al. 2014).

In order to perform model calibration and verification, the dataset of occurred landslides was divided in two parts one used for calibration and one for validation. The dataset separation was based on spatial criteria and on the concept of obtaining the same number of occurred landslide per area. Even though assuming the same spatial parameters for the two areas could be a strong hypothesis, this is justified by the fact that the two areas belong to the same geological features (Conforti et al. 2014).

Results

The 3sVP procedure has been applied in each of its steps on the study area. The optimal model parameter sets resulting from the calibration procedure are presented in Tables 2 and 3 for each OFs (AI, D2PC, and SI) and for model M1 and M2, respectively. The parameters have same order of magnitude across the different OFs and for both the models M1 and M2.

Moreover is clear a compensation effect between parameters such as soil cohesion and friction angle: a slightly underestimation in friction angle results in a slightly overestimation of cohesion and vice-versa. The M1 and M2 performances are presented in the ROC plane for each of the OFs and for calibration and validation period. We invite the reader to notice that, in order to better appreciate the differences between points, we limited the plane to x = (0.2, 0.5) and y = (0.6, 1.0).

In the calibration phase the results of the different OFs are similar for both the models in the sense that moving from D2PC to SI there is and increasing of both TPR (improving) and FPR (deterioration). Although at this stage a clear selection of one OF is not possible yet, the combination of model M2 with D2PC provides the best results. The same consideration lasts in the validation phase: the most performing OF is D2PC for the model M2 and SI for the model M1. In the latter case AI and D2PC seems to provide lower values of the TPR compared to SI. The comparison between the models M1 and M2 suggests that the performances are quite similar in the calibration phase. However, the model M2 systematically provides better TPR and FPR values. This consideration lasts also in the validation phase even if in some cases (M2-AI e M2-D2PC) quite low. Any of the OFs seems to systematically outperform the others and for this

Table 3 Optimal parameter set for model M2 and for the three OFs: average index (AI), distance from perfect classification (D2PC), and success index (SI)

Parameters	AI	D2PC	SI
T [m^2/d]	18.15	28.91	26.36
c [kPa]	12.75	7.65	26.52
ϕ [°]	31.54	33.32	30.33
Q [mm/d]	278.43	188.14	254.75
Gs [-]	2.44	2.4	2.25
p1 [-]	2.95	3.11	3.53
p2 [-]	1.55	1.75	2.18

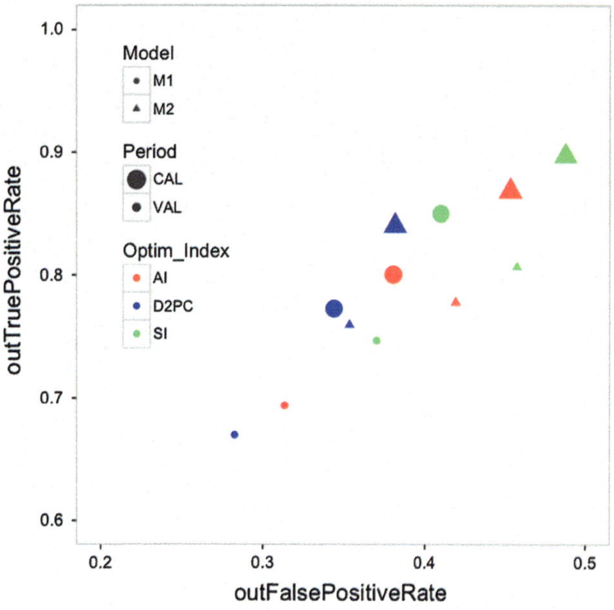

Fig. 2 Models M1 and M2 performances results for calibration and validation period and for the selected OFs: average index (AI), distance to perfect classification (D2PC), and success index (SI)

reason we continue to consider all of them in the next steps (Fig. 2).

The results for second step of the 3SVP are presented in Fig. 3. We used the correlation plot (Murdoch and Chow 1996) where the three MP vectors (MPSI, MPAI, and MPD2PC) are plotted against each other in a matrix form and using an ellipse. The color of the ellipse is scaled according the correlation coefficients between each MP index; the ellipse leans towards the right or the left according the fact that the correlation is positive or negative, respectively; the more prominent is the ellipse's eccentricity the less correlated are two MP vectors. In general for model M2 the MP vectors are more correlated each other respect the model M1. This could be due to the fact that the model structure of model M2 is less flexible that model M1, even thought it has one more parameter. In model M1 H has an independent calibrated value, whereas in model M2 it is strictly related to the slope of the pixel. On the other hand, the correlation between the D2PC and SI indices is higher for model M1 than for model M2, meaning that they provide quite similar model performances. Moreover, the results provided by the second step suggest that the information content provided by SI and D2PC is analogous and that to compare the two models could be sufficient consider ne of them.

Finally, the results of model parameters sensitivity analysis are presented in Figs. 4 and 5 for model M1 and M2, respectively. Each figure represents in terms of boxplot, the variation in the OFs due to a small change of a given parameters. In both the models the D2PC is the most responding and than sensible index to the parameters variation: i.e. changes in parameter variations are reflected in much more evident changes in the model performance. This means that D2PC for our case study was able accommodate eventual parameters responding to these changes instead of providing a "flat" behavior as AI and SI show.

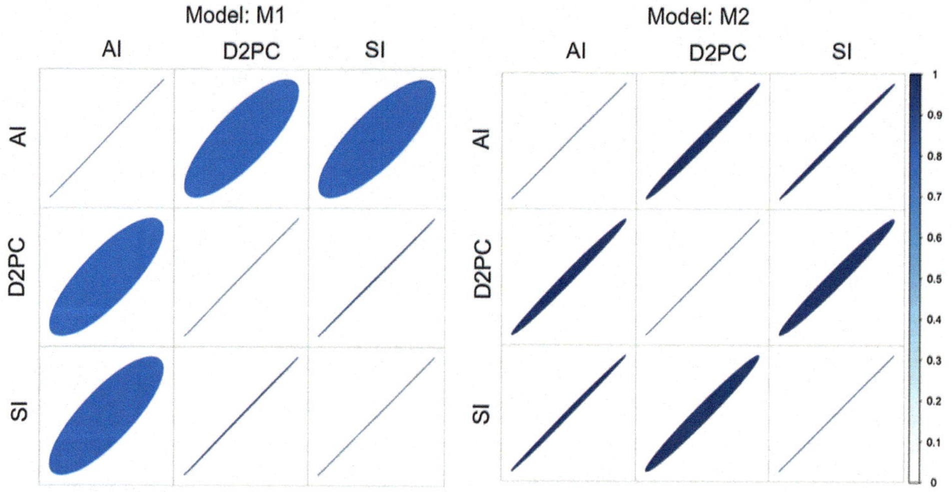

Fig. 3 Correlation plots between model performances indices for the model M1 and M2 and for the three selected objective functions: average index (AI), distance to perfect classification (D2PC), and success index (SI)

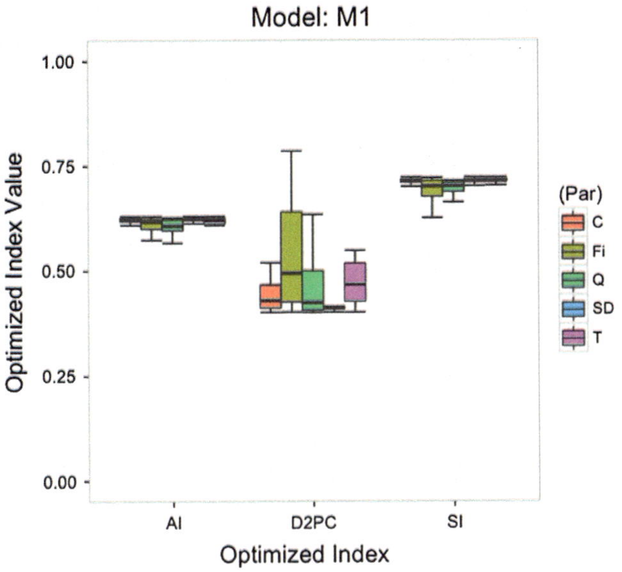

Fig. 4 Model M1 parameters sensitivity analysis

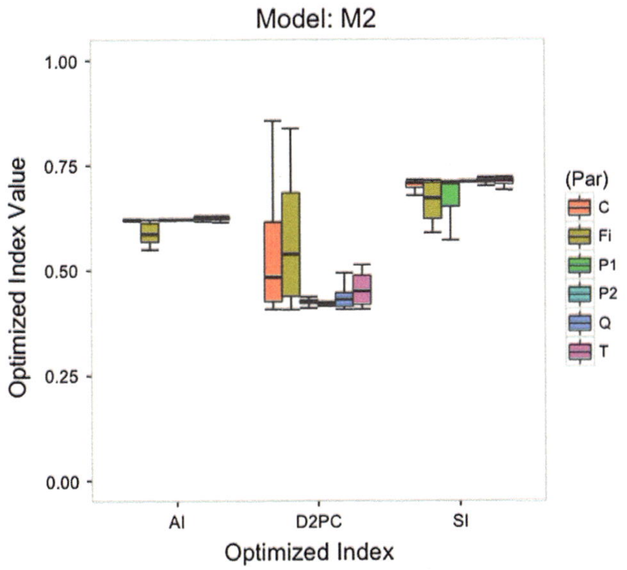

Fig. 5 Model M2 parameters sensitivity analysis

Discussion and Conclusion

In this paper we applied the 3SVP methodology to compare two different landslide susceptibility models. The methodology is organized in three objective steps aimed to individuate the more appropriate OF(s) to quantify the model performance for the analyzed case study. The procedure has been applied in Italy along the highway Salerno-Reggio Calabria and involved 3 steps.

The first step involved the optimization of three OFs, consequently the identification of three optimal parameter sets, and the comparison of the models performances on the ROC plane. In this step the D2PC OF coupled with model M2 provided the best results, even thought in calibration the results provided by different OFs were quite similar.

In the second step has been verified if each OF has is own information content or this is significantly similar to other indices. This is evaluated for each model by computing the correlation between the MP vectors of each OF. This step suggested that D2PC and SI have quite similar information content and thus could be sufficient considering one of them.

Finally the last step provides information about the effects of small variations in model parameters on the OFs. For both the models D2PC shows a higher sensibility compared to AI and SI and thus is the most able to accommodate eventual parameters variations, reflecting them in changes of its values. To conclude, the D2PC is the OF that should be used (i) to calibrate the presented models and (ii) to select the more performing one, which resulted in M2. In addition, the analysis would profit from a model parameter calibration procedure performed by soil classes, in order to take into account of the spatial variability.

Acknowledgements This research was funded by the PON project no. 01_01503 "Integrated Systems for Hydrogeological Risk Monitoring, Early Warning and Mitigation Along the Main Lifelines", CUP B31H11000370005, within the framework of the National Operational Program for "Research and Competitiveness" 2007–2013.

References

Beguería S (2006) Validation and evaluation of predictive models in hazard assessment and risk management. Nat Hazards 37:315–329

Bennett ND, Croke BF, Guariso G, Guillaume JH, Hamilton SH, Jakeman AJ, Marsili-Libelli S, Newham LT, Norton JP, Perrin C, Pierce SA (2013) Characterising performance of environmental models. Environ Model Softw 40:1–20

Beven KJ, Kirkby MJ (1979) A physically based, variable contributing area model of basin hydrology/Un modèle à base physique de zone d'appel variable de l'hydrologie du bassin versant. Hydrol Sci J 24(1):43–69

Brabb EE (1984) Innovative approaches to landslide hazard and risk mapping. In: Proceedings of the 4th International Symposium on Landslides, vol 1. Canadian Geotechnical Society, Toronto, Ontario, Canada, 16–21 September, pp 307–324

Cascini L, Bonnard C, Corominas J, Jibson R, Montero-Olarte J (2005) Landslide hazard and risk zoning for urban planning and development. In: Landslide risk management. Taylor and Francis, London, pp 199–235

Cascini L, Ciurleo M, Di Nocera S (2016) Soil depth reconstruction for the assessment of the susceptibility to shallow landslides in fine-grained slopes. Landslides, 1–13

Conforti M, Pascale S, Robustelli G, Sdao F (2014) Evaluation of prediction capability of the artificial neural networks for mapping landslide susceptibility in the Turbolo River catchment (northern Calabria, Italy). CATENA 113:236–250

Corominas J, Van Westen C, Frattini P, Cascini L, Malet JP, Fotopoulou S, Catani F, Van Den Eeckhaut M, Mavrouli O, Agliardi F, Pitilakis K (2014) Recommendations for the quantitative analysis of landslide risk. Bull Eng Geol Environ 73(2):209–263

Dietrich WE, Bellugi D, Real De Asua R (2001) Validation of the shallow landslide model, SHALSTAB, for forest management. In: Wigmosta MS, Burges SJ (eds) Land use and watersheds: human influence on hydrology and geomorphology in urban and forest areas. American Geophysical Union, Washington, D.C. doi:10.1029/WS002p0195

Formetta G, Capparelli G, Rigon R, Versace P (2014) Physically based landslide susceptibility models with different degree of complexity: calibration and verification. In: Ames DP, Quinn NWT, Rizzoli AE (eds) International Environmental Modelling and Software Society (iEMSs). 7th International Congress on Environmental Modelling and Software, San Diego, CA, USA, 15–19 June 2014. http://www.iemss.org/sites/iemss2014/papers/iemss2014_submission_157.pdf

Formetta G, Capparelli G, Versace P (2015) Evaluating performances of simplified physically based models for landslide susceptibility. Hydrol Earth Syst Sci Discuss 12:13217–13256

Frattini P, Crosta G, Carrara A (2010) Techniques for evaluating the performance of landslide susceptibility models. Eng Geol 111(1):62–72

Goodenough DJ, Rossmann K, Lusted LB (1974) Radiographic applications of receiver operating characteristic (ROC) analysis. Radiology 110:89–95

Guzzetti F, Carrara A, Cardinali M, Reichenbach P (1999) Landslide hazard evaluation: a review of current techniques and their application in a multi-scale study, Central Italy. Geomorphology 31(1):181–216

Guzzetti F, Reichenbach P, Ardizzone F, Cardinali M, Galli M (2006) Estimating the quality of landslide susceptibility models. Geomorphology 81(1):166–184

Heimsath AM, DiBiase RA, Whipple KX (2012) Soil production limits and the transition to bedrock-dominated landscapes. Nat Geosci 5:210–214

Jolliffe IT, Stephenson DB (eds) (2012) Forecast verification: a practitioner's guide in atmospheric science. Wiley, University of Exeter, UK

Kennedy J, Eberhart R(1995) Particle swarm optimization. In: Proceedings of the IEEE International Conference on Neural Networks, 1995, vol 4. IEEE, Perth

Kuriakose SL, Devkota S, Rossiter DG, Jetten VG (2009) Prediction of soil depth using environmental variables in an anthropogenic landscape, a case study in the Western Ghats of Kerala, India. CATENA 79:27–38

Lanni C, Borga M, Rigon R, Tarolli P (2012) Modelling shallow landslide susceptibility by means of a subsurface flow path connectivity index and estimates of soil depth spatial distribution. Hydrol Earth Syst Sci 16(11):3959–3971

Lu N, Godt J (2008) Infinite slope stability under steady unsaturated seepage conditions. Water Resour Res 44:W11404. doi:10.1029/2008WR006976

Montgomery DR, Dietrich WE (1994) A physically based model for the topographic control on shallow landsliding. Water Resour Res 30(4):1153–1171

Murdoch DJ, Chow ED (1996) A graphical display of large correlation matrices. Am Stat 50:178–180

Naranjo JL, van Westen CJ, Soeters R (1994) Evaluating the use of training areas in bivariate statistical landslide hazard analysis: a case study in Colombia. ITC J 3:292–300

O'loughlin EM (1986) Prediction of surface saturation zones in natural catchments by topographic analysis. Water Resour Res 22(5):794–804

Park NW (2011) Application of Dempster-Shafer theory of evidence to GIS-based landslide susceptibility analysis. Environ Earth Sci 62:367–376

Rosso R, Rulli MC, Vannucchi G (2006) A physically based model for the hydrologic control on shallow landsliding. Water Resour Res 42:W06410. doi:10.1029/2005WR004369

Simoni S, Zanotti F, Bertoldi G, Rigon R (2008) Modeling the probability of occurrence of shallow landslides and channelized debris flows using GEOtop-FS. Hydrol Process 22:532–545

Varnes DJ (1984) Landslide hazard zonation: a review of principles and practice. No. 3

Assessing Landslide Dams Evolution: A Methodology Review

Carlo Tacconi Stefanelli, Samuele Segoni, Nicola Casagli, and Filippo Catani

Abstract

In hilly and mountainous regions, landslide dams can be recurring events involving river networks. A landslide dam can form when sliding material reaches the valley floor and closes a riverbed causing the formation of a water basin. Unstable landslide dams may collapse with catastrophic consequences in populated regions because of the resulting destructive flooding wave released. To prevent these consequences, the assessment of landslide dam evolution is a fundamental but not easy task, because of the complex interaction between watercourse and slope dynamics. Several researchers proposed geomorphological indexes to evaluate dam formation and stability for risk assessment purpose. These indexes are usually composed by two or more morphological parameters, characterizing the landslide (e.g. sliding material volume or velocity) and the river (e.g. catchment area or valley width). In this work, a procedure to evaluate landslide dam evolution is applied and reviewed. About 300 obstruction cases occurred in Italy were analyzed with two recently proposed indexes, the Morphological Obstruction Index (MOI) and the Hydromorphological Dam Stability Index (HDSI). The former, which combines the landslide volume and the river width, is used to identify the conditions that lead to the formation of a landslide dam or not. The latter, which combines the landslide volume and a simplified formulation of the stream power (composed by the upstream catchment area and the local slope), allows a near real time evaluation of the stability of a dam after its formation. The two indexes show a good forecasting effectiveness (61% for MOI and 34% for HDSI) and employ easily and quickly available input parameters that can be assessed on a distributed way even over large areas. The indexes can be combined in a convenient procedure to assess, through two subsequent steps, the final stage in which a landslide dam will evolve.

Keywords

Landslide dam • Geomorphological index • Landslide dam stability • Flooding hazard • Landslide • Rivers

C. Tacconi Stefanelli (✉) · S. Segoni · N. Casagli · F. Catani
Department of Earth Sciences, University of Firenze, Via La Pira 4, 50121 Florence, Italy
e-mail: tacconi.carlo@gmail.com

M. Mikoš et al. (eds.), *Advancing Culture of Living with Landslides*,
DOI 10.1007/978-3-319-53483-1_29

Introduction

Landslide dams formation and failure are rather common events in hilly and mountainous regions around the world, causing hazards such as backwater ponding, outburst floods and debris flows on exposed people and properties. Through correct urban planning and flood risk management (Van Herk et al. 2011) the potential damages and correlated significant consequences can be limited. This has stimulated many efforts to assess dams formation and failure using quantitative methods with prevention purpose. Morphological indexes, composed by parameters describing involved elements (the valley, the river, the landslide, the dam and the lake), have been commonly employed to perform such analysis (Swanson et al. 1986; Ermini and Casagli 2003; Dong et al. 2011; Dal Sasso et al. 2014). Regardless their application at local scale, the use of parameters (e.g. discharge, granulometry) that can be accurately estimated only by local survey is troublesome and their assessment over a broad area is difficult (Dong et al. 2011; Dal Sasso et al. 2014). If a large number of landslide dams have to be characterized for prevention or planning activities the employ of parameters that can be easily defined on a distributed way (e.g. morphometrical data via DTM analysis with GIS software) is preferable.

In this paper, a practical methodology to assess landslide dam formation and evolution is presented. This simple method, suitable for large areas and based on the combination of two geomorphological indexes with good prediction effectiveness, is applied to the wider Italian landslide dam database.

Materials

Landslide dams are rather frequent in Italy because its geological and geomorphological characteristics. Tacconi Stefanelli et al. (2015) realized a homogeneous database with a morphometrical characterization of 300 landslide dams collected in Italy (Fig. 1). The inventory, realized mainly by photointerpretation and revision of historical and bibliographical data, characterize each case with a series of information and morphometric attributes taking into account six different aspects: the Localization, the Consequences, the Landslide, the Dam, the Stream and the Lake.

According to their evolution to present, landslide dams are classified in the database in three classes:

- Not Formed: the landslide reduces the riverbed section without forming an upstream basin. The partial damming can evolve with the river path diverted or the landslide toe eroded. It does not cause significant damage.

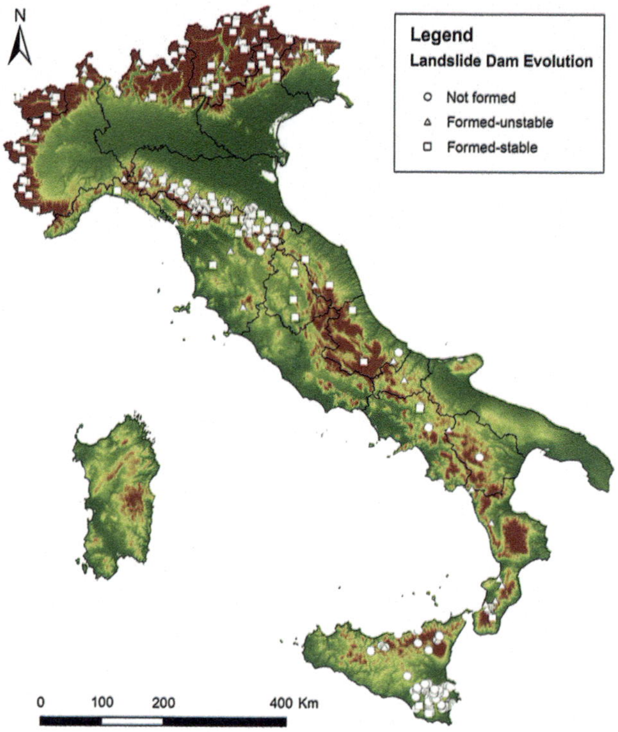

Fig. 1 Landslide dams distribution of Italian, according to three evolution classes (modified from Tacconi Stefanelli et al. 2015)

- Formed-Unstable: the river is completely obstructed by the landslide and upstream, a lake basin is formed. The dam collapses after a variable time (from hours to centuries) and releases a destructive flooding wave threatening life and property, making it the most dangerous case. Dams artificially stabilized or removed for prevention reasons are included in this class as well.
- Formed-Stable: complete obstruction of the river and formation of a dam and an upstream lake, which can be still present or filled by sediments. The dam may be overtopped by water in the past, but it did not suffered total failure producing catastrophic flooding wave.

The frequencies of the three groups are almost equally distributed, as the most frequent class is the formed-stable dams with 39% followed by the not formed dams with 33%, then by the formed-unstable with 28%.

Methodology

The possible evolution process of a dam may be described by the three evolution classes noted above, proposed by Tacconi Stefanelli et al. (2016) in a two consecutive steps flow chart (Fig. 2). At the start of the first step, a landslide involves a riverbed. If the landslide does not realize a

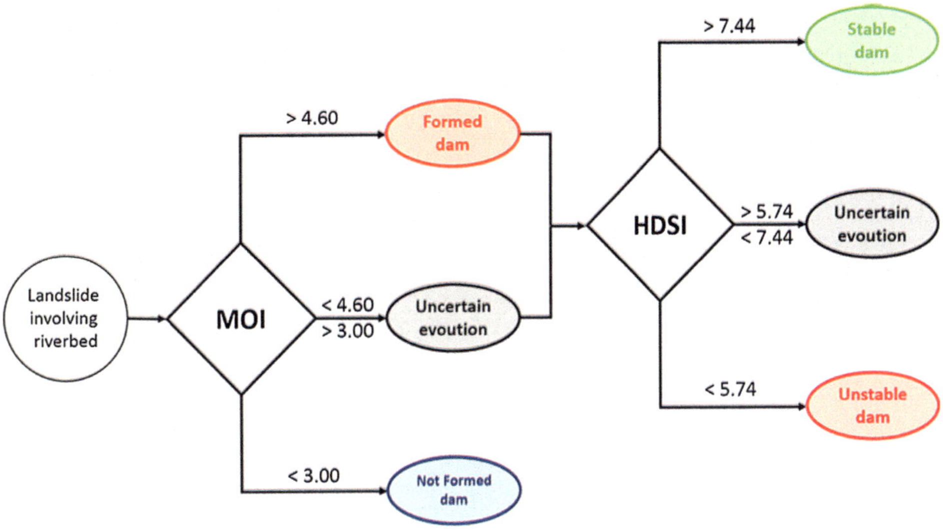

Fig. 2 Flow diagram of the operational procedure for landslide dam formation and stability evaluation (modified from Tacconi Stefanelli et al. 2016)

complete obstruction, the case is classified as "not formed". In the other case, or if the formation is unsure, the flow moves to the second step, where the stability of the formed dam is evaluated. It is classified as "formed unstable", if it can potentially collapse (after an undefined period), or as "formed stable", if it is not going to collapse. The path in the flow process is guided by the assessment of two geomorphological indexes (Fig. 2), first the "Morphological Obstruction Index", MOI:

$$MOI = \log\left(\frac{V_l}{W_v}\right) \quad (1)$$

and than the "Hydromorphological Dam Stability Index", HDSI:

$$HDSI = \log\left(\frac{V_l}{A_b \cdot S}\right) \quad (2)$$

where V_l is the landslide volume (m^3), W_v the dammed valley width (m), Ab the upstream catchment area (km^2) and S the local longitudinal slope of the river. The MOI relates the valley width W_v with the landslide volume V_l able to realize a complete obstruction; the HDSI compare the landslide volume V_l and the erosive capacity of the river, represented by the simplified geomorphological formulation of the stream power $A_b{\cdot}S$, to predict the landslide stability.

Results

The results of the two indexes applied to the Italian landslide dams are shown in Fig. 3 and Table 1. Data belonging to the same evolution classes are grouped on the graphic into evolution domains. The limits a domain is defined according to the extreme values of the evolution class inside it.

For the MOI index the domains are (Fig. 3a): the Non-formation domain (MOI < 3.00), including only landslide not blocking a riverbed, contains 15% of the total; the Uncertain Evolution domain (3.00 < MOI < 4.60), with both formed and not formed dams, contains about 39% of the dataset; the Formation domain (MOI > 4.60), with only landslides that blocked the valley, contains 46% of the dams (see Table 1). Two dashed lines, the lower Non-formation Line and the upper Formation Line, bound the Uncertain Evolution domain that contains the lowest formed and the highest not formed dam.

The HDSI index identify three domains as well in Fig. 3b: the Instability domain (HDSI < 5.74), with only not formed dams; the Uncertain Determination domain (5.74 < HDSI < 7.44), containing both stable and unstable dams; the Stability domain (HDSI > 7.44), with stable dams. The Uncertain Determination domain is encompassed by two dashed lines as well, the upper Stability line and the lower Instability line, comprehend the highest instable and

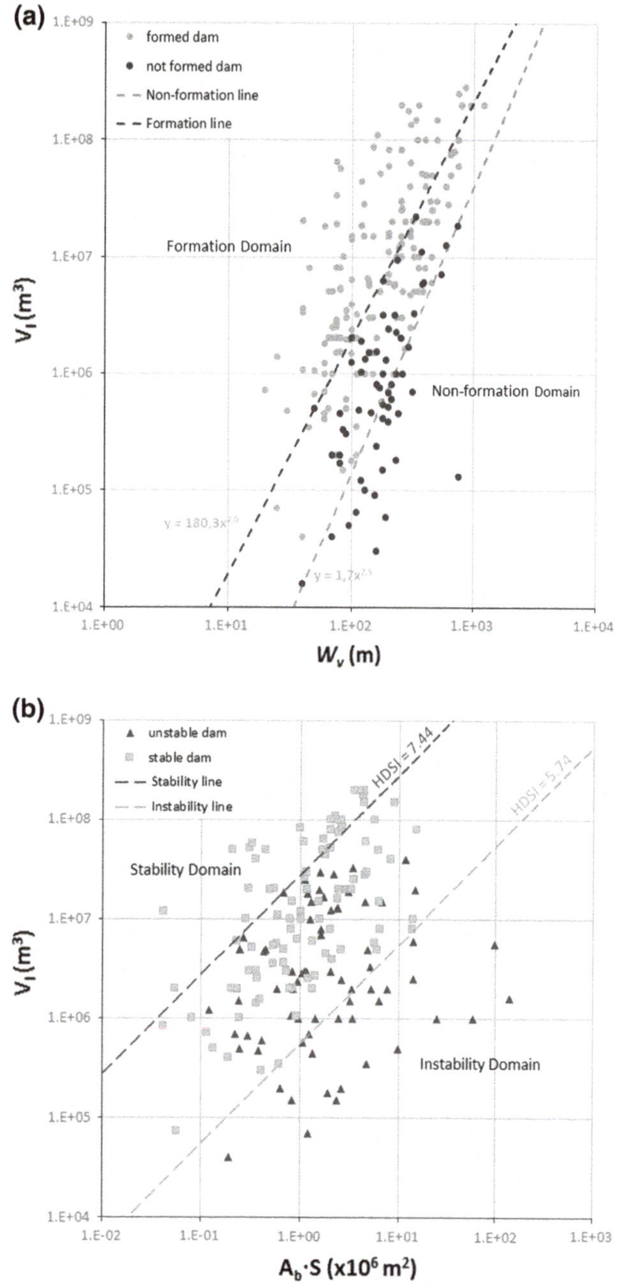

Fig. 3 Bi-logarithmic diagrams of: **a** Morphological Obstruction Index and **b** Hydromorphological Dam Stability Index

the lowest stable dam respectively. This domain is quite extended, with 66% of the total formed dams, while the Stability domain contains 19% of the formed cases, and the Instability domain includes 15% of them (see Table 1).

Table 1 Landslide dams distribution inside MOI and HDSI domains

HDSI domains	Formed dams (n.)	Not formed dams (n.)	% of total
Non-formation	0	34	14.6
Uncertain Evolution	58	33	39.0
Formation	108	0	46.4
MOI domains	**Formed dams (n.)**	**Not formed dams (n.)**	**% of total**
Non-formation	0	34	14.6
Uncertain evolution	58	33	39.0
Formation	108	0	46.4

Conclusion

In this work, a contribution in the rapid damming hazard assessment using practical geomorphological tools has been presented. Two morphological indexes, i.e. Morphological Obstruction Index (MOI) and Hydromorphological Dam Stability Index (HDSI), realized to improve previous methods characteristics (Tacconi Stefanelli et al. 2016) as the easy availability of the input data and the prediction effectiveness, were examined. The MOI allows a reliable analysis and a morphological assessment of landslide ability to block a river, as 61% of the dams in the dataset were correctly classified as formed or not formed. As displayed by the Italian cases, assuming the same valley width, the larger the volume of the landslide, the greater the damming probability. The HDSI correlates landslide volume and a morphological proxy for discharge in order to evaluate the long-term stability of a landslide dam. Even with a wide area of uncertain evolution, 34% of the cases fall in their own class and as the index value increases the general stability of the dam also increases. The high geological and climatic variability in the Italian territory may be responsible for the wide uncertain domain.

A fast operational procedure employing these indexes is presented as a useful tool to carry out a preliminary estimation of landslide dams evolution. This procedure can be employed at basin or smaller scale for forecasting and planning purposes. In the first step, MOI can be used to classify formed and not formed dams. Then, if a case results in a landslide dam, HDSI can be used to discriminate between stable or unstable dams.

References

Dal Sasso SF, Sole A, Pascale S, Sdao F, Bateman Pinzòn A, Medina V (2014) Assessment methodology for the prediction of landslide dam hazard. Nat Hazards Earth Syst Sci 14(3):557–567. doi:10.5194/nhess-14-557-2014

Dong JJ, Tung Y-H, Chen C-C, Liao J-J, Pan Y-W (2011) Logistic regression model for predicting the failure probability of a landslide dam. Eng Geol 117(1):52–61. doi:10.1016/j.enggeo.2010.10.004

Ermini L, Casagli N (2003) Prediction of the behavior of landslide dams using a geomorphical dimensionless index. Earth Surf Proc Land 28:31–47. doi:10.1002/esp.424

Swanson FJ, Oyagi N, Tominaga M (1986) Landslide dams in Japan. In: Schuster RL (ed) Landslide dams: processes risk and mitigation. Geotechnical Special Publication, vol 3. American Society of Civil Engineering, New York, pp 131–145

Tacconi Stefanelli C, Catani F, Casagli N (2015) Geomorphological investigations on landslide dams. Geoenviron Disasters 2(1):1–15. doi:10.1186/s40677-015-0030-9

Tacconi Stefanelli C, Segoni S, Casagli N, Catani F (2016) Geomorphic indexing of landslide dams evolution. Eng Geol 208:1–10

Van Herk S, Zevenbergen C, Ashley R, Rijke J (2011) Learning and action alliances for the integration of flood risk management into urban planning: a new framework from empirical evidence from The Netherlands. Environ Sci Policy 14(5):543–554

4thWLF2017
4th World Landslide Forum
LJUBLJANA SLOVENIA EU

Inventory and Typology of Landslide-Dammed Lakes of the Cordillera Blanca (Peru)

Adam Emmer and Anna Juřicová

Abstract

Despite the fact that landslide-dammed lakes represent less common lake type (n = 23; 2.6% share) in the Cordillera Blanca of Peru, these entities require appropriate scientific attention, because: (i) significantly influence geomorphological processes (erosion-accumulation interactions) at the catchment spatial scale; (ii) act as a natural water reservoirs and balance stream fluctuation on different temporal scales (daily to seasonal); (iii) may represent threat for society (lake outburst flood; LOF). The main objective of this study is to provide inventory of landslide-dammed lakes in the Cordillera Blanca, overview on their typology and discuss their geomorphological significance exemplified by two case studies. Existing, failed and infilled landslide-dammed lakes are simultaneously present in the area of interest. Three sub-types of existing landslide-dammed lakes are distinguished: (i) landslide/rockslide-dammed lakes situated in the main valleys; (ii) debris cone-dammed lakes situated in the main valleys; (iii) lakes situated on landslide bodies irrespective their location. Lakes of sub-types (i) and (ii) reach significant sizes, while lakes of sub-type (iii) do not. The dam formation of lake sub-types (i) and (iii) is usually connected with a single event, while the dams of sub-type (ii) are usually formed by several generations of debris deposition over time. It was shown, that landslide-dammed lakes in the study area are characterized by relatively low mean lake water level elevation (4115 m a.s.l.) and large catchments (in some cases up to 80 km^2), compared to other lake types. Lakes of sub-type (ii) are predominantly situated in central glacierized part of the Cordillera Blanca, while lakes of sub-types (i) and (iii) are situated rather in the already deglaciated piedmont areas, reflecting the conditions and mechanisms of dam formation. Two illustrative examples are, further, studied in detail: rockslide-dammed Lake Purhuay close Huari in Marañon River catchment; debris cone-dammed Lake Jatuncocha in Santa Cruz valley, Santa River catchment.

A. Emmer (✉) · A. Juřicová
Department of Physical Geography and Geoecology, Faculty of Science, Charles University in Prague, Albertov 6, 128 43 Prague, Czech Republic
e-mail: aemmer@seznam.cz

A. Juřicová
e-mail: juricova.an@gmail.com

A. Emmer
Department of the Human Dimensions of Global Change, Global Change Research Institute, The Czech Academy of Sciences, Bělidla 986/4a, 603 00 Brno, Czech Republic

A. Juřicová
Department of Soil Survey, Research Institute for Soil and Water Conservation, Žabovřeská 250, 156 27 Prague, Czech Republic

M. Mikoš et al. (eds.), *Advancing Culture of Living with Landslides*,
DOI 10.1007/978-3-319-53483-1_30

Keywords
Landslide-dammed lake • Natural dam • Landslide dam • Rockslide dam • Lake outburst flood • Cordillera blanca

Introduction

Different types of slope movements occur in the highest Peruvian mountain range—Cordillera Blanca, Andes. These processes claimed thousands fatalities (e.g., Evans et al. 2009) and their occurrence is topically framed by possible role of ongoing climate change and observed glacier ice loss (e.g., Vuille et al. 2008; Huggel et al. 2012), therefore is of great scientific interest. In favourable topographical and hydrological conditions, slope movement may lead to the formation of landslide-dammed lake.

Formation and failure of landslide-dammed lakes is connected to hazardous processes with potential risky consequences: (i) upstream flooding as a result of lake filling by water, and (ii) downstream flooding as a result of sudden water release (lake outburst floods; Costa and Schuster 1988).

Landslide dams and landslide-dammed lakes, however, attracted very little attention compared to other dam/lake types (especially glacial lakes) in the Cordillera Blanca (e.g., Reynolds 2003; Emmer and Vilímek 2013). Only few studies focusing on selected landslide-dammed lakes in relation to remediation works were done (e.g., Zapata 1978).

The main objective of this study is to provide textbook-style overview on existing landslide-dammed lakes in the Cordillera Blanca (defined by the border of NP Huascarán in this work), their inventory, classification and description of selected characteristics.

Existing, Failed and Infilled Landslide-Dammed Lakes of the Cordillera Blanca

Basically, according to the stage of evolution, landslide dams and landslide-dammed lakes of the Cordillera Blanca might be grouped as follow: (i) existing lakes; (ii) failed dams (lakes); and (iii) infilled lakes (lake basins; see below). Groups (ii) and (iii) are referring to the former lakes, which do not exist presently.

Existing Landslide-Dammed Lakes

Emmer et al. (2016) made an inventory of high-mountain lakes of the Cordillera Blanca identifying 882 lakes of which 23 (2.6%) classified as landslide-dammed (current as of 2013). These lakes are further classified according to dam type (see the section Typology of landslide-dammed lakes). Ten lakes (43.5%) are classified as landslide/rockslide-dammed lakes, eight lakes (34.8%) are classified as debris cone-dammed lakes and five lakes (21.7%) are classified as lakes situated on landslide bodies (see Fig. 1).

Existing landslide-dammed lakes are located in elevational range 3486–4706 m a.s.l., while mean elevation is 4115 m a.s.l., which is relatively low compared to other lake types (different types of glacial lakes). Catchments of landslide-dammed lakes may reach an area of several tens of

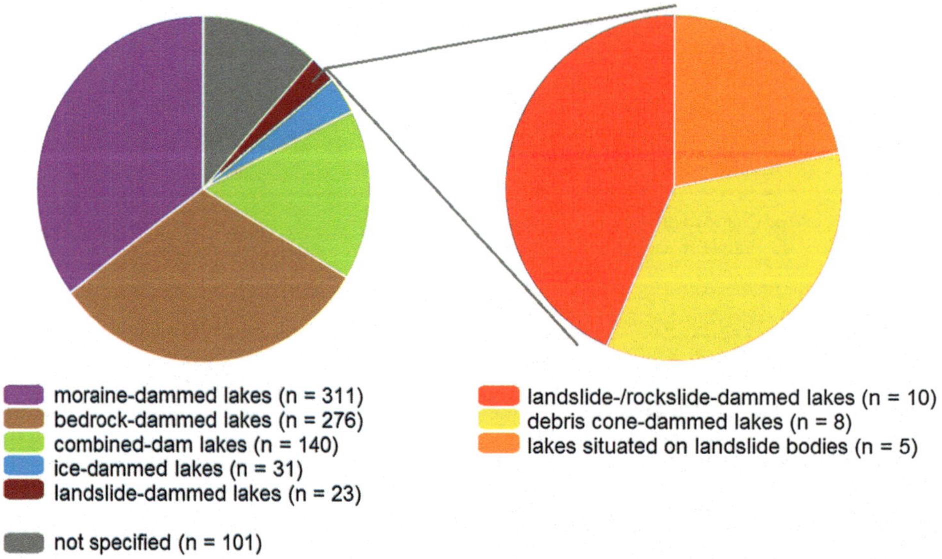

Fig. 1 Typology of lakes of the Cordillera Blanca (partly based on Emmer et al. 2016); landslide-dammed lakes are shown in the *right part* of the figure

Fig. 2 Example of debris cone dam of Lake Yircacocha which failed on 13th December 1941 as a result of a flood wave from *upstream* situated Lake Palcacocha. Breach width is about 150 m and breach depth >25 m. Former lake water level is visible in the *right part* of the figure

km^2 (e.g., Lake Ichiccocha with catchment area about 82.2 km^2) and landslide-dammed lakes may, therefore, influence runoff conditions of large areas (seasonal flow balancing, peak flow mitigation during extreme events, e.g., influence of lakes Jatuncocha and Ichiccocha during 2012 multi-lake outburst flood in Santa Cruz valley; see Mergili et al., in review).

Regarding the characteristics usable in hazard analysis, majority of landslide-dammed lakes (n = 20; 87.0%) has clearly recognizable surface outflow, while three lakes do not, 15 lakes (65.2%) have some lakes situated upstream and 10 lakes (43.5%) have some glaciers in their catchments (Emmer et al. 2016).

Failed Landslide Dams

Landslide-dammed lake may disappear in a catastrophic way by failure of its dam. Dam failure may by triggered by various causes such as dam overtopping or earthquake. Regardless the trigger, dam failure is tied with rapid release of retained water-lake outburst flood (landslide-dammed lake outburst flood, LLOF).

Only two landslide dam failures are documented from the study area since the end of Little Ice Age: (i) dam failure of debris cone-dammed lake Yircacocha (13th December 1941; see Fig. 2); and (ii) dam failure of moraine/debris cone-dammed lake Artizon Bajo (8th February 2012). Both these events were triggered by extremely increased discharge (lake outburst floods originating farther upstream). Released water increased overall volume of the flood and intensified cascading process chain (Mergili et al., in review).

Infilled Landslide-Dammed Lakes

Infilled lakes represent common feature in piedmont as well as mountainous areas of the Cordillera Blanca. Waterlogged ('pampa') areas are found in the bottom of most of the valleys, indicating former existence of lake irrespective the lake type. An example of landslide-dammed lake, which is currently turning into the 'pampa' condition is Lake Ichiccocha in Santa Cruz valley (see Fig. 3).

Lacustrine sediments trapped in lake basin may serve as a valuable source of proxy data for palaeo-climatological reconstructions (e.g., Seltzer and Rodbell 2005). This potential is, however, not fully exploited in the Cordillera Blanca currently and only few studies exist.

Typology of Landslide-Dammed Lakes of the Cordillera Blanca

Three sub-types of existing landslide-dammed lakes are distinguished in the study area according to the type of slope movement and topographical conditions: (i) landslide/rockslide-dammed lakes situated in the main valleys (n = 10); (ii) debris cone-dammed lakes situated in the main valleys (n = 8); (iii) lakes situated on landslide bodies irrespective their location (n = 5).

Fig. 3 Vegetated (*dark brown*) shallow part of Lake Ichiccocha in Santa Cruz valley occupied by cattle during dry season. Photo is taken from debris cone dam of the lake (*upstream view*)

Landslide-/Rockslide-Dammed Lakes

Landslide-/rockslide-dammed lakes situated in main valleys represent the most frequent sub-type of landslide-dammed lakes in the Cordillera Blanca (43.5% share). Three of these lakes reach surface area >100,000 m^2 (Purhuay, Huacrucocha and Tayancocha). The formation of landslide/rockslide dam is usually tied with single event with diverse possibilities of appearance in various topographical, hydrological and geological setting. Large-scale landslides/rockslides in the Cordillera Blanca are typically induced by earthquake (Cluff 1971), permafrost degradation (sensu Haeberli et al. 2016) or heavy rainfalls (Vilímek et al. 2014). Nevertheless, no landslide-/rockslide-dammed lake formation has been directly observed and/or documented.

Debris Cone-Dammed Lakes

Debris cones are common entities in high mountain environment of the Cordillera Blanca. Debris cones are characterized by episodic aggradation (e.g., earthquake-triggered aggradation of the Lake Llanganuco Alto cone dams in 1970; Cluff 1971; Zapata 1978). Debris cone aggradation was recently also linked to climate change (sensu Dietrich and Krautblatter 2016). These are typically built of several generations of debris material transported from upstream parts of the catchment during individual debris flow events. Debris cones are frequently located on the confluence of main and tributary valleys and debris cone-dammed lakes form in the gently-sloped parts of main valleys (thalweg inclination <5°).

Large debris cones often have a diameter of >1000 m, while the slope inclination is typically <25°, reflecting composition and friction angle of debris material. Debris cone-dammed lakes may reach significant areal dimensions >100,000 m^2 (5 lakes in the Cordillera Blanca—Jatuncocha, Llanganuco Alto, Llanganuco Bajo, Gueshguecocha, Querococha, see Fig. 4), while dam height and maximal depth of the lake rarely exceed 20 m (bathymetrical surveying of Unidad de Glaciología y Recursos Hídricos; e.g., UGRH 2015).

Lakes Situated on Landslide Bodies

Specific sub-type of landslide-dammed lakes represent those lakes situated directly on landslides bodies, i.e., lakes filling the depressions in the landslide accumulations regardless its location. Formation of this lake sub-type is, therefore, not interacting with local topography. Lakes situated on landslide bodies do not reach significant dimensions.

Case Studies

Rockslide-Dammed Lake Purhuay

The 130 m deep, over 2 km long Lake Purhuay with an area 850,000 m^2 is situated in the Marañon river catchment at 3580 m a.s.l. in the unnamed valley 7 km northwest of the city Huari. The lake was dammed by the rockslide accumulation 115 × 10^6 m^3 in volume and about 250 m high, that slid from ca. 1000 m high right bank of the valley,

Fig. 4 Debris cone damming the Lake Querococha. Shown cone is about 830 m wide and slope inclination is 12°

composed by Jurassic siliciclastic rocks (see Fig. 5a). Due to presence of a building from Late Intermediate Period (Benozzi and Mazzari 2008), age of this blockage is not less than 1000 years. One smaller (3×10^6 m^3) rockslide was formed at right side of the dam. Until now a considerably part of the dam was eroded by surface outflow, indicating that the dam underwent significant backward erosion (see Fig. 5b).

Debris Cone-Dammed Lake Jatuncocha

Debris cone-dammed Lake Jatuncocha (see Fig. 6a) is situated in the gently-sloped middle part of the Santa Cruz valley at elevation 3870 m a.s.l. under the slopes of the Nevado Artesonraju (6025 m a.s.l.) and Nevado Quitaraju (6040 m a.s.l.). The lake has surface area 450,000 m^2, is 1100 m long, maximum depth is 23 m and the volume of water 7.22×10^6 m^3 (Zapata 1978). Debris cone dam is formed by generations of sediments originating and transported from the hanging tributary valley and north-west slopes of Nevado Artesonraju and its volume was estimated 35×10^6 m^3. The debris of the sediment fans was derives from the recycled older slope moraines by debris flows (Iturrizaga 2008). The lake was equipped by artificial outflow and artificial dam increasing dam freeboard in 1970s.

In 2012, the Santa Cruz valley and the Lake Jatuncocha as well were affected by multi-lake outburst flood. A significant amount of material was deposited in the lake and caused shortening about 80 m and the lake area also slightly decreased. There was also minor damage to the artificial dam of the lake (Fig. 6b) due to increased flow rate. Detailed description of event has been proposed by Emmer et al. (2014) and Mergili et al. (in review).

Fig. 5 Part **a** shows rockslide originating from *right* bant damming Lake Purhuay; part **b** shows *upstream view* on deeply eroded dam

Fig. 6 Part **a** shows view on the lake and debris cone dam of the Lake Jatuncocha; part **b** shows aritifical lake outlet damaged during 2012 event

Conclusions

Landslide-dammed lakes represent minor (n = 23; 2.6% share; Emmer et al. 2016), but important lake type in the Cordillera Blanca of Peru. In this study, three sub-types are distinguished: (i) landslide-/rockslide-dammed lakes; (ii) debris cone-dammed lakes; (iii) lakes situated on landslide bodies. The dam formation of sub-types (i) and (iii) is usually triggered by a single event (single landslide or rockslide), while the dams of sub-type (ii) are usually aggraded gradually. Lakes of sub-type (i) are the most frequent in the study area, while the lakes of sub-type (ii) are most frequently characterized as large lakes with area >100,000 m^2.

Beside the existing lakes, geomorphological evidence of former landslide-dammed lakes is observed in the study area (both infilled lake basins and failed landslide dams), indicating high dynamic of their evolution in time. It was further shown, that landslide-dammed lakes may significantly influence cascade geomorphological processes such as multi-lake outburst floods both in mitigation and amplification way.

Acknowledgements The authors would like to thank Autoridad Nacional del Agua, Huaráz, Peru, for the long-term cooperation in high mountain lakes-related research. Grant Agency of Charles University (GAUK project No. 70 613 and GAUK project No. 730 216), Mobility fund of Charles University and the Ministry of Education, Youth and Sports of CR within the framework of the National Sustainability Programme I (NPU I), Grant No. LO1415 are further acknowledged.

References

Benozzi E, Mazzari L (2008) Los sitios gemelos de Llamacorral y Awilupaccha. In: Convegno Internazionale di Americanistica. Centro Studi Americanistici "Circolo Amerindiano", pp 853–858

Cluff LS (1971) Peru earthquake of May 31, 1970, engineering geology observations. Bull Seismol Soc Am 61:511–533

Costa J, Schuster RL (1988) The formation and failure of natural dams. Geol Soc Am Bull 100(7):1054–1068

Dietrich A, Krautblatter M (2016) Evidence for enhanced debris-flow activity in the Northern Calcareous Alps since the 1980s (Plansee, Austria). Geomorphology, online first, not yet assigned to an issue

Emmer A, Vilímek V (2013) Review article: lake and breach hazard assessment for moraine-dammed lakes: an example from the Cordillera Blanca (Peru). Nat Hazards Earth Syst Sci 13:1551–1565

Emmer A, Vilímek V, Klimeš J, Cochachin A (2014) Glacier retreat, lakes development and associated natural hazards in Cordillera Blanca, Peru. In: Shan W et al (eds) Landslides in cold regions in the context of climate change. Springer, Cham, pp 231–252

Emmer A, Klimeš J, Mergili M, Vilímek V, Cochachin A (2016) 882 lakes of the Cordillera Blanca: an inventory, classification, evolution and assessment of susceptibility to outburst floods. CATENA 147:269–279

Evans SG, Bishop NF, Smoll LF, Murillo PV, Delaney KB, Oliver-Smith A (2009) A re-examination of the mechanism and human impact of catastrophic mass flows originating on Nevado Huascaran, Cordillera Blanca, Peru in 1962 and 1970. Eng Geol 108(1–2):96–118

Haeberli W, Schaub Y, Huggel C (2016) Increasing risks related to landslides from degrading permafrost into new lakes in de-glaciating mountain ranges. Geomorphology, online first, not yet assigned to an issue

Huggel C, Clague JJ, Korup O (2012) Is climate change responsible for changing landslide activity in high mountains? Earth Surf Proc Land 37(1):77–91

Iturrizaga L (2008) Paraglacial landform assemblages in the Hindukush and Karakoram Mountains. Geomorphology 95(1):27–47

Mergili M, Emmer A, Juřicová A, Cochachin A, Fischer J-T, Huggel C, Pudasaini SP (in review) The 2012 multi-lake outburst flood in the Santa Cruz Valley (Cordillera Blanca, Peru): geomorphologic effects and process chain modelling. Water Resour Res, in review

Reynolds JM (2003) Development of glacial hazard and risk minimisation protocols in rural environments: methods of glacial hazard assessment and management in the Cordillera Blanca, Peru. Reynolds Geo-Sciences Ltd., Flintshire

Seltzer GO, Rodbell DT (2005) Delta progradation and Neoglaciation, Laguna Paron, Cordillera Blanca, Peru. J Quat Sci 20(7–8):715–722

UGRH (2015) Consolidado de Actividades Realizadas en el Año 2015 Por la Unidad de Glaciología y Recursos Hídricos. Autoridad Nacional del Agua (ANA), Unidad de Glaciología y Recursos Hídricos (UGRH), Huaráz, 285 p

Vilímek V, Klimeš J, Emmer A, Novotný J (2014) Natural hazards in the Cordillera Blanca of Peru during the time of global climate change. In: Sassa K (ed) Landslide science for a safer geo-environment, vol 1. Springer, Cham, pp 261–266

Vuille M, Francou B, Wagnon P, Juen I, Kaser G, Mark BG, Bradley RS (2008) Climate change and tropical Andean glaciers: past, present and future. Earth Sci Rev 89(3–4):79–96

Zapata ML (1978) Lagunas con obras de seguridad en la Cordillera Blanca. INGEOMIN, Glaciologia y segirudad de lagunas, Huaráz, 9 p

Recommending Rainfall Thresholds for Landslides in Sri Lanka

Udeni P. Nawagamuwa and Lasitha P. Perera

Abstract

Triggering factors for landslides could vary from heavy rainfalls/glacial activities to earthquakes, volcanisms or even vibrations due to nuclear explosions or heavy vehicle movement. However, in Sri Lanka landslides are mostly triggered due to heavy and prolonged rainfall. During last few decades, landslides have occurred with increasing frequency and intensity, causing extensive damage to lives and properties. Therefore, it becomes a necessity to predict and warn landslide hazards before it actually takes place which is still a mammoth task to achieve. Although several researches were done based on daily rainfall data in Sri Lanka, it has been identified that the extreme rainfalls with shorter durations could trigger more disastrous landslides. Hence, recommending hourly rainfall thresholds for landslides is much important in terms of hazard warning and preparedness. In this research an hourly rainfall trend was proposed in order to predict occurrences of landslides in Sri Lanka considering hourly rainfall data from twelve hours before to any disastrous event. Special attention was paid to data from Badulla District where more than 200 people were believed to be dead due to one major landslide in November 2014. The developed relationships are compared with Caine (1980) and observed to be matching to a considerable accuracy. Obtaining hourly rainfall data at exact location was almost impossible with the available rainfall measurement procedure in Sri Lanka. Much finer conclusions could have been made with more accurate data.

Keywords

Landslides • Heavy rainfall • Prediction and early warning • Thresholds

Introduction

Although landslides can be triggered due to many reasons, in Sri Lanka, landslides are mainly triggered due to heavy and prolonged rainfall. Heavy rainfall not only causes water penetration into the subsoil layers, but also affects losing the interlayer cohesion and at the same time increases the weight of the soil. In addition, penetrated water acts as an easy lubricant flowing down slope.

Apart from the damage to life and property, several infrastructural as well as economically important facilities have also been affected, especially water distributary pipes, hydro electricity generating centers, and communication systems. At times, social interests, such as educational and health services, are also severely disrupted. Moreover, frequent landsliding has threatened the destruction to the environment, including the flora and fauna of the areas concerned. Such damage caused to the environment, at times is irreversible and therefore cannot be estimated and perhaps will never be known (Bandara 2005).

U.P. Nawagamuwa (✉) · L.P. Perera
Department of Civil Engineering, University of Moratuwa, Katubedda, Moratuwa, Sri Lanka
e-mail: udeni@uom.lk

L.P. Perera
e-mail: pathum.lasitha@gmail.com

M. Mikoš et al. (eds.), *Advancing Culture of Living with Landslides*,
DOI 10.1007/978-3-319-53483-1_31

In recent times, landslides in Sri Lanka have increased both in frequency and in intensity. During last few years, several hundreds of minor landslides were reported which may reactivate catastrophically in the future. Nawagamuwa et al. (2011) concluded, after a thorough study on regional thresholds, that the recent landslides due to short term high intense rainfalls in the region have caused devastated landslides. Therefore, hourly rainfall data related to landslides would be required and it can be recommended that more regional efforts must be made on this important subject. Therefore, developing a prediction model based on hourly rainfall data is vital in terms of early warning as nowadays most of the recent landslides are triggered due to extreme rainfall events in shorter duration (Crosta 1998; Corominas et al. 2002).

Thresholds have been developed by several researchers, however, the equation established by Caine (1980) considered as a worldwide threshold:

$$\mathbf{I} = \mathbf{14.82\,D^{-0.39}}. \quad (1)$$

where: I is the rainfall intensity (mm h^{-1}), D is duration of rainfall (h)

This threshold applies over time periods of 10 min to 10 days. Most of other relationships were based on daily data and the beauty of this formula is that it is possible to modify the formula to take into consideration areas with high mean annual precipitations by considering the proportion of mean annual precipitation represented by any individual event.

Sri Lankan Thresholds

Bandara (2008) discussed the thresholds prepared after studying number of landslides in Sri Lanka with the correlation of the landslide occurrence and the daily rainfall of the incidence day or day before. According to the studies following generalized thresholds had been decided for the whole landslide prone areas of Sri Lanka.

Type one	Alert Warning—Rainfall exceed 75 mm within 24 h and continuing
Type two	Warning—Rainfall exceed 100 mm within 24 h and continuing
Type three	Evacuation warning—Rainfall exceed 150 mm within 24 h or 75 mm within an hour period and continuing

Those above three types of threshold limits are generally used all over Sri Lanka at present.

Nawagamuwa et al. (2011) proposed a trend lineas expressed in the Eq. (2) discusses the relationship between the normalized rainfall and the number of days till the day of the disaster occurence.

$$\mathbf{Y = 0.1496X - 0.16221} \quad (2)$$

To develop this equation, rainfall data from 6 days priorto the disaster had been considered. However, this too had been a relationship based on daily data and may not be applicable for very recent landsides occurred due to heavy rainfalls within shorter duration.

Data Collection

Major landslides happened in Sri Lanka were selected for the analysis which covers the entire high land of the country. Rainfall data of the exact locations where landslide had taken place was not readily available, hence the nearest collection center was selected with reasonable assumptions and these data was collected from twenty hours prior to the event.

Two types of data were required in this study, i.e., landslide details and hourly rainfall of the particular location where landslide took place.

Collection of Landslide Data

Selecting landslides for the analysis became extremely difficult due to unavailability of proper landslide records. It was even more difficult to collect hourly rainfall data. However they were selected by referring old records of Disaster Management Center of Sri Lanka. But possibility of obtaining hourly rainfall data was verified prior to select a particular landslide.

Collection of Rainfall Data

Hourly rainfall measurements were only available at main weather stations. Hence, for some landslides hourly rainfall data had to be obtained from National Building Research Organization (NBRO) and Maskeliya Plantation PLC. For others hourly rainfall data at main weather stations were used. Main weather station for a particular landslide was chosen based on both the distance from landslide location

Fig. 1 Koslanda, Meeriyabedda landslide

and the rainfall pattern in order to check the validity of such data, daily rainfall was also obtained from the closest sub weather station and compared it with data obtained from main weather station.

Difficulties Faced During Data Collection

Actual hourly rainfall at a required landslide location could not be obtained as they had not been measured there. Those landslide locations where most people dead had not been equipped with any rainfall data collection technology. Although nine landslides were selected for the analysis, due to rainfall data errors such as instrumental errors, only seven were analyzed. Finding exact time of landslide was almost impossible due to unavailability of proper records at particular locations. Such a landslide happened in 2014 in Koslanda, Meeriyabedda, where 37 people were buried alive and 88 families were affected, and its past and present situation is shown in Fig. 1

Data Analysis

Landslides in the area of Palewela, Agalawatta, Watawala, Nuwera Eliya, Matale and Pattipola were considered in this analysis as those locations had a reasonable accurate data. It also clearly shows a huge variation in the above rainfall data during the time of the event. This could be due to several morphological and geological identities of such locations. Hence, those data were normalized to limit their range of variation between zero and one. By considering those normalized values, rainfall threshold is proposed. Table 1 provides the details of those data. Figure 2 shows the cumulative rainfall data variation of above landslides.

Cumulative rainfall data was normalized to limit their variation between zero and one. Two methods were considered; Method 1—Data was normalized with reference to an individual landslide and Method 2—Data was normalized with reference to all landslides considered. Figure 3 provides such information on seven landslides. It was observed that the Method 2 gave the highest goodness of fit (R^2) value and hence it was used for analysis.

These results were plotted to observe the variations between rainfall intensity and duration. As shown in Fig. 4, a graph could be plotted similar to the relationship proposed by Caine (1980). Dotted line is the trend line.

It can be proposed that the rainfall threshold for landslides in Sri Lanka would be,

$$\mathbf{I} = \mathbf{14.02D^{-0.30}} \quad (3)$$

Above threshold is developed with goodness of fit (R^2) of 0.94.

Table 1 Rainfall data of seven previous landslide locations in Sri Lanka

Hours	Palewela	Agalawatte	Watawala 92	Watawala 93	Nuwara eliya	Matale	Pattipola
1	17.5	18.2	21.3	17.9	17.6	18.7	19.1
2	10	9.3	18	7	16	2	1.8
3	0	0	0	0	8	5.5	2.2
4	23.5	0	0.2	0	6.3	4	0.5
5	12.5	0	0.5	0	1.3	6.9	0.1
6	13	0	0.5	0.2	2.2	7	0.4
7	20	0	0.6	0	2.7	16.3	4.2
8	6	5	20	0	6.3	3	7.5
9	7	53	6.2	0.8	6	15	7
10	10.1	14.5	0.3	2	7	11.7	24.5
11	25.6	0	0.2	0.7	6	5.8	3
12	2.4	11.8	0.5	0.1	5	7	5.3

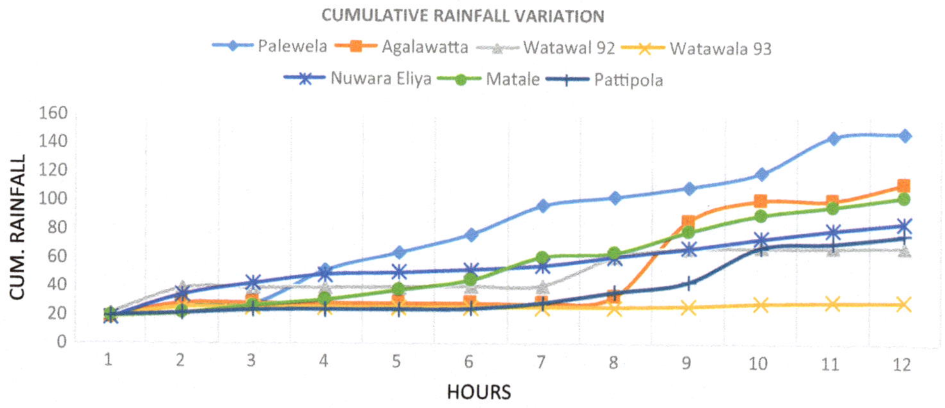

Fig. 2 Cumulative rainfall data variation 12 h prior to the event

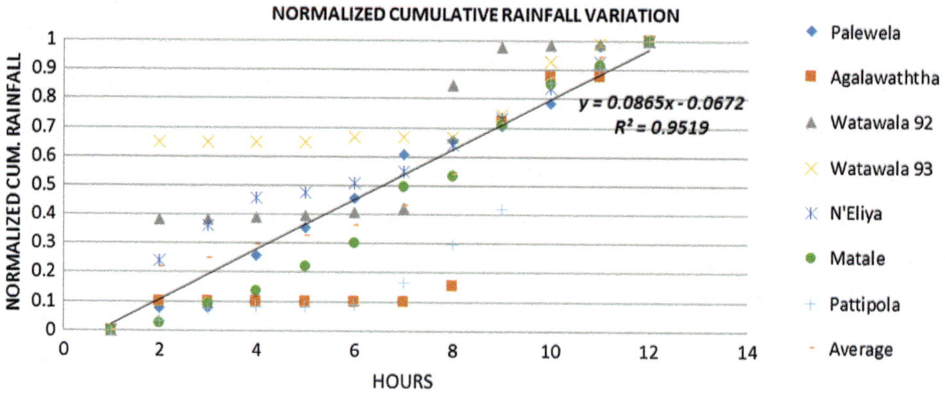

Fig. 3 Normalized data for all landslides

Recent Landslides in Badulla District

In recent past landslides especially in Badulla district have increased both frequency and intensity causing severe damage to human lives as well as their properties. In this study, it was targeted to develop rainfall threshold for Badulla district, however, due to unavailability of proper data in most of the cases, only Koslanda landslide was considered. Figure 5 shows the variations of cumulative rainfall with time and Fig. 6 provides the plot between the intensity and rainfall duration for Badulla District. Dotted line is the trend line.

From the above results, rainfall threshold of landslides for Badulla district could be proposed as,

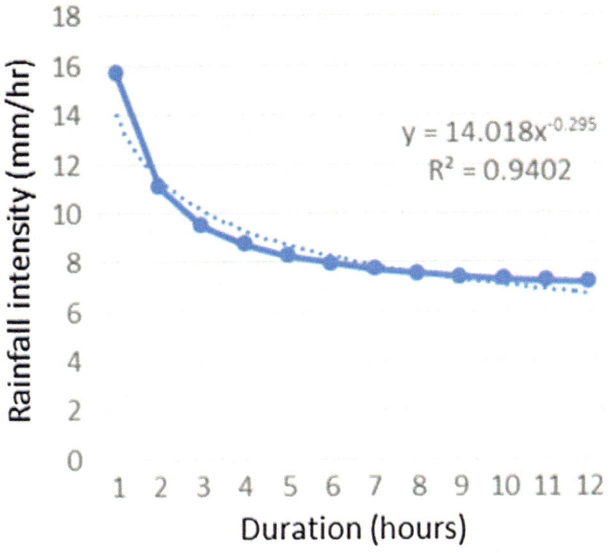

Fig. 4 Rainfall intensity versus duration

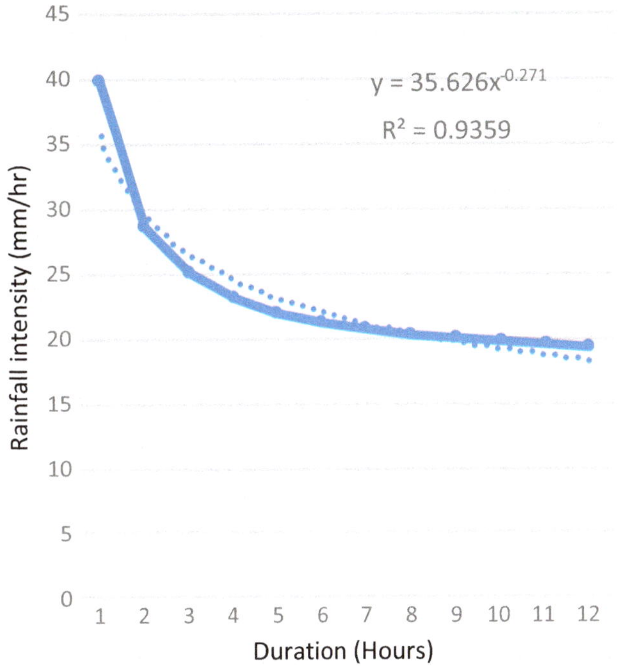

Fig. 6 Rainfall Versus Duration plot for Badulla District

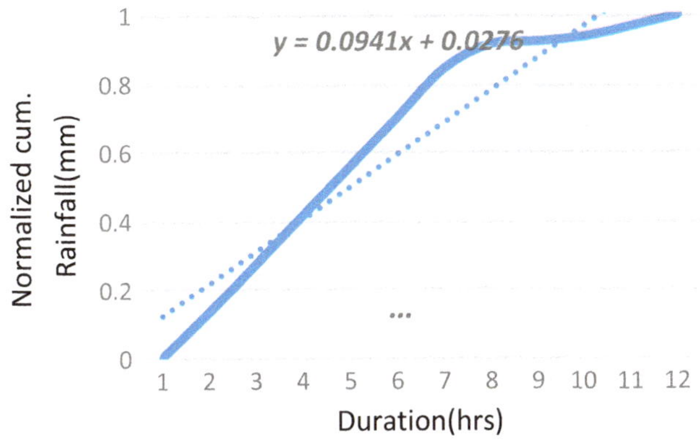

Fig. 5 Cumulative rainfall variation with time for Koslanda, Meeriyabedda landslide

$$I = 35.63D^{-0.27} \tag{4}$$

Above threshold is developed with R^2 value of 0.936.

Figure 7 compares the above developed two thresholds with Caine (1980) worldwide threshold for landslides.

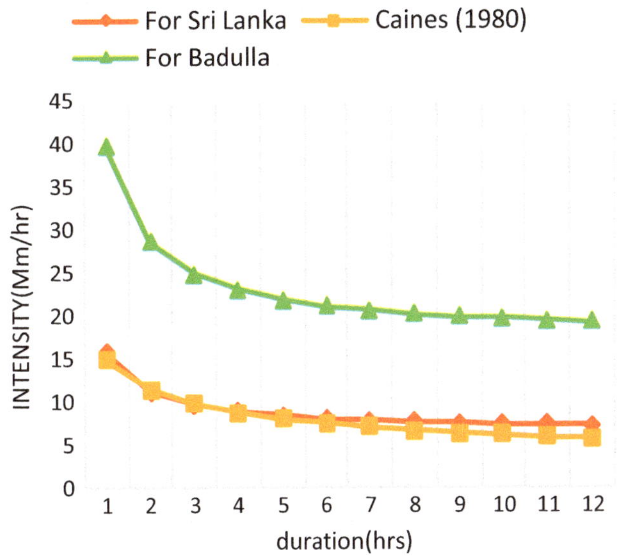

Fig. 7 Comparison of the results with Caine's results

Conclusion

Although developed countries in the world use the latest technology for landslide prediction, Sri Lanka being an underdeveloped country cannot afford such costs. Hence, it is important to conduct research studies for landslide predictions to provide early warnings which will eventually save the lives of innocent people. This study discussed the importance of hourly variation of rainfall data with time and a reasonable relationships were proposed with limited data.

Recommended rainfall threshold for landslides in Sri Lanka is

$$\mathbf{I = 14.02\,D^{-0.30}}$$

Recommended rainfall thresholds for land-slides in Badulla district is

$$\mathbf{I = 35.63\,D^{-0.27}}$$

However, it has to be noted that, much finer conclusions could be made with more findings and data. Rainfall threshold developed for whole country is very much close to that of Caine (1980). But it shows higher rainfall threshold for Badulla District than that of Caine (1980). This research is only based on rainfall data. Further researches need to be done considering geological features as well. It is clear that almost all landslides in Sri Lanka are triggered by rain. Sri Lanka doesn't have a proper hourly rainfall measuring mechanism in all vulnerable sites and due to that reason, conducting researches on this field is still a huge challenge. Therefore, it is strongly recommended to measure hourly rainfall in all sub weather stations or at least in areas where more susceptible to landslides.

Acknowledgements Officials of National Building Research Organization (NBRO) and the Maskeliya Plantations PLC are kindly acknowledged for their cooperation in obtaining data.

References

Bandara RMS (2008) Landslide early warning in Sri Lanka. Regional seminar on experience of geotechnical investigations and mitigation for landslides. Bangkok, Thailand

Bandara RMS (2005) Landslides in Sri Lanka. Vidurawa, vol 22

Caine N (1980) The rainfall intensity-duration control of shallow landslide and debris flow.Geografiska Annaler. Ser A. Phys Geogr 62A:23–27

Corominas J, Moya J, Hurlimann M (2002) Landslide rainfall triggers in the Spanish Eastern Pyrenees. Proceedings of the 4th EGS Plinius conference, Universitat de les Illes Bakesr, Spain

Crosta G (1998) Regionalization of rainfall threshold; an aid landslide hazard evaluation Environ Geol 35: 131–145

Nawagamuwa UP, Rajinder K, Bhasin Kjekstad O (2011) Recommending rainfall threshold values for early warning of landslides in Asian region. In: Proceedings of the 2nd World Landslide Forum, Rome, Italy (Landslide science and practice. Volume 2 early warning, instrumentation and monitoring)

4th WLF 2017
4th World Landslide Forum
LJUBLJANA SLOVENIA EU

Brahmaputra River Bank Failures—Causes and Impact on River Dynamics

Archana Sarkar

Abstract

The Brahmaputra River has been the lifeline of north- eastern India since ages. This mighty trans-Himalayan trans-boundary river runs for 2880 kms through China, India and Bangladesh. The gradient of the Brahmaputra River varies from very steep near the source at the Tibetan plateau (1:385) to very flat in the lower part of Bangladesh (from 1:11,340 to 1:37,700). Geomorphologically, the Brahmaputra basin is very unstable as it is located in a high seismic zone. The Brahmaputra is a large alluvial river with highly variable channel morphology and a high degree of braidedness. The dominant flow is multichannel flow acknowledged to be very complex for mathematical modelling. The problems of flood, erosion and drainage congestion in the Brahmaputra basin are gigantic. The river bank failures are responsible for large scale bank erosion, aggradations and widening of the river channel. This in turn is responsible for lateral channel changes of the Brahmaputra River in many reaches leading to a considerable loss of good fertile land each year. Bank dynamics is also causing shifting of outfalls of its tributaries bringing newer areas under waters. Frequent changes of channel courses and bank erosion with high rate of siltation have also been identified as major threats to the riverine biota. This in turn has a negative impact on the sustainability of the wetlands. Degradation and destruction of the wetlands have considerable impact on the deteriorating flood hazard scenario in the state. This paper outlines the types of river bank failure mechanism and erosion process in the Brahmaputra River. The paper also presents information on the river reaches of Brahmaputra suffering from high bank erosion rates and the impacts of bank erosion on the Brahmaputra basin and people of the region.

Keywords

River bank instability • River bank erosion • Brahmaputra river • Nodal points

The Brahmaputra River System

The Brahmaputra river system is one of the largest in the world, and majestic in multiple aspects: in the volumes of water and sediment that it gathers and passes on, the power with which these flows are routed, and the scale of changes that these powerful flows bring upon the landscape. It is ranked fourth in respect of the average discharge. The annual sediment load of 735 million metric tons, that the river carries, ranks it as the second largest sediment laden river.

The Brahmaputra basin represents a unique physiographic setting vis-à-vis the eastern Himalayas: a powerful monsoon rainfall regime under wet humid conditions, a fragile geologic base and active seismicity. Flowing eastward for 1625 km over the Tibetan plateau, the Brahmaputra, known there as the Tsangpo, enters a deep narrow gorge at Pe (3500 m) and continues southward across the

A. Sarkar (✉)
National Institute of Hydrology, Ministry of Water Resouces
River Development and Ganga Rejuvenation Govt of India
Jal Vigyan Bhawan, Roorkee, 247667, India
e-mail: archana_sarkar@yahoo.com

M. Mikoš et al. (eds.), *Advancing Culture of Living with Landslides*,
DOI 10.1007/978-3-319-53483-1_32

east-west trending ranges of the Himalayas, viz. the Greater Himalayas, Middle Himalayas and sub-Himalayas before debouching onto the Assam plains near Pasighat. Figure 1 shows a view of the upper Brahmaputra basin and the total river length crossing three countries. The different geo-ecological zones have a distinctive assemblage of topographical, geological, climatological and floral characteristics. The gradient of the Brahmaputra River is as steep as 4.3–16.8 m/km in the gorge section upstream of Pasighat, but near Guwahati it is as flat as 0.1 m/km. The dramatic reduction in the slope of the Brahmaputra as it cascades through one of the world's deepest gorges in the Himalayas before flowing into the Assam plains explains the sudden dissipation of the enormous energy locked in it and the resultant unloading of large amounts of sediments in the valley downstream. Two rivers, the Dibang and the Lohit, join the upper course of the Brahmaputra, known as the Dihang (or Siang) river, a little south of Pasighat and the combined flow, hereafter called the Brahmaputra, flows westward through Assam for about 640 km until near Dhubri, where it abruptly turns south and enters Bangladesh.

In the course of its 2880 km journey, the Brahmaputra receives as many as 22 major tributaries in Tibet, 33 in India and three in Bangladesh. Figure 2 shows the Brahmaputra river basin in India along with its major tributaries in north as well as south. The northern and southern tributaries differ considerably in their hydro-geomorphological characteristics owing to different geological, physiographic and climatic conditions. The north bank tributaries generally flow in shallow braided channels, have steep slopes, carry a heavy silt charge and are flashy in character, whereas the south bank tributaries have a flatter gradient, deep meandering channels with beds and banks composed of fine alluvial sediments, marked by a relatively low sediment load. Many of the north bank tributaries are of Himalayan origin fed by glaciers in their upper reaches, e.g., the Subansiri, the Jia Bharali and the

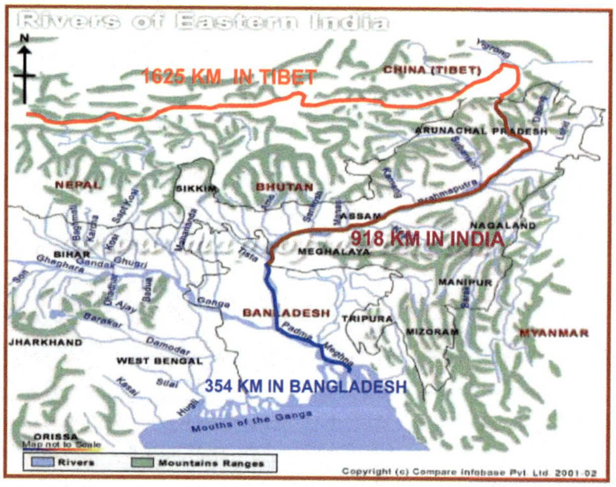

Fig. 1 The Brahmaputra River crossing three countries

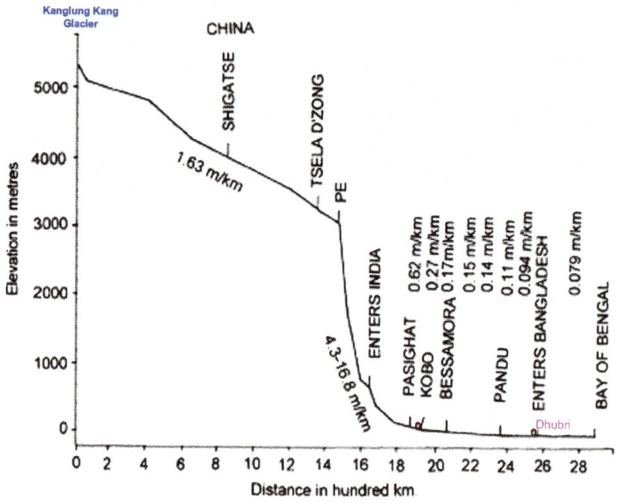

Fig. 2 Longitudinal slope of the Brahmaputra River

Manas. The Debang and the Lohit are two large tributaries emerging from the extreme eastern flank of the Himalayas, while the Jiadhal, the Ranganadi, the Puthimari, the Pagladiya, etc. are major rivers having their sources in the sub-Himalayas, the latter two in Bhutan. Among the south-bank tributaries, Burhidihing originates near the Nagaland-Myanmar border, the Dhansiri and the Dikhow in the Naga hills, the Kopili in the Karbi plateau while the Kulsi and the Krishnai flow from the Meghalaya hills.

The Brahmaputra basin represents a tectono-sedimentary province in India. Due to the colliding Eurasian (Chinese) and Indian tectonic plates, the Brahmaputra valley and its adjoining hill ranges are seismically very unstable. The earthquakes of 1897 and 1950, both of Richter magnitude 8.7, are among the most severe in recorded history. These earthquakes caused extensive landslides and rock falls on hill slopes, subsidence and fissuring in the valley and changes in the course and configuration of several tributary rivers as well as the main course.

The basin is bounded by the Eastern Himalayas on the north and east, the Naga-Patkai ranges on the northeast and Meghalaya Plateau and Mikhir hills on the south. The region can be geologically and tectonically divided into four major zones, viz. the Himalayan folded and Tertiary hills and mountains, the Naga-Patkai ranges, the Meghalaya Plateau and Mikhir hills and the Brahmaputra valley in Assam (Wadia 1975).

The characteristic geologic and tectonic framework coupled with structural complexities has rendered the Brahmaputra basin geomorphologically a most complicated one. As such, the Brahmaputra basin, geomorphologically, is very unstable as it is located in a high seismic zone and is also constituted by alluvial soil. Recent satellite photographs reveal that the Brahmaputra is continuously shifting towards the south and in some places may be migrating at rates as high as 800 m per yr. The shifting of the river in India is more evident in the Assam districts of Dibrugarh, Morigaon and Sonitpur. Lateral migration of the channel is always associated with large scale bank erosion, aggradations and widening of the river channel (Bristow 1987).

Causes of River Bank Instability/Failures

The gradient of the Brahmaputra River varies from very steep near the source at the Tibetan plateau to very flat in the lower part of Bangladesh. The average river gradient in the Tibetan plateau is 1:385 whereas in Bangladesh it varies from 1:11,340 to 1:37,700 as shown in Fig. 2 (NDMA 2011). The length of river shown between the two red points in Fig. 2 is the river length in flood plain area within India. The Brahmaputra River in the valley reach in Assam and in Bangladesh is highly braided and unstable in nature. The bank line is also extremely unstable in many reaches of the river. The river flows into a number of interlacing channels separated by sandy islands. During high stages, the braided channels get submerged to form a vast sheet of moving water. The average river width in the braided portion within valley reach is 8.08 km and varies from 6 km to a maximum of 18 km except in 9 nodal points where the width is restricted within stable banks and flows through deep and narrow channels followed by broad aggrading ones.

The instability of the river is owing to high sediment load, steep slopes (4.3–16.8 m/Km before entering India), transverse (lateral) gradients, high intensity rainfall (as high more than 600 cm annual rainfall in southern slopes of Himalayas) in the region, alluvial nature of the river (as per NBSS and LUP 1994 majority of soils of Assam are inceptisols, entisols & alfisols in a ratio of 45%, 36.6% & 12.3% respectively) and the occurrence of a number of major earthquakes whose epicenters were inside or very close to the basin area and which disturbed the topography and the river system to a great extent; the severest earthquakes occurred in 1762, 1822, 1865, 1869, 1897, 1908, 1937, and 1950 as per available records.

Types of River Bank Failures

The bank of the Brahmaputra River is non-clayey. The composition varies in the proportion of silt and fine sand. The banks can be seen to have two distinct strata. The top one is fine grained and the bottom is coarser. The banks of the Brahmaputra River are made up of silt, sandy material. The clay portion is less than 5%. The inhomogeneous composition of the bank material results in uneven bank slumping. This causes the flow to take a different path and the orientation of the bank-line to the direction of flow also changes. Also, at some localities older alluvium protruding into the river offers significant resistance to the flow and causes changes in hydraulic conditions. The finely divided bank material and the constant change in flow direction produce severe bank caving along the channel. When the flow approaches the bank at an angle, severe under-cutting takes place resulting in slumping of sediments. Slumps are more common along banks composed of clayey silt and silty clay. Quite often, the highly saturated clayey silt will liquefy and tend to flow towards the channel. As the materials flow, the overlying, less-saturated bank sediments tend to slump along well-defined shear planes. Coleman (1969) observed an average of five slumps of bank per mile distance in this river. The bank line retreats under the combination of a favorable fluvial process clearing of the basal material derived out of the slips or other lump failures and the critical soil moisture of the bank. Thorne et al. (1993) assessed that 40% of the eroded bank material in the Brahmaputra River goes as wash load and the rest is incorporated into the bedload. The river swings very often within these bank lines.

There are two prominent types of slumping which cause the bank-line to recede; one operating during flood stage (undercutting), and other during falling stage (flow of highly saturated sediments). However, a large scale slumping of banks is observed during the falling stage. The accumulated water level during the flood stage provides additional support to the bank material as the pore spaces of the loosely bound bank materials are occupied by water and act as a continuous system. With the fall in water level, the support diminishes abruptly and the bank materials are subjected to different degrees and nature of failure. It seems that soil moisture of the bank has a strong relation to its failure in addition to fluvial processes. Undercutting of the banks and the development of overhanging cantilever lead to bank failure. The applied shear on the bank sediments or on natural levee deposits is quite large at high stages. This can entrap voluminous material and cause fluvial failure of the bank. At this stage water is forced into the bank increasing the pore pressure. When the water level in the river falls quickly compared to the bank water condition, the water from the bank moves from the pores. The lateral flow takes away possible sediments into the channel. Caves appear on the banks. The overlaying bank portion fractures into number of blocks. These blocks eventually fail by tilt (Fig. 3).

Shear failure is one of the most effective causes of bankline recession of the Brahmaputra. Majority of the failures occurred as a result of the current undermining of the natural levee deposits, due to which large blocks of natural levee sediments are shearing-off and tilting into the river. The other major cause of shear failure is over-steepening of the bank materials as thalweg of the channel hug banks. In localities where bank materials are slightly cohesive, shear failure of the bank results in a rotated step-like structure

Fig. 3 Bank Erosion in the Brahmaputra River

leading from the top of the bank to the water edge. Most of the shear planes diminish in the slope as they penetrate into sub- surfaces and as a result the blocks are tilted land-ward by rotation. In some areas around Salmora in Majuli Island and in some localities in Kaziranga National Park, the bank is composed of cohesive materials. In such areas banks with a slope approaching 90° and more with over-hangs are observed. This type of over-steepening always enhances the failure of the bank. Fluvial erosion, in turn, is linked to mass-failure processes through the concept of basal endpoint control (Dietrich and Gallinatti 1991). Fluvial erosion of the basal area of the bank can lead to undercutting and subsequent cantilever failure. Equally, a mass failure event supplying sediments to the basal zone will tend to increase bank stability (decreasing bank angle), unless fluvial conditions result in the critical shear stress for removal of the material being exceeded (Lawler 1992). As stream power is observed to be at a peak in the middle reaches of a basin, fluvial erosion will dominate there. With increasing channel depth downstream, Lawler (1992) suggested that there will be a point at which the maximum (or critical) bank height for stability with respect to mass failure is exceeded and mass failure thus dominates erosion downstream of this point (Holden 2003). Different types of shear failure occur during the receding stage of the river. As the water level recedes in the channel, saturated levee materials loose support from the channel side. This results in shearing of small blocks from the saturated bank due to its own weight. However, this type of failure is always observed in a small scale.

A peculiar type of bank failure is observed in some localities around Kaziranga National Park. Here fine-grained overbank deposits with mudcracks are present along the banks. The formation of mudcracks can directly be attributed to subaerial processes, which include wetting and drying of soil and associated desiccation (Green et al. 1999). They weaken the surface of the bank prior to fluvial erosion, thus increasing the efficacy of the latter. The blocks are separated by cracks, detached from the bank. These types of blocky detachment from the bank give it an appearance which resembles the tooth of a saw. The cumulative effect of blocky separation of fine-grained sediments enhances the activities of shearing, which may ultimately lead to large-scale bank failure.

River Bank Erosion in Various Reaches and Tributaries of Brahmaputra River

The entire course of Brahmaputra river in Assam from upstream of Dibrugarh up to the town Dhubri near Bangladesh border for a stretch of around 620 kms along with twenty two major tributaries (Fig. 4) has been studied by Sarkar (2013) using an integrated approach of remote sensing (RS) and geographical information system (GIS). The basic data used in this study are digital satellite images of Indian Remote Sensing (IRS) LISS-I and LISS-III sensor, comprising of scenes for the years 1990, 1997 and 2008. The other collateral data used in the present study are Survey of India (SOI) toposheets at 1:250,000 scale and Landsat ETM images (Path/Row 135/41, 42; 136/42; 137/42; 138/42). The channel configuration of the Brahmaputra river has been mapped for the years 1990, 1997 and 2008 using IRS 1A LISS-I for 1990 and IRS-P6 LISS-III satellite images for 1997 and 2008.

From the above study, it has been observed that in general the river has eroded both the banks throughout its course during the entire study period. For the period of 1990–2008, the total land loss per year excluding forest area is found out to be 62 km^2/year. For the recent period of 1997–2008, the total land loss per year is found to be 72.5 km^2/year which indicates enhanced rate of increase in the land lost due to river erosion in recent years. The study amply suggested that three to four major geological channel control points are present along the Brahmaputra in Assam flood plains which are by and large holding the river around the present alignment. These control points are in the vicinity of

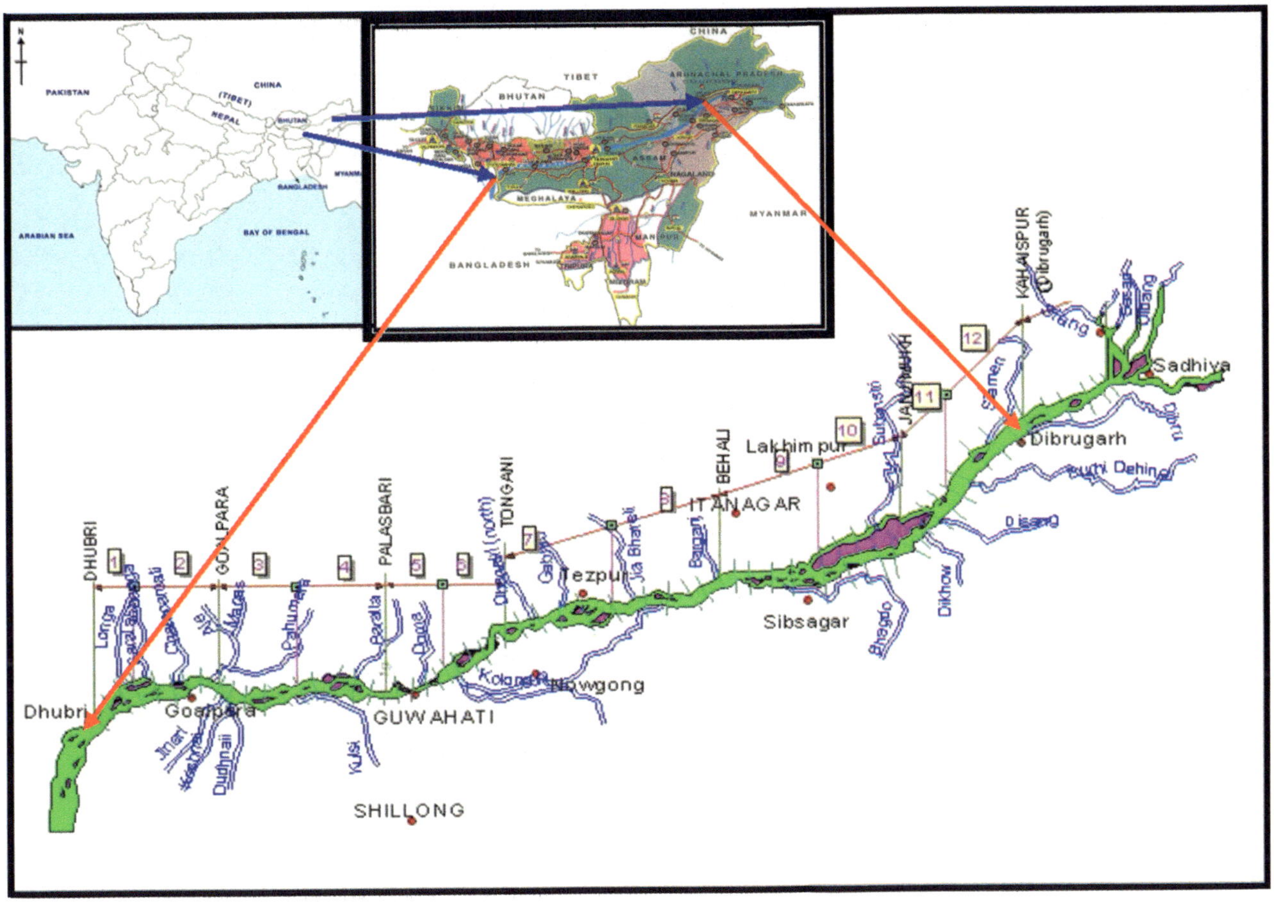

Fig. 4 River reaches and Tributaries for Bank Erosion study in the Brahmaputra River

Jogighopa near Goalpara, Pandu near Guwahati, Tezpur and Bessamora in Majuli. Many reaches along the Brahmaputra river have been perceived as suffering from high erosion that endanger nearby settlements and infra-structure. Figure 5 graphically presents the river bank erosion during 1990–2008 and 1997–2008 period.

Plot of reach-wise land loss (Fig. 5) indicates that downstream of Guwahati (river reach number 4), Brahmaputra River tends to move towards right (north) side whereas in upstream side it tends towards left (south) side keeping the control point at Guwahati invariant. Similar pattern repeats at other control points also.

Based on the study of erosion in various reaches, prioritization of the reaches have been done with respect to the land area lost in 18 years and given in Table 1. The reaches of the main river downstream of Guwahati have shown highest erosion. The stream bank erosion of the Brahmaputra river resulting in land area loss is surmised to be closely linked to the generation of huge runoff and high sediment inflows mainly contributed from the catchment areas receiving heavy precipitation. The highest flood discharge

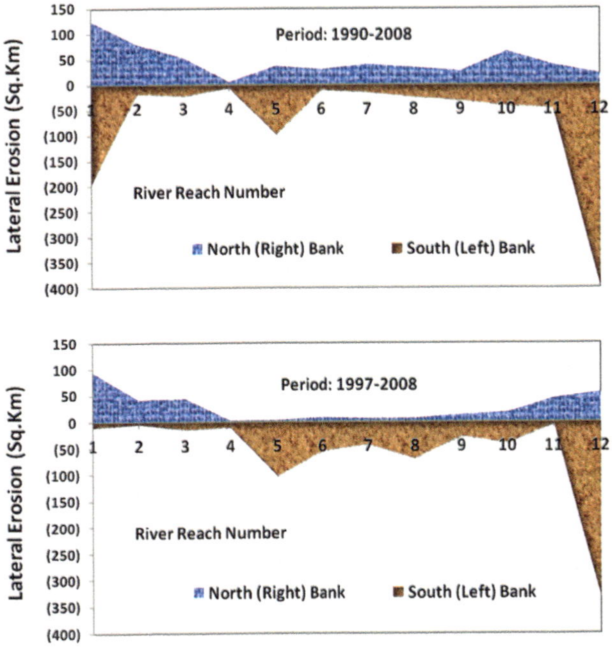

Fig. 5 Reach wise eroded area during 1990–08 & 1997–08

Table 1 Prioritization with respect to land area lost (considering the period 1990–2008)

Left bank		Right bank	
prioritize reaches	Land area lost (km^2)	Prioritize reaches	Land area lost (km^2)
1 (Dhubri)	195.01	1 (Dhubri)	124.46
5 (Morigaon)	99.799	2 (Goalpara)	79.046
11 (Dibrugarh)	47.525	10 (U/s Majuli)	64.273
10 (U/s Majuli)	43.088	3 (Palasbari)	51.970
9 (Majuli)	32.788	7 (Tezpur)	38.758
8 (U/s Tezpur)	26.098	11 (Dibrugarh)	37.896
3 (Palasbari)	23.663	5 (Morigaon)	35.781
2 (Goalpara)	18.411	8 (U/s Tezpur)	32.831
7 (Tezpur)	16.628	6 (Morigaon)	29.057
6 (Morigaon)	11.253	9 (Majuli)	25.562
4 (Guwahati)	6.831	4 (Guwahati)	4.618

Table 2 Area undergone erosion in selected tributaries of Brahmaputra during 18 years (1990–2008)

Trib. No.	Tributary name	Trib. length (km)	Erosion area-left bank (km^2)	Erosion area-right bank (km^2)
N1	Champamati	101.985	42.852	9.194
N2	Aie	131.677	27.118	8.549
N3	Manas	48.927	15.496	1.476
N4	Pahumara	91.432	23.204	4.792
N5	Pagladiya	70.233	19.717	1.839
N6	Bornadi	106.670	18.814	2.250
N7	Dhansiri (N)	89.193	30.509	1.500
N8	Gabharu	57.501	13.671	7.405
N9	JiaBhareli	36.206	8.956	1.974
N10	Borgang	31.502	8.300	9.462
N11	Subansiri	122.906	26.144	29.925
N12	Jiadhol	23.095	10.297	19.554
S1	Jinari	28.790	1.539	6.091
S2	Krishnai	106.792	1.314	15.579
S3	Dudhnoi	34.354	1.306	9.492
S4	Kulsi	80.140	4.390	11.397
S5	KolongKapili	192.511	12.884	23.943
S6	Dhansiri (S)	163.025	7.899	31.557
S7	Jhanji	56.428	4.085	9.667
S8	Dikhow	93.292	9.201	15.734
S9	Disang	197.585	12.306	11.581
S10	BuriDihing	120.406	15.008	17.711

recorded in the Brahmaputra at Pandu (Guwahati, Assam) was of the order of 72,148 m^3/sec (in year 1962), which had a recurrence interval of 100 years (WAPCOS 1993). The river carries about 735 million metric tons of suspended sediment loads annually. Tables 2 and 3 account for the river bank erosion area and erosion rate in the tributaries.

It can be seen from Table 2 that there has been significant bank erosion in most of the tributaries during a period of 18 years. Bank erosion is more prominent in the northern tributaries compared to the southern tributaries. The total land area lost due to river bank erosion during the 18 years period (1990–2008) in twenty two major tributaries of Brahmaputra river system is worked out to be 565.68 km^2

Table 3 Rate of erosion per unit length in selected tributaries of Brahmaputra during 18 years (1990–2008)

Trib. No.	Tributary name	Erosion rate-left bank (km^2/km)	Erosion rate-right bank (km^2/km)
N1	Champamati	0.4202	0.0902
N2	Aie	0.2059	0.0649
N3	Manas	0.3167	0.0302
N4	Pahumara	0.2538	0.0524
N5	Pagladiya	0.2807	0.0262
N6	Bornadi	0.1764	0.0211
N7	Dhansiri (N)	0.3421	0.0168
N8	Gabharu	0.2378	0.1288
N9	JiaBhareli	0.2474	0.0545
N10	Borgang	0.2635	0.3004
N11	Subansiri	0.2127	0.2435
N12	Jiadhol	0.4459	0.8467
S1	Jinari	0.0535	0.2116
S2	Krishnai	0.0123	0.1459
S3	Dudhnoi	0.0380	0.2763
S4	Kulsi	0.0548	0.1422
S5	KolongKapili	0.0669	0.1244
S6	Dhansiri (S)	0.0485	0.1936
S7	Jhanji	0.0724	0.1713
S8	Dikhow	0.0986	0.1687
S9	Disang	0.0623	0.0586
S10	BuriDihing	0.1246	0.1471

giving an annual area loss of 31.43 km^2/year. Out of the total area lost due to erosion in the twenty two tributaries, the north bank tributaries (N1 to N12) exhibit higher area lost which comes out to be 342.99 km^2 during 1990–2008. Total area lost due to erosion in the south bank tributaries (S1 to S10) amounts to 222.68 km^2 during 1990–2008 which is comparatively smaller than the north bank tributaries.

It is evident from Table 3 that the rate of erosion per unit length of the tributaries is highest for the northern tributory namely, Jiadhol which happens to be the smallest in length. Followed by Jiadhol are the Borgang, Subansiri and Gabharu tributaries, also north bank tributaries with high rate of erosion per unit length on both the banks. Besides these four north bank tributaries, other eight north bank tributaries show a high rate of erosion on their left bank, i.e., east bank. It can also be noted that all the ten south bank tributaries show considerably smaller rate of erosion which is more prominent on their right bank, i.e., east bank.

Out of all the twenty two tributaries considered in this study, the Subansiri, a north bank tributary of Brahmaputra exhibits highest bank erosion with highest erodible bank length.

Concluding Remarks

The Brahmaputra basin in India shows marked differences in geology, geomorphology, physiography, climate and soils. The river flows through north east India which is one of the most seismically active regions of the world and witnessed two of the world's most destructive earthquakes. Since the occurrence of the massive earthquake (8.7 magnitude on Richter scale) in Assam in 1950, the stream bank erosion process of the Brahmaputra river and its tributaries displayed heightened momentum. The braided Brahmaputra river in different reaches exhibits differential rate of erosion and deposition.

The inhomogeneity in bank materials and the constant change in flow direction have caused severe undercutting, which enhances the intensity of slumping along the banks. There are two types of bank slumping, undercutting during the flood stage and the flow of highly saturated sediments during the falling stage, through well-defined shear planes.

The river bank erosion study made using the above methodology brings to the fore that Brahmaputra River has been exhibiting sharp increase in fluvial landform changes in recent years causing sizeable land losses. The study amply suggested that three to four major geological channel control points are present along the Brahmaputra in Assam flood plains which are by and large holding the river around the present alignment. These control points are in the vicinity of Jogighopa near Goalpara, Pandu near Guwahati, Tezpur and Bessemora in Majuli. These channel control points usually have well defined and stable hydrographic profiles. Intermittent fanning out and fanning in behavior is displayed between these control points which are being temporally severed in the geological time scale. The channel control points are surmised to be the major co-actors, working in unison with other forcing functions causing severe erosion. Heavy braiding has occurred due to high level dissipation of flow energy when the river emerges downstream of control points (rock outcrops) to more vulnerable sandy and light-cohesive plains. The manifestation of the highest river width of about 20 Km near Mukalamua-Palasbari from constricted width of about 1.2 km near Guwahati is the classic example of the above observation.

As a result of severe bank erosion, many cities including the world's biggest river island, Majuli, are under severe threat. Frequent changes of channel courses and bank erosion with high rate of siltation have been identified as major threats to the riverine biota as they have a great bearing on the faunal composition of the river, some of which belong to the most rare and endangered species. This in turn has a negative impact on the sustainability of the wetlands. Degradation and destruction of the wetlands have considerable impact on the deteriorating flood hazard scenario in the state.

Despite gigantic efforts and colossal expenditure in building embankments, drainage channels and soil conservation, the Brahmaputra continues to wreck havoc through uncontrollable floods year after year. These floods have affected millions of people of the region and millions of hectares of cropland besides claiming a huge number of human lives and innumerable cattle and wildlife causing millions of rupees loss to the region's economy.

However, river bank erosion is also useful for some biota. For example, the coarse sediment generated from erosion of a river bank can provide substrate for fish spawning (Flosi et al. 1998). Also, the irregular river banks generated due to the erosion process become advntageous in providing habitats to the invertebrates, fish and birds (Florsheim et al. 2008).

It is envisaged that the conjunctive use of the information generated out of this study with other ground based data will contribute substantially in more meaningful approach towards planning and execution of means and measures to combat recurring floods and erosion.

References

Bristow CS (1987) Brahmaputra River: channel migration and deposition. Recent fluvial sedimentology. In: Ethridge FG, Flores RM, Harvey MD (eds) Society of economic palaentologists and mineralogists Special publications Vol 39 pp 63–74

Coleman JM (1969) Brahmaputra River: channel processes and sedimentation. J Sedimentry Geology 3(2–3):129–239

Dietrich WE, Gallinatti JD (1991) Field experiment sand measurement programs. Geomorphology J Earth Surf Process Landforms 12:173

Florsheim JL, Mount JF, Chin A (2008) Bank erosion as a desirable attribute of rivers. Bioscience 58(6):519–529

Flosi G, Downie S, Hopelin J, Coey R, Collins B (1998) California salmonid stream habitat restoration manual. Department of Fish and Game, Sacramento California

Green TR, Beavis SG, Dietrich CR, Jakeman AJ (1999) Relating stream bank erosion to instream transport of suspended sediment. J Hydrol Process 13:777

Holden C (2003) Historic island threatened. Random samples. science 300:1368

Lawler DM (1992) Process Dominance. bank erosion systems in lowland floodplain rivers.In: Carling P, Petts GE (eds) Wiley, Chichester pp 117–159

NBSS & LUP (1994) Soils of Assam (India) for land use planning National Beureau of Soil Survey and Land Use Planning report Nagpur, India

NDMA (2011) Study of Brahmaputra River erosion and its control: Phase I. Unpublished report of study conducted by Department. of Water Resources Development and Management I.I.T. Roorkee for National Disaster Management Authority of India, New Delhi

Sarkar A (2013) Runoff and sediment modelling in a part of Brahmaputra river basin. PhD thesis, Indian Institute of Technology, Roorkee, India

Thorne CR, Russell APG, Alam MK (1993) Planform pattern and channel evaluation of Brahmaputra River, Bangladesh. Braided Rivers.In: Best JL, Bristow CS (eds) Geological Society, London Vol 75, pp 257–276 Special Publications

Wadia DN (1975) Geology of India. Tata McGraw Hill Publishers, Delhi, p 506

WAPCOS (1993) Morphological studies of river Brahmaputra. WAPCOS report, New Delhi, India

Downstream Geomorphic Response of the 2014 Mount Polley Tailings Dam Failure, British Columbia

Vanessa Cuervo, Leif Burge, Hawley Beaugrand, Megan Hendershot, and Stephen G. Evans

Abstract

On August 4, 2014, the failure of the Tailings Storage Facility dam at the Mount Polley copper and gold mine in British Columbia (Canada) produced a dynamic and complex geomorphic response downstream. The dam breach was caused by rotational sliding of the embankment due to foundation failure, resulting in sudden loss of containment of water and tailings. The released volume was estimated to be 25 Mm^3. The resulting flow traveled approximately 9 km down the Hazeltine Creek valley over a vertical elevation of 205 m. The August 2014 event is the largest tailings dam failure recorded in Canada and the second largest recorded globally in the last decade. This investigation documents and analyzes the general downstream geomorphic response of the event in the Hazeltine Creek channel and floodplain. It utilizes an integrated approach of geomorphological mapping, topographic analysis, historical aerial photograph analysis, and field surveys to identify and quantify the geomorphic impacts of the event. Overall results indicate that these impacts are significant and comparable to those resulting from extreme debris flow and outburst flood events in mountainous environments in Canada and worldwide.

Keywords

Tailings dam failure • Flow failure • Debris flows • Outburst flood • Geomorphic impacts • Hazard and risk

V. Cuervo (✉) · H. Beaugrand · M. Hendershot
Environment and Geoscience Infrastructure, SNC-Lavalin Inc, 640th 5 Avenue, Calgary, AB T2P 3G4, Canada
e-mail: vanessa.cuervo@snclavalin.com; vacuervo@uwaterloo.ca

H. Beaugrand
e-mail: hawley.beaugrand@snclavalin.com

M. Hendershot
e-mail: megan.hendershot@snclavalin.com

V. Cuervo · S.G. Evans
Department of Earth & Environmental Sciences, University of Waterloo, 200 University Avenue West, Waterloo, ON N2L 3G1, Canada
e-mail: sgevans@uwaterloo.ca

L. Burge
Stantec Consulting Ltd, 400-1620 Dickson Avenue, Kelowna, BC V1Y 9Y2, Canada
e-mail: leif.burge@stantec.com

L. Burge
Geography & Earth and Environmental Science, Okanagan College, 1000 KLO Road, Kelowna, BC V1Y 4X8, Canada

M. Mikoš et al. (eds.), *Advancing Culture of Living with Landslides*,
DOI 10.1007/978-3-319-53483-1_33

Introduction

Breaching of tailings dams presents a safety issue to people and environment. More than 250 dam failures have been recorded in the world since 1937 (ICOLD–UNEP 2001; Rico et al. 2008; WISE Uranium Project 2016). Several of these events have resulted in catastrophic flow failures causing adverse consequences to life, property and the environment (Blight 1997; Davies 2001; Blight and Fourie 2005; Villavicencio et al. 2014). For instance, the Stava disaster in 1985 caused by the failure of two coupled tailings dams resulted in 268 deaths, and significant damage to property in Stava valley near Tesero (Trento) in Italy (Chandler and Tosatti 1996; Sammarco 2004; Luino and De Graff 2012). The Los Frailes tailings dam failure in Aznalcóllar (Spain) in 1998, created a large-scale sulfide tailings slurry that flowed down the Rio Agrio. The spill covered thousands of hectares of farmland in the Rio Agrio and Guadiamar watersheds (Gallart et al. 1999; Benito et al. 2001; Alonso and Gens 2006; WISE Uranium Project 2016). More recently, the failure of the Fundão Dam in Bento Rodrigues (Brazil) in 2015, created a slurry wave that flooded the town of Bento Rodrigues and Paracatu de Baixo, in Mariana, Minas Gerais, and a part of the district of Gesteira, in Barra Longa. The flow destroyed bridges, homes, caused the death of 19 people, and polluted the waters of North Gualaxo River and Rio Doce (WISE Uranium Project 2016; SAMARCO 2016). All these incidents represent examples of flow failures. This type of failure involves breaches of tailings dam that are followed by the release of large volumes of water and tailings previously impounded by the dam and that moves long distances before coming to rest (Blight and Fourie 2005). The processes involved in this type of failure are similar to those occurring during and after a debris flow and glacial outburst flood, where water discharges in excess of the major hydrological probable flood are highly destructive (Hungr et al. 2001). In this paper, we utilize the term flow failure to describe the rapid to extremely rapid surges of fine grain tailings, large woody debris and native material that traveled down Hazeltine Creek Channel following the breach of the Tailings Storage Facility (TSF) at Mount Polley mine on August 4, 2014.

This paper documents and discusses the downstream geomorphic impacts of the flow failure of the Mount Polley tailings dam facility on August 4, 2014. We examine the behavior of the flow in Hazeltine Creek based on the interpretation of field surveys, imagery, and aerial photographs. Both immediate geomorphic impacts from the failure and potentially long-term effects on channel morphology and slope stability are considered. Areas of concern for future erosion and slope instabilities within the impacted area are also identified. Most of the data presented here were collected during the post-event investigation and were presented in the Technical Appendix A of the Post-Event Environmental Impact Assessment Report (PEEIAR) authored by Burge and Cuervo (2015).

Data and Methods

This investigation applied an integrated approach of geomorphological mapping, topographic analysis, time series analysis, and field investigations to identify and quantify the geomorphic impact of the 2014 event at Mount Polley mine. Utilized data sources included: pre- and post-event satellite imagery, post-event airborne LiDAR data, Digital Elevation Models (1 and 15 m resolution), pre- and post-event aerial photographs, helicopter video made by the Cariboo Regional District during failure, and pre-event field data as made available from previous reports, and post-event field data.

Applied methods and techniques included three main stages. The first stage was intended to determine the pre-event hydrologic and geomorphic conditions of the Hazeltine Creek channel and floodplain. Typical cross-sections and dominant geomorphological processes before the event were characterized and mapped from the review of previous technical studies and sequential historical aerial photographs (from 1974, 1996 and 2009). Morphological change over time was examined qualitatively through comparison of features from one period to the next.

The second stage described and reconstructed the event. Two reconnaissance flights were undertaken within a month after the event with the purpose of defining critical areas for field investigation. A field campaign followed to characterize the channel reaches, surficial deposits and bed material. Field surveys included the description of processes such as erosion, aggradation, bank failures, and avulsion that occurred along the reaches.

A third stage involved the evaluation and quantification of flow failure impacts on Hazeltine Creek. Volumes of deposition were estimated through geomorphological and surficial material mapping. Net volume and maximum depth of erosion were estimated based on the difference in elevation between the pre-event floodplain elevation and the post-event surface. Finally, a qualitative slope stability analysis of river banks and valley walls was also conducted.

Physical Settings

The Mount Polley copper and gold mine is located on the Cariboo Plateau in south-central British Columbia, 56 km northeast of Williams Lake. The Tailings Storage Facility (TSF) is located at 52° 31' 7.00"N and 121° 35' 30.98'W (Fig. 1). This region is characterized by low hills and broad valleys within the Quesnel Highlands (Cathro et al. 2003;

Fig. 1 Area of inundation of the August 2014 event and pre-event Hazeltine Creek Channel (*yellow*). UHC = Upper Hazeltine Creek. MHC = Middle Hazeltine Creek. CHC = Canyon Hazeltine Creek. LHC = Lower Hazeltine Creek. **A** Unconfined flow in a wetland area at Upper Hazeltine Creek. Pre-event channel was 7 m wide. **B** Section characterized by a clear gradient shift (3.7%) relative to upstream areas. **C** Creek flowing through the pre-event canyon. Note the presence of cobbles and bedrock in channel bed indicating stability. **D** Low gradient cobble-gravel substrate at Hazeltine Creek Delta

Hashmi et al. 2015). Surficial geology of the area largely reflects glacial processes of the Fraser Glaciation in the Late Wisconsin period (Bichler and Bobrowsky 2003). Till is the dominant surficial material within the area. Extensive blankets of till are encountered displaying massive, matrix supported and poorly sorted diamictons composed of various lithological clasts. Clasts are rounded to subangular in a very stiff to hard sandy silt matrix.

Glaciofluvial sediments are present in both the Hazeltine Creek delta and in wide terraces composed of gravely to cobbly material with a sand matrix located at Middle Hazeltine Creek (Fig. 1). Deposits are clast-supported, rounded to sub-angular and are of mixed lithology. Fluvial deposits transported by Hazeltine Creek were identified on remnants of the pre-event floodplain. Deposits are moderately sorted, sandy to pebbly, and have sub-rounded to rounded clasts. Two different sources of glaciolacustrine sediments were encountered: (1) gray, firm, thinly laminated clay and silty-clay deposits located in the distal zone of Hazeltine Creek delta; and, (2) gray-brown, very stiff (inferred), massive bed exposed at steep slopes in Middle Hazeltine Creek (Burge and Cuervo 2015) (Fig. 1). These deposits correlate with the stratigraphic sequence described by Clague (1991), Bichler and Bobrowsky (2003), and Hashmi et al. (2015). Bedrock slopes are prominent in the Hazeltine Creek Canyon (Fig. 1).

The area inundated by the flow lies within the Hazeltine Creek watershed. This creek is a tributary stream within the Quesnel River watershed in the Cariboo Plateau. Climate in this area is characterized by long warm summers and cool wet winters (Demarchi 2011). Prior to the event, the creek flowed approximately 9 km, draining an area of 112 km^2 into Quesnel Lake at an elevation of 730 m a.s.l. The streams follow a nival hydrologic regime driven by spring snowmelt. Snow is the dominant form of precipitation from mid-September through to late April (based on Horsefly weather station data located 25 km southeast of the TSF, Climate ID 1093598) and remains in storage within the drainage basin until spring melt. As a result, streams experience low flows throughout the winter and peak flows during the spring months as warmer temperatures induce snowmelt. Precipitation falls almost entirely as rain for the rest of the year and contributes directly to stream flow as well as indirectly through rain-on-snow melting events. Low precipitation amounts through the late summer and early fall combined with an exhausted snowpack result in seasonal low flow conditions.

Inundation Area

The inundation area covers approximately 237.4 ha. This area extends from the perimeter of the embankment downstream to Polley Lake and Quesnel Lake. The area inundated by the flow was determined using post-event topographic data derived from LiDAR, a post-event orthophoto (20 cm resolution) and field data. Main criteria to define the area boundaries was based on (1) field observations and, (2) analysis of photographs. Observations and field description of several sites impacted by the flow allowed for the identification of boundaries of tailings, transported mixed material and native deposits. In addition, characteristics of the flow failure were determined from both, aerial and ground post-event photographs. Evidence of geomorphic impacts from the flow included the removal of vegetation, disturbed ground, eroded bedrock, debris jams or wood and boulders with splintered, shattered or broken logs. At areas where visual confirmation was difficult, the delineation of the inundation area was based on morphological features (using slope and elevation contours).

Pre-event Conditions

Previous to the August 2014 event, Hazeltine Creek flowed for 9 km through a well-defined, generally single unconfined channel (Knight Piesold Ltd 2009a; Minnow 2012) (Fig. 1). From Polley Lake to Upper Hazeltine Creek, the channel flowed through a wetland with a channel slope ranging from 0.8 to 1.7%. Downstream, the channel steepened reaching a maximum of 7.3% within the Hazeltine Creek bedrock canyon (Minnow 2007) (Fig. 1C). Downstream of the canyon, the slope decreased as the channel flowed into Quesnel Lake at the Hazeltine Creek delta (Fig. 1D). Analysis of historical aerial photographs (1974, 1996 and 2009) illustrates that overall channel planform remained relatively stable over time. Changes observed between historical aerial photographs occurred in areas of low slope where beaver dams have been reported (Imperial Metals Corporation 1990; Pedersen 1998; Minnow 2007). Depositional areas observed along Hazeltine Creek were mostly thin accumulations in backwaters and small shallow pools, likely depositing and eroding seasonally (Minnow 2012). Channel width, depth, and bed material were highly variable due to woody debris in the channel (Knight Piesold Consulting 2009b). The creek was characterized by dense riparian vegetation (Minnow 2007). Maximum 100-, and 200-year estimated discharges are 3.7 and 4 m^3s^{-1}, respectively (Burge and Cuervo 2015).

The August 2014 Event

On August 4, 2014 the failure of the TSF dam at Mount Polley copper and gold mine in British Columbia (Canada) produced a complex geomorphic response downstream. The Mount Polley Independent Expert Engineering Investigation and Review Panel (2015) concluded that the failure was caused by a dislocation of the embankment due to foundation failure. The breach resulted in the sudden loss of containment of tailings and water. The subsequent flow traveled approximately 9 km down the Hazeltine Creek valley, eroding soil, sediment, and vegetation over a vertical elevation of 205 m. The outflow volume was estimated to be 25 Mm^3 (Golder Ltd. 2015) (Table 1). This volume is one order of magnitude larger than both, the Kimberley, BC, Canada failure in 1948 and the spill in Aznalcóllar, Spain, in 1998 (released volume was 1.1 and 6.8 Mm^3, respectively) (ICOLD–UNEP 2001). Results from a post-event bathymetric survey and volume balance analysis indicate that the largest amount of this material (18.6 ± 1.4 Mm^3) was deposited into Quesnel Lake (Potts et al. 2015). Based on field observation, we believe that the event occurred as a sequence of sediment pulses characterized by increases and decreases in sediment concentration related to sediment supply from the TSF, and entrainment of material from the channel banks and bed. Figure 2 shows tailings flow deposits at the pre-event Hazeltine Creek mouth in Quesnel Lake. At least three flow surges are evident in the photo. Because the majority of the material was deposited into Quesnel Lake, there is uncertainty as to the total number of surges resulting from the flow.

Hazeltine Creek is divided into four sections (upper, middle, canyon and lower Hazeltine Creek) to describe the active processes that took place during and after the event (Fig. 3).

Table 1 Volume estimation for Mount Polley flow failure (Burge and Cuervo 2015)

Estimated volume	Total (Mm^3)
Supernatant water	10.6
Tailings solids	7.3
Interstitial water	6.5
Construction materials	0.6
Total outflow volume	25.0
Net volume of eroded material	0.6–1.7
Total volume of the event	25.6–26.7

Fig. 2 Flow deposits at the pre-event Hazeltine Creek channel. Pre-event surface identified by the presence of roots (*middle right*). TFD Tailings flow deposit. **A** Red layer of sandy tailings. Likely initial stages of the flow where water content is assumed to be higher than solids. **B** Mix of tailings and native material entrained from channel bed and banks. **C** Silty-clay deposits (*Photo* by ARCHER 2015)

Upper Hazeltine Creek (UHC)

Upper Hazeltine Creek represents the area from the TSF dam downstream to the Gavin Lake Road bridge. From the breach site at the TSF, the flow eroded a swath approximately 350 m wide through the formerly forested wetland area and traveled downhill towards Hazeltine Creek and north to Polley Lake (Fig. 3). The portion that traveled towards Polley Lake filled and plugged the Hazeltine Creek outlet with a mixture of tailings, embankment material and eroded native material (Fig. 3A). Thickness of the deposits reached up to 3 m. It was estimated that 0.5–0.6 Mm^3 of material was deposited in this plug area. A second portion of the flow traveled downslope following the pre-event channel and eroding a relatively wide and shallow channel. Vertical erosion by the flow was limited in this area and no major knickpoints occurred (Fig. 3B).

Middle Hazeltine Creek (MHC)

Middle Hazeltine Creek is the most complex section of the flow path. This section is located immediately downstream of Gavin Lake Road and extends to the Canyon (blue section downstream of B in Fig. 3). Within this section, the debris flow travelled rapidly, eroding several meters into native material. Impacts here are mostly erosional driven by an increase in channel slope. Vegetation was completely removed from the path exposing wide terraces of gravely to cobbly material. Degradation of the channel bed is evident by the formation of multiple knickpoint features (Fig. 3C). Two major knickpoints, 6 and 10 m in height, respectively, are located approximately 2.5 and 2.0 km upstream of the Hazeltine Creek Canyon.

The formation of a wide erosion zone is another significant post-event feature. This zone is located immediately downstream of a short bedrock canyon and experienced focused scour and erosion of the valley walls. As a result, slopes are highly unstable (Fig. 3D). Channel aggradation occurred at later stages of the event in the form of thin veneers of tailings (Fig. 3D).

Hazeltine Creek Canyon

The Hazeltine Creek Canyon is the steepest and narrowest section of the flow path (Fig. 3H). The post-event channel flows through a 1.15 km bedrock canyon with slope gradients approaching 6.1%. At this section, the flow likely traveled at extremely rapid velocities, was confined to the valley walls, and eroded virtually all channel sediments, exposing bedrock (Fig. 3E). Maximum erosion depth within the canyon was estimated to exceed 7 m. Deposition of tailings during the event was negligible. Volumes of deposition were estimated to be less than 0.1 Mm^3. Identified deposits consist of thin veneers of sandy tailings and gravel. Main geomorphologic impacts from the flow are related to removal of vegetation, channel widening, and erosion of the channel bed.

Lower Hazeltine Creek

Lower Hazeltine Creek includes the reach downstream of the canyon to Quesnel Lake. From the canyon, the flow continued down Hazeltine Creek channel towards the delta and Quesnel Lake eroding till, fluvial, glaciofluvial and glaciolacustrine material and depositing tailings and large woody debris (LWD) in wide overbank areas. Volumes of deposition in LHC were estimated to be 0.3–0.4 Mm^3. Three important geomorphological processes took place in this section. First, the formation of a large knickpoint incising into till material (Fig. 3F). Second, the abandonment of the pre-event channel and avulsion of Hazeltine Creek, displacing the post-event outlet approximately 400 m north-east of its pre-event location (Fig. 3G). Third, the removal of a large portion of the Quesnel shoreline at the distal zone of the delta (approximately 32,200 m^2 of land was displaced and deposited into Quesnel Lake).

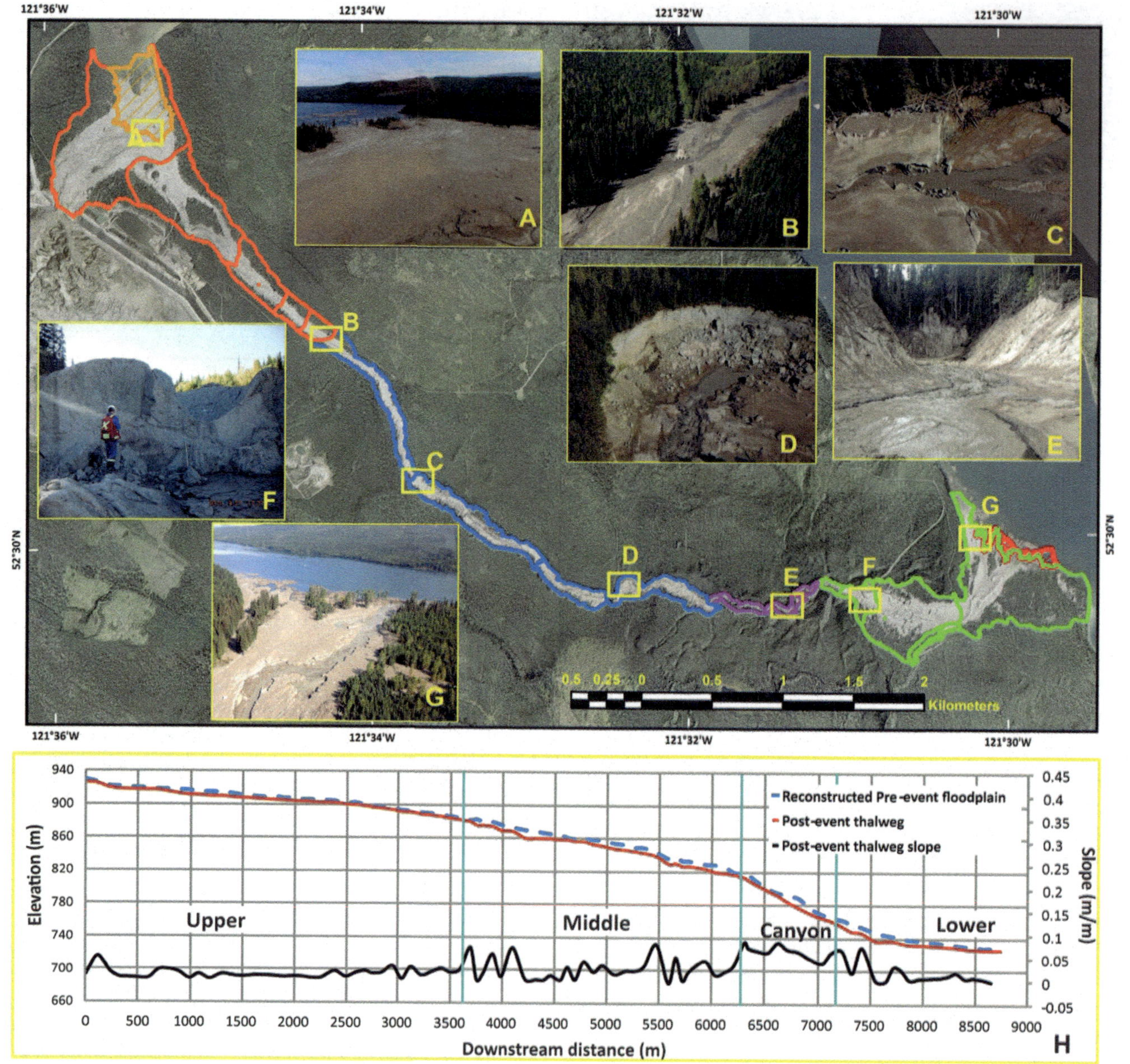

Fig. 3 Highlighted post-event features along the 2014 event flow path. Orange = Upper Hazeltine Creek. Orange dashed = plug area. Blue = Middle Hazeltine Creek. Purple = Canyon. Green = Delta. Red = displaced area **A** Tailings deposits at Polley Lake outlet. **B** Oblique aerial view of Upper Hazeltine Creek. **C** Aerial view of a six-meter-high knickpoint located 2.5 km upstream of the canyon at MHC **D**. Aerial view of a vertical glacio-lacustrine steep slope (40°) displaying fallen soil blocks and topples at a 250 m wide section of the channel. **E** Aerial view of the canyon showing exposed bedrock. **F** Upstream view of the large knickpoint and incision into native material at LHC. The drop in elevation is as much as 4 m. **G** Looking downstream. Channel avulsion at DHC. Pre-event channel was located to the left of the new channel. **H** Post-event long profile displaying several knickpoints in MHC section

Geomorphic Impacts to Hazeltine Creek and Floodplain

The sudden release of tailings and water from the tailings dam produced direct geomorphic impacts on the Hazeltine Creek channel and floodplain. Throughout the event, the channel essentially behaved as the transportation zone of the flow with limited deposition along the flow path and erosion as the dominant mechanism of change. Geomorphic processes were largely controlled by the volume of the flow, characteristics of the material, downstream topography and bed composition. As a result, the post-event geomorphology of the Hazeltine Creek channel and floodplain is dominated by deep erosion zones and cut slopes, thin deposits of

Table 2 Summary of key geomorphic impacts to Hazeltine Creek channel and floodplain

Geomorphic impact	Upper Hazeltine Creek	Middle Hazeltine Creek	Canyon Hazeltine Creek	Lower Hazeltine Creek
Inundated area (ha)	114.2	38.2	4.7	80.3
Channel slope (%)	0.7	2.2	6.1	1.2
Mean width of runout area (m)	292	106	31	248
Deposition of tailings and transported native material	Thick deposits of tailings and construction material obstructed Polley Lake outlet Minor deposition of tailings and LWD in wide, forested overbank areas	Limited deposition. Thin veneers of tailings (thickness of the deposits is less than 0.5 m depth)	Minimal deposition occurred in the final stages of the event in the form of thin veneers of tailings (less than 0.2 m depth) covering the bottom of the canyon	Substantial deposition consisting of a mix of tailings and entrained sediments from the channel bed and banks filling the pre-event channel, and minor deposition of tailings and LWD in wide, forested overbank areas
Changes to channel geometry and bed stability	Widening of the channel in relatively flat areas and localized knickpoints (up to 1 m high)	Several large knickpoints. Largest features are 6 and 10 m high located upstream of the Hazeltine Creek Canyon	Channel widened and eroded the canyon walls. Obliteration of pre-event channel and floodplain by the flow	Large knickpoints within the channel bed and formation of multiple flow channels. Channel avulsion and removal of large portions of the shoreline (distal zone of the delta)
Changes to slope and bank stability	Unstable cut slopes in till material showing signs of continued instability	Active failures on cut glaciolacustrine and till valley slopes and channel banks. Open tension cracks (greater than 10 cm wide) suggesting further activity	Exposed bedrock. Canyon walls are stable under normal conditions	Post-event failures occurred on steep lacustrine banks

tailings and eroded native material, changes to channel geometry and bed stability; and, changes to bank and slope stability. Table 2 summarizes key geomorphic impacts identified at each section.

The aftermath of the 2014 event in the Cariboo region reveals that geomorphic impacts resulting from the failure are very similar to those created by extreme debris flow and outburst flood events. Large erosional features, bank instabilities and widening and deepening of the channel and floodplain have been documented for a number of outburst floods and debris flows resulting from the failure of moraine-dammed and glacier- dammed lakes in Canada and worldwide. For instance, the outburst flood from Ape Lake in 1984 and Nostetuko Lake in 1983, in the southern Coast Mountains of British Columbia (Desloges and Church 1992; Clague and Evans 2000). The glacier outburst floods on September 3, 1977, and August 4, 1985, that modified the Nare Khola, Imja Khola, and Dudh Kosi valleys in Mount Everest Region (Cenderelli and Wohl 2001).

Another important characteristic from the Mount Polley dam failure is that peak discharge was significantly greater than the probable maximum 200-year hydrological flood. Assuming that the maximum outburst volume was 17.4 Mm^3 (estimated volume of tailings and supernatant water released that reached Quesnel Lake), the maximum breach discharge using the empirical formula of Evans (1986) was 5130.67 m^3s^{-1}. This discharge is three orders of magnitude greater than the probable maximum 200-year hydrological flood estimated for Hazeltine Creek. This characteristic is typical for outburst floods and has been widely documented for failures from glacier-dammed and moraine-dammed lakes (Costa and Schuster 1988; Costa 1988; Desloges and Church 1992; O'Connor and Costa 1999; Clague and Evans 1994; Cenderelli 2000; Clague and Evans 2000).

Conclusions

We documented and analyzed the downstream geomorphic response of the flow created by the failure of the Mount Polley TDF. Three types of impacts and resulting geomorphic features are evident within the inundated area: (1) changes to channel geometry and bed stability; (2) changes to bank and slope stability; and, (3) deep erosion and deposition zones. Key factors controlling these impacts on the Hazeltine Creek channel and floodplain were the volume of the flow, characteristics of the material, downstream topography and bed substrate.

Observed geomorphic impacts are comparable to those impacts resulting from historical debris flows and outburst floods events in Canada and other mountainous areas in the world. For this reason, runout of flows and outburst floods from potential tailings dam failures should be an essential component in the development of hazard and risk assessments at mine facilities. Current limitation and uncertainties related to breach modeling make this a challenging task. Understanding the geomorphic response of previous failure events is a first step to the design of better hazard and risk assessment guidelines for tailings dams.

Acknowledgements We thank Imperial Metals for providing pre- and post-event data. The interpretation of the data represents the views of the authors. We also thank Cory McGregor for field work support.

References

Alonso EE, Gens A (2006) Aznalcóllar dam failure. Part 1: Field observations and material properties. Géotechnique 56(3):165–183. doi:10.1680/geot.2006.56.3.165

Benito G, Benito-Calvo A, Gallart F et al (2001) Hydrological and geomorphological criteria to evaluate the dispersion risk of waste sludge generated by the Aznalcóllar mine spill (SW Spain). Environ Geol 40(4):417–428. doi:10.1007/s002540000230

Bichler AJ, Bobrowsky PT (2003) Quaternary geology of the Hydraulic map sheet NTS 93/A12 British Columbia 1:50,000 (British Columbia): NTS 93/A12

Blight GE (1997) Destructive mudflows as a consequence of tailings dyke failures. In: Anonymous Proceedings of the Institution of Civil Engineers–Geotechnical engineering, January 9–18, vol 125. Telford, London, ROYAUME-UNI (1994) (Revue), pp 9–18

Blight GE, Fourie AB (2005) Catastrophe revisited–disastrous flow failures of mine and municipal solid waste. Geotech Geol Eng 23 (3):219–248. doi:10.1007/s10706-004-7067-y

Burge L, Cuervo V (2015) Technical appendices. Appendix A: hydrotechnical and geomorphological impact assessment. In: Golder Associates Ltd. on behalf of Mount Polley Mining Corporation. Post event environmental impact assessment report (PEEIAR), Golder Associates Ltd., Vancouver, BC

Cathro MS, Lane RA, Shives RB et al (2003) Airborne multisensor geophysical surveys in the Central Quesnel Mineral Belt (93A/5, 6, 12). British Columbia Geological Survey Bulletin 185, Victoria, British Columbia

Cenderelli DA, Wohl EE (2001) Peak discharge estimates of glacial-lake outburst floods and "normal" climatic floods in the Mount Everest region, Nepal. Geomorphology 40(1):57–90. doi:http://dx.doi.org/10.1016/S0169-555X(01)00037-X

Chandler RJ, Tosatti G (1996) The Stava tailings dam failure, Italy, July 1985. Int J Rock Mech Min Sci Geomech Abs 33(1):35A

Clague J (1991) Quaternary stratigraphy and history of Quesnel and Cariboo river valleys, British Columbia: implications for placer gold exploration. In: Anonymous current research, Part A, Paper 91-lA edn. Geological survey of Canada, pp 1–5

Clague J, Evans SG (1994) Formation and failure of natural dams in the Canadian Cordillera. Geol Surv Can Bull 464

Clague JJ, Evans SG (2000) A review of catastrophic drainage of moraine-dammed lakes in British Columbia. Quat Sci Rev 19 (17):1763–1783. doi:http://dx.doi.org/10.1016/S0277-3791(00) 00090-1

Costa JE, Schuster RL (1988) Formation and failure of natural dams. Geol Soc Am Bull 100(7):1054–1068

Davies MP (2001) Impounded mine tailings: what are the failures telling us? CIM Distinguished Lecture 2000–2001. The Canadian Mining and Metallurgical Bulletin 94(1052):53–59

Demarchi DA (2011) The British Columbia ecoregion classification, 3rd edn. Ecosystem Information Section. Ministry of Environment, Victoria, British Columbia

Desloges JR, Church M (1992) Geomorphic implications of glacier outburst flooding: Noeick River valley, British Columbia. Can J Earth Sci 29(3):551–564. doi:10.1139/e92-048

Evans SG (1986) The maximum discharge of outburst floods caused by the breaching of man-made and natural dams. Can Geotech J 23 (3):385–387. doi:10.1139/t86-053

Gallart F, Benito G, Martín-Vide JP et al (1999) Fluvial geomorphology and hydrology in the dispersal and fate of pyrite mud particles released by the Aznalcóllar mine tailings spill. Sci Total Environ 242 (1–3):13–26. doi:http://dx.doi.org/10.1016/S0048-9697(99)00373-3

Golder Associates Ltd. on Behalf of Mount Polley Mining Corporation (2015) Post event environmental impact assessment Report (PEEIAR). Vancouver, BC

Hashmi S, Ward BC, Plouffe A et al (2015) Geochemical and mineralogical dispersal in till from the Mount Polley Cu-Au porphyry deposit, central British Columbia, Canada. Geochem: Explor Environ Anal 15(2–3):234

Hungr O, Evans SG, Bovis MJ et al (2001) A review of the classification of landslides of the flow type. Environ Eng Geosci 7 (3):221

ICOLD–UNEP (2001) Tailings dams—risk of dangerous occurrences, lessons learned from practical experiences. Bull 121

Knight Piésold Ltd. (2009a) Assessment of Hazeltine Creek flows. Consultant report prepared for Mount Polley Mining Corporation

Knight Piésold Ltd. (2009b) Recommended maximum discharges from the Mount Polley TSF to Hazeltine Creek. Consultant report prepared for Mount Polley Mining Corporation

Luino F, De Graff JV (2012) The Stava mudflow of 19 July 1985 (Northern Italy): a disaster that effective regulation might have prevented. Nat Hazards Earth Syst Sci 12(4):1029–1044. doi:10.5194/nhess-12-1029-2012

Minnow Environmental Inc. (2007) Hazeltine Creek Habitat Characterization. Consultant report prepared for Mount Polley Mining Corporation

Minnow Environmental Inc. (2012) Mount Polley Mine Supplemental Aquatic Monitoring—2011. Consultant report prepared for Mount Polley Mining Corporation

Mount Polley Independent Expert Engineering Investigation and Review Panel (2015) Report on Mount Polley Tailings Storage Facility Breach

Pedersen R (1998) Overview report–Quesnel river study area–fish habitat assessment procedure. Victoria: Carmanah Research Ltd. Report prepared for Weldwood of Canada Ltd

Potts D, Rogers J, Mathieu J et al (2015) Technical appendices. Appendix B: bathymetry analysis and volume balance. In: Golder associates Ltd. on behalf of Mount Polley Mining Corporation. Post event environmental impact assessment report (PEEIAR), Golder Associates Ltd., Vancouver, BC

Rico M, Benito G, Díez-Herrero A (2008) Floods from tailings dam failures. J Hazard Mater 154(1–3):79–87. doi:10.1016/j.jhazmat.2007.09.110

Rico M, Benito G, Salgueiro AR et al (2008) Reported tailings dam failures: a review of the European incidents in the worldwide context. J Hazard Mater 152(2):846–852. doi:10.1016/j.jhazmat.2007.07.050

SAMARCO (2016) Fundão dam collapse SAMARCO: http://www.samarco.com/en/balanco/

Sammarco O (1999) Impacts of tailings flow slides. Mine Water Environ 18(1):75–80. doi:10.1007/BF02687251

Sammarco O (2004) A tragic disaster caused by the failure of tailings dams leads to the formation of the Stava 1985 Foundation. Mine Water Environ 23(2):91–95. doi:10.1007/s10230-004-0045-z

Villavicencio G, Espinace R, Palma J et al (2014) Failures of sand tailings dams in a highly seismic country. Can Geotech J 51(4):449–464. doi:10.1139/cgj-2013-0142

WISE UP (2016, August 17) Chronology of major tailings dam failures. http://www.wise-uranium.org/mdaf.html. Accessed 18 Aug 2016

The Sediment Production and Transportation in a Mountainous Reservoir Watershed, Southern Taiwan

Chih Ming Tseng, Kuo Jen Chang, and Paolo Tarolli

Abstract

This study examined differences in landslide sediment production and sediment transportation abilities of reservoir watersheds in different geological environments after extreme rainfall. The watershed in this study covered an area of 109 km^2; the upstream river banks of the reservoir contained interbedded shale and faulted shale and had a sandstone dip slope. This paper uses a LiDAR-derived DTM taken in 2005 and 2010 to investigate the landslide sediment production and riverbed erosion and deposition in the watershed. This study also applied the conservation of mass concept to analyze the sediment outflux in the subwatersheds. The research results indicated that although the right bank, which had interbedded shale and a sandstone dip slope, had a substantially greater number of landslides, the sediment production of it was less than that of the left bank, which had numerous deep-seated landslides caused by fault zones. However, affected by the higher sediment production of the left bank and under the same stream power, the left bank subwatersheds also had higher sediment outflux.

Keywords

LiDAR • DTM • Landslide • Reservoir watershed • Sediment yielding • Sediment transport

Introduction

Taiwan is located at the convergent boundary where the Eurasian Plate and Philippine Sea Plate collide, resulting in a terrain environment of frequent earthquakes, geological breakage, and steep slopes; mountainous regions and hilly terrain account for 70% of the land area (CGS 2016). Taiwan receives an average annual rainfall of 2502 mm (WRA 2016. an average of three and a half typhoons annually during the typhoon season from May to September (CWB 2016), during which the rainfall accounts for approximately 80% of the total annual rainfall. Every year, 30% of water resources for the country are supplied by reservoir operations (WRA 2016), and every year concentrated rainfall over mountainous reservoir watersheds during the typhoon season cause landslides and debris flows (Dadson et al. 2004). This heavy rainfall causes large quantities of sediment to be transported into reservoirs, severely affecting the capacity and future water resource management of these reservoirs (Hsu et al. 2016). Therefore, a quantitative analysis of reservoir watershed sediment production from landslides and river sediment transportation characteristics can facilitate inferring the potential quantity of sediment entry into a

C.M. Tseng (✉)
Department of Land Management and Development, Chang Jung Christian University, No.1, Changda Rd, Gueiren District, Tainan City, 71101, Taiwan (R.O.C.)
e-mail: cmtseng@mail.cjcu.edu.tw

K.J. Chang
Department of Civil Engineering, National Taipei University of Technology, No.1, Sec. 3, Zhongxiao E. Rd, Taipei City, 10608, Taiwan (R.O.C.)
e-mail: epidote@ntut.edu.tw

P. Tarolli
Department of Land, Environment, Agriculture and Forestry, University of Padova Agripolis, Viale Dell'Università 16, 35020 Legnaro(PD), Italy
e-mail: paolo.tarolli@unipd.it

M. Mikoš et al. (eds.), *Advancing Culture of Living with Landslides*,
DOI 10.1007/978-3-319-53483-1_34

reservoir. This is a critical topic for the prevention of silting in reservoirs.

In the past, because of the difficulty of performing onsite measurement, obtaining precise, reasonable, and accurate sediment production quantities of watersheds has been almost impossible. The high quality LiDAR (light, detection and ranging) data pre- and post-event providing a unique opportunity for the analysis of the topographical change caused by the typhoon (Tseng et al. 2013). Through a consolidation of digital terrain data of different time periods and a differences comparison and analysis, a rational quantitative assessment of the temporal and spatial trends of the sediment production and transportation as well as of the deposition mass of watershed scales can be performed (Tarolli 2014). The multi-temporal analyses of LiDAR topography such as quantitative landslide-volumetric estimation (Baldo et al. 2009), tracking and evolution of complex active landslides (Ventura et al. 2011), and detection of sediment delivery and process at an active earthflow (DeLong et al. 2012). Accordingly, the goal of this study was to compare the differences of digital terrain generated by LiDAR during two different time periods to analyze sediment production and transportation characteristics of watersheds from different geological environments under the impacts of two typhoon events, each of which brought a total rainfall exceeding 1000 mm. The results can be used as a reference by reservoir management institutions for drafting reservoir watershed sedimentation management strategies.

Study Area

This study examined the Nanhua Reservoir watershed. The Nanhua Reservoir is a critical water resource facility in southern Taiwan that supplies municipal water for Tainan and Kaohsiung; it is the largest reservoir in Taiwan that solely supplies municipal water. Because of Typhoon Kalmaegi in July 2008 (accumulated precipitation: 1113 mm recorded from Guanshan station of Central Weather Bureau) and Typhoon Morakot in August 2009 (accumulated precipitation: 2354 mm at the same station), the reservoir capacity was reduced by approximately 30 million cubic meters, severely affecting its water supply. The Nanhua Reservoir watershed covers a surface area of approximately 10,924 ha, with its terrain mainly inclined from north-east to south-west and the elevation of the watershed ranging between 175 and 1186 m, which is long and narrow in shape (Fig. 1). The inner slope of the watershed primarily ranges between 20° and 40° and accounts for 60% of the entire watershed; the slope at the western part is steeper than that at the eastern part.

The exposed strata of this area are bounded by the Pingxi fault (Fig. 1). The exposed strata on the eastern side of the Pingxi fault, listed in the order from oldest to newest, are Hunghuatzu Formation (Hh) and Changchihkeng Formation (Cc). The exposed strata on the western side of the Pingxi fault, listed in the order from oldest to newest, are Tangenshan Sandstone (Tn), Yenshuikeng Shale (Ys), Ailiaochiao Formation (Al), Maupu Shale (Mp), Choutouchi Formation (Ct), and Peiliao Formation (Pl). Terrace deposits (t) have been discovered on both sides of the Nanhua Reservoir and Houku River. The areas on the left and right banks of the Houku River differ greatly in geological characteristics; the right bank, characterized as having a chutouchi formation, is mainly composed of massive muddy sandstone and interbedded shale and sandstone; the left bank, which is classified as having a peiliao formation, is mainly composed of thick shale. According to the landslide distribution of the entire watershed, the dip slopes and development of landslides are closely correlated. Additionally, rock mass breaking was observed in the areas close to the fault line and fold axis because of the development of cracked surfaces. The area is also prone to large-scale landslides (Lin et al. 2011, 2013).

The geological environments of the left and right banks of the Houku River differ substantially. These differences affect sediment production from landslides and the transportation of sediment by tributaries. Through the aerial photographs from three different time periods, the current study chose the tributary subwatersheds of the Houku River during two severe typhoon events for analysis, during which substantial sediment production from landslides occurred at the upstream slope and the river channel exhibited debris flow or hyperconcentrated flow transportation characteristics such as erosion and widening. Figure 2 details the selection of subwatersheds. Figure 2a illustrates the geomorphic features of the subwatersheds in March 2007 before Typhoon Kalmaegi (July 2008). Figure 2b depicts landslides in multiple areas of the upstream subwatersheds of the tributary in March 2009 after Typhoon Kalmaegi; the river course also shows signs of severe sediment transportation. Figure 2c illustrates evidence of large-scale landslides on the upstream slope surface of the subwatersheds in January 2010 after Typhoon Morakot; the river channel shows a widening trend as a result of sediment transportation. Based on the selection criteria, the study selected 37 subwatersheds on both banks of the Houku River as analysis units for sediment production and transportation characteristics discussion; 16 subwatersheds were selected from the right bank, with the area ranging between 22.2 and 474.8 ha, and 21 subwatersheds were selected from the left bank, with the area ranging between 28.1 and 249.8 ha.

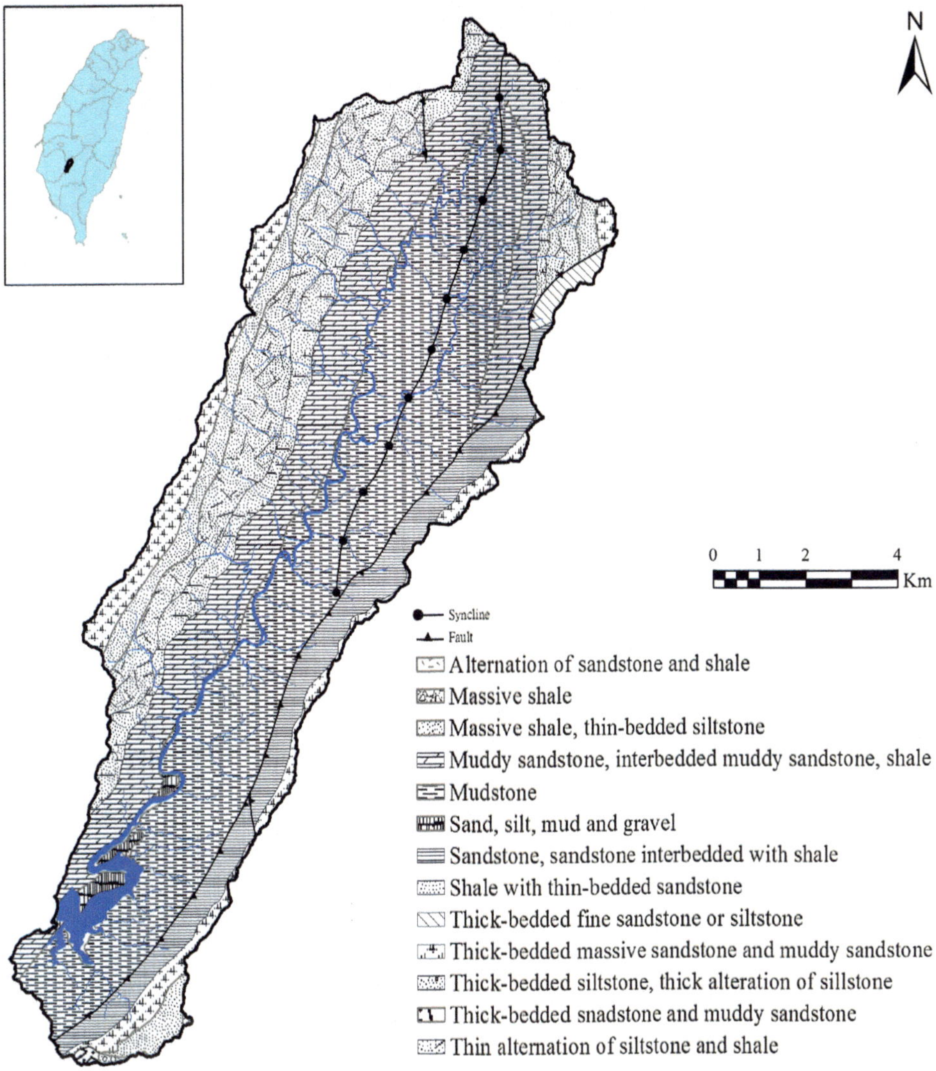

Fig. 1 Geological map of the study area. Geological data from Central Geological Survey

Method and Material

Topography Data

The topography of the study area before and after Typhoon Kalmaegi and Typhoon Morakot is represented by LiDAR data taken in 2005 and 2010. Figure 3 shows shaded relief and aerial photo of a landslide and its connected stream in 2005 and 2010, respectively. Figure 3 shows an evident depletion and deposition of landslide, as well as the deposit along the talweg of the connected stream valley. A 2 m digital terrain model (DTM) was derived from both datasets using a natural neighbor interpolator (Sibson 1981), which has proven useful for geomorphic analysis before (Pirotti and Tarolli 2010). The vertical accuracy of both DTMs, evaluated by a direct comparison of LiDAR and ground differential global positioning system (DGPS) for the study areas, was estimated to be <0.3 m in flat areas (Lin et al. 2013; Tseng et al. 2013, 2015).

Analysis of Sediment Erosion and Deposition

This study used the difference of DTM (DoD) method to perform subtraction on the 2005 and 2010 LiDAR DTMs to obtain the quantitative terrain elevation variance of landslide and river channels (Fig. 4). The erosion and deposition of sediment volumes in both landslides and river beds was calculated by subtracting the 2005 DTM from the 2010 DTM. In calculations the landslide failure surface was divided into 2×2 m cells, and negative values of grid subtraction represent areas that have been eroded. In a similar way, the positive values of subtraction mean areas where debris was deposited. Therefore, for each landslide

(a)March 2007

(b)March 2009

(c)January 2010

Fig. 2 Example of selection of subwatersheds. *Red line* represent the boundary of individual subwatershed

(a)2005

(b)2010

Fig. 3 Shaded relief map and aerial photo of a landslide and its connected stream

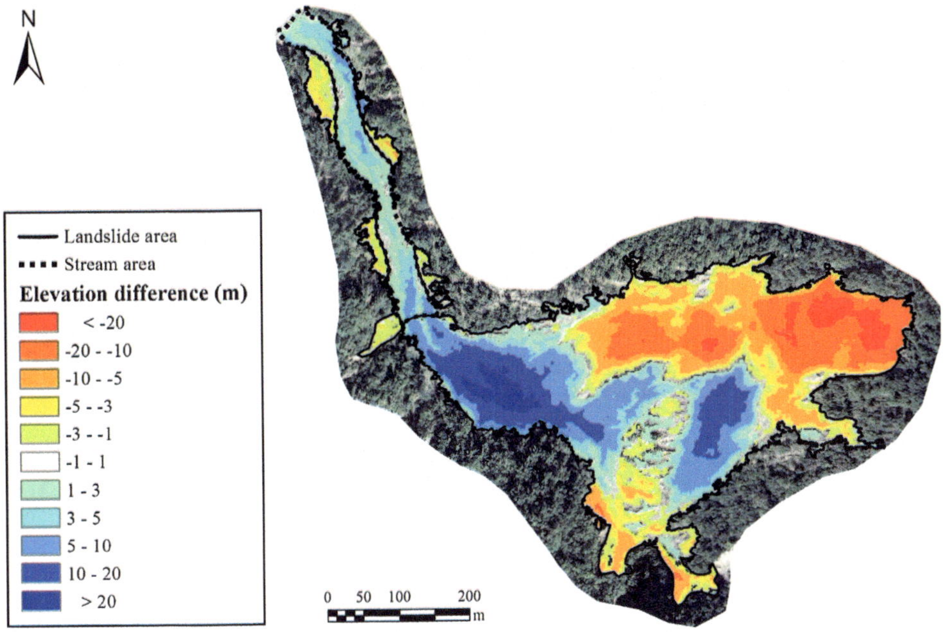

Fig. 4 Example of the difference of DEM. *yellow* and *red color* represent landslide depletion or riverbed erosion, *blue* and *purple color* represent sediment deposited in landslide toe or riverbed deposition

and river section, the summation of the negative and positive values multiplied by the grid area represents the eroded volume V_e or deposited volume V_d. The formula is as follows:

$$V_e(or\ V_d) = A\left(\sum_{i=1}^{n} h_i\right) \quad (1)$$

where n is the total number of grids in the area; A is the area (m^2) of each cell; and h_i is the erosion (negative value) or deposition (positive value) depth (m) of each cell.

Analysis of Sediment Outflux

The Nanhua Reservoir, fed by the Houku River, has a total of 37 subwatershed areas on the left and right banks. Furthermore, debris flow is the primary sediment transportation method of the river during typhoons. According to the debris flow characteristics, the longitudinal sections of the river could largely be classified into three zones: the occurrence zone, flow zone, and accumulation zone. The occurrence zone is characterized by the typical sediment accumulation after landslide occurrence (such as the hillslope toes). The large quantity of water and sediment produced by a landslide gathers in this zone and forms a debris flow. The flow zone is the debris flow transportation section; riverbed scouring and river widening often occur in this area because of debris flow erosion. The accumulation zone is the area of the river where the slope is relatively gentle; the sediment from a debris flow generally cannot continue to be transported past this zone and gradually gets deposited in this area. This area is the section of river generally located at the point before the tributaries joins the main river. The section where the flow zone transforms into the accumulation zone can be considered as a concentration point. The sediment volume pass this concentration point (river section), compared with all the sediment volume of debris flow streams entering the main river, can more reflect the transportation ability of this subwatershed.

Figure 5 illustrates an example of the longitudinal section of stream erosion/deposition depth of a subwatershed on the left bank. At the channel head region within a horizontal distance of 400 m, the scouring depth ranged between 5 and 8 m. As the river slope gradually became gentle, the debris flow erosion and scouring ability was gradually weakened, and the erosion and scouring depth was also greatly reduced; at approximately a 900 m horizontal distance, erosion and deposition depth started to gradually transform from a negative value to a positive value. This study considered the point where the considerable change in erosion and deposition depth value occurred as the downstream endpoint of the primary debris flow transportation section, and defined it as the sediment outflux point; the drainage range of this point was used to analyze the sediment outflux of the 37 subwatersheds. On the basis of the concept of the conservation of mass, the sediment outflux V_o of the subwatershed in the 2005 and 2010 LiDAR digital terrain images was calculated using the following equation:

$$V_o = \sum V_e - \sum V_d \quad (2)$$

where ΣV_e is total sediment production (landslide erosion or riverbed erosion) of the two digital terrains in the drainage

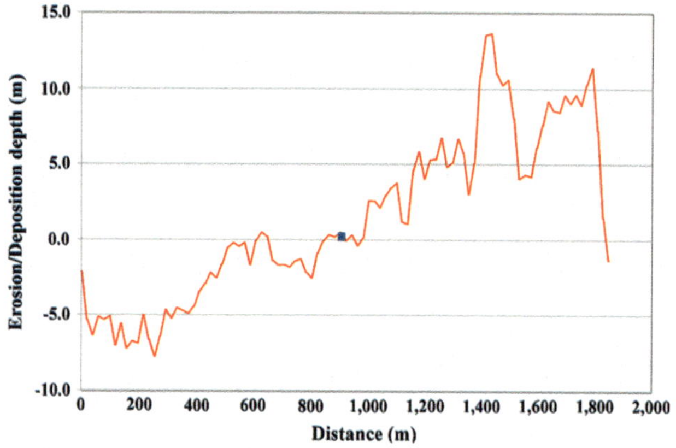

Fig. 5 Example of selecting concentration point of sediment outflux of a subwatershed

range and ΣV_d is the total sediment deposition (landslide deposition or riverbed deposition) of the two digital terrains in the drainage range.

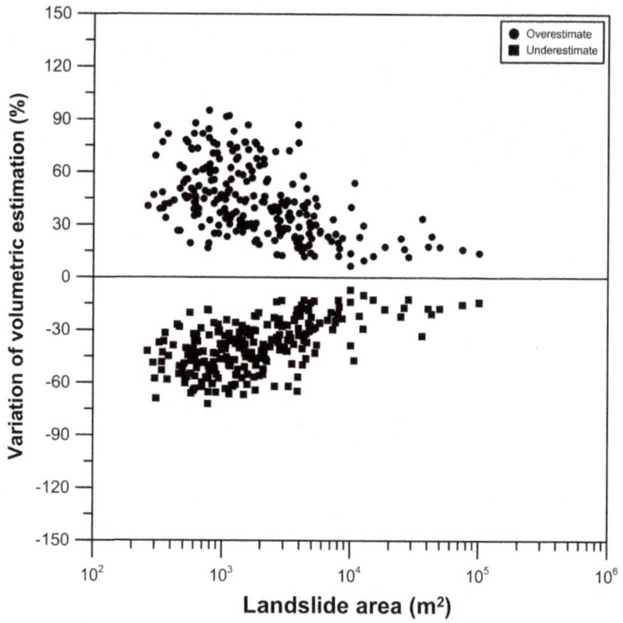

Fig. 6 Variation of volumetric estimation

Discussion

Uncertainty of Volumetric Estimation

The consistency of the LiDAR data was examined to quantify the error in elevation difference between the 2005 and 2010 DTMs. Since the morphology of riverbeds and landslide areas has changed significantly between 2005 and 2010, they are excluded from the validation. The remaining well vegetated areas are treated as undisturbed zone. The variation in elevation difference between these two DTMs is -0.01 ± 1.35 m (mean ± standard deviation) in undisturbed zone, which is therefore a potential measure of error in the elevation differences between the 2005 and 2010 DTM in sediment volume estimations. The error analysis described above is adopted to understand the uncertainty of the volume estimation caused by the error of the LiDAR DTMs. A vertical error value of $\Delta Z = \mu \pm \sigma$ is used to raise or lower the previous DTM, then further obtain the most likely maximum (overestimate) and minimum (underestimate) sediment volume by subtracting the DTM in 254 new landslides that occurred during Typhoon Morakot. For example, the maximum and minimum difference between DTMs is 1.34 m and −1.36 m respectively ($\mu + \sigma = -0.01 + 1.35$ and $\mu - \sigma = -0.01 - 1.35$). Thus, the variation of landslide induced sediment volumes is obtained in Fig. 6, The upper and lower variations represents the most likely maximum and minimum landslide volume estimations based on the DTM vertical error test. The volumetric estimation error is inversely proportional to landslide area. This is because the volume variation caused by vertical error in the total volume is negligible for larger areas. For example, if a landslide area greater than 10^4 m^2, then the absolute errors in estimating volume are only within the range of 20%. Thus, the uncertainties in volumetric calculation are an apparently reliable quantitative estimation technique when using LiDAR-based DTMs.

Sediment Production Characteristics

To understand the differences in sediment production characteristics of the left and right banks of the Houku River that feed into the Nanhua Reservoir, this study analyzed the landslide sediment production characteristics of the 254 new landslides (on the left and right river banks) that occurred during Typhoon Morakot. Table 1 lists the quantitative distribution of the new landslides on the left and right riverbanks; 158 sites were located on the right bank and 96 sites were located on the left bank. However, the total area of

Table 1 Comparison of landslide sites on the left and right banks of the Houku River during Typhoon Morakot

	No. of new landslides	Total area of new landslides (ha)	Total sediment yielding of new landslides (m m^3)
Right bank	158	54	356
Left bank	96	47	270
Total	254	101	626

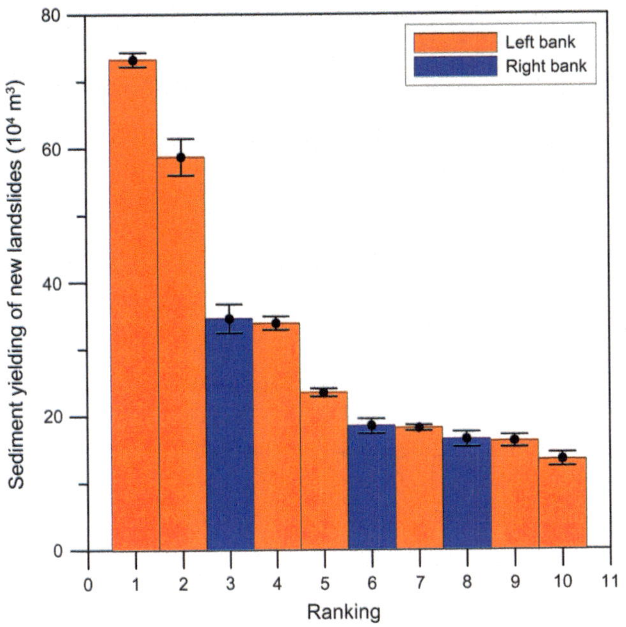

Fig. 7 Top 10 highest producing sites of sediment for new landslides of the study area. The error bar indicates the possible volumetric estimation error

the new landslide sites on the right bank was only larger than that of the left bank by approximately 7 ha; therefore, the scale of the right bank new landslide sites was smaller than that of the left bank sites. The right bank had a little sediment yielding of all new landslide sites totaling 356 million m^3.

Figure 7 depicts the top 10 highest producing sites of sediment for new landslides at the Nanhua Reservoir watershed during Typhoon Morakot. The figure clearly indicates that the left and right banks occupied seven and three positions, respectively. Thus, the left bank was the primary source of new landslide sediments. This analysis shows that the dip slope terrain of the right bank was more prone to landslides, which was proven by the number of new landslide sites and area. However, the dip slope landslide depth is controlled by a structure of interbedded shale and sandstone; not considering large-scale riverbed scouring or curved channel attack slopes, landslides of great depths are uncommon. The development of landslides on the left bank was correlated to the fault strikes distribution; rock mass breaking at the fault caused large-scale landslides; the positions of the large-scale landslides on the left bank conformed to the Pingxi fault strike.

Simonett (1967) proposed that a power relationship existed between landslide surface area A_L and volume of landslide sediment V_{eL}, and could provide inference for a feasible method for calculating landslide sediment production. The power relationship is as follows:

$$V_{eL} = \alpha A_L^{\gamma} \tag{3}$$

where α and γ are the intercept and scaling exponent, respectively. According to Eq. (3), a regression analysis was performed on the 96 left bank new landslides and 158 right bank new landslides. The results are as follows:

$$\text{Left bank}: \quad V_{eL} = 0.2933 A_L^{1.303} \quad R^2 = 0.92 \tag{4}$$

$$\text{Right bank}: \quad V_{eL} = 0.6296 A_L^{1.217} \quad R^2 = 0.90 \tag{5}$$

Equations (4) and (5) revealed that the landslide sediment production on the left bank was greater than that on the right bank. Through use of 10 ha of landslide as an example, the landslide sediment production on the left bank was determined to be 1.25 times that on the right bank.

Sediment Transportation Characteristics

Requirements for debris flow to occur include sufficient water, loose material, and steep slopes. Generally, when assessing susceptibility for debris flow, the two physiographic factors, local slope and drainage area, are often used to reflect the aforementioned three requirements. For example, in past related studies, local slope (S) and root value of drainage area ($A^{0.5}$) have been used to reflect the mobility of slope sediment (Dalla and Marchi 2003; Marchi and Dalla 2005; Borselli et al. 2008). The extended form of the relationship between slope and area uses various power of area ($-\delta$) values to determine potential debris flow sites (Rickenmann and Zimmermann 1993; Zimmermann et al. 1997; Horton et al. 2008); the general equation is as follows:

$$\tan\beta = \alpha A^{-\delta} \tag{6}$$

where $\tan\beta$ is the slope threshold and α is the weight factor and α and δ mainly show the physiographic characteristics of debris flow at various areas. The research results of Rickenmann and Zimmermann (1993) were $\alpha = 0.36$ and $\delta = 0.11$; those of Zimmermann et al. (1997) were $\alpha = 0.32$ and $\delta = 0.2$; and those of Horton et al. (2008) were $\alpha = 0.31$ and $\delta = 0.15$. In Eq. (6), α represents the stream power of a drainage area, and can be considered as the debris flow trigger indicator (D'Agostino and Bertoldi 2014); stream power can also reflect the sediment transportation ability of debris flow.

Ijjasz-Vasquez and Bras (1995) used digital terrain DEMs to obtain the slope and drainage area of a watershed, and proposed four types of zones based on the resultant relationship diagram. In Zone I, the local slope and drainage area are positively correlated, with a diffusion wave power

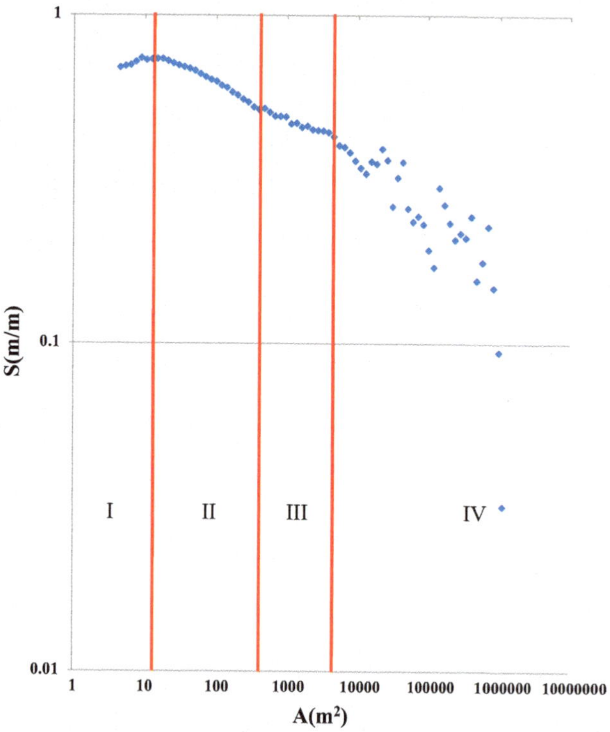

Fig. 8 Example of the division of the relationship of slope and area of a subwatershed

transmission slope being the primary characteristic. In Zone II, the positive slope from Zone I changes to a negative slope, with the terrain characteristic being that of a valley; this area is often the initial formation point of the channel and is called the channel head. Zones II and III consist of debris flow or hyperconcentrated flow. Zone IV is the general silt alluvial channel. Therefore, the current study used a

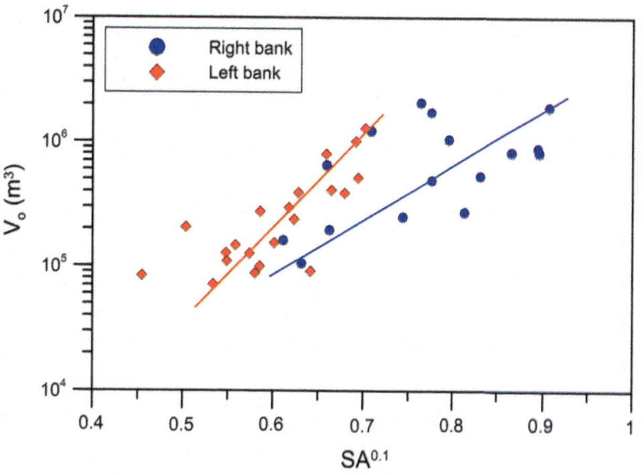

Fig. 9 Relationship between sediment outflux V_o and stream power $SA^{0.1}$ of the 37 subwatersheds

slope and drainage area relationship diagram to determine the parameters δ for Eq. (6); the line gradient of Zone II was used to determine the parameter δ. Figure 8 is an example of the division of the relationship of slope and area of a subwatershed on the left bank. Through a consolidation of the Zone II line gradient of the slope and area relationship diagram of the 37 subwatersheds on the left and right banks, the right bank average line gradient was determined to be −0.0983, and the standard deviation 0.02; the left bank average line gradient was determined to be −0.0933, and the standard deviation 0.03. Therefore, this study used $\delta = 0.1$. Figure 9 illustrates the relationship between sediment outflux V_o and stream power $SA^{0.1}$ of the 37 subwatersheds in the study area. The figure shows that the left bank trend line is steeper than that of the right bank, indicating that under similar stream power, the left bank watersheds have a greater amount of sediment outflux.

Conclusions

The research results are summarized as follows: the number of landslide sites on the Houku River right bank, which primarily has an interbedded shale and sandstone dip slope structure, was markedly higher than that on the left bank, indicating that the dip slope of the right bank was more prone to landslides. However, because of the impact of the fault zone and the numerous deep-seated landslides sites, the left bank had a higher landslide production. In other words, sediment production body is not only controlled by landslide area, but different geological structures have a direct effect on the thickness of the landslide. In the past, direct use of landslide surface slope to infer the thickness of a landslide through an empirical formula, or use of a regressive empirical relationship of landslide volume and area without considering the difference in geological conditions, could cause considerable error in sediment estimation. Additionally, in this study, the sediment transportation ability of the left and right bank subwatersheds differed; if the stream power of the watershed was used to determine the sediment transportation ability, when stream powers were identical, the left bank subwatersheds with greater landslide sediment production had a higher sediment outflux (because they passed the fault zone). Therefore, subwatersheds with deep-seated landslides are the main source of reservoir sediment and more resources should be invested for sediment control accordingly.

Acknowledgements This work is funded by the grant MOST 105-2625-M-309-003 from the Ministry of Science and Technology, Taiwan, ROC.

References

Baldo M, Bicocchi C, Chiocchini U, Giordan D, Lollino G (2009) LIDAR monitoring of mass wasting processes: the radicofani landslide, province of siena, Central Italy. Geomorphology 105(3–4):193–201

Borselli L, Cassi P, Torri D (2008) Prolegomena to sediment and flow connectivity in the Landscape, a GIS and field numerical assessment. Catena 75:268–277. doi:10.1016/j.catena.2008.07.006

Central Geological Survey (CGS) (2016) MOEA, R.O.C. (Taiwan). http://www.moeacgs.gov.tw/english2/index.jsp

Central Weather Bureau (CWB) MOTC, R.O.C. (Taiwan). http://www.cwb.gov.tw/eng/index.htm (October 2016)

Dadson SJ, Hovius N, Chen H, Dade WB, Lin JC, Hsu ML, Lin CW, Horng MJ, Chen TC, Milliman J, Stark CP (2004) Earthquake triggered increase in sediment delivery from an active mountain belt. Geology 32(8):733–736

D'Agostino V, Bertoldi G (2014) On the assessment of the management priority of sediment source areas in a debris-flow catchment. Earth Surf Proc Land 39:656–668. doi:10.1002/esp.3518

Dalla FG, Marchi L (2003) Slope-area relationships and sediment dynamics in two alpine streams. Hydrol Process 17(1):73–87

DeLong SB, Prentice CS, Hilley GE, Ebert Y (2012) Multitemporal ALSM change detection, sediment delivery, and process mapping at an active earthflow. Earth Surf Process Landforms 37:262–272

Horton P, Jaboyedoff M, Bardou E. (2008) Debris flow susceptibility mapping at a regional scale. In Proceedings of 4th Canadian conference on geohazards: from causes to management. Université Laval, Québec, pp 399–406

Hsu S, Tseng C, Lin C (2016) Antecedent bottom conditions of reservoirs as key factors for high turbidity in muddy water caused by storm rainfall. J Hydraul Eng. doi:10.1061/(ASCE)HY.1943-7900.0001241

Ijjasz-Vasquez E, Bras RL (1995) Scaling regimes of local slope versus contributing area in digital elevation models. Geomorphology 12:299–311

Lin CW, Chang WS, Liu SH, Tsai TT, Lee SP, Tsang YC, Shieh CL, Tseng CM (2011) Landslides triggered by the 7 august 2009 typhoon morakot in Southern Taiwan. Eng Geol 123:3–12

Lin CW, Tseng CM, Tseng YH, Fei LY, Hsieh YC, Tarolli P (2013) Recognition of large scale deep-seated landslides in forest areas of Taiwan using high resolution topography. J Asian Earth Sci 62:389–400. doi:10.1016/j.jseaes.2012.10.022

Marchi L, Dalla FG. (2005) GIS morphometric indicators for the analysis of sediment dynamics in mountain basins. Environ Geol 48:218–228. ISSN: 0071–0857

Pirotti F, Tarolli P (2010) Suitability of LiDAR point density and derived landform curvature maps for channel network extraction. Hydrol Process 24:1187–1197. doi:10.1002/hyp.7582

Rickenmann D, Zimmermann M (1993) The 1987 debris flows in Switzerland: documentation and analysis. Geomorphology 8(2–3):175–189

Sibson R (1981) A brief description of natural neighbor interpolation. In: Barnett V (ed) Interpreting multivariate data. Wiley, Chichester, pp 21–36

Tarolli P (2014) High-resolution topography for understanding earth surface processes: opportunities and challenges. Geomorphology 216:295–312

Tseng CM, Lin CW, Stark CP, Liu JK, Fei LY, Hsieh YC (2013) Application of a multi-temporal, LiDAR-derived, digital terrain model in a landslide-volume estimation. Earth Surf Proc Land 38 (13):1587–1601. doi:10.1002/esp.3454

Tseng CM, Lin CW, Fontana GD, Tarolli P (2015) The topographic signature of a major typhoon. Earth Surf Proc Land 40(8):1129–1136. doi:10.1002/esp.3708

Simonett DS (1967) Landslide distribution and earthquakes in the Bewani and Torricelli Mountains, New Guinea. In: Jennings JN, Mabbutt JA (eds) Landform studies from Australia and NewGuinea. Cambridge University Press, Cambridge, pp 64–84

Ventura G, Vilardo G, Terranova C, Sessa EB (2011) Tracking and evolution of complex active landslides by multi-temporal airborne LiDAR data: the montaguto landslide (Southern Italy). Remote Sens Environ 115:3237–3248

Water Resources Agency (WRA) MOEA, R.O.C. (Taiwan). http://eng.wra.gov.tw/lp.asp?ctNode=6299&CtUnit=1177&BaseDSD=43 (October 2016)

Zimmermann M, Mani P, Gamma P (1997) Murganggefahr und Klimaänderung—einGIS-basierter Ansatz. vdf Hochschulverlag AG an der ETH Zürich 162 (in German)

Integration of Geometrical Root System Approximations in Hydromechanical Slope Stability Modelling

Elmar Schmaltz, Rens Van Beek, Thom Bogaard, Stefan Steger, and Thomas Glade

Abstract

Spatially distributed physically based slope stability models are commonly used to assess landslide susceptibility of hillslope environments. Several of these models are able to account for vegetation related effects, such as evapotranspiration, interception and root cohesion, when assessing slope stability. However, particularly spatial information on the subsurface biomass or root systems is usually not represented as detailed as hydropedological and geomechanical parameters. Since roots are known to influence slope stability due to hydrological and mechanical effects, we consider a detailed spatial representation as important to elaborate slope stability by means of physically based models. STARWARS/PROBSTAB, developed by Van Beek (2002), is a spatially distributed and dynamic slope stability model that couples a hydrological (STARWARS) with a geomechanical component (PROBSTAB). The infinite slope-based model is able to integrate a variety of vegetation related parameters, such as evaporation, interception capacity and root cohesion. In this study, we test two different approaches to integrate root cohesion forces into STARWARS/PROBSTAB. Within the first approach, the spatial distribution of root cohesion is directly related to the spatial distribution of land use areas classified as forest. Thus, each pixel within the forest class is defined by a distinct species related root cohesion value where the potential maximum rooting depth is only dependent on the respective species. The second method represents a novel approach that approximates the rooting area based on the location of single tree stems. Maximum rooting distance from the stem, maximum depth and shape of the root system relate to both tree species and external influences such as relief or soil properties. The geometrical

E. Schmaltz (✉) · S. Steger · T. Glade
Faculty of Geosciences, Geography and Astronomy, ENGAGE—Geomorphological Systems and Risk Research, University of Vienna, Vienna, Austria
e-mail: elmar.schmaltz@univie.ac.at

S. Steger
e-mail: stefan.steger@univie.ac.at

T. Glade
e-mail: thomas.glade@univie.ac.at

R. Van Beek
Faculty of Geosciences, Utrecht University, Utrecht, The Netherlands
e-mail: r.vanbeek@uu.nl

T. Bogaard
Faculty of Civil Engineering and Geoscience, Delft University of Technology, Delft, The Netherlands
e-mail: T.A.Bogaard@tudelft.nl

M. Mikoš et al. (eds.), *Advancing Culture of Living with Landslides*,
DOI 10.1007/978-3-319-53483-1_35

cone-shaped approximation of the root system is expected to represent more accurately the area where root cohesion forces are apparent. Possibilities, challenges and limitations of approximating species-related root systems in infinite slope models are discussed.

Keywords

Physical based modelling • Soil reinforcement • Root system approximation • Slope stability

Introduction

The stabilizing effect of vegetation on slopes was addressed in many studies (e.g. Sidle et al. 1985; Sidle and Ochiai 2006; Stokes et al. 2014). Plant roots are known to reinforce hillslopes by mechanically anchoring into the ground and extracting moisture out of the soil mantle (Stokes et al. 2008; Genet et al. 2008; Ghestem et al. 2011; Papathomas-Koehle and Glade 2013). Thereby, the above-ground (organic matter, stem, branches, leafs) and the below-ground biomass (roots) form a soil-plant continuum with the ground the plant is standing on. The negative pressure in the xylem that is induced by evaporation from the leafs, actuate the water extraction of roots from the surrounding soil material. Hence, plants have considerable effects on the hydrological balance of a landscape and are able to significantly increase the stability of the rooted soil mantle (Ghestem et al. 2011).

Several studies tackled the issues that affect the reinforcing potential of vegetation on a slope—in particular of root systems. For instance, Schwarz et al. (2012), Schmidt and Kazda (2002), Pollen and Simon (2005) and Danjon et al. (2002) highlighted the effects of different root systems and architectures on slope stability and on the Factor of Safety (FoS). Greenway (1987), Fan and Su (2008) and Meng et al. (2014) showed that the hydrological influences of forest stands on slopes have direct effects on the mechanical stability. Whereas, Multiple studies all over the globe (Ziemer 1981; Sidle 1991; Montgomery et al. 2000; Sidle and Ochiai 2006; Imaizumi et al. 2008) addressed the effects of tree stand removal and the reduction of stability due to decay of root systems of harvested trees.

Many modelling and simulation approaches attempted to quantify the effect of roots (Danjon et al. 2002; Van Beek et al. 2007; Schwarz et al. 2010, 2012; Thomas and Pollen-Bankhead 2010) and to implement stabilizing forces of roots into the Factor of Safety equation. However, vegetation in general and roots in particular appear to be rather underrepresented in physically based slope stability models (Schmaltz et al. 2016a).

In this study, we highlight the possibilities and drawbacks of a simple root system approximation and its implementation in a well-established hydromechanic slope stability model. Hereto, we compared three different land cover scenarios.

Study Area

The area of investigation is situated at the South-facing slope of the Walserkamm ridge within the Walgau valley in Vorarlberg, Austria. The area is delimited by the creeks Schnifiser Tobel to the East and Montanastbach to the West while ranging in altitude from 625 m.a.s.l. to 1971 m.a.s.l. (Fig. 1). Primarily consolidated morainic material covers the geological underground that is composed of alternating sandstone and claystone layers, partly interrupted by limestones (Friebe 2013). Particularly in the lower slope section between Düns and Montanast and as well in the Pfänder area

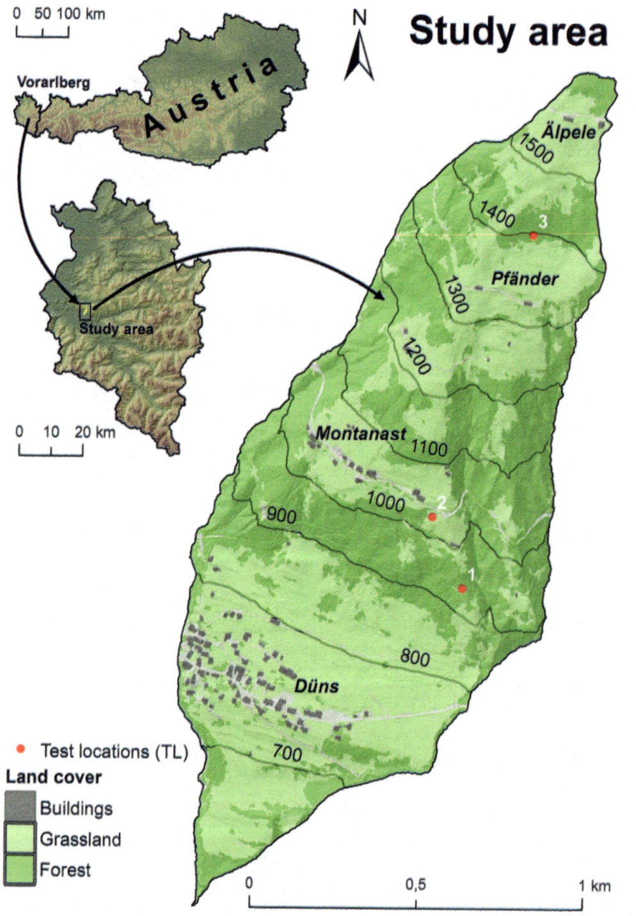

Fig. 1 Location and land cover of the study area

(Fig. 1), marls are widespread. Alpine pastures and timberland depict the land use of the area. A considerable amount of the forests are economically used conifer stands (spruces and firs), which are primarily located in steeper areas of the slopes. However, the species composition of the stands is under continuous change towards a higher diversity of deciduous trees and conifers. In this regard, the forested areas in the lower slope part around Düns and Montanast are mixed stands.

Under consideration of previous studies (Tilch 2014; Ruff and Czurda 2008), several field campaigns have been conducted to investigate the landslide dynamics of the area (Schmaltz et al. 2016a, b). In 1999, 14 landslides were triggered during an intense rainfall event. Although the exact time of failure can not be determined, personal communications with locals and archive data from the Torrent and Avalanche Control (WLV, Wildbach- und Lawinenverbauung) revealed that most of the mapped landslides of that period occurred on the 20th of May 1999.

Data

Digital Terrain Model (DTM)

We used a 1 × 1 m DEM that was obtained from LiDAR flights in 2004 and provided by the Federal State of Vorarlberg. The DEM was resampled to 2.5 m to ensure both, computational feasibility and an adequate level of detail for the approximation of the root systems.

Land Cover Data

Forested areas as well as buildings and sealed areas (roads, parking lots) were digitized, based on a RGB-orthophoto of 2001. It is assumed that the 2001 conditions remain until today. Stands of different tree species were not distinguished. All areas that were not assigned as 'forest' or 'sealed (buildings, roads)' were considered as grassland (as illustrated in Fig. 1).

Climate Data

Meteorological information were obtained from the Austrian Federal Ministry of Agriculture, Forestry, Environment and Water Management. We chose a dataset from November 1998 to October 1999 to cover both, an entire hydrological year and an heavy precipitation event. This heavy rainstorm occurred on the 20th of May 1999 and triggered several landslides in the study area. Former geomorphological studies (Schmaltz et al. 2016a, b) revealed that 14 landslides

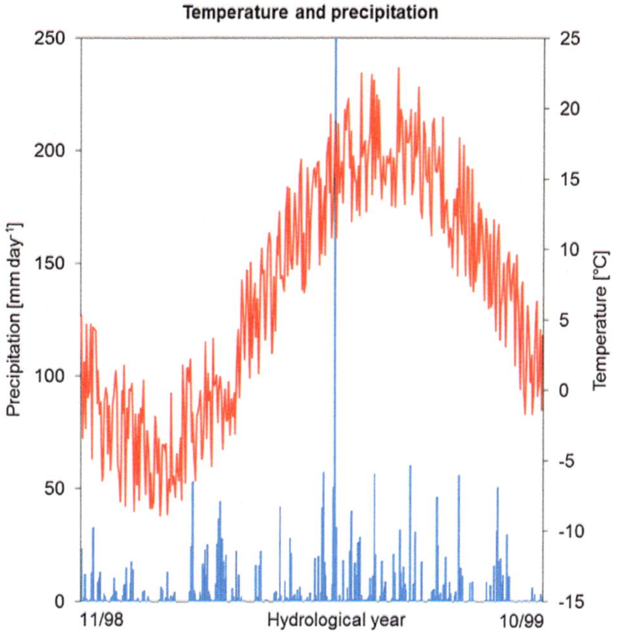

Fig. 2 Rainfall data for the hydrological year of November 1, 1998–October 31, 1999 (365 days). Precipitation data were provided by the Austrian Federal Ministry of Agriculture, Forestry, Environment and Water Management. Synthetic daily temperature was estimated based on monthly temperature information of the study site

with an average scarp area size of ~64 m^2 were triggered within the extent of the investigated area.

We estimated daily temperature data based on monthly precipitation information with a daily standard deviation of the temperature of 2.5 °C. This was performed due to the existent temperature gradient and the high temperature differences along the whole slope section (Fig. 2).

Methodology

Slope Stability Modelling

The probability of failure was assessed using the coupled model STARWARS/PROBSTAB, developed by Van Beek (2002) for translational landslides. The model contains a hydrological component (STARWARS) and a geomechanic module (PROBSTAB). In the hydrological part, the volumetric moisture content for the distinguished soil layers and groundwater level are calculated dynamically and spatially distributed. The geomechanic part uses the simulated hydrological parameters to calculate Factor of Safety (FoS) values and the slope failure probability. STARWARS/PROBSTAB is written in pcrcalc, processed with the pcraster GIS and embedded within the convenient pcraster-python framework. This allows a straightforward manipulation of the model code.

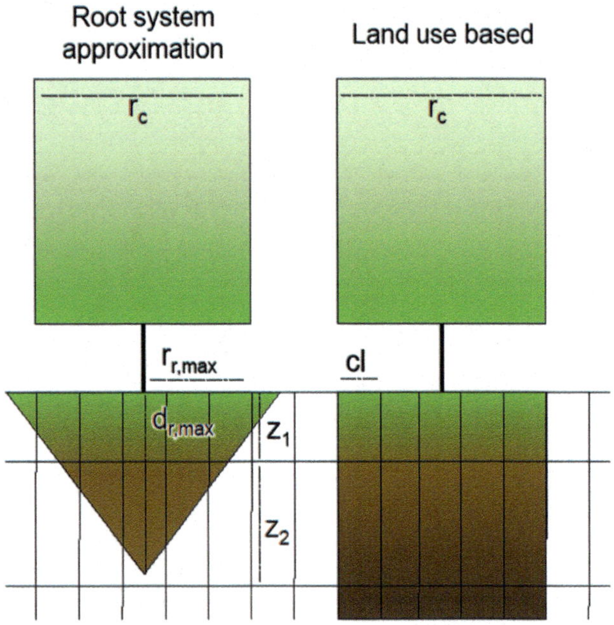

Fig. 3 2D-view of the applied root system approximation scheme in a raster environment. Where r_{c1} [m] is the diameter of the canopy, $r_{r,max}$ [m] is the maximum radius of the root system, $d_{r,max}$ [m] is the maximum depth of the root system, z_1 and z_2 [both m] are depths of soil layer 1 and 2 respectively and cl [m] is the length of a raster cell

No measured data of soil moisture content or groundwater level for validation of the modelling output were available for the chosen period. Thus, we primarily used the probability of failure and changes in volumetric soil moisture contents to compare three different scenarios:

(i) Scenario 1: The area without any vegetation cover.
(ii) Scenario 2: The study area is covered with vegetation—either forest that was assigned as a dense mixed forest (see Table 1) and grassland (land use based).
(iii) Scenario 3: Root system approximation was performed on areas that were assigned as forests. Not-forested areas were considered as sealed or grassland (see Fig. 1).

Schmaltz et al. (2016a) performed similar scenario analyses like those mentioned in bullet point i) and ii) on the different scales, for a greater area (~12 km^2) and lower resolution (10 × 10 m raster). Three test locations were defined to determine both differences in the soil moisture fluctuations of the three distinguished soil layers and the slope failure probability. Therefore, two spots were chosen under a forest cover (location 1 and 3) and in a none forested area (location 2) (see Fig. 8). A landslide was triggered at

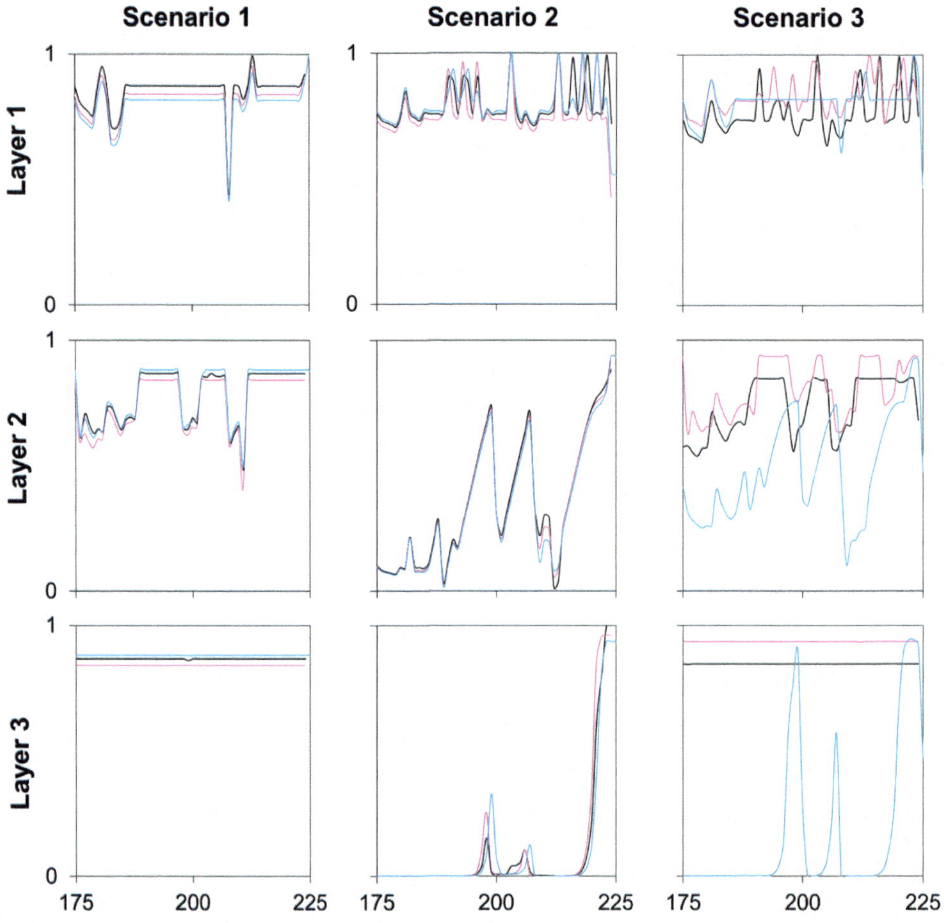

Fig. 4 Changes of volumetric moisture content (y-axis; in % of respective maximum) of layer 1 (*top row*), layer 2 (*mid row*) and layer 3 (*bottom row*) for time steps (x axis; days) 175–225. The colored lines show the different test locations. *Black* location 1, *cyan* location 2, *magenta* location 3

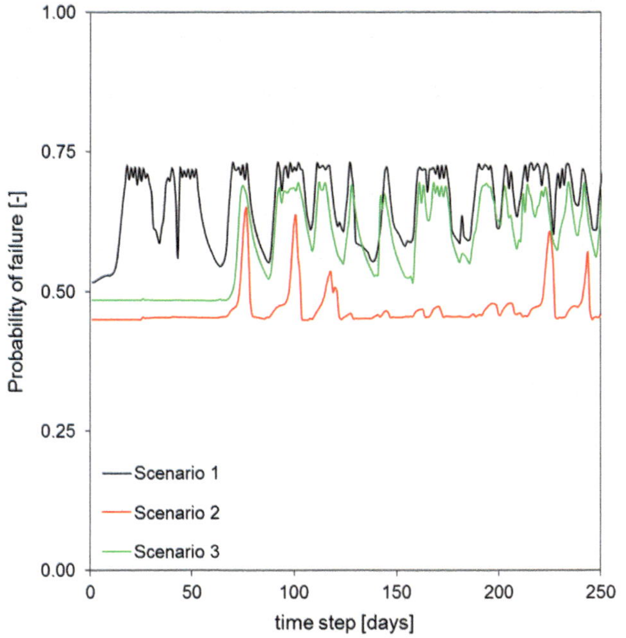

Fig. 5 Slope failure probability (Pf) at test location 1. Location 1 was located in the forest during the modelled hydrological year, however, deforested after a landslide event (c.f. Fig. 8)

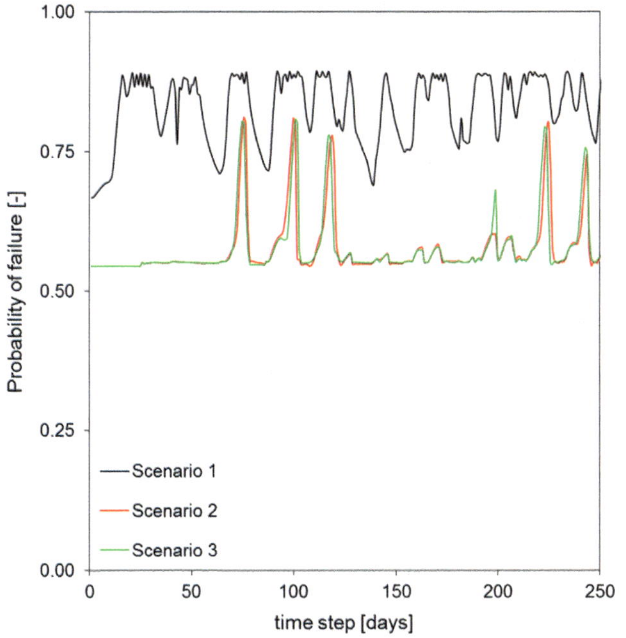

Fig. 6 Slope failure probability (Pf) for a not-forested area (location 2)

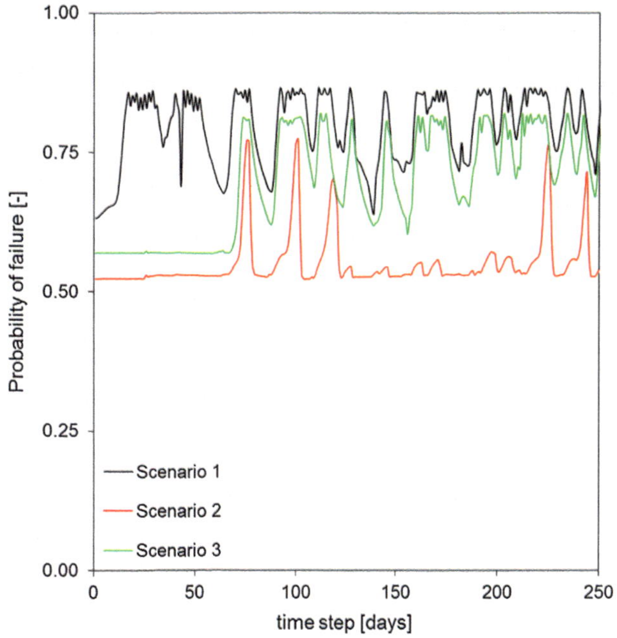

Fig. 7 Slope failure probability (Pf) for a forested area (location 3)

location 1, whereas the other two test locations appeared to remain stable.

Root System Approximation

Within scenario 3 we assumed that roots are distributed around the trunk of a tree with fixed maximum depth and distance. Both, distance and depth depend on tree species, soil type, nutrient and water availability as well as on climatic conditions (Sidle and Ochiai 2006; Thomas and Pollen-Bankhead 2008, Ghestem et al. 2012). For simplification, it is assumed that the maximum rooting distance around the trunk decreases with depth. Thus, the approximated rooted zone forms a cone shaped solid with variable ratios between maximum radius and maximum depth for certain rooting systems (Fig. 3).

Parameterization and Calibration

Soils

To carve out the effect of different vegetation covers, geotechnical soil parameter were kept constant for the whole

Table 1 Input-data of vegetation coverage for the modelled scenarios

Scen.	C_r (kPa)	Int. loss (m)	K_c (–)
1	0	0	0
2	4	4.5	1
3	4	4.5	1

*Scen. = Scenario; c_r = root cohesion; Int. loss = interception loss; K_c = cropfactor

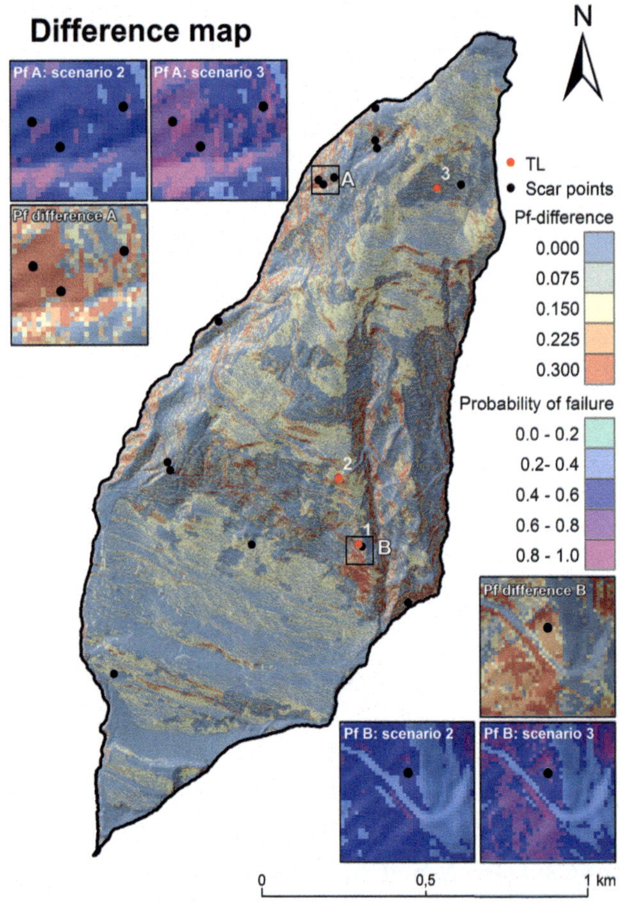

Fig. 8 Difference map between Pf-maps of scenario 3 and 2. Detailed views of Pf-maps and the Pf-difference for two representative areas A and B are provided in the *upper left* and the *lower right* corner of the figure. Moreover, landslides that were triggered 1999 are represented as scar points (*black*). Test locations (TL) that reflect Pf values (c.f. Figs. 5, 6, and 7) and volumetric soil moisture contents of the three soil layers (c.f. Fig. 4) are given as *red dots*

study area. Three different soil layers z_i with distinct geotechnical properties were distinguished. Geotechnical soil parameters (dry bulk density [γ], internal friction angle [Φ] and soil cohesion [c_r] at shear plane) used in the modelling procedure were set to uniform values for all three soil layers in the whole study site: $\gamma = 25.3$ kN m^{-3}, $\Phi = 35°$, $c_r = 2$ kPa.

A simplified version of a generic soil depth prediction model of Pelletier and Rasmussen (2009) was used to estimate soil depth:

$$z = \frac{h_0}{\sqrt{1 + \left|\frac{\partial z}{\partial x}\right|^2}} \ln\left(\frac{P_0}{U}\sqrt{1 + \left|\frac{\partial z}{\partial x}\right|^2}\right) \quad (1)$$

where h_0 is the soil thickness [m] at which bedrock lowering falls of 1/e of its maximum value, P_0/U is the relation between the maximum lowering rate on a flat surface and the rock uplift rate. According to Pelletier and Rasmussen (2009) and Roering (2008) we used cos(θ) for $1/\sqrt{(1 + |\partial z/\partial x|^2)}$ to express the topographic controls in terms of the slope normal coordinate direction z and its derivatives. The ratio P_0/U was neglected and kept as 1. Based on analysis in former studies, h_0 was set to 0.5 m (Heimsath et al. 1999, 2001). For the modelling procedure, we assumed the following percentages of thickness of each layer towards total depth based on field observations: layer 1 = 20% (topsoil), layer 2 and 3 = 2 × 40%.

Vegetation

To determine the effects of root reinforcement solely, the canopy cover of the trees and the maximum canopy interception storage was kept constant for forested areas (Table 1). Distribution of cohesion values within the root systems were not considered. Thus, we assumed a root cohesion value of 4 kPa for both scenarios with vegetation on all areas that are depicted as forested.

Usually, the cropfactor is obtained by considering time intervals between wetting events (e.g. rainfall, snowmelt), evaporation power of the atmosphere and the magnitude of the wetting event. For our study, we set Kc = 1, since this is a recommended value used for mixed and conifer forests (Allen et al. 1998).

The root fraction of a distinct soil layer z_i was calculated according to Eq. 2:

$$rootfrac(z_i) = \frac{\int_0^{z_{\max}} \frac{r_{\max}^2}{z_{\max}^2}\pi x^2 dx - \int_{zi}^{z_{\max}} \frac{r_i^2}{z_{\max}^2}\pi x^2 dx}{\int_0^{z_{\max}} \frac{r_{\max}^2}{z_{\max}^2}\pi x^2 dx} \quad (2)$$

where z_i is the thickness of the soil layer [m], r_{max} [m] and z_{max} [m] are the maximum extent of the root system in lateral and vertical direction, respectively.

Results and Discussion

Figure 4 shows the changes of volumetric moisture content in soil layers 1–3 for all modelled scenarios and test locations. It is clearly recognizable that scenario 1 and 2 show quite similar changes in volumetric moisture content throughout all test locations. Whereas, scenario 3 exhibits patterns of both forests and areas with no vegetation cover.

This might be explained by the fact that the approximated root systems of scenario 3 do not cover as much of the soil column compared to scenario 2. In this regard, it is possible that the cone-shaped root systems do not reach the lithic boundary or even soil layer 3. Therefore, some locations show fluctuations of moisture content that are similar to conditions observed for unvegetated areas—particularly when the deeper soil layers 2 and 3 are considered (see Fig. 4, scenario 3). This moreover effectuates the course of pf-values over all time steps.

Generally, pf-values of the scenarios that include vegetation cover (2 and 3) can be clearly distinguished from those of scenario 1 (without vegetation cover) at all observed locations. Location 1 shows quite diverse reactions of slope failure probability (pf) values in the modelling output (Fig. 5). This forested location shows an immediate reaction on precipitation input for scenario 1 (no vegetation cover) at the first time steps and a resulting increase of pf-values to ~0.82. Scenario 2 and 3 show a significant rise of pf-values from time step 70, which depicts the end of the winter season and thus freezing conditions. However, for all locations, the probability of failure drops rapidly for scenario 2 between time steps 120 and 125 and is evened out until time step 220.

In contrast, the model output of scenario 3 shows a highly erratic pf-curve between time step 70 until the end of the model run at the forested locations 1 and 3.

Pf-curves of scenario 2 and 3 show similar courses for the whole modelled period at the non-forested location 2. However, scenario 3 exhibits a significantly higher peak at time step 200 than scenario 2. Time step 200 represents the high precipitation event at the 20th of May 1999. This finding indicates the effect of different vegetation input parameters on non-vegetated locations and a stronger reaction of the system when root distribution is decreased within the soil column.

The difference map of scenario 2 and 3 shows the spatial discrepancies of slope failure probability distribution between those models that include spatial information on the root system (Fig. 8). The map highlights that particular at areas where actual slope failures occurred, scenario 2 (land cover based approximation) tends to decisively overestimate the stabilizing impact of vegetation (detail views of locations A and B in Fig. 8).

Conclusion

In this study, we showed the effect of geometrical root system approximations on the dynamic hydromechanical slope stability model STARWARS/PROBSTAB. Three different land cover scenarios were compared: (1) the study area without vegetation cover; (2) the root system of forested areas are represented by the raster cell size and the depth of the soil column; (3) the root systems are geometrically approximated as a cone with distinct radius and depth emanating from the location of the tree stem. Soil depth was estimated with a simplified sine-cosine relation derived from generic soil depth prediction model of Pelletier and Rasmussen (2009). Geotechnical parameters as well as surface vegetation input (e.g. canopy coverage, interception capacity, etc.) were considered as constant for the whole modelled area. All scenarios were applied in a small study area in Vorarlberg, Austria.

The results show that the decision on how roots are spatially represented within a physically-based slope stability model (e.g. land cover based vs. cone-shaped approach) affects the modeling outcomes considerably. Scenario 2 and 3 give similar reactions of soil moisture fluctuations and slope failure probability on precipitation input—particularly in not vegetated areas. However, there are significant differences in forested areas recognizable. The model output of scenario 3 shows a much stronger reaction than scenario 2 and appears to align to pf-values of scenario 1 (no vegetation). It is assumed that scenario 3 primarily represents the stabilizing effect of root systems rather in the topsoil layer and thus shows similar patterns like scenario 2. In contrast, deeper soil layers show patterns of no vegetation coverage or a highly reduced vegetation impact respectively. This might be explained by the smaller volume that represents the rooted zone and thus a decreased root fraction for deeper layers of a respective soil column. First visual comparison (as shown in Fig. 8) suggests that, scenario 3 might reproduce slope failures more accurately that actually occurred during the modelled period. Hence, we expect that geometrical root system approximation is able to represent the hydrological and mechanical properties of roots more reliable in physically based models.

However, a drawback of our approach is the lack of quantitative evidence for its reliability due to the absence of measured data for validation. Moreover, tree locations and thus tree stand densities were estimated randomly for the forested areas, which produces a high uncertainty in the accuracy of root fraction calculation. In this context, it is expected that the envisaged inclusion of information based on highly resolved LiDAR data is able to decrease this uncertainty. The implementation of these airborne laser scanning data (ALS) could provide detailed information about vegetation on the surface from which subsurface biomass (e.g. roots) could be estimated more precisely. Moreover, the results outputs of this study open opportunities for better root system approximations (e.g. rotation ellipsoids) or the implementation of hydrological effects (e.g. water uptake capacity).

Acknowledgements We would like to thank the Federal State of Vorarlberg for providing remote sensing data and the Digital Elevation Model. Moreover, we thank our project partners of the BioSLIDE-project, Di Wang, Markus Hollaus and Norbert Pfeifer as well as the Torrent and Avalanche Control in Austria for supporting this study with valuable information and their expertise.

References

Allen RG, Pereira LS, Raes D, Smith M (1998) Crop evapotranspiration—guidelines for computing crop water requirements—FAO irrigation and drainage paper 56. FAO—Food and Agriculture Organization of the United Nations. Rome

Danjon F, Barker DH, Drexhage M, Stokes A (2002) Using three-dimensional plant root architecture in models of shallow-slope stability. Ann Bot 101(8):1281–1293

Fan C-C, Chen Y-W (2008) Role of roots in the shear strength of root-reinforced soils with high moisture content. Ecol Eng 33 (2):157–166

Fan C-C, Chen Y-W (2010) The effect of root architecture on the resistance of root-permeated soils. Ecol Eng 36(6):813–826

Friebe G (2013) Steine und Landschaft—Zur Geologie der Jagdberggemeinden. In: Naturmonographie Jagdberggebeinden. Dornbirn: inatura Erlebnis Naturschau:41–52

Genet M, Kokutse N, Stokes A, Fourcaud T, Cai X, Ji J, Michovski SB (2008) Root reinforcement in plantations of cryptomeria japonica D. Don. effect of tree age and stand structure on slope stability. For Ecol Manage. 256(8):1517–1526

Ghestem M, Sidle RC, Stokes A (2011) The influence of plant root systems on subsurface flow: implications for slope stability. Bioscience 61(11):869–879

Greenway DR (1987) Vegetation and slope stability. In: Anderson MG, Richards KS (eds) slope stability: geotechnical engineering and geomorphology. Wiley and Sons, New York, pp 187–230

Heimsath AM, Dietrich WE, Nishiizumi K, Finkel RC (1999) Cosmogenic nuclides, topography, and the spatial variation of soil depth. Geomorphology 27(1):151–172

Heimsath AM, Dietrich WE, Nishiizumi K, Finkel RC (2001) Stochastic processes of soil production and transport: erosion rates, topographic variation and cosmogenic nuclides in the Oregon Coast range. Earth Surf Process Land 26(5):531–552

Imaizumi F, Sidle RC, Kamei R (2008) Effects of forest harvesting on the occurrence of landslides and debris flows in steep terrain of central Japan. Earth Surf Process Land 33(6):827–840

Meng W, Bogaard T, Van Beek LPH (2014) How the stabilizing effect of vegetation on a slope changes over time: a review. Landslide Sci Safer Geoenvironment 1:363–372

Montgomery DR, Schmidt KM, Greenberg HM, Dietrich WE (2000) Forest clearing and regional landsliding. Geology 28(4):311–314

Papathoma-Köhle M, Glade T (2013) The role of vegetation cover change for landslide hazard and risk. In: Renaud G, Sudmeier-Rieux K, Estrella M (eds) The role of ecosystems in disaster risk reduction. UNU-Press, Tokyo, pp 293–320

Pelletier JD, Rasmussen C (2009) Geomorphologically based predictive mapping of soil thickness in pland watersheds. Water Res Res 45(9):1–15

Pollen N, Simon A (2005) Estimating the mechanical effects of riparian vegetation on stream bank stability using a fiber bundle model. Water Resour Res 41(7):1–11

Roering JJ (2008) How well can hillslope evolution models "explain" topography. Geol Soc Am Bull 120(9–10):1248–1262

Ruff M, Czurda K (2008) Landslide susceptibility analysis with a heuristic approach in the Eastern Alps (Vorarlberg, Austria). Geomorphology 94(3):314–324

Schmaltz E, Steger S, Bell R, Glade T, Van Beek LPH, Bogaard T, Wang D, Hollaus M, Pfeifer N (2016a) Exploring possibilities of including detailed ALS derived biomass information into physically-based slope stability models at regional scale. In: Aversa et al. (eds) Landslides and engineered slopes. Experience, theory and practice,pp 1807–1815

Schmaltz E, Steger S, Bell R, Glade T, Van Beek LPH, Bogaard T, Wang D, Hollaus M, Pfeifer N. (2016b) Evaluation of shallow landslides in the Northern Walgau (Austria) using morphometric analysis techniques. Procedia Earth and Planet Sci. Article in press

Schmid I, Kazda M (2002) Root distribution of Norway spruce in monospecific and mixed stands on different soils. For Ecol Manage 159(1–2):37–47

Schwarz M, Cohen D, Or D (2012) Spatial characterization of root reinforcement at stand scale. Theory and case study Geomorphol 171:190–200

Schwarz M, Lehmann P, Or D (2010) Quantifying lateral root reinforcement in steep slopes—from a bundle of roots to tree stands. Earth Surf Process Land 35(3):354–367

Sidle RC (1991) A conceptual model of changes in root cohesion in response to vegetation management. J Environ Qual 20(1):43–52

Sidle RC, Ochiai H (2006) Landslides: processes, prediction, and land use. Water resources monograph 18, American Geophysical Union, Washington D C

Sidle RC, Pearce AJ, O'Loughlin CL (1985) Hillslope stability and land use. Water Resources monograph 11, American Geophysical Union, Washington D C

Stokes A, Douglas GB, Fourcaud T, Giadrossich F, Gillies C, Hubble T (2014) Ecological mitigation of hillslope instability. Ten key issues facing researchers and practitioners. Plant Soil 377(1–2):1–23

Stokes A, Norris JE, Van Beek LPH, Bogaard T, Cammeraat E, Michovski SB (2008) How vegetation reinforces soil on slopes. In: Norris JE et al. (eds) Slope stability and erosion control: ecotechnological solutions. Springer, pp 65–118

Thomas RE, Pollen-Bankhead N (2010) Modelling root-reinforcement with a fibre-bundle model and Monte Carlo simulation. Ecol Eng 36 (1):47–61

Tilch N (2014) Identifizierung Gravitativer Massenbewegungen mittels multitemporaler Luftbild-auswertung in Vorarlberg und angrenzender Gebiete. Jahrbuch der Geologischen Bundesanstalt 154(1–4):21–39

Van Beek LPH, Wint J, Cammeraat L, Edwards JP (2007) Observation and simulation of root reinforcement on abandoned Mediterranean slopes. In: Stokes A et al. (eds) Eco- and ground bio-engineering: the use of vegetation to improve slope stability. Springer, pp 91–109

Ziemer, RR (1981) The role of vegetation in the stability of forested slopes. In: Proceedings of the International Union of Forestry Research Organizations, XVII World Congress, vol 1. pp 297–308

4th WLF 2017
4th World Landslide Forum
LJUBLJANA SLOVENIA EU

Landslide Deformation Prediction by Numerical Simulation in the Three Gorges, China

Kun Song, Fawu Wang, Yiliang Liu, and Haifeng Huang

Abstract

A few ancient landslides have been reported to be reactivated by the water level fluctuation in the Three Gorges Reservoir, China. On-site monitoring by Global Positioning System (GPS) is one of the most effective and accessible methods to understand the landslide deformation characteristic, which is the dominant factor for evaluating and predicting landslide stability. However, the monitoring points are usually limited and dispersed in landslides. A 3-D numerical simulation model was employed to study the spatial deformation of a landslide in the Three Gorges area, which was established according to its geological condition, and verified by comparisons of actual monitoring information and simulation recording data during a filling-drawdown cycle of the reservoir. The water level fluctuation has caused the landslide deformation, especially during the drawdown. The strong deformation zone exists in the forepart of the landslide, which lies mainly between 140.0 m and 200.0 m (EL.). The reservoir water variation controls the deformation of forepart near the river of the landslide. If the shear strength of slip surface reduces to $c = 22.1$ kPa, $\Phi = 8.7°$, which is resulted by cyclic water filling-drawdown, the failure would occur and almost whole part of the landslide would be in plastic state.

Keywords

Reservoir landslide • Spatial distribution of deformation • Numerical simulation • Prediction

K. Song (✉) · Y. Liu
Hubei Key Laboratory of Disaster Prevention and Reduction, China Three Gorges University, Yichang, 443002, China
e-mail: songkun_ctgu@163.com

Y. Liu
e-mail: lylctgu@126.com

K. Song · F. Wang
Department of Geoscience, Shimane University, Matsue, 690-8504, Japan
e-mail: wangfw@riko.shimane-u.ac.jp

H. Huang
Collaborative Innovation Center for Geo-Hazards and Eco-Environment in Three Gorges Area, Yichang, 443002, China
e-mail: hifn_huang@qq.com

M. Mikoš et al. (eds.), *Advancing Culture of Living with Landslides*,
DOI 10.1007/978-3-319-53483-1_36

Introduction

Landslides are common and prone geo-disasters in reservoir area. Many landslides occurred after initial impoundment of the reservoir, such as the Vajont landslide in 1963 in Italy (Müller et al. 1964; Alonso et al. 2010; Paronuzzi et al. 2013), Qianjiangping landslide in 2003 in the Three Gorges reservoir, China (Wang et al. 2008). Furthermore, a number of ancient landslides have been reactivated and exhibited large deformation due to the reservoir water level fluctuation in the routine operation of the hydraulic projects to satisfy the aim of flood control, power generation, et al. In the Three Gorges reservoir area, such slope movement triggered by water level fluctuation has been observed in large ancient landslides, such as Shuping landslide (Wang et al. 2008), Baishuihe landslide (Lu et al. 2014), Liangshuijing landslide (Wang et al. 2013), and Muyubao landslide. Therefore, there is an urgent need to study deformation characteristics, predict displacement trend and failure mode of landslides under water fluctuation condition, thereby evaluate, predict and prevent the reservoir landslides.

The water level fluctuation has been the focus in previous studies to investigate factors affecting the stability of reservoir landslides. The conventional rigid equilibrium limit method, numerical and physical modeling, and monitoring have been used in reservoir landslides stability and deformation investigation (Müller 1964; Franco and Claudia 2003; Alonso and Pinyol 2010; Song et al. 2011; Yan et al. 2012; Paronuzzi et al. 2013; Zaniboni and Tintin 2014; Tang et al. 2015; Strauhal et al. 2016; Dumperth et al. 2016). Deformation characteristic is the key factor for evaluating and predicting landslide stability, and it can be analyzed with available on-site monitoring data (Wang et al. 2008; Tang et al. 2015). However, the monitoring points are usually limited and dispersed, and they are not sufficient to represent the whole landslide deformation and predict the deformation tendency or even failure mode.

In this paper, in-site monitoring data and 3-D numerical simulation was employed to investigate the spatial deformation characteristic and predict the critical deformation of a reservoir landslide under water level fluctuation in the Three Gorges reservoir, China.

Geological Background of the Landslide

A landslide is located on the right bank of the Yangtze River, in Wushan Town, Chongqing. The landslide is an ancient slide, with a shape likes a tongue. The width is about 350 m and the length is 310 m. Boreholes indicated that the average thickness of landslide is about 30 m with the volume of about 3.25 million m^3. The slide body is composed of silty clay with gravel and underlain by argillaceous sandstone of the Triassic Badong formation (T_{2b}) with the dip direction of 285° and dip of 50°. The potential slip surface is between soil and bedrock. The geological profile is shown in Fig. 1.

Landslide Surface Deformation Monitoring

To monitor the movement of the landslide, the Chinese Ministry of Land and Resource installed six Global Positioning System (GPS) stations (GPS–GPS6 in Fig. 2) along three longitudinal axes, and surveyed monthly from January 2007. Two referenced GPS stations, namely GPS7 and GPS8, were installed outside of the landslide boundary.

Figures 3 and 4 shows the movement data from 2007 to 2009, and more displacements were observed in the front of the slide (GPS4) than in the upper/rear part (GPS3) (Fig. 3).

The displacements of the front showed gradual decrease from left to right, i.e. GPS 2 > GPS 4 > GPS 6 (Fig. 4). The movements increased with the water level fluctuation, especially the drawdown. It indicates that the deformation

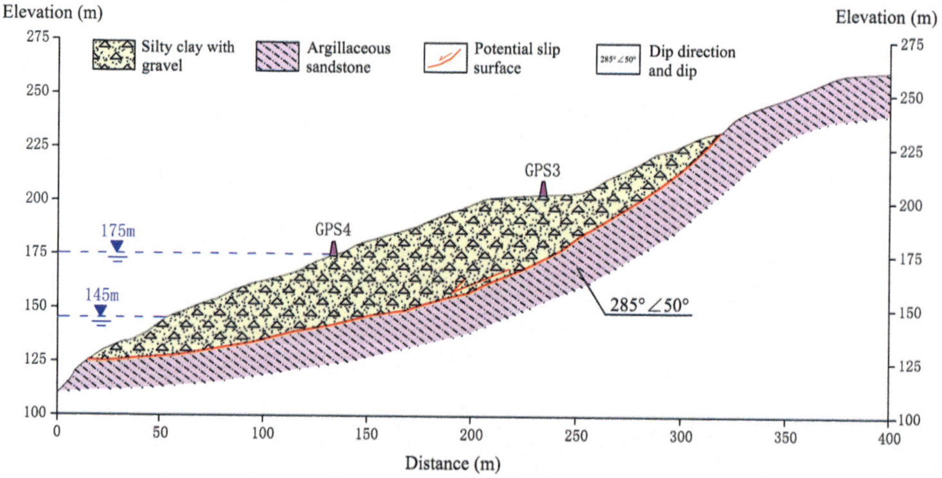

Fig. 1 Engineering geological profile of the landslide

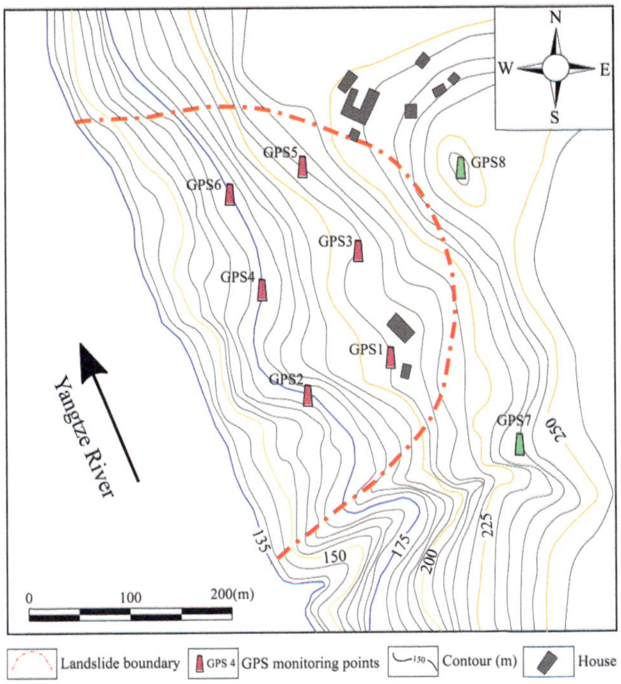

Fig. 2 The landslide plan and surface deformation monitoring system (GPS)

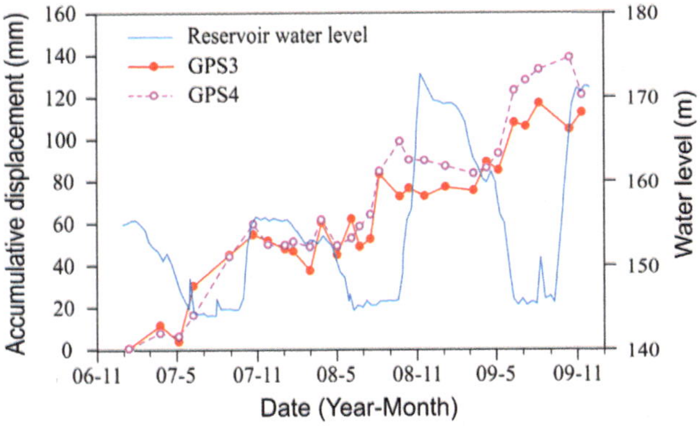

Fig. 3 Accumulative horizontal displacements of points in the middle longitudinal axis and reservoir water level versus time (measured by China Geological Survey)

mode was the moving of forepart resulting adjacent deformation of back part, as the former pulled the latter.

Landslide 3-D Numerical Simulation

Numerical Simulation Model and Parameters

A 3-D numerical simulation model (Fig. 5) of the landslide was established by ANSYS software. The model was meshed by tetrahedral grids, and the element sizes of slide body and bed rock were different to meet the accuracy

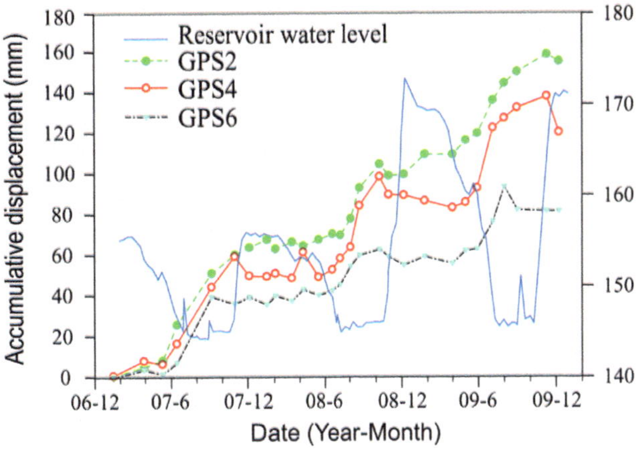

Fig. 4 Accumulative horizontal displacements of points in the front of abscissa axis and reservoir water level versus time (measured by China Geological Survey)

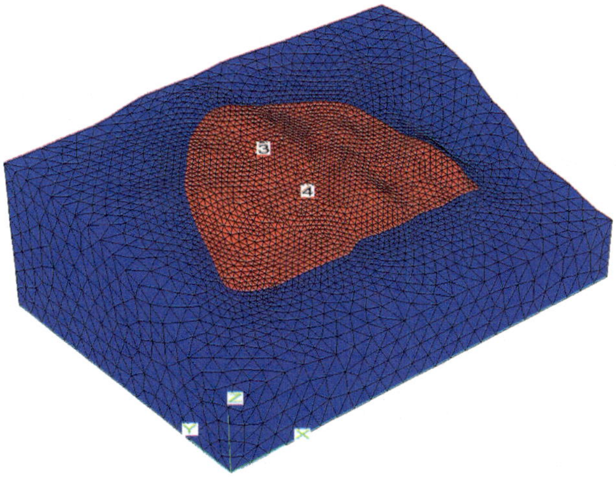

Fig. 5 Landslide 3-D numerical simulation model

requirements. There were 15,961 nodes and 83,050 elements in total in the model.

The hydro-mechanical numerical simulation was conducted by FLAC^{3D} software. The water flow in the landslide is described by Darcy's law, and the pore-water pressure was calculated by transient saturated seepage analysis with reservoir water level variation. The soil or rock of the landslide is elastoplastic and isotropic material, which follow the Mohr-Coulomb strength criterion.

The hydro-mechanical properties are obtained from some previous work by Center of Hydrogeology and Environmental Geology, China Geological Survey, and from literatures for similar materials, which are shown in Table 1. The mechanical properties are bulk unity weight γ, Young's modulus E, Poisson's ratio μ, drained cohesion c', and internal friction angle φ'. The hydrologic parameter is saturated hydraulic coefficient Ks.

Table 1 The physical and mechanical parameters of soil and rock for simulation

Materials	Slide body	Slip surface	Bed rock
γ (kN/m^3)	23.8[a]	–	27.6[a]
c' (kPa)	24.0[a]	26.5[a]	190.0[b]
φ' (°)	18.0[a]	10.4[a]	30.4[b]
E (MPa)	60[a]	–	2000[b]
μ	0.25[a]	–	0.20[b]
Ks (cm/s)	1.8×10^{-3} [a]	–	1.5×10^{-7} [b]

[a]Provided in the survey report by Center of Hydrogeology and Environmental Geology, China Geological Survey
[b]Values of similar materials from literatures

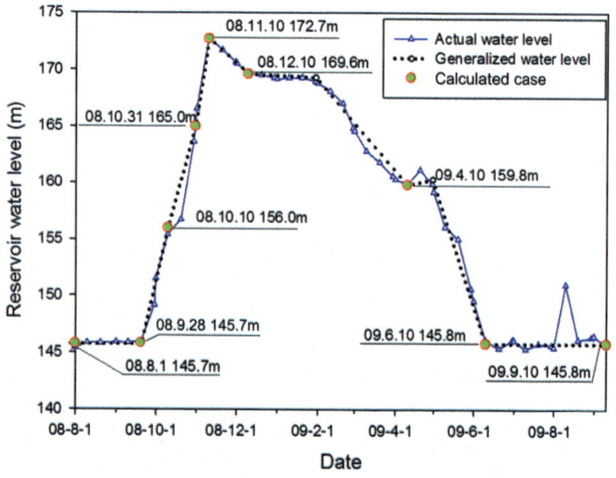

Fig. 6 Reservoir water level fluctuated curve

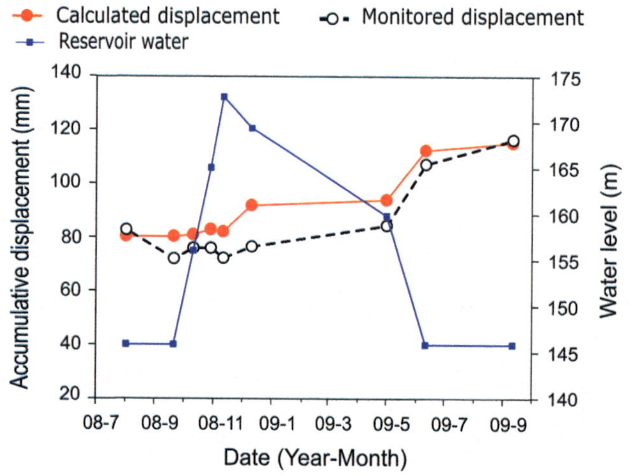

Fig. 7 Comparison of calculated and monitored displacements of point GPS 3 with water level fluctuation

Hydraulic Boundary Condition

Hydraulic boundary condition is basis for seepage simulation in the previous numerical studies of landslide deformation with reservoir water level fluctuation. A filling-drawdon cycle from August 1, 2008 to September 10, 2009 (406 days) was selected as the transient hydraulic boundary condition based on the actual reservoir water level fluctuation data of the Three Gorges Project. The water fluctuation curve was smoothened and simplified as the dot line in Fig. 6 to avoid complex calculation. In the simulation progress, 9 crucial results of simulated cases were saved, which are marked in Fig. 6.

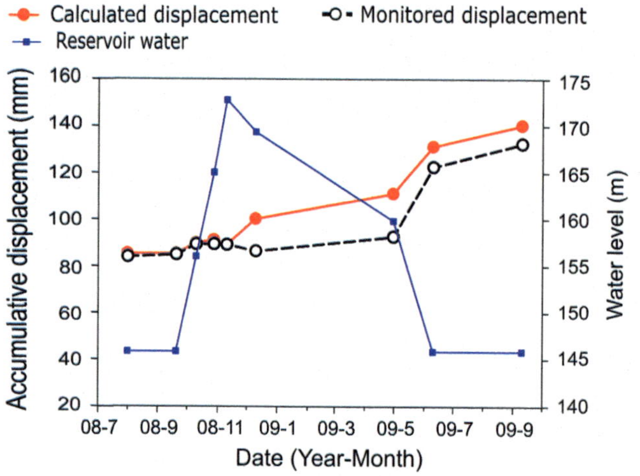

Fig. 8 Comparison of calculated and monitored displacements of point GPS 4 with water level fluctuation

Numerical Model Verification

In the simulation of landslide deformation, displacements of nodes at the same locations as the surface GPS stations were recorded. To verify the numerical simulation, the displacements of monitored and numerical simulation were compared using the generalized water level fluctuation over time in Figs. 7 and 8.

The simulated trends and values of accumulative displacement presented fairly good agreements with GPS monitoring data at points GPS3 and GPS4. It proved that the numerical model was suitable for the landslide deformation

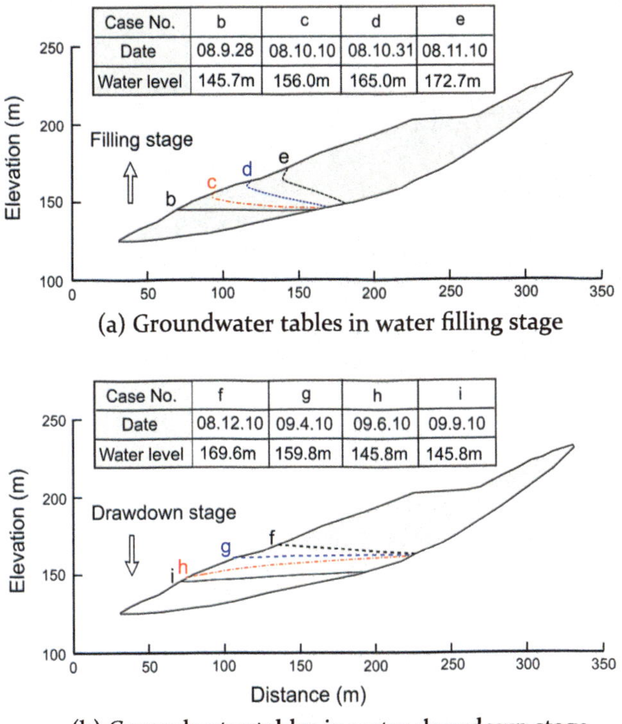
Case No. | b | c | d | e
Date | 08.9.28 | 08.10.10 | 08.10.31 | 08.11.10
Water level | 145.7m | 156.0m | 165.0m | 172.7m
Filling stage
Elevation (m)
(a) Groundwater tables in water filling stage
Case No. | f | g | h | i
Date | 08.12.10 | 09.4.10 | 09.6.10 | 09.9.10
Water level | 169.6m | 159.8m | 145.8m | 145.8m
Drawdown stage
Elevation (m)
Distance (m)
(b) Groundwater tables in water drawdown stage

Fig. 9 Groundwater tables of different cases with reservoir water fluctuation

simulation with reservoir water level fluctuation, and it could be employed for simulation and prediction.

Numerical Simulation Results and Discussion

After the verification, the deformation simulation of the landslide was investigated by the numerical model under the reservoir water level fluctuation condition. Some of the seepage results and mechanical simulation results are shown in Figs. 9 and 10 during the progress.

Some of the seepage results and mechanical simulation results are shown in Figs. 9 and 10 during the progress. When the water level rised to 172.7 m on November 10, 2010 from 145.7 m (August 8, 2008), the displacement contour was presented in Fig. 10a. Then, the displacement contour changed to Fig. 10b after the water level drawdown to 145.8 m on September 10, 2009, which represents a complete filling-drawdown cycle.

In the reservoir filling-drawdown cycle, the strong deformation zone exists in the forepart of the landslide, which lies mainly between 140.0 m and 200.0 m (EL.), and the displacement of the forepart is larger than the back one. It means that the reservoir water variation controls dominantly forepart deformation near the river of the landslide.

Landslide Deformation Prediction

To understand the conditions that may trigger the failure of the landslide and the associated failure regions in the future, the landslide deformation prediction was conducted using the numerical model. As the landslide showed obvious increase of displacement in water drawdown condition, the dominant trigger is the drawdown of reservoir water. Rapid drawdown would cause catastrophic failure of the landslide, but it is a uncommon and rare condition for man-made reservoir expect some catastrophic events. However, the shear strength of the soil would decrease after cyclic water filling-drawdown, and thus, it could result landslide failure.

The shear strength reduction technique was employed in the simulation. It was used as early as 1975 by Zienkiewicz et al. (1975), and has later been widely applied by Dawson et al. (1999), Cheng et al. (2007), Fu and Liao (2010), Song et al. (2012) and others. The displacement of point GPS 4 was recorded in the process, and the result of accumulative displacement and strength reduction factor is presented in Fig. 11. As can be observed, the displacement increases with the strength reduction factor increasing, and there is a inflection of the accumulative displacement curve. It indicates a failure of the slope (Zienkiewicz et al. 1975; Dawson et al. 1999; Cheng et al. 2007; Song et al. 2012). The matching strength reduction factor is 1.20, thus, the shear strength of the soil is c = 22.1 kPa, $\Phi = 8.7°$ in saturated condition. Therefore, when the shear strength of the slip zone soil reduces to c = 22.1 kPa, $\Phi = 8.7°$, the failure would occur and almost whole part of the landslide would be in plastic state (Fig. 12).

Fig. 10 Landslide displacement contour with reservoir water fluctuation

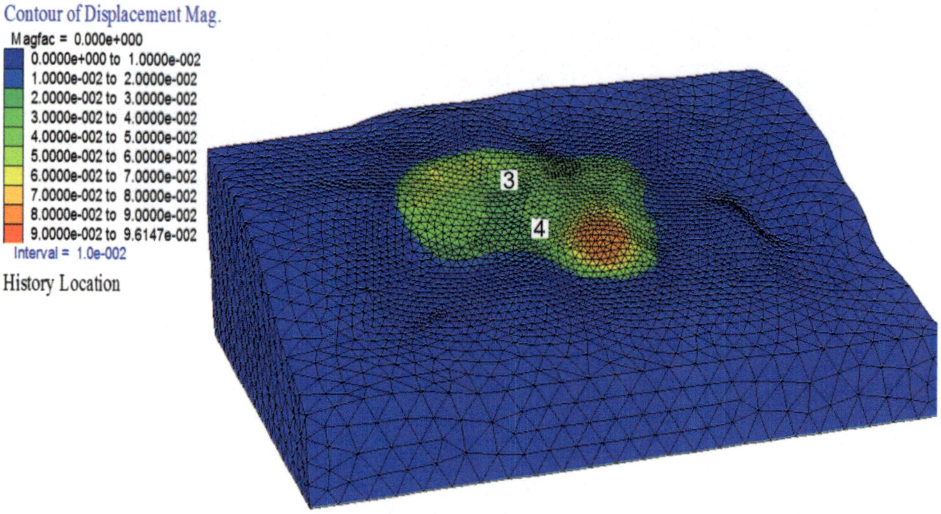

(a) Landslide displacement contour with the reservoir water level rise to 172.7m

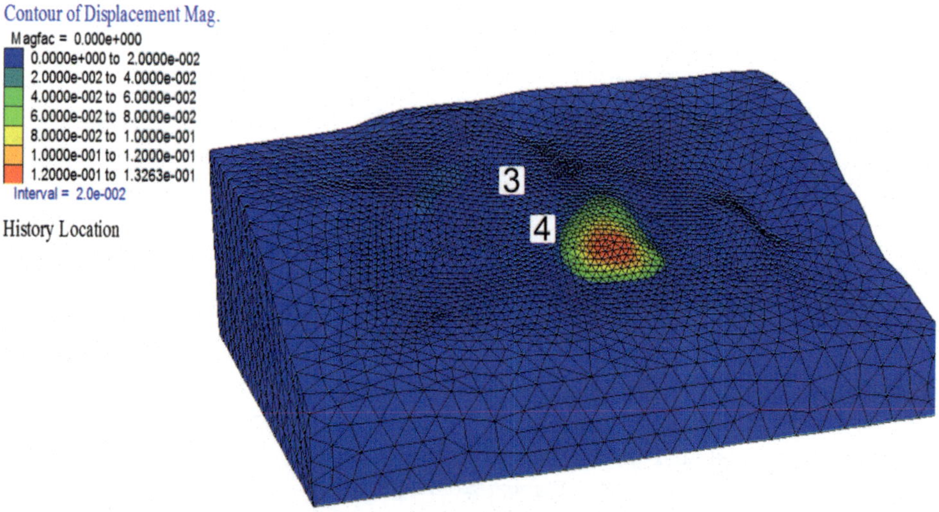

(b) Landslide displacement contour with the reservoir water level drawdown to 145.80m

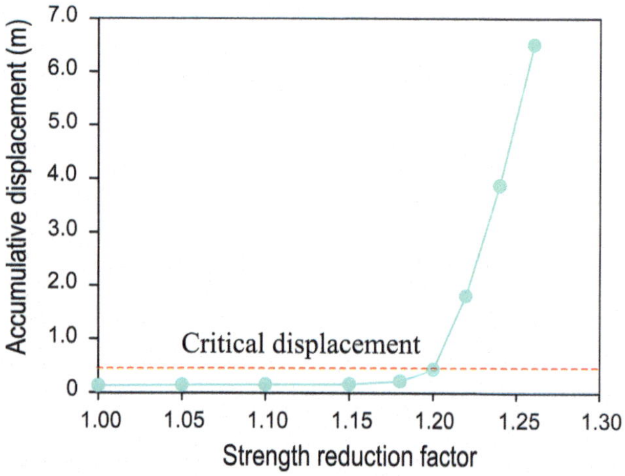

Fig. 11 Comparison of calculated accumulative displacement with strength reduction factor

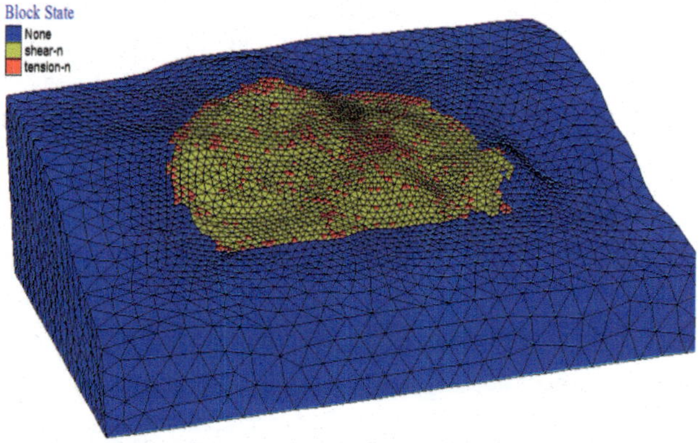

Fig. 12 Landslide failure zone after shear strength reduction

Conclusions

Water level fluctuation is the dominant factor affecting the reservoir landslide deformation and stability. Some limited and dispersed surface GPS points were employed to monitor the deformation of an ancient landslide in the Three Gorges Reservoir, China. The water level fluctuation can induce large landslide deformation, especially during the drawdown, and the forepart has shown the most increase of deformation.

A 3-D numerical simulation model of the landslide was established based on the geological condition, and verified by comparisons of actual monitoring information and simulation recording data of the landslide during a filling-drawdown cycle of the Three Gorges reservoir. The spatial distribution of landslide deformation characteristic was simulated by the verified numerical model with reservoir water level fluctuation. The strong deformation zone exists in the forepart of the landslide, which lies mainly between 140.0 m and 200.0 m (EL.). The reservoir water variation controls dominantly forepart deformation near the river of the landslide.

The landslide displacement increase or failure may be triggered by decrease in shear strength of the soil resulted from cyclic water filling-drawdown. The shear strength reduction technique was used in the numerical simulation for landslide deformation prediction. When the strength reduction factor reach to 1.20, the shear strength of slip zone soil is $c = 22.1$ kPa, $\varphi = 8.7°$, and the failure would occur as almost whole part of the landslide would be in plastic state.

Acknowledgements This study was supported by National Natural Science Foundation of Hubei Province (No. 2015CFB358) and National Natural Science Foundation of China (No. 41372359). The authors are indebted to Prof. Echuan Yan from China University of Geosciences for his valuable suggestion.

References

Alonso EE, Pinyol NM (2010) Criteria for rapid sliding I. A review of Vaiont case. Eng Geol 114(3):198–210

Cheng YM, Lansivaara T, Wei WB (2007) Two-dimensional slope stability analysis by limit equilibrium and strength reduction methods. Comput Geotech 34(3):137–150

Dawson EM, Roth WH, Drescher A (1999) Slope stability analysis by strength reduction. Geotechnique 49(6):835–840

Dumperth C, Rohn J, Fleer A, Xiang W (2016) Local-scale assessment of the displacement pattern of a densely populated landslide, utilizing finite element software and terrestrial radar interferometry: a case study on Huangtupo landslide (P.R. China). Environ Earth Sci 75(10):1–9

Franco M, Claudio V (2003) Neotectonics of the Vajont dam site. Geomorphology 54(1):33–37

Fu W, Liao Y (2010) Non-linear shear strength reduction technique in slope stability calculation. Comput Geotech 37(3):288–298

Lu SQ, Yi QL, Yi W, Zhang GD, He X (2014) Study on dynamic deformation mechanism of landslide in drawdown of reservoir water level-Take Baishuihe landslide in the Three Gorges area for example. J Eng Geol 22(5):869–875

Müller L (1964) The rock slide in the Vaiont valley. Rock Mech Eng Geol 2(1):10–16

Paronuzzi P, Rigo E, Bolla A (2013) Influence of filling–drawdown cycles of the Vajont reservoir on Mt. Toc slope stability. Geomorphol 191:75–93

Song K, Yan EC, Zhu DP, Zhao QY (2011) Base on permeability of landslide and reservoir water change to research variational regularity of landslide stability. Rock and Soil Mech 32(9):2798–2802

Song K, Yan EC, Mao W, Zhang TT (2012) Determination of shear strength reduction factor for generalized Hoek-Brown criterion. Chin J Rock Mechan Eng 31(1):106–112

Song K, Yan EC, Zhang GD, Lu SQ, Yi QL (2015) Effect of hydraulic properties of soil and fluctuation velocity of reservoir water on landslide stability. Environ Earth Sci 74(6):5319–5329

Strauhal T, Loew S, Holzmann M, Zangerl C (2016) Detailed hydrogeological analysis of a deep-seated rockslide at the Gepatsch reservoir (Klasgarten, Austria). Hydrogeol J 24(2):349–371

Tang HM, Li CD, Hu XL, Wang LQ, Criss R, Su AJ, Wu YP, Xiong CR (2015) Deformation response of the Huangtupo landslide to rainfall and the changing levels of the Three Gorges Reservoir. Bull Eng Geol Environ 74(3):933–942

Wang FW, Zhang YM, Huo ZT, Peng XM, Wang SM, Yamasaki S (2008a) Mechanism for the rapid motion of the Qianjiangping landslide during reactivation by the first impoundment of the Three Gorges dam Reservoir, China. Landslides 5(4):379–386

Wang FW, Zhang YM, Huo ZT, Peng XM, Araiba K, Wang GH (2008b) Movement of the Shuping landslide in the first four years after the initial impoundment of the Three Gorges Dam Reservoir, China. Landslides 5(3):321–329

Yan EC, Zhu DP, Song K, Lin YD (2012) Deformation prediction method of typical accumulative landslide in Three Gorges Reservoir based on numerical modeling. J Jilin Univ (Earth Sci) 42(2):422–429

Zaniboni F, Tinti S (2014) Numerical simulations of the 1963 Vajont landslide, Italy: application of 1D Lagrangian modeling. Nat Hazards 70(1):567–592

Zienkiewicz O, Humpheson C, Lewis RW (1975) Associated and non-associated visco-plasticity and plasticity in soil mechanics. Geotechnique 25(4):671–689

Patterns of Development of Abrasion-Landslide Processes on the North-West Coast of the Black Sea

Olena Dragomyretska, Galina Pedan, and Oleksandr Dragomyretskyy

Abstract

The development of abrasion-landslide processes on the north-west coast of the Black Sea depends on several factors: geological structure, slope exposure, neotectonic, climatic and hydrodynamic conditions. However, the nature of this relation is not always obvious and insufficiently studied. The goal is to identify patterns of spatial and temporal development of landslide processes in the north-west coast of the Black Sea. Methods of correlation-regression, and spectral analysis were used. Periods of activation of landslide processes depending on changing climatic and hydrodynamic conditions were determined. On the south-west from the Odessa bay, abrasion process is described by dependence model abrasion indicators from wind wave energy, width of a beach and a precipitation. On the east from the Odessa bay the most proved model is dependence between abrasion indicators, sea level and a precipitation. Beaches have a significant role in the rate of abrasion and landslide processes. Gentle slopes at some sections of the beaches are the result of the predominance of silt and clay components in abraded loess strata of coastal slopes. All identified patterns have significant correlation coefficients ($r > 0.5$). It was found that the landslide processes in relation to climate and hydrodynamic impacts have a delay of 1–2 years, due to the inertia of the coastal geological system.

Keywords

Abrasion • Landslide • Black sea coast • Correlation-regression methods

Introduction

The accumulated factual material on the study of exogenous geological processes on the shores of seas and oceans is summarized in Drannikov (1940), Aksentev (1960), Longinov (1963), Yemelyanova (1972), Voskoboynikov and Likhodedova (1984), Shnyukov et al. (1985), Voskoboynikov and Kozlova (1992). Several solutions to protect in the coastal zone of the north Black Sea region from the Engineering Geodynamics problems are discussed in Rozovskiy et al. (1987); Konikov and Pedan (2003). It has been established that coastal zones dynamics depends on the planetary and cosmic factors (Cherkez 1996; Konikov and Likhodedova 2010). The role of structural and tectonic factors in the development of coastal zone is formulated by Shmuratko (2001). Spatial and temporal modeling of the dynamics of landslide processes and the development of prediction methods is one of the areas of research in the coastal zone (Zelinskiy et al. 1993; Pedan 2001; Dragomyretskyy et al. 2015). Evaluation of slope protection measures and improvement of coast stabilization methods are considered as an urgent task (Cherkez et al. 2006, 2008, 2013; Pedan 2006; Dragomyretska et al. 2013). The dynamics of the beaches on the shores of

O. Dragomyretska (✉)
State Institution "Hydroacoustic Branch of Institute of Geophysics by S.I. Subbotin Name of National Academy of Sciences of Ukraine", 3 Preobrazenska Str, 65082 Odessa, Ukraine
e-mail: alena_dr@mail.ru

G. Pedan · O. Dragomyretskyy
Department of Engineering Geology and Hydrogeology, Odessa I. I. Mechnikov National University, 2 Dvorianskaya Str, 65082 Odessa, Ukraine

M. Mikoš et al. (eds.), *Advancing Culture of Living with Landslides*,
DOI 10.1007/978-3-319-53483-1_37

the seas and oceans and their role in protecting coasts from the destruction have been studied (King 1972; Bascom 1980).

The total length of the north-west Black Sea shoreline (between the delta of Danube river and Crimean peninsula) is approximately 1000 km, 30% of which are represented by actively eroded cliffs, 40%—by retreating accumulative forms and 30%—with stable shores. There formes abrasion-debris, abrasion-landslide and accumulative types of coast. Relative distribution of coast types is not uniform. The accumulative coasts are spread in the southwest. They are represented by typical coastal bars (baymouth barrier) generated during eustatic raise of the sea level during Late Holocene. Their modern development takes place in conditions of sand and shell deficiency. They are retreating accumulative forms.

The typical abrasion-landslide coast is located in the north of the studied territory between the Suhoy and the Berezan limans. Here, 62% of coasts are affected by abrasion and landsliding. Abrasion coastal ledges represent the raised sites of post-Pontic platau. Their height varies from several to 40 m. The typical abrasion-debris coasts are located in the east of the studied territory. Activity of the abrasion-landslide processes on the shores of the Black Sea in the last decades increased due to rising sea level, neotectonic subsidence, as well as human activities.

The development of abrasion-landslide processes depends on several factors: geology, slope exposure, neotectonic, climatic and hydrodynamic conditions. However, the nature of this relationship is not always obvious and has not been studied sufficiently. The aim of our work is to identify patterns of spatial and temporal development of abrasion-landslide processes in the north-west coast of the Black Sea.

Materials and Methods

Origin of the abrasion-landslide processes and conditions of their development is logically-random, so the study was carried out on an integrated methodology. Patterns of the abrasion and landslide processes development both in time and in space were studied using probabilistic and statistical methods utilizing the "STATISTICA" software (Borovikov and Borovikov 1997). Correlation and regression analysis was used, which allows to identify the close connection between the variables and factors to build a regression model

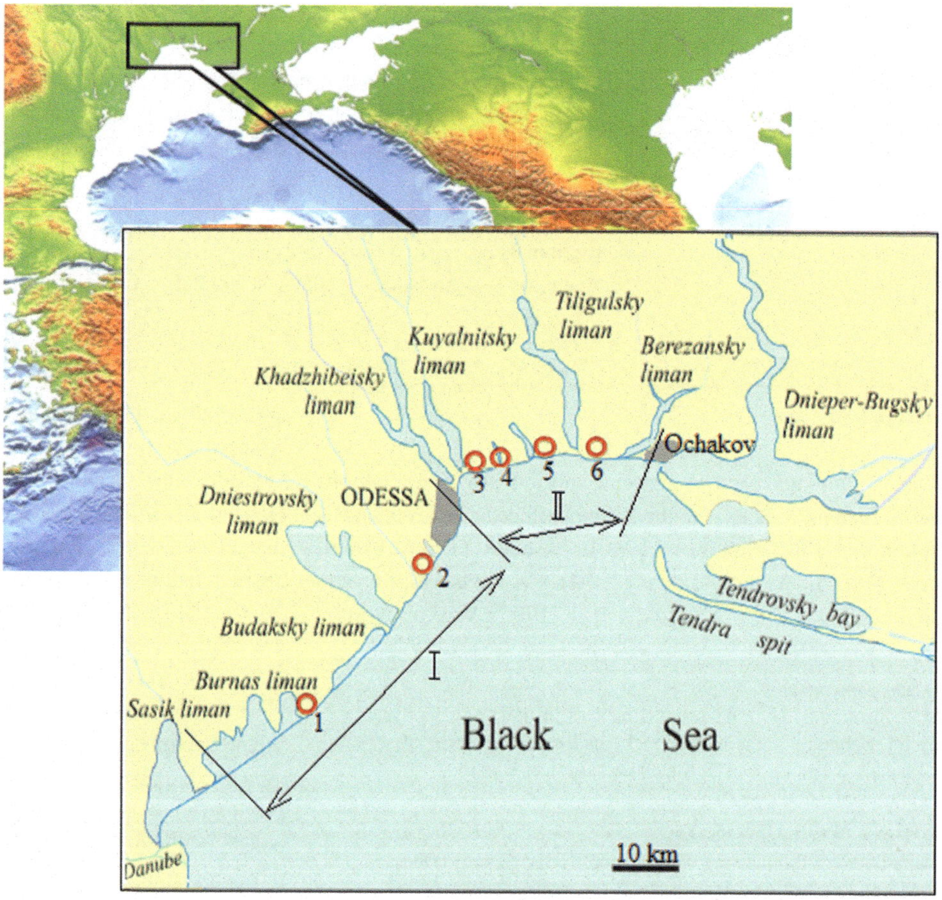

Fig. 1 Map of the north-west Black Sea. ⟵—boundaries of subareas I and II; ○—sites of long-term field of observation; 1—Lebedevka, 2—Sanzheyka, 3—Dofinovka, 4—Grigorievka, 5—Sychavka, 6—Rybakovka

that would adequately reflect the relationship between them (Krumbein and Graybill 1965; Smirnov 1975; Borovikov and Borovikov 1997). To identify patterns of changes in time spectral analysis was used, which allows to identify a periodic component time series, as well as to carry out a procedure for forecasting (Kyunttsel 1978; Sheko 1984).

The following procedures were performed for each time series: (1) determination of the distribution law; (2) check of the consistency of empirical data with a normal law; (3) compilation of histograms, as estimates of the density distribution; (4) reveal of the estimated parameters (mean maximum and minimum values, dispersion, etc.). Long-term field observations of changes in the coastline at 6 experimental sites (Fig. 1) from the Danube Delta to the town of Ochakov during 1976–2000 were analyzed (data of the Odessa National University, "Prichernomor SRGE" geological organization, Odessa). The observational data after 2000 are absent due to termination of the research financing.

The following parameters were used as an activity indicators of the abrasion-landslide processes: linear retreat edge (L_1, m/year), and the slope of the sole (L_2, m/year); the volume of seaworn rocks (Q_1, m^3/linear meters per year), the number of new landslides during the year, the width (S, m), height (H, m) of the beach and the amount of sediment on the beach (Q_2, m^3/lm), vertical and horizontal displacement of soil on the slope (m/year).

Results and Discussion

The Geological Structure and Lithology

Within the studied area the dominated coasts are composed of Quaternary loess mainly, which lie at the base of the red-brown clay (the Danube River—the river Dniester) of the Pliocene-Quaternary age and carbonate rocks of the Miocene (Odessa coast) (Fig. 2) (Zelinskiy 1993). Four main types of rocks and soils are developed in the landslide slopes: Quaternary loess, Pontic limestone, Meotian clay and landslide accumulation of the previous three types. They have various strength and deformation properties. Gradual immersion under the sea level of more and more young deposits takes place to the south-west and the north-east.

Low mechanical strength of these materials facilitates high rate of coastal erosion. The average rate of the coast retreat in different areas is ranging from 0.01 to 3.0 m/year (Fig. 3).

The minimum rate of abrasion is available on the coastal slopes, where the limestone-shell rock is located at the base of cliff. The greatest abrasion occurs on the coastal slopes composed of loess-like loam.

Thus, the geological conditions on the banks of the study area are favorable for the active development of abrasion-landslide processes. Over the past decade the

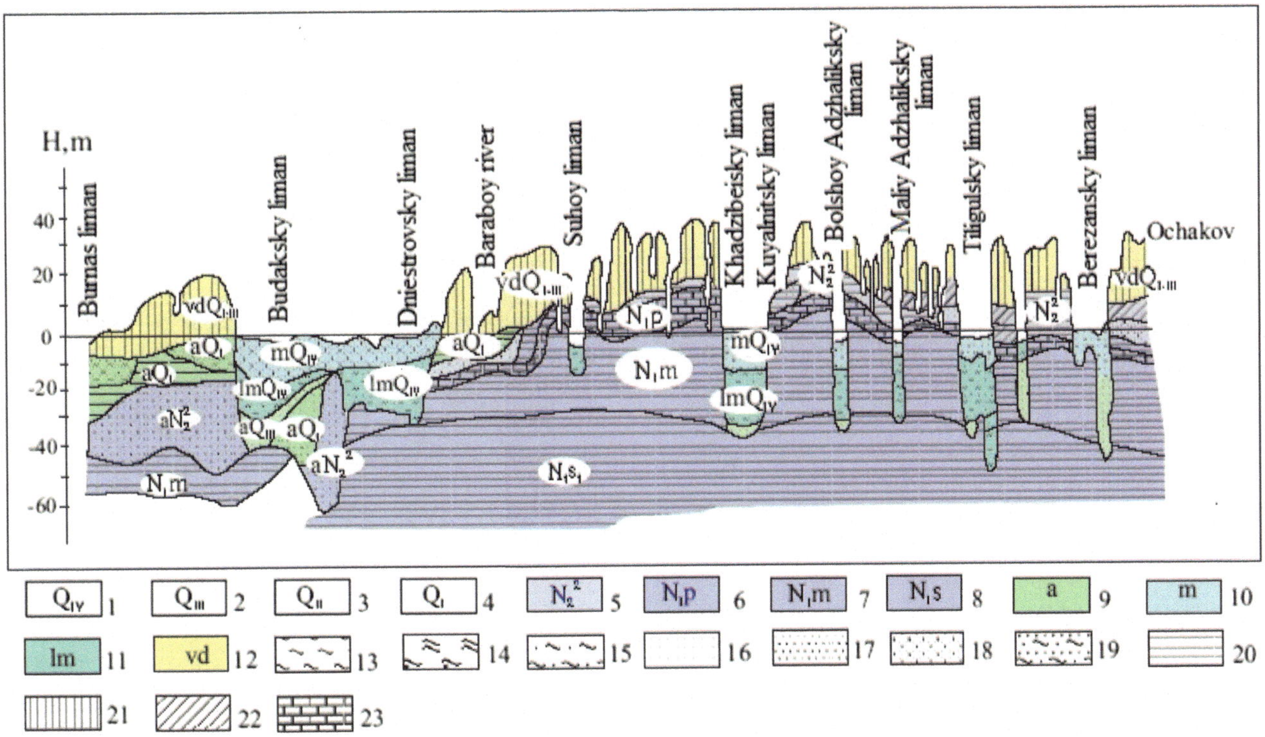

Fig. 2 Geological cross-section of the north-west Black Sea coasts. Stratigraphy: 1—Holocene; 2–4—Upper, Middle and lower Pleistocene; 5–8—Upper and lower Pliocene; 6—Pont, 7—Meotian, 8—Sarmatian. Genetic type of sediments: 9—alluvial, 10—marine, 11—liman, 12—eolian-deluvial. Lithology: 13–15—loamy, clayey and sand silts; 16–17—fine-grained and coarse-grained sands; 18—shells; 19—muddy sand; 20—clay; 21—loess; 22—loam; 23—limestone

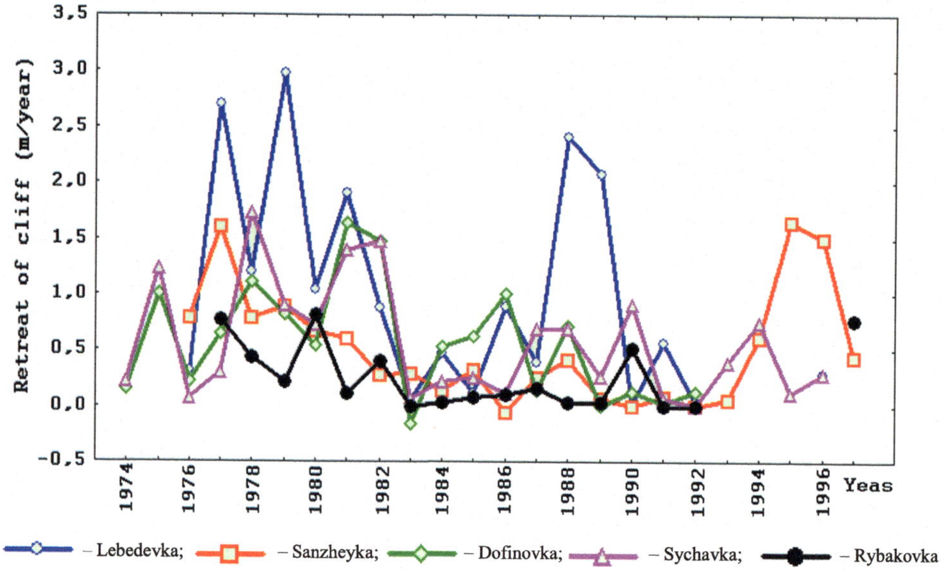

Fig. 3 Cliff dynamics in the experimental sites

shoreline has moved for 100 m towards the land on an average.

Atmospheric Precipitation

Atmospheric precipitation conditions play a significant role in the development of abrasion-landslide processes in the coastal zone, raising the level of groundwater and reducing the stability of the upper part of the slope. Groundwater level regime is characterized by periodic oscillations associated with the frequency of atmospheric precipitation (Yemelyanova 1972). Landslide activity increases sharply during strong rainfall.

The average annual rainfall of 430 mm, and this amount reduces from the south-west to north-east from 437 mm to 376 mm, respectively. The coast is divided into two subareas according to this indicator: 1—the Danube River—Odessa, 2—Odessa—Ochakov. The difference between the minimum and maximum of average annual atmospheric precipitation is about 400 mm, according to the observations during 1867–2010.

Spectral analysis of the time series of atmospheric precipitation revealed a periodic component and the most significant periods: 2–4; 5–6 and 9–12 years. Analysis of the periodicity of atmospheric precipitation allows to predict with some reliability the stages of rising groundwater levels in coastal slopes and activation of landslide processes, based on rather close relationship between these parameters (Fig. 4).

Sea-Level and Wave Regime

The sea level is a significant factor in the consideration of the abrasion of the Black Sea (Longinov 1963). The average speed of sea level rise for the north-western coast of the Black Sea is about 3 mm/yr. The difference between the maximum and minimum values of the average annual sea level is more than 1 m. The main periods of sea-level fluctuations are: 2.4; 3.3–3.6; 4–5; 9–12 years. The maximum average sea level values were observed in 1979, 1980, 1988, 1989, 1997, 1998.

Significant wave height is the most important hydrodynamic factor that affects the mode of abrasion-landslide processes. Wave energy depends directly on the wind conditions in the region. The coast west of Odessa (subarea 1) is bordering the open sea and therefore is under the influence of higher waves than the coast east of Odessa (subarea 2). Increased storm activity of the sea is reflected in the amplitude of sea level fluctuations, which leads to increased abrasion-landslide processes rates. Due to the fact that the waves less than 1.25 m high have little destructive effect or have not any effect on the cliff, in the analysis of spatial and temporal variability we used data on the significant wave height (SWH) exceeding 1.25 m (Fig. 5).

Morphometric Indices of the Beach

Beaches is the only natural protection of the coast from degradation due to the fact that the coastal slops of the

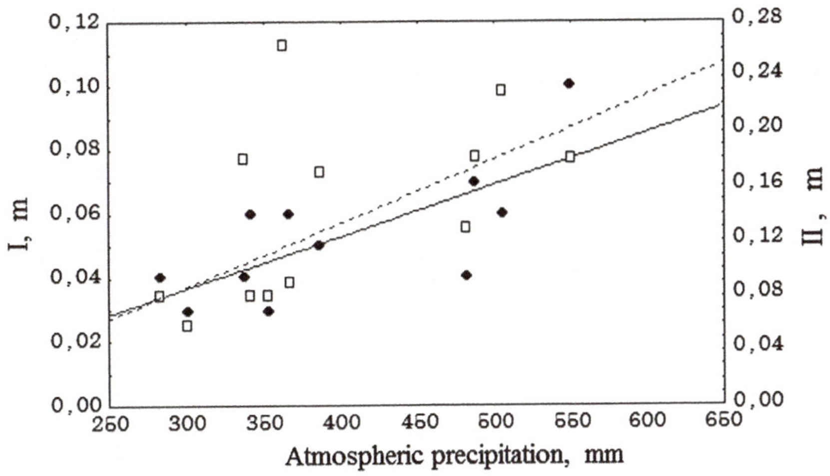

Fig. 4 Relation between atmospheric precipitation and displacements of parts of the landslide slope on the Sychavka site

Displacements of the upper part of the slope – **I** (●); the lower part– **II** (□). Regression line of the upper part ——; the lower part - - - - Correlation coefficient r=0.65; statistical significance p= 0.001

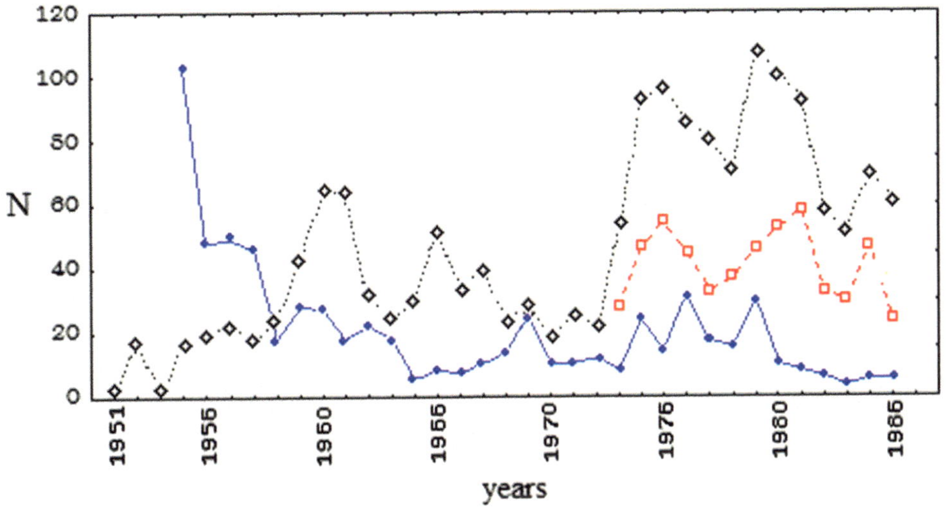

Fig. 5 Rezime of significant wave height (SWH)

N –number of case of SWH on the site: —♦— Lebedevka; - -□- - Sanzheyka; ···◇··· Odessa

north-western part of the Black Sea are composed mainly of loose soils that are easily eroded. Depending on the width of the beach, it fully or partially suppresses waves' energy. Analysis of the data shows that the width of the beaches on this coast is usually insignificant—up to 15–20 m, decreasing to 2–4 m in some areas (Fig. 6).

Changing of the beaches is cyclical with periodicity of 2.0–2.4; 5.5; 7.3 and 9 years. Wind and waves energy has big impact on the width of the beaches as well shore exposure. The study showed that erosion of the slope sharply decreases and becomes minimal when the beach width reaches 20 m, while the volume of deposits on the beach must be equal to more than 30 m^3/m

Features of the Abrasion-Landslide Processes Development

The north-western coast of the Black Sea can be divided into two subareas. They are characterized by:—seacoast forming factors;—regime of abrasion-landslide processes. The subarea 1 is the coast of the Black Sea from the Danube delta to

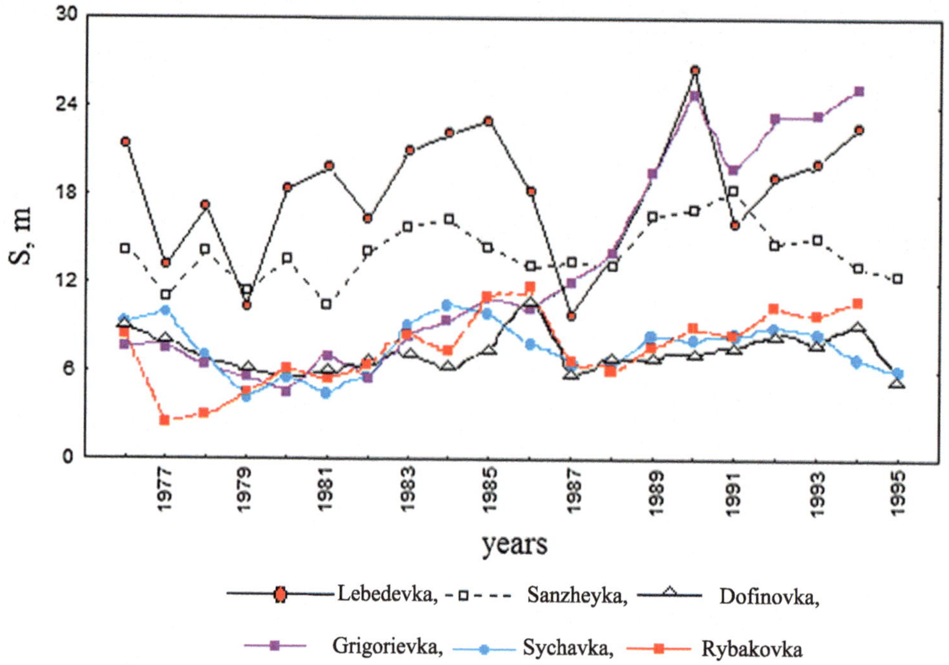

Fig. 6 Changing of the width of the beach (S) on sites

Odessa and the subarea 2—from Odessa to Ochakov (see Fig. 1). The abrasion and landslide processes have multidirectional nature, determined by rapidly changing factors, the relationship between them is shown in Fig. 7 and in Table 1. It should be noted that the reaction of coastal systems to the impact of factors (precipitation, wind wave energy) occurs with 1 year delay. This is taken into consideration in Fig. 7 and Table 1.

Multivariate regression models of coast abrasion have been created for two studied subareas on the basis of statistical processing of long-term observations. The average annual volume of seaworn soils of the abrasion slope Q_1 (t) was chosen as an indicator of the abrasion process. Model is considered accurate in a particular subarea at sufficiently high coincidence of actual and calculated values of the parameter Q_1 (t).

So, the abrasion-landslide process is most closely described by the model that demonstrates dependence between the rate of abrasion indicator and width of a beach, an atmospheric precipitation and wind wave energy on the subarea1 (site Sanzeyka). Model of abrasion-landslide process is shown on the Sanzheyka site as an example (subarea 1):

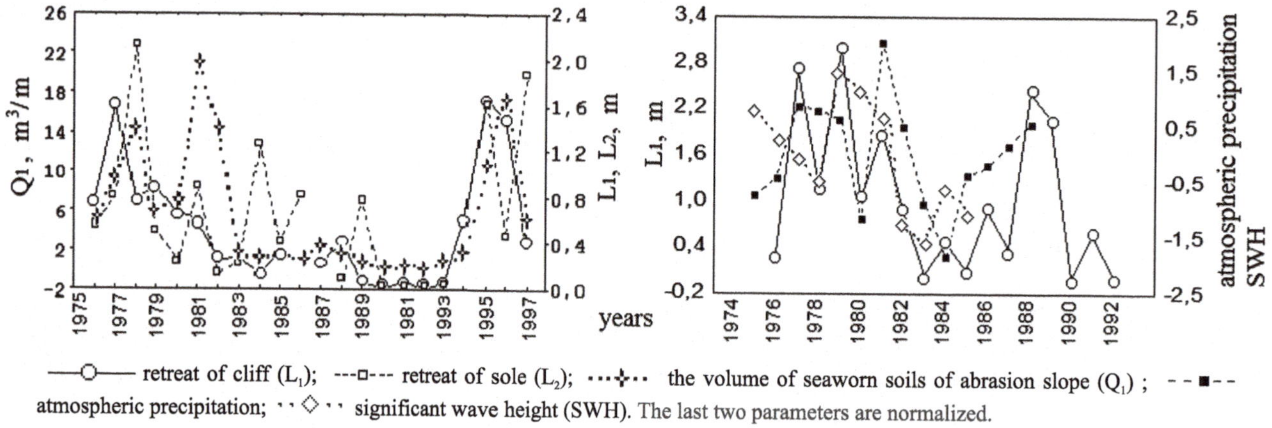

Fig. 7 The total variation of the average annual values of abrasive parameters in the area Sanzheyka (subarea 1)

Table 1 Results of the correlation analysis

Factors	L_1	Q_1
The sum of an atmospheric precipitation for a year, mm. Hydrometeorological station "Ilyichevsk" (1976–1997 years)	$R = 0.60$ $p = 0.015$ $n = 20$	$R = 0.51$ $p = 0.02$ $n = 20$
The sum of an atmospheric precipitation for a year, mm. Hydrometeorological station "Ilyichevsk" (1976–1997 years)		$R = 0.71$ $p = 0.000$ $n = 20$
Wind wave energy		$R = 0.57$ $p = 0.04$ $n = 13$
Width of a beach (S, m)	$R = -0.65$ $p = 0.001$ $n = 22$	$R = -0.62$ $p = 0.002$ $n = 22$

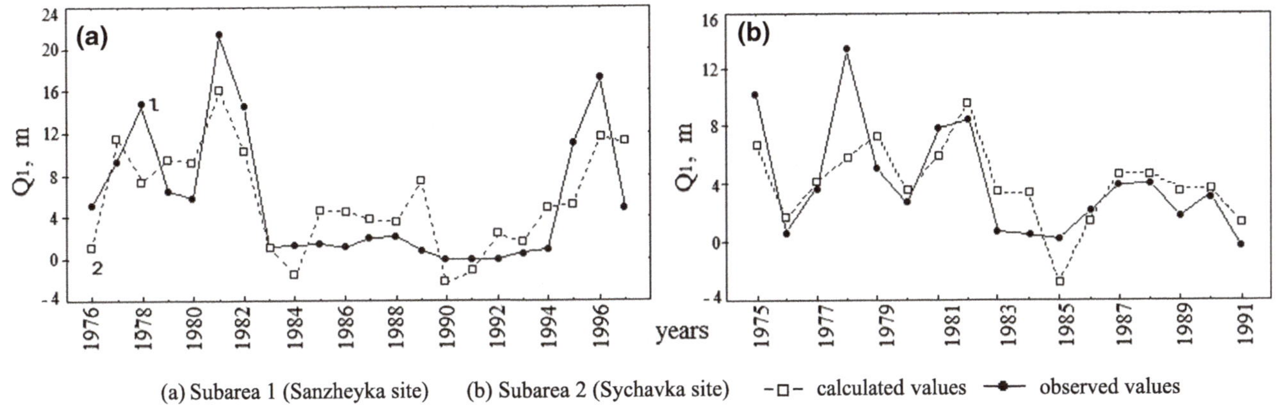

Fig. 8 The correlation between the observed and calculated values of the indicator of the abrasion process (Q_1)

$$Q_1(t) = -1.81 - 1.07\,X_1(t) + 0.0303\,X_2(t-1) + 0.63\,X_3(t-1)$$

$Q_1(t)$.—volume of seaworn rocks (m^3 per 1 linear meter a year); $X_1(t)$.—width of a beach (m); $X_2(t-1)$.—the sum of an atmospheric precipitation for a year (mm). Data are displaced on 1 previous year; $X_3(t-1)$—wind wave energy (dimensionless size). Data are displaced on 1 previous year.

There is a 1–2 year lag between abrasion-landslide indicators and such factors as atmospheric precipitation and wave energy. This results from the inertia of the coastal geological system. The best model for the development of the coast subarea 2 is one:

$$Q_1(t) = -6.99 + 0.591\,X_1(t) + 0.0158\,X_2(t-2) + 0.78\,X_3(t)$$

in which abrasion $Q_1(t)$. depends on the width of the beach $X_1(t)$.; precipitation shifted by 2 years $X_2(t-2)$ and fluctuations of the sea level $X_3(t)$. Sychavka site (subarea 2) is taken as an example.

All discovered patterns have significant correlation coefficients ($r > 0.5$). The correlation between the observed and calculated values is shown on the Fig. 8.

Statistical analysis showed periodicity of activation of abrasion-landslide process 2.3; 3.6; 5.6 and 9 years old, which corresponds to the periods of climatic and hydrological factors.

Conclusions

North-west coast of the Black Sea is a complex geodynamic system and is characterized by a highly dynamic coastline. Abrasion and landslides are the most common of all the exogenous geological process, which both spatial and temporally development are subject to certain regularities, although chance factor is also manifested. Conditions of development and activation of abrasion-landslide processes are the result of the impact of factors on the coastal system, the most significant of which are the geological structure and hydrogeological conditions, sea-level mode, wave conditions, and atmospheric precipitations.

The role of different factors at various sites of coast is not uniform. West from the Odessa bay abrasion process is described by model based on abrasion indicators dependence on wind and wave energy, width of a beach and an atmospheric precipitation. To the east from the Odessa bay, the most proved is model based on dependence between abrasion indicators and an atmospheric precipitation and a sea level.

Beaches have a significant role in the rate of abrasion and landslide processes. The amount of sediment on the beach and its width are determined by the biological characteristics of eroded rocks and sea wave activity. It was found that when the width of the beach is more than 20 m and a volume of sediments on the beach is more than 30 m^3/m abrasion process slows down and becomes insignificant. Low morphometric parameters of some sections of the beaches are the result of the predominance of silt and clay components in abrasion loess strata of coastal slopes.

All identified patterns have significant correlation coefficients ($r > 0.5$). It was found that the abrasion-landslide processes in relation to climate impacts have a delay of 1–2 years, due to the inertia of the coastal geological system. Periods of activity abrasion and landslide processes coincide and identified as equal to 9–12, 6–5 and 2–3 years on the basis of spectral analysis. Patterns of variation of abrasion-landslide indicators (obtained by spectral analysis) confirm the cyclical nature of these processes. Correlation of abrasion and landslide indicators with the periodicity of climatic and hydrological characteristics allows to predict the landslide process. Of course, such a forecast illustrates only the tendency rather reliable, since the obvious impossibility of exact prediction of random fluctuations in the relevant parameters of landslide processes.

References

Aksentev GN (I960) Results of supervision over abrasion activity of the Black Sea at coast of Odessa. In: Proceedings of the Odessa State University, geology and geography, materials on studying of the Odessa landslides, vol 150, No. 7. pp 131–136 (In Russian)

Bascom W (1980) Waves and beashes. Doubleday and Company Inc., New York, 366p

Borovikov VP, Borovikov IP (1997) STATISTICA. Statistical analysis and data processing in a windows environment. Information and Publishing House "Filin". Moscow, 608p. (In Russian)

Cherkez EA (1996) Geological and structural-tectonic factors of landslides formation and development of the north-western Black Sea coast. In: Proceeding of 7 international symposium on landslides, Trondheim. Balkema, Rotterdam, 17–21 June 1996, pp 509–513

Cherkez EA, Dragomyretska OV, Gorokhovich Y (2006) Landslide protection of the historical heritage in Odessa (Ukraine). Landslides 3(4):303–309. doi:10.1007/s10346-006-0058-8

Cherkez EA, Kozlova TV, Shmouratko VI (2008) Geological engineering characteristics of the Primorsky boulevard area in Odessa during construction of the Potyomkin stairs (based on the research of the 1840s historical data). Ecology Environment and protection of life. Kiev. №2: 10–23. (In Russian)

Cherkez EA, Kozlova TV, Shmouratko VI (2013) Engineering geodynamics of landslide slopes of the Odessa sea coast after anti-landslide measures. Bull Odessa Nat Univ, Geogr Geol Sci 18 (1):15–25 (In Russian)

Dragomyretska O, Skipa M, Dragomyretskyy O (2013) Assessment of actual morphodynamic activity of landslide slopes in Odessa. Landslide science and practice. 6. In: Margottini C et al. (eds) Springer-Verlag, Berlin, Heidelberg,pp 318–322. doi:10.1007/978-3-642-31319-6_42

Dragomyretskyy O, Dragomyretska O, Skipa M (2015) Search and assessment of decompression zones in landslide slopes of the north-west coast of the Black Sea (Ukraine). Engineering Geology for Society and Territory. 5. Lollino G. et al. (eds.). Springer International Publishing, Switzerland, pp 811–814. doi:10.1007/978-3-319-09048-1_157

Drannikov AM (1940) General scheme of anti-landslide measures of the coast of Odessa. Odessa. 190p. (In Russian)

King CA (1972) Beaches and coasts. Arnold Publ.Ltd., London. 570p

Konikov EG, Pedan GS (2003) The study, simulation and forecast dynamics of abrasion cliff and abrasion-landslide coasts on the basis method of generalization variables. Bull Odessa Nat Univ, Geogr Geol Sci 8(5):141–149 (In Russian)

Konikov EG, Likhodedova OG (2010) Global and regional factors of level fluctuations of the Black Sea during the last two centuries and forecast its variations as basis for geodynamic model for the coastal zone. Geology and mineral resources of the Word ocean, № 1:84–93. ISSN: 1999-7566. (In Russian)

Krumbein W, Graybill F (1965) An introduction to statistical models in geology. McGraw-Hill, New York, p 396

Kyunttsel VV (1978) The landslide rhythm and some of its manifestations in the USSR. Proc VSEGINGEO 119:28–33 (In Russian)

Loginov VV (1963) Dynamics of the coastal zone of tide-free sea. Publishing House of the USSR Academy of Sciences, Moscow. 379p. (In Russian)

Pedan GS (2001) Study of the coast zone development by area stationery-analogs. Bull Odessa Nat Univ, Geogr Geol Sci 6 (9):144–150 (In Ukrainian)

Pedan GS (2006) Evaluation of effectiveness of anti-landslide measures of the Odessa coast. Ecology environment and protection of life. Kiev. № 2:28–35. (In Russian)

Rozovskiy LB, Zelinskiy IP, Voskoboynikov VM (1987) Engineering-geological forecasts and modeling. Vishcha shkola. Kiev, Odessa. 208p. (In Russian)

Sheko AI (1984) Methods of long-term regional forecasts of exogenous geological processes. VSEGINGEO, Nedra, Moscow. 167p. (In Russian)

Shmouratko VI (2001) Gravitational-resonans exotectogenesis. Astroprint, Odessa. 332p. ISBN: 966-549-576-3. (In Russian)

Shnyukov Ye F, Melnik VI, Inozemtsev Yu I et al. (1985) Geology of shelf of the USSR. Lithology. Naukova dumka, Kiev. 189p. (In Russian)

Smirnov BI (1975) Statistical methods in geology. Vishcha School, Lviv. 122p. (In Russian)

Voskoboynikov VM, Likhodedova OG (1984) Studying and predicting of geological processes on the basis of a method of the generalized variables (by the example of reservoir bank transformation). Eng Geol. №. 1:23–36. (In Russian)

Voskoboynikov VM, Kozlova TV (1992) Use of the geodynamic analysis and method of the generalized variables for estimating and predicting the stability of landslide slopes (by the example of the Northern Black Sea region). Eng Geol. №. 6:34–49. (In Russian)

Yemelyanova EP (1972) Basic patterns of landslide processes. Nedra, Moscow. 310p. (In Russian)

Zelinskiy IP, Korzenevskiy BA, Cherkez EA et al. (1993) Landslides of north-western coast of the Black sea, their study and Prognosis. Naukova dumka, Kiev. 228p. (In Russian)

Part IV

Landslides as Environmental Change Proxies: Looking at the Past

Session Introduction—Landslides as Environmental Change Proxies

Tomáš Pánek

Slope stability reacts sensitively to changes of external factors including tectonic strain, temperature and precipitation regime, permafrost degradation, sea-level oscillations and land-use patterns. Majority of these factors are recently related to global warming and increased human pressure. Despite that outcomes from recent studies supported by numerical modeling suggest indisputable link between global environmental change and changing of landslide frequency and magnitude, there is still a need for empirical data to verify these assumptions. One of the ways to bring new light to the landslide occurrence in the context of environmental changes is looking to the past and evaluation how known paleoenvironmental changes influenced landslide occurrences. Recent progress in dating techniques, monitoring and GIS technologies has brought several new challenges for the determination of the landslide temporal and spatial dynamics.

In this section category, eight papers dealing with several types of topics are presented.

Papers of Hermanns et al. and Schleier et al. concern rockslides in Norway and their temporal response to the retreat of the Scandinavian ice sheet. Results of radiometric dating and numerical modeling suggest short-time response of rock slope failures to glacier withdrawal and different dynamics of rock avalanches which originated before and after the glacier withdrawal.

S. Moreiras also discuss slope instabilities in deglaciated landscape, but in the different time scale. On the example of the Aconcagua region (Central Andes), pronounced mass movements activity within the large moraine complexes suggests permafrost degradation related to recent climate warming.

Although within different regional context, two papers from this section are looking for the link between large paleolandslides and temporal occurrence of strong earthquakes. Based on the cosmogenic radionuclide dating and historical sources, Ivy-Ochs et al. connect voluminous rock avalanche in the Trentino valley (Italy) with medieval earthquake event. Similarly, Nepop and Agatova correlate debris flow deposits with Holocene paleoearthquakes in the seismically active area of the Russian Altai.

Remaining papers present landslides as a result of environmental changes within last several decades. Two papers from the southern Italy (Pisano et al. and Gariano et al.) bring interesting insight how climate and land-use changes affect landslide occurrences and use such data for the prediction of future landslide origins. Study of Ružić et al. deals with sea-wave related increasing slope instability along the abrasion coast of the Krk Island (Croatia) with implications for the coastal hazard assessment.

Although differing in time-scale, character of landscapes and types of studied mass movements, papers in this section underline the importance of historical experience in the study of landslides. Historical analogues could be used for the calibration of numerical models and for the predictions of future landslide scenarios. In concert with increasing datasets of high-resolution paleoenvironmental proxies, dating of landslides will be more and more valuable within the process of hazard and risk assessment.

T. Pánek (✉)
University of Ostrava, Czech RepublicOstrava,
e-mail: tomas.panek@osu.cz

M. Mikoš et al. (eds.), *Advancing Culture of Living with Landslides*,
DOI 10.1007/978-3-319-53483-1_38

Rock-Avalanche Activity in W and S Norway Peaks After the Retreat of the Scandinavian Ice Sheet

Reginald L. Hermanns, Markus Schleier, Martina Böhme, Lars Harald Blikra, John Gosse, Susan Ivy-Ochs, and Paula Hilger

Abstract

We have compiled recently published and unpublished cosmogenic ^{10}Be exposure ages of rock-avalanche deposits and break away scars in western and southern Norway in order to compare those to the retreat of the Scandinavian ice sheet. In total 22 rock-avalanche events were dated by their deposits (19) or break away scars (3). Sampling of rock-avalanche deposits and failure surfaces was not systematic over the region but with few exceptions we sampled all deposits within the same valley. All ages were recently calculated using the CRONUS online calculator and the geochronology ensemble reveal five late Pleistocene events, eight Preboreal events, and nine younger events. The decay of the Scandinavian ice sheet was not spatially synchronous but differed regionally and lasted over several thousand years in places, hence the requirement for widespread dating targets. One rock avalanche (at Innerdalen at 14.1 ka) occurred when ice existed in the valley, which is in agreement with the latest deglacial models. Depositional characteristics of ten (44%) of the rock avalanches suggest ice free conditions although they occurred within the first millennia

R.L. Hermanns (✉) · M. Böhme · P. Hilger
Geological Survey of Norway, Trondheim, Norway
e-mail: reginald.hermanns@ngu.no

M. Böhme
e-mail: martina.bohme@ngu.no

P. Hilger
e-mail: paula.hilger@ngu.no

R.L. Hermanns
Department of Geoscience and Petroleum, Norwegian University of Science and Technology, Trondheim, Norway

M. Schleier
GeoZentrum Nordbayern, University of Erlangen-Nuremberg Erlangen, Erlangen, Germany
e-mail: reginald.hermanns@ngu.no

L.H. Blikra
Norwegian Water and Energy Directorate, Trondheim, Norway
e-mail: lab@nve.no

J. Gosse
Department of Earth Sciences, Dalhousie University, Halifax, Canada
e-mail: John.Gosse@Dal.Ca

S. Ivy-Ochs
ETH Zurich, Institute for Particle Physics, Zurich, Switzerland
e-mail: ivy@phys.ethz.ch

M. Mikoš et al. (eds.), *Advancing Culture of Living with Landslides*,
DOI 10.1007/978-3-319-53483-1_39

following local deglaciation. Five events (22%) occurred between 9 and 7.5 ka at a time when climate was warmer and moister than today. Finally seven events (30%) appear to be relatively evenly distributed throughout the rest of the Holocene. Although limited in number we interpret that the dated events are representative of the temporal distribution of post-ice sheet rock avalanches in western Norway. However, the number of rock avalanches occurring onto the decaying ice sheet is likely underrepresented as those deposits are reworked and can be difficult to distinguish from moraine deposits. Our widespread data reveal a rapid rock slope instability response to the initial local decay of the Scandinavian ice sheet followed by a lower and constant frequency following the climate optimum (ca. 8.5 ka) in the Holocene.

Keywords

Decay of the scandinavian ice sheet • Cosmogenic nuclide • Surface exposure dating • Rock-avalanche deposit • Late pleistocene • Holocene • Preboreal • Fast response

Introduction

Rock avalanches both contribute significantly to the erosion of high relief mountain settings and pose a large threat to societies and infrastructure in mountain environments (e.g. Korup et al. 2007). In particular, the post-glacial landscape evolution of glacially-steepened valleys is controlled by rock avalanches (e.g. Hewitt et al. 2011; Ballantyne et al. 2014) and it is in general agreed that rock slope failures are most frequent after deglaciation (e.g. Cruden and Hu 1993; Ballantyne et al. 2014). However, the deceleration of rock-avalanche activity since deglaciation is difficult to document as it requires a large number of dated events from many deposits which are buried (e.g. Cruden and Hu 1993; Antinao and Gosse 2009; Ballantyne et al. 2014). However the timing of prehistoric rock avalanches and the decay rate of rock avalanche activity after deglaciation are critical factors required to completely evaluate the hazard of rock slope failures that can lead to rock avalanches in the future (e.g. Hermanns et al. 2013).

In Norway, based on historical data, two to five rock avalanches occur per century (Blikra et al. 2006). However the temporal information on prehistoric rock avalanches is limited (Blikra et al. 2006). This is partly because rock avalanches often drop into a fjord or lake and deposits are not preserved onshore (e.g. Hermanns and Longva 2012). Recently a total of 108 rock slope failures, identified and mapped from bathymetric and seismic data around Storfjorden (western Norway), have been dated with a combination of ^{10}Be surface exposure dating, ^{14}C dating of organic material sampled form cores of fjord sediments, and seismic stratigraphy (Böhme et al. 2015). The Storfjorden dataset indicates a rapid response and high frequency of rock avalanching following deglaciation that is followed by a constant frequency since 9 ka.

In this contribution we compile the published data (Table 1) on ^{10}Be ages of rock-avalanche deposits (Hermanns et al. 2011; Böhme et al. 2013, 2015; Schleier et al. 2015, 2016) and report previously unpublished data on ten additional rock avalanche deposits (Table 2). Owing to the page limitations of this paper, and our focus on the temporal distribution of rock avalanches in western and southern Norway, we only provide the sample sites information of the previously unpublished rock avalanches. In the context of the other chronologies, the newly reported ages help provide a more robust relationship with the retreat history of the Scandinavian ice sheet given by Hughes et al. (2016). For an overview of rock avalanche deposits and sample points of previously published ages we refer to the respective publications.

Regional Setting

The landscape of Norway is characterised by an alpine relief with steep slopes, strongly glacially steepened U-shaped valleys reaching below sea level, creating an emergent coastline (10^5 km long) with strongly ramified fjords intruding into the continent for up to 200 km. The bedrock comprises mainly metamorphic rocks of Precambrian to Palaeozoic age that exhibit significant structural fabrics and anisotropies. The bedrock is highly tectonized owing to protracted intense ductile and brittle tectonics acting since Precambrian times over the entire region. In the Quaternary dozens of glaciations covered the landscape with kilometre thick ice caps and ice sheets which deepened the fault-controlled valleys and induced isostatic responses to loading and unloading. The relatively high abundance of tectonically and isostatically induced structures and the overall high steepness of the slopes are two important first

Table 1 Previously published rock-avalanche ages recalculated as given in Schleier et al. (2015)

No.	Rock avalanche	Type of landform	Deposit mean age (0 erosion) (ka ± 1σ ka)	Coefficient of variation (%) (ka)		References
1	Innerdalen old	Isolated boulder patches	14.1 ± 1.4	3.7	0.3	Schleier et al. (2015)
2	Innerdalen young	Rock-avalanche deposit	8.0 ± 0.8	11.8	0.4	Schleier et al. (2015)
3	Innfjordalen 1st	Rock-avalanche deposit	14.3 ± 1.4			Schleier et al. (2016)
4	Innfjordalen 2nd	Rock-avalanche deposit	8.8 ± 0.9			Schleier et al. (2016)
5	Innfjordalen 3rd	Rock-avalanche deposit	1.0 ± 0.4			Schleier et al. (2016)
6	Nokkenibba	Break-away scarp over deposit in fjord	7.1 ± 1.2	3.7	0.3	Böhme et al. (2015)
7	Nakkaneset	Break-away scarp over deposit in fjord	7.2 ± 1.1	4.6	0.3	Böhme et al. (2015)
8	Blåhornet	Break-away scarp (no deposit defined on fjord bottom because lying on much larger older rock avalanche deposit)	2.1 ± 0.3	26.9	0.7	Böhme et al. (2015)
9	Stampa young	Rock-avalanche deposit	3.8 ± 0.4	14.3	0.6	Böhme et al. (2013)
10	Bøydalen	Rock-avalanche deposit	3.3 ± 0.4	6.5	0.2	Hermanns et al. (2011)
11	Øyrahagstolen	Rock-avalanche deposit	9.9 ± 1.1	8.6	0.8	Hermanns et al. (2011)
12	Uraneset	Rock-avalanche deposit	10.5 ± 1.2	7.4	0.8	Hermanns et al. (2011)

Table 2 New rock-avalanche ages of unpublished results calculated as indicated in Schleier et al. (2015)

No.	Sample	Rock-avalanche deposit	Location lat long		Age (0 erosion) (ka ± 1σ ka)	Deposit mean age (ka ± 1σ ka)	Coefficient of variation (%) (ka)	
13	R110803-01	Gråura	62.5667	7.5192	14.9 ± 2.3	14.1 ± 1.9	8.6	1.2
	R110803-02		62.5655	7.5224	13.2 ± 1.4			
14	R100803-01	Skiri 1	62.4297	7.9497	11.7 ± 1.3	11.7 ± 1.3		
15	R100803-04	Skiri 2	62.4302	7.9264	11.4 ± 1.3	11.0 ± 1.3	5.8	0.6
	R100803-06		62.4302	7.9264	10.5 ± 1.2			
16	R160803-01	Svarttinden	62.4125	7.8540	10.0 ± 1.2	8.7 ± 1.1	14.3	1.3
	R160803-02		62.4141	7.8530	7.5 ± 0.8			
	R160803-03		62.4135	7.8680	8.7 ± 1.3			
17	R160803-04	Alstadfjellet	62.3289	7.4889	9.5 ± 0.9	9.3 ± 1.0	3.8	0.4
	R160803-07		62.3291	7.4894	9.0 ± 1.0			
18	R120803-06	Kallen	62.2208	7.4153	11.5 ± 2.1	11.4 ± 2.1	1.9	0.2
	R120803-09		62.2215	7.4167	11.2 ± 2.1			
	R170803-04	Kallen outburst flood	62.2333	7.4161	9.5 ± 0.9	9.5 ± 0.9		
19	R180803-04	Vora a	61.7428	6.5572	12.9 ± 1.9	12.9 ± 1.9		
20	R180803-02	Vora b	61.7408	6.5444	12.1 ± 1.3	12.1 ± 1.3		
21	R180803-07	Vora d	61.7403	6.5331	10.8 ± 1.4	10.8 ± 1.4		
22	R200803-02	Urdbøuri	59.7811	7.7044	12.8 ± 1.6	11.3 ± 1.3	19.5	2.2
	U-01		59.7810	7.7049	9.7 ± 1.0			

order controls on the region's susceptibility to large gravitational rock-slope deformation (e.g. Braathen et al. 2004).

Methods

For the terrestrial cosmogenic nuclide exposure dating (Gosse and Phillips 2001), we sampled the surfaces of 2–3 boulders on each rock avalanche deposit with hammer and chisel, prioritizing large boulders in central parts of the deposit away from any steps in the landscape to avoid boulders that may have toppled after deposition in the rock-avalanche event. At most sites simple lobate morphology indicated that the deposits resulted from single events. This was different at the Innerdalen, Innfjorddalen (Schleier et al. 2015, 2016) and the Vora site where obviously different rock avalanche lobes superimposed each other. At the Innerdalen and Innfjorddalen sites we sampled each lobe independently and dated several samples. The stratigraphic relations of rock avalanche deposits at Vora were published earlier (Aa et al. 2007). At that site we dated a single boulder due to financial restrictions as ages obtained fit with the rock avalanche stratigraphy. We recently calculated and recalculated previously published ages using the online CRONUS calculator at the KU server and determined the age of each landform following calculation details summarized in Schleier et al. (2015). For all sites with multiple dating results uncertainties are reported in the text, figures and tables as Coefficient of Variation in % and in ka but if single ages existed for a deposit as 1 sigma external error. This calculation does not include uncertainties related to inheritance, snow, vegetation cover, and erosion. While the uncertainties related to snow cover, vegetation cover and erosion will similarly effect the obtained ages over western Norway are the uncertainties related to inheritance individual from sample site to sample site and are discussed for the earlier published data in the respective paper and for the new data within this text. Previously published samples were prepared at Dalhousie Geochronology Centre and AMS run at Lawrence Livermore National Laboratory. New results were prepared and run at ETH Zurich, Institute for Particle Physics ca. 11–13 years ago.

Characteristics of Rock-Avalanche Deposits and Sample Sites

Gråura Rock-Avalanche Deposit

The Gråura rock avalanche deposit is a rather atypical deposit as it occurs on a subdued slope on a peninsula only 440 m high (Fig. 1). The deposit however is a continuous boulder field lying adjacent to the scar. The run-out is 310 m over a drop in the terrain of 140 m and has therefore a Fahrböschung of 0.45. Therefore, and based on the lobate form and the absence of any talus cone, it can be classified as a rock-avalanche deposit. Samples were collected from the frontal part and the ^{10}Be ages all fall within the 8.6% coefficient of variation (CoV) about their mean age of 14.9 ka (Table 2).

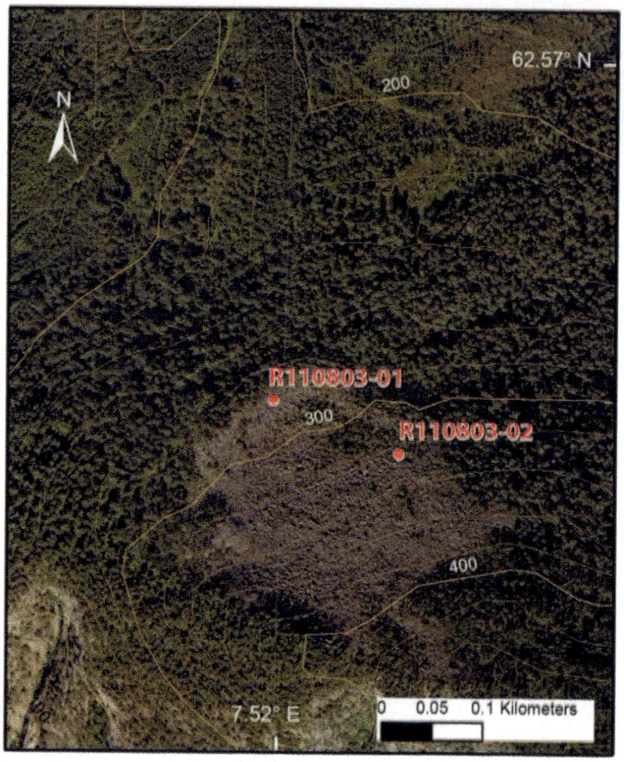

Fig. 1 Distribution of the Gråura rock-avalanche deposit (contours = 20 m)

Skiri Rock-Avalanche Deposits

At the locality of Skiri is the Romsdalen, filled with rock-avalanche deposits over a distance of 2.2 km. In the field they have been interpreted as being two separate deposits although the limit between both could not be distinguished in the field or from aerial imagery (Fig. 2). The ^{10}Be ages of all three boulders (two from one deposit, one from the other) fall within the 5.6% CoV about their mean age of 11.2 ka (Table 2). The valley is steep on both sites and multiple niches along the mountains would accommodate the rock-avalanche volume, however owing to the large spread of deposits concentrating around an eastern and western centre we maintain our interpretation that the deposits were created by at least two independent rock avalanches.

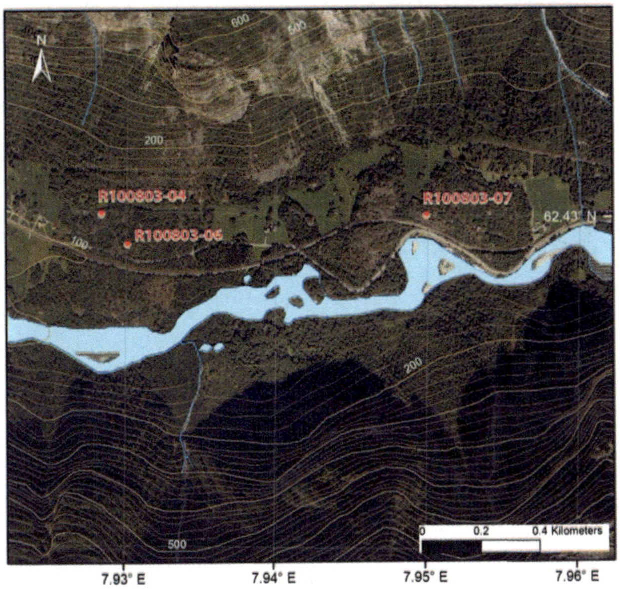

Fig. 2 Distribution of rock-avalanche deposits in Romsdalen at Skiri locality (contours = 20 m)

The Svarttinden Rock-Avalanche Deposits

Svarttinden (1587 m) is characterized by a clear scarp on the E flank. At the foot of the slope on the 1100 m high plateau, a rock-avalanche deposit spreads to the plateau edge and drops into Romsdalen (Fig. 3). In that valley further rock-avalanche deposits exist that are interpreted by Saintot et al. (2012) to belong to a different event that relates to the clear niche along the plateau edge. The deposit is divided by Saintot et al. (2012) into a large event that covers the plateau and partly obstructs the drainage to the SE and a minor deposit of a rock-slope failure that spreads only down to 1200 m and does not represent a rock avalanche. The three ^{10}Be ages fall within the relatively large CoV (14.3%) about their mean age of 8.7 ka (Table 2). However because there was no clear outlier, all three samples were taken to calculate the mean deposit age.

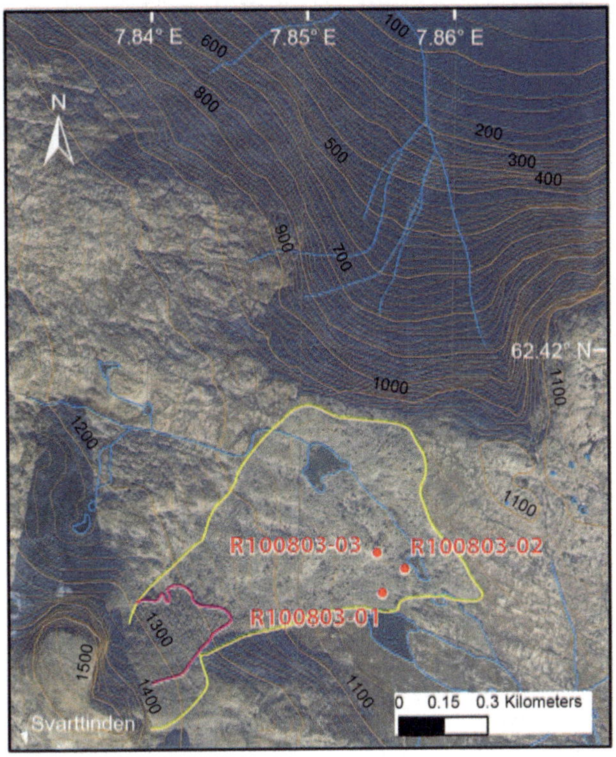

Fig. 3 Distribution of the Svarttinden rock-avalanche deposit after Saintot et al. (2012) (contours = 25 m)

The Alstadfjellet Rock-Avalanche Deposits

Allstadfjellet (1360 m) is characterized by a clear triangular scarp on the S flank. At the foot of that mountain a rock-avalanche deposit spreads over the entire Valldalen (Fig. 4). The mean age of 9.3 ka has a CoV of 3.8% (Table 2).

The Kallen Rock-Avalanche Deposits

The N ridge of Hornet Mountain (1360 m) at the S tip of Tafjord is called Kallen (1360 m). This ridge is characterized by a clear scarp on the N flank, the source of a 0.2 km^3 rock avalanche that dams Onilsavatnet. A large fan interpreted to be an outburst flood event (Hermanns et al. 2009) extends from a clear breach in the rock-avalanche deposit (Fig. 5). Two boulders dated of the rock-avalanche deposit yield a mean age of 11.4 ka with a 1.8% CoV. One boulder on the outburst flood deposit was dated, yielding 9.5 ka, indicating that the outburst flood occurred 2 ka later. This coincides with the 10 ka old fill deposit in Tafjord (Hermanns et al. 2009).

Vora Rock-Avalanche Deposits

Below the north face of Vora mountain (1450 m) a cluster of at least six rock-avalanche deposits was mapped by Aa et al. (2007). These authors dated the rock-avalanche deposits relatively by Schmidt hammer surface weathering tests comparing the rock strength to the rock strength of a Younger Dryas moraine and a modern road cut. Their ages for lobes a, b, c, and d were 10700 ± 260, 9680 ± 440, 9490 ± 500, 9170 ± 280, respectively. We sampled lobe a, b and d for ^{10}Be exposure dating. The ^{10}Be ages we obtained coincide with the established rock-avalanche stratigraphy, however our results suggest that the deposits are 1000–2000 years older (Table 2) (Fig. 6).

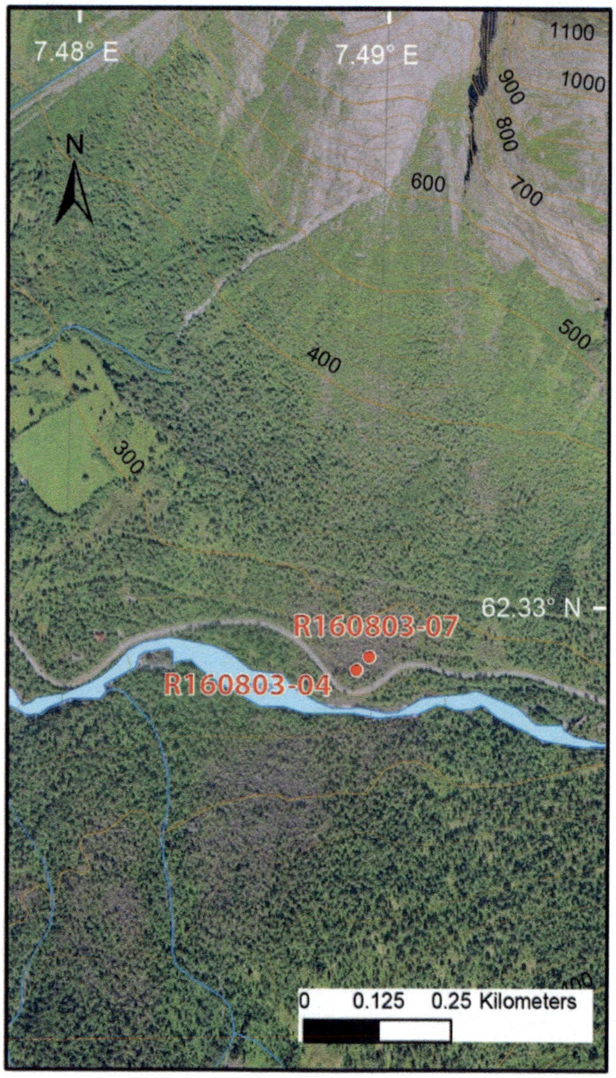

Fig. 4 Distribution of the Alstadfjellet rock-avalanche deposit (contours = 25 m)

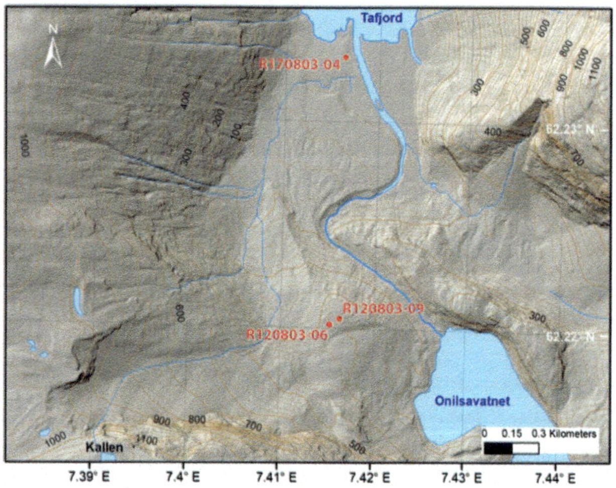

Fig. 5 Distribution of the Kallen rock-avalanche and outburst-flood deposits (contours = 25 m)

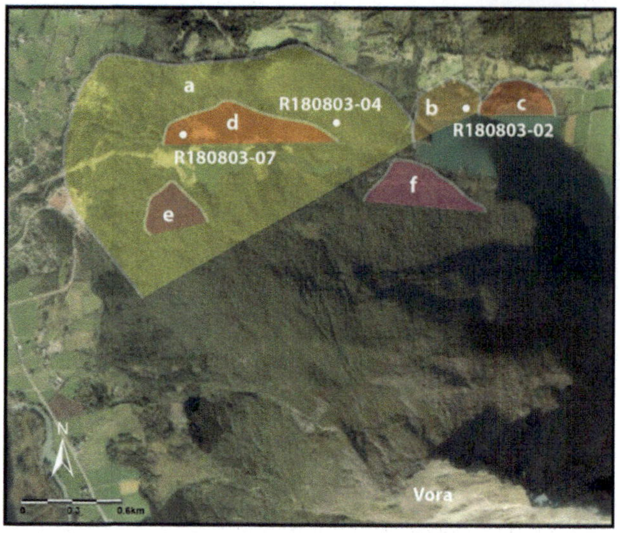

Fig. 6 Distribution of rock-avalanche deposit below Vora mountain. The individual lobes marked in different colors are taken from Aa et al. (2007)

Urdbøuri Rock-Avalanche Deposits

The Urdbøuri rock avalanche has dropped into Totak lake splitting off a bay and forming a new lake Urdbødjønni that has a lake level 5 m higher than Totak lake itself (Fig. 7).

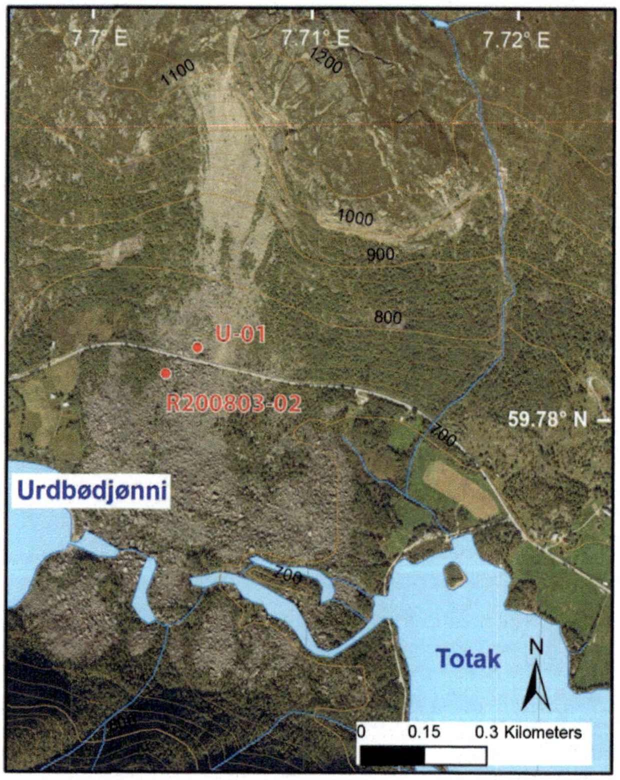

Fig. 7 Distribution of the Urdbøuri rock-avalanche deposit (contours = 25 m)

Samples were taken from the central part of the deposit and ^{10}Be ages obtained do not coincide within 1σ uncertainty limits (Table 2). This deposit is extremely poor in quartz and it was challenging to find adequate samples and we even had to return for a new sample in a later year (U-01) that was spotted while driving by. There are two explanations that might explain the misfit of ages. On the one hand the older sample may contain inherited nuclides of an exposure history prior to failure. On the other hand the younger age might be taken from a boulder that was moved or tilted during road construction and the applied scaling (0°) for surface inclination does not apply and the production on the sample location has been minor resulting in an age too young.

Temporal Spatial Distribution of Rock Avalanches in Relation to the Decay of the Scandinavian Ice Sheet

The temporal pattern of the rock-avalanche distribution is similar to the decay of the Scandinavian ice sheet, which is asynchronous. Oldest deposits occur further N where the

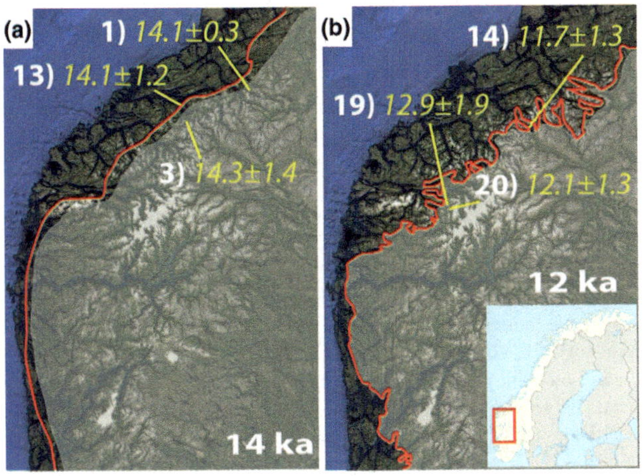

Fig. 8 Comparison of the ages of Late Pleistocene rock-avalanche deposits with the retreat of the Scandinavian Ice sheet for **a** 14 ka, and **b** 12 ka, as given by Hughes et al. (2016). The Scandinavian Ice sheet is shown as the mean likely extend for each time slice in transparent *white* and its maximum possible extend with the *red line*. *Red box* in inset in (**b**) shows the study area in respect to the extent of Norway (Norway is *white*)

ice sheet decayed first and then rock-avalanche activity migrated further south (Figs. 8 and 9). Obvious is that several mean rock-avalanches ages suggest that these events occurred while the ice sheet still existed at that locality.

However such a condition could be shown for the older Innerdalen rock-avalanche deposit only (Schleier et al. 2015). All other rock-avalanche deposits (older Innfjorddalen, the two older Vora, and the Urdbøuri rock-avalanche deposits) do not show any morphological features allowing to interpret that they deposited on ice and thus attest for ice free conditions at the time of occurrence. This suggests a mismatch with the latest ice sheet decay model given by Hughes et al. (2016). Unfortunately, at all these sites only one age determination exists per deposit or the spread of ages is large (ca. 20% of COV) for Urdbøuri. Therefore, it is rather suggested to date those deposits with more samples in order to define a more precise age before reinterpreting the ice decay model. The most striking result of this compilation is that of the total 22 rock avalanches dated 50% (1, 11, 12, 13, 14, 15, 18, 19, 20, 21, and 22) occurred in a close to ice margin environment and within 1 ka after deglaciation (Figs. 8 and 9) and six events (2, 4, 11, 12, 16, 17) close to the climate maximum characterized by warmest temperatures in Norway ($\sim$8.5 ka). This shows that large changes in the slope related to deglaciation is one of the prime conditioning factors for large rock-slope failures. This is similar as in the data set of rock-failure deposits in Storfjorden (Bøhme et al. 2015). The post Boreal rock avalanches (<7.5 ka) distribute relatively evenly in time and space however the total number of 6 is statistically not representative. These results are hence in a way similar to those obtained by CN dating by Ballantyne et al. (2014) for rock slope failures following the decay of the ice sheet of the British Islands suggesting a fast response after the ice sheet decay. However different is that rock slope failures peaked in Norway in the first thousand years while this peak was reached in Scotland not before 1600 years after ice retreat. Different is also that rock-slope failure activity continued in Norway at a constant lower rate throughout the Holocene while CN data on rock slope failure deposits on the British Islands rather suggest an end of activity in the millenia after the Younger Dryas (Ballantyne et al. 2014).

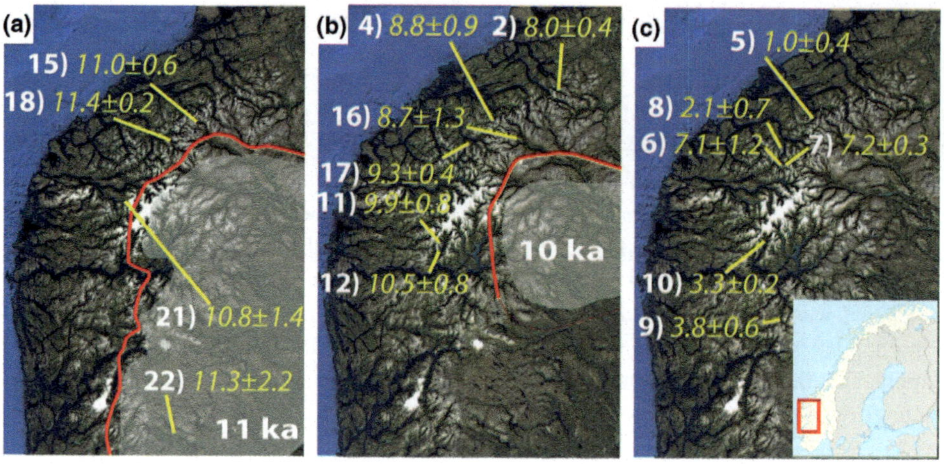

Fig. 9 Comparison of the ages of Holocene rock-avalanche deposits with the retreat of the Scandinacian Ice sheet for **a** 11 ka, **b** 10 ka, and **c** for after the decay of the Scandinavian ice sheet as given by Hughes et al. (2016). The Scandinavian Ice sheet is shown as the mean likely extend for each time slice in transparent *white* and its maximum possible extend with the *red line*. *Red box* in inset in (**c**) shows the study area in respect to the extent of Norway

Acknowledgements Field work and age determination of samples taken in 2003 were financed through the Excellence Centre "International Centre for Geohazards" financed by the Norwegian Research Council. Later samples were taken partly financed through the Norwegian Water Resources and Energy Directorate (NVE) and partly through a Ph.D. thesis by M. Schleier. R.L. Hermanns got funding to write this publication through the NFR-funded CryoWALL project (243784/CLE).

References

Aa AR, Sjåstad J, Sønstegaard E, Blikra LH (2007) Chronology of Holocene rock-avalanche deposits based on Schmidt-hammer relative dating and dust stratigraphy in nearby bog deposits, Vora, inner Nordfjord, Norway. The Holocene 17:955–964

Antinao JL, Gosse J (2009) Large rockslides in the Southern Central Andes of Chile (32–34.5 S): tectonic control and significance for quaternary landscape evolution. Geomorphology 104:117–133

Ballantyne CK, Sandeman GF, Stone JO, Wilson P (2014) Rock-slope failure following Late Pleistocene deglaciation on tectonically stable mountainous terrain. Quatern Sci Rev 86:144–157

Blikra LH, Longva O, Braathen A, Anda E, Dehls JF, Stalsberg K (2006) Rock slope failures in Norwegian fjord areas: examples, spatial distribution and temporal pattern. In: Evans SG, Scarascia Mugnozza G, Strom A, Hermanns RL (eds) Landslides from massive rock slope failure. NATO science series IV: earth and environmental sciences 49. Springer, Dordrecht. pp 475–496

Böhme M, Oppikofer T, Longva O, Jaboyedoff M, Hermanns RL, Derron MH (2015) Analyses of past and present rock slope instabilities in a fjord valley: implications for hazard estimations. Geomorphology 248:464–474

Böhme M, Hermanns RL, Oppikofer T, Fischer L, Bunkholt HSS, Eiken T, Pedrazzini A, Derron M-H, Jaboyedoff M, Blikra LH, Nilsen B (2013) Analyzing complex rock slope deformation at Stampa, western Norway, by integrating geomorphology, kinematics and numerical modelling. Eng Geol 154:116–130

Braathen A, Blikra LH, Berg SS, Karlsen F (2004) Rock-slope failures of Norway, type, geometry deformation mechanisms and stability. Nor Geol Tidsskr 84:67–88

Cruden DM, Hu XQ (1993) Exhaustion and steady state models for predicting landslide hazards in the Canadian rocky mountain. Geomorphology 8:279–285

Gosse JC, Phillips FM (2001) Terrestrial in situ cosmogenic nuclides: theory and application. Quatern Sci Rev 20:1475–1560

Hermanns RL, Oppikofer T, Anda E, Blikra LH, Böhme M, Bunkholt H, Crosta GB, Dahle H, Devoli G, Fischer L, Jaboyedoff M, Loew S, Sætre S, Yugsi Molina F (2013) Hazard and risk classification system for large unstable rock slopes in Norway. In: Genevois R, Prestininzi A (eds) International conference on Vajont—1963–2013. Ital J Eng Geol Environ, Book series 6:245–254

Hermanns RL, Longva O (2012) Rapid rock-slope failures. In: Clague JJ, Stead D (eds) Landslides: types, mechanisms and modeling. Cambridge University Press, Cambridge, UK, pp 59–70

Hermanns RL, Fischer L, Oppikofer T, Böhme M, Dehls JF, Henriksen H, Longva O, and Eiken T (2011) Mapping of unstable and potentially unstable rock slopes in Sogn og Fjordane, NGU rapport 2011.055, 195 p

Hermanns, RL, Blikra LH, Longva O (2009) Relation between rockslide dam and valley morphology and its impact on rockslide dam longevity and control on potential breach development based on examples from Norway and the Andes, Long term behavior of dams.In: Proceedings of the 2nd international conference, Graz, Austria. pp 789–794

Hewitt K, Gosse J, Clague JJ (2011) Rock avalanches and the pace of late quaternary development of river valleys in the Karakoram Himalaya. Geol Soc Am Bull 123:1836–1850

Hughes ALC, Gyllencreutz R, Lohne ØS, Mangerud J, Svendsen JI (2016) The last Eurasian ice sheets—a chronological database and time-slice reconstruction, DATED-1. Boreas 45:1–45

Korup O, Clague JJ, Hermanns RL, Hewitt K, Strom AL, Weidinger JT (2007) Giant landslides, topography, and erosion. Earth Planet Sci Lett 261:578–589

Saintot A, Oppikofer T, Derron M-H (2012) Large gravitational rock slope deformation in Romsdalen valley (Western Norway). Revista de la Asociación Geológica Argentina 69:354–371

Schleier M, Hermanns RL, Rohn J, Gosse J (2015) Diagnostic characteristics and paleodynamics of supraglacial rock avalanches, Innerdalen, Western Norway. Geomorphology 245:23–39

Schleier M, Hermanns RL, Gosse JC, Oppikofer T, Rohn J, Tønnesen JF (2016) Subaqueous rock-avalanche deposits exposed by post-glacial isostatic rebound, Innfjorddalen, Western Norway. Geomorphology. (in press)

The Role of Rainfall and Land Use/Cover Changes in Landslide Occurrence in Calabria, Southern Italy, in the 20th Century

Stefano Luigi Gariano, Olga Petrucci, and Fausto Guzzetti

Abstract

Urbanization in hazardous regions, the abandonment of rural and mountain areas, and changed agricultural and forest practices have increased the impact of landslides through the years. Hence, the changing climate variables, like rainfall, acted and will act on a human-modified landscape. In this work, we analyze the role of rainfall variation and land use/cover change in the occurrence of landslides in Calabria in the period 1921–2010. Combining rainfall and landslide information, we reconstruct and analyze a catalogue of 1466 rainfall events with landslides (i.e., the occurrence of one or more landslide during or immediately after a rainfall event). To investigate the impact of land use/cover changes in the occurrence of landslides, we consider the "Land Use Map" made by the Italian National Research Council and the Italian Touring Club in 1956, and the "CORINE Land Cover" map released in 2000. Since our landslide catalogue is at municipality scale (i.e., for each landslide we known the municipality in which it occurred), we attribute a prevailing land use/cover class to each of the 409 municipalities of Calabria. We split the catalogue in two subsets (1921–1965 and 1966–2010) and correlate the landslides occurred in the first period to the 1956 land use and the landslides occurred in the second period to the 2000 land cover. We find that: (i) the geographical and the temporal distributions of rainfall-induced landslides have changed in the observation period; (ii) land use/cover in Calabria has changed between the two periods, with a huge decrease of arable land and an increase of heterogeneous agricultural areas and forests; (iii) in both periods, most of the landslides occurred in areas characterized by forests and arable land; (iv) in the second period, there was an increase (decrease) of landslides occurred in agricultural areas (arable land).

S.L. Gariano (✉) · F. Guzzetti
Italian National Research Council, Research Institute for Geo-Hydrological Protection, Via Madonna Alta 126, 06127 Perugia, Italy
e-mail: stefano.gariano@irpi.cnr.it

F. Guzzetti
e-mail: fausto.guzzetti@irpi.cnr.it

S.L. Gariano
Department of Physics and Geology, Piazza Università, University of Perugia, 06123 Perugia, Italy

O. Petrucci
Italian National Research Council, Research Institute for Geo-Hydrological Protection, Via Cavour 4/6, 87036 Rende (CS), Italy
e-mail: olga.petrucci@irpi.cnr.it

M. Mikoš et al. (eds.), *Advancing Culture of Living with Landslides*,
DOI 10.1007/978-3-319-53483-1_40

Keywords
Landslide • Rainfall • Climate change • Land use • Land cover • Change • Calabria • Italy

Introduction

Rainfall is the primary trigger of landslides in Italy (Guzzetti et al. 1994). To model and predict the triggering of landslide induced by rainfall, a standard approach, assuming a stationary relationship, consists in a joint analysis of landslides and rainfall records (Guzzetti et al. 2007, 2008). However, rainfall conditions that have resulted in landslides in the past may vary in the future due to climate changes, including variations in rainfall intensity and frequency, which can affect the frequency of rainfall-induced landslides (Crozier 2010). An increase in the frequency and the intensity of extreme rainfall events was observed in several geographical regions in the World (IPCC 2014), including Italy (Brunetti et al. 2002). Temporal variations in the rainfall conditions that result in landslides affect the definition and the application of empirical and physically based models for the prediction of landslide occurrence.

Several authors investigated the impact of climate changes (in particular, rainfall) on the occurrence of landslides at regional scale and on the variation of their frequency and distribution (Jomelli et al. 2004; Jakob and Lambert 2009; Polemio and Petrucci 2010; Polemio and Lonigro 2013; Stoffel et al. 2014; Gariano et al. 2015; Ciabatta et al. 2016; Gariano and Guzzetti 2016). On the other hand, few studies analyzed the role of land use changes on landslide occurrence, frequency, distribution, hazard, and risk (Sidle and Dhakal 2002; van Beek 2002; Glade 2003; van Beek and Van Ash 2004; Alcàntara-Ayala et al. 2006; Beguería 2006; Wasowski et al. 2010; Lonigro et al. 2015).

Land use change is recognized as one of the most important factors influencing the occurrence of rainfall-induced landslides (Glade 2003), and the landslide hazard and risk, playing a relevant role in the distribution of the element at risk (Promper et al. 2014). Land use and land cover may act as predisposing factors of landslide occurrence, and may also control the spatial distribution of landslide consequences (Glade 2003; Beguerìa 2006; Promper et al. 2014). In fact, changes in land use and land cover (e.g., urbanization in hazardous areas, abandonment of rural and mountain areas, changed agricultural and forest practices) have increased the impact of landslides through the years. Hence, the changing climate variables, like rainfall, acted and will act on a landscape that was modified by human actions. Moreover, a change in rainfall regime can alter land cover types and land use, which have consequences both on single slope stability conditions, and on the type, abundance and frequency of landslides (Sidle and Dhakal 2002). This should be always taken into account when analyzing past and future changes in landslide risk.

In this work, we analyze the role of rainfall variation and land use/cover change in the occurrence of landslides in Calabria, southern Italy, in the period 1921–2010.

Materials and Method

Study Area

Calabria, the southernmost part of the Italian Peninsula, extends for 15,080 km^2, with an elevation ranging from sea level to 2260 m a.s.l. The region comprises 409 municipalities ranging in size from 2.4 km^2 to 292.0 km^2 (average 38.4 km^2). Mean annual rainfall is 1150 mm, with the western side of the region rainier than the eastern side. Annual rainfall strictly depends on elevation, with the mountains significantly wetter (>1400 mm) than the coastal plains (<1000 mm). About 70% of the annual rain falls from October to March, and only 10% in the summer. Events with large cumulated rainfall mainly occur between November and January, whereas high intensity events are most common in September and October (Terranova and Gariano 2014).

Catalogue of Rainfall Events with Landslides

We exploited a catalogue of historical landslides occurred in Calabria from January 1921 to December 2010 (Gariano et al. 2015). To find information on landslides, we used different sources, including local and national newspapers, web sites, reports from national and regional agencies and public offices, and post-event field surveys. Moreover, we exploited daily rainfall records obtained from a network of 318 rain gauges, managed by a regional agency, in the same period, to reconstruct almost 500 thousand rainfall events (RE). A RE is defined as a continuous sequence of rainy days (i.e., days with cumulated daily rainfall >0 mm) preceded and followed by at least one dry day (i.e., a day with no measured rainfall). Combining the rainfall and the landslide information, we obtained a catalogue of 1466 rainfall events with landslides (REL), where a REL is the occurrence of one or more landslide during or immediately after a rainfall event, in a distance ≤ 5 km from the location of the rain gauge where the RE is determined. Furthermore, we calculated the duration, D (h), and the cumulated rainfall, E (mm), of all RE associated to the REL.

Land Use and Land Cover Maps

To investigate the evolution of land use and land cover changes in Calabria, we considered two maps: the "Land Use Map" at 1:200,000 scale made by CNR (Italian National Research Council) and TCI (Italian Touring Club) in 1956, and the "CORINE Land Cover" map at 1:100,000 scale released in 2000. The latter was gathered from the Italian National Geoportal (www.pcn.minambiente.it). The first was gathered in printed form, then acquired in digital format and digitalized in GIS environment. The two maps do not refer exactly to the same information. Effectively, land use is defined as a "classification of the territory according to its current and future planned functional dimension or socio-economic purpose", while land cover is defined as "physical and biological cover of the earth's surface including artificial surfaces, agricultural areas, forests, (semi-) natural areas, wetlands, water bodies" (Directive 2007/2/EC of the European Parliament). Moreover, the maps had different legends; therefore, we defined a new common legend composed by 10 fields, mainly based on the third level of the CORINE Land Cover classification and following the works of Falcucci et al. (2007).

Comparing the two maps, several changes in land use/cover could be identified (Table 1). The maximum difference pertains to the "Arable land" with a reduction of more than 3000 km^2 between 1956 and 2000. "Natural grassland" has also experienced a reduction of ca. 1330 km^2. Conversely, the "Heterogeneous agricultural areas" and "Forest" classes have experiences an increase of ca. 2500 and 1650 km^2, respectively. These changes fit a national framework of land use change in Italy in the second part of the 20th Century: Falcucci et al. (2007) found that more than half of the Italian territory changed with respect to land use, from the 1960s to 2000, with an acceleration of the changes in the last decade.

Land Cover Changes and Landslides

Our landslide catalogue is at municipality scale (i.e., for each landslide we known the municipality in which it occurred). Thus, we attributed a prevailing land use class to each of the 409 municipalities of Calabria. In particular, we attributed to each municipality the class covering more than 50% of its area. In case no class reached this value, we selected the class with the highest values of coverage in the territory of the municipality. We split the catalogue in two subsets (1921–1965, P1, and 1966–2010, P2) and correlated the REL occurred in the first period to the 1956 land use map and the REL occurred in the second period to the 2000 land cover map.

Results and Discussion

Main Features of the Landslides in the Catalogue

Figure 1 shows the spatial and temporal distribution of the 1466 REL occurred in Calabria in the period 1921–2010. We acknowledge that our catalogue could not be complete, given that it consider only rainfall-induced landslides that were noticed (and recorded) because they have caused damage to some elements at risk: e.g., public or private properties, or population

The maximum, average, and minimum values of REL in a municipality are 47, 4, and 0, respectively. More than 30

Table 1 Classes of the common legend of land use/cover. Areas included in each class for the Land Use Map (LUM) of 1956 and the CORINE Land Cover (CLC) map of 2000, and related differences are also reported

Code	Description	LUM 1956 area (km^2)	CLC 2000 area (km^2)	Difference (km^2)
100	Artificial surfaces	211.26	459.43	248.17
210	Arable land (permanently irrigated and non-irrigated)	5675.38	24014.36	−3261.02
220	Vineyards, fruit trees, olive groves	2497.76	2434.60	−63.17
230	Pastures	12.91	72.81	59.90
240	Heterogeneous agricultural areas	0.00	2464.13	2464.13
310	Forest	3967.98	5626.32	1658.34
320	Natural grassland	2620.14	1300.21	−1319.93
330	Areas with little or no vegetation	67.30	262.83	195.52
400	Wetlands	0.00	0.00	0.00
500	Water bodies	25.45	33.51	8.07

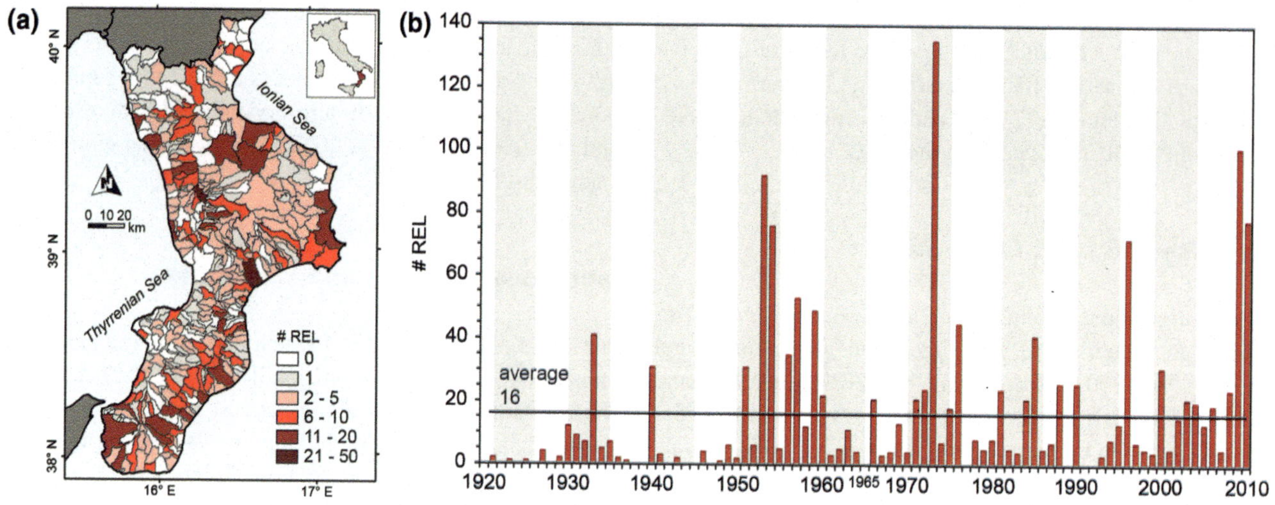

Fig. 1 Number of rainfall events with landslides (#REL) in each municipality (**a**) and per year (**b**), in Calabria in the period 1921–2010. Solid line in (**b**) shows average values for the 90-year period

municipalities experienced 10 or more REL in the investigated period. On average, 16 REL occurred every year with a maximum of 139 REL in 1973. For six years, more than 60 REL were recorded. The month with the largest number of REL was January (337), followed by November (255), and February (254). The average duration of the rainfall events is 7.2 h, while the mean cumulated rainfall is 169 mm.

Figure 2 shows the distribution of REL, per municipality, in the 1921–1965 (P1, Fig. 2a) and 1966–2010 (P2, Fig. 2b) periods. In both periods, the municipality that has experienced the highest number of REL is Catanzaro, with 13 REL in P1 and 34 REL in P2, respectively. It is followed, in the first period, by the municipalities of Rossano and Soverato (both located in the E part of the region) with 8 REL (Fig. 2a). In the second period, the municipalities of Cosenza and Rogliano (both in the N part of the region) have experienced 30 and 17 REL, respectively (Fig. 2b).

In the first period, the municipalities with the highest number of REL were localized mostly in the inner mountainous parts of the region (Fig. 2a). Conversely, in the second period, most municipalities located along the coasts, on hilly landscape, on both the eastern and western sides, have experienced a high number (≥ 6) of REL (Fig. 2b). This result has a double meaning, linked to the fact that our catalogue is composed by damaging landslides, i.e., landslides that affected some elements at risk. Historically, after land reclamation, which occurred in the second part of P1 period, people migrated in Calabria towards coastal sectors, thus increasing the urban pressures in hilly areas (instead than on the mountainous inner sectors), where communication infrastructures could be more easily built (Petrucci and Polemio 2007). This means that, in the 20th century in Calabria, both the landslide hazard and the distribution of the elements at risk—thus, landslide risk—changed.

Changes in Rainfall Events

We found changes in the monthly distribution of REL in our catalogue. In the earlier period (1921–1965) REL were frequent in autumn and winter, mostly from October to February, whereas in the recent period (1966–2010) landslides concentrated in the winter, particularly in December, January, and February. Moreover, we found changes in mean duration and cumulated rainfall of the rainfall events (Table 2). Events that triggered landslides in the recent period were characterized by average and maximum values of cumulated rainfall lower than those occurred in the previous period. We consider this evidence of: (i) the increased propensity of the landscape to generate landslides in the recent period, and (ii) the larger number of vulnerable elements exposed to landslides, in Calabria. No remarkable variations were found in the duration of the rainfall events.

Distribution of Landslides Per Land Use Class

Considering the two periods (1921–1965, P1, and 1966–2010, P2), we correlate 568 REL occurred in the first period to the 1956 land use map and 898 REL occurred in the second period to the 2000 land cover map (Table 3). Figure 3 shows the prevailing land use/cover in the 409 municipalities of Calabria according to the 1956 CNR-TCI Land Use Map (P1, Fig. 3a), and the 2000 CORINE Land Cover map (P2, Fig. 3b).

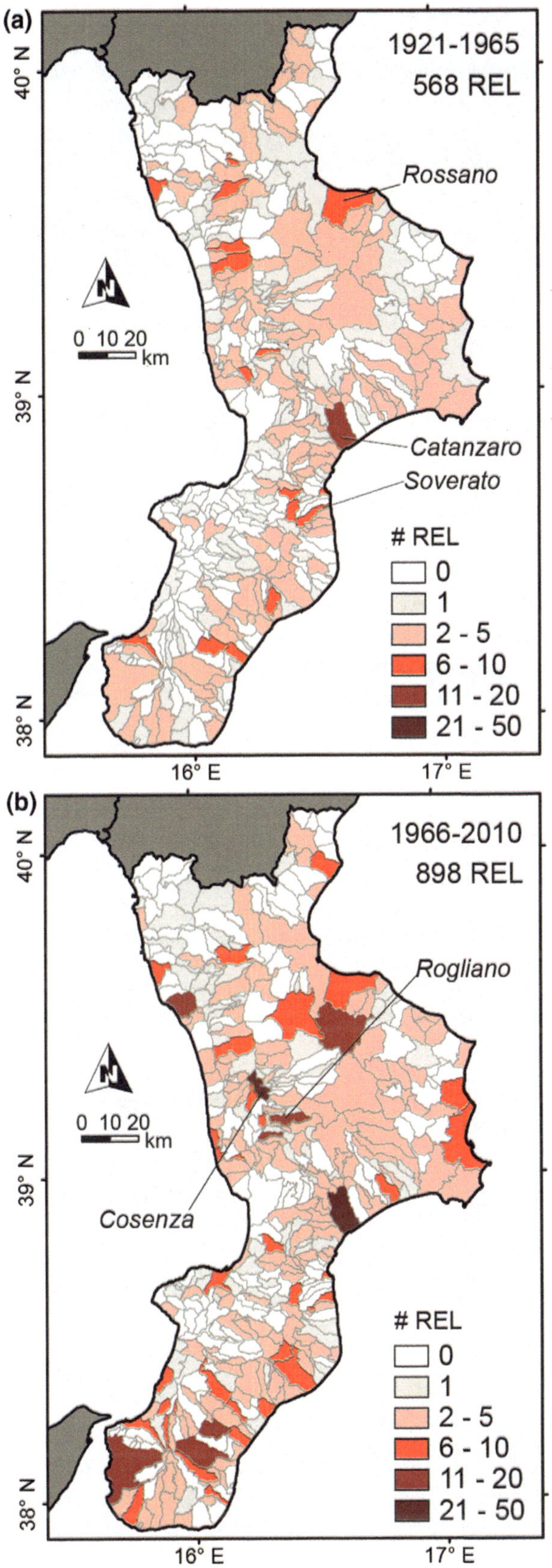

Fig. 2 Number, in classes, of REL per municipality in **a** P1 (1921–1965), and **b** P2 (1966–2010) periods

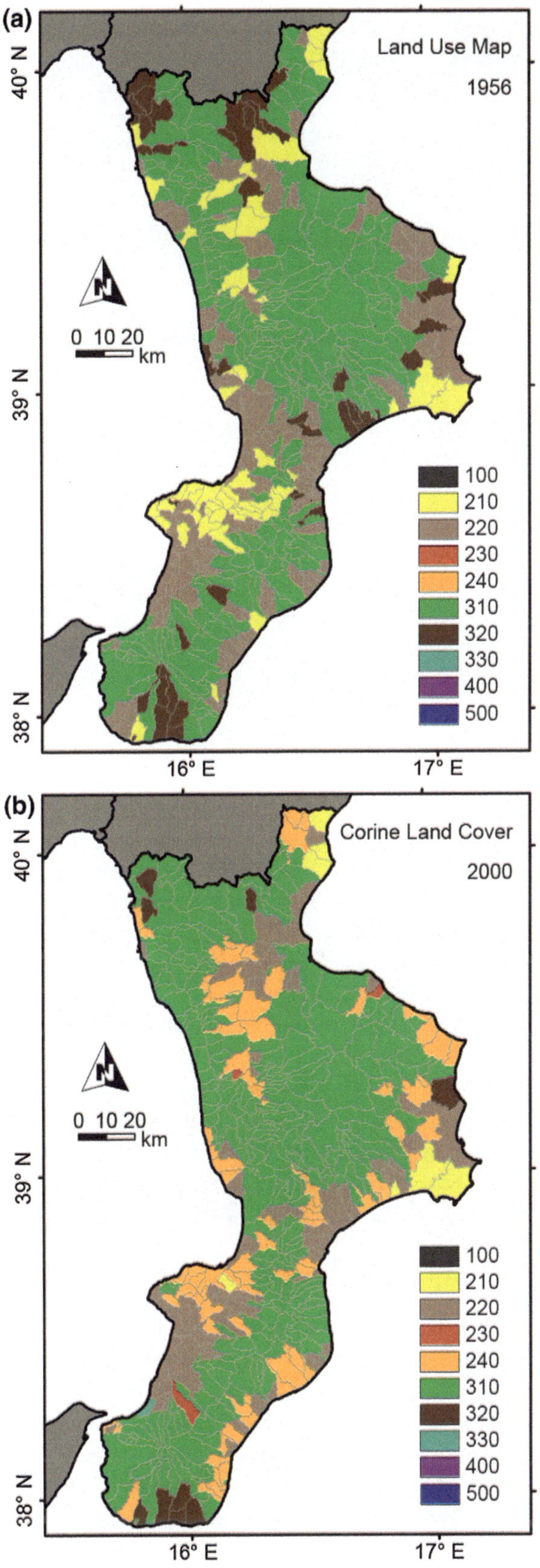

Fig. 3 Prevailing land use/cover in the 409 municipalities of Calabria according to the **a** 1956 CNR-TCI Land Use Map, and the **b** 2000 CORINE Land Cover map. Descriptions of the codes are reported in Table 1

Table 2 Minimum, mean, and maximum values of duration (*D*, in h) and cumulated event rainfall (*E*, in mm) for the rainfall events that triggered landslides in the 1921–1965 (P1) and 1966–2010 (P2) periods

Variable	Period	Min	Mean	Max
D (h)	P1	1.0	7.8	28.0
	P2	1.0	6.5	32.0
E (mm)	P1	11.1	246.4	1504.7
	P2	10.0	176.0	914.5

Table 3 Number of REL (#REL) assigned to the each of 10 classes of land use/cover in the 1921–1965 (P1) and 1966–2010 (P2) periods

Land use/cover code	#REL P1	#REL P2
100	0	0
210	68	18
220	123	159
230	0	3
240	0	182
310	330	513
320	47	18
330	0	5
400	0	0
500	0	0

In the first period, most of the REL occurred in municipalities characterized by "Forest" (330 out of 568 REL, 58%) and "Vineyards, Fruit trees, Olive grove" (123, 22%) prevailing land use/cover classes (Table 3). In the second period, REL are more heterogeneously distributed, with 513 REL (57%) in municipalities with "Forest" prevailing land use/cover, 182 REL (20%) in municipalities with "Heterogeneous agricultural areas", and 159 REL (18%) in municipalities with "Vineyards, Fruit trees, Olive grove" prevailing land use/cover (Table 3). Therefore, in the second period, there was an increase of landslides occurred in municipalities characterized mostly by forested and agricultural areas and a decrease of landslides occurred in arable land.

Conclusions

We reconstructed and analyzed a catalogue of 1466 rainfall events with landslides occurred in Calabria in the period 1921–2010 and we studied the role of rainfall variation and land use/cover change in the occurrence of those events. We split the catalogue in two subsets (1921–1965 and 1966–2010) and correlated the landslides occurred in the first period to the 1956 land use (retrieved from the CNR-TCI Land Use Map released in 1956) and the landslides occurred in the second period to the "CORINE" Land Cover map released in 2000.

We found that the geographical and the temporal (monthly) distributions of the rainfall-induced landslides have changed in the observation period, with a concentration of the events in the winter months. Moreover, we found that land use/cover has changed in Calabria between the two periods, with a huge decrease of arable land and an increase of heterogeneous agricultural areas and forests. In both periods, most of the landslides occurred in areas characterized by forests and arable land; in the second period, there was an increase of landslides occurred in forested areas and a decrease of landslides occurred in arable land. Furthermore, less rain was necessary to trigger landslides in the recent 45-year period than in the previous one. These findings can be used in the correct planning of new settlements in Calabria, also in the light of climatic, environmental, and societal changes

We noted that the observed changes in the number of rainfall-induced landslides in Calabria in the investigated 90-year period are due to changes in the number of the events (a largely natural component) and to changes in the number of the exposed elements (a largely societal component). However, given the complexity of the temporal and spatial variations of rainfall-induced landslides observed in Calabria, we acknowledge that the prediction of possible variations in frequency of such types of phenomena in response to future climatic changes will be difficult and uncertain.

Acknowledgements The "Centro Funzionale Multirischi" of the Agenzia Regionale per l'Ambiente della Calabria (ARPACAL. www.cfd.calabria.it) made available the rainfall data. The CORINE Land Cover Map released in 2000 was acquired from the Italian National Geoportal (www.pcn.minambiente.it). We are grateful to Paola Giostrella who was precious for finding and gathering the CNR-TCI land use map of Calabria.

References

Alcàntara-Ayala I, Esteban-Chavez O, Parrot JF (2006) Landsliding related to land-cover change: a diachronic analysis of hillslope instability distribution in the Sierra Norte, Puebla, Mexico. Catena 65(2):152–165. doi:10.1016/j.catena.2005.11.006

Beguería S (2006) Changes in land cover and shallow landslide activity: a case study in the Spanish Pyrenees. Geomorphology 74 (1–4):196–206. doi:10.1016/j.geomorph.2005.07.018

Brunetti M, Buffoni L, Mangianti F, Maugeri M, Nanni T (2002) Temperature, precipitation and extreme events during the last century in Italy. Glob Planet Change 40(1):141–149. doi:10.1016/s0921-8181(03)00104-8

Ciabatta L, Camici S, Brocca L, Ponziani F, Stelluti F, Berni N, Moramarco T (2016) Assessing the impact of climate-change scenarios on landslide occurrence in Umbria Region, Italy. J Hydrol. doi:10.1016/j.jhydrol.2016.02.007

Crozier MJ (2010) Deciphering the effect of climate change on landslide activity: a review. Geomorphology 124:260–267. doi:10.1016/j.geomorph.2010.04.009

Directive 2007/2/EC of the European Parliament and of the Council. Establishing an Infrastructure for Spatial Information in the European Community (INSPIRE). Official J Eur Union, 14 March 2007. L108/1-L180/14

Falcucci A, Maiorano L, Boitani L (2007) Changes in land-use/land cover patterns in Italy and their implications for biodiversity conservation. Landscape Ecol 22:617–631. doi:10.1007/s10980-006-9056-4

Gariano SL, Guzzetti F (2016) Landslides in a changing climate. Earth Sci Rev. doi:10.1016/j.earscirev.2016.08.011

Gariano SL, Petrucci O, Guzzetti F (2015) Changes in the occurrence of rainfall-induced landslides in Calabria. Southern Italy, in the 20th century. Nat Hazards and Earth Syst Sci 15:2313–2330. doi:10.5194/nhess-15-2313-2015

Glade T (2003) Landslide occurrence as a response to land use change: a review of evidence from New Zealand. Catena. 51(3–4):297–314. doi:10.1016/s0341-8162(02)00170-4

Guzzetti F, Cardinali M, Reichenbach P (1994) The AVI Project: a bibliographical and archive inventory of landslides and floods in Italy. Environ Manage 18:623–633. doi:10.1007/BF02400865

Guzzetti F, Peruccacci S, Rossi M, Stark CP (2007) Rainfall thresholds for the initiation of landslides in central and southern Europe. Meteorol Atmos Phys 98:239–267. doi:10.1007/s00703-007-0262-7

Guzzetti F, Peruccacci S, Rossi M, Stark CP (2008) The rainfall intensity-duration control of shallow landslides and debris flow: an update. Landslides 5:3–17. doi:10.1007/s10346-007-0112-1

IPCC—Intergovernmental Panel on Climate Change (2014) Climate change 2014: synthesis report. Contribution of working groups I. II and III to the fifth assessment report of the intergovernmental panel on climate change. Geneva. Switzerland. 151 p

Jakob M, Lambert S (2009) Climate change effects on landslides along the southwest coast of British Columbia. Geomorphology 107:275–284. doi:10.1016/j.geomorph.2008.12.009

Jomelli V, Pech VP, Chochillon C, Brunstein D (2004) Geomorphic variations of debris flows and recent climatic change in the French Alps. Clim Change 64:77–102. doi:10.1023/B:CLIM.0000024700.35154.44

Lonigro T, Gentile F, Polemio M (2015) The influence of climate variability and land use variations on the occurrence of landslide events (Subappennino Dauno, Southern Italy). Rend Online della Soc Geol Ital 35:192–195. doi:10.3301/ROL.2015.98

Petrucci O, Polemio M (2007) Flood risk mitigation and anthropogenic modifications of a coastal plain in southern Italy: combined effects over the past 150 years. Nat Hazards and Earth Syst Sci 7:361–373. doi:10.5194/nhess-7-361-2007

Polemio M, Lonigro T (2013) Climate variability and landslide occurrence in Apulia (Southern Italy). In: Margottini C, Canuti P, Sassa K (eds) Landslide science and practice. Global environmental change, vol 4. Springer, Berlin, Heidelberg, pp 37–41

Polemio M, Petrucci O (2010) Occurrence of landslide events and the role of climate in the twentieth century in Calabria, southern Italy. Quat J Eng Geol Hydrogeol 43:403–415. doi:10.1144/1470-9236/09-006

Promper C, Puissant A, Malet J-P, Glade T (2014) Analysis of land cover changes in the past and the future as contribution to landslide risk scenarios. Appl Geogr 53:11–19. doi:10.1016/j.apgeog.2014.05.020

Sidle RC, Dhakal AS (2002) Potential effects of environmental change on landslide hazards in forest environments. In: Sidle RC (ed) Environmental change and geomorphic hazards in forests, IUFRO Research Series, vol 9. CABI Publishing, Wallingford, Oxen, UK, pp 123–165

Stoffel M, Tiranti D, Huggel C (2014) Climate change impacts on mass movements—case studies from the European Alps. Sci Total Environ 493:1255–1266. doi:10.1016/j.scitotenv.2014.02.102

Terranova O, Gariano SL (2014) Rainstorms able to induce flash floods in a Mediterranean-climate region (Calabria. southern Italy). Nat Hazards and Earth System Sci 14:2423–2434. doi:10.5194/nhess-14-2423-2014

van Beek LPH (2002) Assessment of the influence of changes in land use and climate on landslide activity in a Mediterranean environment. Netherlands Geo-Graphical Studies NGS 294, Utrecht. 360 p

van Beek LPH, Van Asch TWJ (2004) Regional assessment of the effects of land-use change on landslide hazard by means of physically based modelling. Nat Hazards 31(1):289–304. doi:10.1023/B:NHAZ.0000020267.39691.39

Wasowski J, Lamanna C, Casarano D (2010) Influence of land-use change and precipitation patterns on landslide activity in the Daunia Apennines, Italy. Quat J Eng Geol Hydrogeol 43:387–401. doi:10.1144/1470-9236/08-101

Geomorphology and Age of Large Rock Avalanches in Trentino (Italy): Castelpietra

Susan Ivy-Ochs, Silvana Martin, Paolo Campedel, Kristina Hippe, Christof Vockenhuber, Gabriele Carugati, Manuel Rigo, Daria Pasqual, and Alfio Viganò

Abstract

Within a project aimed at understanding past catastrophic rock slope failure in the Trentino Province of Italy, we studied the Castelpietra landslide. Castelpietra encompasses a main blocky deposit, with an area of 1.2 km^2, which is buried on the upper side by more recent rockfall debris. The release area is the Cengio Rosso rock wall, which is comprised of Dolomia Principale and overlying Calcari Grigi Group dolomitized limestones. ^{36}Cl exposure dates from two boulders in the main blocky deposit indicate that the landslide occurred at 1060 ± 270 AD (950 ± 270 yr ago). The close coincidence in time of the Castelpietra event with several events that lie within a maximum distance of 20 km, including Kas at Marroche di Dro, Prà da Lago and Varini (at Lavini di Marco) landslides, strongly suggests a seismic trigger. Based on historical seismicity compilations, we have identified the "Middle Adige Earthquake" at 1046 AD as the most likely candidate. Its epicenter lies right in the middle of the spatial distribution of the discussed landslides.

S. Ivy-Ochs (✉) · K. Hippe · C. Vockenhuber
Laboratory of Ion Beam Physics, ETH-Honggerberg, 8093 Zurich, Switzerland
e-mail: ivy@phys.ethz.ch

K. Hippe
e-mail: hippe@phys.ethz.ch

C. Vockenhuber
e-mail: vockenhuber@phys.ethz.ch

S. Martin · M. Rigo · D. Pasqual
Dipartimento Di Geoscienze, Università Di Padova, 35133 Padua, Italy
e-mail: silvana.martin@unipd.it

M. Rigo
e-mail: manuel.rigo@unipd.it

D. Pasqual
e-mail: daria.pasqual@unipd.it

P. Campedel · A. Viganò
Servizio Geologico Della Provincia Autonoma Di Trento, 38121 Trento, Italy
e-mail: paolo.campedel@provincia.tn.it

A. Viganò
e-mail: alfio.vigano@provincia.tn.it

G. Carugati
Dipartimento Di Scienza E Alta Tecnologia, 22100 Como, Italy
e-mail: gabriele.carugati@uninsubria.it

M. Mikoš et al. (eds.), *Advancing Culture of Living with Landslides*,
DOI 10.1007/978-3-319-53483-1_41

Keywords
Cosmogenic ^{36}Cl • Rock avalanche • Historical seismicity

Introduction

As part of a project aimed at understanding causes of past catastrophic rock slope failures in the Trentino Province of Italy, we are studying several of the largest rock avalanches. Increasing population density in narrow mountain valleys in the Alps makes understanding of such factors especially crucial. A key piece of information vital for disentangling triggering mechanisms, dominantly climate or dominantly seismic, is the timing of events. Before the availability of isotopic dating, it was generally accepted that the largest Alpine landslides occurred within a few millennia after deglaciation 18,000 years ago. Indeed, many of the large landslides were interpreted to have come down when glaciers still occupied the main valleys. As more and more landslides were isotopically dated, it became apparent that most had actually occurred during the Holocene. In the Alps, two main periods of increased landslide activity have been recognized: (1) the early Holocene ca. 10–9 kyr and (2) the middle Holocene ca. 5–4 kyr (Prager et al. 2008 and references therein). For northern Italy, an additional period of enhanced landslide activity during the last 1500 yr was recognized and suggested to have been related to neotectonics (Galadini et al. 2001; Sauro and Zampieri 2001; Martin et al. 2014).

To reconstruct sequences of events at each site, we combine detailed field mapping with LiDAR imagery interpretation and sedimentological study. Cosmogenic nuclide surface exposure dating, combined with archeological and historical data, provides time control. By studying individual sites in detail, we assemble a database that allows informed examination of magnitude-frequency patterns in the past. Clusters in time may then be interpreted as rock slope failure being either climatically driven (high porewater pressures related to periods of extreme precipitation) or seismically triggered. Herein, we present recent results from the Castelpietra landslide (Fig. 1), which is located in the middle Adige Valley about 15 km south of Trento (Fig. 2). The role of earthquakes in rock failure is gauged by comparing temporal patterns of dated rock slope failures lying in close proximity to Castelpietra with databases of historical seismicity and recent fault activity.

Fig. 1 Castelpietra landslide deposits with Cengio Rosso head scarp and lower scarp below. Note 2009 rockfall deposits

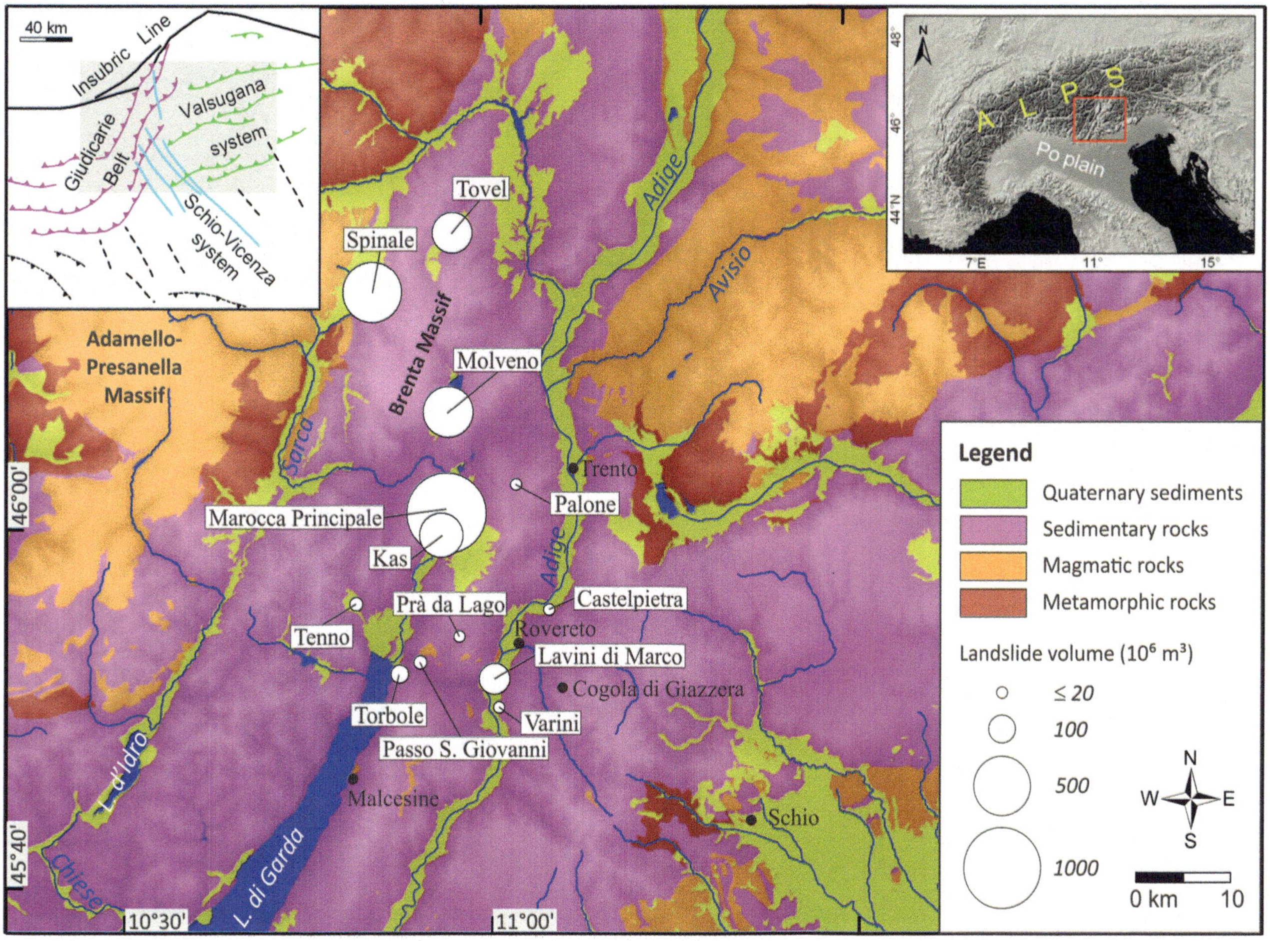

Fig. 2 Simplified geological map of the Sarca and Adige Valleys (data from the Geological Survey of Trento, CARG project database). *Circle sizes* are proportional to landslide volumes. *Upper left* inset shows major fault systems (Viganò et al. 2015 and references therein); *grey shade* indicates extent of main map

Castelpietra Geology and Geomorphology

Castelpietra deposits lie at the foot of a reddish rock wall known as Cengio Rosso (920 m a.s.l.), which is the main detachment zone (Fig. 1). The wall is 300 m high and 2 km long and composed of Dolomia Principale (upper Carnian—Rhaetian) overlain by Gruppo dei Calcari Grigi (Monte Zugna Formation of Hettangian—Sinemurian and Calcare Oolitico di Loppio Formation of Sinemurian ages) (Fig. 3). Bedding planes dip 30–45° to the W and WNW, forming a dip slope. A second detachment scarp is located beneath the Cengio Rosso, at about 450 m a.s.l. It is 70 m high and 800 m long, and composed only of Dolomia Principale. The lower scarp is buried along the southern end by rockfall debris. The NE-SW oriented Cengio Rosso head scarp belongs to the set of near-vertical fracture planes commonly observed along the Adige Valley lateral slopes. It is cut by NNE-trending fractures (Fig. 3), E-W-oriented open fractures perpendicular to the wall (Fig. 1), and by NW-SE fault planes related to the Schio-Vicenza system. The latter are as well exemplified by the two NNW-oriented narrow valleys visible at the center bottom of Fig. 3.

The Castelpietra deposits cover about 1.2 km^2, between 400 and 200 m a.s.l. The main blocky deposits on the Adige River plain are buried on the eastern side by large rockfall debris cones (Fig. 3). The suggestion of this giant scree apron, and the presence of an upper and lower scarp indicate repeated gravitational activity of varying magnitude after deposition of the main valley floor blocky deposit. Boulders on the surface of the main deposit range up to 20 m in diameter and are dominated by Dolomia Principale and Calgari Grigi dolomitized limestones. Boulder surfaces weather white to pink due to the strongly recrystallized dolomitic composition. No rillenkarren or other karst

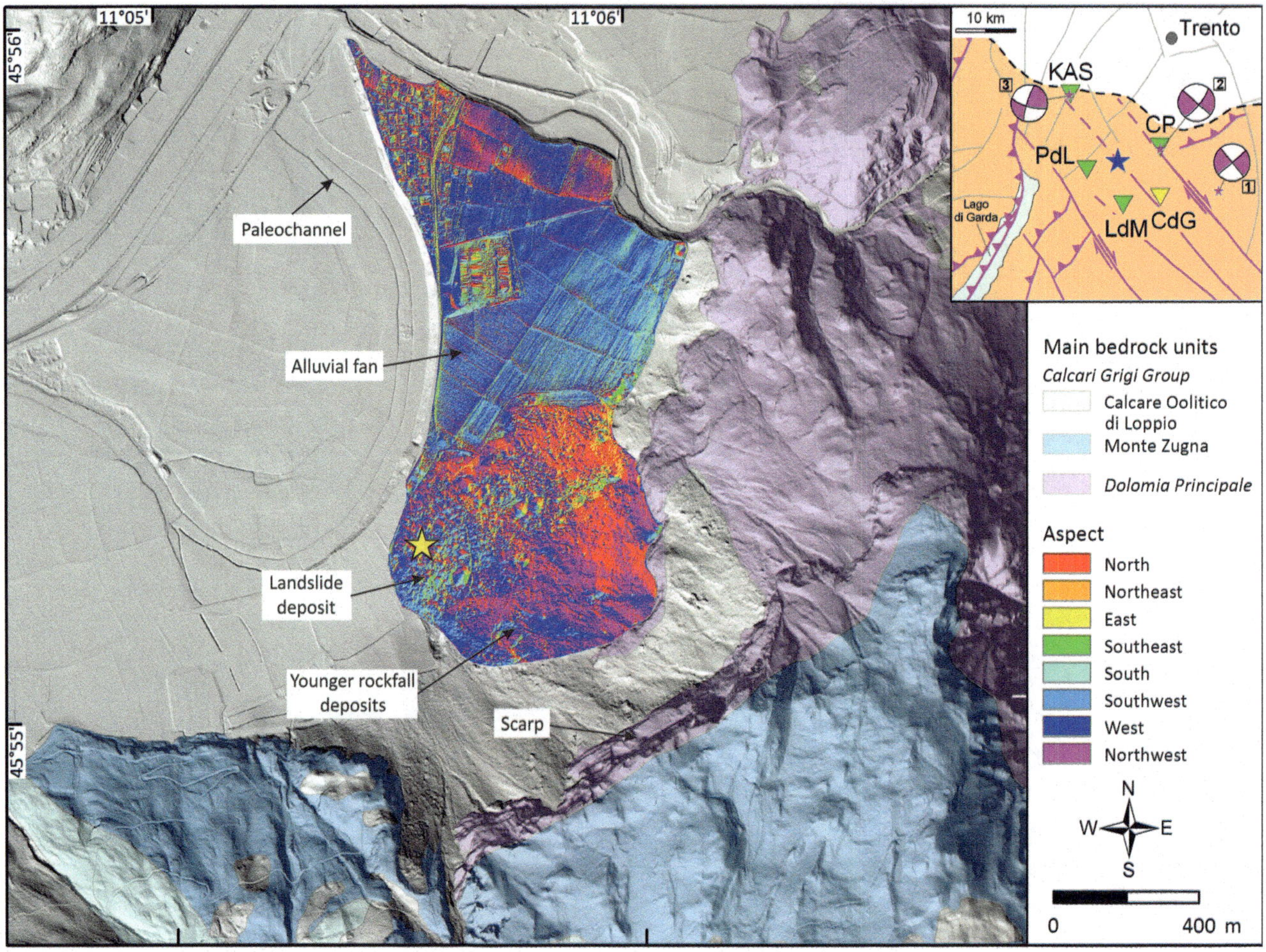

Fig. 3 Hillshade map (sun azimuth 225°) of Castelpietra; landslide and fan deposits shown with aspect map. *Yellow star* indicates location of dated boulders (CP1, CP2, see Table 2). Bedrock data from the Geological Survey of Trento, CARG project database (modified). *Inset map* Seismotectonic sketch, *orange shading* indicates high-seismicity area, major faults shown in *gray*, active faults in *magenta*, *green triangles* indicate landslides discussed in text (LdM Lavini di Marco, PdL Prà da Lago), *yellow triangle* is Cogola di Giazzera (CdG) cave. 1046 AD earthquake epicenter from Guidoboni and Comastri (2005) is shown as *blue star*. Three recent earthquake epicenters (chronologically listed) from Viganò et al. (2015) shown as *magenta stars*, with focal mechanisms (see text)

dissolution structures were visible on boulder surfaces. The travel distance of debris of the blocky valley floor deposit is 0.7 km; a steep Fahrboschung angle of 42° was calculated by Abele (1974). Paleochannels (the Adige course before correction in 1850 and 1860 AD) visible in Fig. 3 show that the river meandered quite close to the toe of the landslide but was not deflected by the event.

Age of the Castelpietra Landslide

Cosmogenic ^{36}Cl Exposure Dating

The tops of two boulders within the main Castelpietra blocky deposits were sampled for ^{36}Cl surface exposure dating, with two samples taken from each boulder (A and B designations in Table 1). Thin sections show that the two boulders we sampled are strongly dolomitized limestone (Calcari Grigi). Total Cl and ^{36}Cl were determined at the ETH AMS facility of the Laboratory for Ion Beam Physics (LIP) with the 6MV tandem accelerator. The ^{36}Cl/Cl ratios of the samples were normalized to the ETH internal standard K382/4N with a value of ^{36}Cl/Cl = 17.36 × 10^{-12}, which is calibrated against the primary ^{36}Cl standard KNSTD5000 (Christl et al. 2013). Full process chemistry blanks (4.2 ± 0.8 × 10^{-15}) were subtracted from measured sample ratios. We calculated surface exposure ages with the LIP ETH in-house MATLAB code based on the parameters presented in Alfimov and Ivy-Ochs (2009 and references therein). We used a production rate of 54.0 ± 3.5 ^{36}Cl atoms (g Ca)$^{-1}$ yr^{-1}, which encompasses a muon contribution at the rock surface of 9.6%; and a value of 760 ± 150 neutrons (g air)$^{-1}$ yr^{-1}. These values are in excellent agreement with those recently published by Marrero et al. (2016).

Table 1 Site data for the ^{36}Cl exposure dated rock samples at Castelpietra

Sample No.	Boulder height (m)	Lat. (°N)	Long. (°E)	Elev. (m)	Thick. (cm)	Shield.
CP1A	6	45.920	11.093	207	2	0.968
CP1B	6	45.920	11.093	207	4	0.952
CP2A	3	45.920	11.093	206	2.5	0.960
CP2B	3	45.920	11.093	206	2	0.912

Table 2 Measurement data and ^{36}Cl exposure ages for the two Castelpietra boulders

Sample No.	CaO (wt%)	Cl (ppm)	^{36}Cl (10^4 atoms/g)	Exposure age (years)	Exposure age (years) boulder mean
CP1A	33.67	21.8 ± 0.5	1.04 ± 0.35	640 ± 210	820 ± 260
CP1B	33.43	36.6 ± 0.2	2.01 ± 0.49	1180 ± 290	
CP2A	34.27	34.4 ± 0.5	1.78 ± 0.53	1040 ± 310	1070 ± 230
CP2B	33.53	45.5 ± 0.1	1.92 ± 0.57	1110 ± 330	

Production rates were scaled to the latitude, longitude, and altitude of the sites based on Stone (2000). No correction was made for karst weathering of the boulder surfaces. Stated errors of the exposure ages include both analytical uncertainties and those of the production rates. An error-weighted mean was calculated from the two samples (e.g. CP1A, CP1B) from each boulder (Table 2). For the age of the deposit as a whole a mean (non-weighted) of 950 ± 270 yr ago was determined. The uncertainty of the mean is at the 1σ level and includes the uncertainties of the individual boulder means (boulder CP1 and CP2).

Historical Constraints

Castelpietra is one of the numerous castles present in the Adige Valley along the main road (Via Claudia Augusta), which went from Verona to northern countries since Roman Imperial time. The road was located on the east bank of the river, not far from the castle. The castle itself (Castelpietra) was built sometime before 1206 AD, as it is mentioned in a document of that time describing "Castel Pietra" as belonging to the Castel Beseno family (Castel Beseno is another castle located few km north of Castelpietra). The location is in the narrowest segment of this stretch of the Adige River valley where a wall could control passage. The structures were built right on landslide debris; the oldest tower is on top of a landslide mega boulder. Therefore, landslide boulders were present before 1206 AD. Roman coins attributed to the Augustus Imperial age (about 27 BC), and younger, were discovered in the area (Roberti 1925), but the exact location where the coins were found is unknown.

Several smaller events from Cengio Rosso were recorded during historic times although the details are obscure. Dante Alighieri (1314) in Canto XII of Hell in the Divine Comedy, titled "The Landslide and the Minotaur," wrote about a rockslide event (Gorfer 1994). For some authors, the landslide described by Dante was that of Castelpietra. By contrast, for Giuseppe Tellani, failure occurred at Castelpietra 1310 AD, after Dante Alighieri's visit a few years earlier (Gorfer 1994). For Tellani the landslide described by Dante was that of Marco ("Slavini di Marco"), especially as he referred to a sliding down of rock slabs. The Castelpietra castle was apparently threatened by a third historical rockslide with detachment from Cengio Rosso, as described by Trapp in 1506 (Gorfer 1994). The extent of that deposit was not recorded. In February 2009 (also in 2016), several small rockfalls fell from Cengio Rosso as visible in Fig. 1.

Discussion and Conclusions

The blocky debris at Castelpietra was deposited 950 ± 270 yr ago. Considering the sampling year of 2009, this corresponds to 1060 ± 270 AD. At Lavini di Marco, which lies about 10 km downstream from Castelpietra, ^{36}Cl surface exposure ages of 1600 ± 100 yr and 1400 ± 100 yr were obtained from bedrock bedding plane detachment surfaces in the upper reaches. From the steep head scarp itself an exposure age of 800 ± 200 yr was found (Martin et al. 2014). These ages correspond to 510 ± 100 AD and 1210 ± 200 AD, respectively, for the youngest events at Lavini di Marco. The Varini deposits comprise the southernmost independent slide at Lavini di Marco site. The age of a palesol buried during the Varini event is 565—970 AD based on radiocarbon (Martin et al. 2014 and references therein). Recently, we presented ^{36}Cl exposure ages of 1080 ± 160 yr (930 ± 160 AD) for the Kas event (volume 300 10^6 m^3) in the Sarca Valley (Ivy-Ochs et al. submitted). Kas lies about 15 km from Castelpietra and 17 km from Lavini di Marco (Fig. 2). Finally, in the Cogola di Giazzera cave located about 6 km southeast of Rovereto, Frisia et al. (2005) dated the last episode of stalagmite failure to 1060 ± 70 AD and propose linkage to a major seismic event. Moreover, they suggest that the nearby Prà da Lago rockslide (Fig. 2), which has been dated to 1020–1260 AD based on radiocarbon (Castellarin et al. 2005), was related to the same event. The remarkable coincidence in time of the several events which all lie within a maximum distance of 20 km from each other strongly suggests a seismic origin.

Based on compilations of historical seismicity (Catalogue of strong earthquakes in Italy, CFTI4Med by Guidoboni et al. 2007; Parametric Catalogue of Italian Earthquakes, CPTI15 by Rovida et al. 2016), we therefore searched for an event within this region during the time period in question. We highlight here the two strongest earthquakes that struck southern Trentino and the central Po plain around 1000 AD. They are the "Middle Adige Valley" event dated to 1046 AD (estimated epicentral intensity IX MCS; Guidoboni and Comastri 2005; Guidoboni et al. 2007), and the "Verona" earthquake, which occurred in 1117 AD (estimated epicentral intensity IX MCS; Guidoboni et al. 2007). According to a contemporary source more than 30 castles were partly destroyed during the 1046 AD event (Guidoboni and Comastri 2005). Both earthquakes are comparable in age with the Castelpietra event at 1060 ± 270 AD dated here and the Kas deposits of the Marocche di Dro rock avalanche complex at 930 ± 160 AD.

Although there are significant uncertainties associated with the epicentral coordinates of the 1046 AD earthquake, its location (Guidoboni and Comastri 2005) is considerably closer to the Castelpietra and Kas sites (about 6 and 11 km of epicentral distance, respectively) than the 1117 AD earthquake (about 72 and 80 km, respectively) (Fig. 3 inset). Based on an epicentral intensity of IX MCS (Guidoboni et al. 2007) and a distance of about 10 km, the Peak Ground Acceleration (PGA) for the 1046 AD event can reasonably be considered to have exceeded the threshold proposed by Guo et al. (2015) for triggering a landslide on a slope. In contrast, a much greater distance from the seismic source (the distance between Castelpietra and Kas and the 1117 AD epicenter is more than six times greater) would have yielded considerably lower PGA values. In light of the non-linear decrease of co-seismic ground acceleration with distance (Boore et al. 2014 and references therein), we can therefore suggest a possible correlation between the 1046 AD seismic event and both the Castelpietra and Kas landslides, and possibly the Varini and Prà da Lago events as well.

Principal present-day sources of seismicity include faults belonging to the Schio-Vicenza and Southern Giudicarie systems. The three strongest instrumentally recorded earthquakes in the study area occurred on 13 September 1989, 26 April 1999, and 16 June 2000 (Viganò et al. 2015) (chronologically numbered locations shown in Fig. 3 inset). The 1989 event, with local magnitude of 4.7 and focal depth of 9.1 km, testifies that the Schio-Vicenza fault system is locally active and can generate earthquakes able to produce strong shaking. The 1999 event (local magnitude 3.4, focal depth 1.5 km) was just under Castelpietra (at very shallow depth). The 2000 event (local magnitude 3.2, focal depth 5.3 km) was located right under Marocche (also at quite shallow depth). The latter two cases emphasize the likelihood of progressive deforming and damaging of rock slopes (fatigue process; cf. Gishig et al. 2015) by local seismic activity as a key preparatory factor in the study region. Moreover, all three events have almost the same focal mechanisms (Fig. 3 inset, cf. Viganò et al. 2015) indicating similar kinematics.

Faults of the Schio-Vicenza system most probably provided the active trigger and facilitated seismic wave transmission to landslide sites during seismic activity. Repeated seismic shaking both related to or prior to the 1046 AD event weakened the rock. Based on the prevalence of active faults at all four sites discussed here a scenario, which involves rock fatigue due to numerous events, and trigger right at the 1046 AD event(s), is likely.

Acknowledgements Funding from the Geological Survey of Trento Province and the University of Padova (Progetto di ricerca di Ateneo 2014, CPDA140511) allowed this research. We thank the Geological Survey of Trento Province for access to the DEM. The AMS group at Ion Beam Physics provided support of fieldwork, labwork, and AMS measurements.

References

Abele G (1974) Bergsturze in den Alpen. Ihre Verbreitung, Morphologie und Folgeerscheinungen. Wiss. Alpenvereinshefte 25, Munchen

Alfimov V, Ivy-Ochs S (2009) How well do we understand production of ^{36}Cl in limestone and dolomite? Quat Geochr 4:462–474

Alighieri D (1314) Divina Commedia, Inferno, XII, 4–9

Boore DM, Stewart JP, Seyhan E, Atkinson GM (2014) NGA-West2 equations for predicting PGA, PGV, and 5% damped PSA for shallow crustal earthquakes. Earthq Spectra 30:1057–1085

Castellarin A, Picotti V, Cantelli L et al. (2005) Note illustrative della carta geologica d'Italia, Foglio no. 80. Riva del Garda

Christl M, Vockenhuber C, Kubik PW, Wacker L, Lachner J, Alfimov V, Synal H-A (2013) The ETH Zurich AMS facilities: performance parameters and reference materials. Nucl Instrum Methods Phys Res B 294:29–38

Frisia S, Borsato A, Richards DA, Miorandi R, Davanzo S (2005) Variazioni climatiche ed eventi sismici negli ultimi 4500 anni nel Trentino meridionale da una stalagmite della Cogola Grande di Giazzera. Studi Trent Sci Nat Acta Geol 82:205–223

Galadini F, Galli P, Cittadini A, Giaccio B (2001) Late Quaternary fault movements in the Mt. Baldo-Lessini Mts. sector of the Southalpine area (northern Italy). Neth J Geosci 80:187–208

Gischig V, Preisig G, Eberhardt E (2015) Numerical investigation of seismically induced rock mass fatigue as a mechanism contributing to the progressive failure of deep-seated landslides. Rock Mech Rock Eng 49:2457–2478

Gorfer A (1994) I castelli di Rovereto e della Vallagarina. Arti grafiche Saturnia

Guidoboni E, Comastri A (2005) Catalogue of earthquakes and tsunamis in the Mediterranean area from the 11th to the 15th century. Istituto Nazionale di Geofisica e Vulcanologia, Storia Geofisica Ambiente, Bologna, Italy

Guidoboni E, Ferrari G, Mariotti D, Comastri A, Tarabusi G, Valensise G (2007) CFTI4Med, catalogue of strong earthquakes in Italy (461 B.C.–1997) and Mediterranean area (760 B.C.–1500). An advanced laboratory of historical seismology. INGV-SGA. http://storing.ingv.it/cfti4med/

Guo D, He C, Xu C, Hamada M (2015) Analysis of the relations between slope failure distribution and seismic ground motion during the 2008 Wenchuan earthquake. Soil Dyn Earthq Eng 72:99–107

Ivy-Ochs S, Martin S, Campedel P, Hippe K, Alfimov V, Vockenhuber C, Andreotti E, Carugati G, Pasqual D, Rigo M, Viganò A Geomorphology and age of the Marocche di Dro rock avalanches (Trentino, Italy). Quat Sci Rev (submitted)

Marrero SM, Phillips FM, Caffee MW, Gosse JC (2016) CRONUS-Earth cosmogenic ^{36}Cl calibration. Quat Geochron 31:199–219

Martin S, Campedel P, Ivy-Ochs S, Viganò A, Alfimov V, Vockenhuber C, Andreotti E, Carugati G, Pasqual D, Rigo M (2014) Lavini di Marco (Trentino, Italy): ^{36}CI exposure dating of a polyphase rock avalanche. Quat Geochron 19:106–116

Prager C, Zangerl C, Patzelt G, Brandner R (2008) Age distribution of fossil landslides in the Tyrol (Austria) and its surrounding areas. Nat Hazard Earth Syst Sci 8:377–407

Roberti G (1925) Monete di accertata provenienza trentina nel Museo Nazionale di Trento. Studi Trentini di Scienze Storiche 6:307–316

Rovida A, Locati M, Camassi R, Lolli B, Gasperini P (eds) (2016) CPTI15, the 2015 version of the parametric catalogue of Italian earthquakes. INGV. doi:10.6092/INGV.IT-CPTI15

Sauro U, Zampieri D (2001) Evidence of recent surface faulting and surface rupture in the Fore-Alps of Veneto and Trentino (NE Italy). Geomorphology 40:169–184

Stone JO (2000) Air pressure and cosmogenic isotope production. J Geophys Res: Solid Earth 105(B10):23753–23759

Viganò A, Scafidi D, Ranalli G, Martin S, Della Vedova B, Spallarossa D (2015) Earthquake relocations, crustal rheology, and active deformation in the central—eastern Alps (N Italy). Tectonophysics 661:81–98

Coupled Slope Collapse—Cryogenic Processes in Deglaciated Valleys of the Aconcagua Region, Central Andes

Stella Maris Moreiras

Abstract

This paper presents coupled geomorphological processes such as glacial advances, gravitational collapses, and solifluction engaged to the environment climate changes. Complex landslides with a puzzling classification were identified by a landslide inventory of the Aconcagua Park involving the highest peak of South America (Aconcagua peak 6958 m a.s.l.). These deformed deposits were interpreted as gravitational collapsed moraines occurred after the Holocene–Pleistocene ice retreat on these Andean valleys. The stabilized huge masses began to be partially remobilized by solifluction phenomena generating protalus ramparts. At present well developed debris rock glaciers are established at the top landslide surfaces. This finding confirms glacial/interglacial cycles in the Central Andes are related to glacial advances supported by preserved moraines and gravitational collapses caused by ice loss during glacial retreat. However, the occurrence of cryogenic processes after collapse could evidence a periglacial environment restoration linked to a colder period. Therefore, available debris/sediments infilling deglaciated valleys will be mainly mobilized by glaciers, slope collapses or periglacial processes depending on the climate environment conditions.

Keywords

Complex landslides • Permafrost degradation • Debris rock glaciers • Paleo-climate

Introduction

In the Argentinean Andes (32°S) largest complex landslides have been commonly associated with warm-wet periods (Espizúa 2005; Fauqué et al. 2009a), even though local evidences of such interstadials are lacking. Moreiras (2006) suggested that although the rock avalanches clustered in Cordon del Plata range had a seismic origin, probable they occurred in a wetter interstadial previous to Uspallata glaciation. But geological record of this wetter period has not been found in the Mendoza River valley. As well, rock avalanches generated from the Tolosa mount were associated with a Holocene interstadial which is not certainly proved by local climate proxies (Rosas et al. 2008). Hence at present, a climatic cause of palaeolandslides is weakly supported in the Central Andes as local records of glacial—interglacial periods seem to be vague (Moreiras et al. 2015, 2016).

Current investigations in the Argentinean Andes (32ºS) reinterpret several moraines as collapses. The Holocene deposit described as Horcones moraine (Espizúa 1993) is presently considered as the distal facies of a rock avalanche sourced in the south face of Mount Aconcagua (6958 m a.s.l.). This chaotic mass travelled 30 km until arrived to the Cuevas River where it generated a palaeolake (Fauqué et al. 2009a). Likewise, occurrence of MIS 2 locally represented by Penitentes glaciation is controversial. The outwash of this drift has

S.M. Moreiras (✉)
CONICET- IANIGLA (CCT Mendoza), Av Ruiz Leal S/N. Capital. Parque Gral San Martin, 5500 Mendoza, Argentina
e-mail: moreiras@mendoza-conicet.gob.ar

S.M. Moreiras
Fac. de Cs Agrarias. Univ. Nacional de Cuyo. a. Brown S/N, Chacras de Coria, Mendoza, Argentina

M. Mikoš et al. (eds.), *Advancing Culture of Living with Landslides*,
DOI 10.1007/978-3-319-53483-1_42

been recently re-interpreted as a debris flow resulting from the El Abuelo—Mario Ardito rock avalanche. Therefore, reinterpretation of the genesis of chaotic deposits along Mendoza River valley argued both Horcones and Penitentes glaciations (Fauqué et al. 2009b; Hermanns et al. 2015). Thus, paleoclimate correlation established by Espizúa (1993) with glaciations of the Aconcagua valley (Caviedes 1972) is out-of-date as Portillo and Salto del Soldado were reclassified as landslides (Antinao and Gosse 2009; Moreiras and Sepúlveda 2015). Here, the study of the gravitational collapsed moraines occurred in deglaciated valleys of the Aconcagua Park may prove the questionable link between landslides and paleoclimate conditions in the Central Andes.

Study Area

The study area involves the Aconcagua Park (32° 36′–32° 48′ S, 69° 50′–70° 04′ W) covering an area of

710,000 km^2 (Fig. 1). This reserve was founded in 1983 due to presence of highest peak of the Andes, Aconcagua mount (6958 m a.s.l.). Such peak is a main icon for climbers around the World resulting in an international tourist attraction. This region corresponds to the Main Cordillera. This morphostructural unit is characterized by Jurassic, Cretaceous, and Oligocene–Miocene continental volcaniclastic rocks intruded by Miocene–Pliocene dikes. The basement is constituted by Carboniferous rocks. Structurally, this region corresponds to the Aconcagua fold and thrust tectonic belt (AFTB) integrated by westward-verging reverse slip that experienced combinations of compressional, extensional, and shear tectonics throughout the Late Cenozoic (Ramos 1996).

The Aconcagua Park comprehends U-shaped valleys of Las Vacas and Horcones rivers. This high mountain landscape was modelled by Pleistocene glaciations (Espizúa 1999, 2004). Tundra climate predominates between 2700 and 4100 m a.s.l. Permanent ice covers the highest altitudes were mean monthly temperature is lower than 0 °C (Koeppen 1948). Mainly precipitation falls as snow during winter season. Mean annual precipitation is 500 mm. Strong violent winds flow in this region with speed over 200 km/h.

Methods

Landslides began to be identified recently in this sector of the Central Andes. The earliest description of megalandslides along the valleys of Las Cuevas and Mendoza rivers comes from a geomorphological study in the 60's by Salomón (1969). Later, this kind of processes was recognised in the Chilean Andes (Abele 1984) by reinterpretation of ancient moraines of the Aconcagua and Yeso valleys (Caviedes 1972; Marangunic and Thiele 1971). Following these ideas, several studies reinterpreting the glacial origin of Quaternary deposits were performed. At the same time revision of the world global glaciations epochs continues. In this research we used satellite images interpretation and field checking for identifying landslide deposits in the Aconcagua region. A geomorphological map was carried out by GIS tools including digitalization of different landforms. This geomorphological map of Aconcagua park (1:125,000) involves 852.5 km^2, surrounding the valleys of Las Vacas, Horcones, Santa María and Cruz de Caña rivers. As well a landslide inventory map was obtained including classification of this type of processes (WP/WLI 1993). A statistical analysis lets to estimate areas of identified deposits. We recognized different types of landslides, but greatest ones are complex gravitational collapses associated with relict lateral moraines.

Landslide Inventory

Landslides coves at least 7% of the studied region. These landforms affect a total extension of 67.3 km^2 representing (7.8%) of the reserve. This spatial distribution coincides partially with discontinuous permafrost environment over 3.200 m above sea level. However, landslides stay at a lower altitude of continuous permafrost evidenced by the presence of uncovered glaciers. Due to an intense cryogenic weathering, talus rockfall (105) are very common processes. Nevertheless, complex landslides cover a more extensive area (30.2 km^2), while slides cover 17.1 km^2. Greater landslides correspond to complex collapses with more than a type of movement. They use to involve debris material of relict moraines and under laying rock outcroppings. Maybe, the most impacting landslide is the just mentioned Horcones rock avalanche (2.5 km^2) (Pereyra 1995; Fauqué et al. 2009a; Hermanns et al. 2015).

Contrary to our expectations, few rockslides were distinguished along the Vacas River; even though they are very common in the Horcones valley. Instead, enormous collapsed mass of relict lateral moraines were found. They were interpreted as gravitational collapsed moraines probably generated after Late Pleistocene ice retreat on these deglaciated valleys. These extremely huge landforms commonly dammed valleys. Wide up-stream alluvial plains evidence these paleo natural dammings even finer lake sediments are rarely found.

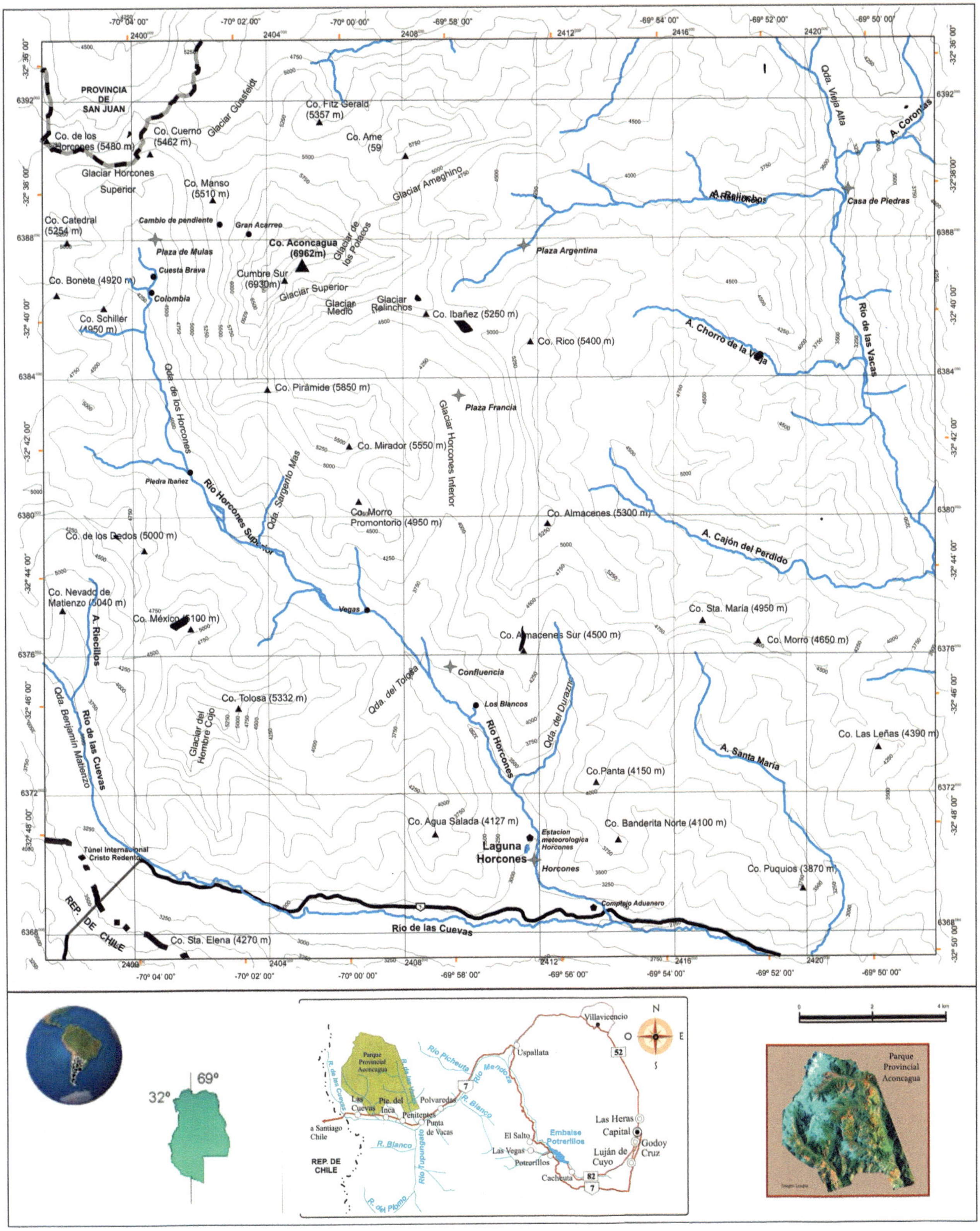

Fig. 1 Location of study area with main localities of the Aconcagua park

Gravitational Collapsed Moraines—Cryogenic Processes

Prominent debris rock glaciers were identified in source areas or main scarps of landslides classified as gravitational collapsed moraines. Load of debris material from the crown of these complex landslides is being remobilized by cryogenic rock glaciers. These particular ice bodies are generated by a type of permafrost reptation known as solifluction phenomenon. This cryogenic process used to generate marked tongues and ridges growing until generate protalus ramparts.

Our observations imply that the glacial origin deposit collapsed after ice retreat during interglacial warmer conditions, but later the same material is mobilized by solifluction during colder conditions until today. The permafrost degradation conditioning moraine instability was followed by a colder period during which periglacial environment was restored at this altitude (3.200 m a.s.l.).

The debuttressing process, maybe combined with land isostatic rebounding, after Late Pleistocene glacier mass retreat was proposed as the main conditioning factor of catastrophic geological processes in Las Cuevas valley, as it was pointed out for landslides in the Argentinean Patagonia (Gonzalez Diaz 2003, 2005). Abele (1974) remarked that greater permafrost degradation and thawing during the Late Pleistocene-Holocene have strongly influenced on slope instability in the Central Andes. However, the existence of debris rock glaciers has not been interpreted as a signal of cooling during the Holocene. Age of this type of debris rock glaciers was estimated on 3000 years BP in Blanco River basin (Moreiras et al. 2016); whereas a Late Pleistocene-Holocene age was established from moraines (Espizúa 1999) that collapsed lately.

Fig. 2 Complex landslides in the Aconcagua Park: **a** Rock avalanche coming from the Tolosa peak, **b** Deep complex landslide in Pampa de Leñas, **c** Gravitational collapsed moraines in Casa de Piedra (Vacas River), and **d** Gravitational collapsed moraine along the Vieja alta valley

Conclusions

Huge gravitational collapsed moraines were identified in the Vacas River valley, Aconcagua Park. Their location on upper catchment of deglaciated valleys and their relative proximity to present glaciers suggest a link to glacial/cryogenic processes.

The debuttressing of main valleys after ice retreat during Late Pleistocene-Holocene should conditioned slope instability in these high steep mountain ranges. However, the incidence of permafrost degradation during interglacial warm periods could mainly force instability of these saturated instable moraines. These huge masses of debris material represent available sediments for being mobilized by cryogenic processes or debris rock glaciers in the periglacial environment.

A coupled system between slope collapses—cryogenic processes occurred in these ice abandoned valleys of the Aconcagua region, Central Andes (32°). Quaternary regional studies (Espizúa 1993, 1999, 2004; Espizúa and Bigazzi 1998) suggest glacial advances occurred in the Late Pleistocene, being moraine collapses a later phenomenon mainly related to permafrost degradation during an interglacial period. Newer protalus ramparts engendering debris-rock glaciers in the Aconcagua Park should be established during a Holocene colder period (Fig. 2).

Acknowledgements The research was funded by SECTYP 2011-2013 leader by Moreiras.

References

Abele G (1974) Bergstürze in den Alpen: the Verbreitung, Morphologie und Folgeerscheinun-gen. Wissenschaftliche Alpenvereinshefte, Heft 25, Munchen

Abele G (1984) Derrumbes de montaña y morenas en los Andes chilenos. Revista de Geografía Norte Grande 11:17–30

Antinao JL, Gosse J (2009) Large rockslides in the Southern Central Andes of Chile (32–34.5°S): tectonic control and significance for Quaternary landscape evolution. Geomorphol 104:117–133

Caviedes C (1972) Geomorfología del Cuaternario del valle de Aconcagua, Chile Central. Freiburger Geographische Hefte 11:153

Espizúa LE (1993) Quaternary Glaciations in the Rio Mendoza Valley, Argentine Andes. Quat Res 40:150–162

Espizúa LE (1999) Chronology of late Pleistocene glacier advances in the Río Mendoza Valley, Argentina. Glob and Planet Change 22: 193–200

Espizúa LE (2004) Pleistocene glaciations in the Mendoza Andes. In: Ehlers J, Gibbard PL (eds) Quaternary glaciations extent and chronology, Part III. Elsevier, 69–73

Espizúa LE (2005) Megadeslizamientos pleistocénicos en el valle del río Mendoza, Argentina. In: Proceedings of XVI Congreso geológico Argentino, La Plata, vol 3, pp 477–482

Espizúa LE, Bigazzi G (1998) Fission-track dating of the punta de vacas glaciation in the rio Mendoza valley, Argentina. Quat Sci Rev 17:755–760

Fauqué L, Hermanns RL, Wilson CGJ (2009a) Mass removal in the Andean region. Revista de la Asociacion Geologica Argentina 65 (4):687

Fauqué L, Hermanns R, Hewitt K, Wilson C, Baumann V, Lagorio SY Di Tomasso I (2009b) Mega-landslide in the South face of Aconcagua mount and its relation with deposits associated with a Pleistocene glaciation. Revista de la Asociación Geológica Argentina 65(4):691–712.c

González Díaz EF (2003) El englazamiento en la región de la caldera de Caviahue-Copahue (Provincia del Neuquén): Su reinterpretación. Revista de la Asociación Geológica Argentina 58(3): 356–366 (Buenos Aires)

González Díaz EF (2005) Geomorfología de la región del volcán Copahue y sus adyacencias (centro-oeste del Neuquén). Revista de la Asociación Geológica Argentina, 60(1):072–087 (Buenos Aires)

Hermanns R, Fauqué F, Wilson C (2015) 36Cl terrestrial cosmogenic nuclide dating suggests Late Pleistocene to Early Holocene mass movements on the south face of Aconcagua mountain and in the Las Cuevas-Horcones valleys, Central Andes, Argentina. In: Geodynamic Processes in the Andes of Central Chile and Argentina, Special Publication, Geological Society of London, 399. doi:10.1144/SP399.19

Koeppen W (1948) Climatología. Fond de Cult Econ, México

Marangunic C, Thiele R (1971) Procedencia y determinaciones gravimétricas de espesor de la morrena de la Laguna Negra, Provincia de Santiago. Comunicaciones 38:25

Moreiras SM (2006) Chronology of a Pleistocene rock avalanche probable linked to neotectonic, Cordon del Plata (Central Andes), Mendoza—Argentina. Quat Int 148(1):138–148

Moreiras SM, Hermanns R, Fauqué L (2015) Cosmogenic dating of rock avalanches constraining quaternary stratigraphy and regional neotectonics in the Argentine Central Andes (32°S). Quat Sci Rev 112:45–58

Moreiras SM, Lenzano MG, Riveros N (2008) Inventario de procesos de remoción en masa en el Parque provincial Aconcagua, provincia de Mendoza—Argentina. Multiequina Lat Am J Nat Resour 17:129–146

Moreiras SM, Páez MS, Lauro C, Jeanneret P (2016) First cosmogenic ages for glacial deposits from the Plata range (33°S): New inferences for Quaternary landscape evolution in the Central Andes. Quaternary International. Special volume. Interactions between Quaternary climatic forcing, tectonics and volcanism along some different tectonic settings of South America Quaternary International (in press). doi:10.1016/j.quaint.2016.08.041

Moreiras SM, Sepúlveda SA (2015) Megalandslides in the Andes of Central Chile and Argentina (32–43°S) and potential hazards. In: Geodynamic processes in the Andes of Central Chile and Argentina. Special Publication 399, Geol Soc of London 399:329–344

Pereyra FX (1995) Esquema geomorfológico del sector norte del valle del río Las Cuevas, entre Puente del Inca y Las Cuevas, Prov. de Mendoza. Revista de la Asociación Geológica, Argentina 50(1–4):103–110

Ramos VA (1996) Evolución tectónica de la alta cordillera de San Juan y Mendoza. En Ramos VA (ed) Geología de la región del Aconcagua, provincias de San Juan y Mendoza. Subsecretaría de Minería de la Nación, Dirección Nacional del Servicio Geológico, Anales 24(12):447–460. (Buenos Aires)

Rosas M, Wilson C, Hermanns H, Fauqué L, Baumann V (2008) Avalanchas de rocas de las Cuevas una evidencia de la desestabilización de las laderas como consecuencia del cambio climático del Pleistoceno superior. Proccedings of the XVII Congreso Geológico Argentino, Jujuy: 313–314

Salomón JN (1969) El alto valle del río Mendoza. Estudio de geomorfología. Boletín de Estudios Geográficos, vol XVI, N° 62:1–50

WP/WLI (1993) Multilingual landslide glossary. Bi-Tech Publishers, Richmond, British Columbia Canada, 59

Coastal Erosion and Instability Phenomena on the Coast of Krk Island (NE Adriatic Sea)

4th WLF 2017
4th World Landslide Forum
LJUBLJANA SLOVENIA EU

Igor Ružić, Čedomir Benac, and Sanja Dugonjić Jovančević

Abstract

This paper presents the influence of the marine erosion on slope stability in the south eastern coastal area of the Krk Island (north-eastern part of the Adriatic Sea). The bedrock (Paleogene marls and flysch) is occasionally covered with talus breccia from Quaternary period. The coast is strongly exposed to wave attack and thereby to marine erosion. Comparison of few orthophoto map generations shows significant coastal retreat during the fifty-year period. This phenomenon has been a fundamental trigger off different instability phenomena. The type of instabilities is a consequence of local geological fabric and resistivity of rock mass to marine erosion. In the investigated area, rock falls and slumps prevail in cliffs formed in talus breccias. Extremely high tides from 2008 and 2012 have caused significant coastal erosion. This is obviously an indicator of the possible higher hazard degree caused by the sea-level rise.

Keywords

Coast • Slope stability • Marine erosion • Sea level rise • Adriatic sea

Introduction

On the global scale, coastal zones are one of the most dynamic and rapidly changing environments. These changes are strongly induced by anthropogenic pressure in synergy with natural coastal processes and the global climate changes. Coastal risks at particular coastal locations depends on the global mean sea-level rise and its regional deviation (Masselink and Hughes 2003).

The Adriatic Sea is one of the several basins within the Mediterranean in which coasts are geomorphologically distinctively different. Eastern coast of the Adriatic Sea is locus typicus of "Dalmatian type coast", a type of drowned mountainous coast consisting of parallel fold ranges, resulting in channels parallel and normal to the general coastline and for this reason strongly opposed to the sandy coast that is found along the western part of the Adriatic Sea (Pikelj and Juračić 2013). Therefore, north eastern coast of Kvarner area and island chains Cres-Lošinj and Krk-Rab-Pag have "Dinaric" direction of elongation NW-SE (Fig. 1).

Recent shape of carbonate rocky coast in the Kvarner area is primarily a consequence of drowning of karst relief due to the sea-level rise. In less resistant flysch rocks depressions and bays were formed. In this area marine erosion caused by wave impact is not very pronounced due to its sheltered position. Therefore, coastal shape is primarily formed due to coastal geological fabric. Very steep carbonate coasts with scars up to 100 m high are tectonically formed (Juračić et al. 2009).

The great part of the coast in the Kvarner area has a low risk of wave induced marine erosion and it has low vulnerability to predicted sea-level change. But, marine erosion is locally expressed in some zones. Coastal zone near Stara

I. Ružić · Č. Benac · S.D. Jovančević (✉)
Faculty of Civil Engineering, University of Rijeka, Radmile Matejčić 3, 51000 Rijeka, Croatia
e-mail: sanja.dugonjic@uniri.hr

I. Ružić
e-mail: igor.ruzic@uniri.hr

Č. Benac
e-mail: cedomir.benac@uniri.hr

M. Mikoš et al. (eds.), *Advancing Culture of Living with Landslides*,
DOI 10.1007/978-3-319-53483-1_43

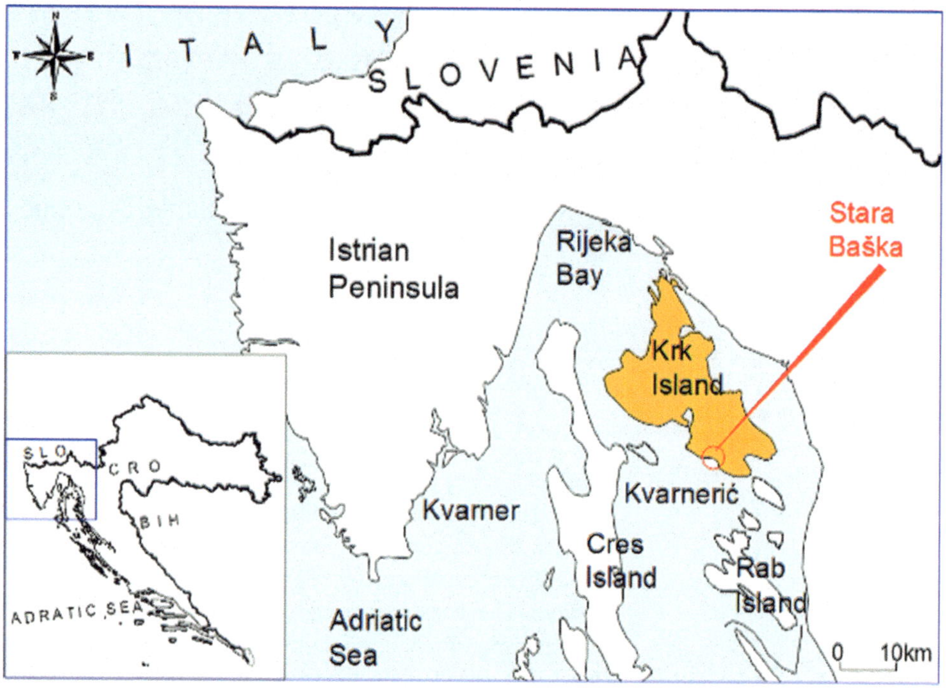

Fig. 1 Study area with the location of Stara Baška

Baška settlement in south eastern part of the Island of Krk is exposed to a significant coastal erosion (Fig. 1). This paper presents the interrelation between marine erosion, coastal changes, cliff retreat and slope stability.

Geological, Geomorphological and Meteorological Settings

The elevations around Stara Baška settlement are formed in Upper Cretaceous and Paleogene limestone, dolomitic limestone and dolomitic breccias (Benac et al. 2013). This is typical bare karstic landscape. Outcrops of Paleogene siliciclastic rocks mass, consisting of marls, siltstones and sandstones, are visible sporadically in the narrow coastal zone (Fig. 2). Despite bare karst in surrounding altitudes, a large part of this rock mass is covered by Quaternary talus breccia.

Mean sea level recorded at the nearest tidal gauge in Bakar is 0.15 m (vertical datum of the Republic of Croatia—CVD). This is a micro-tidal environment with a tidal range between 30 and 35 cm (Ružić et al. 2015). Kvarnerić aquatorium is a relatively enclosed channel part of the Adriatic Sea (Fig. 1). Most frequent are low and moderate winds, with silent periods, while the storm winds (speed over 30 m/s) are rare (Leder et al. 1998).

The stagnation of the sea-level occurred in the last 5000 years, after rapid Late Pleistocene–Holocene rise in the Adriatic Sea (Surić 2009). Near the mean sea level the erosional processes are reduced, while marine erosion increases. Those are favourable conditions for a cliff cut development in talus breccias and gravel beaches formation from eroded cliff material in investigated area near Stara Baška settlement.

North eastern wind bora has the highest frequency and reaches maximum speeds (>30 m/s), but waves are moderate due to small fetches. Winds from the southeast (SE) generate highest waves that have the biggest influence on coastal erosion. Figure 3 shows SWAN wave model results (Booij et al. 1999). Model simulations were generated on Kvarner bay bathymetry, SE wind direction and constant wind speed 26.5 m/s (50 years return period). Rab Island shelters the part of Krk Island coastal area of from SE wind waves. The significant wave height near investigated area are approximately 3 m.

The storm surge occurring during SE winds, can increase the water level up to 1.18 m CVD as recorded in Bakar on December 1st 2008. The combination of waves from SE and storm surge has major influence on coastal processes in the area of Stara Baška.

As a result of climate changes, accelerated sea level rising is expected in the Mediterranean Sea. Estimated sea level rise could be 62 ± 14 cm till the end of the 21st century (Orlić and Pasarić 2013).

Coastal Erosion and Instabilities Phenomena

Cliffs and marine terrace are dominant geomorphological forms in the investigated zone. The cliffs are formed in talus breccias. Their height is between 5 and 6 m. The top layer at the subsurface, of about 1 m thickness is completely weathered

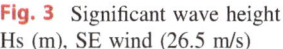

Fig. 2 Simplified engineering geological map of the investigated area: *1* carbonate rocks (Upper Cretaceous and Palaeogene limestones), *2* siliciclastic rock mass (Palaeogene marls and flysch), *3* periodical surface water flow, *4* partially canalized water flow, *5* cliffs

Fig. 3 Significant wave height Hs (m), SE wind (26.5 m/s)

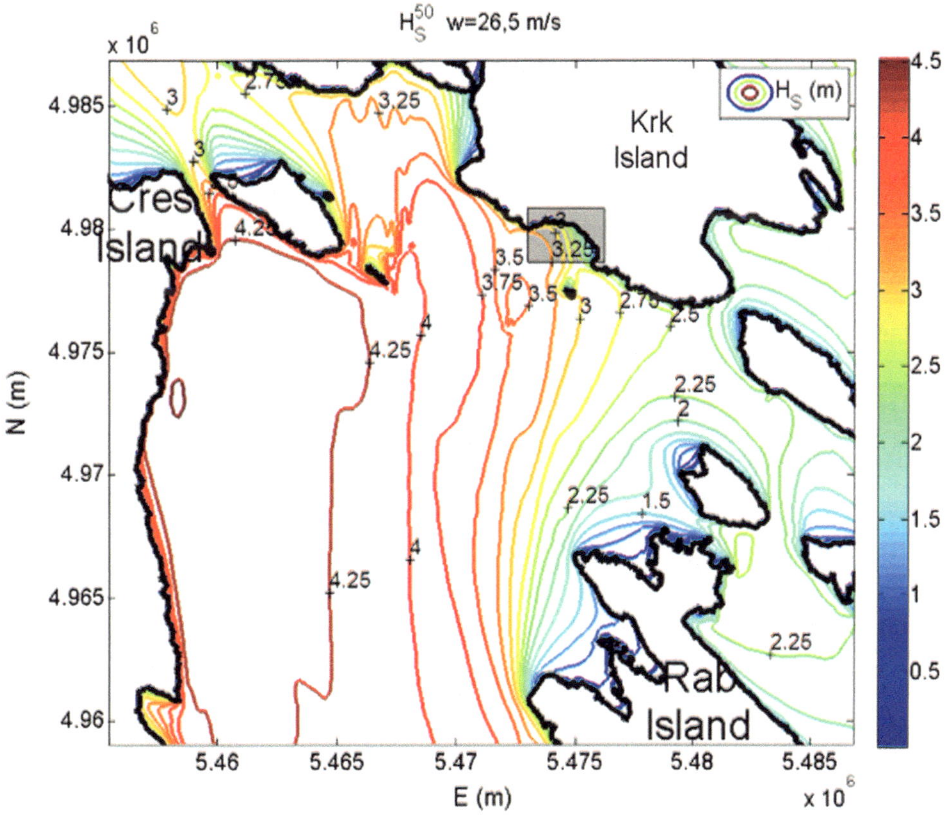

Fig. 4 The cliff formed in talus breccias and gravelly beach (*photo* Č. Benac)

and has residual soil features. The breccias have pronounced horizontal stratification and the joints were not visible. Horizontal layers consist of angular limestone fragments from few millimeters to blocks greater than 50 cm in size, with prevailing fragments of 1–4 cm. The matrix is reddish silt-sandy cement with a different $CaCO_3$ component. This degree of calcification has a great influence on the erodibility as well as on the breccias strength parameters (Fig. 4).

Gravelly beaches are formed on the cliff toes. The beach sediment average grain size is between 2 and 20 mm. Beach bodies are partially covered by large blocks.

Evidence of Coastal Changes

Because of this complex geological and meteorological setting, the coastline at Stara Baška is earmarked by a high variety of features including pocket beaches, headlands, cliffs, cliff undercuts, rock piles and landslides. Comparing two orthorectified aerial photographs from 1966 and 2004, cliffs have retreated by up to 5 m. Estimated horizontal accuracy is up to 0.50 m due to an old areal images distortion (Ružić et al. 2014, 2015).

In this paper the changes in the cliff-top position were identified by a comparison of ortho-rectified photos from 2002 and 2011, evidencing cliff retreat up to 2 m (Fig. 5). Estimated horizontal accuracy is up to 0.25 m due to an old areal images distortion. These cliffs are partially very unstable and have significantly retreated during the past decade. The cliffs face erosion is permanent phenomenon. Eroded coarse grained material from cliffs provides nourishment of the beach bodies. Despite that, the collapse in rock mass is periodical event, and partially eroded blocks of breccia are situated between beaches (Fig. 5).

Figure 5 shows beach coastline changes and cliff retreat. Beach coastline changes are just approximation because the sea levels during the areal images recording were not noted. It can cause significant errors in the assessment of the beach coastline. The analyses of historical aerial images have shown that the breccias are prone to rapid cliff recession (Ružić et al. 2014, 2015). Furthermore, large rocky blocks, can be found at the toe of the cliffs, indicating earlier local block falls. These cliffs are very unstable and have significantly retreated during the past five decades. Some of the island's important infrastructure, such as the tourist camp in Stara Baška, is threatened by these cliff collapses. Hence, a prediction of this type of geohazard is extremely valuable for future spatial planning and development on this coastline.

Stability Analyses

The cliffs formed in talus breccias are very unstable in the coastal area, near Stara Baška. Collapsed rocky blocks are often found on the beaches fronting the cliffs. A combined method for the stability analysis of the coastal cliffs that

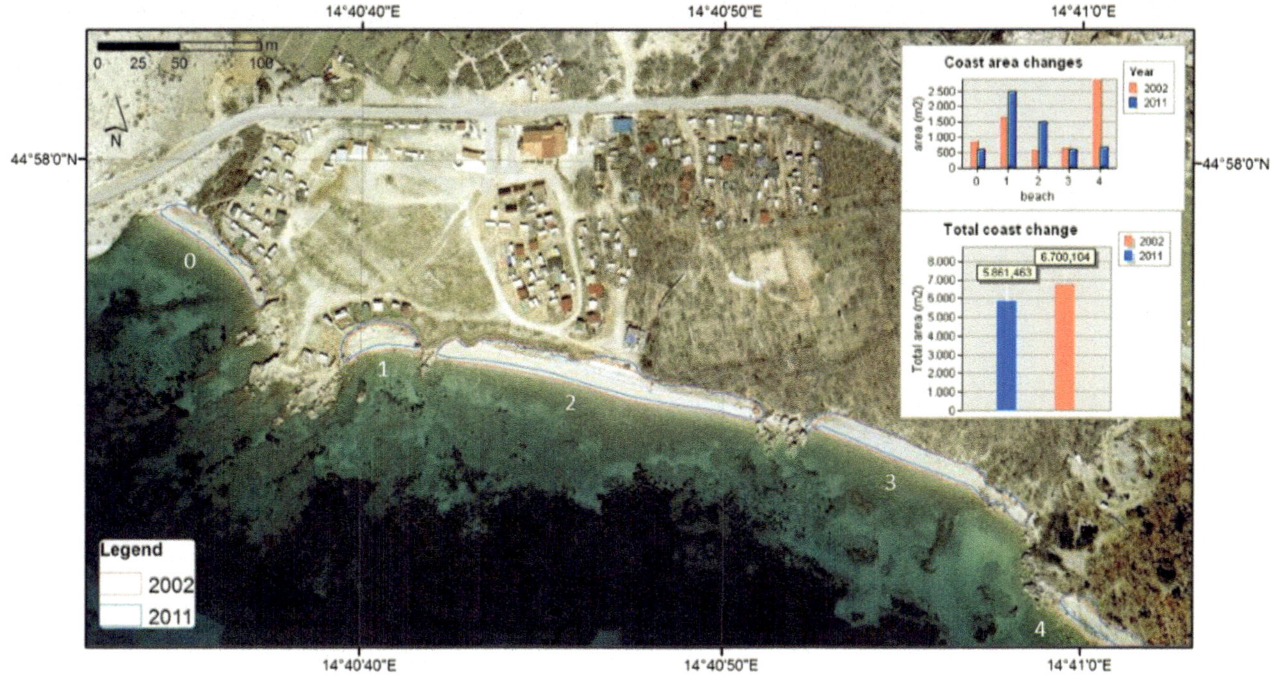

Fig. 5 The beach and cliff changes between 2002 (*red polygon*) and 2011 (*blue polygon*)

incorporates the cantilever-beam model was tested. The cantilever-beam model (Kogure et al. 2006) has been used for stability analyses of coastal cliffs with the presence of notches in a number of studies. The model is based on the assumption that the cliff-overhanging material above the notch acts as cantilever-beam load. The tensile stresses, which are distributed on the upper section, and the compressive stress, distributed on the lower section of the cliff face, are derived from the model (Timoshenko and Gere 1972). The cliff's cross-section is calculated by multiplying the average cliff height by the horizontal distance between the retreat point of the notch and the averaged location of the cliff face.

To calculate the stresses, input parameters such as the rock density and the geometry of the cliff face were required. The detailed cliff geometry was derived from the 3D point cloud delivered using structure from motion (SfM) photogrammetry. It provided precise cliff geometrical parameters and the spatial distribution of the stresses along the cliff face (Ružić et al. 2015).

On the other hand, it was difficult to determine the geotechnical parameters in lithologically very heterogeneous rocks like breccia. A correlation between the heterogeneous breccias compressive and tensile strengths cannot be unambiguously defined. The stability analyses are based on laboratory-derived data. The material bulk density in dry conditions is 24.5, and 25.3 kN/m^3 in wet conditions. The uniaxial compressive strengths are UCS = 29.21–70.21 MPa in dry conditions and UCS = 19.82–62.10 MPa in wet conditions. The estimated Geological Strength Index is GSI = 25–35 (Marinos and Hoek 2000). Since the subsurface top layer, with soil features, of about 1 m thickness, is completely weathered, and the rest of the cliff material has been subject to different mechanical and chemical processes, the disturbance factor of the material was defined as D = 0.7 (Fig. 6).

Strength parameters of the cliff material were defined in Rocscience, RocData using the described input parameters defining the Hoek-Brown criterion (Hoek and Brown 1991; Hoek 1996; Marinos and Hoek 2000), and are shown on Fig. 7. The corresponding Mohr Coulomb strength parameters are $\Phi = 49°$ and c = 47 kPa.

Kogure et al. (2006) assessed a cliff failure process based on three stages: (1) a developing of notch forms as a column like a cantilever beam; (2) this column generates tension cracks on the top surface of the cliff due to the tensile stress arising from its own weight; and (3) a vertical extension of the cracks causes bending failure of this column-shaped mass, which results in a toppling mode of failure. The cross-section geometry was used to calculate parameters such as the cliff area, the cliff overhanging material weight and the centre of gravity. These are then used in the cantilever-beam model for an estimation of the cliff's maximum bending stress ($\sigma_{T\text{-max}}$). Figure 8 illustrates the distribution of stresses for a cliff notch near cliff collapse shown on Fig. 6, which may be considered as a 3D phenomenon.

It is shown that this combined method can identify locations with maximum stresses. The strength of the tensile stresses and their distribution result in a non-uniform cliff

Fig. 6 The recent cliff collapse (*photo* Č. Benac)

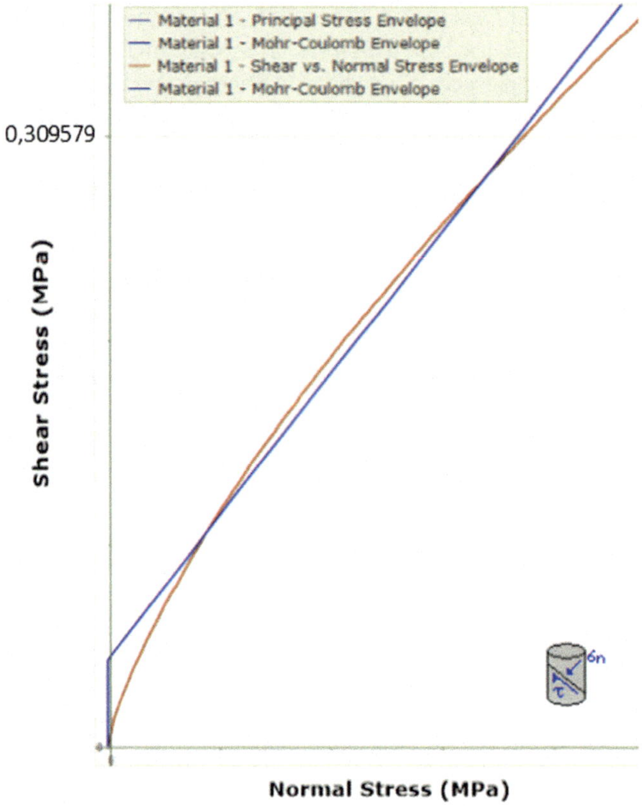

Material 1	
Hoek Brown Classification	
intact uniaxial compressive strength	35 MPa
GSI	25
mi	20
disturbance factor	0.7
intact modulus	17500 MPa
modulus ratio	500
Hoek Brown Criterion	
mb	0.325
s	1.903e-005
a	0.531
Failure Envelope Range	
application	slopes
sig3max	0.129 MPa
unit weight	0.025 MN/m3
slope height	5.5 m
Mohr Coulomb Fit	
cohesion	0.047 MPa
friction angle	49.163 deg
Rock Mass Parameters	
tensile strength	-0.002 MPa
uniaxial compressive strength	0.109 MPa
global strength	2.231 MPa
modulus of deformation	528.923 MPa

Fig. 7 Strength criterion of the cliffs material from Rocscience RocData

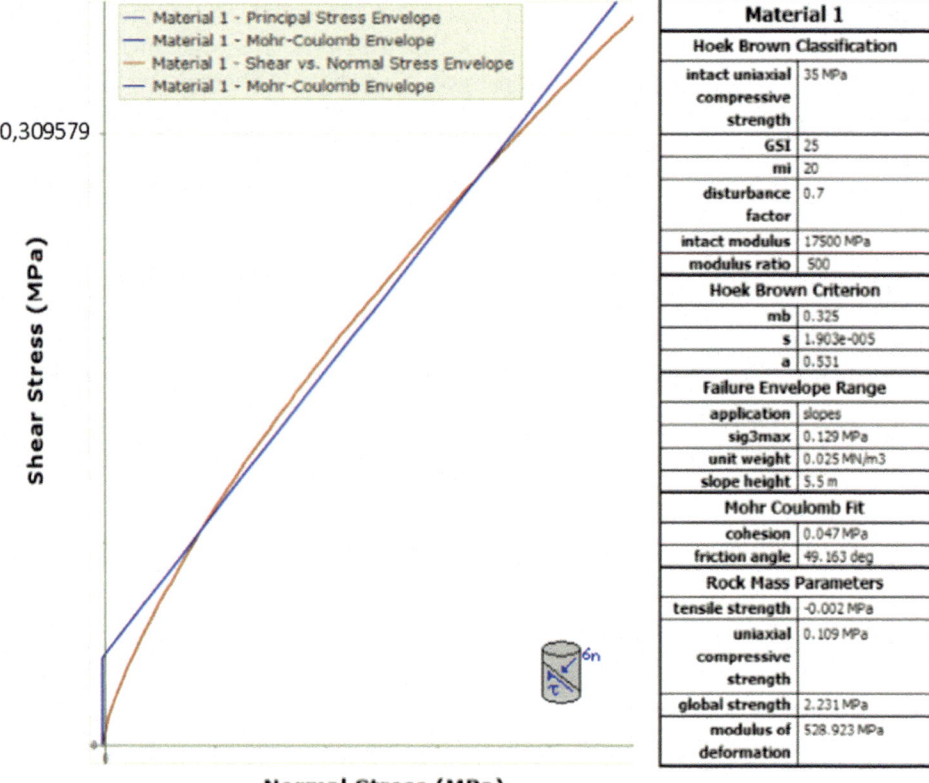

recession, also seen from the analysis of the historical aerial photographs (Ružić et al. 2015). Assuming that the cliff collapse is solely the result of the notch length and amount of material overhanging the notch, the proposed combined method could be useful for estimating the stresses in the cliff face and its stability. A top layer of material which has no tensile bearing capacity can significantly increase tensile stresses (Fig. 8).

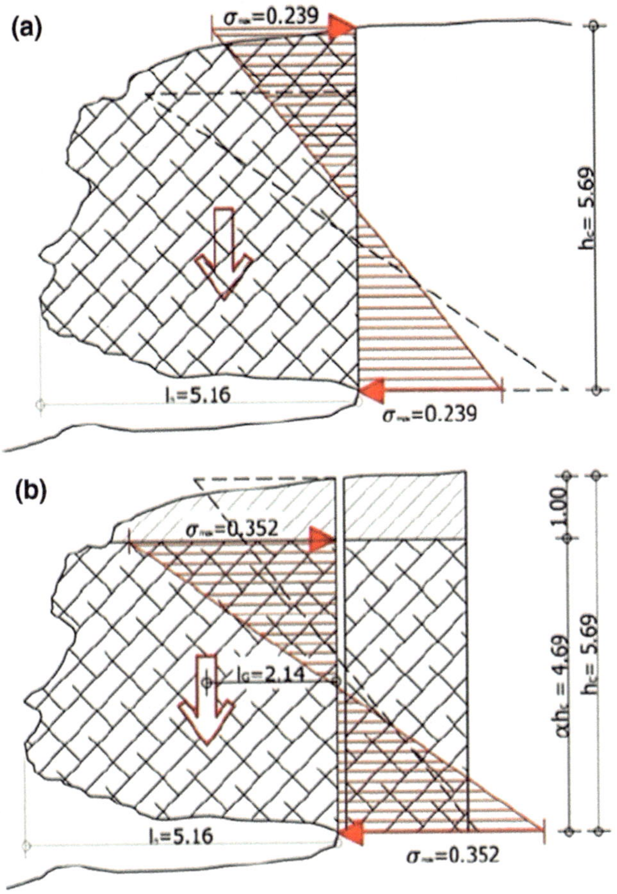

Fig. 8 Stress distribution inside a cliff with no tension crack (**a**) and with a soil layer (**b**) (according: Ružić et al. 2015)

Conclusions

The coastal area around Stara Baška settlement on the south eastern part of the Krk Island has a specific geological fabric. These are breccias of varying lithification degrees and marls bedrock. This part of the coast is more exposed to marine erosion caused by wave action and storm surges. Namely, the investigated part of the coast is exposed to the destructive action of storm waves from the south east to south west direction.

Due to the interaction of the weathering and gravity, the erosion of the cliffs is a permanent phenomenon. The nourishment of beach bodies is mostly provided from carbonate coarse fragments. These beaches have a natural role in protecting the coast from the intense destructive wave's action.

The collapse of blocks up to 100 m^3 from cliffs happens periodically. The partial remains of eroded blocks evidence past rock falls, which are potentially very dangerous events, whereas a large number of people are present on the beaches during the summer period.

From the comparison of orthorectified aerial photographs, recorded in 2002 and 2011, a few meters cliff retreat is visible. At the same time changes at the beach surfaces cannot be defined due to the possibly different sea levels during the areal images record. The reason for the accelerated erosion of the cliffs can be more frequent extreme high sea level with a combination of stormy winds from the SE direction in the Kvarner area.

Since the tourist camp is situated near the investigated location, it will be necessary to ensure the stability of cliffs by adequate geotechnical and hydro operations. However, this could result in reduction of beach nourishment what could also require ensuring of the beach bodies hydrodynamic stability.

References

Benac Č, Juračić M, Matičec D, Ružić I, Pikelj K (2013) Fluviokarst vs. karst: examples from the Krk Island, Northern Adriatic Croatia. Geomorphol 184:64–73

Booij N, Ris RC, Holthuijsen LH (1999) A third-generation wave model for coastal regions, Part 1, Model description and validation. J Geophys Res 104(C4):7649–7666

Hoek E (1996) Strength of rock and rock masses. ISRM News J 2(2):4–16

Hoek E, Brown ET (1991) Practical estimates of rock mass strength. Int J Rock Mech Min Sci 34(8):1165–1187

Juračić M, Benac Č, Pikelj K, Ilić S (2009) Comparison of the vulnerability of limestone (karst) and siliciclastic coasts (example from the Kvarner area, NE Adriatic, Croatia). Geomorphol 107(1–2):90–99

Kogure T, Aoki H, Maekade A, Hirose T, Matsukura Y (2006) Effect of the development of notches and tension cracks on instability of limestone coastal cliffs in the Ryukyus, Japan. Geomorphology 80 (3–4):236–244

Leder N, Smirčić A, Vilibić I (1998) Extreme values of surface wave heights in the Northern Adriatic. Geophys 15:1–13

Marinos P, Hoek E (2000) GSI: a geologically friendly tool for rock mass strength estimation. In: Proceedings of the Geoengineering. Melbourne, pp 1422–1446

Masselink G, Hughes MG (2003) Introduction to coastal processes and geomorphology. Arnold, London, p 354

Orlić M, Pasarić Z (2013) Semi-empirical versus process-based sea-level projections for the twenty-first century. Nat Clim Change 3:735–738

Pikelj K, Juračić M (2013) Eastern Adriatic Coast (EAC): geomorphology and coastal vulnerability of a karstic coast. J. Coast Res 29 (4):944–957

Ružić I, Marović I, Benac Č, Ilic S (2014) Coastal cliff geometry derived from structure-from-motion photogrammetry at Stara Baška, Krk Island, Croatia. Geo-Mar Lett 34:555–565

Ružić I, Benac Č, Marović I, Ilić S (2015) Stability assessment of coastal cliffs using digital imagery. Acta Geotech Slov 2(12): 25–35

Surić M (2009) Reconstructing sea-level changes on the Eastern Adriatic Sea (Croatia)—an overview. Geoadria 14(2):181–199

Timoshenko SP, Gere JM (1972) Mechanics of materials. van Nostrand Reinhold Co., New York

Rock Avalanches in a Changing Landscape Following the Melt Down of the Scandinavian Ice Sheet, Norway

Markus Schleier, Reginald L. Hermanns, and Joachim Rohn

Abstract

Rock avalanches can form complex deposits for which the interpretation can be challenging, especially if they occur in valleys affected by other 'fast' geological processes, such as, glaciations or isostatic rebound. The mountains of western Norway enable to study rock avalanches in such a complex geological setting. Within the two valleys of Innerdalen and Innfjorddalen (~70 km afar), several rock avalanches occurred since the Late Pleistocene. The rock avalanches in Innerdalen have volumes of 31×10^6 and 23×10^6 m^3 and yielded terrestrial cosmogenic nuclide ^{10}Be ages of 14.1 ± 0.4 and 7.97 ± 0.94 ka, while those in Innfjorddalen have volumes of 15.1×10^6, 5.4×10^6 and 0.3×10^6 m^3 and yielded ages of 14.3 ± 1.4, 8.79 ± 0.92 ka and 1611–12 CE, respectively. Although being of nearly similar age, the rock avalanches in both valleys occurred under different environmental settings associated with the decay of the Scandinavian ice sheet. One of which fell onto a retreating valley glacier, partly depositing as supraglacial debris (Innerdalen), while the contemporaneous one fell into a fjord, partly forming a subaqueous deposit which is today exposed due to post-glacial isostatic uplift (Innfjorddalen). The younger rock avalanches fell into the ice-free valleys onto the older rock-avalanche deposits. All of the observed rock avalanches are preserved in rock-boulder deposits distributed on the valley floor and its slopes showing a variety of geomorphological features and landforms, which are diagnostic for their paleodynamics. Numerical runout modeling using DAN3D supports the landform interpretations, which are further confirmed by ^{10}Be ages of the rock-avalanche deposits. The presented description of rock-avalanche deposits can enable a better identification and interpretation of similar deposits in other mountain areas and contributes to the knowledge of Quaternary landscape evolution in western Norway, such as, ice-sheet thickness and post-glacial isostatic rebound.

M. Schleier (✉) · J. Rohn
GeoZentrum Nordbayern, University of Erlangen-Nuremberg
Erlangen, Erlangen, Germany
e-mail: markus.schleier@fau.de

R.L. Hermanns
Geological Survey of Norway, Trondheim, Norway

R.L. Hermanns
Department of Geology and Mineral Resources Engineering, Norwegian University of Science and Technology, Trondheim, Norway

M. Mikoš et al. (eds.), *Advancing Culture of Living with Landslides*,
DOI 10.1007/978-3-319-53483-1_44

Keywords
Rock avalanche • Supraglacial rock avalanche • Subaqueous rock avalanche • Post-glacial isostatic rebound • Innerdalen and Innfjorddalen • Western Norway

Introduction

Rock avalanches and associated deposits significantly impact landscape evolution in high mountain areas, by, for instance, affecting sediment discharge, forming rock-avalanche dams or interacting with glaciers (e.g., Shulmeister et al. 2009; Evans et al. 2011; Hewitt et al. 2011; and references therein). Therefore, their deposits are often difficult to identify and interpret.

Because of its geological history, the west Norwegian mountains are appropriate to study those complex interactions of rock avalanches and the environment. On the one side, this area is intensively affected by Pleistocene glaciations with associated effects, such as, large ice shields and post-glacial isostatic uplift, and neotectonics (Sollid and Sørbel 1979; Fjeldskaar et al. 2000; Hughes et al. 2016). On the other side, rock-slope failures and rock avalanches are quite common phenomena in this area (e.g., Blikra et al. 2006; Hermanns et al. 2012; Böhme et al. 2015; and references therein).

Moreover, as recent studies show (Böhme et al. 2015; Hermanns et al. 2016, submitted), spatial and temporal distribution of past rock slope failures can be used as additional evidence for the interpretation of Quaternary landscape evolution, or can confirm those interpretations, such as, regional deglaciation history.

This contribution contains a case study on paleodynamics of rock-avalanches in western Norway, which combines the main findings of our studies in the two valleys of Innerdalen and Innfjorddalen (Schleier et al. 2015, 2016). It puts them into a more regional context to highlight their implications for landscape evolution following the melt down of the Scandinavian ice sheet.

Regional Setting

The two valleys of Innerdalen and Innfjorddalen are located in western Norway within a distance of about 70 km (Fig. 1).

With simplified geology, the disclosed lithologies are chiefly high metamorphic gneisses of the West Norwegian Gneiss Region that are characterized by a strong foliation and various joint sets and tectonic faults (Tveten et al. 1998).

The area was intensively affected by the Pleistocene glaciations, including the Last Glacial Maximum and the Younger Dryas, and the following deglaciation (Hughes et al. 2016; Sollid and Sørbel 1979). The landscape is characterized by steep mountain topography with over-steepened slopes and deep fjords. It was further affected by an intense post-glacial isostatic uplift of up to locally two hundred meters (Hansen et al. 2014) and associated relative sea level drop. The landmass shows a current apparent uplift rate of 2–3 mm $year^{-1}$, although probably not only of post-glacial origin (Fjeldskaar et al. 2000; Dehls et al. 2000).

Materials and Methods

The presented results are based upon the works of Schleier et al. (2015, 2016), in which the methodologies are described in detail (see also references therein). The basically applied methods include the following:

(1) Geomorphological field mapping to determine spatial and relative temporal distribution as well as morphometry of surficial deposits, with focus on rock-boulder deposits. Field mapping was supported by remote sensing data, such as, orthophotos and digital elevation models.
(2) Surface-exposure dating using the in situ produced terrestrial cosmogenic nuclide ^{10}Be (Gosse and Phillips 2001) to determine temporal distribution by calculating absolute ages of the identified rock-boulder deposits in order to confirm landform interpretations.
(3) Numerical dynamic runout modeling using the code DAN3D (McDougall and Hungr 2004) to study the dynamics of rock avalanches and to test hypotheses of paleodynamics of rock avalanches in order to support landform interpretations.

Distribution and Characteristics of Rock-Avalanche Deposits

In Innerdalen and Innfjorddalen Valley, several rock avalanches are identified by their deposits, which, in both cases, failed repeatedly from the same source area since Late Pleistocene. These rock avalanches are preserved on the valley floors and their slopes in deposits formed by rock-avalanche debris showing a carapace of angular rock

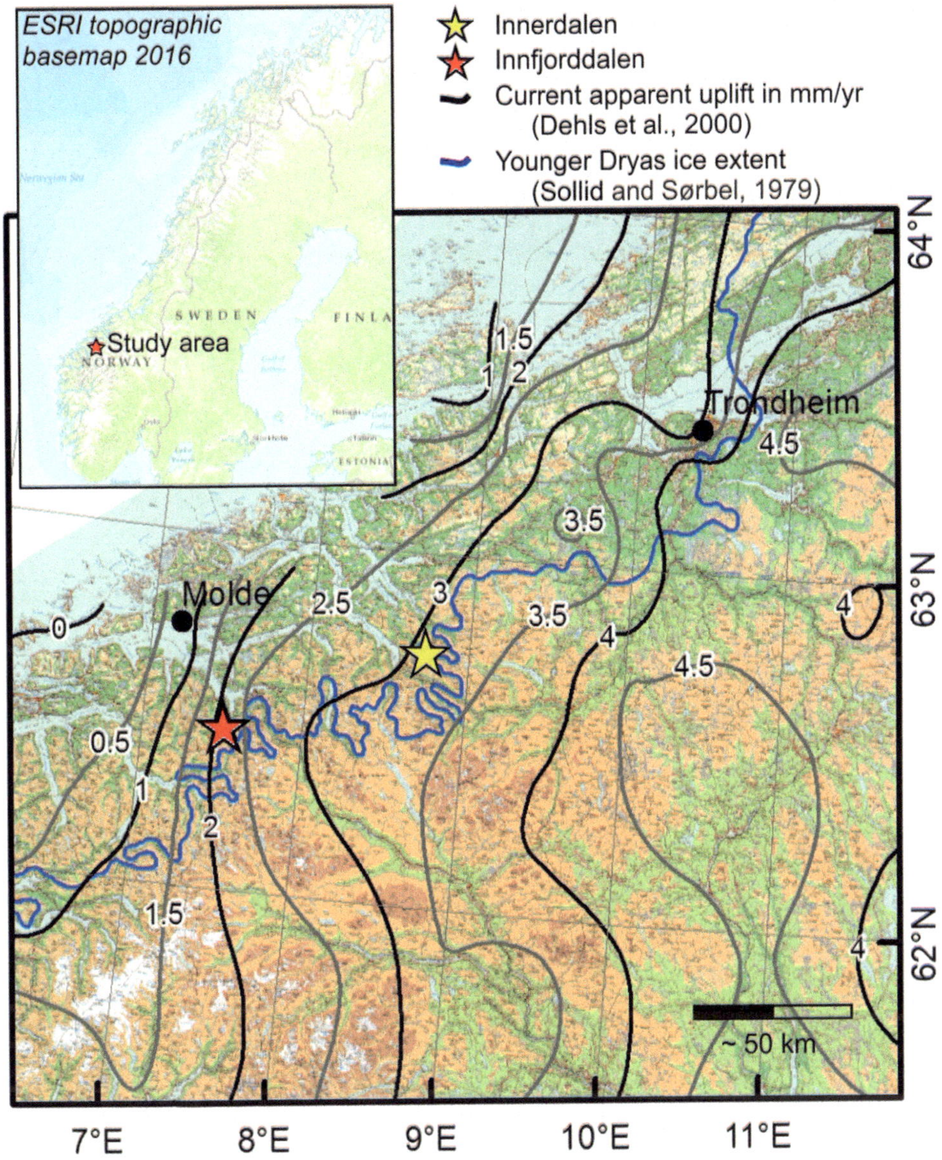

Fig. 1 Location of the study areas of Innerdalen and Innfjorddalen (after Schleier et al. 2016)

boulders, several meters up to tens of meters large. Besides this, some of these deposits show other characteristics typical for rock avalanches with high mobility, such as, lobes, longitudinal ridges and runup on the opposite slope. However, some of the rock avalanches exhibit some special geomorphological features and landforms.

Rock Avalanches in Innerdalen

Two rock avalanches occurred in Innerdalen that can be determined by spatial and temporal distribution of their deposits (Fig. 2). The characteristics of the rock-avalanche deposits are summarized in Table 1.

One rock avalanche with a volume of about 31×10^6 m^3 is preserved in discontinuous deposits composed of rock-avalanche debris. These deposits show special geomorphological features, such as, marginal moraine ridges and isolated concentric hills behind them, isolated boulder patches about 300–350 m high above the valley floor, and large valley-parallel ridges. This event is dated to about 14.1 ± 0.4 by means of ^{10}Be.

Another rock avalanche with a volume of about 23×10^6 m^3 is preserved in a continuous deposit of rock-avalanche debris, which has lobes, longitudinal ridges and a runup on the opposite slope. The rock avalanche formed a natural dam and impounded a lake. The deposit yields a ^{10}Be age of 7.97 ± 0.94 ka.

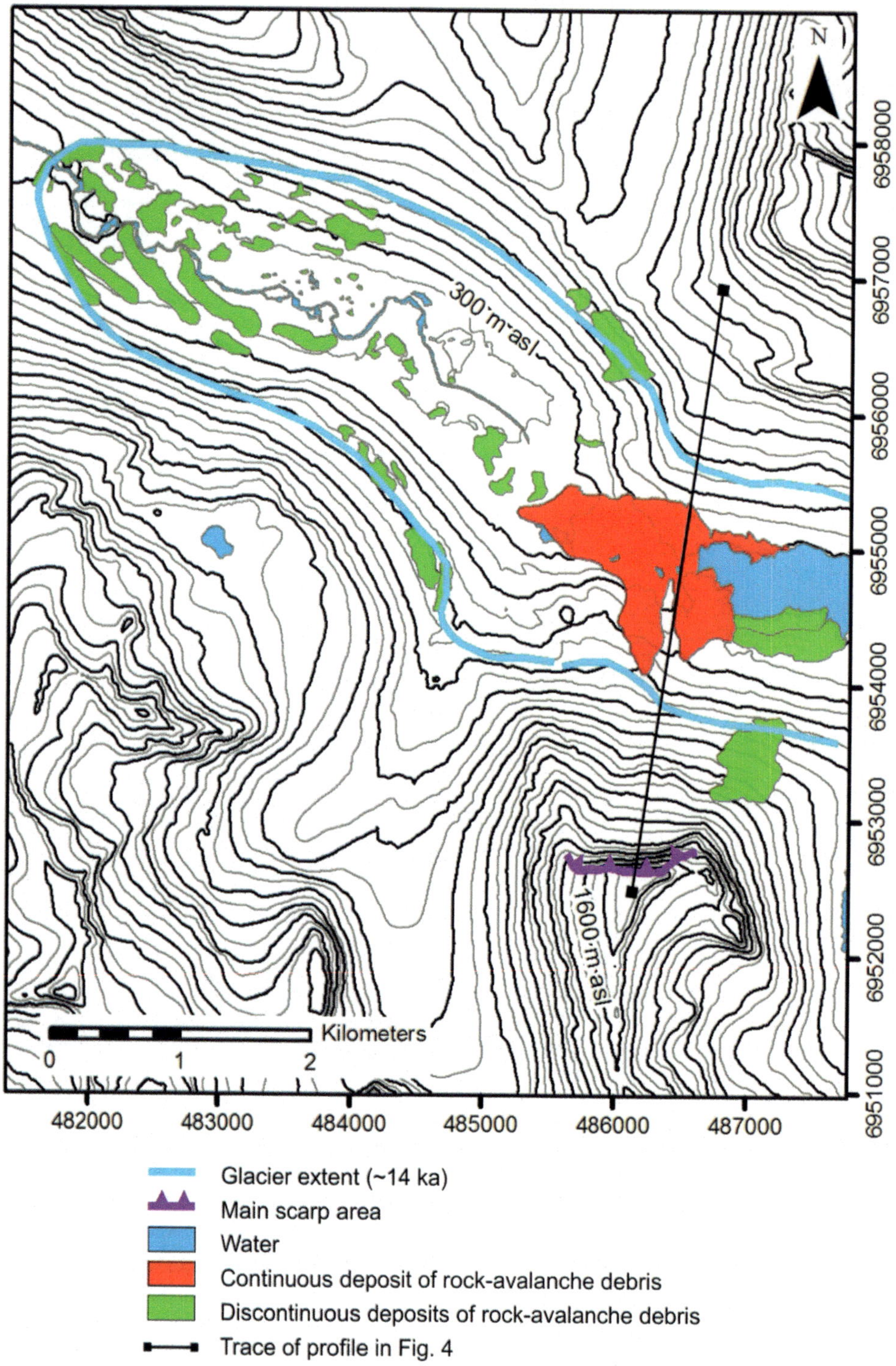

Fig. 2 Distribution of rock-avalanche deposits in Innerdalen (contours = 50 m)

Rock Avalanches in Innfjorddalen

Three rock avalanches occurred in Innfjorddalen that can be determined by spatial and temporal distribution of their deposits (Fig. 3). The rock-avalanche deposits, of which the characteristics are summarized in Table 1, rather form a stratified succession.

The oldest rock avalanche with a volume of about 15.1×10^6 m^3 is preserved in discontinuous deposits composed of rock-avalanche debris. These deposits show special geomorphological features, such as, a large continuous deposit with lobes, longitudinal ridges and a runup on the opposite slope, and a deposit formed by isolated concentric hills detached from the main deposit by deformed valley-fill

Table 1 Characteristics of rock avalanches in Innerdalen and Innfjorddalen

Timescale	Characteristics	Rock avalanches in Innerdalen	Rock avalanches in Innfjorddalen
Late Pleistocene	**Age (ka)**	**14.1 ± 0.4**	**14.3 ± 1.4**
	Volume (10^6 m^3)	31.5	15.1
	Paleodynamics	(partly) supraglacial	(partly) subaqueous
	Deposit extent (10^3 m^2)	1517	1437
	Drop height (m)	n.n. (1240)	1340 (1307)
	Runout distance (m)	n.n. (5280)	4188 (4143)
	Fahrböschung (°)	n.n. (13.2)	17.7 (17.5)
	Travel angle (°)	n.n. (16.5)	n.n. (27.3)
	Runup height (m)	n.n.	100
	Runup velocity (m s^{-1})	n.n.	44
Holocene	**Age (ka)**	**7.97 ± 0.94**	**8.79 ± 0.92**
	Volume (10^6 m^3)	23.5	5.4
	Paleodynamics	Subaerial	Subaerial
	Deposit extent (10^3 m^2)	1121	523
	Drop height (m)	1491 (1493)	1292 (1263)
	Runout distance (m)	3695 (3510)	2841 (2882)
	Fahrböschung (°)	22.0 (23.0)	24.5 (23.7)
	Travel angle (°)	n.n. (26.6)	n.n. (29.1)
	Runup height (m)	80	n.n.
	Runup velocity (m s^{-1})	40	n.n.
Historic	**Age**	**n.n.**	**1611–12 CE**
	Volume (10^6 m^3)	n.n.	0.3
	Paleodynamics	n.n.	Subaerial
	Deposit extent (10^3 m^2)	n.n.	44
	Drop height (m)	n.n.	1280 (1267)
	Runout distance (m)	n.n.	2390 (2321)
	Fahrböschung (°)	n.n.	28.2 (28.6)
	Travel angle (°)	n.n.	n.n. (29.6)

The values of height, distance, Fahrböschung and travel angle presented in parentheses are derived by DAN3D modeling, additionally to those derived by field mapping

sediments. The main part of the deposit yields a ^{10}Be age of 14.3 ± 1.4 ka.

The second rock avalanche with a volume of about 5.4 × 10^6 m^3 is preserved in a continuous deposit of rock-avalanche debris that also shows lobes and longitudinal ridges, and that formed a natural dam and impounded a lake. It is dated to 8.79 ± 0.92 ka by means of ^{10}Be.

The third rock avalanche is much smaller with a volume of about 0.3 × 10^6 m^3 and is correlated to an event of 1611–12 CE, which is recorded in historic documents.

Paleodynamics of Rock Avalanches in Relation to the Decay of the Scandinavian Ice Sheet

We interpret paleodynamics of the rock avalanches based on spatial and temporal distribution of the observed deposits of rock-avalanche debris and regional environmental conditions at time of occurrence, such as the extent of the ice sheet or the relative sea level. Numerical runout modeling supports the landform interpretation by reproducing the deposits using typical modeling parameters.

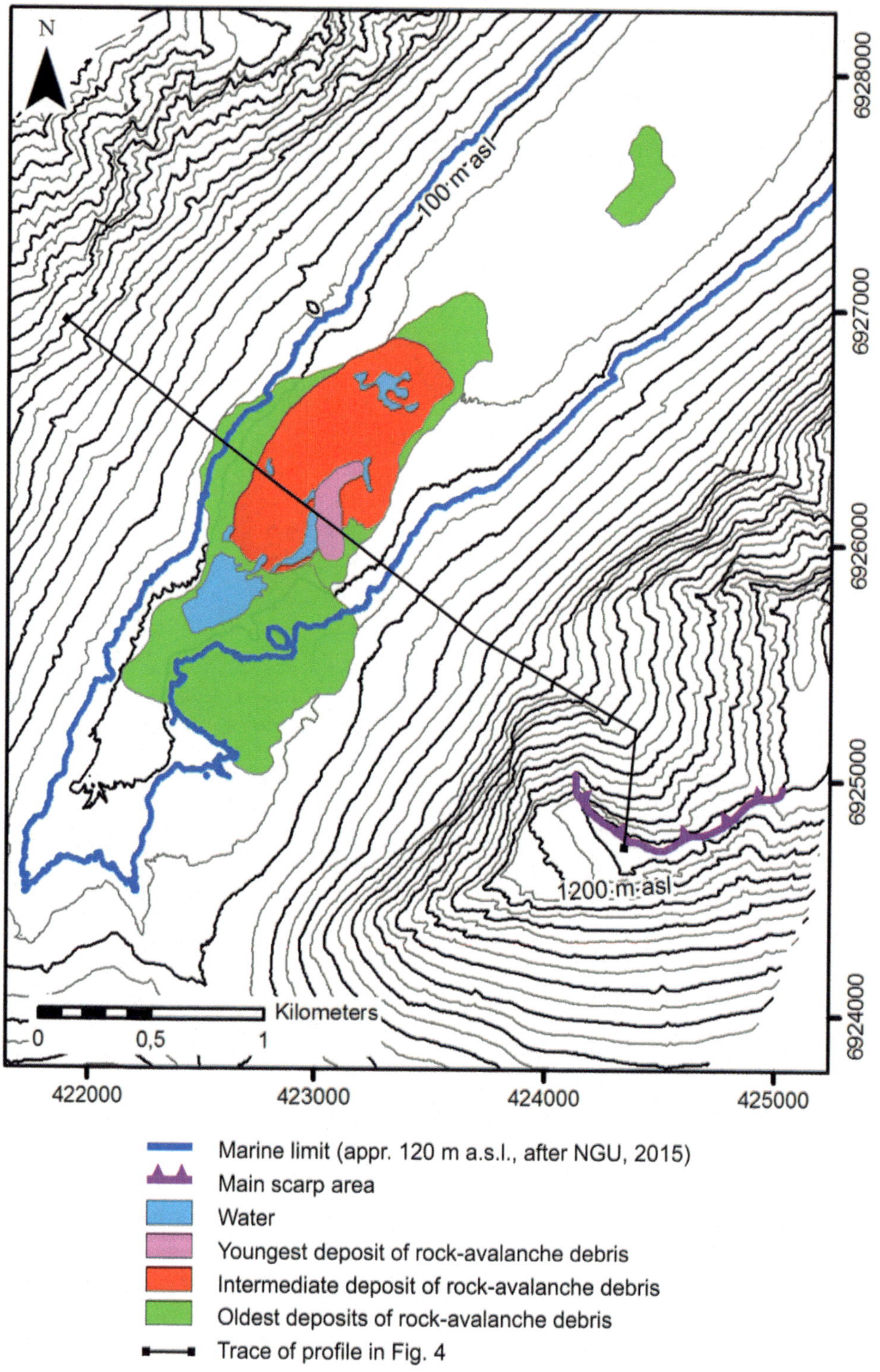

Fig. 3 Distribution of rock-avalanche deposits in Innfjorddalen (contours = 50 m)

Late Pleistocene Rock Avalanches

The oldest rock avalanches in both valleys (~14 ka) occurred under paraglacial conditions under a relatively cold climate. Although being of similar age and climatic conditions, they show different paleodynamics that can be linked to the local environmental setting (Fig. 4a).

We interpret that the rock avalanche in Innerdalen fell onto a glacial ice body lying in the valley and that it was, at least partly, deposited as supraglacial debris below the

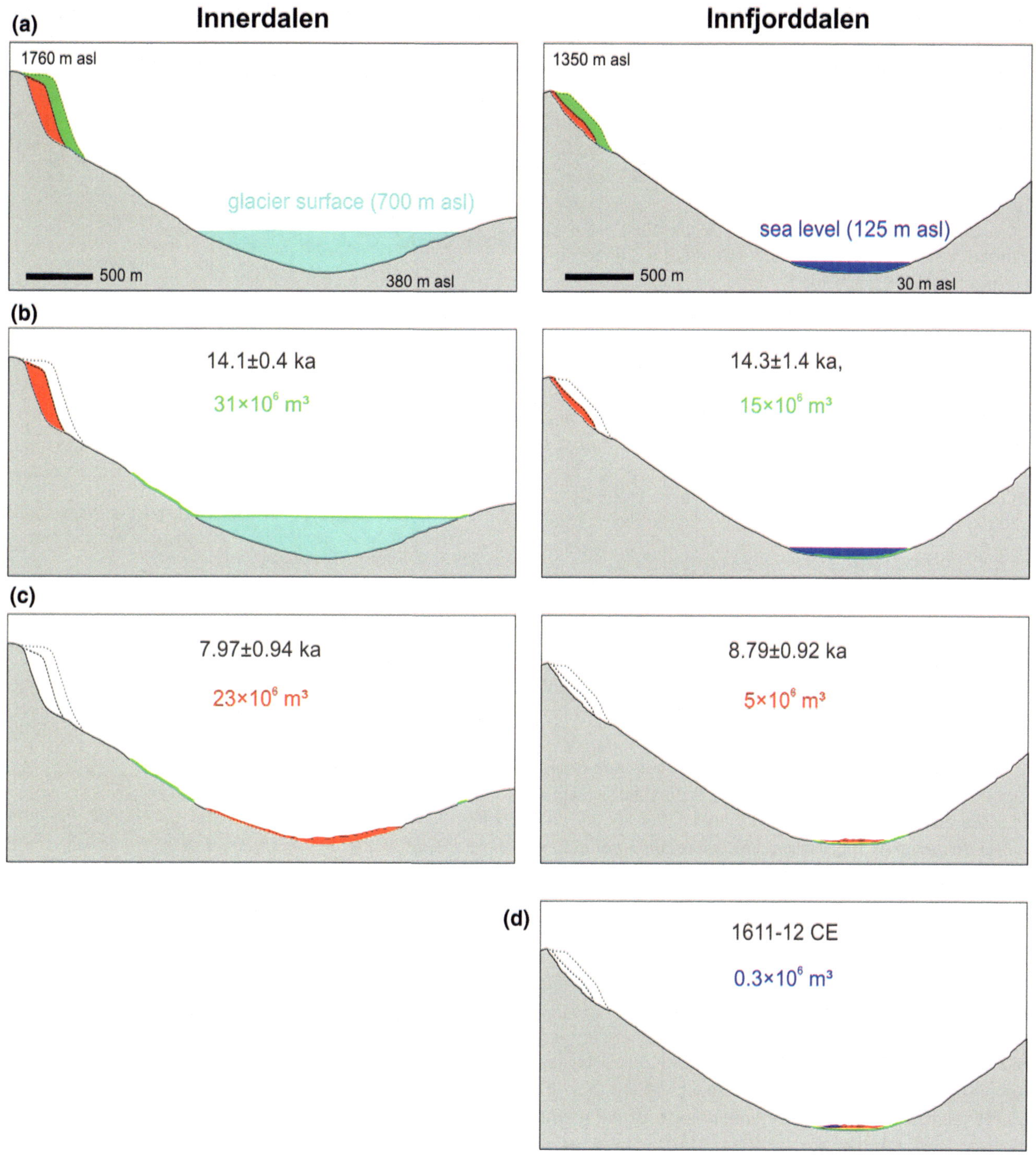

Fig. 4 Topographic profile and conceptual model of rock avalanches in Innerdalen and Innfjorddalen following the melt down of the Scandinavian ice sheet, showing pre- and post-failure conditions of the source and the deposits of **a**, **b** the Late Pleistocene rock avalanches, **c** the Holocene rock avalanches, and **d** additionally for Innfjorddalen, the historic rock avalanche. Traces of the profiles are indicated in Figs. 2 and 3

paleo-equilibrium-line altitude (Fig. 4b). Glacial and supraglacial modification, re-transportation and re-deposition can explain the distribution in discontinuous deposits and the landforms observed today (Figs. 2 and 4c).

In contrast, the approximately contemporaneous rock avalanche in Innfjorddalen fell into the water body of the former fjord, and was, at least partly, deposited below the former sea level (Fig. 4b). This subaqueous propagation and deposition over/on water-saturated sediments and the later following post-glacial isostatic uplift can explain the distribution in discontinuous deposits and the landforms observed today (Figs. 3 and 4c).

Holocene and Historic Rock Avalanches

In both of the valleys a second rock avalanche ($\sim$8.4 ka) occurred within the Boreal (9–7.5 ka) under a climate warmer and moister than today.

We interpret quite similar paleodynamics as both rock avalanches fell into an ice-free valley but, at least partly, onto older rock-avalanche deposits (Fig. 4c). Both deposits created a natural dam, which impounded a lake, and which is rather stable since its origin (Figs. 2 and 3).

Additionally, a much smaller rock avalanche occurred in historic times in Innfjorddalen (1611–12 CE), which again fell onto older rock-avalanche deposits (Figs. 3 and 4d).

Implications for Landscape Evolution

Similar ages of the rock avalanches could imply a similar triggering mechanism, for instance, a climatic control or seismicity, however the dataset is much too small to provide a reliable interpretation. Thus it has to be checked and compared with more regional database.

Nevertheless, implications for regional landscape evolution following the melt down of the Scandinavian ice sheet are more clear.

The findings in Innerdalen, i.e., especially spatial and temporal distribution of rock-avalanche deposits reveal an ice thickness in the valley of about 300–350 m at about 14.1 ka (Figs. 2 and 4). Schleier et al. (2016) further interpret a local glacial advance induced by the supraglacial rock-avalanche deposit and interpret the following deglaciation to be rather stagnant ice decay.

The findings in Innfjorddalen reveal that the valley was ice free but in the condition of a shallow fjord at 14.3 ka, and that post glacial uplift was in total about 120 m after this time (Figs. 3 and 4). This observation fits with the marine limit modeled for this area (NGU 2015). Schleier et al. (2016) additionally interpret the amount of uplift after 3.9 ka to be about 15 m, and interpret an observed significant ^{10}Be age discrepancy (about 10.4 ka) of previously contemporaneous subaqueous rock-avalanche deposits to be caused by the post-glacial uplift.

Our observations of ice extent in both valleys fit with the ice-margin data presented in Hughes et al. (2016) and contribute to a more regional dataset presented by Hermanns et al. (2016, this issue) to trace the decay of the ice shield.

Acknowledgements We thank the Norwegian Water Resources and Energy Directorate (NVE) for financial support for field work under the project "Rock avalanche mapping in Møre og Romsdal County". R.L. Hermanns got funding to contribute to this publication through the NFR-funded CryoWALL project (243,784/CLE). We acknowledge J. C. Gosse, G. Yang and S. Zimmerman for their contribution for ^{10}Be dating, and O. Hungr for providing the DAN3D code. This contribution is based on two sub-chapters of the senior author's doctoral thesis.

References

Blikra LH, Longva O, Braathen A, Anda E, Dehls JF, Stalsberg K (2006) Rock slope failures in Norwegian fjord areas: examples, spatial distribution and temporal pattern. In: Evans SG, Scarascia Mugnozza G, Strom A, Hermanns RL (eds) Landslides from massive rock slope failure. NATO Science Series IV: Earth and Environmental Sciences 49. Springer, Dordrecht, pp 475–496

Böhme M, Oppikofer T, Longva O, Jaboyedoff M, Hermanns RL, Derron MH (2015) Analyses of past and present rock slope instabilities in a fjord valley: implications for hazard estimations. Geomorphology 248:464–474

Dehls JF, Olesen O, Bungum H, Hicks EC, Lindholm CD, Riis F (2000) Neotectonic map: Norway and adjacent Areas. Geological Survey of Norway, Trondheim

Evans SG, Delaney KB, Hermanns RL, Strom A, Scarascia-Mugnozza G (2011) The formation and behaviour of natural and artificial rockslide dams; implications for engineering performance and hazard management. In: Evans SG, Hermanns RL, Strom A, Scarascia-Mugnozza G (eds) Natural and artificial rockslide dams. Lecture Notes in Earth Sciences 133. Springer, Berlin/Heidelberg, pp 1–75

Fjeldskaar W, Lindholm C, Dehls JF, Fjeldskaar I (2000) Postglacial uplift, neotectonics and seismicity in Fennoscandia. Quat Sci Rev 19:1413–1422

Gosse JC, Phillips FM (2001) Terrestrial in situ cosmogenic nuclides: theory and application. Quat Sci Rev 20:1475–1560

Hansen L, Høgaas F, Sveian H, Olsen L, Rindstad BI (2014) Quaternary geology as a basis for landslide susceptibility assessment in fine-grained, marine deposits, onshore Norway. In: L'Heureux J-S, Locat A, Leroueil S, Demers D, Locat J (eds) Landslides in sensitive clays. Springer, Netherlands, pp 369–381

Hermanns RL, Hansen L, Sletten K, Böhme M, Bunkholt HSS, Dehls JF, Eilertsen RS, Fischer L, L'Heureux J-S, Høgaas F, Nordahl B, Oppikofer T, Rubensdotter L, Solberg I-L, Stalsberg K, Yugsi Molina FX (2012) Systematic geological mapping for landslide understanding in the Norwegian context. In: Eberhardt E, Froese C, Turner K, Leroueil S (eds) Landslides and engineered slopes: protecting society through improved understanding. Taylor & Francis Group, London, pp 265–271

Hermanns RL, Schleier M, Gosse JC (2016) Ages of rock-avalanche deposits allow tracing the decay of the scandinavian ice sheet. In: 32nd Nordic geological winter meeting, Helsinki, 13–15 Jan 2016

Hermanns RL, Schleier M, Böhme M, Blikra LH, Gosse JC, Ivy-Ochs S (submitted) Rock-avalanche activity in western and southern Norway peaks in the first millennia after the retread of the scandinavian ice sheet. In: Proceedings of the 4th World Landslide Forum (this issue)

Hewitt K, Gosse J, Clague JJ (2011) Rock avalanches and the pace of late Quaternary development of river valleys in the Karakoram Himalaya. Geol Soc Am Bull 123:1836–1850

Hughes ALC, Gyllencreutz R, Lohne ØS, Mangerud J, Svendsen JI (2016) The last Eurasian ice sheets—a chronological database and time-slice reconstruction, DATED-1. Boreas 45:1–45

McDougall S, Hungr O (2004) A model for the analysis of rapid landslide motion across three-dimensional terrain. Can Geotech J 41:1084–1097

NGU (2015) Online database: Nasjonal løsmassedatabase—Standard-kart: marin grense. Norges geologiske undersøkelse. http://geo.ngu.no/kart/losmasse/. Last accessed: 13 Jan 2015

Schleier M, Hermanns RL, Rohn J, Gosse J (2015) Diagnostic characteristics and paleodynamics of supraglacial rock avalanches, Innerdalen, Western Norway. Geomorphology 245:23–39

Schleier M, Hermanns RL, Gosse JC, Oppikofer T, Rohn J, Tønnesen JF (2016) Subaqueous rock-avalanche deposits exposed by post-glacial isostatic rebound, Innfjorddalen, Western Norway. Geomorphology (in press)

Shulmeister J, Davies TR, Evans DJA, Hyatt OM, Tovar DS (2009) Catastrophic landslides, glacier behaviour and moraine formation—A view from an active platemargin. Quat Sci Rev 28: 1085–1096

Sollid JL, Sørbel L (1979) Deglaciation of Western Central Norway. Boreas 8:233–239

Tveten E, Lutro O, Thorsnes T (1998) Geologisk Kart over Norge, Berggrunnskart Ålesund, 1:250,000. Geological Survey of Norway, Trondheim

Multi-Temporal Landslide Susceptibility Maps and Future Scenarios for Expected Land Cover Changes (Southern Apennines, Italy)

Luca Pisano, Veronica Zumpano, Žiga Malek, Mihai Micu, Carmen Maria Rosskopf, and Mario Parise

Abstract

Human activities, including extensive land use practices, such as deforestation and intensive cultivation, may severely affect the landscape, and have caused important changes to the extent of natural forests during the last century in Southern Italy. Such changes had a strong influence on the frequency of occurrence of natural hazards, including landslides. Being one of the most significant control factors of slope movements, any variation in land cover pattern may determine changes in landslide distribution. The study area is the Rivo Basin which is located in Molise (Southern Apennines of Italy), a region severely affected by landslides. We prepared multi-temporal land cover and landslide inventory maps, aimed at developing different susceptibility maps to evaluate the effects of land cover changes in the predisposition to landslides. Based on the observed land cover trends in the study area, we simulated future scenarios of land cover in the attempt to assess potential future changes in landslide distribution and susceptibility. By investigating the relationship between the spatial pattern and distribution of past land cover settings and location factors (as elevation, slope, distance to settlements), we were able to calibrate a land cover change model to simulate future scenarios. The obtained results give important information both regarding the impact of past trends of land cover changes on landslide occurrence and possible future directions. They could be useful to provide insights toward a better land management for the study area, as well as for similar landslide-prone environments in Southern Italy, contributing to establish good practices for future landslide risk mitigation.

L. Pisano (✉) · M. Parise
CNR-IRPI, Via Amendola 122-I, 70126 Bari, Italy
e-mail: L.Pisano@ba.irpi.cnr.it

M. Parise
e-mail: M.Parise@ba.irpi.cnr.it

L. Pisano · C.M. Rosskopf
Department of Biosciences and Territory, University of Molise, Contrada Fonte Lappone, 86090 Pesche (IS), Italy
e-mail: rosskopf@unimol.it

V. Zumpano · M. Micu
Institute of Geography, Romanian Academy, 12 Dimitrie Racovita, Bucharest, Romania
e-mail: zumpanoveronica@gmail.com

M. Micu
e-mail: Mikkutu@yahoo.com

Ž. Malek
Environmental Geography Group, Faculty of Earth and Life Sciences, Vrije Universiteit Amsterdam, De Boelelaan 1085, 1081HV Amsterdam, The Netherlands
e-mail: ziga.malek@gmail.com

M. Mikoš et al. (eds.), *Advancing Culture of Living with Landslides*,
DOI 10.1007/978-3-319-53483-1_45

Keywords
Landslides • Multi-temporal susceptibility • Land cover changes • Future scenarios • Molise

Introduction

Spatial and temporal distribution of landslides occurrence is strongly influenced by natural and human factors. One of the most important factors is the land cover that controls not only the spatial distribution of landslides, but also their effects and impacts on elements at risk. Therefore, landslide hazard and risk should be considered through a dynamic approach, by taking into account global and local changes (such as those regarding the land cover).

Land cover is a highly dynamic factor, it changes due to natural and anthropogenic actions such as economic development, population growth or decrease and land management (Promper et al. 2014). It can be considered also one of the most rapid drivers of global change (Slaymaker et al. 2009) and may be subjected to rapid modifications (Reichenbach et al. 2014).

In Italy, due to the profound social and economic rearrangements in a post-war environment, many significant changes have occurred after World War II. The most significant ones concerned industrialization and urbanization with extensive abandonment of hilly and mountainous rural areas (Antrop 2004; Mazzoleni et al. 2004). In other sectors, the increase of agricultural and breeding lands in lowland areas was observed (Falcucci et al. 2007). This made it possible to meet the growing demand of goods by the increasing population, in order to maximize food production and reduce the related costs (Falcucci et al. 2007; Pelorosso et al. 2011; Bracchetti et al. 2012).

In this study, we analyze local changes in land cover and landslide distribution (in terms of where and how) for a small catchment, the Rivo basin, located in the Molise region of Southern Italy.

The attention toward understanding the influence of land cover changes on mass movement distribution is rising in the research community. There are for instance recent studies in different regions of the world analyzing the effects of past land management and land cover changes on slope stability (Bruschi et al. 2013; Guns and Vanacker 2014; Reichenbach et al. 2014). Furthermore, several works address future projections as well (Promper et al. 2014).

The aim of this study is to capture the types and distribution of past land cover changes for two time intervals (1954–1981, 1981–2007) and their effects on slope instability. With this purpose, three landslide inventory maps for these two time spans have been produced and used to obtain landslide susceptibility maps by means of the Spatial Multi-Criteria Evaluation (SMCE) approach. The obtained maps have been compared in order to detect the location and severity of the effects of land cover changes on landslide predisposition. Furthermore, we developed a spatially explicit land cover change allocation model, to simulate a plausible scenario for 2030. Based on historic trends, these projected land cover changes are used to understand future effects on landslide susceptibility.

Understanding the past and future land cover changes in the study area could give important information, useful for decision-makers especially in those areal sectors where multi-temporal analysis has never been carried out. Furthermore, given the influence of land cover changes on the landslide hazard and the distribution of elements at risk as well, a multi-temporal analysis can produce suitable data for land-management strategies and risk reduction measures.

Study Area

The Rivo basin is located along the Adriatic side of the Molise region, in a hilly-low mountain sector, characterized by elevations ranging from 230 to 980 m a.s.l. It has an extension of 82 km^2 and is part of the Trigno River catchment which crosses most of Molise region in NE-SW direction, perpendicular to the Adriatic coastline (Fig. 1). The Rivo catchment is subject to a warm temperate climate (Aucelli et al. 2006) and characterized by monthly temperatures between 8 and 30 °C, with mean annual precipitations ranging from 600 to 850 mm and mainly concentrated between October and February.

From a geological standpoint, the Rivo basin is located in the sheet 393 "Trivento" of the Geological Map of Italy in scale 1:50,000 (ISPRA). The outcropping geological formations consist mainly of clays and marly clays, sands and sandstones, marls and marly limestones and, locally, of limestones and limestone breccias (Aucelli et al. 2001).

The variety and distribution of these lithologies are related to the tectonic setting of the area which also strongly influences its morphology. In general, harder lithologies like marly limestones, limestones and limestone breccia give origin to the highest peaks emerging from the surrounding gentle hilly landscape, which dominates the area and is mainly shaped in the weaker clays (Borgomeo et al. 2014).

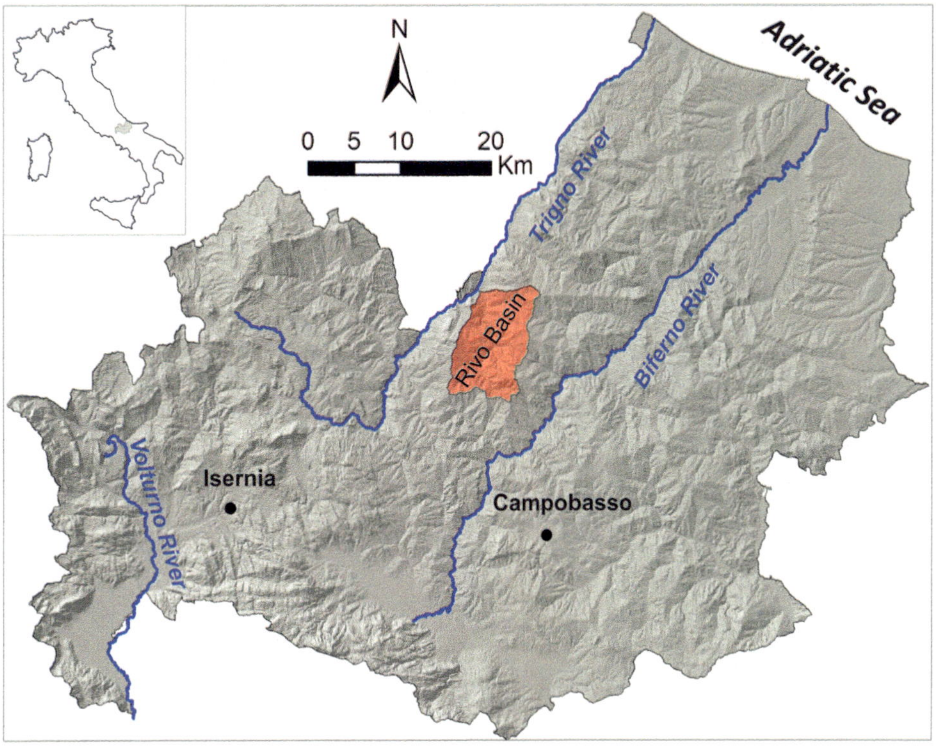

Fig. 1 Study area location in the Molise region (Italy)

The slopes formed by weak lithologies with predominantly clay component are widely affected by landsliding and other mass movements. Landslides are mainly represented by earthflows and complex earth-slide-earthflows, and are primarily triggered by rainfall (Rosskopf and Aucelli 2007; Borgomeo et al. 2014). These movements are typically located along slopes that are directly connected to river incisions and, therefore, often affected by undercutting. They are mostly dormant and relatively shallow (maximum depth 4 m).

Data and Methods

The work is structured in three main parts, namely the landslide and land cover mapping for the selected past years, the related landslide susceptibility analysis, and the evaluation of future scenarios (land cover and related landslide susceptibility projection).

Multi-Temporal Land Cover and Landslide Mapping

The analysis of land cover changes was carried out through the following steps: 1—aerial photo interpretation and production of homogenous land cover maps; 2—analysis of land cover changes and landscape dynamics through the use of parametric indices; 3—landscape pattern stability analysis based on land cover type, by means of the evaluation of rates, sizes and shapes of the various changing patches.

The analyses were conducted using the GIS software ESRI Arcgis 10.0 to investigate a total time interval of 53 years (1954–2007) divided into two distinct periods, 1954–1981 and 1981–2007, respectively.

As concerns the years 1954 and 1981, the aerial photographs from the Italian Geographic Military Institute (IGMI) (http://www.igmi.org/) were scanned in digital format with a resolution of 800 dpi. For the year 2007, an integration of the 2003 aerial photographs and the 2007 orthophotos was used. The aerial photographs were ortho-rectified using the ERDAS LPS software. The ortho-rectification was based on a DEM (Digital Elevation Model) with 5 × 5 m cells. In order to obtain a good accuracy, for each aerial photograph at least 12 GCPs (Ground control point) together with 20 Tie points (points of connection) and 3 control points (check points) were detected.

Land cover mapping was operated at scale 1:5000, choosing a Minimum Map Units (MMU) of 50 × 50 m (Bracchetti et al. 2012) with each cell having a size equal to 2500 m^2, in order to facilitate the photo-interpretation. The distinguished land cover classes are settlements, cropland, forest land, grassland, wetland, shrubland and bare land (modified after the IPCC-GPG-LULUCF classification, see Marchetti et al. 2012). Choice of the classes was essentially linked to two factors: (1) the difficulty to distinguish the sub-classes in low-quality aerial photo sets, and (2) more sub-classes with similar contribution to the slope stability are combined into a single class.

As regards landslides, one of the key elements of the work was to detect with the maximum possible accuracy only the landslides developed in the selected period of time and environment (i.e. land cover type). Therefore, only the landslides that showed evidence of recent movement, such as clear scarp and deposit zones, and related change in the vegetation pattern, were considered. These landslides were mapped by contouring the entire body and classified according to Cruden and Varnes (1996).

The multi-temporal landslide database was built by using a digital stereoscopic visualization (digital photogrammetry) that simplifies the acquisition of features which can be stored directly in a GIS database, and reduces the acquisition time and associated errors (Ardizzone et al. 2013).

Multi-Temporal Landslide Susceptibility

The multi-temporal landslide susceptibility analysis was carried out using the Spatial Multi-Criteria Evaluation (SMCE) approach. This is an expert based, semi quantitative approach integrated in the open source GIS software ILWIS (ITC 2007). The SMCE approach is based on the conceptual model of the Analytical Hierarchical Process (AHP) developed by Saaty (1980). The SMCE application helps users to perform multi-criteria assessment for a defined problem. The model is organized by using a criteria tree whose root represents the problem defined by the user, whilst its branches are the sub-goals or alternatives. The user is guided towards a criterion weighting and standardizing process by the software tool in which the criteria are spatially defined by maps, and the final output is a composite index map which is calculated by adding up the performance of all cell values of the different criteria for the particular alternative (Castellanos Abella and van Westen 2007).

The use of this approach was driven by the aim to obtain comparable final maps through an equable methodological approach between the past landslide susceptibility and the future 2030 scenario, hence, a no-landslide site dependent method was required to perform the analysis. Another convenience of this method is represented by the fact that the expert plays a key role in the weighting procedure, and the variables weights can be maintained constant in each performance.

For this application, seven predisposing factors have been selected. The geomorphological conditions of the study area are represented by slope gradient, internal relief, aspect and altitude; these factors are coupled to the distances to roads, lithology, and land cover. In addition, in order to evaluate the possible contribution of rainfall in the observed changes of the multi-temporal susceptibility, rainfall data coming from the Trivento raingauge station have been preliminarily analyzed.

The SMCE was performed for each investigated year starting with the definition of the problem and the set up of the criteria tree. The factors were standardized using the direct ranking method and the factor's classes applying the maximum standardization method based on the landslides density for each class.

To validate the susceptibility maps, the area under curve (AUC) of the receiver operating characteristic (ROC) curve was used. This method is based on obtaining a curve by plotting the true positive rate (sensitive) against the false positive rate (1-specificity) at various threshold settings (Pourghasemi et al. 2012).

Land Cover and Landslide Susceptibility Scenario for 2030

A spatially explicit land cover allocation model in Dinamica EGO, a raster based GIS platform (Soares Filho et al. 2002), was developed. The model was divided into two parts: spatial land cover allocation and non-spatial scenario demand (amount of land cover).

First, the spatial part was calibrated. We performed an analysis of spatial drivers on past land cover changes. Using the Bayesian method weights of evidence (WoE), we analyzed the significance of the following location factors: distance to settlements, distance to roads, elevation, slope, aspect and curvature. Afterwards, a cellular automata (CA) allocation algorithm with the observed long-term land cover trends (1954–2007) was calibrated. CA models are often used in detailed scale land change studies, as they can successfully simulate decision making. For example, future urban expansion is the more likely situation to occur near existing settlements. The model was trained with the observed landscape metrics of land cover changes, i.e. the mean size of new urban, forest, cropland, etc. patches. In this way it was possible to capture a more realistic spatial pattern of the potential future changes.

The non-spatial scenario demand part was also based on the observed changes, with slight modification. Future forest land, grassland, cropland, shrubland and settlement changes all followed the past trends. Bare areas were considered stagnant. Although there had been an increase in cropland between 1954 and 2007, we allowed the model to simulate a decrease in cropland until 2030. This finds its reason in the recent ongoing forest expansion which represented the most significant process in terms of spatial extent. For this reason forests at the expense of cropland areas, as they were the most suitable for expansion. Although the non-spatial part is an extrapolation of past trends, we wanted to analyze the potential consequences of the continuation of the observed land cover change processes.

The landslide susceptibility scenario was obtained using the abovementioned methodology for the other time intervals. In this case, the density scores used as class weights for the geomorphological factors were the same as before, while for the roads and the land cover the mathematical projections of the landslide densities for each class were used.

Results

Land Cover and Landslide Distribution Maps

From the observation of the obtained maps, it was possible to catch the land cover distribution in 1954, 1981 and 2007, respectively (Fig. 2). The land cover in 1954 was composed by croplands, occupying more than 50% of the whole area, followed by forest land and grassland with approximately 21 and 14%, respectively.

The comparison of the maps highlighted that over the entire analyzed period (1954–2007) major changes concern the expansion of forest land and the reduction of grassland, shrubland and bare land.

The relative increment of forest land consists in more than 11%, whilst grassland and bare land are reduced by two-thirds. The histogram in Fig. 2 shows that the relative increments in forest land occurred rather steadily during the entire analyzed period, while the reduction of grassland and shrubland was greater during the second period (1981–2007). Other land cover classes such as settlements, wetland and bare land had a very small extension in 1954 and their overall changes were very small; nonetheless the relative increase of settlements was about ten times greater, whilst wetland decreased of about two-thirds. Analyzing the

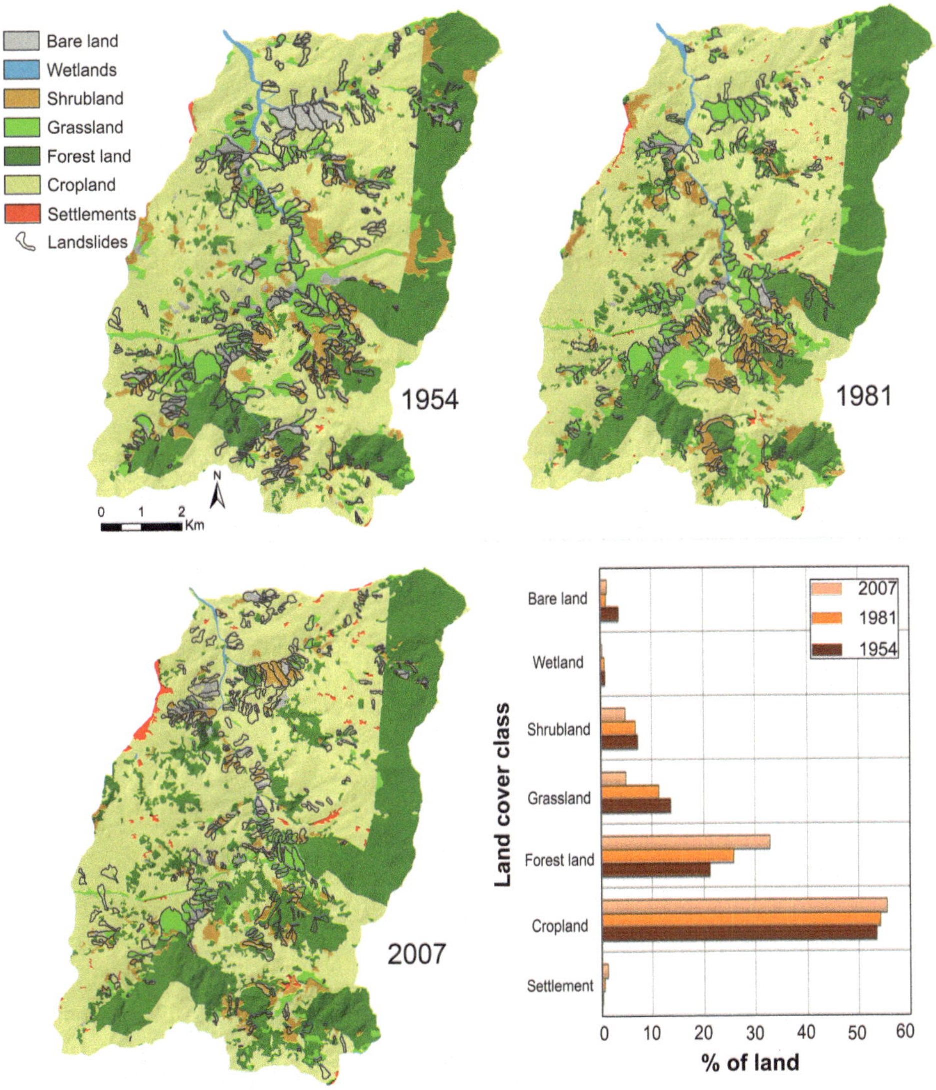

Fig. 2 Multi-temporal land cover maps (1954, 1981 and 2007). Extent (in percentage) of land cover classes during each year is reported in the histogram

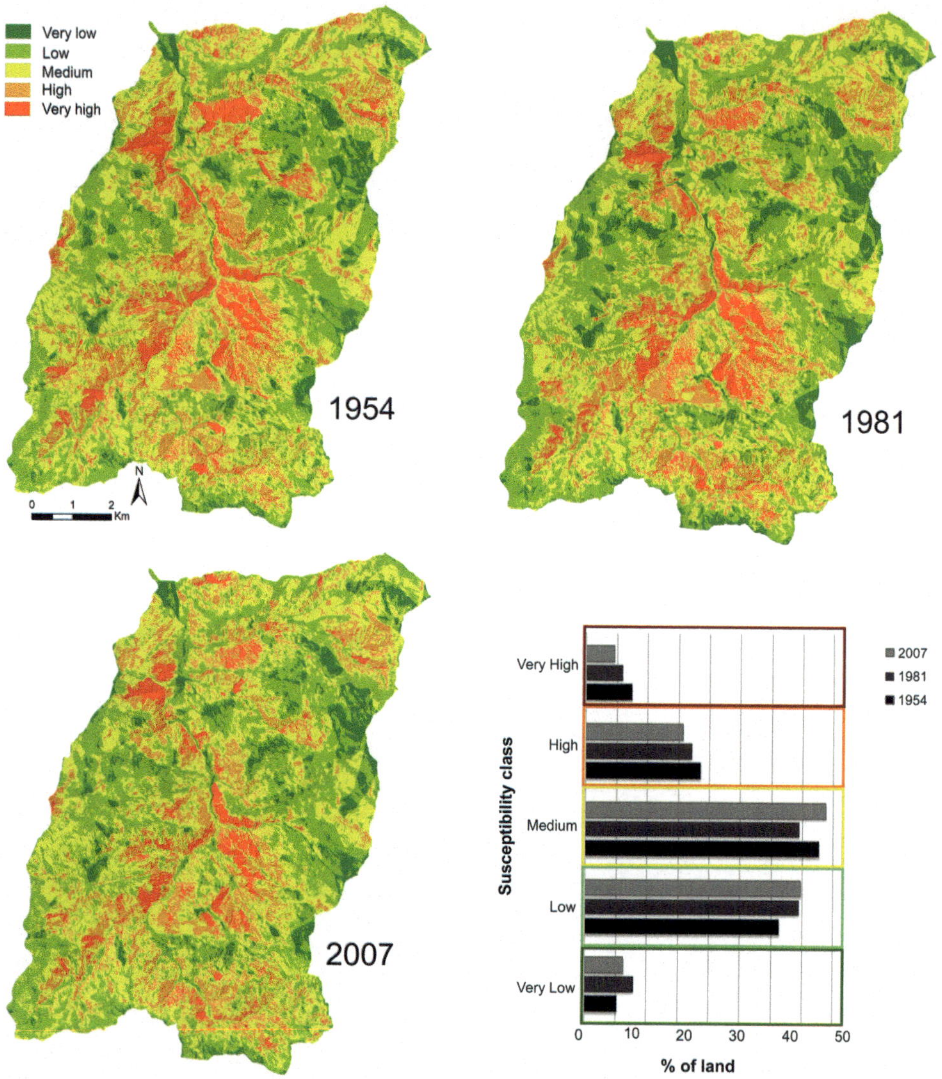

Fig. 3 Multi-temporal landslide susceptibility maps obtained for 1954, 1981, 2007. In the histogram the variation of the susceptibility class in each year

transition of land patches from one class to another that resulted in the above mentioned total changes, some interesting information emerged. In the first period (1954–1981), the most evident transitions were from grassland to cropland and from shrubland to forest land. From 1981 to 2007, major transitions occurred from shrubland to forest land, from grassland to cropland and from grassland to shrubland.

The areal extent of mapped landslides allowed us to determine that in 1954, 12.8% of the study area was covered by landslides, and this percentage reduced to 8.8% in 1981, and to 8.6% in 2007. For what concerns the relationship between landslide density and land cover typologies, the analysis showed that the highest landslide densities characterized the categories bare land (50%), grassland (25%) and shrubland (15%) for year 1954, with similar values for the other two years investigated, whilst the remaining land cover types were characterized by densities lower than 5%.

In the presented analysis the comparison of landslide inventories did not take into account the possible influence of rainfall on landslide distribution over time and space. This is due to the fact that the analysis of rainfall data for the time range analyzed at the Trivento weather station did not reveal any significant variations in annual rainfall of the antecedent years with respect to those analyzed, or during the wet season (October–March) before it.

Multi-Temporal Landslide Susceptibility Maps

The obtained susceptibility maps (1954, 1981, 2007) have been reclassified in five classes from very low to very high, using the following class limits: 0.35, 0.45, 0.55, 0.65, >0.65.

They show satisfying AUC ROC values, namely 0.78 for 1954, 0.83 for 1981 and 0.82 for 2007.

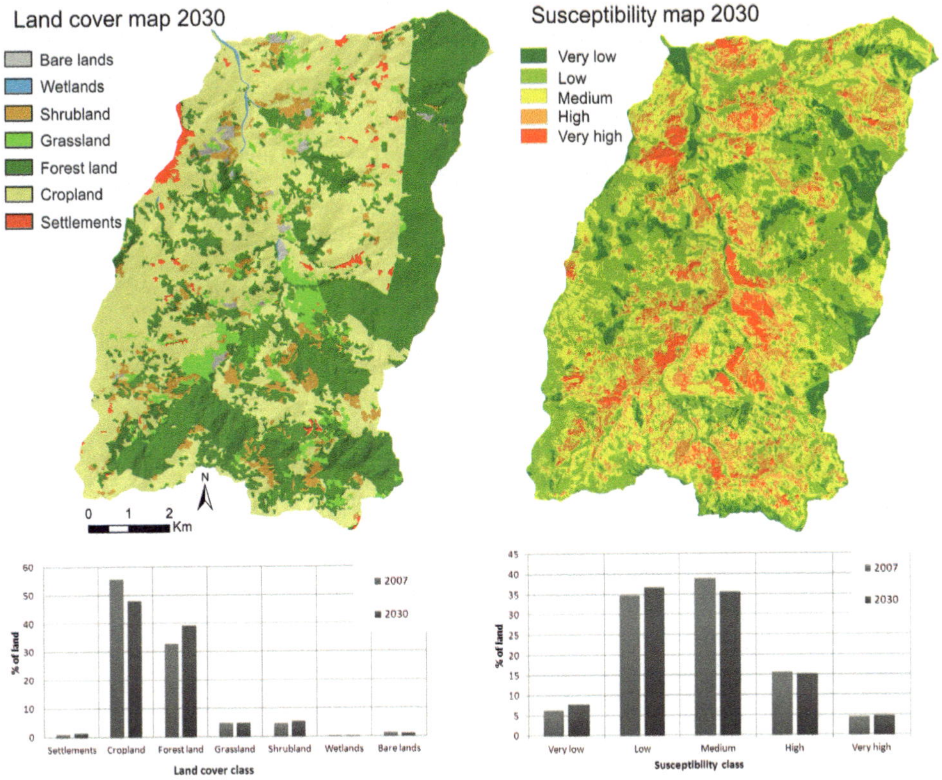

Fig. 4 Land cover and landslide susceptibility scenario for 2030. Histograms show the changes with respect to the correspondent map of 2007

The susceptibility maps (Fig. 3) show a decrease in the extension of the highest susceptibility classes (very high and high), and a correspondent increase of the lowest ones (very low and low). In fact, summing up the high and very high classes of each analyzed year, it is possible to appreciate a decrease of their extension with respect to the total area, from 25.9% in 1954 to 23.03% in 1981, and 20.19% in 2007.

In particular, between 1954 and 1981 the area that shows a decrease in susceptibility is 19.96%, whilst 7.62% shows an increase. Combining the susceptibility changes with the previously analyzed land cover transitions, the obtained information suggests for the first time interval (1954–1981) that the most outstanding decrease of susceptibility has occurred where the land cover changed from grassland to cropland (~5% of the total area).

Another important decrease occurred in correspondence of the transition from shrubland to forest land (~2%). Other changes close to 1% are recorded at the change from bare land to grassland or to shrubland. Conversely, an increment is mainly registered at the transition from cropland to grassland and from cropland to shrubland (respectively 2.7 and 1.3%).

For period 1981–2007, a decrease in the extension of the highest susceptibility classes of about 3% is registered. Most of the changes occurred in correspondence of the transition of grassland to cropland and of shrubland and grassland to forest land.

Land Cover and Landslide Susceptibility Scenario for 2030

The multi-temporal assessment of the influence of land cover on slope stability (addressed here as landslide susceptibility) was completed with a scenario projection for 2030.

The simulated land cover map for 2030 shows only minor changes. This is likely due to the short time period analyzed, but also to the fact that the developed scenario does not take into account any major changes in the study area due to land management or extreme natural events as forest fires. Despite this, it was possible to observe a decrease of cropland and an increase of forest land (Fig. 4).

The obtained land cover map was used as input data for the susceptibility analysis. The output map is indicating an increase of 3.5% in the low and very low classes which corresponds to decrements of 3.18% in the medium and high classes with respect to the entire study area. The main increases occurred where the scenario has modeled the transition from cropland to shrubland or to grassland. On the other hand, a decrease in susceptibility was registered where shrubland and cropland changed to forest land.

Discussion and Conclusion

We used multi-temporal information to investigate the effects of land cover changes on the spatial distribution of landslides in a small catchment in the southern Apennines of Italy. The analysis performed in the Rivo catchment covers a period of 53 years between 1954 and 2007; moreover, a possible land cover change scenario for 2030 was analyzed.

In this study, multi-temporal landslide susceptibility maps have been produced for 1954, 1981, 2007 and 2030 and investigated in order to highlight possible differences due to land cover changes. The comparison of the maps reveals a general decrease of the highest susceptibility classes for the past time interval, attributable primarily to the conversion of grassland and bare land respectively to cropland and to grassland and shrubland, and, secondly, to the change of shrubland to cropland and forest land. It is therefore possible to assume that the forest has a stabilization effect, together with the cropland; while the first one is well documented in literature (Schwarz et al. 2010; Caviezel et al. 2014) for cropland areas there might be some controversial issues. Despite the numerous previous studies sustaining the negative effects of cropland on slope stability (Begueria 2006; Wasowski et al. 2014), in our case the transition from land cover types that do not foresee any human action (or land abandonment), such as from bare land or grassland to cropland, is showing a positive effect on the spatial distribution and extent of landslides. In fact, for the analyzed catchment it is possible to state that the presence of cultivations requires a correct water control resulting in a much desired maintenance action of the land, possibly resulting in a reduction of slope instability. Increase in slope stability is confirmed even in a minor extent in the 2030 scenario where an increment of forest land is foreseen.

The performed study can be an explanatory example of what may also be observed in other sectors of the Italian Apennine with similar geo-environmental and rural structures. The information obtained can be helpful to lead toward improved land management if taken in consideration by stakeholders during land planning actions.

References

Antrop M (2004) Landscape change and the urbanization process in Europe. Landscape and Urban Plann 67(1–4):9–26

Ardizzone F, Fiorucci F, Santangelo M, Cardinali M, Mondini AC, Rossi M, Reichenbach P, Guzzetti F (2013) Very-high resolution stereoscopic satellite images for landslide mapping. Landslide Sci Pract 1:95–101. doi:10.1007/978-3-642-31325-7_12

Aucelli PPC, Cinque A, Rosskopf CM (2001) Geomorphological map of the Trigno basin (Italy) explanatory notes. Geogr Fis Dinam Quat 24:3–12

Aucelli PPC, De Angelis A, Colombo C, Palombo G, Scarciglia F, Rosskopf CM (2006) La stazione sperimentale per la misura dell'erosione del suolo di Morgiapietravalle (Molise, Italia): primi risultati sperimentali. Proceedings of the final workshop of the project "water erosion in Mediterranean environment: direct and indirect assessment in test areas and catchments". Brigati, Genova, pp 87–104

Beguería S (2006) Changes in land cover and shallow landslide activity: a case study in the Spanish Pyrenees. Geomorphology 74 (1–4):196–206. doi:10.1016/j.geomorph.2005.07.018

Borgomeo E, Hebditch KV, Whittaker AC, Lonergan L (2014) Characterising the spatial distribution, frequency and geomorphic controls on landslide occurrence, Molise, Italy. Geomorphology 226:148–161. doi:10.1016/j.geomorph.2014.08.004

Bracchetti L, Carotenuto L, Catorci A (2012) Land-cover changes in a remote area of central Apennines (Italy) and management directions. Landscape and Urban Plann 104(2):157–170. doi:10.1016/j.landurbplan.2011.09.005

Bruschi VM, Bonachea J, Remondo J, Gómez-Arozamena J, Rivas V, Barbieri M, Capocchi S, Soldati M, Cendrero A (2013) Land management versus natural factors in land instability: some examples in northern Spain. Environ Manag 52(2):398–416

Castellanos Abella EA, Van Westen CJ (2007) Generation of a landslide risk index map for Cuba using spatial multi-criteria evaluation. Landslides 4(4):311–325. doi:10.1007/s10346-007-0087-y

Caviezel C, Hunziker M, Schaffner M, Kuhn NJ (2014) Soil-vegetation interaction on slopes with bush encroachment in the central Alps—adapting slope stability measurements to shifting process domains. Earth Surf Proc Land 39(January):509–521. doi:10.1002/esp.3513

Cruden DM, Varnes DJ (1996) Landslide types and processes. In: Turner KA, Schuster RL (eds) Landslides: investigation and mitigation, transport research board special report, vol 247. National Academy Press, Washington DC, pp 36–75

Falcucci A, Maiorano L, Boitani L (2007) Changes in land-use/land-cover patterns in Italy and their implications for biodiversity conservation. Landscape Ecol 22(4):617–631

Guns M, Vanacker V (2014). Shifts in landslide frequency–area distribution after forest conversion in the tropical Andes. Anthropocene, 1–11. doi:10.1016/j.ancene.2014.08.001

ITC (2007) ILWIS 3.3 Academic. Available at https://www.itc.nl/ilwis/downloads/ilwis33.asp

Marchetti M, Bertani R, Corona P, Valentini R (2012) Changes of forest coverage and land uses as assessed by the inventory of land uses in Italy. Forest@—Rivista Di Selvicoltura Ed Ecologia Forestale, 9(4), 170–184. doi:10.3832/efor0696-009

Mazzoleni S, Di Pasquale G, Mulligan M, Di Martino P, Rego F (2004) Recent dynamics of the mediterranean vegetation and landscape. Wiley & Sons Ltd, Chichester, UK

Pelorosso R, Della Chiesa S, Tappeiner U, Leone A, Rocchini D (2011) Stability analysis for defining management strategies in abandoned mountain landscapes of the Mediterranean basin. Landscape and Urban Plann 103(3–4):335–346. doi:10.1016/j.landurbplan.2011.08.007

Pourghasemi HR, Pradhan B, Gokceoglu C (2012) Application of fuzzy logic and analytical hierarchy process (AHP) to landslide susceptibility mapping at Haraz watershed, Iran. Nat Hazards 63:965–996

Promper C, Puissant A, Malet JP, Glade T (2014) Analysis of land cover changes in the past and the future as contribution to landslide risk scenarios. Appl Geogr 53:11–19. doi:10.1016/j.apgeog.2014.05.020

Reichenbach P, Busca C, Mondini AC, Rossi M (2014) The influence of land use change on landslide susceptibility zonation: the briga catchment test site (Messina, Italy). Environ Manage 54:1372–1384. doi:10.1007/s00267-014-0357-0

Rosskopf CM, Aucelli PPC (2007) Analisi del dissesto da frana in Molise. In: Rapporto sulle frane in Italia. Il progetto IFFI—Metodologia, risultati e rapporti regionali. Rapporti APAT vol. 78, pp 493–508
Saaty TL (1980) The analytical hierarchy process. McGraw Hill, New York, p 350
Schwarz M, Lehmann P, Or D (2010) Quantifying lateral root reinforcement in steep slopes—from a bundle of roots to tree stands. Earth Surf Process Land 35:354–367
Slaymaker O, Spencer T, Embleton-Hamann C (2009) Geomorphology and global environmental change. Cambridge University Press, Cambridge
Soares-Filho BS, Coutinho Cerqueira G, Lopes Pennachin C (2002) DINAMICA—A stochastic cellular automata model designed to simulate the landscape dynamics in an Amazonian colonization frontier. Ecol Model 154(3):217–235. doi:10.1016/S0304-3800(02)00059-5
Wasowski J, Lagreca MD, Lamanna C (2014) Land-use change and shallow landsliding: a case history from the apennine mountains, Italy. In: Sassa K, Canuti P, Yin Y (eds) Landslide science for a safer geoenvironment, vol 1. The International Programme on Landslides (IPL). Cham, Springer International Publishing, pp 267–272. doi:10.1007/978-3-319-04999-1_36

4thWLF2017 4th World Landslide Forum LJUBLJANA SLOVENIA EU

Holocene Seismically and Climatically Driven Mass Wasting Processes in Boguty Valley, Russian Alta

R.K. Nepop and A.R. Agatova

Abstract

Study of post Last Glacial relief evolution in the southeastern part of Russian Altai (SE Altai) revealed that debris flows were widely presented in the region. After degradation of the last Pleistocene glaciation and drying giant ice-dammed lakes seismic excitation was one of the main factors that controlled intensification of slope activity. Geological and geomorphological researches of 2011–2015 in the Boguty river valley, SE Altai, were focused on studying geomorphic processes and reconstructing landscape evolution within the eastern periphery of the Chuya intermountain depression and Chikhachev range during the late Pleistocene-Holocene. Multidisciplinary approach, including determination of bio-composition of plant remnants from deposits of different genesis, tree species composition from charcoal fragments, micromorphological studies of surface and buried soils as well as litho-stratigraphic and pedogenetic analysis were applied for paleoclimatic and paleoenvironmental reconstructions. Geochronological investigations included radiocarbon dating of fossil soils, buried peats, lacustrine gyttjas and charcoal fragments. Within area of investigations seven sections were studied and 18 radiocarbon age estimations, covered time interval of the last 14 ka, were obtained. Obtained results indicate the tectonic origin of tributary valley of the Boguty river, which was settled along one of the faults in the late Pleistocene. Periods of tectonic quiescence and slope stability were repeatedly interrupted by periods of intensification of slope processes. Debris flows took place here about 7000, between 2800 and 1000, 650, 250 cal. years BP and could be triggered by both climatic and seismic events.

Keywords

Mass wasting processes • Debris flows • Radiocarbon analysis • Holocene • Russian Altai

Introduction

Relief development and landscape evolution within the high mountain provinces of Central Asia are usually controlled by complex combination of endogenous and exogenous factors. Cenozoic formation of Russian Altai was accompanied by uplifting of peripheral parts of intermountain depressions as well as by deepening of main valleys, which separate mountain ranges. These valleys of tectonic origin usually follow regional faults. As a result of tectonic uplifting, unconsolidated pre-Pleistocene sediments had been brought to the day surface. Later together with the extensive Pleistocene glacial deposits, they were affected by exogenous processes. Slope processes, including those triggered by ongoing tectonic movements, fluvial erosion, draining of ice- and moraine-dammed lakes transported sediments to lower hypsometrical levels. After drying Pleistocene

R.K. Nepop (✉) · A.R. Agatova
V.S. Sobolev Institute of Geology and Mineralogy SB RAS, Ak. Koptyuga av., 3, Novosibirsk, 630090, Russia
e-mail: agatr@mail.ru

R.K. Nepop · A.R. Agatova
Ural Federal University, Mira Str. 19, Yekaterinburg, 620002, Russia

M. Mikoš et al. (eds.), *Advancing Culture of Living with Landslides*,
DOI 10.1007/978-3-319-53483-1_46

ice-dammed lakes, seismotectonic activity and climate variations are the main triggers that controlled these processes in the Holocene (Agatova et al. 2016).

Studying post-glacier dynamics of landscape evolution within the high mountainous southeastern part of Russian Altai (SE Altai), which is characterized by arid climate, indicate that slope mass wasting processes play an important role among all surface processes. Debris flows were quite common nature phenomenon here. Earthquakes, as well as climatic factors (rapid snow melting, permafrost degradation etc.) served as one of the main triggers of slope processes in this high mountainous area with strongly dissected topography. Seismic excitations affected slope stability and prepare slope material for further destruction and removing. Presence of outcrops with organic material (such as fossil soils, buried peats, lacustrine gyttjas and charcoal fragments) in debris flow channels gives an opportunity to reconstruct the Holocene chronology of slope processes intensifications.

This paper presents the results of our multidisciplinary investigations of the Holocene landscape evolution and paleogeographical reconstructions within eastern periphery of Chuya intermountain depression, which is the largest one in Russian Altai.

Methods

A multidisciplinary approach was used to reconstruct the landscape evolution of the studied area.

Detailed geomorphological investigations and process analyses were based on interpretation of aerial photographs, Landsat-TM images, topographic maps (scale 1:50,000), and field investigations including mapping of landforms and deposits of different genesis.

Selected exposures in the Boguty River valley were studied to examine the sediments and landforms associations. Suitable material from key locations was sampled for determining the radiocarbon chronology of the late Pleistocene deglaciation, the Holocene hydrological system transformation, slope processes intensification, and the timing of other major geomorphic processes.

The bio-composition of plant remnants from peat and boggy deposits, tree species composition from charcoal fragments, micromorphological studies of surface and buried soils as well as litho-stratigraphic and morphological pedogenetic descriptions for the soil-sedimentary sequences were determined for palaeolandscape and paleoclimate reconstructions, revealing environmental conditions and sedimentation patterns.

The spatial distribution of in situ archaeological sites and their affiliation to a certain archaeological culture was used for timing and analyzing the patterns of Holocene hydrological system transformation, estimating ages (*terminus ante quem*) of terraces and alluvial fans.

Study Area

The Altai Mountains are the northern part of the Central Asia collision belt. They stretch northwest more than 1500 km across the borders of Mongolia, China, Kazakhstan, and Russia, and form a wedge shape narrowest in the southeast and widest in the northwest. The elevation increases in the opposite direction from 400 m a.s.l. to about 4000 m a.s.l.

The SE Altai is the highest part of the Russian Altai and includes the Chuya-Kurai system of intermountain depressions, located at about 1750–2000 and 1500–1600 m a.s.l. correspondingly, and framing ridges with the altitudes about 3500–4200 m a.s.l. Deeply incised valleys represent a system of imbedded troughs including up to three generations. The ice-sheet mid-Pleistocene and the piedmont late Pleistocene glaciations greatly affected the SE Altai landscapes (Devyatkin 1965). Smaller Holocene mountain valley glaciers mainly occupied the heads of troughs.

The mean annual precipitation is less than 200 mm in the floor of intermountain depressions, increasing with altitude. At the same time the mean annual precipitation near the snow line decreases along the W-E axis from 2000 mm (in the western part of the Katun range) to less than 500 mm in the Chikhachev range near the Mongolian border (Narozny and Osipov 1999) due to decreasing of moisture transfer from the Atlantic Ocean.

The SE Altai is the most seismically active part of Russian Altai. High regional seismicity is evidenced by paleoruptures and numerous large late Pleistocene-Holocene earthquake induced landslides (Nepop and Agatova 2008). Recently it was supported by the 2003 Chuya earthquake ($M_S = 7.3$). Strongly dissected topography and widely distributed unconsolidated Cenozoic sediments provide favorable conditions to produce earthquake induced landslides in the region.

The SE Altai is one of the areas where extensive ice-dammed lakes were formed in intermountain depressions throughout the Pleistocene. The Kuray-Chuya system of intermountain depressions is well-known worldwide due to the giant Pleistocene ice-dammed palaeolakes with their cataclysmic runoff into the Arctic Ocean along the Ob River (Rudoy and Baker 1993), which led to significant landscape changes in the drainage valley network over hundreds of kilometers. The water surface of that lakes could reach a maximum altitude of about 2250 m a.s.l., but even well expressed in topography level of 2100 m a.s.l. implies a lake volume of 607 km^3 and area 2630 km^2 (Herget 2005). According to exposure ages obtained from dropstones in the former lake basin, the most recent ice-dam outburst flood occurred at 15,000 ± 1800 BP (Reuther et al. 2006).

Area examined in this paper is located within the western slope of the Chikhachev range (Fig. 1), which is stretched southward more than 100 km. Southern part of the range with altitudes up to 3550–3700 m a.s.l. forms eastern periphery of the Chuya depression including its small branch—Boguty basin. Stair-step morphology of the range slopes indicates fault-block deformations of pre-orogenic peneplain surface. Remnants of this surface are intensively reworked by glacial processes and affected by cryogenic weathering. Some stairs were most likely developed as a result of glacial planation. Alpine topography characterizes the axial part of the range. Moraine deposits completely fill the piedmont part of the basin up to 2250–2230 m a.s.l.

The source of the Boguty river is located on the western slope of the Chikhachev range in its crest area. Complexes of terminal and side moraines, moraine-dammed and thermokarst lakes are presented in the Boguty depression from confluence of Left and Right Boguty rivers down to the mouth of the Boguty river. Numerous seismic ruptures on the trough's slopes serve as an evidence for the high seismic activity of the region during Post Glacial epoch.

The northern slope of the Boguty river valley near the Kok-Kul lake (Fig. 2) is a key location for landscape reconstructions, as well as for chronological reconstruction of slope processes intensifications. Valley slope of southwestern exposure in this place is covered by moraine sediments with large boulders and blocks, and is affected by solifluction. Slope inclination is about 16°–18° and altitude gradient—up to 150 m. Sliding of moraine cover is revealed down from 2550 m a.s.l. This altitude is significantly higher than the maximal water level of giant ice dammed lakes, which were developed in the Kurai-Chuya system of intermountain depressions in Pleistocene. It is also high above the water level of local Holocene moraine dammed lakes that occupied the Boguty basin during the degradation of Last Glaciation. Thus, drying of these reservoirs could not affect processes of slope material sliding in this part of the Boguty depression. The petrographic composition of the rocks (predominance of granite boulders in moraine structure) indicates transportation of debris material from the head of the Boguty valley, i.e. formation of moraine cover here was not controlled by the nearest glacier activity (that one, which occupied the nearby peaks).

Several faults of different ranks are located within the study site and point to a possible seismic trigger of sliding. One of the young ruptures goes along the main scarp of this sliding area and also supports this thesis. Seismic origin of studied transversal nameless valley—right tributary of the Boguty river (Fig. 2), is evidenced by its location on the extension of one of the secondary faults, which is involved in the formation of block structure of the western slope of the Chikhachev range. This tributary valley also contrasts sharply by its depth and slopes steepness. Loose sediments fill chaplet-shaped widenings of the valley in places where it crosses sliding ramparts and depressions. Nowadays these sediments are cut by younger creek channel. Repeated powerful debris flows run along this tributary into the Boguty valley. Earthquakes, which are quite frequent in this part of Russian Altai (Agatova and Nepop 2016), together with the seasonal wetting as a result of snow melting and permafrost degradation, served as a main trigger of mass wasting processes.

The lower part of alluvial fan/debris cone is confined to 2400 m a.s.l.—the maximal water level of the paleo-Kok-Kul lake, which is well defined in topography. Further lowering of the moraine dammed lake level down to 2394 m a.s.l. and then to the modern level about 2387 m a.s.l. defined changes of local erosion basis.

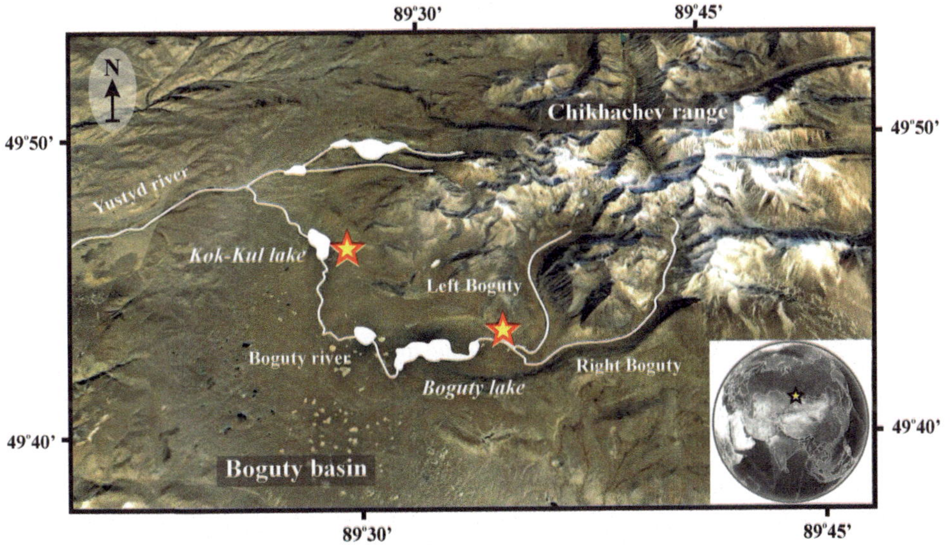

Fig. 1 Studied area. Stars show site locations in the mouth of nameless tributary near Kok-Kul lake (Fig. 2) and in the head of the Boguty valley

Fig. 2 Northern slope of the Boguty river valley near the Kok-Kul lake and transversal *right* tributary valley, where numerous debris flows occurred during the Holocene. Numbers indicate: *1* the channel of the side debris flow; *2–7* location of studied outcrops. *Dashed line* shows the water level 2400 m a.s.l. of the paleo-Kok-Kul lake

Results and Interpretations

Multidisciplinary investigations in the Boguty river valley, conducted in 2011–2015, were focused on studying relief forming processes and reconstruction of landscape evolution within the eastern periphery of Chuya intermountain depression and Chikhachev range during the late Pleistocene-Holocene. Analysis of topography and newly discovered outcrops revealed that after degradation of last glaciation repeated debris flows took place in the basin of the Boguty river.

In the context of this investigation generally seven sections were studied and eighteen radiocarbon age estimations were obtained within the Boguty river basin. Three sections are located in the upper part of the debris flow channel (tributary valley of the Boguty river). Three more outcrops were studied in its mouth, including the section of buried peats in the coastal outcrop of the Kok-Kul lake. Radiocarbon ages of fossil peats from this section allowed us to estimate the time of lake draining to its modern level. Additionally, alluvial/colluvial cone with the fossil soil was studied in the head of the Boguty river. Radiocarbon dating and analyses of sediments from these sections suggest some chronological stages of landscape evolution and intensifications of slope activity in the Boguty basin, which were controlled by climate changes and seismicity.

Radiocarbon age (IGAN 4098) of the oldest peat laid above moraine deposits on the bottom of the tributary valley (Fig. 2, number 5) indicate that at least by 14 ka BP the Sartan (late Würm) glaciers in this part of Russian Altai were either completely degraded or retreated above 2500 m.a.s.l. Studied tectonic valley had been already formed but it was shallow and boggy. Tree and bush vegetation remnants preserved in the peat layers points to a more warm and humid climate in this phase. There is no tree vegetation within the western slope of the Chikhachev range nowadays.

About 11 ka BP (SOAN 9368) there is a period of tectonic quiescence, stabilization of slope processes, and soils formation at 2470 m.a.s.l. (SOAN 9368) (Fig. 2, number 4).

Abundance of charcoal in paleosoil layers demonstrates widespread distribution of larch (*Larix sibirica* Ledeb) in this part of the SE Altai about 8.5–8.0 ka BP (SOAN 8674, 9366-2; IGAN 4089) and repeated forest fires, which led to tree deaths. Radiocarbon ages of fossil soils with charcoal fragments provide evidence of prolonged period of tectonic quiescence and stabilization of slopes 8.2–7.8 ka BP (SOAN 9366-1, 9369; IGAN 4091) (Fig. 2, number 1). In all cases charcoal fragments are 300–400 years older than inclosing soils. This fact could be explained by later beginning of soil formation and longer duration of pedogenic process.

Removal and transportation of large boulders and blocks by debris flow about 7.0 ka BP was reconstructed by analysis

Fig. 3 Section of deposits exhumed in coastal outcrop of right nameless tributary of the Boguty river at 2470 m a.s.l. (Fig. 2, number 4). Calibrated (2σ) radiocarbon dates are presented. *1* contemporary and fossil soils; *2* alluvial deposits; *3* debris flow deposits with fragments of fossil soils; *4* alluvial deposits affected by solifluction similar in structure to those in pack 2

of sedimentation sequences in the outcrop at 2470 m a.s.l. (Fig. 2, number 4). Paleosoil layer was "wrapped" over the granite boulder in debris flow deposits (Fig. 3). By this time the water level of the Kok-Kul lake was already close to its modern one, which is evidenced by radiocarbon age (IGAN 4823) of peat layer covering lacustrine sediments at the bottom of coastal outcrop near the mouth of studied tributary (Fig. 2, number 1). Thus debris cone was developed on the drying bottom of the Boguty river valley.

Soil and peat layer overlying redeposited moraine sediments within the tributary valley was formed at 2500 m a.s.l. by 3.5 ka BP. At that time the slope processes intensification occurred and caused peat burying within the right slope of this tributary valley. About 2.9–2.8 ka BP (SOAN 9370, 9371) a small local lake occupies this place (Fig. 2, number 5), which is evidenced by peaty lacustrine gyttjas. The presence of sands and gravels in the upstream part of studied outcrop indicates periodic drainage of this lake. Later the tributary valley was significantly deepened, probably as a result of debris flow activity, and creek channel moved to the left slope of the valley. About 1.0 ka BP (SOAN 8675) peat and later soil was developed above sliding slope deposits on the surface of preserved fragment of the right-bank terrace.

Intensification of slope processes, which caused redeposition of fossil soil with charcoal fragments at 2470 m a.s.l., occurred also between 7.0 and 2.7 ka BP (Fig. 2, number 4, Fig. 3). About 2.7 ka BP another soil horizon developed above alluvial deposits affected by solifluction.

About 650 years BP (SOAN 9373) soil horizon, developed on the dried bottom of the Kok-Kul lake in the distal part of debris cone within studied tributary valley at 2396 m a.s.l. (Fig. 2, number 6), was covered by a layer of alternating loams, fine sands, gruss and coarse sands. This fact argues for powerful removal of debris material from the tributary valley at that time.

The time of last powerful debris flows in the area (about 250 years BP) is determined by the age of fossil soil in the mouth of the tributary valley at 2460 m a.s.l. (Fig. 2, number 2). This soil is interlying between older debris flow deposits reworked by water and the bouldery deposits of younger debris flow. Deposits of the last debris flow are spread over a large area, which can be seen in numerous coastal outcrops downstream the creek channel (Fig. 2, number 3). This date (about 250 years BP) coincides with the radiocarbon age (IGAN 4087) of thin fossil soil layer in the sediments of alluvial/colluvial cone in the head of the Boguty river. On the one hand, this fossil soil provides evidence for the short period of surface stabilization, and on the other, marks abrupt intensification of slope activity, which led to its burial. Taking into account wide areal distribution of surface effects of slope processes intensification in the region and coinciding the period of this activity with the 1761 Great Mongolian earthquake, it could be suggested that intensive ground motion during the 1761 strong seismic event served as a possible trigger of one of the recent powerful debris flows in the Boguty river valley.

Conclusion

Geomorphological and geochronological analyses indicate that after degradation of the last Pleistocene glaciation mass wasting processes played an important role in landscape development within the Boguty basin (eastern branch of Chuya intermountain depression). By 14 cal. ka BP moraine sediments here were reworked by slope processes and nameless tributary valley of tectonic origin near the Kok-Kul lake was formed.

Radiocarbon dating of peat layers, fossil soils, lacustrine gyttjas and charcoal fragments from outcrops within this tributary valley and in the head of the Boguty basin suggest some chronological stages of landscape evolution and time intervals of intensification of slope activity, which were controlled by climate changes and seismicity. Eighteen radiocarbon dates indicate periods of tectonic quiescence and slope stability, which were repeatedly interrupted by powerful mass wasting processes. Debris flows took place here about 7000, between 2800 and 1000, 650, and 250 cal. years BP and could be triggered by both climatic and seismic events.

Acknowledgements O.N. Uspenskaya (RSRIV RAAS) and V.S. Myglan (SFU, Krasnoyarsk) are kindly thanked for determination of bio-composition of plant remnants and tree species. We would like to thank also M.A. Bronnikova (IG RAN) for morphological pedogenetic analysis. Radiocarbon analysis was carried out by I.Yu. Ovchinnikov (IGM SBRAN) and E.P. Zazovskaya (IG RAN).The study was partly funded by Russian Foundation for Basic Researches (grants 15-05-06028 and 16-05-01035).

References

Agatova A, Nepop R (2016) Dating strong prehistoric earthquakes and estimating their recurrence interval applying radiocarbon analysis and Dendroseismological approach—case study from SE Altai (Russia). Int J Geohazards Environ 2(3):131–149

Agatova AR, Nepop RK, Bronnikova MA, Yu Slyusarenko I, Orlova LA (2016) Human occupation of South Eastern Altai highlands (Russia) in the context of environmental changes. Archaeological and Anthropological Sciences 8:419–440

Devyatkin EV (1965) Cenozoic deposits and neotectonics of South-eastern Altai [Kajnozojskie otlozheniya i neotektonika Jugovostochnogo Altaya]. USSR Acad Sci Publisher, Moscow (in Russian)

Herget J (2005) Reconstruction of Pleistocene ice-dammed lake outburst floods in Altai-mountains, Siberia. In: Geological Society of America, Special Publication, p 386

Narozny YuK, Osipov AV (1999) Oroclimatic conditions of the Central Altai glaciations [Oroclimaticheskie uslovija oledenenija Central'nogo Altaja]. News Russ Geogr Soc 131(3):49–57 (in Russian)

Nepop RK, Agatova AR (2008) Estimating magnitudes of prehistoric earthquakes from landslide data: first experience in southeastern Altai. Russ Geol Geophys 49(2):144–151

Reuther A, Herget J, Ivy-Ochs S, Borodavko P, Kubik PW, Heine K (2006) Constraining the timing of the most recent cataclysmic flood event from ice dammed lakes in the Russian Altai Mountains, Siberia, using cosmogenic in situ ^{10}Be. Geology 34:913–916

Rudoy AN, Baker VR (1993) Sedimentary effects of cataclysmic late Pleistocene glacial outburst flooding, Altay Mountains, Siberia. Sed Geol 85:53–62

Part V
Student Session

Stress-Strain Modelling to Investigate the Internal Damage of Rock Slopes with a Bi-Planar Failure

Alberto Bolla and Paolo Paronuzzi

Abstract

The bi-planar failure, sometimes referred to as "bi-linear failure", is a particular type of rupture of rock slopes that occurs when a steep rock joint intersects a discontinuity having a lower inclination and that daylights at the rock face. The bi-planar configuration requires, differently from other well-known failure types (such as planar, wedge and circular failures), a considerable inner deformation and/or rock fracturing to make the block movement and the subsequent collapse possible. In the present paper, a forward analysis has been performed on a high natural rock slope (height = 150 m) made up of stratified limestone and characterised by a bi-planar sliding surface. The slope stability has been investigated adopting a 2D finite difference analysis (*FDA*). Two specific failure mechanisms (1 and 2) have been identified, based on the different strength parameters assumed in the models. In failure mechanism 1, a combination of internal shear and tensile fracturing occurs so as to form a deep, curvilinear rupture surface that links the two pre-existing planar surfaces. The block kinematism is an en-block roto-translation that, in turn, causes additional internal fracturing to accommodate deformation. In failure mechanism 2, a large shear band with obsequent dip enucleates within the unstable block, thus subdividing it into two main sub-blocks with different kinematisms. Model results demonstrate that bi-planar rock slope failures are associated with internal block damage that can also determine possible inner block splitting and differential movements between the secondary blocks. Stress-strain modelling is a very effective study approach that can be used to understand the key role played by rock fracturing and inner deformation occurring during the long preparatory phase that precedes the final collapse.

Keywords

Rock slope • Bi-planar (bi-linear) failure • Limestone • Failure mechanism • Internal block damage • Block splitting • Stress-strain analysis

A. Bolla (✉) · P. Paronuzzi
Università degli Studi di Udine, Dipartimento Politecnico di Ingegneria e Architettura, via Cotonificio 114, 33100 Udine, Italy
e-mail: alberto.bolla@uniud.it

P. Paronuzzi
e-mail: paolo.paronuzzi@uniud.it

M. Mikoš et al. (eds.), *Advancing Culture of Living with Landslides*,
DOI 10.1007/978-3-319-53483-1_47

Introduction

The bi-planar failure, sometimes referred to as "bi-linear failure", is a particular type of rupture of rock slopes that occurs when a steep rock joint intersects a discontinuity having a lower inclination and that daylights at the rock face. This type of failure is rarely studied, but there are many examples of actual rockslides denoting this geometrical configuration, especially for sedimentary rock masses characterised by bedding planes and affected by faults or by very persistent tectonic fractures. Bi-planar failures are most commonly encountered in shallow mine and engineered slopes (Stead and Eberhardt 1997; Corkum and Martin 2004; Fisher 2009; Alejano et al. 2011; Havaej et al. 2014), whereas they have been more rarely reported for high natural rock slopes (Eberhardt et al. 2004). Bi-linear rupture surfaces can characterise ancient huge landslides that filled valley floors after catastrophic slope collapses (Paronuzzi and Bolla 2015; Song et al. 2015; Chen et al. 2016). However, the failure mechanism of bi-planar failures often differs between mine and natural slope cases (Fisher 2009).

The bi-planar configuration can be determined by fully persistent discontinuities (O–P and P–Q in Fig. 1a) that enable sliding of the rock block to occur. In more complex cases, the lower sliding surface may be not persistent (O'–P in Fig. 1b) or even absent (Fig. 1c). When fully persistent discontinuities are lacking, a more complex failure mechanism has to occur through damage involving an 'intact' rock mass at the base of the unstable block (O–O' and O–P in Figs. 1b, c). In these circumstances, the slope topography is of key importance since strong changes in the slope inclination are associated with localised zones of stress concentration (grey circles in Fig. 1). High stress concentrations are responsible for fracture initiation, propagation and coalescence.

The bi-planar failure requires, differently from other well-known failure types (such as planar, wedge and circular failures), a considerable inner deformation and/or rock fracturing to make the block movement and the subsequent collapse possible. Mencl (1966) and Kvapil and Clews (1979) proposed the development of a Prandtl's prism transition zone or an active-passive block mechanism to explain slope failure controlled by a bi-planar surface. In the active-passive wedge mechanism, the forces from the upper block drive the lower block outward. To account for the importance of internal shearing, Sarma (1979) formulated a limit equilibrium method with inclined slices. Martin and Kaiser (1984) showed that internal shear and rock mass dilation are necessary to accommodate sliding along a basal rupture surface. Several authors emphasised the role of the rock mass strength degradation through newly-formed shear/tensile fractures and internal deformation that characterise rock slope failures (e.g. Stead et al. 2006; Havaej and Stead 2016; Paronuzzi et al. 2016).

In the present paper, a forward analysis has been performed on a high natural rock slope made up of stratified limestone and characterised by a bi-planar sliding surface. The slope stability has been investigated adopting a 2D finite difference analysis (*FDA*).

The Investigated Bi-Planar Failure

The site under investigation is located in the Carnic Alps, in the Province of Udine (Friuli Venezia Giulia Region, Italy). From a morphological point of view, the unstable slope is characterised by a large rock pyramid that is bounded at its western limit by a vertically incised stream cut and that culminates with a small terrace at its top (Fig. 2). The slope has the shape of a rock spur that forms a morphological relief with respect to the continuous strip of steep rock cliffs resting behind. The rock spur has a scarp height of more than 150 m and is covered at its toe by a coarse loose deposit that was originated by antecedent rockfalls.

A detailed field survey has been performed in order to determine the structural arrangement of the rock masses. The

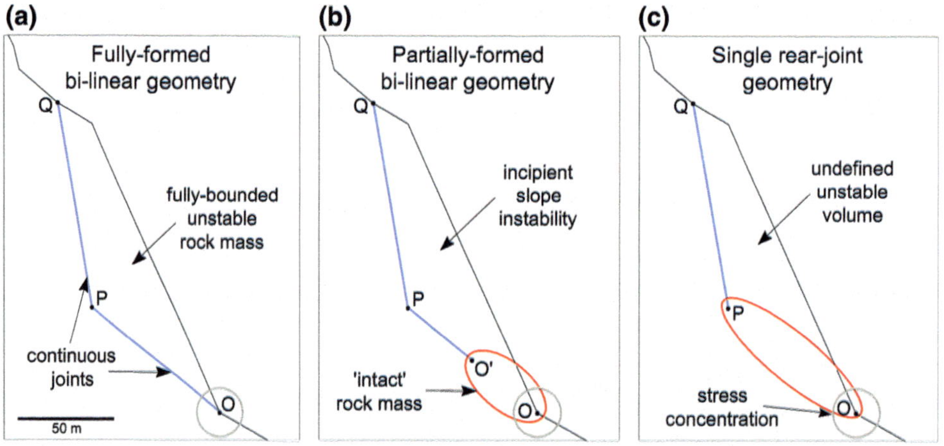

Fig. 1 Different geometrical configurations for bi-planar failures in high natural rock slopes

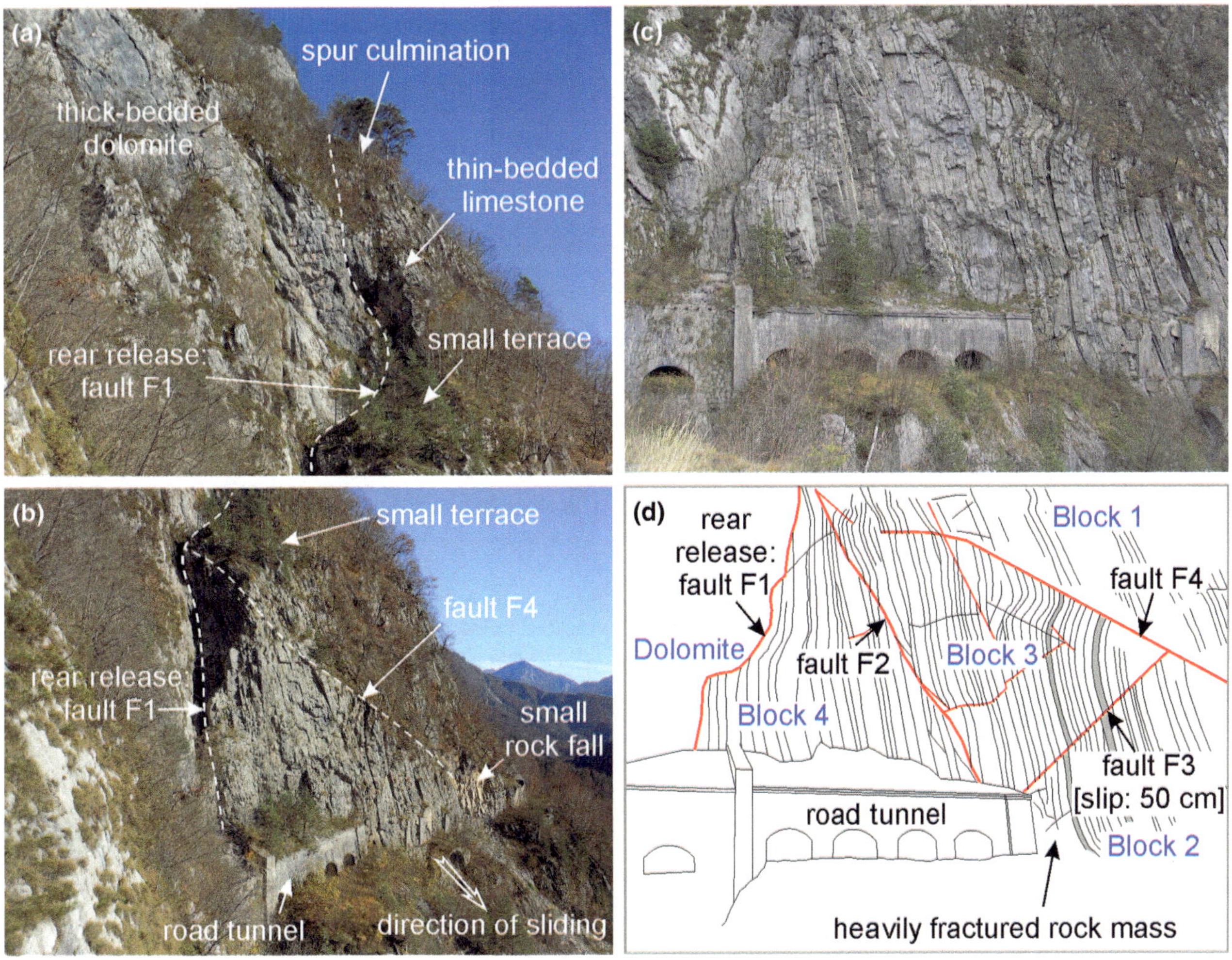

Fig. 2 **a** and **b** The investigated limestone slope characterised by a bi-planar failure. **c** and **d** Geomechanical sketch showing the major discontinuities affecting the slope, subdividing it into sub-blocks

rock spur is made up of highly-tilted thinly-stratified limestone (bed thickness: 10–30 cm, on average), whereas the rear and upper rock scarps are formed by thick-bedded dolomite (bed thickness: 1–3 m). The contact between the two lithostratigraphic units coincides with a high-angle (80°) fault that bounds the rock spur at its back and represents a possible rear detachment surface (F1 in Fig. 2). The rock mass is crossed by major discontinuities (faults) that subdivide the whole rock spur into many secondary blocks with different structural conditions (Figs. 2c, d). On the whole, these faults are characterised by high persistence and size (length = 20–100 m), low waviness and extremely low roughness (*JRC* = 1–2). The fault surfaces show a highly variable weathering degree, emphasised by the results of the Schmidt hammer tests carried out (*JCS* = 60–90 MPa). Some infilling mainly characterised by crushed rocks has been locally identified along the faults.

The fault F4 outcrops at the slope face with an apparent dip angle ranging from 30° to 45°, and intersects the fault F1 upward (Figs. 2b, d). This fault completely delimits an upper block (Block 1) that is characterised by a considerable damage associated with folded and fractured limestone layers. The fault F3 has an obsequent dip with a fault slip of about 50 cm, and separates two blocks with a very different damage. The upper block (Block 3) has a moderate degree of fracturing, with closed joints and some folded layers (Fig. 2). On the contrary, the lower block (Block 2) is highly fractured, with closely-spaced and open joints that isolate small blocks (even rotated) and that sometimes are filled with loose soil material (Fig. 2d). A small rockfall (volume = 200 m^3) recently occurred in this frontal portion of the slope (Fig. 2b). The fault F2 (apparent dip angle of 50°) intersects upward the rear fault F1 but terminates downward in correspondence to the Block 2 without daylighting at the slope toe (Fig. 2). The lower block (Block 4) is characterised by a lower degree of fracturing with closed unfilled joints and undeformed rock layers. Finally, two large and persistent faults (F5 and F6) that are sub-parallel to the fault F1 occur within the rock mass behind the limestone spur. As a result of the possible collapse of the rock spur, the slope geometry would change drastically, thus destabilizing a rear block made up of dolomite (Dolomite in Fig. 2d). The faults F5 and F6 can act as rear releases of an unstable slope portion that can originate a retrogressive failure.

The overall mechanical behaviour of the slope is mainly governed by the kinematism of the internal blocks, which in turn, depends on the geometry and mechanical properties of the major discontinuities that delimit the blocks. The stratification joints have scarce influence on the mechanical behaviour of the rock spur. In fact, the mean dip direction of the bedding is oblique when compared to the potential sliding direction of the slide. Field evidence, especially the ascertained slip along the fault surfaces, the strong fracturing of the rock mass at the slope toe and the occurrence of previous rockfalls, testifies that the rock spur has a stability condition very close to the critical one.

On the basis of the geological evidence and of the mechanical interpretation, three possible scenarios of slope collapse have been identified: (1) a slope failure that involves the upper part of the rock spur (volume = 100,000 m^3) and is characterised by a bi-linear rupture surface that is fully formed by the intersection of the rear fault F1 and the basal fault F4; (2) a larger rockslide (volume = 225,000 m^3) characterised by a bi-linear rupture surface almost entirely formed by the faults F1 and F2 and involving a resisting part of rock mass at the slope toe; and (3) a retrogressive failure involving a slope portion behind the rock spur (volume = 190,000 m^3) in which only the rear joints are present (high-angle faults F5 and F6). Each scenario identified corresponds with one of the three reference geometries discussed in the introductory section (Fig. 1).

Stress-Strain Modelling

The investigated slope has been analysed with some 2D mechanical simulations by using the explicit finite difference code FLAC 6.0 (Itasca, 2008). In the stress-strain modelling, the three cases of bi-planar failure previously described have been analysed (Fig. 3).

Model Setting and Input Parameters

The stress-strain analyses have been performed considering a reference cross section that is parallel to the potential sliding direction (Fig. 2b) and that passes through the spur culmination. A finite difference grid with quadrilateral elements of variable size (0.5–20 m) has been generated. No bottom vertical and horizontal displacements and no lateral

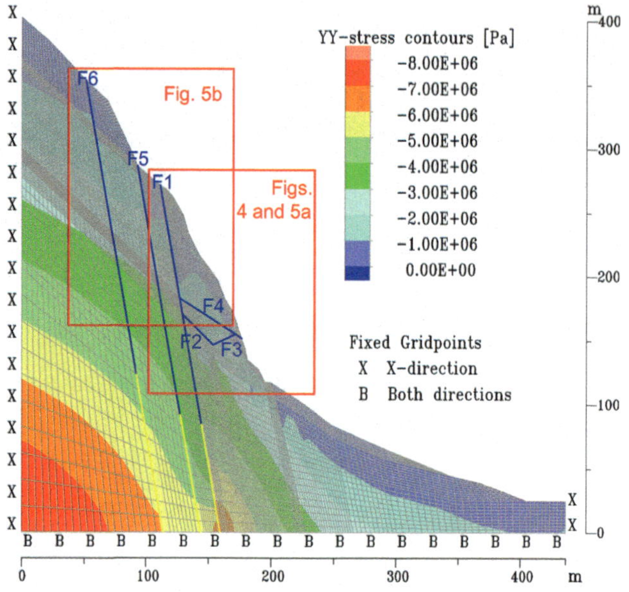

Fig. 3 Initial stresses and boundary conditions of the 2D model used to analyse various bi-planar failures. The interfaces simulating the main discontinuities crossing the slope (*blue lines*) are also shown

horizontal displacements were adopted as boundary conditions (Fig. 3).

Five mechanical units have been assumed in the model. These units correspond to the various blocks (Blocks 1–4 and Dolomite) previously identified through the field analysis (Fig. 2d). The geomechanical parameters of the blocks are listed in Table 1. A Mohr-Coulomb failure criterion has been adopted for the shear strength of the involved units. The M-C strength parameters (cohesion c and friction angle φ) as well as the Young's modulus E of the different blocks have been obtained through a field characterisation based on the Hoek-Brown criterion. The parameters for the intact rock and the rock mass are the following: σ_{ci} = 80 and 130 MPa; GSI = 25–60, respectively. A constant value equal to 9 has been assumed for m_i. The values of the equivalent cohesion and friction angle have been calculated through the linearization of the curvilinear Hoek-Brown failure envelopes for the 5 blocks, considering the actual slope height (H = 50–100 m). The friction angle φ differs very slightly among the various blocks (not more than 3°), thus a same value of φ = 50° has been assumed. This assumption had no influence on the modelling results. In some models, a lower value of the friction angle for the blocks (φ = 40°) has also been considered.

The faults crossing the slope (F1–F6) have been modelled as interfaces separating adjacent blocks. Because of the uncertainties concerning the actual strength properties of these faults, various values of their cohesion and friction angle have been considered in the simulations. These values vary considering the type of material involved (limestone),

Table 1 Geomechanical parameters adopted in the stress-strain analyses

Unit	Unit weight γ (kN/m^3)	Young's modulus E (MPa)	Normal stiffness k_n (Pa/m)	Shear stiffness k_s (Pa/m)	Friction angle φ (°)	Cohesion c (kPa)	Tensile strength σ_t (kPa)
Block 1	25.0	2500	–	–	40, 50	500	500
Block 2	25.0	1500	–	–	40, 50	400	400
Block 3	25.5	4000	–	–	40, 50	600	600
Block 4	25.5	6000	–	–	40, 50	800	800
Dolomite	26.0	14,000	–	–	40, 50	1000	1000
Interfaces	–	–	1.0E10	1.0E9	25–40	0–300	0

the high persistence and the low waviness of the faults as well as the possible presence of localised intact rock bridges (Table 1). The normal (kn) and shear (ks) stiffnesses of the interfaces have been assumed accordingly to physical properties of limestone joints (Bandis et al. 1983). Since no in situ stress data is available for high rock slope cases, a coefficient of earth pressure at rest (K_O) equal to 0.5 has been assumed. The slope has only been analysed in dry conditions.

Modelling Results

In some preliminary simulations, the stability condition of the slope has been investigated through a strength reduction analysis. The strength reduction factor (SRF) of the slope has been calculated assuming different combinations of the M-C strength parameters of the blocks and of the interfaces (Table 1). The SRF values vary between 1.00 and 1.15. These low values confirm that the slope is in a critical stability condition, as demonstrated by the considerable rock mass damage observed in the field.

After this preliminary slope stability evaluation, some stress-strain analyses have been performed with the aim of exploring the failure mechanism associated with the three joint configuration of bi-planar failures. In these analyses, various combinations of the mechanical parameters (friction angle of the blocks, cohesion and friction angle of the interfaces) have been assumed. The critical stage associated with the slope failure has been reached through a decrease in the cohesion values of the limestone blocks. A total of about eighty simulations have been performed. Two basic slope failure mechanisms have been identified.

Modelling outcomes show that, in order to enable slope displacements and the final collapse, internal block damage has to occur. Figures 4 and 5 show localised shear and tensile failures occurring within the slope before the collapse for the three cases analysed. When analysing a fully-formed bi-linear geometry (Fig. 4), the stress-strain models highlight two specific failure mechanisms (1 and 2), based on the different strength parameters assumed. In failure mechanism 1 (Model A1–A3 in Fig. 4), a combination of internal shear and tensile fracturing occurs so as to form a deep, curvilinear rupture surface that links the two pre-existing planar surfaces.

The block kinematism is an en-block roto-translation that, in turn, causes additional internal fracturing to accommodate deformation, especially tensile failures in the upper rear part of the unstable slope. In failure mechanism 2 (Model B1–B3 in Fig. 4), a large shear band with obsequent dip enucleates within the unstable block, thus subdividing it into two main sub-blocks. These secondary blocks have different kinematisms, being the displacement vectors sub-parallel to the basal sliding surface for the lower block (dip = 32°) and sub-parallel to the rear sliding surface for the upper one (dip = 80°).

When analysing the partially-formed geometry of the bi-linear rupture surface (Fig. 5a), the stress-strain models show that shear and tensile failures firstly occur at the slope toe. These internal ruptures grow upwards to form a continuous surface that joins the angular point of the rock face (point O in Fig. 1) with the lower extremity of the pre-existing fault F2 (Models C1 and D1 in Fig. 5a). After this first stage, a strong shear fracturing occurs within the unstable block, starting from the zone of intersection between the two pre-existing faults F1 and F2. The phenomenon of internal growing damage becomes very similar to the one characterising the fully-formed bi-linear geometry (Fig. 4). In fact, a curvilinear rupture surface enucleates in failure mechanism 1 (Model C3), whereas a clear obsequent shear band is formed in failure mechanism 2, well-representing the active-passive wedge mechanism (Model D3).

The case of single rear-joint geometry has been analysed considering the possible retrogressive slope failure involving the Dolomite block (Fig. 5b). In this case, shear failures occur in the zone of slope angle change at the toe and progressively propagate upward. A new rupture surface

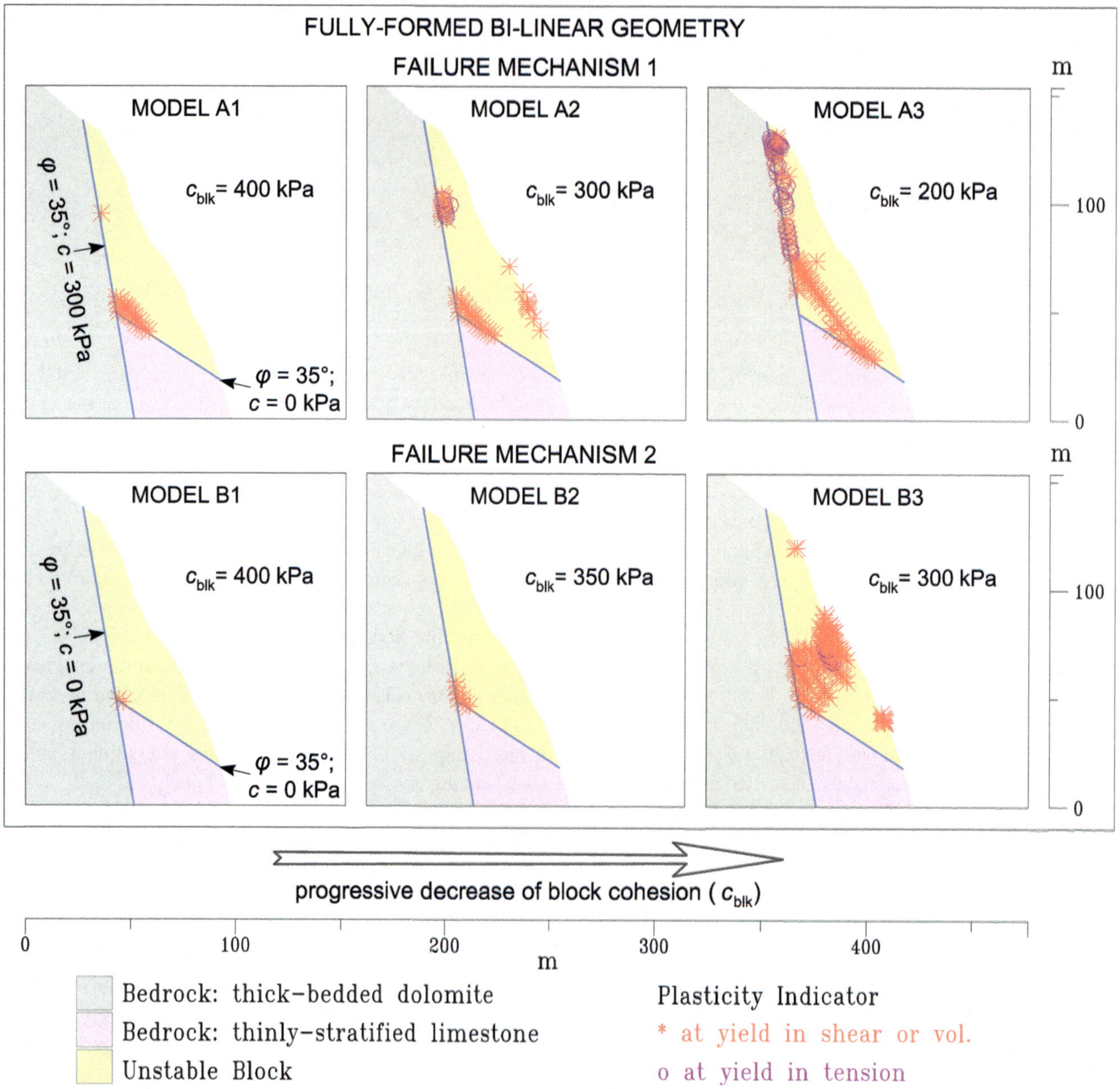

Fig. 4 Patterns of localised internal ruptures associated with the fully-formed bi-linear geometry

enucleates up to join the faults F5 and F6, destabilizing firstly a frontal block (Model E1), and subsequently a rear block (Model E2). The slope collapse is reached as a consequence of strong shear fracturing occurring in the deep zone of intersection between the fault F6 and the newly-formed rupture surface (Model E3). The final failure surface has a less evident curvilinear shape when compared to the previous cases analysed. However, the internal damage mechanism can be associated with the previously identified failure mechanism 1. The stress-strain analyses did not show evidence of a failure mechanism characterised by an internal subdivision of the unstable blocks. Nevertheless, an internal block subdivision can reasonably occur to enable the slope collapse with a single rear-joint geometry.

Model outcomes show that the failure mechanism 1 (curvilinear rupture surface) occurs when lower values of the strength parameters for the faults located at the base of the unstable blocks are assumed. In this case, higher slope displacements are enabled, favouring an overall roto-translation of the blocks. On the contrary, the failure mechanism 2

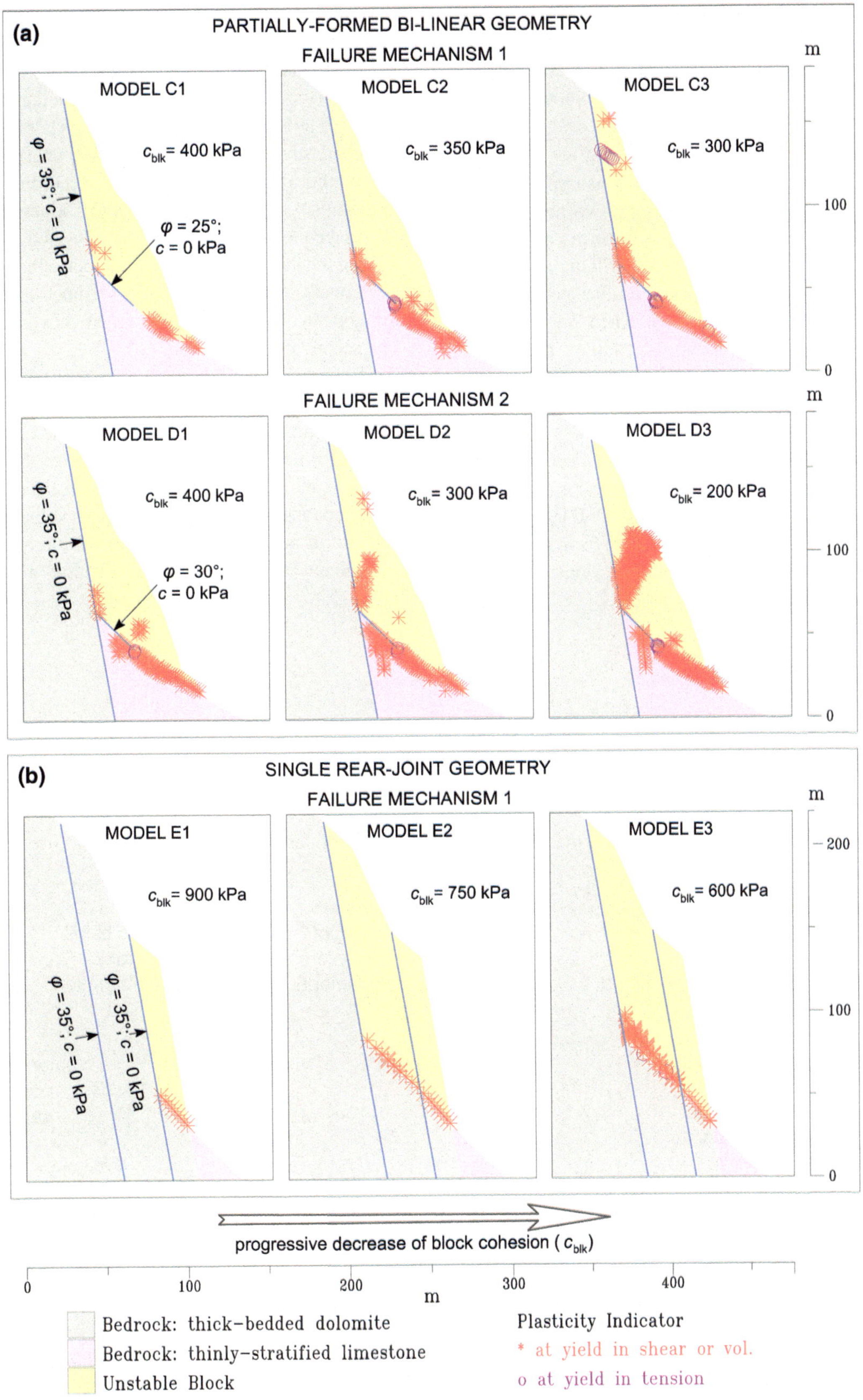

Fig. 5 Localised internal ruptures associated with the partially-formed **a** and the single rear-joint **b** geometry

(obsequent internal shear band) occurs when higher strength values for the basal faults are adopted. As a consequence of lower slope displacements, the unstable block tends to internally subdivide to accommodate deformation.

Conclusions

Bi-planar, or bi-linear, failures in high natural rock slopes occur when a rock mass slides along two persistent discontinuities, the upper sliding surface being more inclined than the lower one that daylights at the slope face. When the lower sliding surface does not daylight at the slope face (Fig. 6a), or even when only the upper sliding surface is present (Fig. 6b), failure must initiate through shearing of the rock mass, especially in coincidence of high stress concentrations at the slope toe (point O in Fig. 6).

Shear and secondary tensile ruptures occur and propagate up to enucleate a continuous shear surface at the base of the unstable block (O–O' and O–P in Figs. 6a, b, respectively). As a result, the fully-formed bi-planar surface allows for shear displacements along both planes (Fig. 6c). However, the slope collapse is only enabled through inner block damage in the form of the enucleation of shear and tensile ruptures that originate internal secondary failure surfaces and/or damage zones. Two possible failure mechanisms can occur, depending on the different strength ratios between the unstable block and the delimiting discontinuities. According to failure mechanism 1 a roto-translational kinematism occurs along a curvilinear failure surface (Fig. 6d). According to failure mechanism 2, an internal block subdivision occurs as a result of the formation of an inner shear band with obsequent dip, causing differential movements between the secondary blocks (Fig. 6e).

To comprehensively study bi-planar failures involving rock slopes we must consider the interaction between pre-existing discontinuities, rock mass strength and newly-formed fractures. The analysis of the deformation process and of the progressive failure that precede the slope collapse is a geomechanical issue that is difficult to capture in actual rockslides and for this reason is rarely considered in the stability evaluation of rock slopes.

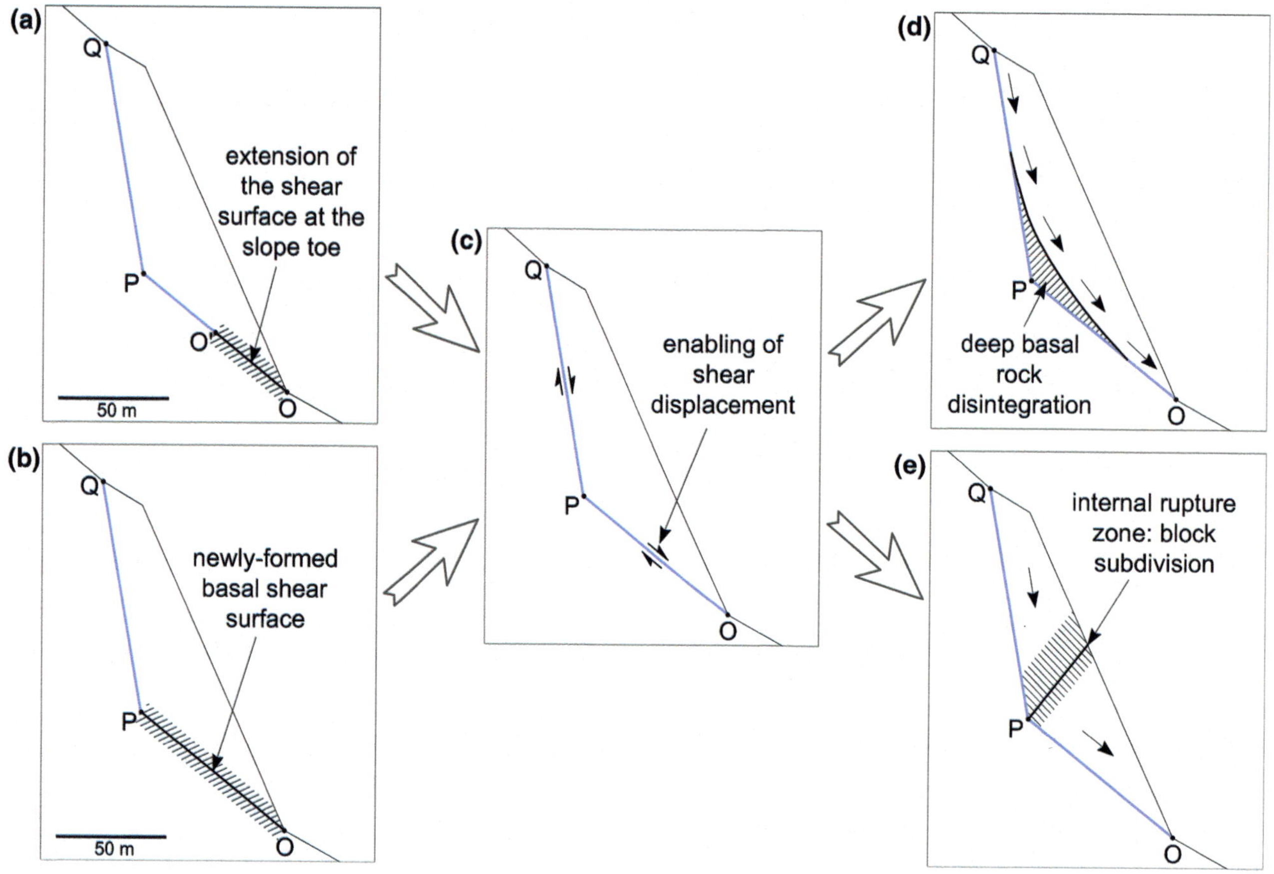

Fig. 6 Possible failure mechanisms associated with the bi-planar geometry

These decisive mechanical processes involving the inner rock mass remain, in most cases, invisible to the eyes of the technicians. Stress-strain modelling is a very effective study approach that can be used to understand the key role played by rock fracturing and inner deformation that occur during the long preparatory phase that precedes the final collapse.

References

Alejano LR, Ferrero AM, Ramírez-Oyanguren P, Álvarez Fernández MI (2011) Comparison of limit-equilibrium, numerical and physical models of wall slope stability. Int J Rock Mech Min Sci 48:16–26

Bandis S, Lumsden A, Barton N (1983) Fundamentals of rock joint deformation. Int J Rock Mech Min Sci Geomech Abstr 20:249–268

Chen XP, Zhu HH, Huang JW, Liu D (2016) Stability analysis of an ancient landslide considering shear strength reduction behaviour of slip zone soil. Landslides 13:173–181

Corkum AG, Martin CD (2004) Analysis of a rock slide stabilized with a toe-berm: a case study in British Columbia, Canada. Int J Rock Mech Min Sci 41:1109–1121

Eberhardt E, Stead D, Coggan JS (2004) Numerical analysis of initiation and progressive failure in natural rock slopes-the 1991 Randa rockslide. Int J Rock Mech Min Sci 41:69–87

Fisher BR (2009) Improved characterization and analysis of bi-planar dip slope failures to limit model and parameter uncertainty in the determination of setback distances. Ph.D. Thesis, University of British Columbia, Vancouver, Canada

Havaej M, Stead D (2016) Investigating the role of kinematics and damage in the failure of rock slopes. Comput Geotech 78:181–193

Havaej M, Stead D, Eberhardt E, Fisher BR (2014) Characterization of bi-planar and ploughing failure mechanisms in footwall slopes using numerical modelling. Eng Geol 178:109–120

Kvapil R, Clews M (1979) An examination of the Prandtl mechanism in large dimension slope failures. Trans Inst Min Metall A 88: A1–A5

Martin CD, Kaiser PK (1984) Analysis of rock slopes with internal dilation. Can Geotech J 21:605–620

Mencl V (1966) Mechanics of landslides with noncircular slip surfaces with special reference to Vaiont slide. Géotechnique 16(4):329–337

Paronuzzi P, Bolla A (2015) Gravity-induced rock mass damage related to large en masse rockslides: evidence from Vajont. Geomorphology 234:28–53

Paronuzzi P, Bolla A, Rigo E (2016) Brittle and Ductile behavior in deep-seated landslides: learning from the Vajont experience. Rock Mech Rock Eng 49:2389–2411

Sarma SK (1979) Stability analysis of embankments and slopes. J Geotech Eng Div, ASCE 105:1511–1534

Song K, Yan E, Zhang G, Lu S, Yi Q (2015) Effect of hydraulic properties of soil and fluctuation velocity of reservoir water on landslide stability. Environ Earth Sci 74:5319–5329

Stead D, Eberhardt E (1997) Developments in the analysis of footwall slopes in surface coal mining. Eng Geol 46:41–61

Stead D, Eberhardt E, Coggan JS (2006) Developments in the characterization of complex rock slope deformation and failure using numerical modelling techniques. Eng Geol 83:217–235

Slope Mass Assessment of Road Cut Rock Slopes Along Karnprayag to Narainbagarh Highway in Garhwal Himalayas, India

Saroj Kumar Lenka, Soumya Darshan Panda, Debi Prasanna Kanungo, and R. Anbalagan

Abstract

Slope instability is a major problem in hilly terrains. Stability assessment of road cut rock slopes is of paramount importance for planning and construction of infrastructures in hilly terrain. Slope Mass Rating (SMR) technique developed by Romana (1985) is a geomechanical method to assess the stability of rock slopes, that in principle uses basic Rock Mass Rating (Bieniawski 1979, 1989) and geometrical relationship between slope and rock discontinuities. In the present study, 39 rock slopes along Gopeswar-Almora road from Karnprayag to Narainbagarh in Chamoli district of Garhwal Himalayas have been studied in detail for their slope mass assessment using the mentioned technique. Based on the field observations, quartzite is the dominant lithotype with maximum of three sets of discontinuities in the study area. Meta Volcanic and Meta Sedimentary rocks are also present, but within narrow patches along the road, with phyllitic and schistose rocks. The parameters pertaining to Rock Mass Rating (RMR) and Slope Mass Rating (SMR) techniques were collected from field and laboratory studies for all the rock slopes. The basic RMR values of quarzitic rocks range between 50 and 88 whereas the SMR values vary from 07 to 84 depending upon the geometrical relationship between orientation of Slope face and discontinuities. For most of the schistose and phyllitic rocks basic RMR values are observed to be below 50 and the SMR values lie between 0 and 38. Based on the SMR values, these cut slopes were categorized into five different failure potential classes (Romana 1985). From the results, it is inferred that out of total 39 rock slopes, six slopes are completely unstable, 17 slopes are unstable, another six slopes are partially stable, nine slopes are stable, and only one slope is completely stable. From the kinematic analysis of slope face in relation to discontinuities, it was found that planar mode of failure is the most

S.K. Lenka (✉) · S.D. Panda · R. Anbalagan
Department of Earth Sciences, Indian Institute of Technology Roorkee, Roorkee, India
e-mail: sarojlenka29@gmail.com

S.D. Panda
e-mail: soumya.goodlu@gmail.com

R. Anbalagan
e-mail: anbaiitr@gmail.com

D.P. Kanungo
Geotechnical Engineering Group, CSIR-Central Building Research Institute (CBRI), Roorkee, India
e-mail: debi.kanungo@gmail.com

M. Mikoš et al. (eds.), *Advancing Culture of Living with Landslides*,
DOI 10.1007/978-3-319-53483-1_48

predominant type followed by wedge mode of failure in these rock slopes. Out of total slopes, 19 slopes are prone to planar failure, 8 slopes are prone to wedge failure, 3 slopes are more likely to fail by planar mode but has significant wedge failure component, 1 slope is prone to topple failure and 8 stable slopes.

Keywords

Slope mass assessment • Rock slopes • Discontinuities • Basic RMR • SMR • Kinematic analysis

Introduction

In the Himalayan hilly terrains, road networks are the only means of communication. Construction of these roads through blasting and excavation without proper knowledge of structure, lithology and rock mass strength often produce unstable slopes along the road, which on failure creates a lot of problem in flow of local traffic as well as pilgrimage to the holy shrines of Badrinath and Kedarnath as well as a loss to life and property. The present study area is a 30 km stretch along Gopeawar-Almora highway of Chamoli district of Uttrakhand from Karnprayag (lat.30°15′40″N/long.79°12′ 14″E) to Narainbaragh (30°07′43″N/long.79°03′12″E) which lies within Inner Lesser Himalayas and is bounded by North Almora Thrust (NAT) in south & Main Central Thrust (MCT) in north (Bhattacharya 1990). The studied rock masses are dominantly low to medium grade Beringag Quartzite of Mesoproterozoic age (Valdiya 1980), along with Quarzitic Phyllites, Phyllites and Schists which are exposed in narrow patches along the road section.

Method of Study

For Slope Stability assessment, Rock Mass rating (RMR) technique developed by Bieniawski (1973, 1979, 1989) and Slope Mass Rating (SMR) technique developed by Romana (1985) were followed.

Rock Mass Rating (RMR)

It is a geomechanical classification that describes quality of a rock mass. The RMR in principle uses five parameters to evaluate basic RMR value (RMR_{basic}). These parameters include: (i) Uniaxial Compressive Strength (UCS) of the Rock mass, (ii) Rock Quality designation (RQD), (iii) Spacing of discontinuities, (iv) Condition of discontinuities, (v) Ground water condition.

To obtain UCS, point load testing method was followed to obtain Point load Strength Index (PSI) by using the following formula along with the specified sample dimensions given by Bureau of Indian Standards (IS 8764:1998).

$$\mathrm{IL}\,(50) = \mathrm{P}/\left\{(\mathrm{D_W})^{0.75} \times (\mathrm{De})^{0.5}\right\}$$

where: IL (50): PSI of sample in kgf/cm^2, P: Peak Load at failure in kgf, D_W: Cross-section area of the sample in cm^2, De: Standard size of lump = 5 cm.

RQD gives information about the quality of a drilled core. Mathematically RQD is given by:

$$\mathrm{RQD} = \frac{\text{Total length of individual intact cores of length more than 10 cm}}{\text{Total length of recovered core}} \times 100$$

In the field, RQD was calculated using Palmstrom (2005) equation, RQD = 110–2.5 Jv, where Jv is volumetric joint count. Another method was also worked out to calculate RQD in the field for road cut rock slopes. If a vertical hole would have been drilled through the rock formation, the recovered core should contain all the discontinuities exposed on the road cut slope face as visible in the photograph (Fig. 1). The red line on the photograph represents a 1 m drill length and B, C, D & E represents the core lengths to be considered for calculating the RQD.

Spacing of discontinuities, condition of discontinuities and water conditions are either directly measured or indirectly inferred from field observations. All the above mentioned parameters for different rock slopes were observed and assessed based on field and laboratory investigations and rated according to Bieniawski (1989) for obtaining the RMR_{basic}.

Slope Mass Rating (SMR)

SMR was developed by Romana (1985) for stability assessment of rock slopes. SMR technique uses RMR_{basic} with due consideration for four adjustment ratings, out of which first three adjustment ratings (F_1, F_2 & F_3) are dependent on the geometrical relationship between attitudes of slope face and discontinuities and the fourth one (F_4) is dependent on the nature of blasting carried out on the slope face.

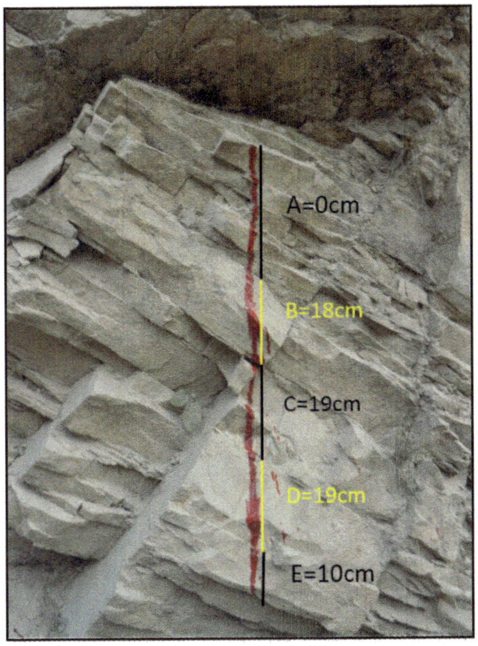

Total length of the Red line = 100 cm (1 m)
Volumetric joint count (J_v) = 17
RQD = 110-2.5J_v = 67.5 (As per Palmstrom, 2005)
Total length rock portion ≥10cm (B+C+D+E) =66cm.
RQD= 66% (As per proposed method for road cut rock slopes)

Fig. 1 Field Photograph Showing calculation method of RQD in field

$$SMR = RMRbasic + F_1 \cdot F_2 \cdot F_3 + F_4$$

where, F_1: measure of parallelism between slope face (α_s) and joint plane (α_j) or line of intersection between two joint planes {$F1 = (1 - \text{Sin } A)^2$}; F2: depends on the dip of the joint plane or plunge of the line of intersection of two joint planes {F2 = tan B}; F3: depends on the difference of dip of slope (β_s) and dip of joint plane (β_j) or plunge of the line of intersection of two joint planes; F4: Depends on the nature of blasting; A = |αj-αs| or |αi-αs|; B = βj. The ratings or the adjustment factors for different slopes with planar, wedge and toppling type of failures are computed from the tables given by Romana (1985) and Anbalagan et al. (1992). The SMR value, in general, ranges from 0 to 100. According to the SMR value, the rock slopes are classified into five potential failure classes as per Romana (1985).

Evaluation of Data

The collected field data is thoroughly analyzed to obtain RMR, SMR and stability condition of all the 48 rock slopes. Furthermore Kinematic analysis is also performed using stereo plot to infer the most probable type of failure. An example of data evaluation is illustrated in Fig. 2 and Table 1.

Results and Discussion

Stability condition of the slopes are assessed and classified into five potential failure classes based on their SMR values. SMR values vary widely ranging from 0 to 75.5. Table 2 summarizes the RMR, SMR, stability condition, and principal failure types of the slopes. It is also inferred that the RMR_{basic} values show a range from 29 to 88. It is also evident that a number of slopes with higher RMR values have significantly low SMR values. This is because of unfavorable discontinuity conditions, that makes the slope

Fig. 2 Field Photograph and stereo plot of rock slope no. 21

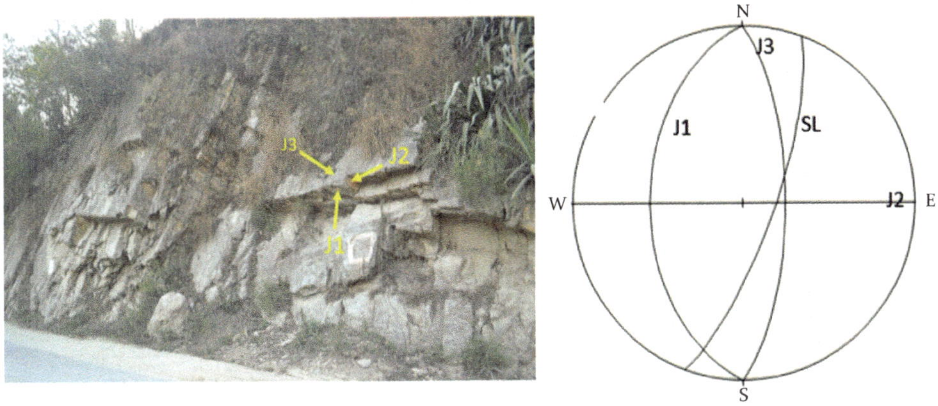

Table 1 RMR and SMR ratings for rock slope no. 21

Parameters		Value	Rating	Adjustment Factors				
PSI		7.959	12	Corrections	F_1	F_2	F_3	F_4
RQD		85	17	Values	20	70	−5	Normal blasting
Spacing of discontinuities		60–200 mm	6	Rating	0.7	1.0	−50	0
Condition of discontinuities	Persistence	3–10	2	SMR = RMR_{basic} + (F1.F2.F3) + F4 = 37				
	Separation	None	6	Most probable failure type: Planar along J_3				
	Infilling	None	6	Slope description: Bad condition				
	Roughness	Rough	5	Stability Condition: Unstable				
	Weathering	Moderate	3	Location: 30°12′05.7″N/79°16′39.6″E				
Ground water condition		Dry	15	Attitudes: Slope: 75°/N110° J1: 45°/N270°				
RMR_{basic}			72	J2:90°/N180° J3: 70°/N90°				

Table 2 RMR, SMR, stability condition, and principal failure types of the slopes

Slope No.	RMR	SMR	Rock type	Slope condition	Stability condition	Failure type
01	77	35	Blocky quartzite	Bad	Unstable	Planar
02	88	37	Massive quartzite	Bad	Unstable	Planar
03	75	33	Quartzite	Bad	Unstable	wedge
04	65	29.3	Quartzite	Bad	Unstable	Wedge
05	29	00	Phyllite	Very bad	Completely unstable	Planar
06	83	75.5	Blocky quartzite	Good	Stable	Planar
07	86	35	Blocky quartzite	Bad	Unstable	Wedge
08	39	30	Phyllite	Bad	Unstable	Planar
09	41	32	Schist	Bad	Unstable	Wedge
10	84	84	Massive quartzite	Very good	Completely stable	Stable
11	38	38	Schist	Bad	Unstable	Wedge
12	38	38	Schist	Bad	Unstable	Wedge
13	67	07	Blocky quartzite	Very bad	Completely unstable	Planar/some wedge
14	76	38.5	Blocky quartzite	Bad	Unstable	Planar
15	73	73	Quartzite	Good	Stable	Stable
16	76	34	Blocky quartzite	Bad	Unstable	Wedge
17	47	27	Quartzite	Bad	Unstable	Planar
18	50	08	Quartzite	Very bad	Completely unstable	Planar
19	60	39.67	Blocky quartzite	Bad	Unstable	Planar
20	74	39	Blocky quartzite	Bad	Unstable	Planar/some wedge
21	72	37	Blocky quartzite	Bad	Unstable	Planar
22	86	78.5	Quartzite	Good	Stable	Stable with some Wedge
23	58	07	Phyllite	Very bad	Completely unstable	Planar
24	74	65	Quartzite	Good	Stable	Wedge
25	81	73.5	Blocky quartzite	Good	Stable	Planar
26	79	54	Blocky quartzite	Normal	Partially stable	Topple
27	64	22	Quartzite	Bad	Unstable	Wedge
28	64	22	Quartzite	Bad	Unstable	wedge
29	72	63	Blocky quartzite	Good	Stable	wedge
30	86	68.5	Quartzite	Good	Stable	Stable with some Topple

(continued)

Table 2 (continued)

Slope No.	RMR	SMR	Rock type	Slope condition	Stability condition	Failure type
31	74	65	Blocky quartzite	Good	Stable	Planar
32	76	56.8	Quartzite	Normal	Partially stable	Planar
33	81	39	Massive quartzite	Bad	Unstable	Planar
34	83	41	Quartzite	Normal	Partially stable	Planar
35	81	21	Quartzite	Bad	Unstable	Planar
36	78	30	Massive quartzite	Bad	Unstable	Planar
37	72	36.3	Quartzite	Bad	Unstable	Planar
38	79	35.5	Quartzite	Bad	Unstable	Planar
39	81	21	Blocky quartzite	Bad	Unstable	Wedge
40	81	46	Blocky quartzite	Normal	Partially stable	Planar
41	81	59.75	Quartzite	Normal	Partially stable	Topple
42	70	34	Blocky quartzite	Bad	Unstable	Planar
43	83	32	Massive quartzite	Bad	Unstable	Wedge
44	81	57	Quartzite	Normal	Partially stable	Planar
45	71	49.75	Massive quartzite	Normal	Partially stable	Topple
46	71	21	Blocky quartzite	Bad	Unstable	Planar
47	78	18	Blocky quartzite	Very bad	Completely Unstable	Wedge
48	76	51	Massive quartzite	Normal	Partially stable	Topple

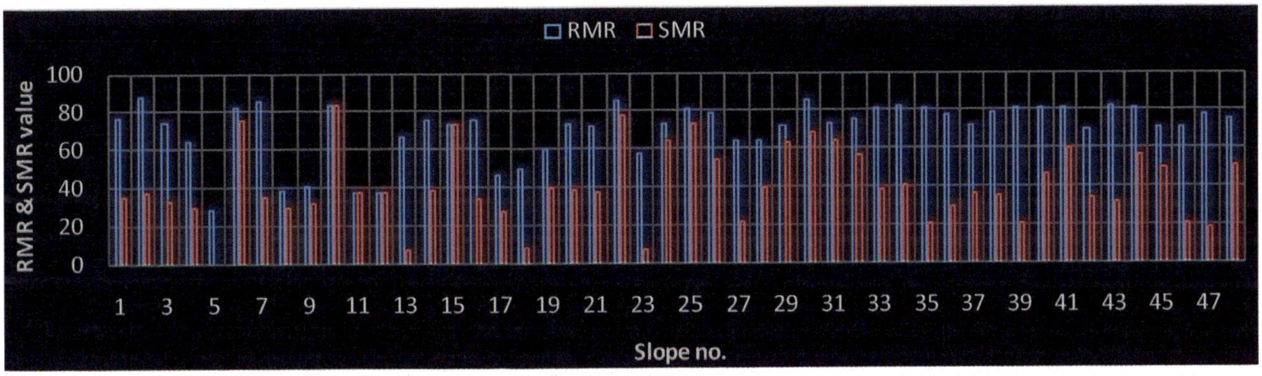

Fig. 3 Graphical representation comparing RMR and SMR values of 48 slopes

with good quality of rock mass unstable and prone to failure. Figure 3 compares the RMR and SMR values for corresponding slopes.

The studied slopes are prone to different types of failure; planar failure being the dominant mode of failure followed by wedge and toppling failures. Out of 48 slopes, only two slopes are completely stable without having any type of failure component. Figure 4 depict the number of slopes prone to different types of failures.

The Google Earth (R) images (Figs. 5 and 6) depict the spatial location of different slopes along with their stability condition.

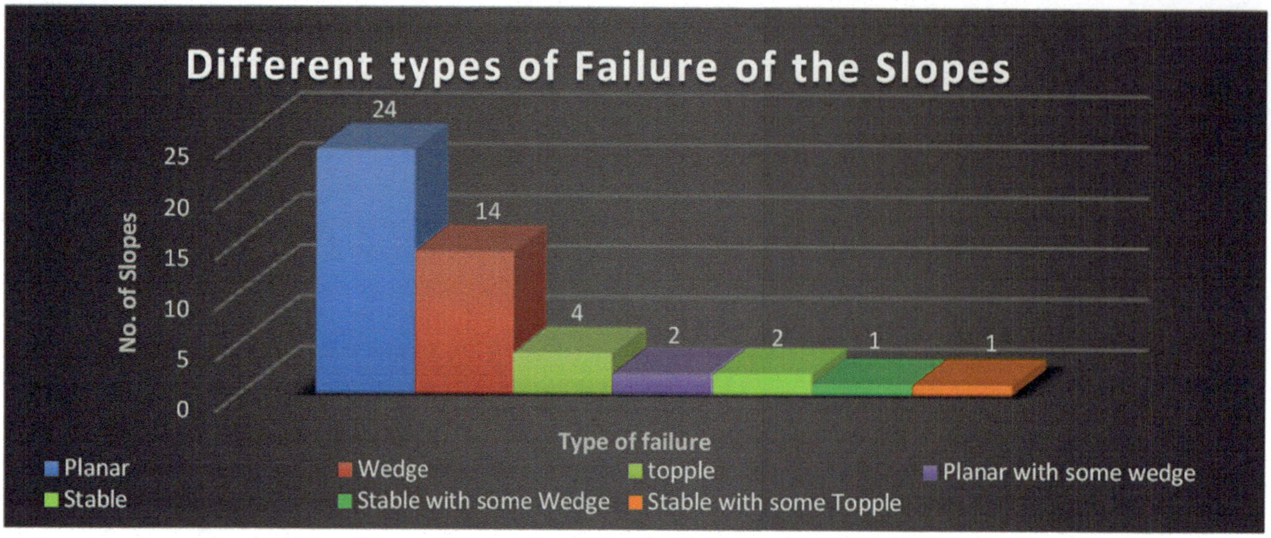

Fig. 4 Scenario of different types of failure in rock slopes

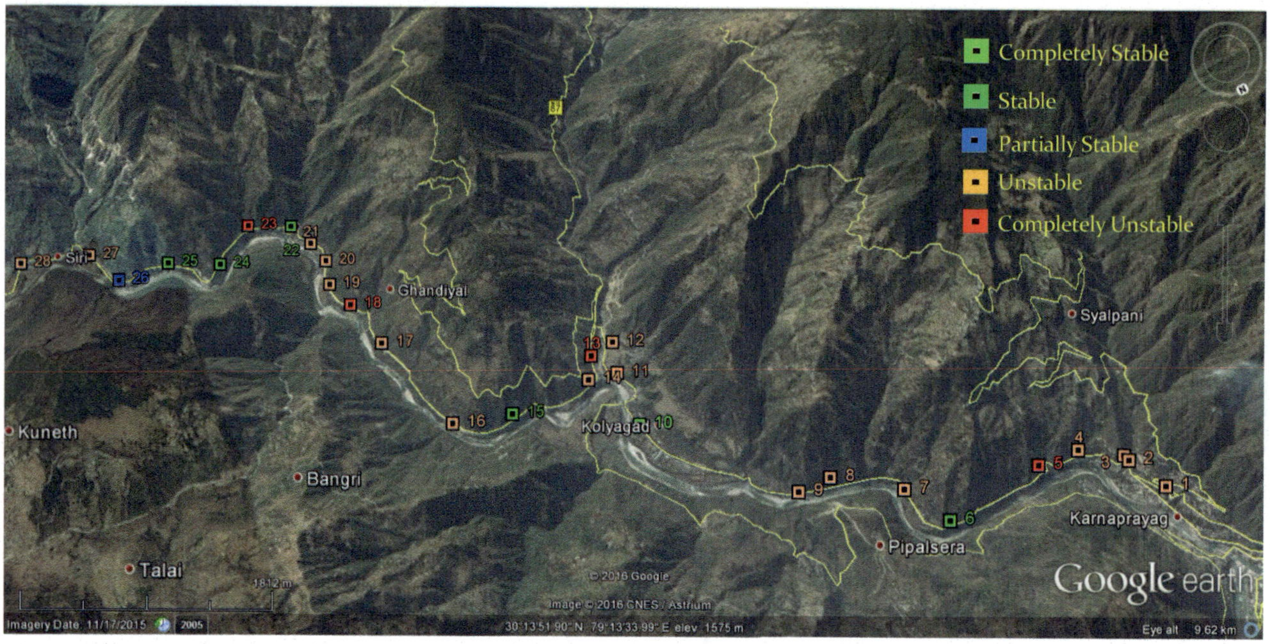

Fig. 5 Spatial Location of slopes (slope nos. 1–28) and their stability conditions

Fig. 6 Spatial Location of slopes (slope nos. 29–48) and their stability conditions

Conclusions

The SMR is a very significant and powerful technique for characterization of rock mass especially for slope stability assessment. It is one of the most comprehensive and widely used method for slope stability assessment. The present study is conducted for stability assessment of 48 road cut rock slopes in hilly regions of Garhwal Himalayas along Gopeswar-Almora Highway, and carefully studied applying RMR and SMR techniques. The study area is composed of low to medium grade meta-sedimentary rocks, dominantly Quartzite with three to four sets of discontinuities. From the results, it is inferred that out of total 48 rock slopes, five slopes are completely unstable, 26 slopes are unstable, eight slopes are partially stable, another eight slopes are stable, and only one slope is completely stable. Thus, majority of the slopes are unstable and prone to some kind of failure either wedge or planar, the latter being more likely to occur. These slopes require immediate reinforcement to avoid slope failure.

Acknowledgements The third author is thankful to the Director, CSIR-Central Building Research Institute, Roorkee, India for their permission to publish this paper.

References

Anbalagan R, Sharma S, Raghuvanshi TK (1992) Rock mass stability evaluation using modified SMR approach. In: Proceedings of 6th National Symposium on Rock Mechanics, 2009. Bangalore, India, pp 258–268

Bhattacharya AR (1990) Crustal processes associated with the Main Central Thrust of the Himalaya. Zeitschr. Geowissen, 18(4):293–299

Bieniawski ZT (1973) Engineering classification of jointed rock masses. Trans S Afr Inst Civ Eng 15:335–344

Bieniawski ZT (1979) The geomechanical classification in rock engineering applications. In: Proceedings of 4th International Congress Rock Mechanics, 1979. Montreux, Balkema, Rotterdam, vol 2, pp 41–48

Bieniawski Z T, (1989) Engineering rock mass classifications. Wiley-Interscience, New York: ISBN 0-471-60172-1, 251p

IS 8764 (1998) Method of determination of point load strength index of rocks [CED 48:Rock Mechanics]

Palmstrom A (2005) Measurements of and correlations between block size and rock quality designation (RQD). Tunn Undergr Space Technol 20:362–377

Romana M (1985) New adjustment ratings for application of Bieniawski classification to slopes. In: Proceedings of International Symposium on Role of Rock Mechanics, 1985. Zacatecas, pp 49–53

Valdiya KS (1980) Stratigraphy of the Lesser Himalaya: Synthesis. In: Valdiya KS and Bhatia SB (eds) Stratigraphy and Correlations Of The Lesser Himalayan Formations. Hindustan Publishing Corporation, Delhi, pp 283–298

Towards Decentralized Landslide Disaster Risk Governance in Uganda

Sowedi Masaba, N. David Mungai, Moses Isabirye, and Haroonah Nsubuga

Abstract

Decentralized governance is critical to reducing disaster risks. This paper evaluates the decentralized landslide disaster risk governance in Uganda. Primary data were collected through household surveys and key informant interviews conducted in the landslide disaster prone Mount Elgon district of Bududa, Eastern Uganda. Secondary data were collected through document review. Primary and secondary data were analyzed using descriptive statistics and content analysis. The study findings reveal that in Uganda, landslide disaster risk reduction is perceived as a shared responsibility between different actors and involves wider stakeholder participation that has enhanced resource mobilization. Coordination of landslide disaster risk reduction has also been streamlined. Decentralized landslide disaster risk governance however, faces several challenges, including; financial and human resource constraints, political interference, corruption, uncooperative constituents and lack of an enabling sectoral law. Decentralized governance should therefore be upscaled to achieve landslide disaster risk reduction. Future research should focus on mapping key actors and institutions using Social Network Analysis to enable better resource allocation for landslide disaster risk reduction in the Country.

Keywords

Decentralization • Disaster risk reduction • Governance • Landslides • Uganda

S. Masaba (✉) · N. David Mungai
Wangari Maathai Institute for Peace and Environmental Studies, University of Nairobi, P.O BOX 30197 00100, Nairobi, 254, Kenya
e-mail: sowedimasaba@gmail.com

N. David Mungai
e-mail: mungaidavid@uonbi.ac.ke

M. Isabirye
Natural Resource Economics, Busitema University, P.O BOX 236, Tororo, Uganda
e-mail: isabiryemoses@yahoo.com

H. Nsubuga
Institute of Continuing Education, Zanzibar University, P.O BOX 2440, Zanzibar, Tanzania
e-mail: nsubugaharoonah@yahoo.com

M. Mikoš et al. (eds.), *Advancing Culture of Living with Landslides*,
DOI 10.1007/978-3-319-53483-1_49

Introduction

Globally, climate change related disasters are increasing in frequency, severity and impact. Between the year 2003 and 2013, the number of disaster events increased by 26% worldwide, affecting 2.9 billion people, causing 1.2 million deaths and economic loss exceeding US$1.7 trillion. In Africa, about 1700 disaster events were recorded between 1980 and 2008, affecting more than 319 million people, killing over 708,000 and causing economic loss exceeding US$24 billion (Anderson 2013; EM-DAT Database; McEntire 2001; Millennium Ecosystem Assessment 2005; Munich Re 2014; Palliyaguru et al. 2014; UNDP 2007; UNISDR 2013a; Walhastrom 2013). Disasters threaten development in Africa with Uganda listed among the 11 countries most at risk of disaster induced poverty in the world (Shepherd et al. 2013).

In Uganda, at least 61 disaster events occurred between 1980 and 2010 affecting 4.9 million people, killing more than 2200 and causing economic loss exceeding US$72.6 million. Disasters affect more than 200,000 people annually and between 2000 and 2005, about 66% of the households experienced at least one type of disaster in Uganda (Akera 2012; EM-DAT Database; National Planning Authority 2010; Office of the Prime Minister 2010). Landslides kill more people (14%) than any other socio-natural disaster in Uganda and affect 4% of the population. The Country has experienced enormous losses due to landslides since 1933, including: 1903 deaths; 427,658 people affected; 2487 houses destroyed; and 53 educational centers, 2350 hectares of crops and 80,535 meters of roads damaged. One such disaster was the March 1, 2010 landslide which was ranked among the top ten disasters by number of deaths in the world. The landslide killed 388 and affected at least 8500 people in the Mount Elgon District of Bududa in Eastern Uganda (EM-DAT Database; DesInventar 2015; Doocy et al. 2013; Misanya 2011; Office of the Prime Minister 2010; Terry 2011; Vlaeminck et al. 2015).

Disaster risk governance, through decentralization of risk management functions, powers and resources can enhance disaster risk reduction (Ahrens and Rudolph 2006; Kahn 2005; Lemos and Agrawal 2006; Office of the Prime Minister 2010; Tierney 2012). This paper evaluates the decentralized disaster risk governance system in Uganda, with specific reference to landslides.

Materials and Methods

Study Area Description

Bududa district is situated on the South Western slopes of Mount Elgon in Eastern Uganda, along the Kenya border (Fig. 1). The district lies between latitude 2°49′N and 2°55′N, and longitude 34°15′E and 34°34′E, and covers a total land area of about 274 km^2. The area receives very high annual rainfall (above 1500 mm) and is characterized by high altitude and steep slopes ranging between 1250 and 2850 m above sea level. Although Mount Elgon national park covers about 40% of the district, the area has fertile volcanic soils and subsistence farming is the main economic activity. With the exception of Central Bukigai zone which is a carbonatite hill, the study area is dominated by "problem soils" i.e. where slope failure can occur even without human intervention (Claessens et al. 2007, 2013; Bududa District Local Government 2007; Kitutu 2010; Mugaga 2011; Cox 2013; Shilaku J., personal communication, January 11, 2015).

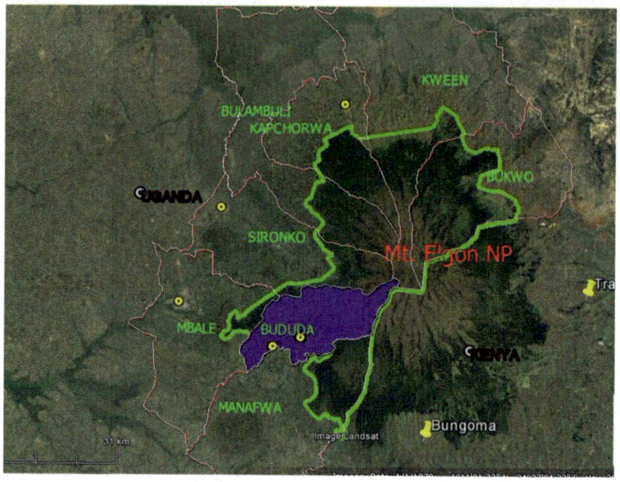

Fig. 1 Map showing the location of Bududa District in Uganda

Bududa is a highly populated and predominantly rural district. Between 2002 and 2014, the population grew by 72% from 123,103 to 211,683. At 4.5%, the annual population growth rate is very high and far above the national average of 3%. The population density is also very high (>450 persons per km^2) far above the national average of 123. The average household size is 5.7 people. The population is relatively homogeneous and traditional, with a predominant household population of 99.8% and the Bagisu or Bamasaba constitute the major ethnic group (99%). In terms of administrative units, Bududa District has 15 Sub-counties, one Town council, 36 Parishes and 336 Villages (Bududa District Local Government 2007; Cox 2013; UBOS 2009, 2013, 2014).

Research Design and Sampling

The study adopted a mixed method approach involving household surveys, key informant interviews and document review, and employed both qualitative and quantitative approaches. Such a mixed method approach is superior to a single method because it enhances data quality through triangulation, facilitation and complementarities (Lassa 2010; Oso and Onen 2008; Palliyaguru et al. 2014; Russel 2002; Were et al. 2013).

For the household surveys, the target population was all the 37,028 households in Bududa district. The sample size was 300 households, and determined statistically (Russell 2002). The study used various sampling techniques. Purposive sampling was used to select Bududa district as the study area. Bududa district was selected because it experiences the highest number of landslide disasters in the country (DesInventar 2015). Stratified random sampling was used to select the sample Sub-counties of Bukigai, Bushika and Bukalasi on the basis of low, medium and high landslide disaster risk respectively (Cox 2013). Simple random sampling was used to select the sample parishes of Bunamubi, Bufutsa and Bundesi while systematic random sampling was used to select the sample households. Such randomization enhances data validity and reliability since it reduces the effects of extraneous variables (Oso and Onen 2008).

Data Collection and Analysis

Primary data were collected using questionnaires and key informant interviews. The choice of data collection methods was guided by the study objective and nature of data to be collected. The objective of the research was to assess the institutional capacity for landslide disaster risk reduction in the Mount Elgon region of Uganda. The research was therefore mainly concerned with views, opinions, perceptions, feelings and attitudes, and such information can best be collected using questionnaires and key informant interviews. To enhance data validity and reliability, the questionnaires were pretested before final use (Russel 2002; Oso and Onen 2008). Primary data were collected from household heads or their representatives and staff of key disaster risk reduction agencies. Secondary data was collected through document analysis, including review of government of Uganda disaster risk reduction policy documents. Both primary and secondary data were analyzed using descriptive statistics and content analysis.

Results

Benefits of Decentralized Landslide Disaster Risk Governance

The study findings reveal that through decentralization, landslide disaster risk governance in Uganda has been streamlined. The governance framework for disaster risk reduction in the Country is derived from the 1995 Constitution, Local Governments Act of 1997 and National Policy for Disaster Preparedness and Management of 2010. The institutional framework aims at establishing an efficient mechanism for integrating disaster preparedness and management into the socio-economic development planning process at national and local government levels. It also establishes a national disaster preparedness and management institutional structure and defines its functions.

At the national level, cabinet is the chief policy making organ of government and advises the President on disaster risk reduction matters. The Ministerial Policy Committee of cabinet which handles cross sectoral matters is responsible for policy formulation, oversight and mainstreaming disaster preparedness and management in the governance of the country. The Ministry for Disaster Preparedness, Management and Refugees in the office of the Prime Minister is the lead agency that coordinates disaster preparedness and management issues in the Country. The National Platform for Disaster Management is in charge of implementing disaster risk reduction policy while the National Emergency Coordination and Operations Centre (NECOC) is responsible for coordinating emergency response.

At the sub national level, the decentralized governance structure provides for City, District, Municipal and Town disaster policy committees which offer policy direction while the respective management committees implement disaster policy. The governance structure also provides for the District Emergency Coordination and Operations Centre (DECOC) which is in charge of coordinating emergency response. The Sub County Disaster Management Committees are in charge of implementing disaster policy at that level while Village Disaster Management Committees are the lowest units of disaster policy implementation in the country. Coordination between the different landslide disaster risk reduction agencies has therefore been streamlined.

Decentralized governance has also enhanced ownership of landslide disaster risk reduction since it is perceived a shared responsibility between the different actors. The majority of respondents (47%) attributed the responsibility of landslide disaster risk reduction to both government and households, 16% attributed it to households, 15% attributed it to the local community, 14% attributed it to government while 7% attributed it to others, including government and local community, God, and civil society, particularly the Uganda Red Cross Society. For the case of government and households, most of the respondents felt that the chairpersons of village local councils (LC1) and household heads should take the greatest responsibility for landslide disaster risk reduction.

The polycentric and multi-scale nature of decentralized disaster risk governance has enabled wider stakeholder participation. The majority of respondents (36%) perceived both governmental and non-governmental organizations as key actors in landslide disaster risk reduction. Whereas 23% of respondents viewed non-governmental organizations as the most important players, 23% indicated that governmental agencies were the key actors. Among the governmental agencies and programmes, the National Agricultural Advisory Services (NAADS) and Northern Uganda Social Action Fund (NUSAF) were perceived to have made a major contribution to landslide disaster risk reduction. However, among the non-governmental organizations, the majority of respondents viewed the civil society as a key player compared to private sector and international organisations. Among civil society actors, the Uganda Red Cross Society was perceived by the local community as the leading organization for landslide disaster risk reduction in the Country. Overall, the majority of respondents (78%) perceived the actor organisations effective.

Decentralized governance has also enabled better resource mobilization and several landslide disaster risk reduction activities have been implemented using both governmental and non-governmental resources. For example, construction of houses for resettling the March 2010 landslide disaster victims in Kiryandongo District was done using funds from the Government of Uganda and UN Habitat. The majority respondents (36%) indicated that the actor organisations contributed resources for pre-disaster mitigation, emergency response and post-disaster recovery. However, 32% of the respondents perceived the organisations as contributing resources for emergency response while 17% felt that the organisations focused on pre-disaster mitigation. Emergency response mainly focused on provision food, medical care, shelter, clothing, education, retrieving dead bodies buried by landslides, and relocation of affected households while pre-disaster mitigation focused on provision of agricultural inputs, tree seedlings and sensitisation farmers about sustainable land use practices.

Challenges Facing Decentralized Landslide Disaster Risk Reduction

Despite the above-mentioned benefits, decentralized landslide disaster risk reduction in Uganda faces several challenges, including: limited financial and human resources, political interference, corruption, lack of a sectoral law, and non-cooperative constituents.

Effective implementation of landslide disaster risk reduction policy requires adequate and dedicated financial resources, including contingency funds. Most of the key informants indicated that decentralized disaster risk reduction is inadequately funded. At the national level, the National Disaster Preparedness and Management Fund has not been put in place, and currently, both Government of Uganda and Bududa District Local Government spend less than 1% of their annual budgets on disaster risk reduction interventions. This has affected implementation of pre-disaster mitigation measures, emergency response, and post-disaster recovery. During the financial year 2012/2013, only 50% of the planned houses for resettling landslide victims were constructed in Kiryandongo District by the Central Government. Due to financial constraints, Bududa district has not yet established its own Emergency Coordination Centre and still relies on the neighbouring Mbale district to coordinate emergency response. Several other landslide risk reduction policy measures including; gazetting of landslide disaster prone areas and prohibiting settlement in such risky areas, and resettling of people living in landslide prone areas have also not been effectively implemented. The majority of key informants also indicated that emergency logistics and equipment were inadequate, and no significant post-disaster recovery interventions have been undertaken in Bududa District since the March 2010 landslide disaster.

The human resource capacity for landslide disaster risk reduction is also limited. Although the District disaster policy and technical committees have been put in place, most of the members have not been trained and do not have good knowledge of landslide disaster risk reduction. The study findings further reveal that most of the Sub-county disaster management committees have either not been established or are not effective. For those that have been put in place, most of the committee members do not know their mandate. The District, Sub county and Village disaster management committees do not have volunteer teams that can assist during emergency response and rely on the Uganda Red Cross Society for volunteers. Besides, the district does not have its own environmental police to enforce land use regulations that are essential for landslide risk reduction and relies on the neighbouring Mbale district. Most of the Village local councils (LC1s) that are mandated to serve as Village disaster management committees are not aware of the mandate. Most of the Village disaster management committees are therefore non-functional, and the majority of respondents who are potential members were neither aware of the existence (89%) nor could confirm their membership (98%) to the committees.

Political interference and corruption were also identified as key bottlenecks to decentralized landslide disaster risk reduction. The polycentric and multi-scale nature disaster risk governance, characterized by disaster policy and management committees that are embedded in central and local government structures has subjected landslide disaster risk reduction to undue political interference. Most of the respondents indicated that failure of the March 2010 landslide disaster resettlement programme to Kiryandongo and Bulambuli districts was partly due to some local politicians who de-campaigned it. Corruption was also cited as another serious challenge, particularly in the distribution of emergency relief items. Several respondents indicated that during the March 2010 landslide disaster, some non-victims were registered as beneficiaries of emergency relief, leaving out bonafide victims, and some of the emergency relief items were sold by local leaders.

Effective disaster risk reduction depends on the cooperation of local communities. Although resettlement of populations at risk appears to be the most feasible landslide disaster risk reduction policy measure, most of the respondents did not support its implementation due to strong social-cultural ties with their ancestral lands, and fear of the harsh bio-physical conditions and poor services in the destination areas. For example, only 40% of the March 2010 landslide disaster affected families accepted to relocate to Kiryandongo District in Central Uganda and most of them have since returned to Bududa. The proposed resettlement to Bunambutye lowlands in Bulambuli District has also been contested by the potential beneficiaries and local leaders on account of being a swampy area.

Decentralized landslide disaster risk reduction has also been hampered by lack of a sectoral law on Disaster Preparedness and Management. The study findings reveal that the proposed National Disaster Preparedness and Management Act, and related regulations have not been developed by the Central Government. Bududa district local government has also not yet passed any by-laws to deal with landslide disaster risk reduction. In absence of a sectoral law and regulations, Uganda currently relies on the National Environment Act of 1995 and National Environment (Hilly and Mountainous Areas Management) Regulations of 2000 which do not adequately address the issue of landslide disaster risk reduction. A sectoral law, regulations and by-laws are therefore needed to enforce the provisions of the National Policy for Disaster Preparedness and Management 2010.

Discussion and Conclusions

One key feature of decentralized risk governance in Uganda is its polycentric and multiscale nature that has enabled wider stakeholder participation, enhanced resource mobilization and coordination of disaster risk reduction interventions. There is however, limited private sector participation in landslide disaster risk reduction activities. Decentralized risk governance also faces several challenges; including financial and human resource constraints, political interference, corruption, uncooperative constituents and lack of enabling laws and regulations. As noted by Maes et al. (2015), lack of enabling laws and regulations affects implementation of landslide risk reduction measures. Inspite of the decentralized governance, several landslide disaster risk reduction functions, powers and resources remain centralized at the national level thus undermining the functionality of local level disaster management committees.

Although decentralized landslide disaster risk governance has yielded several benefits, it is faced with several challenges. To achieve landslide disaster risk reduction, the identified bottlenecks should be addressed and decentralized risk governance upscaled. An enabling law and regulations to operationalize the National Disaster Preparedness and Management Policy of 2010 should be put in place. Local level disaster management committees should also be made functional and enabled to undertake their landslide disaster risk reduction mandate. Future research should focus on mapping key actors and institutions using Social Network Analysis to enable better resource allocation for landslide disaster risk reduction in the Country.

Acknowledgements The authors wish to thank Prossy Kanzila, Julius Kutosi, Simon Mugalu and John Sekajugo for the support during field data collection, analysis and maps. Funding for the study was provided by Busitema University through the ADB-HEST Project. We also acknowledge the cooperation of all household heads and local leaders in Bududa district during the study.

References

Ahrens J, Rudolph PM (2006) The importance of governance in risk reduction and disaster management. J Contingencies Crisis Manag 14(4):207–220

Akera S (2012) Towards implementation of the Uganda National Disaster Preparedness and Management Policy. In: ISDR, Disaster risk reduction in Africa. Special issue on drought 2012. United Nations International Strategy for Disaster Reduction Regional Office for Africa, Nairobi, Kenya

Anderson MG (2013) landslide risk reduction in developing countries: perceptions, successes and future risks for capacity building. Landslide Sci Pract 247–256

Bududa District Local Government (2007) 2002 Population and housing census analytical report. Bududa district

Claessens L, Knapen A, Kitutu MG, Poesen J, Deckers JA (2007) Modeling landslide hazard, soil redistribution and sediment yield on the Uganda foot slopes of Mount Elgon. Geomorphology 90:23–35

Claessens L, Kitutu MG, Poesen J, Deckers JA (2013) Landslide hazard assessment on the Ugandan foot slopes of Mount Elgon: the worst is yet to come. Landslide Sci Pract 527–531

Cox J (2013) Landslide risk in Mount Elgon region, Eastern Uganda: local perceptions in a DPSIR framework. M.Sc. dissertation, KU Leuven, Belgium

DesInventar: The national disaster loss database-http://desinventar.net (last accessed: 31 Dec 2015)

Doocy S, Russell E, Gorokhovich Y, Kirsch T (2013) Disaster preparedness and humanitarian response in flood and landslide-affected communities in Eastern Uganda. Disaster Prev Manage 22 (4):326–339

EM-DAT Database, http://www.emdat.be/ (last accessed: 18 June 2014)

Kahn ME (2005) The death toll from natural disasters: the role of income, geography and institutions. Rev Econ Stat 87(2): 271–284

Kitutu KMG (2010) Landslide occurrences in the hilly areas of Bududa District in Eastern Uganda and their causes. Ph.D. thesis, Makerere University, Kampala, Uganda

Lassa J (2010) Institutional vulnerability and governance of disaster risk reduction: Macro, meso and micro scale assessment (With case studies from Indonesia). Ph.D. thesis, University of Bonn, Germany

Lemos MC, Agrawal A (2006) Environmental governance. Annu Rev Environ Resour 31:297–325

Maes J, Kervyn M, Vranken L, Dewitte O, Vanmaercke M, Mertens K, Jacobs L, Poesen J (2015) Landslide risk reduction strategies: an inventory for the global South. EGU General Assembly, vol 17, EGU2015-5988

McEntire DA (2001) Triggering agents, vulnerabilities and disaster reduction: towards a holistic paradigm. Disaster Prev Manage 10 (3):189–196

Millennium Ecosystem Assessment (2005) Ecosystems and human well-being: Synthesis. Island Press, Washington, DC

Misanya D (2011) The role of community based knowledge and local structures in disaster management. A case study of landslide occurrences in Nametsi parish of Bukalasi sub county in Bududa District, Eastern Uganda. Masters' thesis, University of Agder, Norway

Mugaga F (2011) Land use change, landslide occurrence and livelihood strategies on Mount Elgon slopes, Eastern Uganda. Ph.D. thesis, Nelson Mandela Metropolitan University, South Africa

Munich Re (2014) NatCatservice: http://www.unisdr.org/2005/wcdr/thematic-sessions/presentations/session2-8/munichre.pdf (last accessed 24 June 2014)

National Planning Authority (2010) National development plan 2010/11-2014/15. Kampala, Uganda

Office of the Prime Minister (2010) The national policy for disaster preparedness and management. Kampala, Uganda

Oso WY, Onen D (2008) A general guide to writing a research proposal and report. A handbook for beginning researchers. 2nd edn. Makerere University Printery, Kampala, Uganda

Palliyaguru R, Amaratunga D, Baldry D (2014) Constructing a holistic approach to disaster risk reduction: the significance of focusing on vulnerability reduction. Disasters 38(1):45–61

Russel HB (2002) Research methods in anthropology. Qualitative and quantitative methods. 3rd edn. Altamira Press, USA

Shepherd A, Mitchell T, Lewis K, Lenhardt A, Jones L, Scott L, Muir-Wood R (2013) The geography of poverty, disasters and climate extremes in 2013. Overseas Development Institute, UK

Terry G, (2011) Climate, change and insecurity: views from a Gisu Hillside. Ph.D. thesis, University of East Anglia, UK

Tierney K (2012) Disaster governance: social, political, and economic dimensions. Annu Rev Environ Resour 37:341–363

Uganda Bureau of Statistics (UBOS) (2014) National population and housing census 2014. provisional results. November 2014. Revised Edition. http://www.ubos.org/onlinefiles/uploads/ubos/NPHC/NPHC%202014%20PROVISIONAL%20RESULTS%20REPORT.pdf (last accessed 14 Dec 2015)

Uganda Bureau of Statistics (UBOS) (2013) Statistical abstract. August 2103. Uganda Bureau of Statistics, Kampala, Uganda

Uganda Bureau of Statistics (UBOS) (2009) Profiles of the higher local governments. May 2009. Kampala, Uganda

United Nations Development Programme (UNDP) (2007) Human Development report 2007/2008. Fighting climate change: human solidarity in a divided world. UNDP, New York; USA. ISBN 978-0-230-54704-9

United Nations International Strategy for Disaster Reduction (UNISDR) (2013a) Disaster impacts/2000-2013 http://www.preventionweb.net/files/31737_20130312disaster20002012copy.pdf (last accessed: 24 June 2014)

Vlaeminck P, Maertens M, Isabirye M, Vanderhpydonks F, Poesen J, Deckers J, Vranken L (2015) Coping with landslide risk through preventive resettlement. Designing optimal strategies through choice experiments for the Mount Elgon region, Uganda. Bioeconomics Working Paper Series. Working Paper 2015/4. http://ees.kuleuven.be/bioecon/ (last accessed: 15 Dec 2015)

Walhastrom M (2013) Progress and challenges in global disaster risk reduction. Int J Disaster Risk Sci 4(1):48–50

Were NA, Isabirye M, Poesen J, Maertens M, Deckers J, Mathijs E (2013) Decentralized governance of Wetland resources in the Lake Victoria basin of Uganda. Nat Resour 4(1):55–64

Automatic Landslides Mapping in the Principal Component Domain

Kamila Pawłuszek and Andrzej Borkowski

Abstract

The availability of digital elevation model (DEM) delivered by airborne laser scanning (ALS) opens new horizons in the geomorphological research, especially in the landslide studies. This detailed geomorphological information allows for mapping of landslide affected areas using DEM data only. In order to map landslide areas in the automatic manner using machine learning classification algorithms and only DEM, generation of several DEM derivatives is needed. These first and second order derivatives provide information about specific properties of the terrain. However, involving a set of topographic features in the machine learning process increases significantly time of computations. Moreover, the topographic features are correlated since they are generated using the same DEM. The objective of this study is an in-depth exploration of the topographic information provided by the DEM data as well as the reduction of the computational time while the automatic landslide mapping. For this reason, a set of DEM derivatives have been generated and transformed into the principal component domain. The Principal Component Analysis (PCA) is a procedure that converts the set of correlated features into a set of linearly uncorrelated components using the orthogonal transformation. For the automatic landslide detection, the support vector machine (SVM) algorithm was used. The achieved results were compared with the existing landslide inventory map and overall accuracy and kappa coefficient were calculated. For the non-reduced original topographic model, we received 73% of overall accuracy. For the PCA-reduced models, accuracy parameters are not significantly worse. For instance, using only 7 principal components, which provide 90% of the total variability of the original topographic features, we received the overall accuracy of 72% while the computation time was reduced.

Keywords

Landslide inventory mapping • DEM-derivatives • Principal component analysis • Support vector machine

K. Pawłuszek (✉) · A. Borkowski
Institute of Geodesy and Geoinformatics, Wroclaw University of Environmental and Life Sciences, Grunwaldzka 53, 50-357 Wroclaw, Poland
e-mail: kamila.pawluszek@igig.up.wroc.pl

A. Borkowski
e-mail: andrzej.borkowski@igig.up.wroc.pl

M. Mikoš et al. (eds.), *Advancing Culture of Living with Landslides*,
DOI 10.1007/978-3-319-53483-1_50

Introduction

Landslides are significant, costly and damaging natural hazard (Moosavi et al. 2014). Worldwide, considerable money are spent in order to mitigate landslides implications (Mahalingam et al. 2016). For that purposes, landslide susceptibility and hazard mapping are crucial to effective anticipation of future landslide occurrence and improving preservation in that landslide-prone regions (Moosavi et al. 2014). According to the geomorphologist main principal "the past and present are keys to the future", it was assumed that landslide are probable to appear in the same areas where historical landslide have already occurred (Varnes and the IAEG Commission on Landslides and other Mass-Movements 1984; Guzzetti et al. 2012). Therefore, landslide inventory maps are indispensable requirement for reliable hazards and risk assessment (Carrara and Merenda 1976; Van Den Eeckhaut and Hervás 2012; Guzzetti et al. 2012; Martha et al. 2013).

Landslide inventory maps are often created using conventional methods involving geomorphological field reconnaissance and the visual interpretation of stereoscopic aerial photographs (Guzzetti et al. 2012). These methods are time-consuming, resource-intensive and require a team of experienced people (Borkowski et al. 2011; Martha et al. 2013). Galli et al. (2008) evaluated that production of landslide inventory map required on average one month per interpreter to cover a 100 km^2 area in Umbria region, Italy. Based on Guzzetti et al. (2012), it is a misconception that landslide mapping by field reconnaissance is more accurate than landslides mapping using remote sensing techniques (e.g. photographs, satellite images, very high resolution DEMs). In field investigations, it is not straightforward to identify landslide's boundaries, which are often obscure or fuzzy. Especially within the hardly accessible areas, where topography is hummocky and where vegetation is tall and/or dense (Borkowski et al. 2011; Guzzetti et al. 2012). Moreover, the possibility to map landslide boundaries in field is limited in case of complex landslide, where an overview of the entire landslide scope is limited as a consequence of the local perspective and the landslide's size (Santangelo et al. 2010; Guzzetti et al. 2012). Thus, the perspective created by a distant view from remote sensing techniques is advantageous.

In the last decade, a development in remote sensing techniques opened new horizons in the geomorphological research, especially in the landslide studies. Among these techniques, airborne laser scanning (ALS) is characterized by its usefulness and effectiveness in providing detailed information about surface topography (Tarolli 2014). The ALS technology makes it possible to acquire detailed information about terrain topography shaped by landslides activity, particularly in the forested areas (Borkowski et al. 2011). This detailed geomorphological information allows for mapping of landslide affected areas using ALS-derived digital elevation model (DEM) only. Unfortunately, only few researchers made attempt to automate the process of landslide mapping using DEM (Van Den Eeckhaut et al. 2012; Chen et al. 2014). Various machine learning methods combined with diverse DEM-derivatives can be utilized for automatic landslide mapping. However, combining many DEM-derivatives with sophisticated classifiers from machine learning theory (e.g. Support Vector Machine (SVM), Artificial Neural Network or Random Forests) are often computationally time consuming.

McKean and Roering (2004) used local variability of unit vector orientations of slope and aspect to outline landslide spatial extent. Chen et al. (2014) applied extended set of DEM-derived features and random forests algorithm for semi-automatic landslide mapping and overall accuracy of 78.24% was achieved. Van Den Eeckhaut et al. (2012) proposed another approach based on object-oriented analysis (OOA). The results obtained show that OOA using DEM-derivatives allows them identifying up to 70% of the landslide bodies.

The objective of this study is an in-depth exploration of the topographic information provided by the ALS data as well as the reduction of the computational effort while the automatic landslide detection. For this reason, DEM and 19 DEM-derivatives have been generated and SVM classification was applied. In order to reduce redundant information provided by the original dataset, it is transformed into the principal component domain. The set of principal components is utilized then for classification. Several combinations of subsequent principal components were involved into automatic landslide detection and for each combination accuracy measures were determined. In this manner, the increasing accuracy of mapping is correlated with increasing number of principal components used for SVM classification. Thus, this research should answer the question how many principal components have to be taken in consideration to receive acceptable result of the classification.

Study Area

The proposed approach was tested on a study area located in the central part of the Outer West Carpathians in Poland and is covering an area of ca 3 km^2 (Fig. 1). The study area is affected by more than 50 landslides. The geographical location of study area is 49°44′N latitude and 20°43′E longitude. The elevation within the study side ranges from

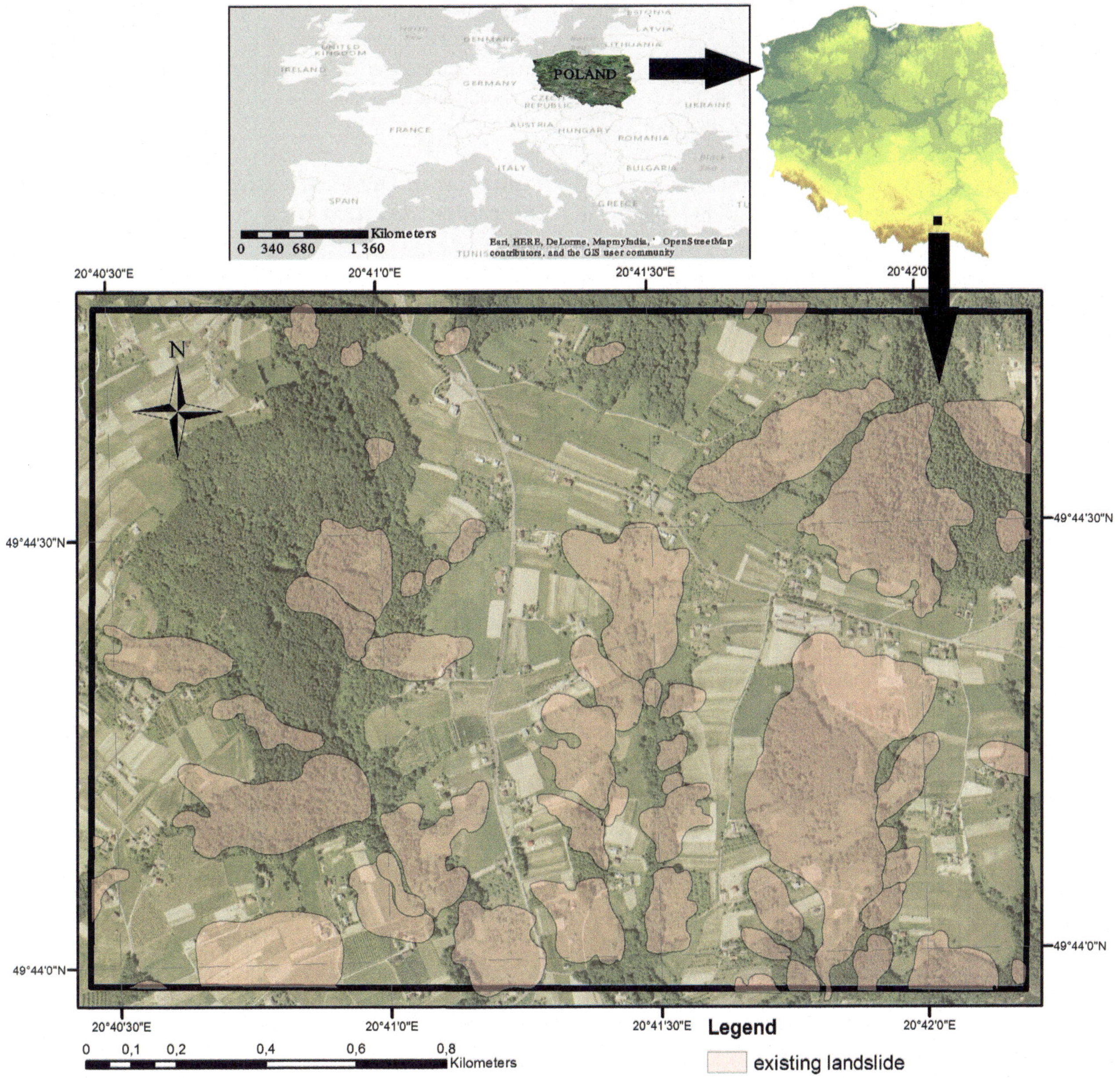

Fig. 1 The study area with the existing landslides (*pink polygons*) and ortophoto (Pawluszek and Borkowski 2016)

267.48 to 477.7 m. From geological perspective, the study areas is located on the Ciężkowice Foothills (Starkel 1972).

From the hydrological context, the annual mean precipitaion within the study area achive 800 mm in 1981–2010 (Woźniak 2013). The main landslide triggering factors within the study area are abundant rainfalls, level water fluctuation in the Rożnów Lake and flysch rock types (Borkowski et al. 2011). This area was chosen because of its landslides frequency over the past few years. Most of the landslides located within the study area are: translational, rotational and combined rock-debris slides or debris slides (Gorczyca et al. 2013).

Data Used

ALS Data Specification and DEM Generation

Point cloud with resolution of 4–6 points/m^2 was obtained from airborne laser scanning in the framework of the ISOK project and the nominal height accuracy of 15 cm (Pawłuszek et al. 2014). The DEM was generated with resolution 1 m × 1 m using natural neighbor interpolation. According to Pawłuszek et al. (2014) the height component accuracy of the ISOK data does not exceed 23 cm for forested areas.

Landslide Inventory Map

Having considered the negative effects of landslide occurrences and activities, Polish Geological Institute created "Landslide Counteracting System" called SOPO (Borkowski et al. 2011; Wójcik et al. 2015). This system aims to produce landslide inventory maps of all existing landslides in Poland and map landslide-prone areas. Moreover, within this system landslides are differentiated into the active, inactive or active periodically. Furthermore, areas with high risk of landslide activity were mapped. All of this data are stored in one database, which is available online and is free of charge (Wójcik et al. 2015).

Methodology

DEM-Derivatives

ALS-based DEM and its derivatives generation is the first step in our methodology (Fig. 2). DEM provides topographical information about the nature of the surface. Applying different mathematical operation on the DEM allows enhancing different morphological information contained in the DEM. These mathematical operations are based on window-moving calculations, where the central pixel value is changed by new value resulting from mathematical calculations of neighboring cells. These resulting calculations are called first and second-order derivatives and are implemented in many GIS software. ArcGIS is suitable for the analysis of geomorphological calculation because of the variety of tools to process raster digital models. Therefore, 19 diverse DEM-derivatives were produced within ArcGIS. Table 1 depicts the DEM-derivatives and their main characteristics and references.

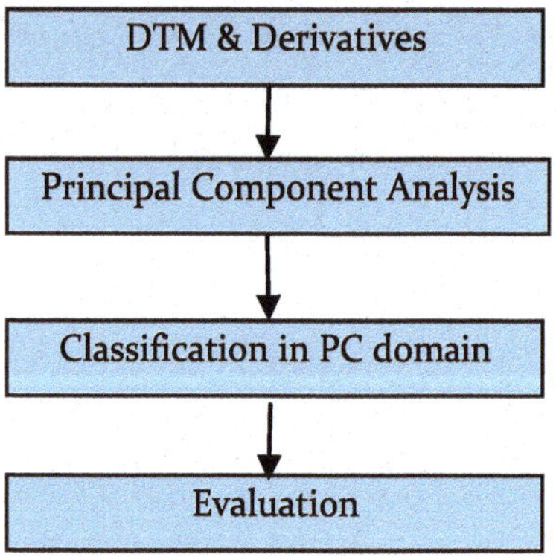

Fig. 2 Methodology flowchart

Reduction of DEM-Derivatives in Principal Component Domain

Principal Component Analysis (PCA) is popular variables-reduction tool. It allows replacing dataset by smaller number of variables. This tools aims to extract the important information from the extended set of inter-correlated variables and represent it by new orthogonal space as a principal component (PC) domain (Abdi and Williams 2010). Other applications of PCA include de-noising signals, blind source separation and data compression.

In practice, PCA uses a vector space transform to reduce the dimensionality of big data sets. By applying mathematical projection, the original data set, which involves many variables, can often be characterized by just a few variables (called principal components – PCs).

Having considered that, PCA allows the user to spot trends, patterns and outliers in the data, easier than would have been possible without implementation of the PCA (Richardson 2009).

The transformation of the initial dataset into the PC domain is the second step of the methodology that is implemented in this research (Fig. 2).

Support Vector Machine Classification

SVM classification was introduced by Vapnik in 1995 as a new classification tool from machine learning theory (Vapnik 2013). SVM classification main idea is to find a separating hyperplane that divide two classes of data in new space. The transformation of dataset into new space is performed using kernel function. The main principal is to divide this data and to have the largest distance from the hyper-surface to the nearest points of the specific class. The objective of an SVM is to search for an n-dimensional hyperplane that differentiates the two classes with maximum gap (Peng et al. 2014). In this research, SVM classification was performed in ENVI software using four degree of kernel polynomial function, bias term equal to three and kernel

Table 1 DEM-derivatives and their main characteristic and references (Pawłuszek and Borkowski 2016)

DTM-derivatives	Information and references
Slope	Spatial Analyst Toolbox in ArcGIS (Van Westen et al. 2008)
Standard deviation of shaded relief [std_of_shadedrelief]	– calculated by moving standard deviation filter using 3 × 3 kernel size—Spatial Analyst Toolbox in ArcGIS
Openness	Difference between original DEM and DEM_{ki}, where DEM_{ki} is interpolated DEM of moving average filter with 9 × 9 kernel size (Van Den Eeckhaut et al. 2012)—Raster Calculator in ArcGIS
Topographic roughness	GIS Geomorphometry & Gradient Metrics toolbox by Evans et al. (2015)
Contour density	– calculated based on each 20 cm contour line per circle with radius of 3 m—implemented using Python in ArcGIS™
Area solar radiation [ASR]	ASR represents the solar energy for a given pixel and for specific date—Spatial analyst toolbox in ArcGIS™
Morphological gradient [gradient]	– represent difference between the dilation and erosion of DEM-image—implemented using Python in ArcGIS™
Topographic position index [TPI]	Implemented by Land Facet Corridor Designer by Janness et al. (2013)
Skewness	– represents the asymmetry of the probability distribution. Calculated by moving skewness filter with 3 × 3 kernel size
Curvature	Spatial analyst toolbox in ArcGIS™
Integrated moisture index [IMI]	Implemented by Geomorphometry & Gradient Metrics toolbox by Evans et al. (2015)
Stream power index [SPI]	– describes potential flow erosion at the specific location of the surface Spatial analyst toolbox and Raster Calculator in ArcGIS™
compound topographic index [CTI]	Implemented by Geomorphometry & Gradient Metrics toolbox by Evans et al. (2015)
Semivariogram	Moving semivariogram index filter using 9 × 9 kernel size. Implemented by Python in ArcGIS™
Slope/aspect transformation [slope/aspect]	Implemented by Geomorphometry & Gradient Metrics toolbox by Evans et al. (2015)
Difference between DEM and polynomial surface fitted into DEM [offset]	Calculated using raster calculator and interpolation function in ArcGIS™
Deviation from trend	Calculated using GIS Geomorphometry & GradientMetrics toolbox by Evans et al. (2015)
Standard deviation from slope [std_slope]	– calculated by moving standard deviation filter using 3 × 3 pixel kernel—ArcGIS™
Multiple shaded relief	Calculated using Spatial analyst toolbox in ArcGIS™ (Eeckhaut et al. 2007)

bandwidth $\gamma = p^{-1}$, where p is the number of DEM-derivatives (Hsu et al. 2003). According to the applied methodology, the SVM classification was performed on set of the PCs (Fig. 2). A subset covering 24.5% of the whole study area was utilized for machine learning. The remaining part covering 75.5% of the study area was used for evaluation of the classification.

Results

As the example, Fig. 3 shows the results of automatic landslide mapping. The figure presents the identification results based on seven PCs only. It was observed that the classification results for various number of principal component can vary slightly.

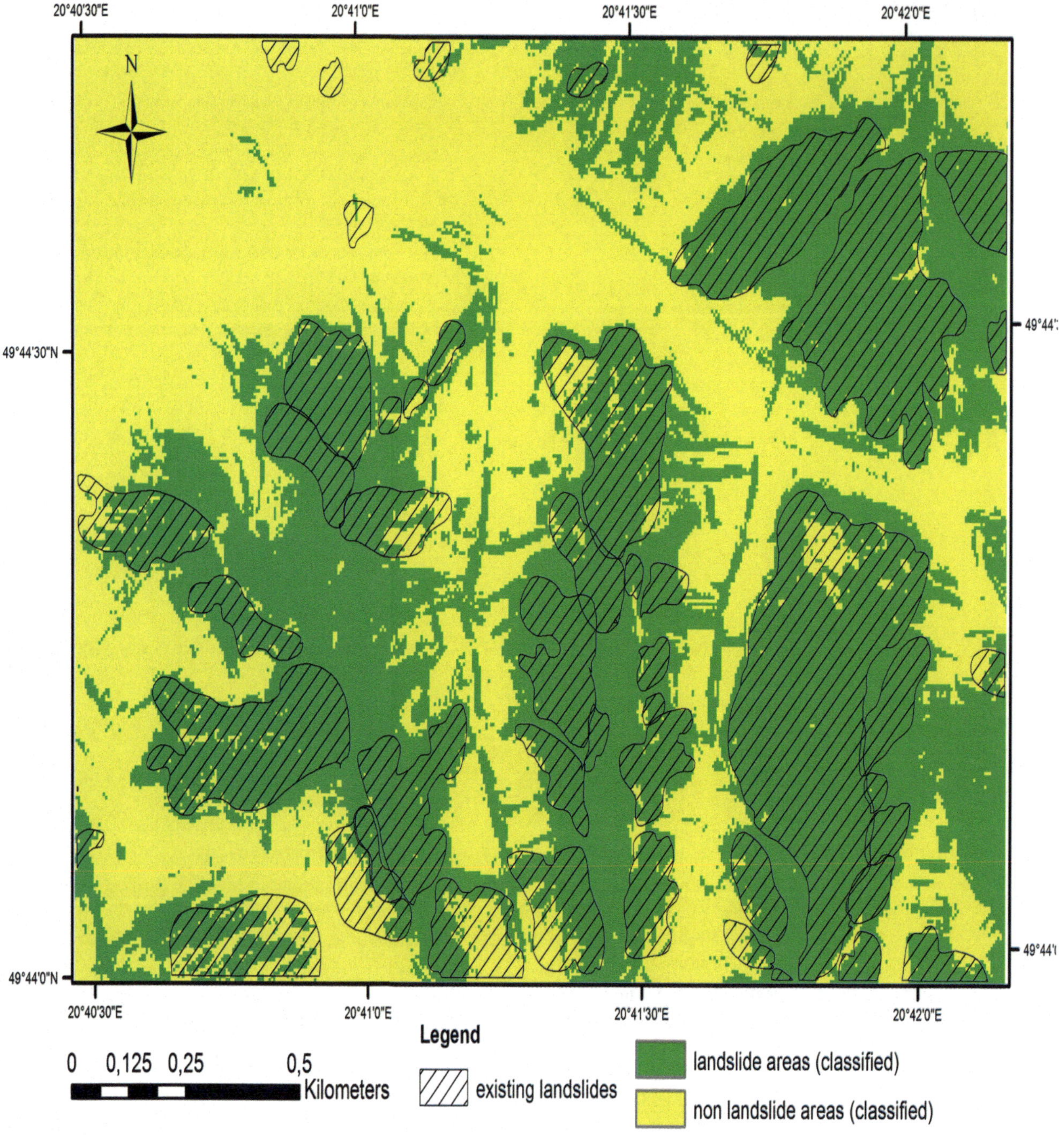

Fig. 3 Example of landslide detection based on seven PCs and existing landslide

Analyzing Fig. 4 it can be seen that accumulative eigenvalue increases as the number of subsequent principal component increases, but it increases significantly only for few first components. Adding further components, especially from the last part of the principal component spectrum, does not change accumulative eigenvalue significantly (Fig. 4). Figure 5 presents the overall accuracy in reference to the number of principal components taken into account. Based on the PCA accuracy chart (Fig. 5), it can be seen that not all information contain in the full set of DEM-derivatives provides valuable information in order to increase classification accuracy. Only

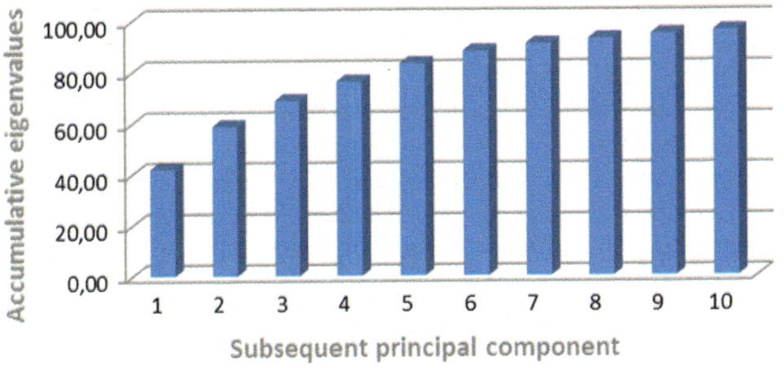

Fig. 4 Accumulative eigenvalues in subsequent principal component

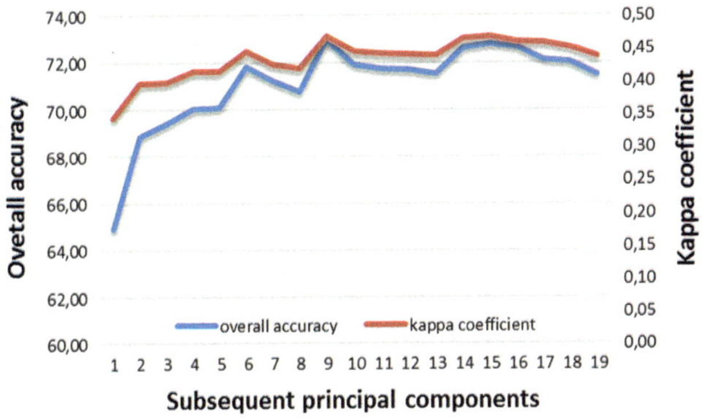

Fig. 5 Accuracy assessment in reference to subsequent principal components

few first principal components have an important impact on accuracy increasing of the landslide mapping. For instance ninth, eleventh, twelfth PCs decrease slightly classification accuracy. The best accuracy was achieved using ten principal components (72.95%) with 95.98% of total eigenvalues while for original dataset (all original DEM derivatives) overall accuracy is 73.37%. Additionally, the Kappa coefficients in relation to the number of principal components taken into account are visualized in Fig. 5.

In order to identify the most relevant variables and intercorrelated variables, loading plots were performed. The loading plots reveal the relationships between variables in the space of the first two components and are widely used to interpret the significance of variables (Crosta et al. 2013). Those located close to the orgin of coordinate system (low eigenvalue coefficient) are characterized as not significant. Based on the Fig. 6a and b, it can be seen that among the topographic indicators, multiple shaded relief, offset, elevation, ASR, standard deviation of shaded relief,topographic roughness and alope/aspect transformation have the largest eigenvector coefficients. Therefore they are the most significant variables.

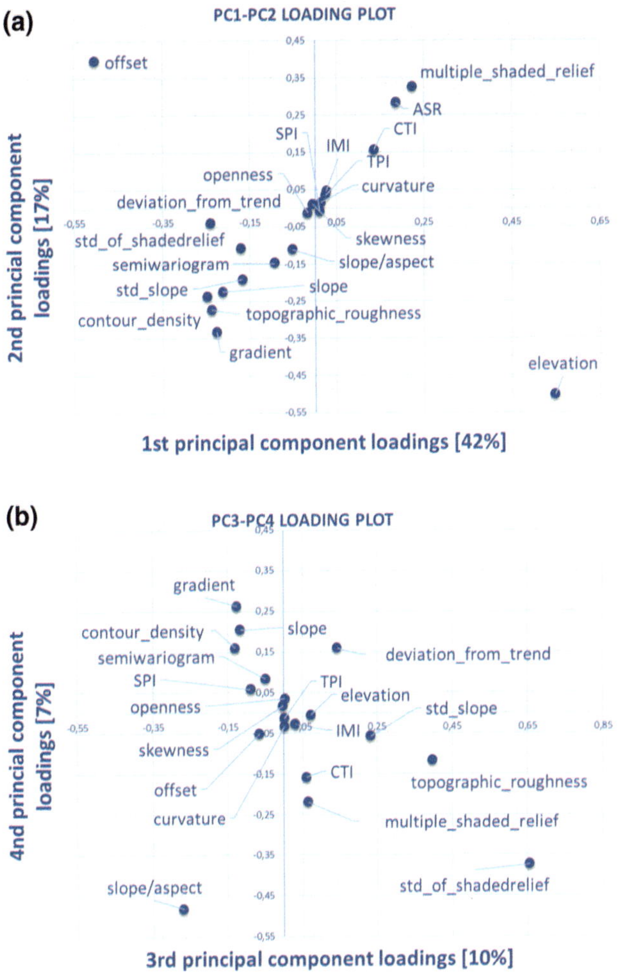

Fig. 6 Principal component loadings plot for **a** PC1-PA2 and **b** PC3-PC4

Summary and Conclusions

The objective of this study was deep exploration of the topographic information provided by the ALS data, while the computational effort should be reduced. For this reason, DEM and 19 DEM-derivatives have been generated and SVM classification was applied. Original dataset including DEM and 19 diverse DEM-derivatives were transformed into PCA domain in order to reduce secondary information and to reduce noise contained within this dataset.

Based on received results, it can be concluded that almost the same accuracy can be achieved by using ten principal components instead of extended set of DEM-derivatives. Applying PCA reduction of the variables, taken into account during the classification process, makes it possible to reduce calculation effort. For 10 principal components computational time was 10 min while for original dataset it was 25 min.

Authors will continue this research in further studies taking into account several test sites representing different landslide types. Nevertheless, the proposed approach, which combines the DEM-derivatives, PCA and the SVM algorithm, can be used for automatic landslide mapping, especially in hardly accessible areas covered by forest.

References

Abdi H, Williams LJ (2010) Principal component analysis. Wiley Interdisc Rev Comput Stat 2(4):433–459. doi:10.1002/wics.101

Borkowski A, Perski Z, Wojciechowski T, Jóźków G, Wójcik A (2011) Landslides mapping in Rożnów Lake vicinity, Poland using airborne laser scanning data. Acta Geodynamica et Geomaterialia 8(3):325–333

Carrara A, Merenda L (1976) Landslide inventory in northern Calabria, southern Italy. Geol Soc Am Bull 87(8):1153–1162. doi:10.1130/0016-7606(1976)87<1153:liincs>2.0.co;2

Chen W, Li X, Wang Y, Chen G, Liu S (2014) Forested landslide detection using Lidar data and the random forest algorithm: a case study of the Three Gorges, China. Remote Sens Environ 152:291–301. doi:10.1016/j.rse.2014.07.004

Crosta GB, Frattini P, Agliardi F (2013) Deep seated gravitational slope deformations in the European Alps. Tectonophysics 605:13–33. doi:10.1016/j.tecto.2013.04.028

Eeckhaut M, Poesen J, Verstraeten G, Vanacker V, Nyssen J, Moeyersons J, Van Beek L, Vandekerckhove L (2007) Use of LIDAR-derived images for mapping old landslides under forest. Earth Surf Proc Land 32(5):754–769. doi:10.1002/esp.1417

Evans J, Oakleaf J, Cushman S, Theobald D (2015) An ArcGIS Toolbox for surface gradient and geomorphometric modeling, version 2.0-0

Galli M, Ardizzone F, Cardinali M, Guzzetti F, Reichenbach P (2008) Comparing landslide inventory maps. Geomorphology 94(3):268–289. doi:10.1016/j.geomorph.2006.09.023

Gorczyca E, Wrońska-Wałach D, Długosz M (2013) Landslide hazards in the Polish Flysch Carpathians: example of Łososina Dolna Commune. In Geomorphological impacts of extreme weather, Springer Netherlands, pp 237–250. doi:10.1007/978-94-007-6301-2_15

Guzzetti F, Mondini AC, Cardinali M, Fiorucci F, Santangelo M, Chang KT (2012) Landslide inventory maps: new tools for an old problem. Earth Sci Rev 112(1):42–66. doi:10.1016/j.earscirev.2012.02.001

Hsu CW, Chang CC, Lin CJ (2003) A practical guide to support vector classification. Available at: http://www.csie.ntu.edu.tw/~cjlin/papers/guide/guide.pdf (last access: 26 Sept 2016)

Jenness J, Brost B, Beier P (2013) Land facet corridor designer: topographic position index tools. Available at: http://www.jennessent.com/downloads/Land_Facet_Tools.pdf (last access: 26 Sept 2016)

Mahalingam R, Olsen MJ, O'Banion MS (2016) Evaluation of landslide susceptibility mapping techniques using lidar-derived conditioning factors (Oregon case study). Geomatics Nat Hazards Risk 1–24. doi:10.1080/19475705.2016.1172520

Martha TR, van Westen CJ, Kerle N, Jetten V, Kumar KV (2013) Landslide hazard and risk assessment using semi-automatically created landslide inventories. Geomorphology 184:139–150. doi:10.1016/j.geomorph.2012.12.001

McKean J, Roering J (2004) Objective landslide detection and surface morphology mapping using high resolution airborne laser altimetry. Geomorphology 57(3):331–351. doi:10.1016/s0169-555x(03)00164-8

Moosavi V, Talebi A, Shirmohammadi B (2014) Producing a landslide inventory map using pixel-based and object-oriented approaches optimized by Taguchi method. Geomorphology 204:646–656. doi:10.1016/j.geomorph.2013.09.012

Pawłuszek K, Borkowski A (2016) Landslides identification using airborne laser scanning data derived topographic terrain attributes and support vector machine classification. Int Arch Photogramm Remote Sens Spatial Inf Sci XLI-B8:145–149. doi:10.5194/isprs-archives-XLI-B8-145-2016

Pawłuszek K, Ziaja M, Borkowski A (2014) Accuracy assessment of the height component of the airborne laser scanning data collected in the ISOK system for the Widawa River Valley (in Polish). Acta Scientiarum Polonorum. Geodesia et Descriptio Terrarum 13(3–4). Available at: http://yadda.icm.edu.pl/yadda/element/bwmeta1.element.baztech-a1b0f405-cffd-4d91-b1e2-33e49dd0eb50

Peng L, Niu R, Huang B, Wu X, Zhao Y, Ye R (2014) Landslide susceptibility mapping based on rough set theory and support vector machines: a case of the Three Gorges area, China. Geomorphology 204:287–301. doi:10.1016/j.geomorph.2013.08.013

Richardson M (2009) Principal component analysis. URL: http://www.dsc.ufcg.edu.br/~hmg/disciplinas/posgraduacao/rn-copin-2014.3/material/SignalProcPCA.pdf (last access: 26 Sept 2016)

Santangelo Á, Cardinali Á, Rossi Á, Mondini AC, Guzzetti F (2010) Remote landslide mapping using a laser rangefinder binocular and GPS. Nat Hazards Earth Syst Sci 10(12):2539–2546. doi:10.5194/nhess-10-2539-2010

Starkel L (1972) An outline of the relief of the polish Carpathians and its importance for human management. Problemy Zagospodarowania Ziem Górskich 10:75–150 (in Polish)

Tarolli P (2014) High-resolution topography for understanding earth surface processes: opportunities and challenges. Geomorphology 216:295–312. doi:10.1016/j.geomorph.2014.03.008

Van Den Eeckhaut M, Hervás J (2012) State of the art of national landslide databases in Europe and their potential for assessing landslide susceptibility, hazard and risk. Geomorphology 139:545–558. doi:10.1016/j.geomorph.2011.12.006

Van Den Eeckhaut M, Kerle N, Poesen J, Hervás J (2012) Object-oriented identification of forested landslides with derivatives of single pulse LiDAR data. Geomorphology 173:30–42. doi:10.1016/j.geomorph.2012.05.024

Van Westen CJ, Castellanos E, Kuriakose SL (2008) Spatial data for landslide susceptibility, hazard, and vulnerability assessment: an overview. Eng Geol 102(3):112–131. doi:10.1016/j.enggeo.2008.03.010

Vapnik V (2013) The nature of statistical learning theory. Springer Science & Business Media. doi:10.1007/978-1-4757-2440-0

Varnes DJ (1984) IAEG Commission on Landslides and other Mass-Movements—Landslide hazard zonation: a review of principles and practice. The UNESCO Press, Paris, pp 1–63. doi:10.1007/bf02594720

Wójcik A, Wojciechowski T, Wódka M, Krzysiek U (2015) Landslide inventory map of landslide in Gródek nad Dunajcem in the scale of 1: 10000. Municipality of Łososina dolna, district: nowosądecki, province: małopolskie. URL: http://geoportal.pgi.gov.pl/portal/page/sopo (access on 6 July 2016)

Woźniak A (2013) Precipitation in the polish Carpathian Mountains in 2010 compared to the period 1881–2010. Prace Geograficzne 133:35–48 (in Polish)

Geological Aspects of Landslides in Volcanic Rocks in a Geothermal Area (Kamojang Indonesia)

I. Putu Krishna Wijaya, Christian Zangel, Wolfgang Straka, and Franz Ottner

Abstract

This Kamojang area is an Indonesian geothermal field that produces electricity since 1983. As many other geothermal fields in Indonesia, Kamojang area is located in high relief volcanic terrain where landslides frequently occur. Hydrothermal alteration of volcanic rocks is an important geological process which reduce slope stability due to the reduction of the rock mass strength properties. Landslides in a geothermal area are hazards that can adversely affect roads, pipelines, as well as injection and geothermal wells. The aim of this study is to enhance the understanding of landslide processes in highly weathered and hydrothermal altered volcanic rock masses based on field, laboratory and numerical modelling studies. Exemplarily for this geological situation, the study site of the geothermal area of Kamojang in Indonesia was chosen. This article presents an overview of the study area and some preliminary results from a data compilation study and a field survey during this summer.

Keywords

Landslide • Volcanic rock • Hydrothermal alteration • Clay minerals

Introduction

Volcanic regions often exhibit significant landslide hazard. Typically, such areas are characterized by mountainous topography, weakened volcanic rocks due to hydrothermal alteration processes, and substantial seismic activities (Voight 1992). Kamojang area is an Indonesian geothermal field that has been producing electricity since 1983 and is located in West Java, at an altitude of 1500 m above sea level about 40 km southeast of the city of Bandung (Fig. 1, Utami and Browne 1999). Kamojang is a vapor-dominated system hosted in a caldera surrounded by volcanic ranges (Raharjo et al. 2010; Suryadarma et al. 2005). This area contains numerous thermal manifestations e.g. mud pools, fumaroles, hot springs, mud pots, and steam vents (Fig. 1, Mulyanto et al. 2010). As many other geothermal fields in Indonesia, Kamojang area is located in high relief volcanic terrain where landslides frequently occur (Barbano et al. 2014; Hürlimann et al. 2000; Mitchell et al. 2015; Moon and Simpson 2001; Shuzui 2001).

Hydrothermal fluids passing through the host rock are changing its composition by adding, removing, or redistributing components under certain temperature, pressure and chemical conditions and generate hydrothermal altered rocks. This change from rock to soil-like material has a considerable impact on the slope stability. In addition, weathering effects further reduce the strength properties of the slope and slope failure can occur, especially during high precipitation events (Day 1996; Pioquinto et al. 2010; Frolova et al. 2014).

I.P.K. Wijaya (✉) · C. Zangel · W. Straka · F. Ottner
Institute of Applied Geology, BOKU-WIEN, Vienna, Austria
e-mail: krishna_wijaya@yahoo.co.id

C. Zangel
e-mail: christian.j.zangerl@boku.ac.at

W. Straka
e-mail: wolfgang.straka@boku.ac.at

F. Ottner
e-mail: franz.ottner@boku.ac.at

M. Mikoš et al. (eds.), *Advancing Culture of Living with Landslides*,
DOI 10.1007/978-3-319-53483-1_51

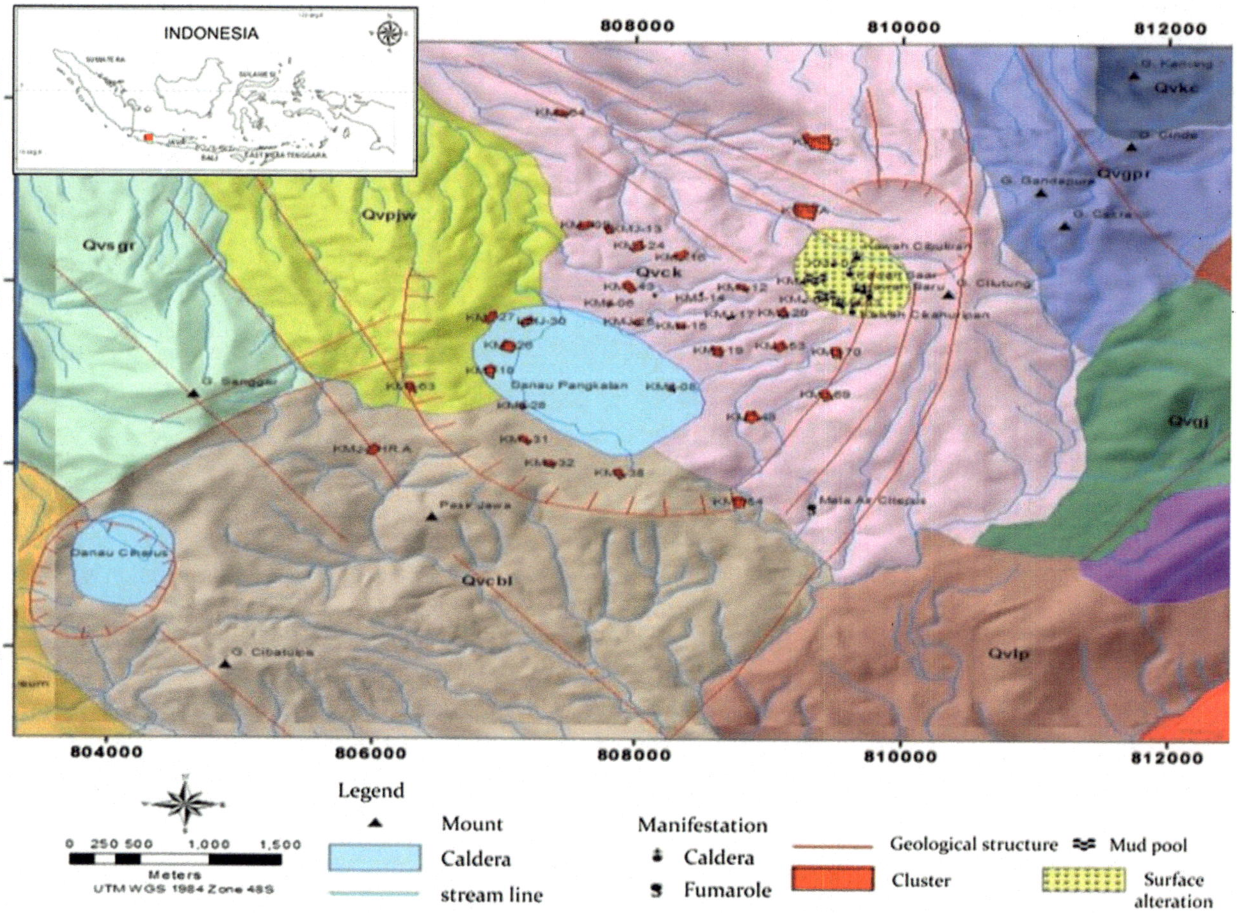

Age (million years)			Lithology Units	Age (million years)		Lithology Units
Quartenary	Post caldera	0.025 ± 0,1	Qvgt : Alteration	Pre- Caldera	0,425 ± 0,015	Qvpjw : Andesite Pasir Jawa
			Qvms : Basaltic Andesite Mt. Masigit			Qplv : Basaltic Andesite Legokpulus
			Qvck : Basaltic Andesite Mt. Cakra			Qvkt : Andesite porphyry Mt. Tampomas
			Qvgj : Andesite Mt. Gajah		1,02 ± 0,02	Qvcbl : Hornblende Andesite Mt.Cibautipis
			Qvgpr : Pyroxene Andesite Mt. Cibereum			Qvsgr : Pyroclastic Mt. Sanggar
			Qvck : Pyroxene Andesite Mt. Kancing			Qvcbr : Pyroxene Andesite Mt. Cibereum
						Qvrkt : Basalt Mt. Rakutak

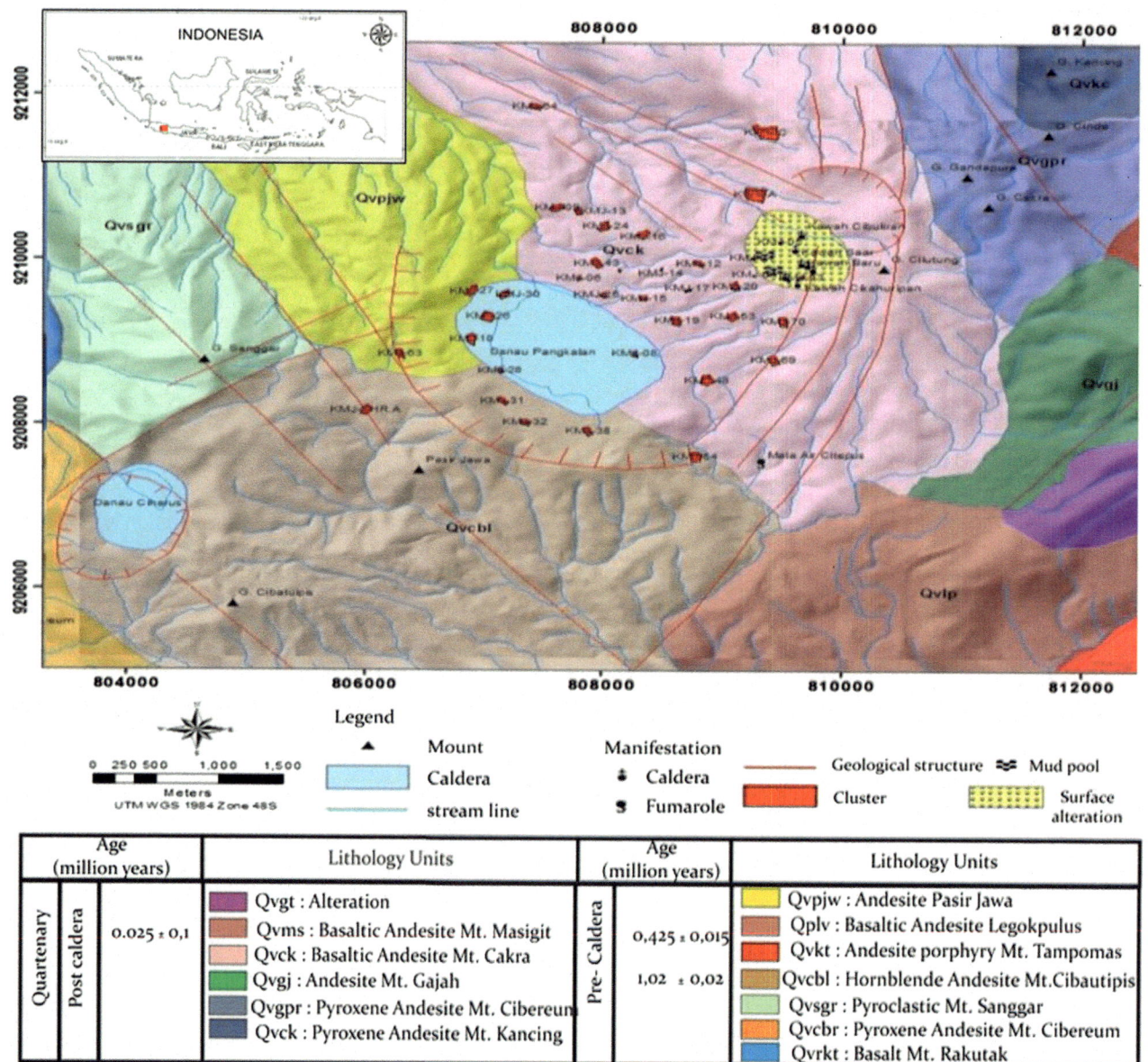

Fig. 1 Regional geological map of study area, modified after (Pertamina 2013). Insert map is the location of study area (*red box*), modified after (Sofyan et al. 2010)

Hydrothermal alteration lead to the formation of clay minerals e.g. kaolinite, halloysite, and smectite. Clays of the smectite group are produced by the transformation of volcanic glass under the action of thermal fluids. As suggested by Frolova et al. (2014) newly formed minerals from the smectite group cause a significant decrease of the strength properties, especially at water saturated conditions. Thus, high rainfall and storm events can trigger landslides in such geological environment (Shen et al. 2016; Robbins 2016). Most research topics in Kamojang geothermal area that have been published are related with geothermal systems (e.g. Raharjo et al. 2010; Zuhro et al. 2005). There are no publications focusing on landslides in this area, although several landslides may have adversely affected roads, pipelines and wells (Pioquinto 2010; Shangyao and Tian 1986).

The aim of this preliminary study is to compile pre-existing data and to acquire new data about the geological and geomechanical situation of landslide events in Kamojang area and to increase the understanding of landslide mechanisms in hydrothermal altered volcanic rocks.

Geological Situation of the Study Area

Kamojang geothermal area is situated along the axis of Sunda Volcanic Arc. This Arc-trench system represents the convergent boundary between Eurasian Plate on the north and Indo-Australian plate on the south. The orientation of plate convergent in Jawa is approximately N-S, therefore it shows that the subduction is nearly perpendicular to the arc front in central Java but increasingly oblique towards Sumatra (Mulyanto et al. 2010).

The Kamojang geothermal area lies in a 15 km long and 4.5 km wide in volcanic chain (Rakutak-Guntur). Utami and Brownie (1999) suggest that the volcanoes in this chain erupted sequentially from the WSW to ENE. Hantono et al. (1996) suggest that this volcanic chain is constituted by following succession of volcanic complexes: Mt. Rakutak, Ciharus complex, Pangkalan complex, Gandapuro complex, Mt. Masigit and Mt. Guntur (Sudarman and Hocstein 1983), (Fig. 1). Erosion stage of these volcanoes have progressively grown from the WSW to ENE thus Gunung Rakutak is the oldest and G. Guntur is the youngest.

This volcanic chain has been affected by a large NW-SE graben, 6 km wide extending from Ciharus to Kamojang (Fig. 1). The depression can be clearly observed in the field and is divided into several sub-structures such as horst and graben structures. The magmatic axis and the NW-SE depression are affected by normal faults oriented approximately N-S. Two structures have been distinguished: the first is a graben structures of 500 m width which traces through the part of Ciharus, and the second is a fault bundle which extent into the eastern part of Kamojang area where surface manifestations e.g. fumarole, hot springs, mud pool occur (Fig. 1). Most of the major surface thermal manifestations occur at the edge of Citepus fault (Mulyanto et al. 2010). Most of the volcanic complexes exhibit circular structures. (Hantono et al. 1996).

Based on a regional geological map, Pertamina (2013) Kamojang area consist of 17 lithological units that can be separated into two main groups; i.e. the pre-caldera and post-caldera lithology units (Fig. 1). The pre-Caldera units mainly consist lava basaltic-andesite and pyroclastic rocks. These units can be divided into 9 lithological sub-units; (a) Basalt Mt. Rakutak (Qvrkt), (b) Basalt Dodog (Qvdd), (c) Pyroxene Andesite of Mt. Cibeureum (Qvcbr), (d) Pyroclastic of Mt. Sanggar (Qvsgr), (e) Pyroxene Andesite of Mt. Cibauti (Qvcbi), (f) Andesite Porphyry of Mt. Katomas (Qvkt), (g) Andesite-Basaltic Legokpulus (Qvlp), (h) Andesit-Basaltik Mt. Putri (Qvp) on the northen part this units altered into kaolinite, chlorite by hydrothermal process. (i) Andesit Pasir Jawa mainly consist of intercalation of lava and pyroclastic rocks.

Post-Caldera lithology units are composed of 8 lithologies sub-units; (a) Pyroxene Andesite Mt. Kancing (Qvkc), (b) Andesite-Basaltic Mt. Batususun (Qvbs), (c) Andesite-Basaltic Mt. Gandapura (Qvgp), (d) Andesite Mt. Gajah (Qvgj), (e) Andesite-Basaltic Mt. Cakra (Qvck), (f) Andesite-Basaltic Mt. Masigit (Qvms), (g) hydrothermal Alteration units (Qvaltr); Hydrothermal activity altered volcanic rocks into the (clay) minerals; kaolinite, montmorilonite, chlorite, alunite and gypsum. Hydrothermal alteration in widely observed on the top of Mt. Cakra, Mt. Masigit, Mt. Guntur, Mt. Gandapura and Mt. Putri. Alluvium (Qval); composed of lapili, lahar, andesite-basaltic rocks (gravel-boulder size, subangular, poorly sorted).

The Kamojang Geothermal field is now, vapor-dominated but the hydrothermal minerals show that the rock-altering fluid were dominantly liquid (Utami 2000; Utami and Brownie 1999). There are two hydrothermal mineral assemblages present, those produced by "acid" and "neutral" pH fluids. The "acid" mineral assemblage that occupies shallow levels (down to 100–300 m) consists of kaolin, smectite, alunite, quartz, cristobalite, and pyrite. This clay minerals especially smectite have a high swelling index (Kariuki et al. 2004).

Most of lithology in study area belongs to (Qvck) formation, which consist of andesite-basaltic of Mt. Cakra products. Another small part belongs to (Qvcbi) formation that consists of hornblende andesite as a result of Mt. Cibautipitis eruption. Most of landslide locations are near faults and the rim structure (Fig. 2).

Landslide Characteristics

In the study area 28 landslide events characterized by different geomorphological and geological situations were mapped (Fig. 2). This article will primarily focus on the largest landslide in the study area, termed as site no. 28 (Fig. 3). On April 6th, 2013 a large landslide occurred in the southeastern part of the study area within highly weathered and hydrothermally altered hornblende andesite. This landslide was located in the upper part of one geothermal injection well platform. The slope where the landslide occurred is close to the NE-SW striking Pateungteng fault system. The occurrence of altered rocks and hot springs as well as fumaroles near the fault trace indicate a hydrothermal overprint of this area. Hydrothermal alteration led to the transformation of the primary mineral composition into a clay-rich mineral composition. Based on investigations from Pertamina (2013), the clay mineralogy is composed of kaolinite, halloysite, smectite, and mixed layer minerals (Fig. 4).

Based on the landslide classification system developed by Cruden and Varnes (1996), the movement type of the landslide in study area can be characterized as a complex landslide. Concerning the type of material, the landslide is an earth slump-earth flow. According to the more recent

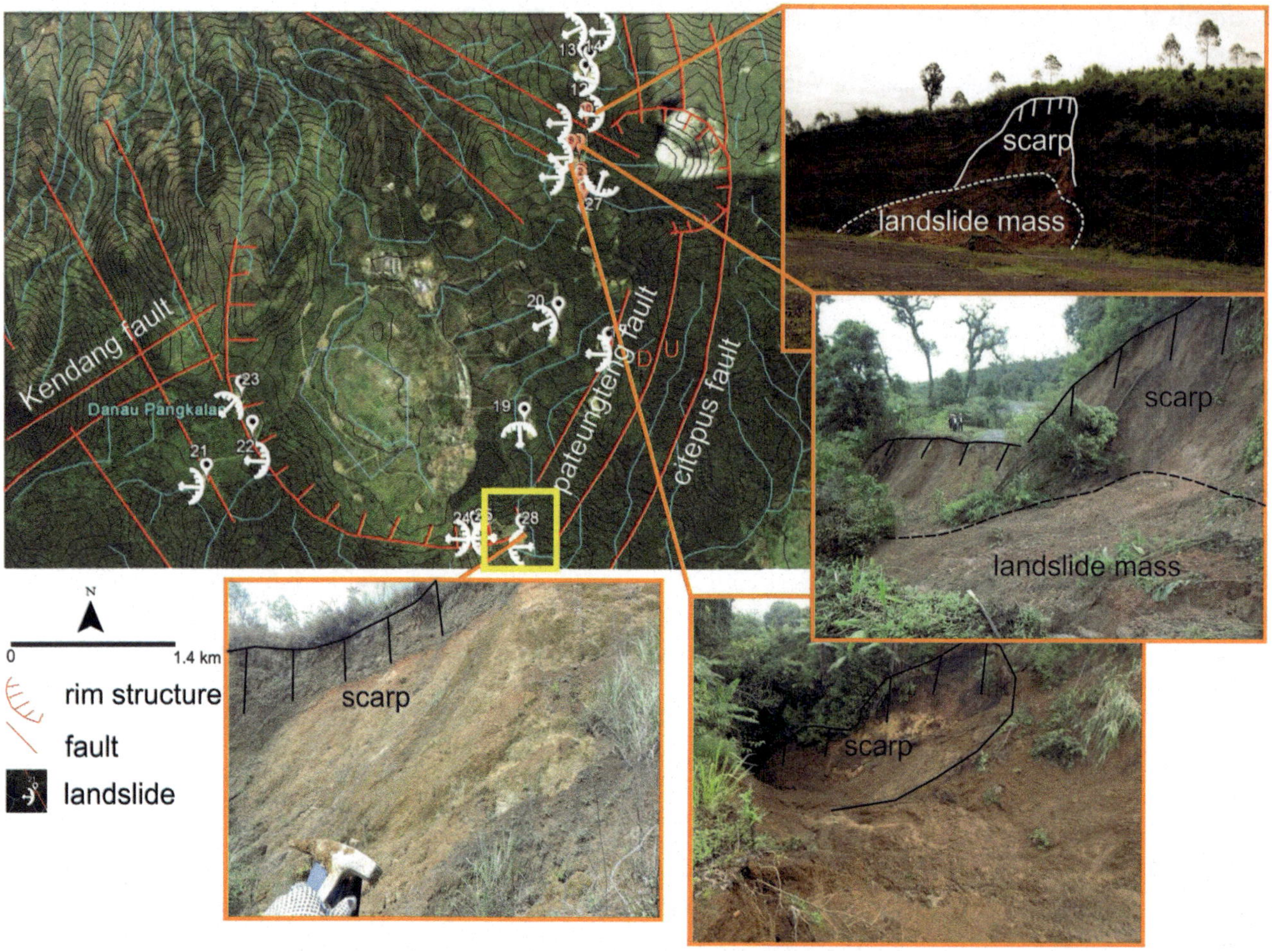

Fig. 2 Landslide inventory map of study area. The *yellow box* shows the location of the landslide of site 28, which is also explained in more detail in Fig. 3

classification system from Hungr et al. (2014) the landslide can be categorized as a rotational slide-earthflow characterized by a circular failure plane and a transition into an earthflow. Including both, the source and accumulation area the landslide has a total length of 330 m, a width of 90 m and a maximum thickness of 20 m. Concerning the main scarp a width of 30 m, a length of 110 m, a mean thickness of 10 m, and a failure volume of 33,000 m^3 is estimated. Between the main scarp and the deposition area a runout and elevation distance of 330 and 70 m was measured. Based on this a runout travel angle of 12° can be determined.

The basal shear zone was formed preliminary in the hydrothermal altered andesitic rock mass (Fig. 5). Within the sliding surface the clay-rich soil was sheared and the mineral grains were aligned along the plane in the direction of shear. Thus distinct slickenside surfaces were formed. The sliding surface and the slickenside striations in the main scarp area dip with 39° towards east. The plunge direction of the slickenside striation was in accordance with the movement direction of the landslide i.e. the slide slid in east direction and then changed into an earthflow with southeast movements along a stream channel (Fig. 5).

Generally, rainfall is one of the main landslide triggers especially in tropical high-precipitation regions (Peruccacci et al. 2012). The study area West Java is characterized by high annual precipitation rates. Monthly and daily rainfall data between 2010 and 2016 obtained from the online BMKG (data access on May 2nd 2016) rainfall database were analyzed. Data show that rainfall is concentrated during the wet season between December and April where more than 500 mm per month were measured. The landslide occurred during April 2013 where the cumulative rainfall for that month was 317 mm (Fig. 6). Although the landslide occurred after several rainfall events an instantaneous cause-effect relationship cannot be proven. This is mainly because the months before the failure event are characterized

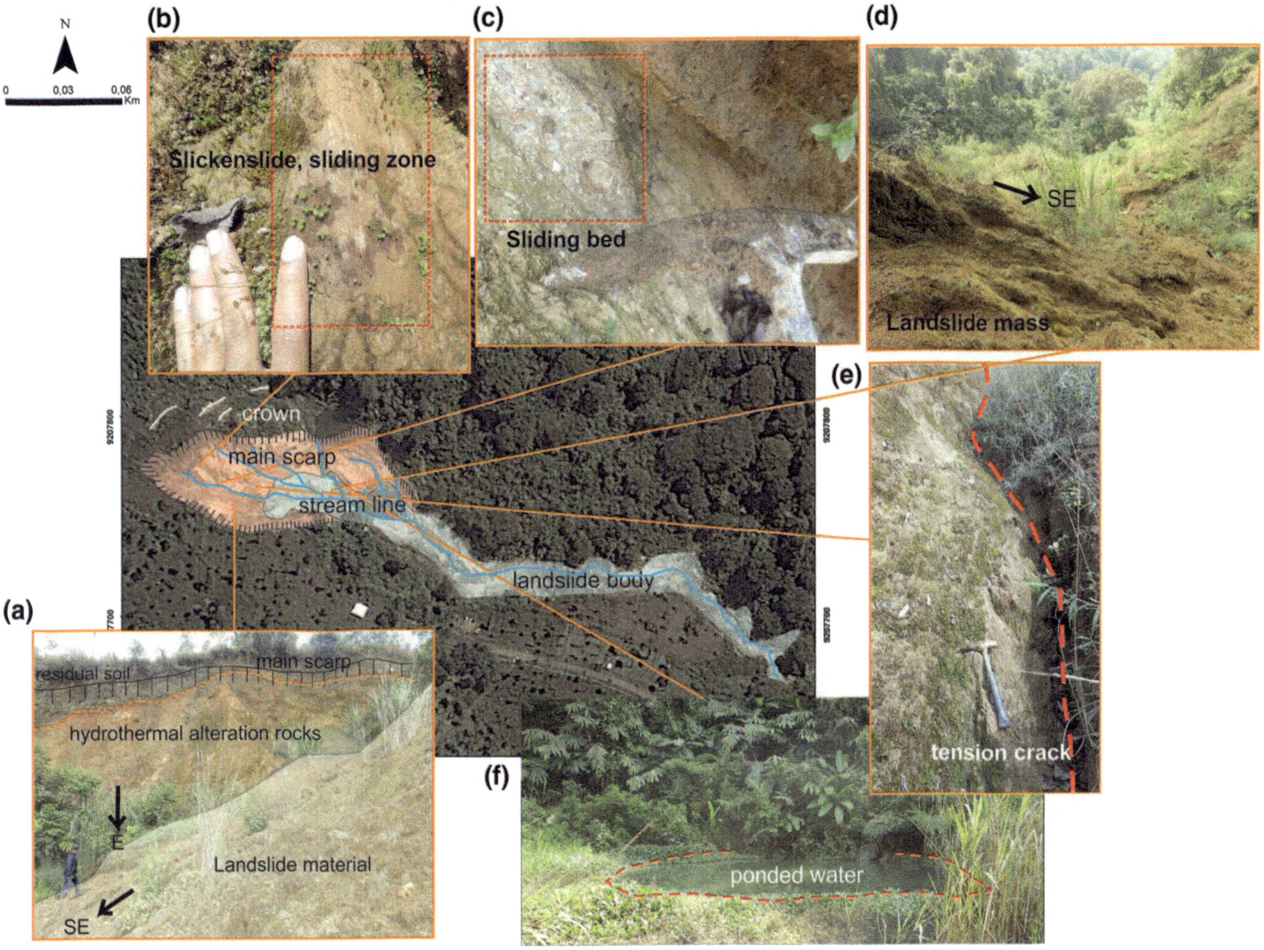

Fig. 3 **a** Geometry of landslide no. 28 on the west scarp. **b** Slickenside striations on the sliding plane in boundary between bedrock and clay alteration. **c** Basal sliding zone. **d** Landslide material moved in E direction then change into SE direction. **e** Tension crack. **f** Ponded water

by a higher precipitation. In addition, the cumulative rainfall during the months January, February and March was higher than in April during the year 2013. Nevertheless, a day before and during the landslide event was raining (Fig. 7).

Lithology and Mineralogy

The northern part of the study area is built by products of the Mt. Cakra eruptions (i.e. a post-caldera event), which are basaltic andesites. In the southern part of the study area volcanic products of the Mt Cibatuipis eruptions which belongs to older pre-caldera event were deposited as hornblende andesites.

Lithologically, the study area is characterized by intercalations of lava flows and pyroclastic depositions. The basaltic andesite lava flow rocks show a dark grey color and a porphyro-aphanitic texture (McPhie et al. 1993). Phenocrysts of plagioclase, hornblende, pyroxene and other mafic minerals are embedded within a microcrystalline groundmass (Fig. 8a). The pyroclastic deposits are composed of andesitic rocks and show a porphyritic texture with phenocrysts of plagioclase and pyroxene. Sometimes poikilitic intergrowth textures can be observed (Fig. 8b). The groundmass consists of plagioclase, pyroxene, and small amounts of volcanic glass. Magnetite was found as inclusions.

The andesitic rock is jointed and altered by hydrothermal processes. A transformation into clayey, silty and sandy soil material took place. Based on XRD analyses the clay mineral assemblage of the soil consists mainly of kaolinite, halloysite, smectite, and mixed layer clay minerals. According to Corbett and Leach (1998), those mineral assemblages can be categorized as an argillic and advanced argillic alteration type. Whereas kaolinite is an alteration product of plagioclase, montmorillonite (smectite) is formed by alteration of amphiboles and plagioclases.

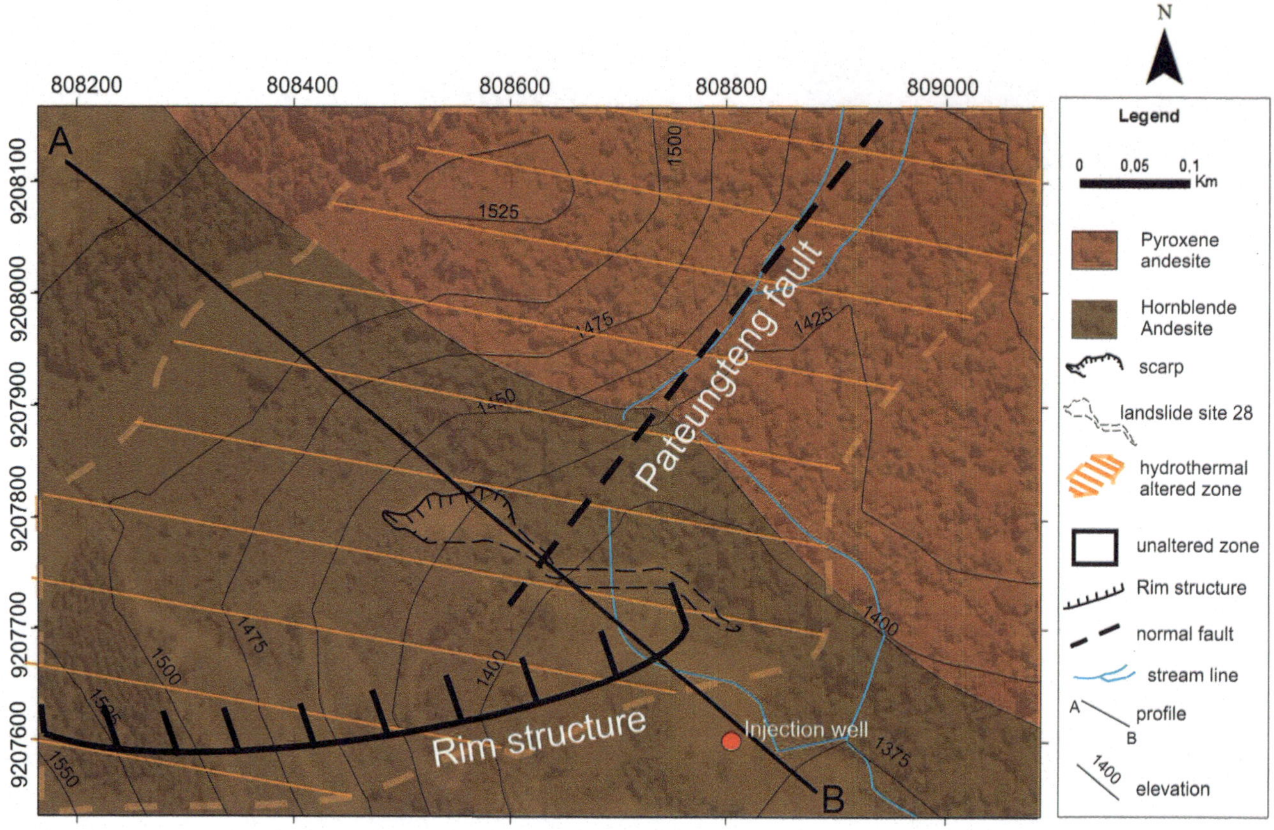

Fig. 4 Schematic geological map of study area

Fig. 5 Geological section A-B

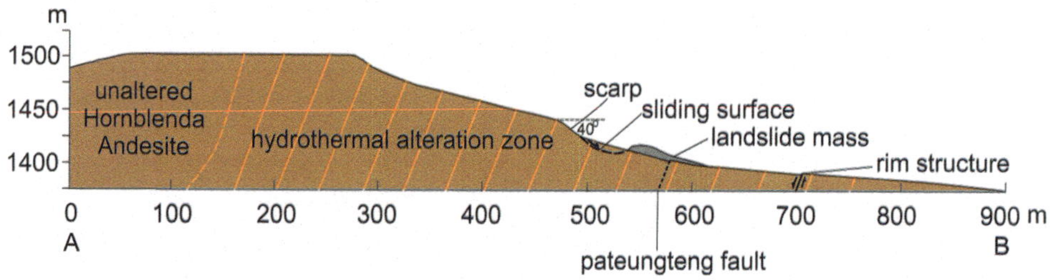

Fig. 6 Rainfall data from January–December 2013

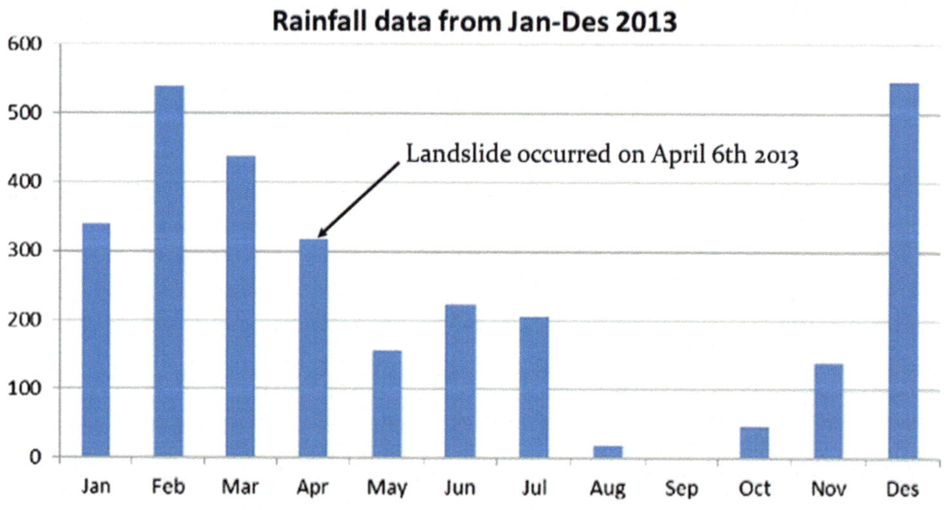

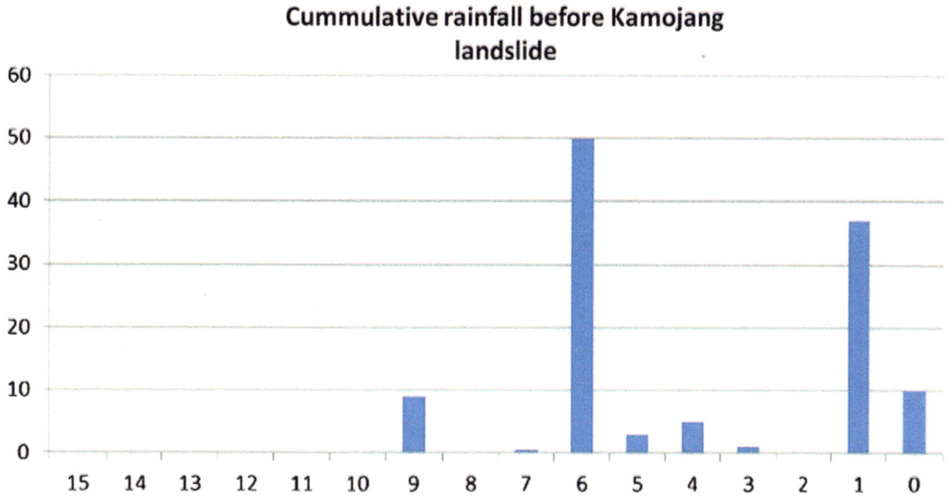

Fig. 7 Cummulative rainfall before Kamojang Landslide (April 6th 2013)

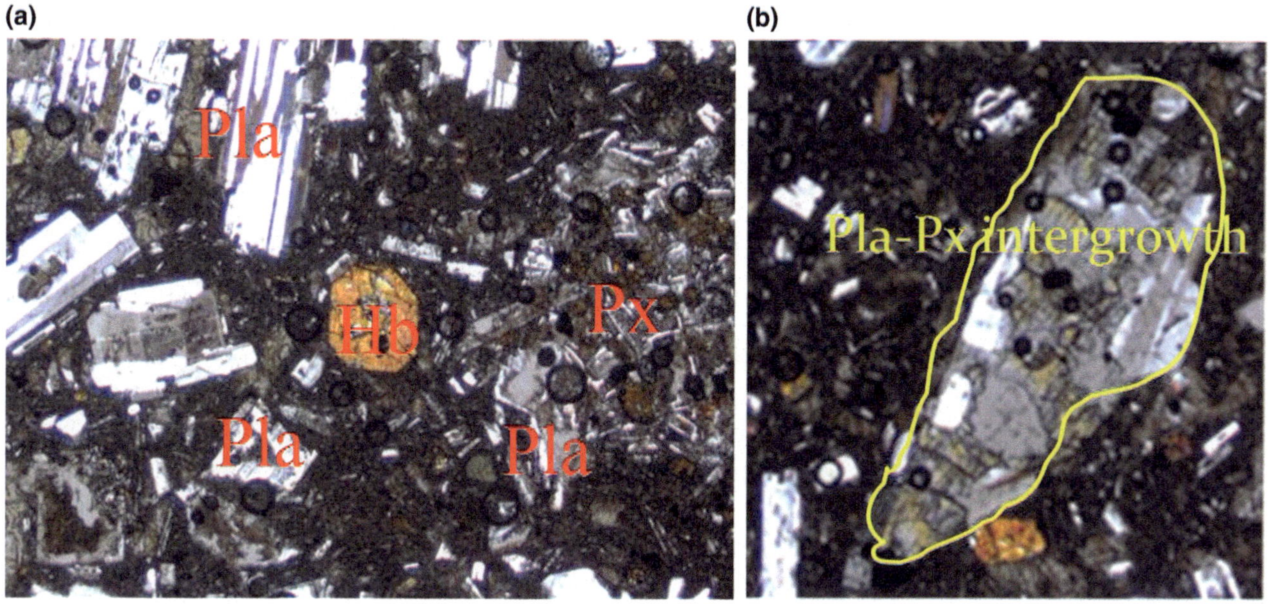

Fig. 8 **a** Petrography analysis of sample KAM STA-2. **b** Plagioclase and pyroxene showing intergrowth structure (poikilitic)

Geotechnical Properties

The geothermal company Pertamina (2013) studied landslides in the same area with similar geological conditions (see landslide no. 3 and 12 in Fig. 2). In the framework of this study geotechnical properties were presented. Based on the Unified Soil Classification System (USCS) the soil in surrounding of the landslide no. 3 and 12 can be subdivided into two layers. A layer of clayey/silty sand is on top of a layer of inorganic silt. The thickness of the clayey sand soil layer varies between 1 and 8 m. For the top layer sieve analyses show a sand grain fraction in the range between 50 and 77% and a clay/silt fraction in the range between 16 and 49%. According to the USCS, this soil could be categorized as poorly graded clayey sand or silty sand with internal friction angle between 33 and 41°. Cohesion obtained from direct shear test are in range between 3 and 22 kPa. The top layer has a variable state of compaction from loose to dense sand. Based on visual criteria the inorganic silt layer can be subdivided into a yellowish and reddish silt. Depending on the water content, this layer can reach a state of compaction from medium to stiff or very soft. The inorganic silt layer is composed of 50–86% silt grain fraction. A liquid limit and plasticity index of about 50% and respectively of 17% was found. The silt layer has a consistency ranging between medium and stiff. Based on a direct shear test a cohesion of 19 kPa and a friction angle of 34° was obtained.

Discussion and Conclusion

Intensive hydrothermal alteration along the Pateungteng fault and interlinked discontinuities and weathering processes changed the mineralogy of andesitic lava and pyroclastic breccia. The newly formed clay minerals characterized by a high swelling indexes led to a considerable change in the strength properties and slope stability. The observed mineral assemblage based on kaolinite, halloysite, trydimite, smectite and mixed layer minerals show an argillic and advance argillic alteration type. Most probably, the high amount of precipitation, especially during the period from December to April may have favored the formation of the landslide. The low runout travel angle of only 12° suggest a high water content of the landslide material during the event. The landslides in Kamojang geothermal area are similar to other landslides cases in geothermal area in the world which are located in volcanic rocks (Frolova et al. 2014; Kugaenko and Saltykov 2010; Kiryukhin et al. 2010; Pioquinto et al. 2010) (Fig. 9).

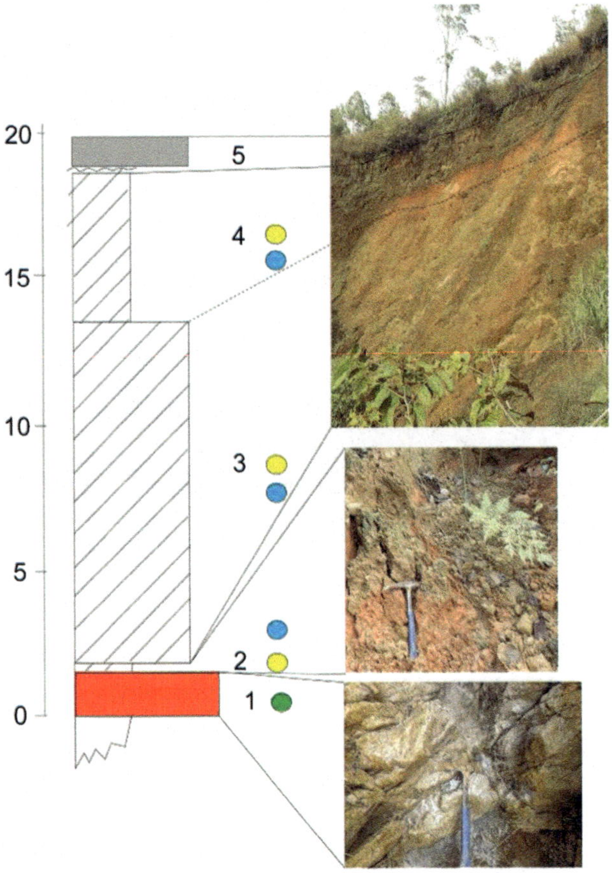

Fig. 9 Stratigraphy column of landslide. *1* Bedrock consist of hornblende andesite that slightly altered due to hydrothermal activity. *2* Reddish clay alteration. *3* Yellowish silt-fine sand alteration. *4* Reddish clay alteration. *5* Weathered residual soil. The sliding zone is between *1* and *2*. *Green*, *yellow*, *blue* dot are petrography, XRD, and geotechnics sampling

Acknowledgements We would like to thanks PT Pertamina Geothermal Energy for the permission to study landslide events in the Kamojang area.

References

Barbano MS, Pappalardo G, Pirrotta C, Mineo, S (2014) Landslide triggers along volcanic rock slopes in eastern Sicily (Italy). Nat Hazards 73:1587–1607. doi:10.1007/s11069-014-1160-1. 24 May 2016

Corbett GJ, Leach TM (1998) SW Pacific Rim Au/Cu Systems; structure, alteration and mineralization. Special publication 6, Society of Economic Geologist, p 238

Cruden DM, Varnes DJ (1996) Landslide investigation and mitigation: landslide types and processes. In: Turner AK, Schuster RL (eds) Special report 247. Transportation Research Board, USA

Day SJ (1996) Hydrothermal pore fluid pressure and the stability of porous permeable volcanoes. Geol soc London 110:77–92

Frolova J, Ladygin V, Rychagov S, Zukhubaya D (2014) Effects of hydrothermal alterations on physical and mechanical properties of rocks in the Kuril-Kamchatka island arc. Eng Geol 183:80–95

Hantono D, Mulyono A, Hasibuan A (1996) Structural control is a strategy for exploitation well at kamojang geothermal field west java, Indonesia. Proceeding 21 workshop on geothermal resources engineering, California

Hase H, Hasimoto T, Sakanaka S, Kanda W, Tanaka Y (2005) Hydrothermal system beneath Aso Volcano as inferred from self-potential mapping and resistivity structure. J Volcanol Geptthermal Res 143:259–277

Hungr O, Leroueli S, Picarelli L (2014) The varnes classification of landslide types, an update. Landslide 11:167–194

Hürlimann M, Garcia-Piera JO, Ledesma A (2000) Causes and mobility of large volcanic landslides: application to Tenerife, Canary Islands. J Volcanol Geoth Res 103:121–134

Kariuki PC, Woldai T, van der Meer, FD (2004) Effectiveness of spectroscopy in identification of swelling indicator clay minerals. Int J Remote Sens 25(2):455–469

Kavzoglu T, Kutlug E, Sahin, Colkesen I (2015) Selecting optimal conditioning factors in shallow translational landslide susceptibility mapping using genetic algorithm. Eng Geol 192:101–112

Kiryukhin AV, Rychkova TV, Droznin VA, Chernykh EV, Puzankov MY, Vergasova LP (2010) Geysers Valley hydrothermal system (Kamchatka): recent changes related to landslide of June 3 2007. Proceedings world geothermal congress 2010

Kugaenko Y, Saltykov V (2010) Kamchatkan Valley of the geysers: geodynamic processes, seismicity, landsliding. Proceedings world geothermal congress 2010

McPhie J, Doyle M, Allen RJ (1993) Volcanic Textures. centre of deposit & exploration studies. University of Tazmania, Tazmania

Mitchell TM, Smith SAF, Anders MH, Di Toro G, Nielsen S, Cavallo A, Beard AD (2015) Catastrophic emplacement of giant landslide aided by thermal decomposition; Heart mountain, Wyoming. Earth Planet Sci lett. 411:199–207

Moon V, Simpson CJ (2001) Large scale mass wasting in ancient volcanic materials. Eng Geol 64:41–64

Mulyanto Y, Nani, Agus A, Zuhro Y, Ahmad (2010) Surface thermal manifestation monitoring Kamojang geothermal field. Proceeding world geothermal congress 2010 2:1–4

Pertamina (2013) Laporan Utama Konsultan Mitigasi Potensi Bencana dan Evaluasi Kestabilan serta Perencanaan Perkuatan Lereng di Proyek dan Area PT Pertamina Geothermal Energy Area Kamojang, Unpublished

Peruccacci S, Brunetti MS, Luciani S, Vennari C, Guzzetti F (2012) Lithological and seasonal control on rainfall thresholds for the possible initiation of landslides in central Italy. Geomorphology 139–140:79–90

Pioquinto WPC, Caranto JA, Bayrante LF (2010) Mitigating a deep-seated landslide hazard- the case of 105 Mahiao Slide area, Leyte geothermal production field, Philippines. Proceedings world geothermal congress, 2010

Raharjo IB, Maris V, Wannamaker PE, Chapman DS (2010) Resistivity structures of Lahendong and Kamojang Geothermal systems revealed from 3-D Magnetotelluric Inversions, a comparative study. Proceedings world geothermal congress, Indonesia, 1999

Robbins JC (2016) A probabilistic approach for assessing landslide-triggering event rainfall in Papua New Guinea, using TRMM satellite precipitation estimates. J Hydrol 541:296–309

Shen P, Zhang LM, Zhu H (2016) Rainfall infiltration in a landslide soil deposit: Importance of inverse particle segregation. Eng Geol 205:116–132

Shuzui H (2001) Process of slip surface development and formation of slip surface clay in landslides in tertiary volcanic rocks, Japan. Eng Geol 61:199–219

Sofyan Y, Daud Y, Kamah Y, Ehara S (2010) Sustainable geothermal utilization deduced from mass balance estimation, A case study of Kamojang geothermal field, Indonesia. Proceedings world geothermal congress, 2010

Sudarman S, Hochstein NP (1983) Geophysical structure of the Kamojang Geothermal field. 5th NZ Geothermal Workshop, New Zealand, 225–230

Suryadarma, Azimuddin T, Dwikorianto T, Fauzi A (2005) The Kamojang geothermal field: 25 Years operation. Proceedings world geothermal congress, Turkey, 2005

Shangyao H, Tian T (1986) Study of environmental impact in geothermal development and utilization. Proceedings of the 7th Asian Geothermal Symposium, 2006, July 25–26

Utami P, Brownie PRL (1999) Subsurface hydrothermal alteration in the Kamojang geothermal field, West Java, Indonesia. Proceedings 24th workshop on geothermal reservoir engineering, Stanford University, Stanford, California, Sgp-Tr-162

Utami P (2000) Characteristics of the Kamojang geothermal reservoir (West Java) as revealed by its hydrothermal alteration mineralogy. Proceedings world geothermal congress 2000, Kyushu - Tohoku, Japan, pp 1921–1926

Voight B (1992) Causes of Landslides: Conventional factors and special considerations for geothermal sites and volcanic regions. Geoth Resour Counc Trans 16:529–533

Adaptive Learning Techniques for Landslide Forecasting and the Validation in a Real World Deployment

T. Hemalatha, Maneesha Vinodini Ramesh, and Venkat P. Rangan

Abstract

A forecasting algorithm using Support Vector Regression (SVR) used to forecast potential landslides in Munnar region of Western Ghats, India (10.0892 N, 77.0597 E) is presented in this paper. Forecasting for the possibility of landslide is accomplished by forecasting the pore-water pressure (PWP) 24 h ahead of time, at different locations and across soil layers under the ground at varying depths, and computing Factor of Safety (FoS) of the slope. It is done by learning from the real-time sensor data gathered from Amrita University's Wireless Sensor Network (WSN) system deployed in Western Ghats for monitoring and early warning of landslides. We use two variations of SVR, SVR-Historic and SVR-Adaptive. SVR-Historic algorithm is trained with the data from July 2011 to December 2015 and tested for the period from January to November 2016. SVR-Adaptive algorithm is adaptively trained from July-2011 onwards and tested for the period from January to November 2016. PWP and the computed FoS from both the algorithms are compared with the actual PWP and FoS data and the Mean Square Error (MSE) for the SVR-Historic model is found to be 48.726 and 0.002 whereas the MSE for SVR-Adaptive model is found to be 12.438 and 0.0007 respectively. The PWP and the computed FoS from both the algorithms are tested for correlation using Pearson's correlation test, with 95% confidence interval and the coefficients for PWP is found to be 0.804 and 0.959 respectively with *p*-value of 2.2e−16, whereas for FoS it is 0.802 and 0.955 with *p*-value of 2.2e−16. The confidence intervals for PWP and FoS from both the models is 0.763 to 0.839 and 0.950 to 0.969 respectively. Among the two forecasting models, SVR-Adaptive model performs better with a low MSE of 12.438 and 0.0007 in forecasting PWP and the computed FoS values respectively and correlates with the real-time data ~95% of the times. Application of this forecasting algorithm in real-world can thus provide 24 h extra time for early warning which is a boon for government and public to prepare for landslides after early warnings.

T. Hemalatha (✉) · M.V. Ramesh · V.P. Rangan
Amrita Center for Wireless Networks and Applications,
Amrita School of Engineering, Amrirapuri Campus,
Amrita Vishwa Vidyapeetham, Amrita University, Kollam,
690525, India
e-mail: hemalathat@am.amrita.edu

M.V. Ramesh
e-mail: maneesha@amrita.edu

V.P. Rangan
e-mail: venkat@amrita.edu

M. Mikoš et al. (eds.), *Advancing Culture of Living with Landslides*,
DOI 10.1007/978-3-319-53483-1_52

Keywords
Learning techniques • Support vector regression • Forecasting methods • Early warning system

Introduction

In the pursuit of forecasting potential landslides and other natural disasters, government and other research organizations all over the world have gathered humungous amount of data by virtue of different technology such as remote sensing, synthetic aperture radar, wireless sensor networks, etc. Even though a lot of forecasting systems are developed using this massive data gathered, providing sufficient early warnings for landslides is still a challenge. In the context of forecasting and early warning for landslides, the techniques used are rainfall threshold based methods (Gabet et al. 2004; Segoni et al. 2015), time series methods (Dore 2003; Loew et al. 2015), Electrical Resistive Tomography based methods (Dostál et al. 2014), Interferometric Synthetic Aperture Radar-based methods (Herrera et al. 2009; Bozzano et al. 2011), infinite slope stability model based methods (Crozier 1999; Chae et al. 2015), soil water index based methods etc (Brocca et al. 2012). Every method has its own advantages and limitations. Rainfall threshold based methods are well established and work well for a regional scale landslide alert and it has its own limitation for a site-specific alert. In the context of landslides, time series based methods are well established to forecast seasonal changes like monsoon precipitation and has its limitations if the data is discontinuous, non-stationary and if it has complex seasonal behavior. For instance, any real-time slope monitoring data will be discontinuous for several reasons and not seasonal necessarily. Also, most of the forecasting models are parametric in nature and makes strong assumptions about the data distribution and its underlying model (Schmidt et al. 2008). In a real world scenario, the data recorded from a real-world process usually does not follow any standard distribution and hence a distribution-free learning needs to be done in order to discover the underlying hidden behavior of the real world process. The learning technique needs to make very few or no assumptions about the data, which is the drawback of the current learning techniques. Hence, non-parametric methods, which makes fewer or no assumptions about the data are more conducive to discover the underlying hidden behavior of the environment. Therefore, in this work, adaptive non-parametric methods are used. SVR-Historic and SVR-Adaptive are the two models presented in this paper. SVR-Historic model learns the PWP changes in the soil from the historic data to forecast the future PWP. SVR-Adaptive model learns the changes happening in the PWP of the soil from the historic data and real-time data and forecast the future PWP of the slope. With the onset of monsoon rainfall, dynamic changes in rainfall rate will trigger variability in soil properties at different heterogeneous soil layers, infiltration rate, pore pressure at different depths, groundwater table, etc. This variability in the environment and slope is captured by various sensors and the data streaming in real time is used for learning the variabilities happening in real-time in the environment. Among all these sensed information, the information about PWP at different soil layers is considered vital for the following four reasons: (1) PWP change happens slowly over time, whereas the parameters like displacement changes and crack occurrence are observed within a fraction of seconds at a later time. There is always enough time to early warn when the threshold limits of piezometer sensors are overcome; (2) the slope accounts for a slide when the soil loses its cohesion and PWP is indirectly proportional to soil cohesion; (3) piezometers deployed at different soil layers sense the PWP on these layers, and this information can be used for understanding the pressure gradient under the soil and localizing the vulnerable layers and slip surfaces.; (4) PWP at different location serves as input for the FoS calculations along with other soil and slope parameters. Therefore, it is highly necessary to understand PWP behavior with respect to rainfall and forecasting its value helps in forecasting the FoS of the slope ahead of time. Forecasting PWP and FoS 24 h ahead provides 24 h extra time for the government and public to prepare for landslides.

Study Area and Amrita's WSN System

The study area is a small town 'Munnar' (10.0892 N, 77.0597 E) located in the landslide prone Western Ghats mountain region of Kerala in India. Rainfall varies from 3500–5500 mm every year with a maximum during the south-west monsoon months June, July & August respectively. The other monsoon is the North-east monsoon during the months of October, November and December. Idduki district, where Munnar is located is a landslide hotspot in the Western Ghats and every year landslides are very common during monsoon season, (Kuriakose et al. 2009; Lukose Kuriakose et al. 2010; Vijith and Madhu 2008).

Amrita University's WSN based landslide monitoring system is deployed in the town of Munnar (Ramesh 2014a, b; Ramesh and Vasudevan 2012), which is densely inhabited with shops, residents, schools and colleges. The study area has experienced two great historical landslides (Ramesh 2014a, b). The WSN based system is the world's first comprehensive landslide monitoring and early warning system of its kind, consisting of heterogeneous sensors distributed spatially at different locations and across varying soil layers at different depths (Ramesh 2014a, b; U.S. Patent No. 8,692,668 B2). The heterogeneous sensors include meteorological sensors like rain gauge, geological sensors like moisture sensor, piezometer, vibration sensors like geophone and movement sensors like strain gauge, and tilt meter. These sensors sense the vital parameters like precipitation rate, soil moisture, PWP, ground vibration, slope movements respectively, and wirelessly transmit the data in

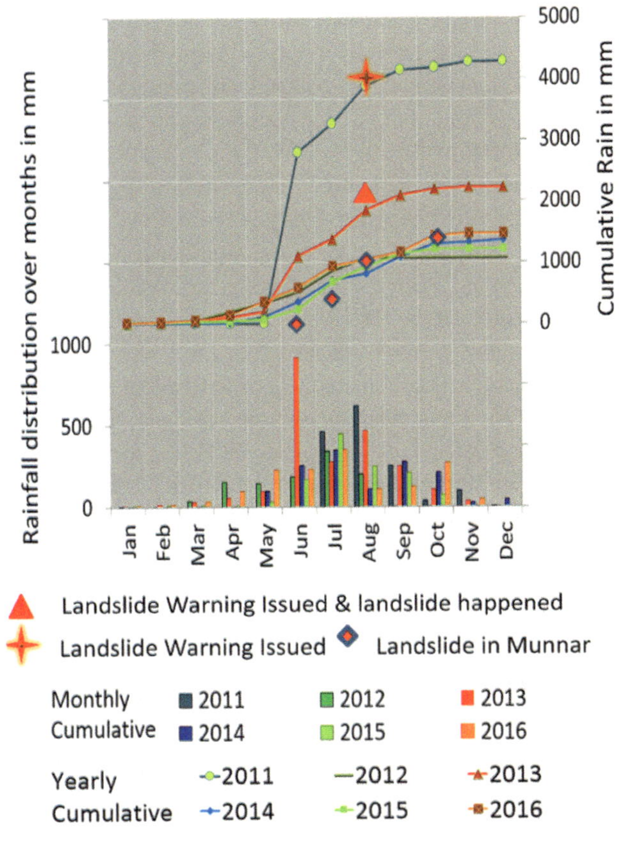

Fig. 1 Annual rainfall distribution and Cumulative rain over years from 2011 to 2016 in the deployment site Munnar. Legends are given for landslide warning issued and landslide happened in the deployment site. The other legend with 'Landslide in Munnar' corresponds to the landslides in the Munnar location far from the deployment site

real-time to the data management center in Amrita University. The deployment of the system started in 2006, and the data is been collected since 2009, the complete set of data for all sensors is available since 2011. Figure 1 shows the monthly rainfall distribution and cumulative rainfall distribution for the years 2011–2016, measured by the rain gauge in Munnar. The historical data and the real-time data are analyzed and early warnings are issued. The early warnings are issued at three levels, they are 'Early', 'Intermediate' and 'Imminent'. The forecasting techniques discussed in this paper, will provide 24 h extra time for issuing the Intermediate level warning. So far the system has given three warnings, and they are in July-2009, August-2011 and August-2013. The time of warnings are highlighted in Fig. 1. Intermediate level warning was issued in July-2009 and early level warning was issued in August-2011 and both of them were conditional warnings for landslides with a higher rainfall rate for the next two days. There was little or no rainfall for the next two days after warnings, and naturally the slope rendered to stable conditions slowly. In August-2013, when the rainfall intensity crossed the threshold (Caine 1980) the early level of warning for landslides was issued and landslides happened at several locations in and around Munnar. After the intermediate level warnings in August-2013, a landslide happened in the very near vicinity of 150 meters from the deployment site, which actually validated the successful working of our system. There was no death toll and no major economic loss happened during August-2013 landslide, since the public was prepared for landslides.

Real World Data and the Need for a Distribution-Free Learning

To understand and model the behavior of PWP over time, we tried finding out the statistical distribution pattern of the data. We used six years of data of piezometer sensors from 2011 to 2016, since the sample size of our data is relatively high, we were able to easily reject well-known standard distributions. To visualize the distribution of piezometer sensor data with respect to other standard distributions we plotted the square of skewness vs kurtosis of the observed piezometer sensor data and all standard distributions. This plot is popularly known as Cullen and Frey graph and is shown in Fig. 2. To account for the uncertainty of the estimated Kurtosis and Skewness of the PWP data 1000 bootstrap samples are drawn and compared with other standard distributions. The figure shows that the observed data do not follow any distribution. Therefore, in order to model the PWP behavior and variations under the soil, we preferred to use non-parametric methods, which do an assumption and distribution-free learning.

SVR for Assumption Free and Distribution-Free Learning

Support Vector Regression is a non-parametric machine learning algorithm, which had proved to provide excellent results in many benchmark datasets (Russell et al. 2003); (Soman et al. 2009). SVR algorithm makes no assumption on the data and learns from the training data provided to it. The training data is created in such a manner, that it contains the relation between rainfall and PWP. The SVR formulation models the underlying relation as a regression function in the terms of kernel function and few other tuned parameters. The SVR formulation is discussed in detail in the sections below.

Relation Between Rainfall and PWP

PWP variability at any layer of soil is predominantly because of two factors, they are the rainfall condition and ground water table. Rain is the primary known factor and ground water table is the secondary unknown factor. So we used rain conditions alone as the primary independent factor to determine the dependent factor PWP. Figure 3 shows the time series plot of the daily cumulative rain from July-2011 to November-2016 and the PWP variations from January-2011 to November-2016. Data from the piezometer at location 6 and 14 m (Ramesh and Vasudevan 2012) is used for analysis throughout this paper. Rain gauge deployment was completed

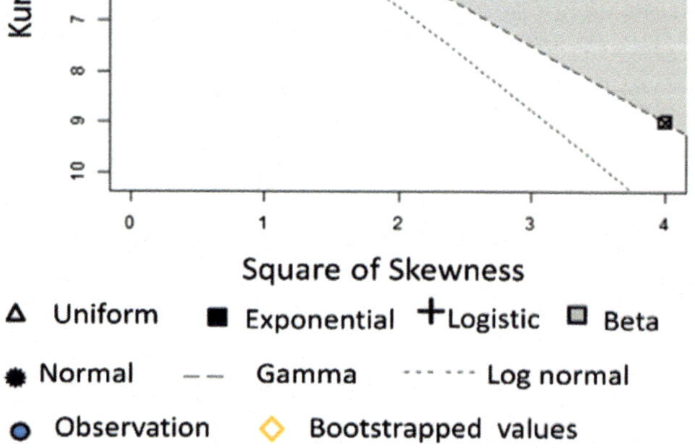

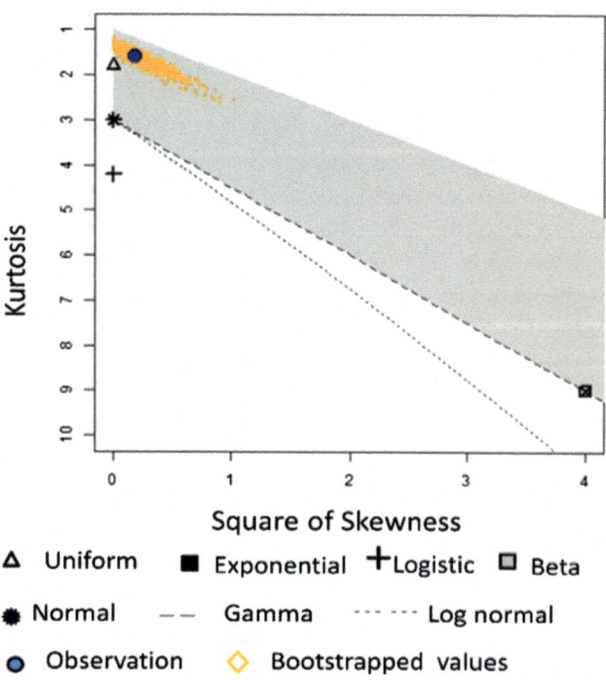

Fig. 2 Cullen and Frey graph to visualize the distribution of *PWP* data

in June-2011 and therefore we have the rain data from July-2011. From Fig. 3, it can be seen that both PWP variation and rainfall data has peaks and valleys. Peaks are nothing but the instances where rainfall data and PWP variation are higher. Valleys are the instances when there is no rainfall or minimum PWP. Peaks and valleys of both the data do not occur at the same instant of time. There is a time lag between the peaks and valleys of rainfall data and PWP variation. This clearly indicates that after receiving ample amount of rainfall, there is a time lag for the PWP to build up.

In other words, the PWP is dependent upon the antecedent rainfall. The PWP values are at the minimum during the months of March, April and May. April and May being the pre-monsoon period, the PWP continues to be the same. This is because of two reasons, (1) maximum temperature is noticed during these months and there is very little or no rainfall, from January to March; (2) Because of high temperature and no rainfall, there will be less water between soil pores. With the start of pre-monsoon rains, the water percolates the soil pores and there is enough space between the pores for the water to percolate. Once when the soil pores are completely filled with water, the soil reaches its saturation state and pressure builds up in the pores suddenly due to increasing rainfall rate from monsoon rains. After heavy rainfall in June, the pressure in the soil pores starts increasing and reaches a maximum by the end of July and August. With the reduction in rainfall rate the pore-water pressure also gets reduced slowly. The decline of PWP starts from October and continues until March. Another inference from Fig. 3 is that, there is very little or no rain during the months of December, January & February, but still there is significant PWP values. On the contrary, there are rain during the months of April, May and June, but the PWP values are significantly low. The above inference clearly indicates that, the PWP behavior is dependent on the previous months or antecedent rainfall. To forecast PWP, as a function of rain, the relation between antecedent rainfall conditions and PWP has to be learned initially. To accomplish that, we created linear models between different antecedent rainfall conditions AR_n and PWP. In AR_n, n refers to the 'number of days' of 'Antecedent Rainfall' AR. Antecedent rainfall conditions starting from 1 day to 180 days were calculated and linear models were created between different antecedent rainfall conditions. To evaluate the goodness of linear models, R-Squared correlation coefficient and Mean Square Error (MSE) are used as the metrics. The goodness of linear model results are shown in Table 1 and Fig. 4. From Table 1 and its corresponding plot in Fig. 4, it is found that 130 days of antecedent rainfall has a higher R-Squared correlation of 0.71 with less mean square error of 39.02. Hence it can be interpreted that, 130 days of antecedent rainfall AR_{130} can better explain the PWP at that instant. For creating SVR models we have used PWP and AR_{130}.

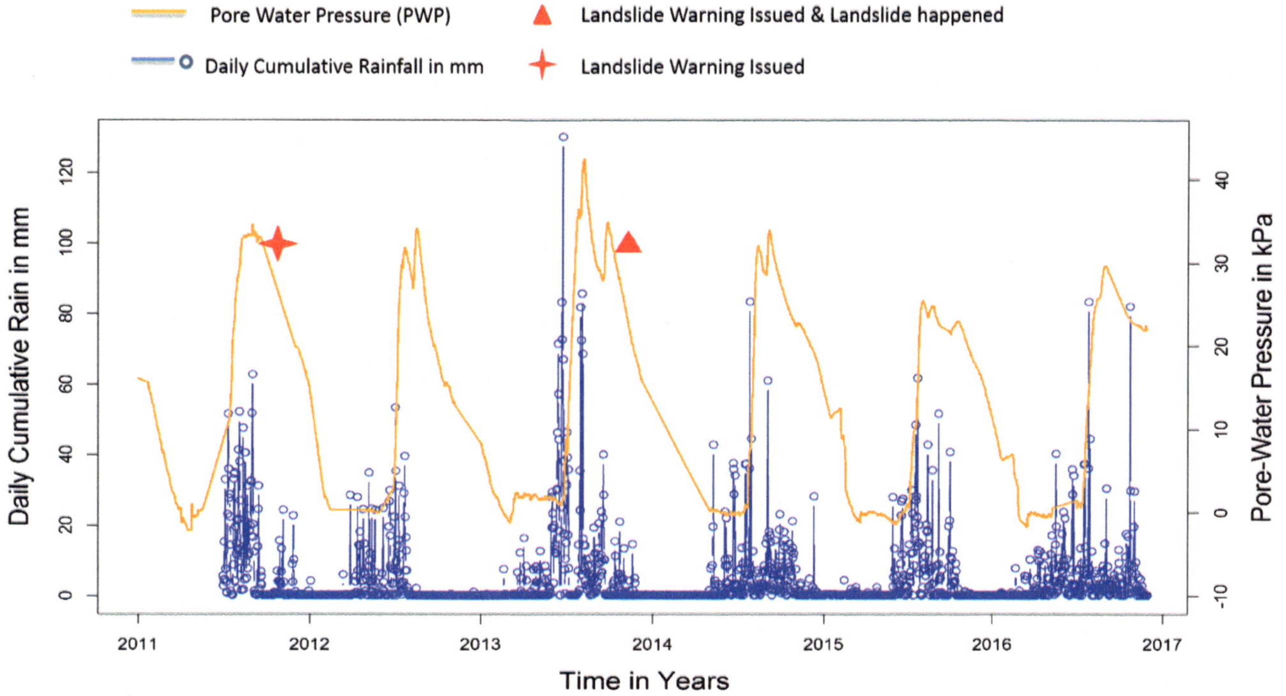

Fig. 3 Plot of Daily cumulative rainfall and PWP at a depth of 14 m in location L6 (Ramesh and Vasudevan 2012) from 2011 to 2016

Table 1 Goodness of linear model between AR_n and PWP results

R_n	R-Square	MSE	AR_n	R-Square	MSE
AR_1	0.011	131.62	AR_{90}	0.602	54.53
AR_5	0.028	129.65	AR_{100}	0.642	49.05
AR_{10}	0.055	126.30	AR_{110}	0.675	44.39
AR_{20}	0.117	118.50	AR_{120}	0.700	40.94
AR_{30}	0.196	108.38	AR_{130}	0.713	39.02
AR_{40}	0.281	97.35	AR_{140}	0.713	39.14
AR_{50}	0.364	86.45	AR_{150}	0.701	40.69
AR_{60}	0.441	76.37	AR_{160}	0.678	43.87
AR_{70}	0.507	67.53	AR_{170}	0.646	48.36

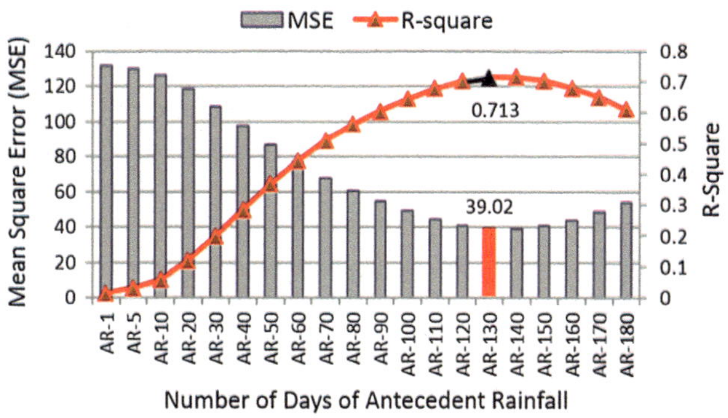

Fig. 4 Plot of the goodness of linear model results

SVR Formulation

C-SVR and Radial basis kernel function gives good results in modeling PWP variations compared to other SVR and kernel functions and so we have used the same for implementation in this paper. Radial basis function kernel for a multiple input training data is shown below.

$$\phi(x_i, x_j) = \exp\left(-\sigma||x_i - x_j||^2\right) \tag{1}$$

where σ is a positive parameter, x_i, x_j are the input of the training data. Radial basis kernel maps the training data to an infinite dimensional space, therefore even the complex functions in the original input dimensional space become simpler in infinite dimensional space. The formulation of C-SVR, with ε insensitive loss function, is given below

$$\begin{aligned} & Minimize \\ & \frac{1}{2}||w||^2 + C\sum_{i=1}^{l}\left(\xi_i + \xi_i^*\right) \\ & Subject\ to \\ & y_i - \langle w^T\phi(x_i)\rangle - \gamma \le \varepsilon + \xi_i \\ & \langle w^T\phi(x_i)\rangle + \gamma - y_i \le \varepsilon + \xi_i^* \\ & \xi_i^*, \xi_i \ge 0, \end{aligned} \quad (2)$$

where, w—weight vectors to be learned from the training data, y_i—output variable in the training data, i.e. PWP values that develop 24 h later, γ—scalar quantity, generally known as a bias term, $\phi(x_i)$—Radial basis kernel function of input training data, ε—small tolerable error, ξ_i and ξ_i^*—error values greater than ε, generally known as slack variables. The regression function learned from the above formulation is of the form

$$f(x) = w^T\phi(x) + \gamma. \quad (3)$$

A-insensitive loss function for an SV regression is that while learning the regression function $f(x)$ any error, less than ε from the actual target value (i.e. PWP) y_i is tolerated. Any error value greater than ε from the actual PWP value y_i is represented using slack variables ξ_i and ξ_i^*. In C-SVR, the amount of error that can be tolerated is decided while creating the model, and with respect to the model in this paper the value of ε is 0.01 kPa. The objective of the above formulation in [2] is to minimize $\frac{1}{2}||w||^2$ and the sum of errors ξ_i and ξ_i^* which deviates larger than ε. The role of $\frac{1}{2}||w||^2$ in the objective function is to achieve generalization and avoid the problem of over-fitting in the learned regression function $f(x)$. C is a constant, and the proper choice of C is essential for good generalization power. The parameter C and ε in C-SVR and the parameter σ of radial basis kernel is fine-tuned, to create a good regression model from C-SVR formulation. The fine-tuned values of C, ε and σ are 1, 0.01 and 10 respectively.

SVR Forecast Models from the Historical Data (SVR-Historic)

Two regression models are created from the SVR formulation discussed in the previous section, they are SVR-Historic and SVR-Adaptive. SVR-Historic model learns from the historic data, whereas the SVR-Adaptive model learns from both the historic data and the real-time streaming data. Both the models are discussed in the sub-sections below. Forecasting PWP 24 h ahead is accomplished by training the SVR with future PWP values. The SVR model is trained for AR_{130} antecedent rainfall conditions from the current real-time and the PWP that developed 24 h later. Therefore, to the learned model, if the current AR_{130} value is given, from the learned knowledge, the SVR model forecasts the PWP that is expected to develop 24 h later.

SVR—Historic

The SVR model is trained using the data from July-2011 to December-2015. Radial basis function kernel is generated from the input training data and the weight vectors and bias terms are learned by minimizing the error variables ξ_i and ξ_i^*. The regression function thus learned is given as $f(x) = 358.34\,\phi(x) + 0.454$, with w as 358.34 and γ as 0.454, and $\phi(x)$ is the kernel function computed out of the training data. To assess the quality of the trained result, 10 fold cross validation is done on the training data, which resulted in a total Mean Squared Error of 41.7 and r-squared correlation coefficient as 0.718. The trained model is then used to forecast the period 2016-January to 2016-November. Forecast results are shown in Fig. 5. Forecasted pore pressure values are compared with the actual pore pressure values, which arrives 24 h later from the forecasted time. The results of the same are discussed in the section Results Discussion.

SVR—Adaptive

Changes happening in the slope are not necessarily seasonal all the time, due to several factors like stabilization or unstabilization in certain regions of the slope, variation in ground water table, previous years wilting point, the amount of pre-monsoon and post-monsoon rainfall received etc. For instance, in Fig. 3, two peaks for PWP are noticed for the years, 2012–2015. Whereas for the year 2011 and 2016, there is only one peak noticed. Therefore, a model created from the historical data alone will not be sufficient to cater to the changes happening in the slope. Hence we modified the SVR formulation to adapt itself in real-time along with the streaming real-time data to efficiently forecast the future behavior of the slope. By adapting means, along with the historical data, the SVR algorithm learns the real-time data and updates the kernel function $\phi(x)$, weight w and γ in real-time, thereby the learned model has the knowledge from the historical data, and the current state of the environment from the real-time data. This knowledge helps in improving the forecast accuracy. Using the historical data from July-2011 to December-2015, and real-time data from January-2016 onwards, the period of January-2016 to

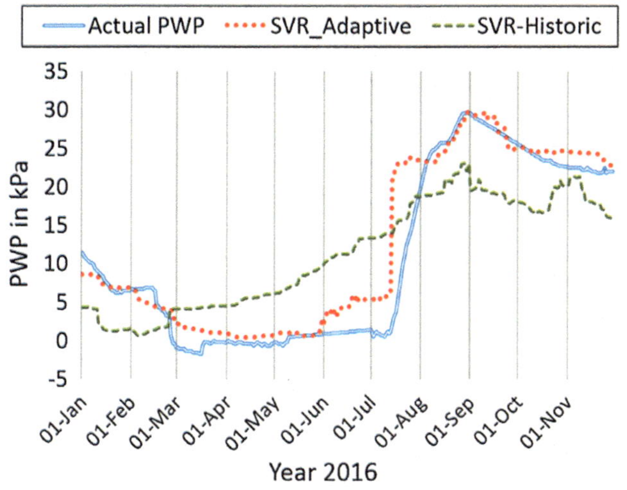

Fig. 5 Plot of actual *PWP* values and the forecasted *PWP* values

November-2016 is forecasted. The forecast results are shown in Fig. 5 and the results are discussed 'Results Discussion' section.

Factor of Safety (FoS) and Amrita's Early Warning System (EWS)

The landslide mechanism in Munnar mainly depends on the rainfall duration, rainfall intensity, pore-water pressure distribution underneath the soil, ground water table and soil properties. After extensive study of different slope stability models, we chose Iverson's model (Iverson 2000), for computing the FoS. We have used PWP values instead of the simulated pressure values in Iverson's model to calculate the FoS of the slope. FoS of the slope forms one of the input for our Early warning System (EWS). Our EWS consists of three levels of warnings, Early, Intermediate and Imminent. 'Early level' of early-warning' is issued when the rainfall intensity crosses a threshold (Caine 1980). Since the early level warning is based on rainfall threshold, it's a regional level warning, and it is applicable for Munnar area wide. 'Intermediate' level warning is issued, when the Factor of Safety (FoS) value of the slope is less than 1. 'Immediate' level of early warnings are issued when significant displacements are noticed from movement sensor or when crack occurrences are observed in Geophones. Intermediate and Immediate level warning is a site specific warning and is applicable to the deployment site and the areas in the near vicinity of the deployment site. After the deployment of our system, the people wait for the Intermediate warning and then vacate themselves and their belongings. There is limited time for the people to vacate after the Intermediate level warnings. The forecasted PWP and FoS values gives 24 h extra time for issuing the Intermediate level warning. Actual

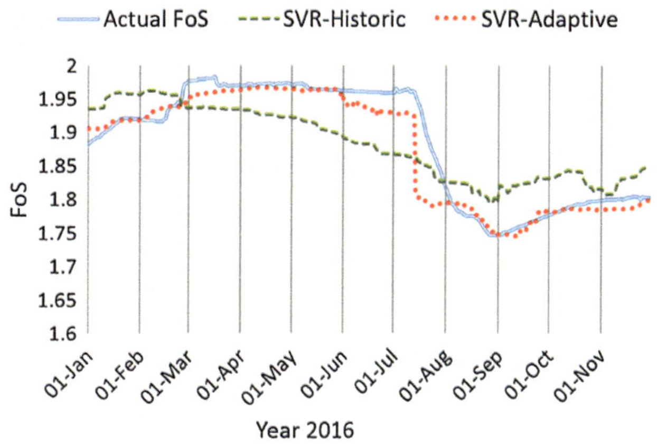

Fig. 6 Plot of actual FoS values and the forecasted FoS values

FoS values and the forecasted values are shown in Fig. 6. Results are discussed in the next section.

Results Discussion

Forecasted PWP, FoS results from the SVR-Historic and SVR-Adaptive models are compared with the actual PWP, FoS values. The results are shown in Tables 2 and 3. We have used MSE and Pearson's correlation test as the metrics to compare the forecasted values with the actual values. From Table 2, for PWP values, it can be seen that, the SVR-Adaptive model has a lesser MSE of 12.438 compared to the MSE of SVR-Historic, which is 48.726. From Table 3, for FoS values, it can be seen that, the SVR-Adaptive model has a lesser MSE of 0.0007 compared to the MSE of SVR-Historic, which is 0.002.

Pearson's test for correlation coefficient is performed to compare the correlation between actual PWP, FoS values and forecasted PWP, FoS values from SVR-Historic and SVR-Adaptive models. From Table 2, for PWP values, the correlation coefficient for SVR-Adaptive has a higher value of 0.959 with a relatively smaller confidence interval of 0.950–0.969 when compared to the SVR-Historic which has a correlation coefficient of 0.804 and a confidence interval of 0.763–0.839. Similarly from Table 3, for FoS values, the correlation coefficient for SVR-Adaptive has a higher value of 0.955 with a relatively smaller confidence interval of 0.950–0.969 when compared to the SVR-Historic which has a correlation coefficient of 0.804 and a confidence interval of 0.763–0.839. For both PWP and FoS values, SVR-Adaptive model has a higher correlation coefficient and lesser MSE, from which we can interpret that SVR-Adaptive model performs better than the SVR-Historic model. The null hypothesis of the Pearson's correlation test states that, "True correlation does not exist between the actual PWP, FoS values and the forecasted PWP, FoS values". For the PWP

Table 2 Comparison of forecasted *PWP* values from SVR-Historic and SVR-Adaptive with actual *PWP* values

Model compared with real-time PWP data	MSE	Correlation coefficient	*p*-value	95% Confidence Interval
SVR-Historic	48.726	0.804	2.2e −16	0.763, 0.839
SVR-Adaptive	12.438	0.959	2.2e −16	0.950, 0.969

Table 3 Comparison of FoS values computed from the forecasted *PWP* with actual FoS values

Model compared with real-time FoS data	MSE	Correlation coefficient	*p*-value	95% Confidence Interval
SVR-Historic	0.002	0.804	2.2e −16	0.763, 0.839
SVR- Adaptive	0.0007	0.955	2.2e −16	0.950, 0.969

and FoS forecast from both the models the *p*-value is found to be 2.2e−16 which is very low, therefore the null hypothesis of true correlation equal to zero is rejected in both the models. From Fig. 5 it can be seen that the SVR-Historic model forecasts the two peaks that was observed in the historic data, whereas the SVR-adaptive model adapts to the changes in real time and the peak was linearized. Figures 5 and 6 shows that, SVR-Historic model gives a generalized forecast and SVR-Adaptive model learns from the real-time data and adapts to the changes happening in real-time.

Conclusion and Future Work

Rainfall distribution and cumulative rainfall for the years 2011-2016 are shown in Fig. 1, along with the landslide warnings and landslide incidences in the deployment site and in other Munnar locations far from the deployment site. In this paper, we have shown from Fig. 2 that the PWP data does not follow any standard distribution, so we chose a nonparametric method 'Support Vector Regression' to learn and forecast the PWP data 24 h ahead of time. From Fig. 3, it can be clearly understood that the PWP build up is due to antecedent rainfall conditions. To understand the relation between AR_n and PWP, linear models are created and found that AR_{130} can better explain PWP variations and training data is created using the same. Two models SVR-Historic and SVR-Adaptive are presented in this paper. SVR-Historic learns and forecasts from the historical data alone. SVR-Adaptive learns from the historical data and the real-time data and forecasts the PWP in real-time. The kernel function $\varphi(x)$, weight w and γ also changes in real-time, thereby adapting to the changes happening in real-time. SVR-Historic model is trained from July-2011 to December-2015 and tested for January-2016 to November-2016. SVR-Adaptive model is adaptively trained from July-2011 and tested for the period from January-2016 to November-2016. From the MSE values and correlation coefficient values of both the models for forecasting PWP and FoS, we can conclude that SVR-Adaptive model performs better than the SVR-Historic model. PWP and FoS forecast results from the SVR-Adaptive model correlates the actual PWP and FoS values approximately 95% of the times. So this model can be used to know PWP and FoS values with 95% accuracy 24 h ahead, which will help the government and public with extra time for landslide preparedness. As a future work we would like to explore other nonparametric methods and compare them with SVR-Adaptive method and also perform probabilistic forecasts, and include weather forecast for improving the forecasting accuracy.

Acknowledgements The authors would like to express their immense gratitude to Satguru Sri Mata Amritanandamayi Devi, the chancellor of Amrita University for her constant support and guidance in all the research activities. This work is partly funded by Ministry of Earth Sciences (MoES), Government of India under the project titled "Advancing Integrated Wireless Sensor Networks for Real-time monitoring and detection of Disasters" and partly funded by Amrita University. We wish to express our gratitude to Prof Balaji Hariharan and Ramesh Guntha for their valuable suggestions.

References

Bozzano F, Cipriani I, Mazzanti P, Prestininzi A (2011) Displacement patterns of a landslide affected by human activities: insights from ground-based InSAR monitoring. Nat Hazards 59(3):1377–1396

Brocca L, Ponziani F, Melone F, Moramarco T, Berni N, Wagner W (2012) Improving landslide movement forecasting using ASCAT-derived soil moisture data. In: EGU general assembly conference abstracts, vol 14, p 2307

Caine N (1980) The rainfall intensity: duration control of shallow landslides and debris flows. Geografiska Ann Ser A Phys Geography, 23–27

Chae BG, Lee JH, Park HJ, Choi J (2015) A method for predicting the factor of safety of an infinite slope based on the depth ratio of the wetting front induced by rainfall infiltration. Nat Hazards Earth Sys Sci 15(8):1835–1849

Crozier MJ (1999) Prediction of rainfall-triggered landslides: A test of the antecedent water status model. Earth Surf Proc Land 24(9):825–833

Dore MH (2003) Forecasting the conditional probabilities of natural disasters in Canada as a guide for disaster preparedness. Nat Hazards 28(2–3):249–269

Dostál I, Putiška R, Kušnirák D (2014) Determination of shear surface of landslides using electrical resistivity tomography. Contrib Geophys Geodesy 44(2):133–147

Gabet EJ, Burbank DW, Putkonen JK, Pratt-Sitaula BA, Ojha T (2004) Rainfall thresholds for landsliding in the Himalayas of Nepal. Geomorphol 63(3):131–143

Herrera G, Fernández-Merodo JA, Mulas J, Pastor M, Luzi G, Monserrat O (2009) A landslide forecasting model using ground based SAR data: The Portalet case study. Eng Geol 105(3):220–230

Iverson RM (2000) Landslide triggering by rain infiltration. Water Resour Res 36(7):1897–1910

Kuriakose SL, Sankar G, Muraleedharan C (2009) History of landslide susceptibility and a chorology of landslide-prone areas in the Western Ghats of Kerala. India Environ Geol 57(7):1553–1568

Loew S, Gschwind S, Gischig V, Keller-Signer A, Valenti G (2015). Monitoring and early warning of the 2012 Preonzo catastrophic rockslope failure. Landslides, 1–14

Lukose Kuriakose S, Sankar G, Muraleedharan C (2010) Landslide fatalities in the Western Ghats of Kerala, India. In: EGU general assembly conference abstracts, vol 12, p 8645

Ramesh MV (2014a) Design, development, and deployment of a wireless sensor network for detection of landslides. Ad Hoc Netw 13:2–18

Ramesh MV (2014) U.S. Patent No. 8,692,668. U.S. Patent and Trademark Office, Washington, DC

Ramesh MV, Vasudevan N (2012) The deployment of deep-earth sensor probes for landslide detection. Landslides 9(4):457–474

Russell SJ, Norvig P, Canny JF, Malik JM, Edwards DD (2003) Artificial intelligence: a modern approach, vol 2. Prentice hall, Upper Saddle River

Schmidt J, Turek G, Clark MP, Uddstrom M, Dymond JR (2008) Probabilistic forecasting of shallow, rainfall-triggered landslides using real-time numerical weather predictions. Nat Hazards Earth Sys Sci 8(2):349–357

Segoni S, Battistini A, Rossi G, Rosi A, Lagomarsino D, Catani F, Moretti S, Casagli N (2015) Technical note: an operational landslide early warning system at regional scale based on space–time-variable rainfall thresholds. Nat Hazards Earth Sys Sci, 15(4):853–861

Soman KP, Loganathan R, Ajay V (2009) Machine learning with SVM and other kernel methods. PHI Learning Pvt. Ltd, 477. ISBN:978-81-203-3435-9

Vijith H, Madhu G (2008) Estimating potential landslide sites of an upland sub-watershed in Western Ghat's of Kerala (India) through frequency ratio and GIS. Environ Geol 55(7):1397–1405

Influence of Mixture Composition in the Collapse of Soil Columns

Lorenzo Brezzi, Fabio Gabrieli, Simonetta Cola, and Isabella Onofrio

Abstract

The collapse and consequent spreading of a column of granular or cohesive material is a simple experiment used by many research groups to study the rheology of the soils and for calibrating numerical propagation models. This paper deals with the results of a comprehensive experimental program carried out with mixtures of sand, kaolin and water: the main aim of the program is to know and understand how the mixture composition influences the collapse and run-out mechanism. In particular, the run-out length and the profile of the final deposit are the two fundamental characteristics taken into consideration to distinguish each test and to find a relation with the mixture composition. Four percentages of kaolin and water are considered for the experiments and different amounts of sand are added to these matrices. The main aim is the comprehension of the role of the coarser granular material in a cohesive collapsing mass. Finally, the dependency of the final runout on the aspect ratio of the initial column is discussed.

Keywords

Slump test • Collapse • Kaolin • Granular mixtures

Introduction

The prediction of the runout length of flow-like landslides is relevant in the context of hazard assessment and mitigation. Forecasting the possible scenarios may help the public subjects to rank the warning situations and to adopt the most appropriate solutions.

Several numerical models are available to evaluate the mobility of the soil masses and their possible evolution during the propagation phase. Some of them are purely empirical (Corominas 1996; Scheidl et al. 2013) or analytical (Hungr et al. 1984; Rickenmann 2005); others, like SPH (Pastor et al. 2009) or MPM (Soga et al. 2016), are numerical methods developed with particular attention to the treatment of large deformations typical of these problems.

The numerical models generally consider the soil masses like an equivalent fluid described by a specific rheological model (Bingham, Voellmy, etc.) (Körner 1976). The calibration of mechanical parameters specific for adopted model results the most complex and critical phase of the forecasting process since a wrong estimation of the parameters may provide not realistic scenarios (Iverson 1997).

Typically, the calibration is conducted by back-analysis of previous events (Brezzi et al. 2015; Pirulli and Sorbino 2008). More rarely the parameters are estimated on the base of tests with viscometer (Major and Pearson 1992; Phillips 1988), flume tests (Iverson 2015) or collapse tests (Lagrée et al. 2011). Generally, digital elevation models or

L. Brezzi (✉) · F. Gabrieli · S. Cola · I. Onofrio
Department ICEA, University of Padua, Via Ognissanti 39, 35129 Padua, Italy
e-mail: lorenzo.brezzi@dicea.unipd.it

F. Gabrieli
e-mail: fabio.gabrieli@unipd.it

S. Cola
e-mail: simonetta.cola@unipd.it

I. Onofrio
e-mail: isabella.onofrio@studenti.unipd.it

M. Mikoš et al. (eds.), *Advancing Culture of Living with Landslides*,
DOI 10.1007/978-3-319-53483-1_53

geological surveys are used to obtain the topographical and geometrical data of the past events. It is a very rare case to have kinematic data from real cases. The difficulties related to find out some kinematic measurements of real flow-like landslides involve the most common use of data related to the shape of the final collapse.

Two main features, i.e. the grain size composition and the water content, control the mechanical behavior of the involved material and, consequently, the triggering of phenomenon and its evolution along the collapsing path upon its deposit. A higher value of water content fluidizes the material moving its viscosity to the water-like values. Instead, a larger amount of coarser particles in the mixture moves its rheology from cohesive to frictional (Major and Pearson 1992).

In order to understand the effect of different percentages of the constitutive materials on the rheological properties, some binary mixtures composed by kaolin, water and sand were studied performing several column collapse tests.

The work of Lajeunesse et al. (2004) and Lube et al. (2005) with granular materials inspired these tests. Other authors considered wet granular materials (Gabrieli et al. 2013; Artoni et al. 2013) while we used slump tests of a cohesive matrix. As stated by Lube et al. (2004), the semplicity of the test conditions allows to vary the initial parameters of the system, investigating how the final deposit is directly dependent on the initial geometry of the mass and on the material composition.

Fig. 1 Layout of the collapse test apparatus

Experimental Strategy

Tests Procedure

Collapse tests were performed using a transparent Perspex cylinder (internal diameter D = 5.8 cm). The cylinder was initially filled with the specific matrix up to reach the prescribed height. A thin layer of Vaseline was used to seal the bottom border of the cylinder and to avoid the leak of material. After that, the cylinder was uplifted by means of a wire and a wheel-pulley system loaded with a weight (Fig. 1). A high-speed camera having a frontal view of the collapsing sample captured several frames of the process.

Lateral profiles of the deposits extracted from the final frame of the video were used for an estimation of the collapsed volumes according to the hypothesis of axial-symmetric conditions. This assumption allows the calculation of volumes by computing the solid of revolution of the 2D profiles.

To this aim, the image distortions and rotations were firstly reduced by using some fixed optical targets mounted on the external frame.

Then the final shape of the deposit profile was extracted using an in-house Matlab code (Fig. 2). We used three values of color thresholds in RGB components to separate the background (a black backdrop) and the foreground objects (lighter column of material) in the images. Once we had calibrated the right segmentation thresholds, we extracted the binary masks from all the final frames of each video and we recorded the shape of the deposits.

Finally, the coordinates of perimeter were scaled and converted from pixels to meters.

The material preparation was conducted as follow: the components were initially weighted and inserted in different containers. Then, they were mixed ensuring the same blending time to be used. After that, the mixture was poured in the cylinder, up to reach the prescribed filling height. To verify the effective amount of material used in the collapse test, the weight of the mass was measured before and after having filled the Plexiglas cylinder.

To ensure that the lifting velocity of the cylinder remains almost constant in all the tests, also the load on the wheel-pulley system and the drop height were controlled. The weight has been raised 20 cm above the level at which the rope is under tension. In that way, the falling mass was able to transfer to the cylinder an initial lifting velocity, which was almost the same in all the tests. This condition ensured the impulsivity of the collapse tests.

Tests Summary

Three types of tests were performed. Initially the repeatability of the tests was investigated to verify the reliability of the data obtained. The mixture used is the WK2 (Table 1) and four repetitions of the same collapse test were carried out. In this

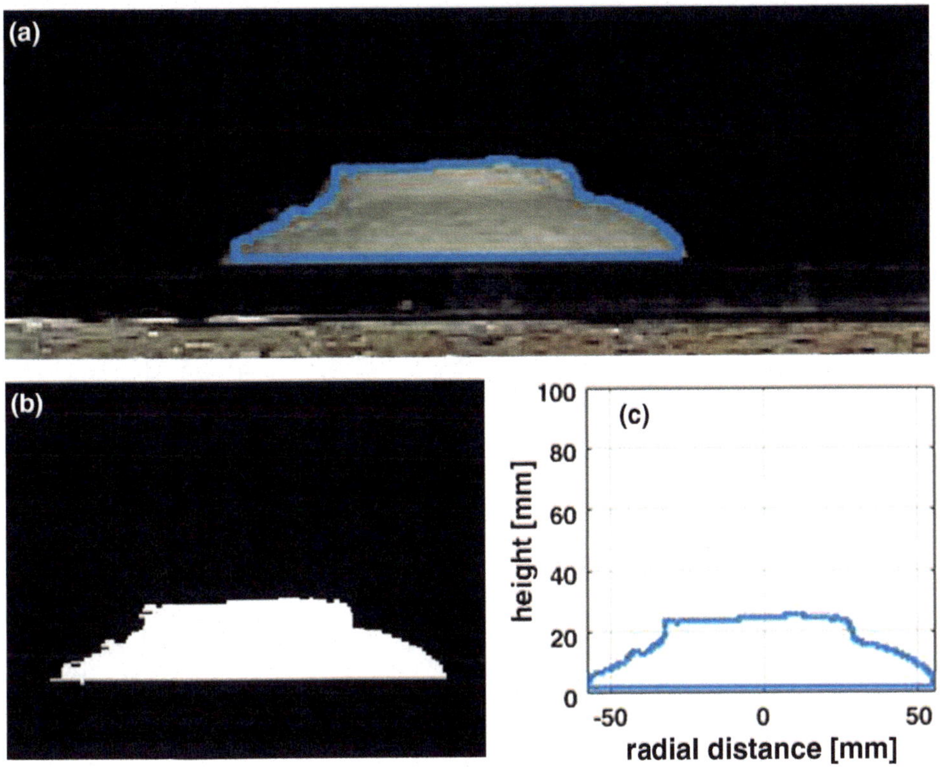

Fig. 2 Example of image processing steps for the final frame: **a** color threshold segmentation, **b** extraction of the binary mask with seed, **c** coordinates units conversion

Table 1 List of investigated mixtures

Basal mixture	Kaolin content (%)	Density (kg/m^3)	Sand addition (%)
WK_1	56.1	1455.1	10–40
WK_2	53.3	1422.8	10–50
WK_3	50.5	1391.9	10–50
WK_4	47.7	1362.3	10–60

phase, some minor details of the test procedure were verified and improved to obtain a high repeatability of the results.

Secondly, some tests are repeated with the same matrix to analyze the thixotropic hardening effect. These investigations mainly focused on the effects of the elapsed time (resting time) between the mixing and the collapsing phases.

Finally, 20 tests were performed with 4 different water-kaolin matrices and adding a various amount of sand, quantified by volume (Table 1). The percentage of granular material included in the collapsing mass ranged between 10 and 60%, depending on the workability of the basal kaolin-water matrix.

Experimental Collapsing Tests

Test Repeatability

The critical issues considered are related to the material preparation, the duration of each preparation phase, the lifting phase, and the measurement phase.

With regard to the material preparation, we observed that the degree of homogeneity decreases with the sand content: the reason is related to the difficulties in mixing a sandy sample with a sticky water-kaolin matrix. On the other hand, for liquid-like matrices, the sand tends to sediment at the bottom of the column and the sample decrease its homogeneity with time. For these reason the initial condition must be checked and, particularly, the preparation time has to be maintained fixed.

Even the lifting phase plays an important role in the collapsing behavior of the specimen: the higher the initial velocity of the cylinder, the lower the uncertainty of the test. In order to reduce the variability of the measurement phase, the camera and the lamps should be fixed to the rigid frame of the collapse test device.

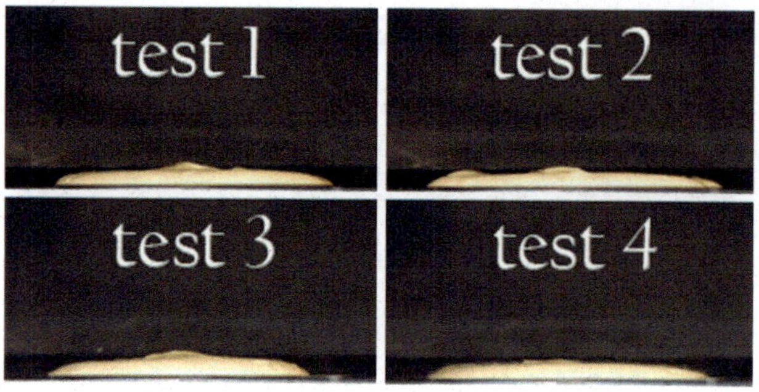

Fig. 3 Final frames of the repeatability tests

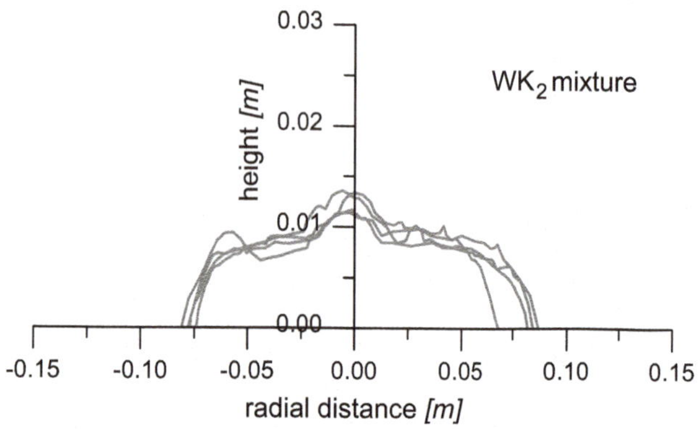

Fig. 4 Final profiles for the repeatability test

To measure the degree of repeatability we compared the final profiles, the runout values and the stopping time of four tests conducted on the WK2 mixture in the same conditions. In Fig. 3 the final frames of each test are reported.

Extracting and overlying the scaled final profiles it is possible to observe the level of variability, which appears quite low (Fig. 4).

Thixotropic Hardening Effect

Many cohesive materials exhibit an increase of viscosity, stiffness and strength with time, which is called thixotropic hardening effect and that should not be confused with the effect of the evaporation of the liquid phase. This phenomenon is particularly evident in high plasticity clays (montmorillonite and illite) with water content close to the liquid limit (Seng and Tanaka 2012) and, thanks to this property, slurries of these clays are widely used in many geotechnical applications, such as the stabilization of excavation faces in low-cohesion soils. The mixtures of kaolin and water here used exhibit this behavior also, which becomes evident after few minutes of resting time of the mixture and reflects on the kinematics of the collapse tests and on the final profile of deposits.

In order to study this effect, 3 mixtures were prepared with the same contents of kaolin and water. The first sample was tested immediately after the mixing phase, while the second and the third after 30 min and 1 h respectively.

In Fig. 5, the final deposits are compared. It can be noted that the highest viscosity of the aged mixture strongly affects the height and the final runout length of the profiles. Figure 6 shows the final runouts of the three experiments (0.18, 0.17 and 0.14 m, respectively) plotted respect the delay between the preparation of the mixture and the execution of the test, evidencing the thixotropic hardening effect.

The time-dependent behavior of the mixture affects also the duration of each collapse test (i.e. the time between the beginning of the lifting phase and the stop of the collapsed mass). For the first sample, the duration extracted from the video frames was 0.243 s; the second mixture (30 min of resting time) takes 0.256 s to stop its movement, while, in the last test (60 min of resting time) the mass moved slower and requested 0.276 s to complete the collapse. From the analysis of the movies, it is possible to determine also the mean horizontal velocity of the front of the mass. After a first phase of acceleration, in fact, the velocity reaches a steady state value and then starts to decrease, gradually decelerating to zero with the final deposition. As reported in Fig. 6, the mean velocities of these tests were 0.75, 0.70 and 0.49 m/s, respectively.

Another observed phenomenon is a partial separation of the solid phase from water occurring in the time between the preparation of the mixture and the test: as the solid grains moves towards the base of the recipient a thin layer of water formed on the top. As result, as soon as the test starts, the water runs out from the cylinder, while the denser cohesive mixture shows a more viscous response.

These tests underline how the material preparation plays a fundamental role in the results of laboratory tests with slurry mixtures and how it is very important to operate with a

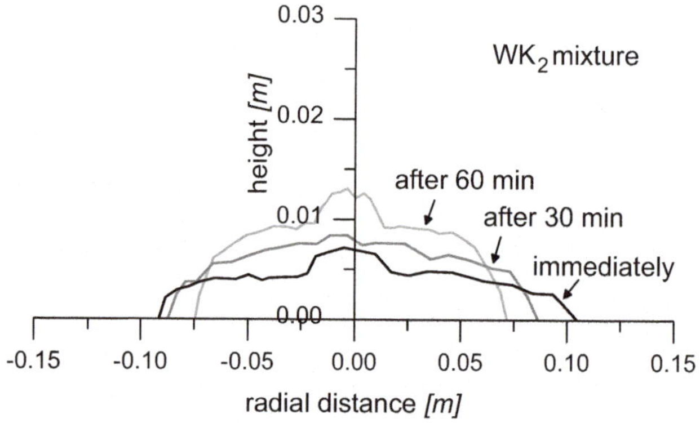

Fig. 5 Final profiles for different resting time

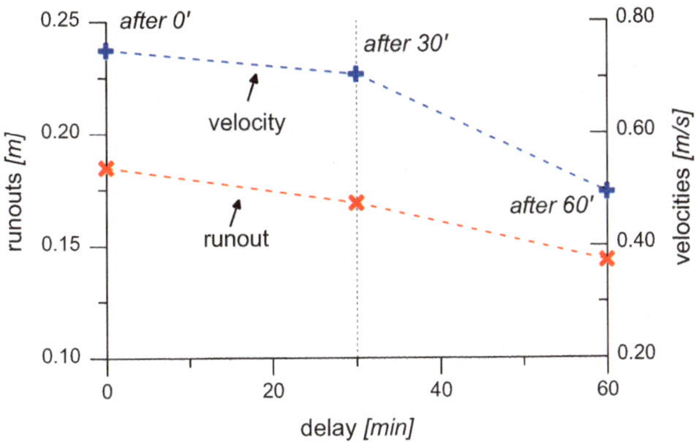

Fig. 6 Runouts and velocities variation due to thixotropic hardening effects

standard procedure in order to record experimental results comparable among them.

Effect of the Basal Matrix for Constant Sand Content

Figure 7 collected the final runout R_f of all the performed tests in function of the sand content of the mixture. The final runout R_f is normalized by the initial radius R_i to obtain a relative measurement.

It is evident that adding sand to a cohesive matrix without altering the water content of the matrix, the mixture results more stiff and the runout distance decrease. On the other hand, the effect is not regular, i.e. it is not possible to find a unique trend in the variation of runout distance with the sand amount.

In order to better exam the experimental results two different approaches could be used. If we intersect the four dashed curves with a horizontal line, it is possible to compare how the same runout may be obtained working with different materials. Conversely, when vertical lines are considered, we are keeping a constant content of sand and we investigate how the inclusion of the same granular contents influenced the behavior of different cohesive matrices.

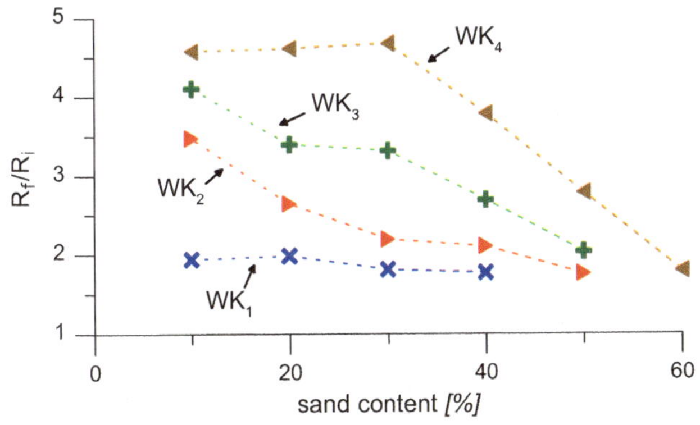

Fig. 7 Runout values for different kaolin and sand content

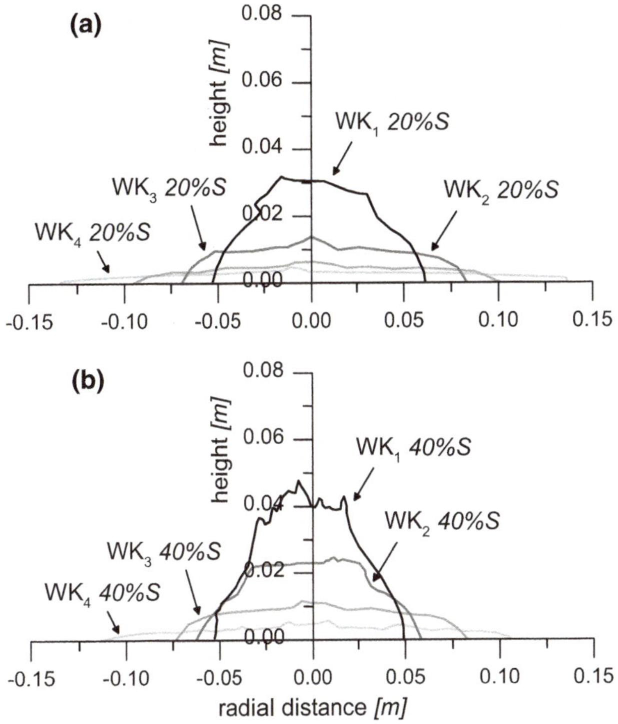

Fig. 8 Final profiles with different fine matrix water content and **a** a 20% or **b** a 40% of sand

Adopting the second approach, two sets of tests are extracted: a first group with a sand quantity equal to 40%, the second with 20% of sand content.

In Fig. 8a, b the final profiles of these tests are reported for comparison. It is evident that the deposits of specimens are strongly dependent on the basal mixture (water-kaolin) composing the mass.

When the decrease of viscosity due to a reduction of the finer material is not compensated by an increase of the granular one, in fact, the runout significantly changes.

Different Matrices with the Same Runout

Typically, in real cases, the final runout values are used for back-analysis and calibration of rheological parameters. However, as it can be noted in Fig. 7, different mixtures may provide similar values of runout and similar shapes of deposits.

To highlight this feature, we can consider in Fig. 7 tests showing the similar runout values (as previously said, this could be obtained drawing a horizontal line intersecting at least three curves in the plot). For instance, we report in Fig. 9 the tests giving a runout around 0.78 m, corresponding to a runout ratio R_f/R_i = 2.7. It is interesting to

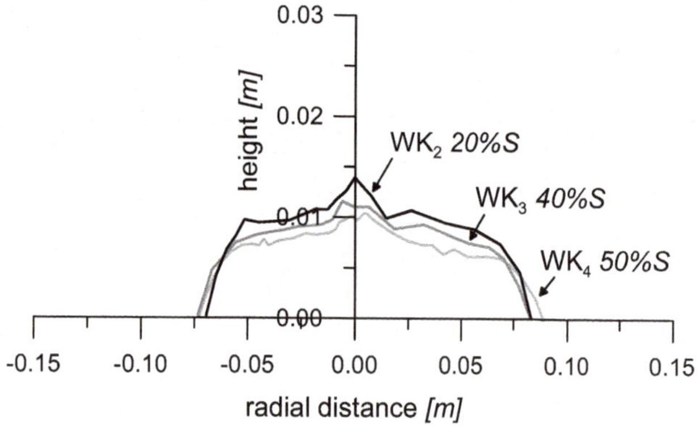

Fig. 9 Final profiles of collapse tests with similar runout values but with different mixture compositions

notice how a small decrease in the kaolin content has to be compensated by a more large increase of sand to reach the same runout value. Passing from WK_2 to Wk_4, in fact, the kaolin content goes from 33.5 to 28.7% while the sand content should more than double, from 20 to 50%.

Effect of the Initial Aspect Ratio

As in the works of Lajeunesse et al. (2004) and Lube et al. (2005), we considered the effect of different aspect ratio (namely $a = H_i/R_i$) on the final runout. Both the authors considered several granular materials and they observed two different trends in the diagram R_f/R_i vs a, with a threshold for $a = 0.74$. The trends are ruled by the equivalence between the volume of the initial cylinder shape and the volume of the final shape assuming a maximum slope equal to the material friction angle.

In the graph of Fig. 10, looking at the rescaled radius R_f/R_i, the two trends and the experimental data by

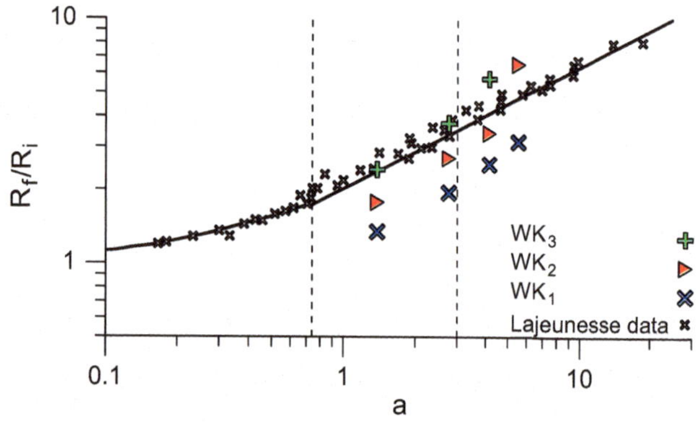

Fig. 10 Rescaled radius R_f/R_i of the deposits as a function of the mass aspect ratio a

Lajeunesse et al. are reported. For $a < 0.74$ the final deposits have the shape of a truncated cone of height H_f/H_i and of angle close to the repose angle of the beads. Using the mass conservation, Lajeunesse et al. obtained the relationship:

$$a < 0.74 \quad \frac{R_f}{R_i} = \frac{1}{2\mu_r}\left(a + \sqrt{4\mu_r^2 - \frac{a^2}{3}}\right) \qquad (1)$$

When $a > 0.74$, a conical volume can roughly approximate the deposits, so the relationship between the rescaled radius and the aspect ratio becomes:

$$a > 0.74 \quad \frac{R_f}{R_i} = \sqrt{\frac{3a}{0.74}} \qquad (2)$$

It is important to underline that these results are achieved working with granular material. In the same figure, the results obtained working with our cohesive materials are also included. All the experiments have been performed with $a > 0.74$. An increase of the normalized runout value with the aspect ratio is observed for all the matrixes with a trend that is similar to the trend found for granular materials. However, the data do not follow a single curve.

The distance between the curve proposed by Lajeunesse et al. and our data may be viewed as the degree of similarity of the final shape of our deposits with a regular cone. In our case, the initial column collapses becoming an irregular taller cylinder, bulged at the bottom.

The final shape depends mainly on the composition of the mixture: the WK1 material is very dense and a cone cannot approximate the collapsed profile (see for instance the final profile of tests WK1 in Fig. 8a, b). At higher water content, the final deposit becomes more regular, without a horizontal plateau on the top.

Conclusions

The composition of the debris and muds is a key aspect in the evaluation of their mobility and in the calibration of propagation models for hazard assessment. This aspect was investigated in laboratory with reference to simple collapse tests with binary mixtures.

The finer material (kaolin) mixed with water constitutes the basal matrix and mainly controls the kinematic of the flow and the final profiles of the deposits. The sand content plays a minor role in comparison to the kaolin content and its effect is less important for dense and viscous basal mixture (high kaolin content). The final shape of deposit and the runout are not sufficient to distinguish the initial composition of the binary matrix and especially of the sand content. In this case, the analysis of kinematic data probably may help in recognizing the differences.

However, determining the sand content of the deposit and the final shape and the runout of the deposits, in the hypotheses of no material entrainment, no drainage and homogeneity of the mass, it may be possible to estimate the water content of the matrix by a back-analysis of the propagation phase.

References

Artoni R, Santomaso AC, Gabrieli F, Tono D, Cola S (2013) Collapse of quasi-two-dimensional wet granular columns. Phys Rev E 87 (3):032205

Brezzi L, Bossi G, Gabrieli F, Marcato G, Pastor M, Cola S (2015) A new data assimilation procedure to develop a debris flow run-out model. *Landslides*, 1–14. doi:10.1007/s10346-015-0625-y

Corominas J (1996) The angle of reach as a mobility index for small and large landslides. Can Geotech J 33(2):260–271

Gabrieli F, Artoni R, Santomaso A, Cola S (2013) Discrete particle simulations and experiments on the collapse of wet granular columns. Phys Fluids 25(10):103303

Hungr O, Morgan GC, Kellerhals R (1984) Quantitative analysis of debris torrent hazards for design of remedial measures. Can Geotech J 21(4):663–677

Iverson RM (1997) The physics of debris flows. Rev Geophys 35(3):245–296

Iverson RM (2015) Scaling and design of landslide and debris-flow experiments. Geomorphology 244:9–20

Körner HJ (1976) Reichweite und Geschwindigkeit von Bergstürzen und Fließschneelawinen. Rock Mech 8(4):225–256

Lagrée PY, Staron L, Popinet S (2011) The granular column collapse as a continuum: validity of a two-dimensional Navier-Stokes model with a μ (i)-rheology. J Fluid Mech 686:378–408

Lajeunesse E, Mangeney-Castelnau A, Vilotte JP (2004) Spreading of a granular mass on a horizontal plane. Phys Fluids 16(7):2371–2381

Lube G, Huppert HE, Sparks RSJ, Hallworth MA (2004) Axisymmetric collapses of granular columns. J Fluid Mech 508:175–199

Lube G, Huppert HE, Sparks RSJ, Freundt A (2005) Collapses of two-dimensional granular columns. Phys Rev E 72(4):041301

Major JJ, Pierson TC (1992) Debris flow rheology: Experimental analysis of fine-grained slurries. Water Resour Res 28(3):841–857

Pastor M, Haddad B, Sorbino G, Cuomo S, Drempetic V (2009) A depth-integrated, coupled SPH model for flow-like landslides and related phenomena. Int J Numer Anal Met Geomech 33(2):143–172

Phillips CJ (1988) Rheological investigations of debris flow materials. *PhD Thesis*. Univ of Canterbury

Pirulli M, Sorbino G (2008) Assessing potential debris flow runout: a comparison of two simulation models. Nat Hazard Earth Sys Science 8:961–971

Rickenmann D (2005) Runout prediction methods. Debris-flow hazards and related phenomena. Springer, Berlin Heidelberg, pp 305–324

Scheidl C, Rickenmann D, McArdell BW (2013) Runout prediction of debris flows and similar mass movements. In: *Landslide Science and Practice*. Springer, Berlin Heidelberg, pp 221–229

Seng S, Tanaka H (2012) Properties of very soft clays: A study of thixotropic hardening and behavior under low consolidation pressure. Soils Found 52(2):335–345

Soga K, Alonso E, Yerro A, Kumar K, Bandara S (2016) Trends in large-deformation analysis of landslide mass movements with particular emphasis on the material point method. Géotechnique 66(3):248–273

New Thoughts for Impact Force Estimation on Flexible Barriers

Daoyuan Tan, Jianhua Yin, Jieqiong Qin, and Zhuohui Zhu

Abstract

Flexible barriers have received increasing attention in debris flow control because they are more economical and easier to install when compared with rigid barriers. However, in the design of a flexible barrier, the debris impact force is difficult to estimate, even if sophisticated numerical analysis is employed. In this paper, suggestions for simplified impact force estimations are given. At first, the existing approaches to estimate the impact force for impermeable rigid barriers are modified to cater for the case of a flexible barrier. We consider that there are two key characteristics of flexible barriers when compared with rigid barriers: flexibility and permeability. Flexibility exemplifies itself in a longer duration of impact. A simple spring-mass system is used to represent the interaction of the debris flow and barrier and observed impact times are considered. It is deduced that the impact force on a flexible barrier should be less than half of that for a rigid barrier, both being impacted by the same/similar debris flow. Furthermore, for a ring net which is impacted by a debris flow of substantial mass and velocity, it is considered that the impact load is proportional to the elastic deformation of the flexible barrier in the direction of flow. Impact force calculated using the preceding assertion has been compared with the impact force in published results, and a satisfactory comparison is found. Large-scale experiments are proposed so that the validity of the above assertions can be ascertained. Permeability, the other key characteristic of a flexible barrier, can also influence the impact force as less force will be imposed on the barrier if less debris mass is retained by the barrier. Large-scale experiments are also proposed to investigate the relationship between the barrier net opening size and the debris impact force. Besides, existing approaches for estimating debris

D. Tan · J. Yin (✉) · J. Qin · Z. Zhu
Department of Civil and Environmental Engineering, The Hong Kong Polytechnic University, Hung Hom, Kowloon, Hong Kong, China
e-mail: cejhyin@polyu.edu.hk

D. Tan
e-mail: 14900443r@connect.polyu.hk

J. Qin
e-mail: jieqiong.qin@connect.polyu.hk

Z. Zhu
e-mail: zhuo-hui.zhu@connect.polyu.hk

M. Mikoš et al. (eds.), *Advancing Culture of Living with Landslides*,
DOI 10.1007/978-3-319-53483-1_54

flow loading on impermeable rigid barriers are reviewed and improved by introducing a drag force which can impede the impact force. Then the largest force combination during the impact process cannot be simply determined as the largest dynamic loading or the largest earth pressure loading, and it can only be decided by calculating the largest force of all three stages.

Keywords
Debris flow • Impact force • Flexible barrier • Flexibility • Permeability

Introduction

Based on experimental and theoretical researches, different models have been developed to estimate the debris flow impact forces on flexible barriers.

Yifru (2014) classified impact force models into hydraulic, solid body collision, shock wave theory collision, and pure empirical models, while Cui et al. (2015) improved the hydrodynamic model by introducing the Froude number. The hydraulic models can be further separated into hydrostatic, hydrodynamic and mixed models.

However, the interaction between debris flow and a flexible barrier is complex and is usually not considered in detail by the present impact force calculation methods. For example, the interaction between debris flow and barrier lasts from several seconds to more than one minute, and the loading on the barrier changes with the deposition of debris materials. These facts have been verified by experiments and videos of small and large scale experiments have been made (Ashwood and Hungr 2016; Vagnon and Segalini 2016; Wendeler 2016; Yu 1992; Kaiheng et al. 2006). Apparently, a single model cannot reflect the whole complex interaction process. In this paper, the barrier loadings in different stages of the debris flow impact are considered.

Unlike rigid barriers, which have been applied in debris flow mitigation for many years, flexible barriers were originally used for rock fall mitigation. Over the past years, flexible rockfall barriers have occasionally been hit by debris flows and landslides in Hong Kong and other countries and districts. It is known that some of the flexible barriers were able to stop and retain a certain scale of landslide debris (Segalini et al. 2014). Flexible barriers, compared with rigid barriers, have two key characteristics which enable them to resist/mitigate debris flows with more kinetic energy: flexibility and permeability. This paper provides directions for impact force estimation by considering both characteristics through the introduction of parameters which can be used to relate the impact force on a rigid barrier to that on a flexible barrier. A method to calculate impact loading using the deflection of a flexible barrier is also proposed.

Capitalizing on the Key Characteristics of Flexible Barriers

Many methods and equations have been propounded to estimate the impact force on rigid barriers, and both small and large-scale experiments have been conducted to verify those methods. Load cells can also be attached to the impact surface of a rigid barrier to measuring the impact pressures directly.

In this paper, parameters based on the key characteristics of flexible barriers (i.e. flexibility and permeability) as compared with rigid barriers are proposed. Then possible relationships between the impact loadings of rigid and flexible barriers can be investigated. The relationships, when found, can be used in the determination of impact loadings on a flexible barrier.

Flexibility

A simple spring-mass model is proposed to describe the interaction between debris flow and barriers. The model consists of a moving mass 'm', representing the debris flow, and an elastic contact spring 'k', which represents the elastic deformation and activated resisting the force of the structure, and the interaction process can be viewed as a mass with a certain velocity impacting a spring. Different factor 'k' of equivalent spring is applied to quantify the flexibility difference between rigid and flexible barriers under impact loading (Bischoff et al. 1990). In this interaction system, energy loss in the process of collision is ignored, and the simplified mechanical model is presented in Fig. 1. To the spring-mass system, the time of finishing the oscillation should be:

$$T_0 = 2\pi \cdot \sqrt{\frac{m}{k}} \tag{1}$$

The impact time is the time taken for the velocity of the mass to decrease to 0 units, which is a quarter of the period T_0. So the impact time is:

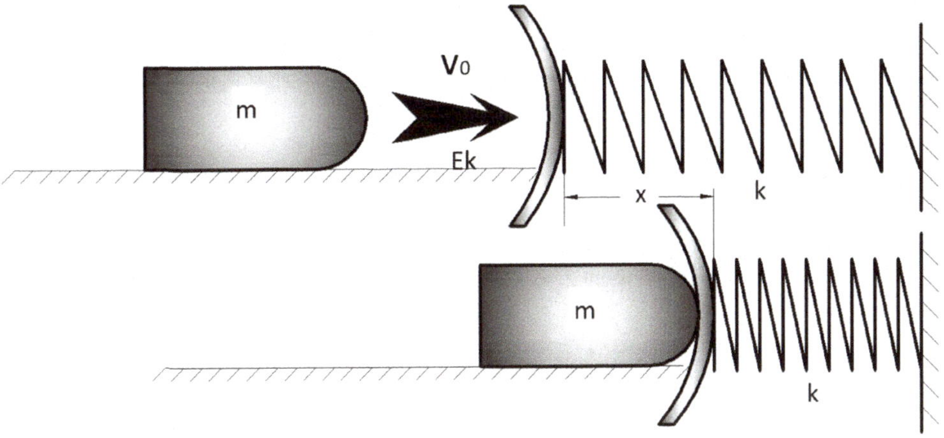

Fig. 1 Simplified interaction process between debris flow and barriers

$$t_{imp} = \frac{1}{4}T_0 = \frac{1}{2}\pi \cdot \sqrt{\frac{m}{k}} \tag{2}$$

In the impact period, the kinetic energy of the mass m is transformed into elastic potential energy:

$$\frac{1}{2}k \cdot x^2 = E_k, \text{ Hence, } x = \sqrt{\frac{E_k}{k}}, F_{max} = kx = \sqrt{k \cdot E_k} \tag{3}$$

Considering two types of barriers with different k and different debris retaining rate ω, relationship of their impact time and maximum impact force should be:

$$\frac{t_{imp1}}{t_{imp2}} = \frac{\frac{1}{2}\pi \cdot \sqrt{\frac{\omega \cdot m}{k_1}}}{\frac{1}{2}\pi \cdot \sqrt{\frac{m}{k_2}}} = \sqrt{\frac{\omega \cdot k_2}{k_1}}, \ E_k = \frac{1}{2} \cdot m \cdot v^2$$
$$\frac{F_{max1}}{F_{max2}} = \frac{\sqrt{k_1 \cdot E_k}}{\sqrt{k_2 \cdot E_k}} = \frac{\sqrt{k_1 \cdot \frac{1}{2}\omega \cdot m \cdot v^2}}{\sqrt{k_2 \cdot \frac{1}{2} \cdot m \cdot v^2}} = \sqrt{\frac{k_1 \cdot \omega}{k_2}} = \frac{t_{imp2}}{t_{imp1}} \cdot \omega \tag{4}$$

In the above equations, T_0 is the oscillation period of the spring-mass system; m is the mass of debris flow; k_1, k_2 are constant factors of springs representing the two types of barrier; t_{imp1}, t_{imp2} are the impact periods of the two types of barrier; x is the equivalent deformation of the spring-mass system when kinetic energy of debris flow was absorbed; E_k is the kinetic energy of the impacting mass; F_{max1}, F_{max2} are the largest impact forces of two types of barrier in the impact period; while, ω is the debris soil retaining rate of the flexible barrier considering its permeability, and the retaining rate of the rigid barrier is reagarded as 1.

The difference of impact time between flexible barriers and rigid barriers are 2∼4 s compared with 0.5 s from records of some experiments, when the impact of the first debris front is considered (Wendeler et al. 2006; Cui et al. 2015; Vagnon and Segalini 2016; DeNatale et al. 1999; Hu et al. 2006; Ishikawa et al. 2008; Wu et al. 2016). From the above equations, the impact force on a flexible barrier should be less than half of that on a rigid barrier under the impact of the same debris flow. The reduction factor can be better ascertained if more experiments can be conducted.

A Simplified Method to Estimate the Impact Force on a Flexible Barrier

A method which extended the elastic assumption was presented in this paper to estimate the impact force on the flexible barrier in the large-scale experiments. In this method, impact force on the flexible barrier is simplified into the 2D situation, which can be considered as a concentrated force at the center of the cable with both tied ends, and the simplified mechanical model can be seen from Fig. 2. In this calculation, the cable is assumed to be able to sustain tensile force only. Even though friction between rings and stiffness of steel wire may impart some bending stiffness to the ring net, the value is relatively small compared with its axial stiffness.

Both elastic strain and plastic strain occur in the impact process. Results of single ring's behavior tests and tensile loading test on ring nets showed that the stress-strain curve

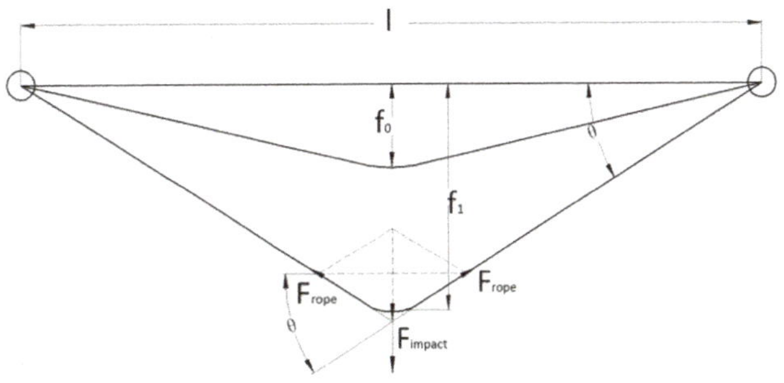

Fig. 2 Deformation f in the flow direction and force analysis of the cable representing the flexible barrier

of ropes is non-linear on initial loading (Denlinger and Iverson 2001). More specifically, conclusions made by this paper based on single ring tension tests presented that the ring first appears soft with a high plastic dissipation because of its deformation (from a circle to an oval). But after the initial stage, the apparent stiffness increases very quickly as the steel strands which make up the ring have high elastic modulus and yield strength.

Because of the length of impact time and the value of impact force, it is assumed that elastic strain dominates at the moment of largest impact force. The cable representing the flexible barrier performs as a cable with a constant elastic modulus (E) and cross-sectional area (A). Experiments can disclose the relationship between rope tensile force and impact force by measuring the largest deformation of cable in the moving direction of debris flow. The slack because of the initial loose state and deformation of rings should be acquired in advance and excluded in the equation because that deformation is not elastic. The equations are presented as follows:

The length of cable before the elastic deformation is:

$$s_0 = (1+\beta) \cdot l \tag{5}$$

The length of cable at the end of impact process is:

$$s_1 = 2\sqrt{f_1^2 + \left(\frac{l}{2}\right)^2} \tag{6}$$

According to the force situation of cable:

$$F_{rope} = A \cdot E \cdot \frac{s_1 - s_0}{s_0} \tag{7}$$

$$F_{impact} = 2F_{rope} \cdot \sin\theta = 4A \cdot E \cdot \frac{s_1 - s_0}{s_0} \cdot \frac{f_1}{s_1} \tag{8}$$

In the above equations, s_0, s_1 are lengths of the equivalent cable before and after the elastic deformation; α is the coefficient of initial slack of cable due to the loose state of the cable; f_1 is the deflection of the equivalent cable in the direction of impact; l is the span of the flexible barrier; F_{rope} is the tensile force of the cable; F_{impact} is the largest impact force in the moving direction of the debris flow; θ is the angle of rope at the moment of largest impact force.

Data from observation of WSL 2006 tests were acquired to determine the combination $A \cdot E$. In that test, $F_{rope} \approx 190\,\text{kN}$, $l = 10.4\,\text{m}$, $f_1 = 2.5\,\text{m}$ (Wendeler 2016). However, the value of initial slack coefficient was not measured. In that case, the initial slack coefficient β of rope before elastic deformation was assumed to be 0.033 based on the initial slack $l/30$ assumed by Wendeler (2016). Calculated from above data, $A \cdot E = 2388kN$. So the equation can be rewritten to:

$$F_{impact} = 2F_{rope} \cdot \sin\theta_1 = 9552(\text{kN}) \cdot \frac{s_1 - s_0}{s_0} \cdot \frac{f_1}{s_1} \tag{9}$$

This equation can be used to estimate the impact force on the flexible barrier by measuring its maximum deflection during the filling process.

Results of large scale tests were used to assess the reliability of this equation (DeNatale et al. 1999; Ashwood and Hungr 2016). In 1999, some large-scale experiments were conducted by USGS. Data of that tests were observed by Vagnon and Segalini (2016), which were: $l = 9.1m$, $f_1 = 1.5m$. However, the initial slack coefficient '$\beta = 0$' was assumed because that the net is pre-tensioned in this test. Based on those above data, the impact force calculated by equation (9) was 79 kN, which was similar to the predicted result of the numerical model proposed in that paper (74kN), since the impact force was not measured in the tests conducted by DeNatale et al. (1999). The above assessment proved that this simplified equation can predict reasonable impact force value by measuring deflections of flexible barrier during the process of impact, which is much easier to acquire.

However, more data is needed to calibrate the combination '$A \cdot E$'. Meanwhile, the initial slack coefficient β also needs consideration before calculation.

Permeability

Permeability is another key characteristic of a flexible barrier. Wendeler and Volkwein (2015) explored the relationship between d_{90} (= size of sieve through which 90% of the debris material will pass through) and the amount of retained debris material by conducting small-scale tests. They found that when the mesh size of flexible barrier is equal to or smaller than d_{90} of debris materials, almost all the debris materials are stopped by the barrier. It can also be found from photos of retained volumes that when the mesh size of the barrier is changed to 4 cm and 6 cm, only 80% and 60% volume respectively of the debris flow material was trapped. Considering these results, a parameter $d = d_{90}/d_{mesh}$ is applied to amend the impact force because smaller force will impose on flexible barriers if less debris materials are trapped. However, the relationship between the parameter d and the impact force needs more data from large-scale experiments.

Impact Force Calculation of Different Stages

First Thrust (Usually with Largest Dynamic Force)

$$F_1 = \alpha \cdot \rho \cdot v_f^2 \cdot A \tag{10}$$

It is widely recognized that the first impact of debris flow is usually the largest dynamic loading in the whole impact process. In Eq. (10), α is the coefficient for dynamic impact, ρ is the density of debris flow, v_f is the velocity of debris flow, and A is the area of impact.

The first impact can be explained as the impact on the barrier by a continuous homogeneous fluid with a certain volume and a uniform velocity. The dynamic pressure on the barrier is assumed to be uniformly distributed.

The empirical coefficient α depends on the fluid flow properties, gradation characteristics of the debris and the mechanical properties of the barrier. Values of the coefficient were estimated based on laboratory tests and field observations (Cui et al. 2015). However, a model that can be applied to all types of flow or some sort of debris flow is still lacking. Table 1 showed some values of the empirical coefficient in hydrodynamic models introduced by some researchers.

As showed in the table, the values of coefficient α range from 0.45 to 5 based on different debris flow measurements.

There is no doubt that the hope for a constant number to represent this coefficient is unrealistic. Cui et al. (2015) introduced the Froude Number which describes the kinetic component ratio between horizontal and vertical directions to be a component of the coefficient because the impact force of debris flow is mainly determined by the kinetic component in the horizontal direction. Finally, an equation for the dynamic pressure p based on 155 sets of data (based on field observations and miniaturized flume tests) has been presented:

$$\mathrm{p} = 5.3\mathrm{Fr}^{-1.5} \cdot \rho \cdot v_f^2 \tag{11}$$

However, the above equation has not considered the effects of flow properties such as viscosity, debris gradation characteristics, and the flexibility of barrier.

Deposition Process (with Largest Static Force)

$$F_2 = \frac{1}{2} \cdot \rho \cdot g \cdot K_a \cdot \left(H_{max}^2 - h_f^2\right) \cdot B + \alpha \cdot \rho \cdot v_f^2 \cdot A - \rho \cdot g \cdot h_f \cdot \left(\tan\varphi + \frac{v^2}{h \cdot \varepsilon}\right) \cdot \frac{\left(H_{max} - h_f\right)}{\sin\theta} \cdot \cos\theta \cdot B \tag{12}$$

At this stage, the flexible barrier has already retained a certain volume of debris, and both dynamic and static loadings are imposed on the flexible barrier simultaneously. Drag force from deposited debris materials weakened the kinetic energy of flowing debris. Under this circumstance, the largest loading in this stage appeared at the time when the last front impacted the barrier for the largest earth pressure. In Eq. (12), g is the gravitational acceleration; K_a is active lateral earth pressure coefficient; H_{max} is the final height of debris deposited behind the barrier; h_f is the depth of debris flow; B is the width of debris flow;φ and ε are the effective friction angle of deposited debris and turbulence coefficient; θ is the slope angle of deposited debris.

In the drag force estimation, Kwan and Cheung (2012) proposed a combination $(\tan\varphi + \frac{v^2}{h \cdot \varepsilon})$ to represent the coefficient of drag force. In this combination, the Voellmy fluid model originally for snow avalanches was applied, which contains both a friction term $\tan\varphi$ and a turbulence term $\frac{v^2}{h \cdot \varepsilon}$. Since this model was originally applied in snow avalanche

Table 1 Values of the empirical coefficient α

Source	Value of α	Description
Zhang (1993)	3–5	Estimated with the data measured in Jiangjia Ravine, China, characterized by viscous debris flow
Watanabe and Ikeya (1981)	2.0	Estimated with the data measured in Nojiri Ravine, Japan, characterized by volcanic debris flow (mudflow)
Hungr et al. (1984)	1.5	Back analysis of the stony type debris flows in British Columbia, Canada
Mizuyama (1979)	1.0	Derived from jet impact theory, applied in Yakedake, Nigorisawa Urakawa debris flows in Japan
Armanini (1997)	0.45–2.2	Estimated with the data measured in laboratory experiments, the material is a mixture of PVC and water with densities 1800–1300 kg/m^3
Kwan and Cheung (2012)	2.0	Consider the interaction characteristics of flexible barrier
Lo (2000)	3.0	A comprehensive collection of field measurements of debris impact pressure in China gives a range from 2 to 4

calculation, its applicability in debris flow needs further verification. Moreover, a modified Bingham model has been presented to describe the rheological behaviour of mud flow (Jeong 2014). In Bingham model, basal shear resistance (τ) can be calculated by the following equation (Yifru 2014):

$$\tau^3 + 3 \cdot \left(\frac{\tau_y}{2} + \frac{\mu_B}{2} \cdot v\right) \cdot \tau^2 - \frac{\tau_y}{2} = 0 \tag{13}$$

where τ is basal shear resistance; τ_y is the apparent yield stress; v is the velocity of debris flow; μ_B is the Bingham viscosity which is a constant after yielding.

Regarding viscosity measurement, de Campos and Galindo (2016) presented a simple and direct way to measure the Bingham viscosity (μ_B) of debris flows whose results can fit well with data in the literature.

Largest Drag Force (with Overflow)

$$F_3 = \frac{1}{2} \cdot \rho \cdot g \cdot K_a \cdot H_{barrier}^2 \cdot B + K_a \cdot \rho \cdot g \cdot h_f + \rho \cdot g \cdot h_f \cdot \tan \varphi_e \cdot \frac{H_{barrier}}{\sin \theta} \cdot \cos \theta \cdot B \tag{14}$$

At this stage, the flexible barrier has been filled by retained debris. The loadings consist of active earth pressure derived from both retained debris and surcharge pressure from the weight of the overflowing debris, and the drag force from the overflowing debris. $H_{barrier}$ is the height of the flexible barrier which may reduce in the impact process.

Conclusions

In this paper, flexibility and permeability are proposed to be the two key characteristics through which we can modify the impact force estimation methods which were originally developed for rigid barriers. Regarding flexibility, a simple spring-mass theory has been presented to compare impact force on flexible and rigid barriers by comparing their impact time difference. Besides, a simplified method to estimate the impact force on flexible barriers based on the deflection of the flexible barrier when absorbing the kinetic energy of debris flow is proposed. It is shown that the impact force value back calculated by this equation can match well with published results of numerical simulation. Permeability, the other key characteristic of a flexible barrier, can also influence the impact force as less force will be imposed on the barrier if less debris mass is retained by the barrier. Moreover, the impact process has been divided into 3 stages, and the maximum loading can be chosen from values calculated by three equations based on impact force estimation methods on rigid barriers. In the future, large-scale experiments are proposed so that the validity of the above assertions can be ascertained. Large-scale experiments are also proposed to investigate the relationship between the barrier net mesh size and the debris impact force.

Acknowledgements The authors acknowledge financial supports from Collaborative Research Fund (CRF) of Research Grants Council of Hong Kong Special Administrative Region Government of China ((PolyU12/CRF/13E, A/C: E-RB09) and supports from The Hong Kong Polytechnic University (A/C: 1-ZVEH). Finally, I would like to offer my special thanks to Dr. H.C. Mark Chan, who offered so much help in his valuable comments and suggestions to improve the quality of the paper.

References

Armanini A (1997) On the dynamic impact of debris flows. Recent developments on debris flows. Springer, Berlin

Ashwood W, Hungr O (2016) Estimating the total resisting force in a flexible barrier impacted by a granular avalanche using physical and numerical modeling. Can Geotech J

Bischoff PH, Perry SH, Eibl J (1990) Contact force calculations with a simple spring-mass model for hard impact: a case study using polystyrene aggregate concrete. Int J Impact Eng 9:317–325

Cui P, Zeng C, Lei Y (2015) Experimental analysis on the impact force of viscous debris flow. Earth Surf Process Land

de Campos T, Galindo M (2016) Evaluation of the viscosity of tropical soils for debris flow analysis: a new approach. Géotechnique 1–13

DeNatale JS, Iverson RM, Major JJ, Lahusen RG, Fiegel G, Duffy JD (1999) Experimental testing of flexible barriers for containment of debris flows. US Geol Surv Open File Rep 99(205):38

Denlinger RP, Iverson RM (2001) Flow of variably fluidized granular masses across three-dimensional terrain: 2. numerical predictions and experimental tests. J Geophys Res: Solid Earth 106:553–566

Hu K, Wei F, Hong Y, Li X (2006) Field measurement of impact force of debris flow. Yanshilixue Yu Gongcheng Xuebao/Chin J Rock Mech Eng 25:2813–2819

Hungr O, Morgan G, Kellerhals R (1984) Quantitative analysis of debris torrent hazards for design of remedial measures. Can Geotech J 21:663–677

IshikawaN, Inoue R, Hayashi K, Hasegawa Y, Mizuyama T (2008) Experimental approach on measurement of impulsive fluid force using debris flow model, na

Jeong SW (2014) The effect of grain size on the viscosity and yield stress of fine-grained sediments. J Mt Sci 11:31–40

Kaiheng H, Fangqiang W, Yong H, Xiaoyu L (2006) Field measurement of impact force of debris flow. Chin JRock Mech Eng 1

Kwan JSH, Cheung RWM (2012) Suggestions on design approaches for flexible debris-resisting barriers. Geo discussion note

Lo DOK (2000) Review of natural terrain landslide debris-resisting barrier design. GEO-Report No. 104, Geotechn Eng Off, Hong Kong Special Administrative Region

Mizuyama T (1979) Computational method and some considerations on impulsive force of debris flow acting on sabo dams. J Jpn Soc Erosion Control Eng 112:40–43

Segalini A, Brighenti R, Umili G (2014) A simplified analytical model for the design of flexible barriers against debris flows. Landslide science for a safer geoenvironment. Springer, Berlin

Vagnon F, Segalini A (2016) Debris flow impact estimation on a rigid barrier

Watanabe M, Ikeya H (1981) Investigation and analysis of volcanic mudflows on Mt. Sakurajima, Japan. Erosion and sediment transport measurement

Wendeler C (2016) Debris-flow protection systems for mountain torrents. In: Steffen K (ed) WSL Berichte. Swiss Federal Institute for Forest, Snow and Landscape Research WSL CH-8903 Birmensdorf, Swiss Federal Institute for Forest, Snow and Landscape Research WSL

Wendeler C, McArdell B, Rickenmann D, Volkwein A, Roth A, Denk M (2006) Field testing and numerical modeling of flexible debris flow barriers. Proc Int Conf on Phys Modell Geotech, Honkong

Wendeler C, Volkwein A (2015) Laboratory tests for the optimization of mesh size for flexible debris-flow barriers. Nat Hazards Earth Syst Sci 15:2597–2604

Wu F, Fan Y, Liang L, Wang C (2016) Numerical simulation of dry granular flow impacting a rigid wall using the discrete element method. PLoS ONE 11:e0160756

Yifru AL (2014) Assessment of rheological models for run-out distance modeling of sensitive clay slides, Focusing on Voellmy Rheology

Yu FC (1992) A study on the impact forces of debris flow. Proc. NSC, Part A: Phys Sci Eng 16, 7

Zhang S (1993) A comprehensive approach to the observation and prevention of debris flows in China. Nat Hazards 7:1–23

A Check-Dam to Measure Debris Flow-Structure Interactions in the Gadria Torrent

Georg Nagl and Johannes Hübl

Abstract

To design technical mitigation structure against debris flows in torrents it is important to define realistic design impact loads. Presented impact forces were mostly derived from back calculations of past events and yielded a rough estimation. The majority of scientific publications of impact force are based on small scale experiments in laboratories, but the transfer to real scale problems is limited due to scaling issues. Monitoring in real scale of debris flows is necessary to understand the process and the apparent phenomena. The Gadria valley in the Autonomous Province of Bozen-Bolzano is one of the rare areas were debris flows frequently occur. Therefore the Gadria torrent is already equipped with a monitoring station to provide data to analyze the occurring debris flows. To measure real scale impact forces of a debris flow and the variables that are necessary to calculate and understand the impact process and the debris flow/structure/ground interaction, a special monitoring check dam is designed and will be built in September 2016. An arrangement of 38 sensors will measure the impact forces on the check dam, the acceleration of the construction and the interaction with the ground. Also the pore water pressures, weights and heights of the debris flows will be recorded. This arrangement of sensors should help to understand the debris flow structure interaction to facilitate the calibration of numerical models and to improve guidelines for national standards.

Keywords

Debris flows • Monitoring • Impact force

Introduction

Alpine regions are exposed to different mass wasting processes, for example debris flows, landslides and rock fall. Debris flows are high mobile gravity driven mixtures of soil, rock and water. The high velocity and the ability to carry large boulders endanger human lives and infrastructures. For the design of mitigation measures a design impact pressure is required. Due to the enormous destructive power of impact forces of debris flows, realistic real scale data are seldom. Worldwide just a few investigations were made to quantify (Hu et al. 2011; Suwa et al. 1973; Zhang and Yuan 1984; Wendeler et al. 2007). From rough estimations, which were made by back-calculations of destroyed structures to small scale experiments in the laboratories, the process of impact mechanism of a debris flow is not fully understood yet (Hungr et al. 1984; Scheidl et al. 2013).

Monitoring plays an important role in debris flow research for data gathering to understand these hazardous natural processes. Many areas around the world have monitoring sites, for example in China (Zhang 1993), in Japan (Suwa et al. 2011), Italy (Marchi et al. 2002), and in Austria (Hübl et al. 2013). But there is none with a permanent and

G. Nagl (✉) · J. Hübl
University of Natural Resources and Life Sciences, Institute of Mountain Risk Engineering, Peter-Jordan-Straße 82, 1190 Vienna, Austria
e-mail: georg.nagl@boku.ac.at

J. Hübl
e-mail: johannes.huebl@boku.ac.at

M. Mikoš et al. (eds.), *Advancing Culture of Living with Landslides*,
DOI 10.1007/978-3-319-53483-1_55

continuous impact force measuring system. For that reason, an automated monitoring station is built in September 2016 in an already well-instrumented test site in the Gadria valley in the Autonomous Province of Bolzano in Italy, to observe various characteristics of debris flows and the mechanism of impacts of debris flows. Among the most commonly measured parameters like bulk velocity, flow depth and mean discharge, the monitoring system will observe the direct impact forces of the debris flow, the interaction of foundation/soil, basal normal and shear force and also the pore water pressure. This work aims to describe the new monitoring barrier for impact force analysis and the interaction of debris flow/barrier/foundation.

Study Site

The Gadria torrent is located in Autonomous Province of Bozen-Bolzano in Italy, in the Vinschgau valley. The catchment areas of 6.3 km^2 exhibit an altitude difference of 1551 m, from 2,945 m to 1,394 m a.s.l., see Table 1. The high frequencies of occurrence of debris flows are granted through the steep topography, high deformed metamorphic rock and thick glacio fluvial deposits. The already existing monitoring installation consists of rain gauges, radar sensors, geophones, video cameras, piezometers and soil moisture probes. At the end of the straight reach a retention basin is situated. A slit check dam ensures the passage of water and fine sediment of the retention basin (Fig. 1). A detailed description of the existing study site can be found in Comiti et al. (2014, Arattano et al. (2015).

Concept

To design active measures, like debris flow breakers, an engineering task is to find solutions and specifications for all necessary subjects. Starting from the estimation of the flow behavior, to the determination of design load and load distribution to the generated stresses inside the reinforced concrete structures to finally transmit the impact energy into the ground. At all stages big uncertainties are given. The basis for proper structural design is a comprehensive approach of all range of subjects. At all stages, specific parameters for description must be identified to choose the best instrumentation to illuminate the different connections, see Fig. 2.

Stage 1. Flow process parameter (Flow behavior):
Stony debris flows and liquid mudflows show different flow behavior, and exert different impact pressures. Some of these essential parameters for specifications are velocity, density and flow height. These parameters can easily be measured with common and often used instrumentation, like force plates, ultra-sonic measurer and video cameras (Berti et al. 2000).

Stage 2. Impact load parameter:
A difficult task faced by the engineer is the accurate estimation of the loads that might be applied by a debris flow impact. Depending on the used impact model, different parameters are needed. A rough classification can be made by hydraulic and solid collisions models. The impact force mechanism is influenced by the composition of a debris flow. The fluid phase attaches a surface pressure while the coarse particle exerts point loads (Zhang and Yuan 1984; Zhang 1993). Fluid pressure attains values up to 100 kN/m^2, the estimated grain impact loads reaches values an order of magnitude higher. For this reason, also the composition of the debris flow is needed to be observed via video analysis and with debris sampling. To assist the engineer in estimating the magnitudes of impact forces records have to be assembled.

Stage 3. Structure parameter:
For the unknown model the nonparametric identification is useful. The identification of the system is based on comparison of system response to a given load of a debris flow impact. These loads are the reason for structural damage due to deterioration. The damage generated by each load at a microscopic level cause failure by an accumulation of damage. This can be monitored by resistive strain gauges, which measure local material deformation.

Stage 4. Earth parameter:
A correct assessment of stresses in the soil is necessary for a founded structure. In high alpine regions it is often impossible to describe accurately the ground.

Table 1 Main parameters of the Gadria study site

Catchment area (km^2)	6.4
Minimum elevation (m) a.s.l.	1394
Maximum elevation (m) a.s.l	2945
Average slope (%)	79.1
Mean annual precipitation (mm)	480

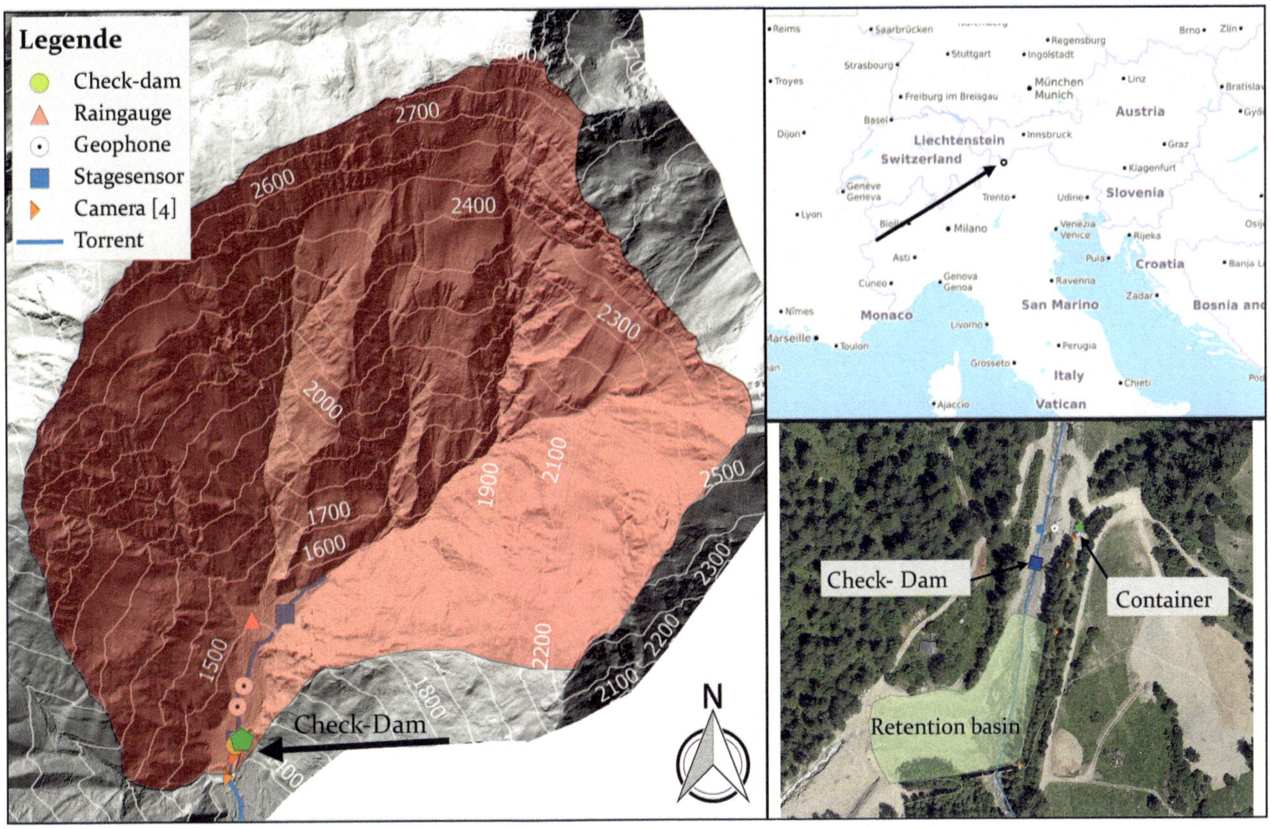

Fig. 1 Location map of the Gadria catchment

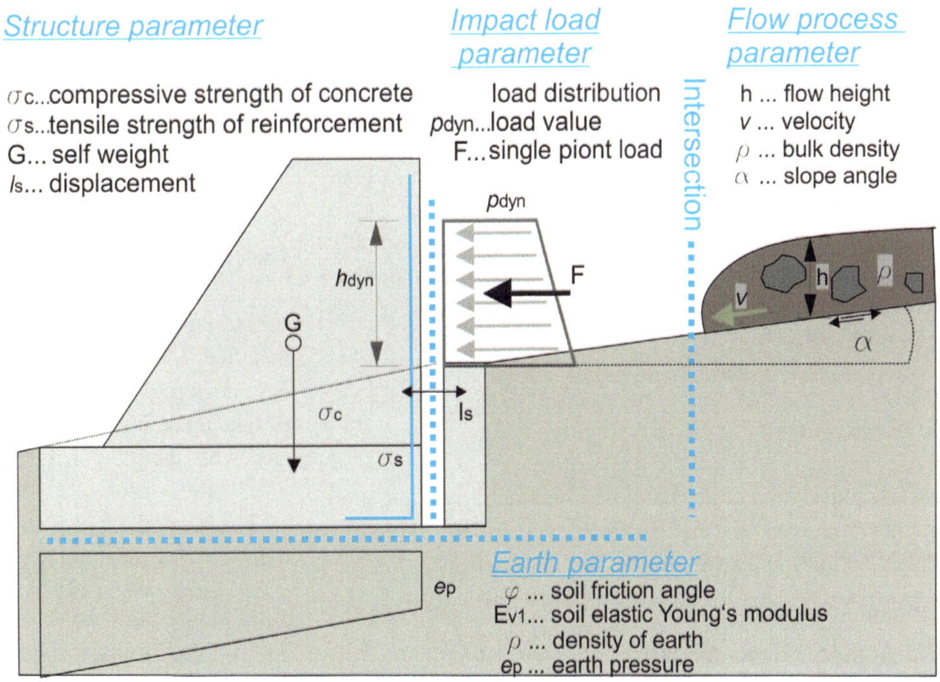

Fig. 2 Schematic concept of measurable parameters to of all stages

Table 2 Overview of the sensor equipment

Sensor	Number of	Range	Unit
Load cells	14	2000 kN	(kN)
Strain gauges	8		(N/mm^2)
Earth pressure sensor	9	400 kPa	(kPa)
Force plates	2	400 kN	(kN)
Pore-water pressure	2	2 bar	(bar)
3D-accelerometer	1	400 g	(m/s^2)
Ultra-sonic measurer	1	0–10 m	(m)
Displacement sensors	2	0–0.25 m	(m)
Video camera	1	25 Pic/sec	

The most reliable source of information for soil/structure interaction is usually obtained by field observations with instruments, e.g.: earth pressure cells with direct contact to the soil (Table 2).

Instrumentation of the Check-Dam

Check-Dam

The check-dam consists of a 1.4 m thick foundation with a length of 6 m and a width of 5 m. In the center of the foundation there is a single wall, 1 m thick and 4 m in height. Upstream a steel construction on the wall, equipped with load cells, faces the debris flow impact.

Load Cells

To measure the forces, 14 load cells with 1–2 MN load capacity are arranged alternatively on the front of the check-dam, see Fig. 3a. This allows imaging the spatial impact distribution. Each load cell is protected with a steel plate against the debris, and to measure the undamped impact energy. Each sensor can be easily replaced on suspicion of defect.

Strain Gauges

Stresses inside the concrete will be measured by eight strain gauges mounted on the reinforcements in two levels of height to track the stress inside the construction.

3D-Accelerations Sensor

With the acceleration and vibrations of the construction, which will be detected by a 3D-accelerometer, the interaction with the structure can be analyzed.

Displacement Sensors

Two displacement sensors are installed on two sides of the barrier to measure possible displacements.

Ultra-Sonic Measurer

To observe the flow height and to calculate the density, an ultra-sonic measurer is installed on the top of the structure.

Cameras

A camera system will observe the debris flow event and deliver an opportunity to verify the flow height and to estimate the bulk- and surface velocity and also to estimate the boulder size. The system is equipped with an infra-red spotlight to make night shots possible.

Force Plates with Pore Water Pressure Sensors

In the upstream transverse structures, 2 force plates are installed. Each device is mounted with four measuring pins. One device is in front of the structure to measure the weight of the impacting debris flow in order to determine the density with the flow height which will be measured by an ultra-sonic sensor.

The second force plate beside the check-dam is able to measure the shear in two directions and the normal basal stress as shown in Fig. 4. Both devices can also gauge the basal pore pressure with an installed pressure transducer, as can be seen in Fig. 5.

The pressure transducer is an adapted version of a previous type which Kaitna et al. (2016) used in laboratory experiments. The pressure transducer is mounted on an oil-filled reservoir. The hole facing the debris flow material

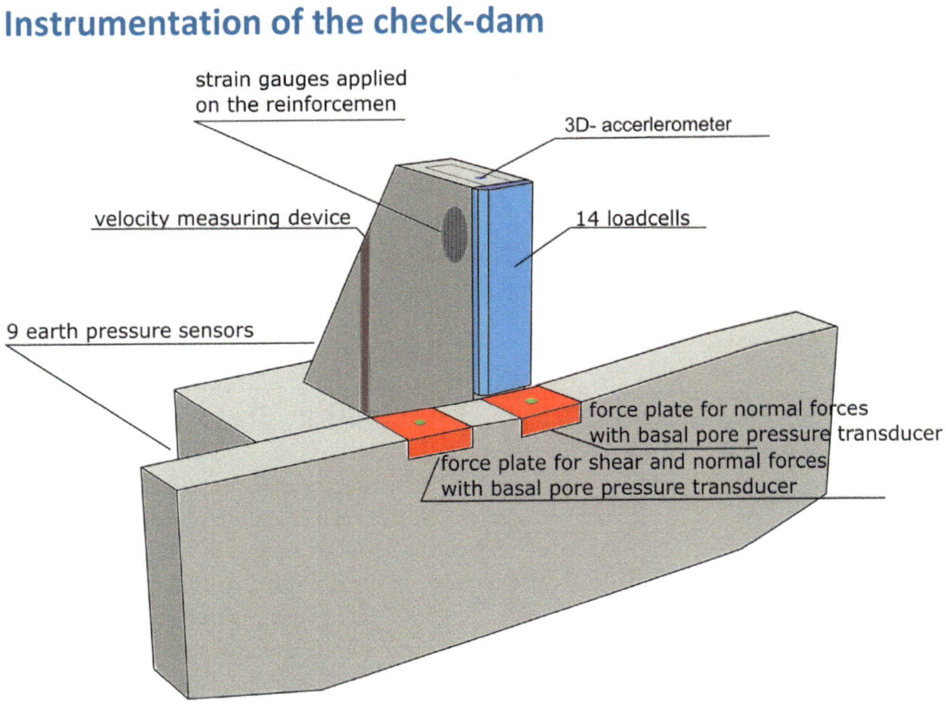

Fig. 3 Schematic concept of the monitoring structures

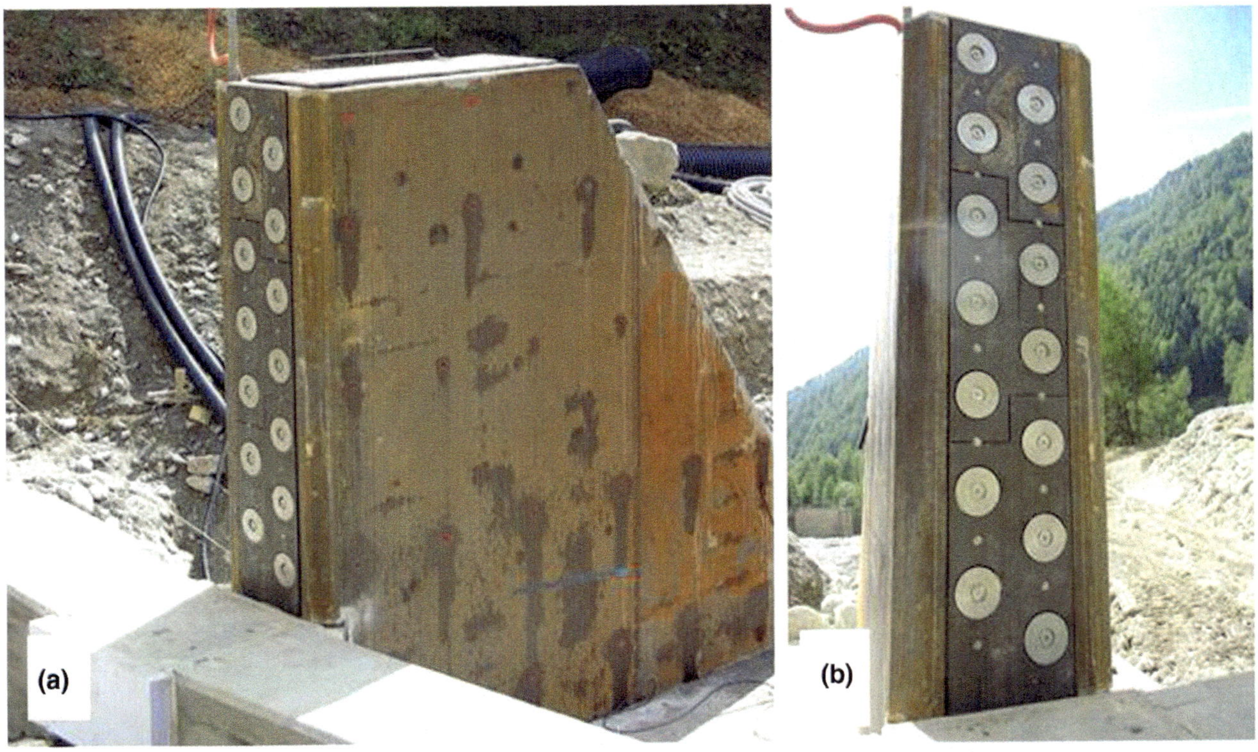

Fig. 4 Front of the check-dam with 14 load cells (**a**); construction of the impact measuring instrumentation (**b**)

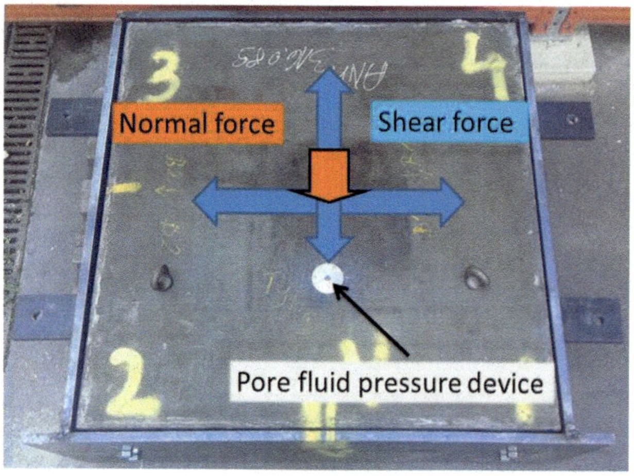

Fig. 5 Pore water pressure sensor mounted on a force plate

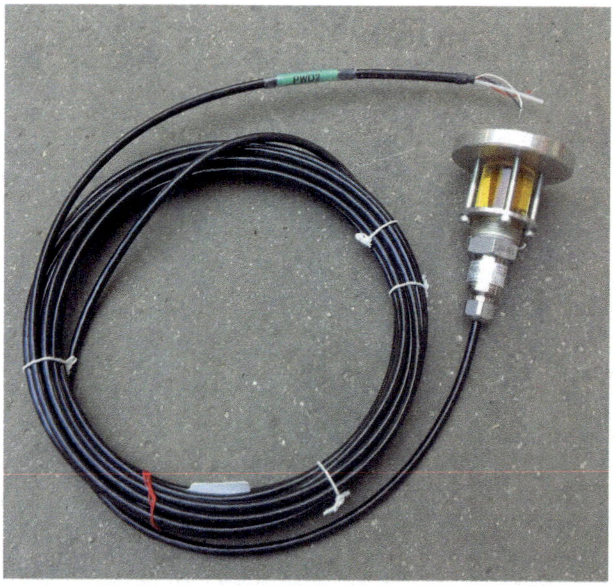

Fig. 6 Pore water pressure sensor with an oil-filled reservoir

is closed by a synthetic foil. Two steel meshes of different sizes protect the foil against penetration of coarse material (Fig. 6).

Measurement System

The data acquisition system consists of a MGC + which will be contained inside the barrier which is furthermore connected to a control PC inside the monitoring container. The monitoring container is placed on the left side of the torrent. The system will be fully automated and can be remotely controlled.

At the event mode, data storage will be triggered by reaching a threshold of geophones or ultra-sonic measurer or of the load cells. Internal, the system always measures with full measuring frequency, but only in case of a trigger signal the data will be saved and stored in the computer. At starting the event mode the system will inform serval institutes by a message.

The load cells and also the force plates will have a measuring frequency of 2400 Hz. For some parameters, a high resolution of the measuring frequency isn´t needed, for example the ultra-sonic measurer and the earth pressure sensors they will be measured with a lower sampling rate of 400 Hz.

Earth Pressure Sensor

Nine earth pressure sensors (Geokon 3500) below the foundation will describe the interaction with the ground during a debris flow event and will give a surface distribution of exert earth pressure. Each sensor consists of two circular steel plates, welded together and filled between with de-aired oil. The corresponding oil pressure will be measured by a pressure transducer (Fig. 7).

Power Supply

The whole system will be connected to main electricity (AC). In case of a blackout the system will be buffered by a

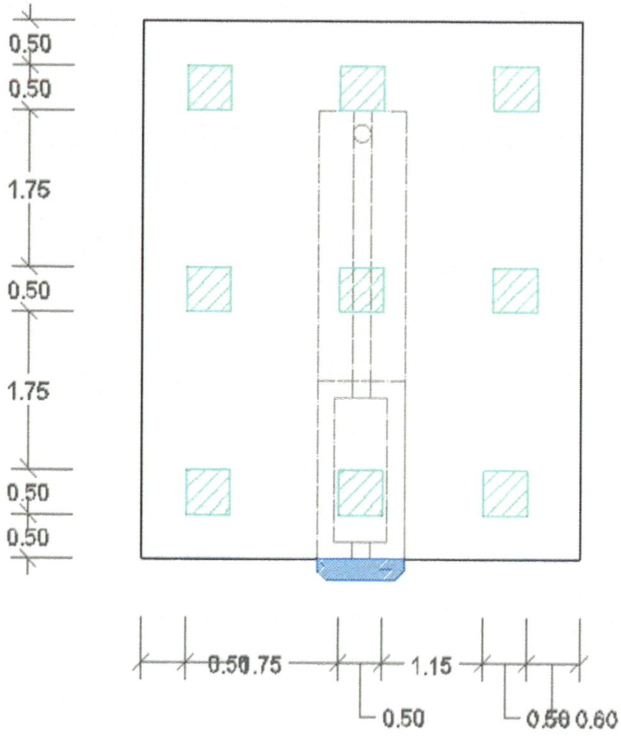

Fig. 7 Nine earth pressure sensors will be placed under the foundation to observe the interaction structure/soil

backup power supply inside the check-dam. All setups can be changed and adapted by remote control.

Conclusion

Numerus investigations carried out around the world to quantify impact forces of debris flows, starting from many small scale experiments to just a few real scale measurements, but a number of question remains.

In this paper, a monitoring structure has been introduced to measure debris flow impact directly in the field. The setup of monitoring arrangement will help to investigate the evolution of the impact dynamics to determine the essential parameters for designing purpose and help to improve the understanding of the impact mechanism and the structure interaction.

References

Arattano M, Coviello V, Cavalli M, Comiti F, Macconi P, Theule J, Crema S (2015) Brief communication: a new testing field for debris flow warning systems. Nat Hazards Earth Syst Sci 15(7):1545–1549. doi:10.5194/nhess-15-1545-2015

Berti M, Genevois R, LaHusen R, Simoni A, Tecca PR (2000) Debris flow monitoring in the Acquabona watershed on the Dolomites (Italian Alps). Phys Chem Earth, Part B: Hydrol, Oceans and Atmos 25(9):707–715

Comiti F, Marchi L, Macconi P, Arattano M, Bertoldi G, Borga M. et al. (2014) A new monitoring station for debris flows in the European Alps: first observations in the Gadria basin. Nat Hazards 73(3):1175–1198. doi:10.1007/s11069-014-1088-5

Hu Kaiheng, Wei Fangqiang, Li Yong (2011) Real-time measurement and preliminary analysis of debris-flow impact force at Jiangjia Ravine, China. Earth Surf Process Landforms 36(9):1268–1278. doi:10.1002/esp.2155

Hübl J, Schimmel A, Kogelnig A, Suriñach E, Vilajosana I, Mcardell BW (2013) A review on acoustic moni toring of debris flow. Int J SAFE 3(1):105–115. doi:10.2495/SAFE-V3-N2-105-115

Hungr O, Morgan GC, Kellerhals R (1984) Quantitative analysis of debris torrent hazards for design of remedial measures. Can Geotech J.21(4):663–677. doi:10.1139/t84-073

Kaitna R, Palucis MC, Yohannes B, Hill KM, Dietrich WE (2016) Effects of coarse grain size distribution and fine particle content on pore fluid pressure and shear behavior in experimental debris flows. J Geophys Res Earth Surf 121(2):415–441. doi:10.1002/2015JF003725

Marchi L, Arattano M, Deganutti AM (2002) Ten years of debris-flow monitoring in the Moscardo Torrent (Italian Alps). Geomorphology 46(1–2):1–17. doi:10.1016/S0169-555X(01)00162-3

Scheidl C, Chiari M, Kaitna R, Müllegger M, Krawtschuk A, Zimmermann T, Proske D (2013) Analysing debris-flow impact models, based on a small scale modelling approach. Surv Geophys 34(1):121–140. doi:10.1007/s10712-012-9199-6

Suwa H, Okano K, Kanno T (2011) Forty years of debris flow monitoring at kamikamihorizawa creek, mount yakedake, Japan. Ital J Eng Geol Environ. Checked on 24 June 2016

Suwa H, Okuda S, Yokoyama K (1973) Observation system on rocky mudflow. Bulletin Disaster Prevention Institute, Kyoto University, vol 23. Parts 3–4 No. 213

Wendeler C, Volkwein A, Roth A, Denk M, Wartmann S (2007) Field measurements and numerical modelling of flexible debris flow barriers. Debris-Flow Hazards Mitig Mech Predict Assess. Millpress, Rotterdam, pp 681–687

Zhang S (1993) A comprehensive approach to the observation and prevention of debris flows in China. Nat Hazards 7(1):1–23

Zhang S, Yuan J (1984) Impact forces of debris flow and its detection. In Memoirs of Lanzhou Institut of Glaziology and Cryopedology Chinese Academy of Science 1984 (04). Checked on 18 Aug 2016

4th WLF 2017 4th World Landslide Forum LJUBLJANA SLOVENIA EU

Detail Study of the Aratozawa Large-Scale Landslide in Miyagi Prefecture, Japan

Hendy Setiawan, Kyoji Sassa, Kaoru Takara, and Hiroshi Fukuoka

Abstract

The deep large-scale landslide near Aratozawa Dam in Miyagi Prefecture of Japan was occurred in 2008 and still the initiation mechanism and motion behavior were not explained in detail up to now. This paper aims to report briefly the detail study of the Aratozawa landslide. We conducted several experiments to test the Aratozawa samples using the newest version of the undrained dynamic loading ring shear apparatus. As reported by Sassa et al. (2014), the ring shear apparatus was designed with the single central axis-based for the normal stress loading system, with the normal stress and pore pressure measurement capacities of up to 3.0 MPa. The friction coefficient, shear displacement at the start and end of shear strength reduction, mobilized friction angle and steady state shear resistance of the Aratozawa samples were obtained from the ring shear tests. Experiments results implied that the shear strength reduction in progress of shear displacement of the Aratozawa samples was caused not only by the earthquake but also by factor of the initial pore pressure (Setiawan et al. 2014, 2016). Further analysis has been conducted by occupying shear parameters of soil failure resulted from experiment as a critical inputs for the LS-RAPID geotechnical simulation. LS-RAPID landslide simulation model is used to observe the overall process of landslide phenomena started from initiation to moving process. The Aratozawa landslide was successfully simulated using LS-RAPID model which involves the pore pressure increase, seismic loading, and landslide volume enlargement during traveling process. However, factor of the reservoir and its relation to the groundwater and bedrock is still needed to analyze in further.

H. Setiawan (✉) · K. Takara
Disaster Prevention Research Institute (DPRI), Kyoto University, Uji, 611-0011, Japan
e-mail: hendy@flood.dpri.kyoto-u.ac.jp

K. Takara
e-mail: takara.kaoru.7v@kyoto-u.ac.jp

K. Sassa
International Consortium on Landslides (ICL), Kyoto, 606-8226, Japan
e-mail: sassa@iclhq.org

H. Fukuoka
Research Institute for Natural Hazards and Disaster Recovery, Niigata University, Niigata, 950-2181, Japan
e-mail: fukuoka@cc.niigata-u.ac.jp

M. Mikoš et al. (eds.), *Advancing Culture of Living with Landslides*,
DOI 10.1007/978-3-319-53483-1_56

Keywords
Aratozawa landslide • Shear strength reduction • Pore pressure • LS-RAPID model • Ring shear tests

Introduction

Reservoir landslides often occurred due to the water impoundment and or significant earthquakes. An antecedent factor such as continuous fractures, prolonged heavy rainfall, weathered bedrock layers and reactivated landslide zone might also contributed to the scale of the landslides. Most notable landslide reservoir in the past was reported when about 270 million m^3 of mass flows down to the Vaiont Reservoir in northern Italy. It was occurred due to the water impounding, but the Vaiont landslide is an example of landslide reservoir that the long-term of slow slope deformation contributes (Ghirotti 2012). The Qianjiangping landslide at the Qinggan-he River, tributary of the Yangtze River of China was occurred in 2003 due to the first impoundment of the Three Gorges Reservoir. Although the creep failure took place before the event, water rising of the Yangtze River has a main role for triggering rapid motion of the Qianjiangping landslide (Wang et al. 2008).

Hazard assessment efforts for the reservoir landslides are very important to reduce its effect to the reservoir capacity, dam infrastructure performance and damage as well as human casualties and economic loss. Even though not frequently happened, reservoir landslides poses serious problems, such as in the Vaiont where thousands people were dead because of the tsunami wave-generated by huge landslide that overtopped the Vaiont dam, mostly at the town of Langarone (Ghirotti 2012; Genevois and Ghirotti 2005). While in the Qianjiangping landslide, hundreds of houses were destroyed and hectares of paddy field and several factories near the Qinggan-he River were damaged (Wang et al. 2004).

Another example of reservoir landslides was took place near the Aratozawa rock filled-dam in Miyagi Prefecture of Japan in 2008, where the Iwate-Miyagi Nairiku earthquake with the magnitude of 7.2 hit the area of Mount Kurikoma (Fig. 1).

This paper aims to report briefly the detail study of the Aratozawa landslide, including explanation of the initiation mechanism and motion behavior based on laboratory experiment through ring shear apparatus and simulation by LS-RAPID model.

The 2008 Aratozawa Large-Scale Landslide

Geological Features

The rock-filled Aratozawa dam located at the southeastern part of Mount Kurikoma, Ohu Mountains in the region of Tohoku, near the border of Miyagi and Iwate Prefecture. Functioned as flood control and irrigation, the Aratozawa

Fig. 1 The deep large-scale Aratozawa landslide (photo by H. Fukuoka, *bottom left* directing to the reservoir)

reservoir started to operate in 1998, which can store the amount of water of up to 14,130,000 m^3 with the catchment area of 20.4 km^2 where Nihasama River flows from Mount Kurikoma.

The large-scale landslide at the upstream part of Aratozawa reservoir was occurred soon after the Iwate-Miyagi Nairiku earthquake on 14 June 2008. A reverse fault earthquake with the depth of 8 km beneath the Ohu Mountains had caused the translational block glide at the Aratozawa area, about 14 km southwestern part from the epicenter. The total landslide mass was about 67 million m^3, depth of more than 100 m, with very gentle gradient, moved about 320 m to the southeast direction close to the reservoir, and part of the landslide mass (approximately 1.5 million m^3) was moved down to the reservoir as the second blocks failure (Miyagi et al. 2008, 2011; Kazama et al. 2012)

The Aratozawa area mainly consists of Tertiary and Quaternary pyroclastic rocks, e.g. sedimentary and welded tuff (Kazama et al. 2012). Large-scale and deep landslide in 2008 was located on the landslide topography where several deep landslides were formed in the past (Yagi et al. 2009; Miyagi et al. 2011). Thus, we agreed that this landslide was reactivated and made new landslide topography in the Aratozawa area. The geological structure of the Aratozawa was explained by Moriya et al. (2010) where the slope consists of volcano debris deposit, welded tuff, a thin layers of sandy clay, silty clay and mudstone, massive pumice tuff, alternating beds of sandstone and silty mudstone and lapilli tuff, in descending order. Based on the depth of landslide, the suspected sliding surface might formed in the layers of alternating beds of sandstone and silty mudstone.

Site Investigation and Sampling

Volcanic activities of Mount Kurikoma are mostly influences the development of soil structure in the Aratozawa area. The upper soil layers were young and formed in a Quaternary period. We found significant cracks at the upper part of head scarp during site investigation in 2013 (Fig. 2), nearly 5 years after the event. The cracks were still active and monitored with pairs of extensometers during reconstruction work of the Aratozawa slope by authorities. Such distinctive features might similarly exist before the event in 2008, indicated that the upper layer of Aratozawa area, particularly the volcano debris rocks and deposit, is not well-intact and prone to form fractures that contribute to the landslide topography in the past.

The volcanic tuff materials that we found squeezed out within the major landslide blocks at the collapse zone are very fine (like talc powder), loose, weathered and very crushable when it is in wet condition (almost saturated) (Fig. 3). Meanwhile, at the toe part of the landslide, the gullies were easily formed where the surface water drained and eroded the tuff materials.

We found difficulties to obtain enough samples from boring log to be tested in the laboratory. Therefore, to explain the initiation mechanism and motion behavior of this landslide, we took samples of squeezed tuff material within major landslide blocks and at the toe part of landslide (Setiawan et al. 2016). From those two samples, we do the tests using the newest version of the undrained dynamic loading ring shear apparatus to obtain soil parameters necessary to generate Aratozawa landslide simulation in the LS-RAPID model.

Ring Shear Tests Results

The Undrained Dynamic Loading Ring Shear Apparatus

A high stress undrained dynamic loading ring shear apparatus was developed by the International Consortium on Landslides, namely the ICL-2. As the upgrade version of the previous ring shear apparatus DPRI versions (Sassa et al. 2004), the ICL-2 applied the single central axis for normal loading system. The normal stress is produced by giving a control signal to two loading piston system through servo motor, then pulled the central axis and pushed down the loading plate to the samples (Sassa et al. 2014). The loaded normal stress of the samples which measured by vertical load cell is used as the feedback signal to the servo-amplifier that control the value of servo-motor of the normal stress. The connection between control signal of servo motor, load cells or sensors and feedback signal of servo amplifier are also used similarly in the system of shear stress, gap control as well as pore pressure control (Fig. 4). With all these features, the ICL-2 is capable to perform undrained tests for the Aratozawa landslide e.g. undrained speed control test, cyclic shear test, pore pressure control and earthquake loading test. Working preparation for the Aratozawa sample test using ring shear apparatus are explained already in Setiawan et al. (2014, 2016).

Shear Parameters for Simulation

In the ring shear tests, peak friction angle (ϕ_p) is obtained from the arctangent of shear resistance at the peak state before failure divided by the total normal stress (σ) that applied in the samples. After failure, the mobilized friction angle during motion (ϕ_m) in the shear strength reduction stage is calculated similarly along the failure line. Afterwards, the shear resistance in a steady state condition (τ_{ss}) is reached in the progress of long shear displacement when

Fig. 2 Continuous cracks at upper part of the head scarp, monitored by the extensometers

Fig. 3 Volcanic tuff materials within landslide blocks at the collapse zone of the Aratozawa landslide

there are no further decrease of shear strength and no further increase of pore-water pressure that both observed from the sliding surface zone of soil samples within the ring shear box (Sassa et al. 2004, 2010). As a result, the apparent friction angle (ϕ_a) act in the sliding surface is then calculated based on the arctangent of the steady state shear resistance divided by the total normal stress (σ) that applied in the samples.

All tested samples of the Aratozawa landslide were set in a normal consolidation. We gave the value of 3000 kPa with rate 2.0 kPa/s in a drained condition. Following consolidation,

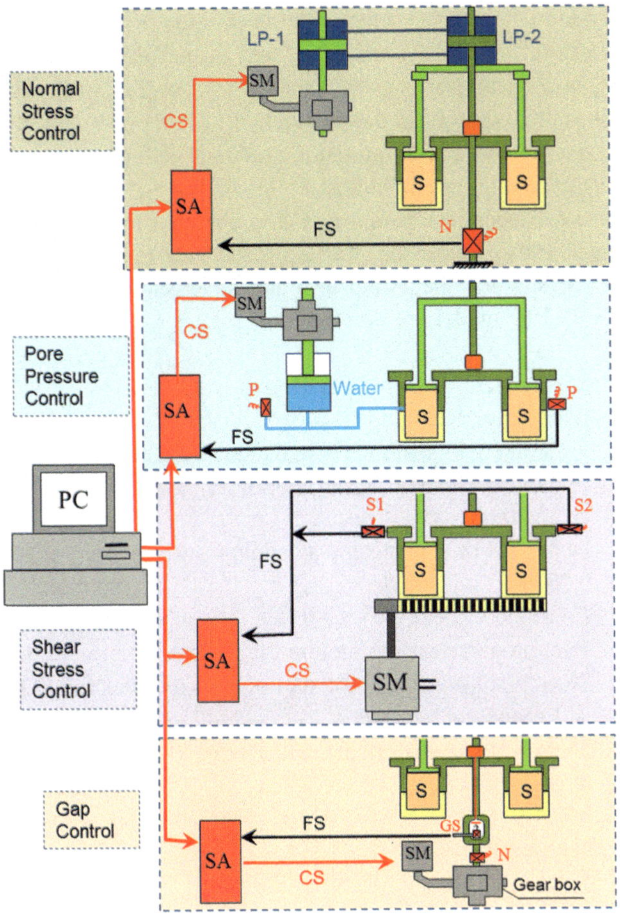

Fig. 4 Servo-control system of the undrained dynamic loading ring shear apparatus ICL-2 (Sassa et al. 2014), PC—Computer system; S—Sample; N—Load cell for normal stress; S1,S2—Load cell for shear resistance; GS—Gap sensor; FS—Feedback signal; CS—Control signal; PC—Computer; SM—Servo motor; LP—Loading piston; SA—Servo amplifier

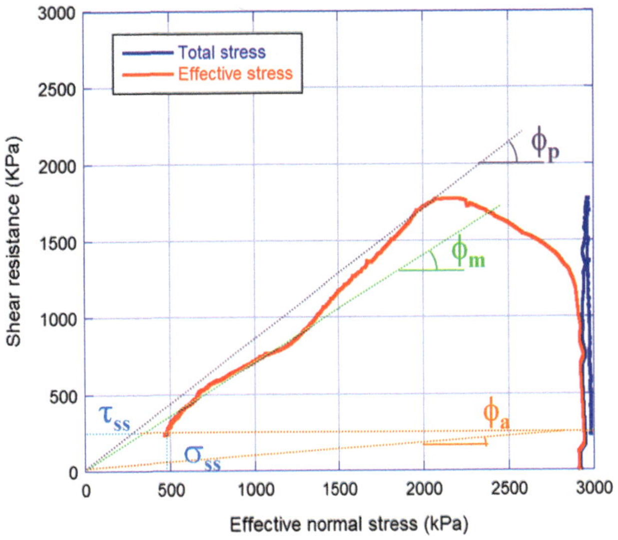

Fig. 5 The undrained speed control test for the volcanic tuff at the collapse zone of Aratozawa landslide

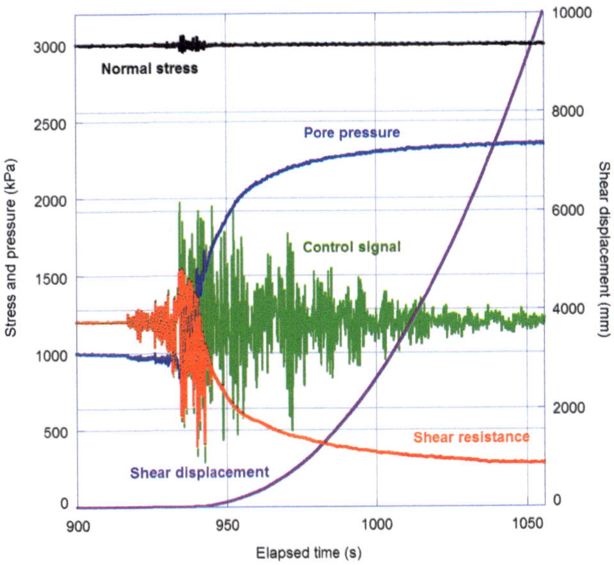

Fig. 6 The earthquake loading test with predefined pore pressure for the sample from collapse zone

we applied undrained speed control test to obtain the values of steady state shear resistance (τ_{ss}), peak friction angle (ϕ_p), mobilized friction angle (ϕ_m) and apparent friction angle (ϕ_a) (Fig. 5). Because the volcanic tuff samples has a low rate of pore pressure dissipation, then the undrained speed control test was appropriate to monitor the generation of pore pressure (instead of the monotonic stress control test that might appropriate for the sandy soils to obtain the shear parameters).

For the earthquake loading test, the 2008 Iwate-Miyagi Nairiku earthquake record of NS direction on MYG004 at Tsukidate measurement station was used with the maximum acceleration of 740 gal. The pore pressure was generated of only 270 kPa with very small shear displacements if the earthquake loading has applied directly to the samples in undrained condition, while the critical pore pressure ratio reached 0.61–0.63 in the normal stress of 3000 kPa (Setiawan et al. 2016).

Based on the results above, we combined the loading of initial pore pressure and earthquake that represent the initiation mechanism of Aratozawa landslide (Fig. 6). The test conditions are: volcanic tuff from collapse zone, 3000 kPa of normal stress, 1100 kPa of static shear stress and 20° of the gradient. For the translational movement of the block glide in a gentle sliding surface, we simulated the undrained dynamic loading test for the sample from toe part, with 3000 kPa of normal stress, 500 kPa of static shear stress and 9° of the gradient (Fig. 7).

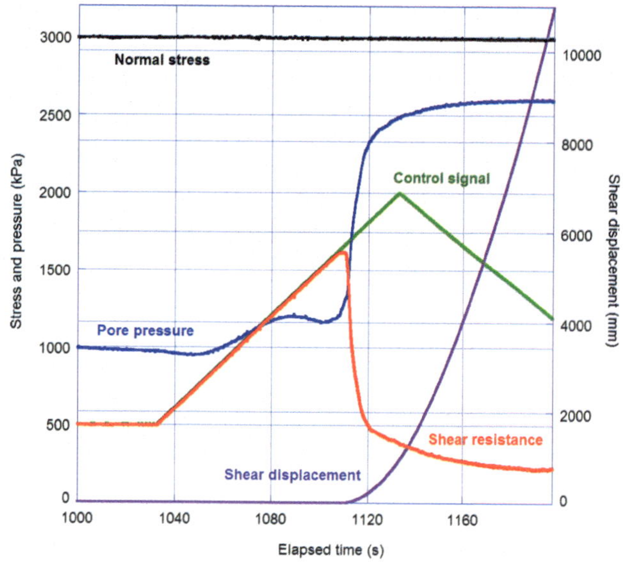

Fig. 7 The undrained dynamic loading test with predefined pore pressure for the sample from toe part

Rubber Edge Friction and Corrected Values

The values of shear resistance after failure in Figs. 6 and 7 are including the rubber edge friction that appears during the shearing process. Due to the high normal stress of 3000 kPa and shearing, the rubber edge friction for shear displacement of more than 3 m was measured of about 101–110 kPa. Thus, the shear resistance in a steady state condition is less than 120 kPa for earthquake loading test in Fig. 6 and less than 60 kPa for undrained shear dynamic loading test in Fig. 7. By neglecting the delineation of the shear strength parameters at each landslide area e.g. initiation and moving blocks area, we took the value of steady state shear resistance (τ_{ss}) approximately of 80 kPa due to the earthquake and dynamic loading for whole landslide area.

Landslide Simulation Modeling Results

The LS-RAPID Model

The LS-RAPID landslide simulation was used to clarify the initiation mechanism and motion of the Aratozawa landslide which triggered by earthquake. The model is generated by considering a vertical imaginary column with the acting forces within a moving landslide mass (Sassa et al. 2010, 2012). The failure initiation, pore pressure increase, seismic loading, moving process and the deposition stage of landslides can be simulated by the LS-RAPID model. The LS-RAPID model has already performed for the case of Leyte landslide in the Philippines (Sassa et al. 2010), the Senoumi submarine megaslide in Suruga Bay of Japan (Sassa et al. 2012) and the past megaslide of Unzen Mayuyama in Nagasaki of Japan (Sassa et al. 2014). The cohesion expressed in the unit weight (h_c), the pore pressure ratio (r_u), the lateral pressure ratio (k) and the apparent friction coefficient mobilized at the sliding surface (ϕ_a) are the key dependent parameters that obtained from the ring shear tests. Based on the ring shear tests results, the parameters for the Aratozawa landslide simulation using LS-RAPID model are described as follows:

- Steady state shear resistance (τ_{ss}) is 80 kPa;
- Lateral pressure ratio ($k = \sigma_h/\sigma_v$) is 0.50;
- Friction coefficient inside landslide mass ($\tan \phi_i$) is 0.50;
- Friction coefficient during motion at sliding surface ($\tan \phi_m$) is 0.66;
- Peak friction coefficient at sliding surface ($\tan \phi_p$) is 0.78;
- Cohesion at peak (c_p) is 80 kPa (assumption);
- Friction angle during motion (ϕ_m) is 33.4°;
- Shear displacement at the start of strength reduction (D_L) is 3 mm;
- Shear displacement at the end of strength reduction (D_U) is 3000 mm;
- Rate of excess pore pressure generation (B_{ss}) ranged between 0.60-0.95;
- Total unit weight of the mass (γ_t) is 20 kN/m^3 (assumption); and
- Excess pore pressure ratio in the fractured zone (r_u) is 0.33 (as initial pore pressure ratio before failure).

Aratozawa Landslide Simulation Results: Progressive Motion After Initiation Due to the Earthquake

We conducted Aratozawa landslide simulation by LS-RAPID Version 2.11 with the initial pore pressure ratio of 0.33 (normal stress of 3000 kPa), and followed by the Iwate-Miyagi earthquake loading, as shown in Fig. 8. The initial slope failure was occurred when the main shock of the earthquake arrived (Fig. 8[2]). But the surface failure was appeared not exactly at the head scarp. By checking with the geological cross section of the Aratozawa landslide (Miyagi et al. 2011), the initial surface failure was took place at the edge of volcanic deposit and welded tuff layers, disconnected with the pumice tuff layer. The progressive motion was appeared when the earthquake has ceased and causing the undulated forms of the main landslide blocks (Fig. 8[3]), this represents the dynamic loading shown in Fig. 7. The main landslide block remained near the toe part and some of them are facing to the reservoir (Fig. 8[4]) and possibly

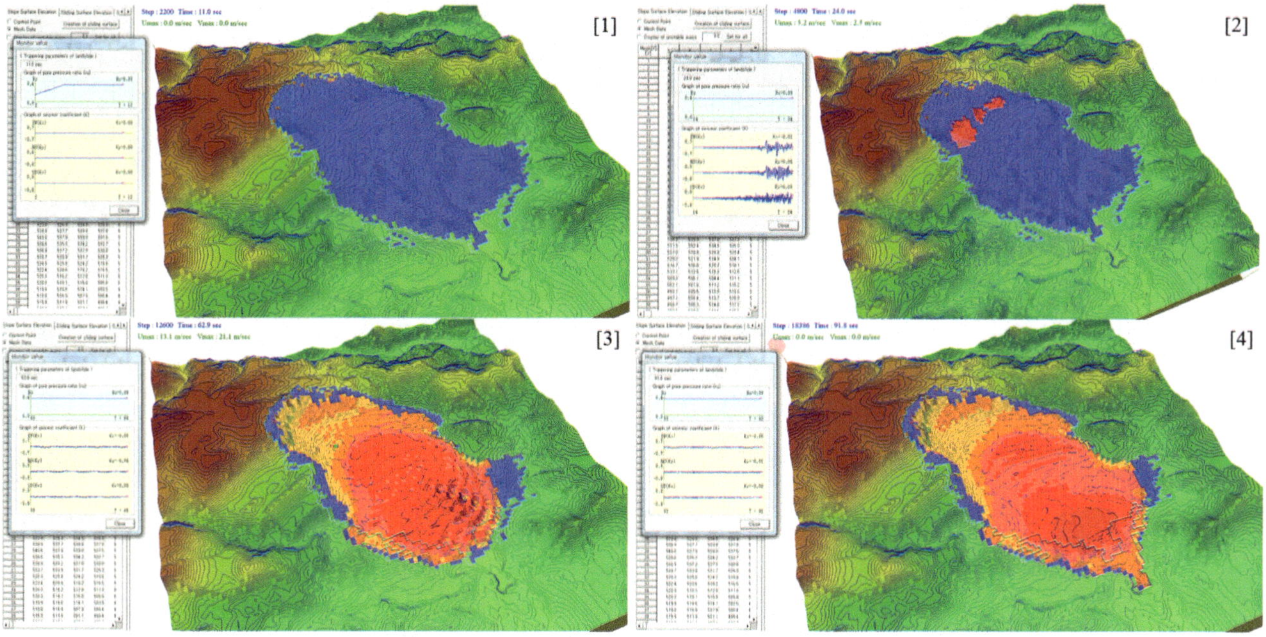

Fig. 8 The Aratozawa landslide simulation using LS-RAPID model: *1* Initial stage, when the pore pressure ratio increased up to 0.33. *2* Initial slope failure, when earthquake wave applied. *3* The progressive motion after earthquake wave was decelerated. *4* Final deposition stage

cause the second failure down to the reservoir. The LS-RAPID simulation results for the Aratozawa large-scale landslide are fit enough with geomorphology analysis that reported by Miyagi et al. (2011).

Conclusion

Detail study of the Aratozawa large-scale landslide was carried out by using a high stress undrained dynamic loading ring shear apparatus of ICL-2. Samples from collapse zone are occupied for the earthquake loading test with predefined pore pressure as the initiation mechanism, while samples from toe part focusing for the undrained dynamic loading that represent the translational movement of the block glide. The shear parameters and corrected values were obtained for the LS-RAPID simulation. The Aratozawa landslide was successfully simulated using the LS-RAPID model which involves the pore pressure increase, seismic loading, and the landslide volume enlargement during traveling process. However, factor of the reservoir and its relation to the groundwater and bedrock is still needed to analyze in further.

Acknowledgements The assistance from Japan Conservation Engineering and Japan Forestry Agency is gratefully acknowledged. The authors would like to thank Kawanami Akiko from Ministry of Agriculture, Forestry and Fisheries of Japan for her kind assistance during site investigation. This landslide research is supported by the Global Survivability Studies Program of Kyoto University (A Leading Graduate School Program of MEXT). The undrained ring shear apparatus ICL-2 version is developed by the International Consortium on Landslides (ICL) as a part of SATREPS (Science and Technology Research Partnership for Sustainable Development Program of the Government of Japan) for Vietnam project in 2012–2017.

References

Genevois R, Ghirotti M (2005) The 1963 Vaiont landslide. G Geol Appl 1:41–52

Ghirotti M (2012) The 1963 Vaiont landslide, Italy. In: Clague JJ, Stead D (eds) Landslides: types, mechanisms and modeling. Cambridge University Press, 359p

Kazama M, Kataoka S, Uzuoka R (2012) Volcanic mountain area disaster caused by the Iwate-Miyagi Nairiku earthquake of 2008, Japan. Soils Found 52(1):168–184

Miyagi T, Kasai F, Yamashina S (2008) Huge landslide triggered by earthquake at the Aratozawa Dam area, Tohoku, Japan. Proceedings of the First World Landslide Forum. ICL, ISDR, Tokyo, pp 421–424

Miyagi T, Yamashina S, Esaka F, Abe S (2011) Massive landslide triggered by 2008 Iwate-Miyagi inland earthquake in the Aratozawa Dam area, Tohoku, Japan. Landslides 8:99–108

Moriya H, Abe S, Ogita S, Higaki D (2010) Structure of the large-scale landslide at the upstream area of Aratozawa dam induced by the Iwate-Miyagi earthquake in 2008. J Jpn Landslide Soc 47(2):77–83 (in Japanese)

Sassa K, Fukuoka H, Wang GH, Ishikawa N (2004) Undrained dynamic-loading ring shear apparatus and its application to landslide dynamics. Landslides 1:7–19

Sassa K, Nagai O, Solidum R, Yamazaki Y, Ohta H (2010) An integrated model simulating the initiation and motion of earthquake and rain induced rapid landslides and its application to the 2006 Leyte landslide. Landslides 7:219–236

Sassa K, He B, Miyagi T, Strasser M, Konagai K, Ostric M, Setiawan H, Takara K, Nagai O, Yamashiki Y, Tutumi S (2012) A hypothesis of the Senoumi submarine megaslide in Suruga Bay in Japan—based on the undrained dynamic-loading ring shear tests and computer simulation. Landslides 9:439–455

Sassa K, Dang K, He B, Takara K, Inoue K, Nagai O (2014) A new high-stress undrained ring-shear apparatus and its application to the 1792 Unzen-Mayuyama megaslide in Japan. Landslides 11: 827–842

Setiawan H, Sassa K, Takara K, Miyagi T, Fukuoka H, He B (2014) The simulation of a deep large-scale landslide near Aratozawa Dam using a 3.0 MPa undrained dynamic loading ring shear apparatus. In: Sassa K, Canuti P, Yin Y (eds) Landslide science for a safer geoenvironment. Springer International Publishing, pp 459–465

Setiawan H, Sassa K, Takara K, Miyagi T, Fukuoka H (2016) Initial pore pressure ratio in the earthquake triggered large-scale landslide near Aratozawa Dam in Miyagi Prefecture, Japan. Proc Earth Planet Sci 16:61–70

Wang F, Zhang Y, Huo Z (2004) The July 14, 2003 Qianjiangping landslide, Three Gorges Reservoir, China. Landslides 1:157–162

Wang F, Zhang Y, Huo Z, Peng X, Wang S, Yamasaki S (2008) Mechanism for the rapid motion of the Qianjiangping landslide during reactivation by the first impoundment of the Three Gorges Dam Reservoir, China. Landslides 5:379–386

Yagi H, Sato G, Higaki D, Yamamoto M, Yamasaki T (2009) Distribution and characteristics of landslides induced by the Iwate-Miyagi earthquake in 2008 in Tohoku District, Northeast Japan. Landslides 6:335–344

Identification of Rock Fall Prone Areas on the Steep Slopes Above the Town of Omiš, Croatia

Marin Sečanj, Snježana Mihalić Arbanas, Branko Kordić, Martin Krkač, and Sanja Bernat Gazibara

Abstract

The aim of this paper was identification of rock fall prone areas above the historical town of Omiš, located at the Adriatic coast in Croatia. Unstable areas were identified by kinematic analysis performed based on relative orientations of discontinuities and slope face. Input data was extracted from the surface model created from the high-resolution point cloud. The town of Omiš is threatened by rock falls, because of its specific location just at the toe of Mt. Omiška Dinara. Rock fall risk is even higher due to rich cultural and historical heritage of the town. Collection of spatial data was performed by Time of Flight and phase-shift terrestrial laser scanners in order to derivate high resolution point cloud necessary for derivation of surface model. Split-FX software was used to extract discontinuity surfaces were semi-automatically from the point cloud data. Spatial kinematic analysis was performed for each triangle of TIN surface model of the investigated slopes to identify locations of possible instability mechanism. From the results of the spatial kinematic analysis, the most critical parts of the slope have identified for planar and wedge failure and flexural and block toppling. Verification of identified rock fall areas was performed by visual inspection of hazardous blocks at the surface model. Identified rock fall prone areas, unstable blocks and probable instability mechanisms on the steep slopes above the town Omiš, present the input data for risk reduction by efficient design of countermeasures.

Keywords

Rock fall • Point cloud • Split-FX • Discontinuity extraction • Kinematic analysis • Omiš (Croatia)

M. Sečanj (✉) · S. Mihalić Arbanas · M. Krkač · S. Bernat Gazibara
Faculty of Mining, Geology and Petroleum Engineering, University of Zagreb, Pierottijeva 6, 10000 Zagreb, Croatia
e-mail: marin.secanj@oblak.rgn.hr

S. Mihalić Arbanas
e-mail: smihalic@rgn.hr

M. Krkač
e-mail: mkrkac@rgn.hr

S. Bernat Gazibara
e-mail: sbernat@rgn.hr

B. Kordić
Faculty of Geodesy, University of Zagreb, Kačićeva 26, 10000 Zagreb, Croatia
e-mail: bkordic@geof.hr

M. Mikoš et al. (eds.), *Advancing Culture of Living with Landslides*,
DOI 10.1007/978-3-319-53483-1_57

Introduction

Rock fall is a common phenomenon on the steep slopes along the Croatian part of the Adriatic Coast. They are usually caused by unfavorable characteristics of the rock mass, weathering in combination with heavy rainfall and artificial influences (Arbanas et al. 2012). Located just at the toe of the steep slopes, historical town of Omiš is frequently threatened by rock falls. To reduce the risk of injury or death for people and to prevent damage of historical landmarks, settlements and infrastructure, it was necessary to identify potential rock fall prone areas on the slopes and to conduct detail geotechnical investigation that are necessary for adequate design of countermeasures.

For the collection of all relevant discontinuity parameters necessary for the identification of rock fall prone areas, traditional surveys require direct access to the rock face. Employing remote sensing techniques is necessary to rapidly obtain information from the inaccessible areas on the slopes. During the last years, many authors have been working on semi-automatic and automatic extraction of rock mass structural data from remotely acquired high resolution point clouds (Slob et al. 2005; Jaboyedoff et al. 2007; Gigli and Casagli 2011; Lato and Vöge 2012; Cacciari and Futai 2016) and some of them have been working on identification of rock fall source areas from surface models created from the point cloud (Günther et al. 2012; Gigli et al. 2012; Fanti et al. 2013). By these approaches, rock mass structural data can be quickly obtained, along with the identification of potentially instable rock blocks, with the relatively high accuracy.

Study Area

The historical town of Omiš, with the population nearly to 7000, is situated on the coast of Adriatic Sea, at the mouth of the river Cetina. The town is a popular tourist destination because of its rich cultural and historical heritage, such as the fortresses which were built in the 13th century and used by the infamous Omiš corsairs. Due to the specific location, just at the toe of the steep slopes of Mt. Omiška Dinara (Fig. 1), the Omiš area is frequently threatened by rock fall and during the last years there were numerous reports of damaged buildings, local roads and other material goods.

Omiška Dinara is a 15 km long mountain with the highest peak at 865 m a.s.l. It is a part of a large nappe system and it is represented as an overturned anticline striking NW–SE (Marinčić et al. 1976) which is a result of compressional tectonics occurred since Cretaceous to Miocene. The core of the anticline is built of Senonian rudist limestones, while the limbs of the anticline are built of Eocene breccia, limestones and flysch (Marinčić et al. 1977).

Fig. 1 View of the southern slopes of Omiška Dinara Mt. from the old town center

In the Omiš area, geological contacts between Cretaceous and Paleogene deposits are usually along steep reverse faults, striking E–W, with the tectonic transport top to south. Complexity of the geological-structural setting, caused by faulting and folding in the area, led to the formation of numerous discontinuities in rock mass. Progressive weathering of discontinuities led to the formation of unstable rock blocks with unfavorable orientation that are prone to rock falls. Moreover, according to Herak et al. (2011), wider Omiš area belongs to one of the seismically most active parts of Croatia, which definitively adds to the rock fall potential in the area.

Slopes above the town of Omiš cover the total area of around 0.15 km^2, with highest peak at around 300 m a.s.l. The residential area is mostly located beneath the southern slopes from which rock falls most often occur. Investigated slopes are very steep, with an average dip mostly over 60°, only locally transected by natural berms in the relief (Fig. 2). Numerous sets of discontinuities with unfavorable orientation and potentially unstable rock blocks were determined by field reconnaissance. Due to high fracturing, detachments of rock blocks of various dimensions were determined, which are related to wedge failure and toppling.

Methodology

Detailed engineering geological mapping was carried out on the limited parts of the rock mass which were accessible. Because of inaccessibility to the entire rock mass, the

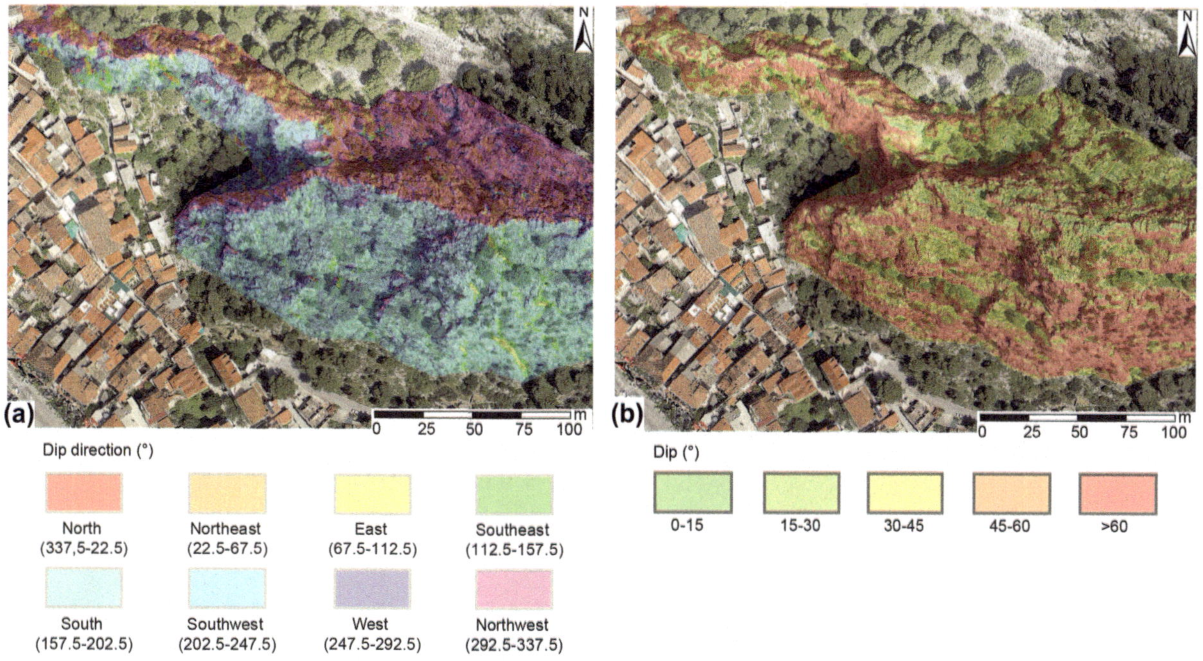

Fig. 2 Distribution of investigated rock mass surface dip direction (**a**) and dip values (**b**) extracted from DEM with 1 × 1 m resolution

mapping of the entire rock mass and extraction of relevant geometrical properties was carried out by remote sensing on a high resolution point cloud.

In order to derivate high resolution point cloud, field surveying and collection of spatial data was performed by Time of Flight (TOF) and phase-shift terrestrial laser scanners (TLS). Phase-shift laser scanner was used in the center of town, while surveying of the area at the greater distances above the old town was performed with TOF laser scanner. Unmanned aircraft system (UAS) photogrammetry method (Haala et al. 2011) was used for measuring of inaccessible areas as well as those which cannot be measured with TLS. Unique georeferenced point cloud was obtained based on the registration and vegetation filtering. Point cloud was composed of about 6 million points with average point resolution of 2 cm or 125,000 points per square meter. Triangulated irregular network (TIN) and digital elevation model (DEM) with 1 × 1 m resolution were derivate from the point cloud.

Point cloud processing and extraction of structural data, including discontinuity orientation and spacing, was performed by Split-FX software (Split Engineering LCC 2007). From the point cloud a triangulated surface, i.e., mesh surface was created. To construct a high quality mesh surface, it was necessary to adjust instrument view, i.e., scan line direction to be perpendicular to the rock face, because it was not set properly prior to the point cloud processing. Patches which represent discontinuity planes, were automatically generated by grouping neighboring mesh triangles which fit the flatness criteria. and afterward by fitting a plane through the points bounded by the grouped triangles (Fig. 3).

It was found that best results were obtained when the mesh surface was created from 10,000 triangles per meter and minimum patch size was set to 15 mesh triangles with the maximum patch neighbor angle of 10°. For valid data representation, noisy patches were excluded and all unrecognized discontinuities inserted manually on to the mesh surface. For those discontinuities that were only visible as traces on the rock face, best fitting plane were assigned and their orientation extracted. In the end, the total amount of 1449 discontinuity planes was extracted. The orientation of discontinuities was presented on a stereonet plot as poles and with cluster analysis 7 discontinuity sets were determined (Fig. 4). Mean orientation values of discontinuities sets are listed in Table 1.

Rock fall prone areas on the investigated slopes were identified by kinematic analysis because of its capability to determine if and where a specific instability mechanism is kinematically possible considering the geometry of the slope and discontinuities. Instability mechanism investigated by this method are plane and wedge failure (Hoek and Bray 1981) and flexural and block toppling (Goodman and Bray 1976; Hudson and Harrison 1997).

Fig. 3 Automatically generated patches plotted on the point cloud in Split-FX software

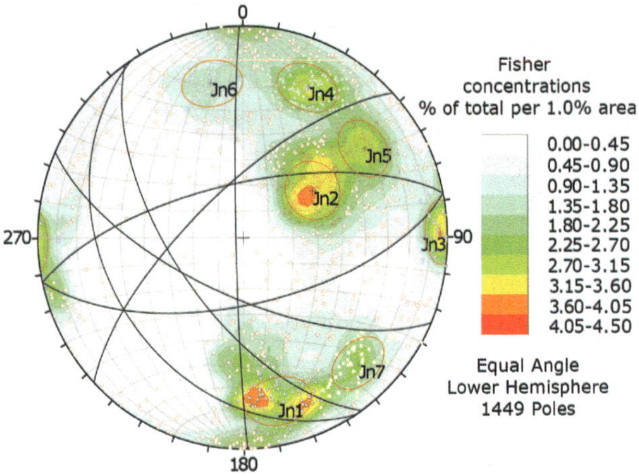

Fig. 4 Stereographic projection of discontinuity poles and mean orientation of discontinuity sets extracted with the Split-FX software

For each instability mechanism, Kinematic Hazard Index, introduced by Casagli and Pini (1993), was calculated by counting poles, discontinuities and their intersections falling into critical areas within the stereographic projection. The Kinematic Hazard Index is calculated as follows:

Table 1 Discontinuity sets mean orientation and standard deviation extracted from the semi-automatic analysis in Split-FX software

Discontinuity set	α (°)	β (°)	σ (°)
Jn1	348	69	10.4
Jn2	237	33	8.4
Jn3	269	87	10.3
Jn4	204	67	9.7
Jn5	235	62	9.6
Jn6	168	64	9.7
Jn7	317	71	9.2

$C_{pf} = 100 \times (N_{pf}/N)$ for plane failure,
$C_{wf} = 100 \times (I_{wf}/I)$ for wedge failure,
$C_{bt} = 100 \times (N_{bt}/N) \times (I_{bt}/I)$ for block toppling and
$C_{ft} = 100 \times (N_{ft}/N)$ for flexural toppling,

where N_{pf}, N_{bt} and N_{ft} are number of poles satisfying plane failure, block toppling and flexural toppling conditions, I_{wf} and I_{bt} are number of intersections satisfying wedge failure and block toppling conditions, while N and I are total number of poles and intersections respectively.

The input parameters for the analysis are the slope dip and dip direction (Fig. 2), the discontinuity surface orientations (obtained from the point cloud in Split-FX software) and discontinuity friction angle value of 38°, which was assigned according to Nikolić (2015).

Spatial kinematic analysis are performed for each triangle or raster cell of the surface model. Similar approach was used by several authors (Gigli et al. 2012; Fanti et al. 2013; Sdao et al. 2013). For this purpose, from the point cloud TIN surface was created using minimum thinning method. This method was used in order to reduce the effect of vegetation leftover after point cloud filtration. For every triangle of TIN surface model, geometrical properties of slope and discontinuities were assigned and spatial kinematic analysis was performed.

It is important to notice that this analysis only take into account those discontinuities that were extracted from the point cloud with Split-FX and does not consider minor or irregular fractures and so for underestimating the probability of instabilities in those areas.

Results

The results of spatial kinematic analysis for each mechanism investigated are presented in Figs. 5 and 6. The investigated slopes show high probability of occurrence of kinematic instability mechanisms with the maximal kinematic index up to 40% which is related to wedge failure (Fig. 5b). The areas with higher kinematic index are generally located at the

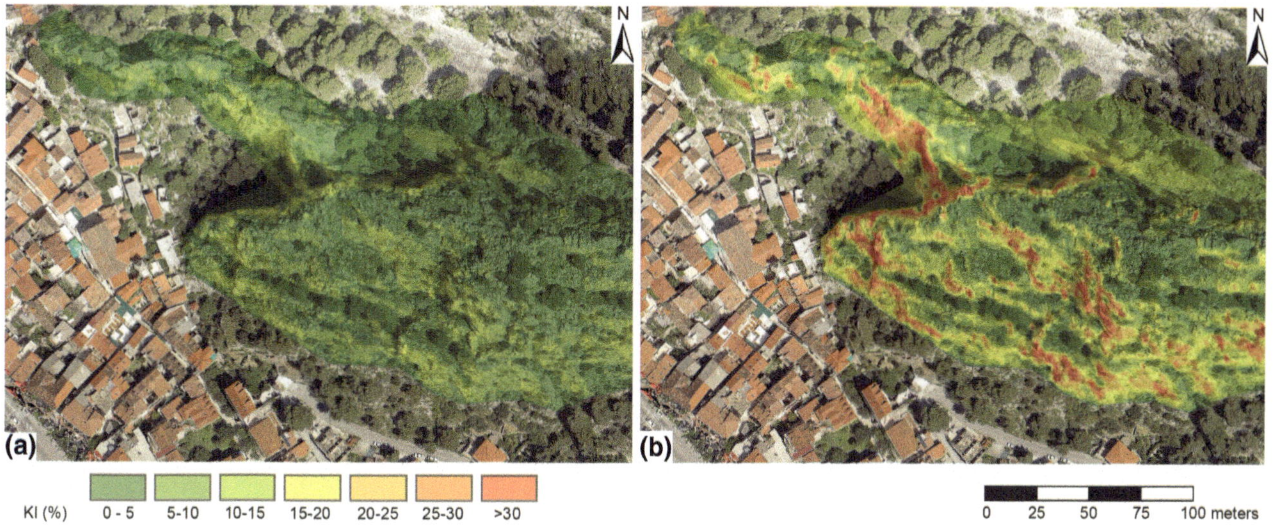

Fig. 5 Results of the kinematic analysis. The higher kinematic hazard indexes (KI) are, the higher probability that the investigated instability mechanism will take place: **a** planar failure and **b** wedge failure

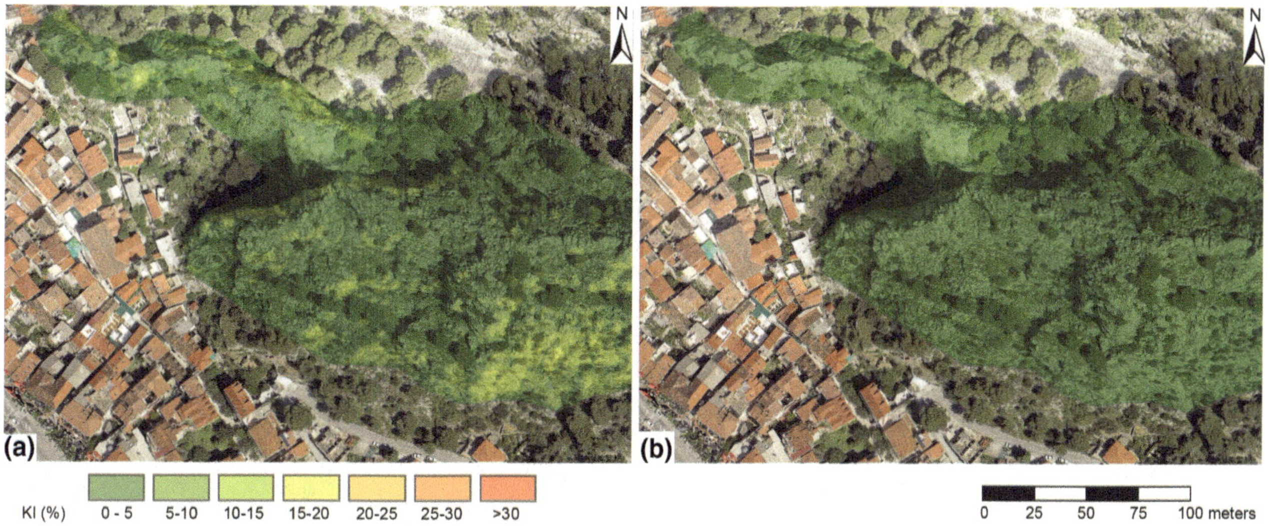

Fig. 6 Results of the kinematic analysis. The higher kinematic hazard indexes (KI) are, the higher probability that the investigated instability mechanism will take place: **a** flexural toppling and **b** block toppling

steepest parts of the slope. The maximal kinematic index calculated for planar failure is around 18% (Fig. 5a), while for flexural toppling is around 19% (Fig. 6a). The kinematic analysis has shown that block toppling is the mechanism that is least likely to happen, with the maximal kinematic index around 3% (Fig. 6b). The probability of instabilities is expressed by the color scale, varying from green to red as the kinematic hazard index increases. The areas with high kinematic hazard index were grouped and identified as rock fall prone areas, especially prone to wedge failure.

By comparing the location of identified rock fall prone areas with the correspondent area on the point cloud, potentially hazardous rock blocks were determined. Some of them are located directly beneath old historical landmarks and above the buildings of the historic center and pose a significant threat (Fig. 7). A closer view of these blocks show the presence of discontinuities which define wedge geometry. With future weathering or seismic shock, a probable detachment of the blocks by wedge failure could occur.

Fig. 7 Example of identified areas prone to wedge failure with the photographs of potentially hazardous blocks

Discussion and Conclusion

The main advantage of the approach presented in this paper are rapidity and objectivity of the analysis. The remote sensing technique allows quick retrieval of a high resolution point cloud of the rock slopes and provides information for unreachable parts of the slope. Semi-automatic discontinuity surface extraction, along with spatial kinematic analysis represent fast and objective method for the identification of rock fall prone areas with the information whether a certain instability mechanism is feasible. The results of this analysis were also confirmed by field observation (Fig. 7).

On the other hand, there are some drawbacks in the presented approach. Because of quick retrieval of point cloud data, accuracy of some parts of created mesh surface is questionable because of the shadow areas. For the better final results, shadow areas should be rescanned and the analysis reapplied to those areas. Also, the spatial kinematic analysis was applied on the 2.5D surface and not on a true 3D surface which should provide better results as described in Fanti et al. (2013) and Gigli et al. (2012). The reason for this was to maintain simplicity and rapidity of the analysis and to reduce the effect of vegetation leftover after point cloud filtration without using any specialized software other than GIS.

Identifying rock fall prone areas, localizing potentially hazardous and probable instability mechanisms on the steep slope above the historical town of Omiš present the input data for risk reduction by efficient design of countermeasures.

References

Arbanas Ž, Grošić M, Udovič D, Mihalić S (2012) Rockfall hazard analyses and rockfall protection along the Adriatic coast of Croatia. J Civ Eng Arch David Publishing Company 6(3):344–355

Cacciari PP, Futai MM (2016) Mapping and characterization of rock discontinuities in a tunnel using 3D terrestrial laser scanning. Bull Eng Geol Env 75:223–227

Casagli N, Pini G (1993) Analisi cinematica della stabilità in versanti naturali e fronti di scavo in roccia. In: Proceedings 3° Convegno Nazionale dei Giovani Ricercatori in Geologia Applicata, Potenza (in Italian)

Fanti R, Gigli G, Lombardi L, Tapete D, Canuti P (2013) Terrestrial laser scanning for rockfall stability analysis in the cultural heritage site of Pitigliano (Italy). Landslides. 10:409–420

Gigli G, Casagli N (2011) Semi-automatic extraction of rock mass structural data from high resolution LIDAR point clouds. Int J Rock Mech Min Sci 48(2):187–198

Gigli G, Frodella W, Mugnai F, Tapete D, Cigna F, Fanti R, Intrieri E, Lombardi L (2012) Instability mechanisms affecting cultural heritage sites in the Maltese Archipelago. Nat Hazards Earth Syst Sci 12:1883–1903

Goodman RE, Bray JW (1976) Toppling of rock slopes. In: Proceedings of special conference on rock engineering for foundations and slopes, vol 2. ASCE, Boulder, Colorado, pp 201–234

Günther A, Wienhofer J, Konietzky H (2012) Automated mapping of rock slope geometry, kinematics and stability with RSS-GIS. Nat Hazards 61:29–49

Haala N, Cramer M, Weimer F, Trittler M (2011) Performance test on UAV-based photogrammetric data collection. In: Proceedings of the international conference on Unmanned Aerial Vehicle in Geomatics (UAV-g), 14–16 Sept 2011, vol XXXVIII-1/C22

Herak M, Allegretti I, Herak D, Ivančić I, Kuk V, Marić K, Markušić S, Sović I (2011) Karta potresnih područja Republike Hrvatske M 1:800.000, Geofizički zavod PMF-a, Zagreb (in Croatian)

Hoek E, Bray JW (1981) Rock slope engineering. (Revised 3rd edn). Institution of Mining and Metallurgy, London, pp 257–250

Hudson JA, Harrison JP (1997) Engineering rock mechanics: an introduction to the principles and applications. In: Pergamon (ed). Oxford

Jaboyedoff M, Metzger R, Oppikofer T, Couture R, Derron MH, Locat J, Turmel D (2007) New insight techniques to analyze rock-slope relief using DEM and 3D-imaging cloud points: COLTOP-3D software. In: Proceedings of 1st Canada—U.S. rock mechanics symposium, Vancouver, 27–31 May 2007, pp 61–68

Lato MJ, Vöge M (2012) Automated mapping of rock discontinuities in 3D lidar and photogrammetry models. Int J Rock Mech Min Sci 54:150–158

Marinčić S, Korolija B, Majcen Ž (1976) Osnovna geološka karta SFRJ 1:100.000. List Omiš K33-32.– Institut za geološka istraživanja Zagreb (1968–1969), Savezni geol. zavod, Beograd (in Croatian)

Marinčić S, Korolija B, Mamužić B, Magaš N, Majcen Ž, Brkić M, Benček Đ (1977) Osnovna geološka karta SFRJ 1:100.000. Tumač za list Omiš K33-32.– Institut za geološka istraživanja Zagreb (1969), Savezni geol. zavod, Beograd, pp. 21–35 (in Croatian)

Nikolić M (2015) Rock mechanics, failure phenomena with pre-existing cracks and internal fluid flow through cracks. Ph.D. thesis, Faculty of Civil Engineering, Architecture and Geodesy, University of Split, Split, Croatia

Sdao F, Lioi DS, Pascale S, Caniani D, Mancini IM (2013) Landslide susceptibility assessment by using a neuro-fuzzy model: a case study in the Rupestrian heritage rich area of Matera. Nat Hazards Earth Syst Sci 13:397–407

Slob S, Hack R, Van Knapen B, Turner K, Kemeny J (2005) A method for automated discontinuity analysis of rock slopes with 3D laser scanning. Transport Res Rec 1913:187–208

Split Engineering LCC (2007) Split-FX V 2.4, Tucson, AZ

Automatic Detection of Sediment-Related Disasters Based on Seismic and Infrasound Signals

Andreas Schimmel and Johannes Hübl

Abstract

The automatic detection of sediment related disasters like landslides, debris flows and debris floods, gets increasing importance for hazard mitigation and early warning. Past studies showed that such processes induce characteristic seismic signals and acoustic signals in the infrasonic spectrum which can be used for event detection. So already many studies has been done on signal processing and detection methods based on seismic or infrasound sensors. But up to date no system has been developed which uses a combination of both technologies for an automatic detection of debris flows, debris floods or landslides. This work aims to develop a system which is based on one infrasound and one seismic sensor to detect sediment related processes with high accuracy in real time directly at the sensor site. The developed system compose of one geophone, one infrasound sensor and a microcontroller where a specially developed detection algorithm is executed. Further work tries to get out more information of the seismic and infrasound signals to enable an automatic identification of the process type and the magnitude of an event. Currently the system is installed on several test sites in Austria, Switzerland and Italy and these tests show promising results.

Keywords

Infrasound signals • Seismic signals • Early warning system • Debris flow • Debris flood

Introduction

Automatic detection of alpine mass movements is an important tool for protecting people and property in the fast socio-economic developing mountain areas.

Alpine mass movements like debris flows and debris floods induce, by the collision of stones and by the friction of the flow to the channel, waves in the low-frequency infrasonic spectrum (e.g. Chou et al. 2007, 2010; Kogelnig et al. 2010) and characteristically seismic waves (e.g. Huang et al. 2007; Burtin et al. 2014, 2016; Arattano 2003). These infrasound and seismic waves produced by the mass movement can be used for detecting events before a surge passes the sensor location and to monitor mass movements from a remote location unaffected by the process.

There have already been several approaches for automatic detection of debris flows based on seismic signals (e.g. Arattano et al. 2014; Coviello et al. 2015) and also infrasound signals are commonly used for detecting avalanches (e.g. Ulivieri et al. 2012; Marchetti et al. 2015) or debris flows (e.g. Zhang et al. 2004). Seismic and infrasound waves have different advantages and disadvantages. As an example, infrasound signals may be disturbed by strong noise due to the wind but show low signal damping at local distances. Seismic signals may suffer a higher attenuation depending on the geology but show lower disturbance due to weather and wind. So a combination of both technologies can

A. Schimmel (✉) · J. Hübl
Institute of Mountain Risk Engineering, University of Natural Resources and Life Sciences, Vienna, Austria
e-mail: andreas.schimmel@boku.ac.at

J. Hübl
e-mail: johannes.huebl@boku.ac.at

M. Mikoš et al. (eds.), *Advancing Culture of Living with Landslides*,
DOI 10.1007/978-3-319-53483-1_58

increase detection probability and reduce false alarms (e.g. Suriñach et al. 2009; Kogelnig 2012; Hübl et al. 2013).

But up to date no system has been designed which uses a combination of seismic and infrasound signals for an automatic detection of sediment related disasters of different types.

So this work aims to develop a reliable automatic detection system for alpine mass movements, which is based on one infrasound and one seismic sensor and can detect different processes in real time directly at the sensor site. In a further step, the infrasound and seismic signals will be used to identify the process type and to get an estimation of the event-magnitude.

Event Detection System

System Setup

This approach for a detection system is based on one infrasound sensor, one geophone and a microcontroller which is used as datalogger and where a detection algorithm is executed (Schimmel and Hübl 2014, 2015). The advantage of this setup is, that it can offer a low-cost and easy to install solution for a warning system for different applications. Future application of such a system could be the protection of traffic lines by controlling a light signal, or for protecting construction sites inside torrents, like the cleaning up of a basin after an event.

Currently we have three different infrasound sensors in operation and two different types of geophones:

One infrasound sensor used for this system is the Chaparral Physics Model 24 with a resolution of 2 V/Pa and a frequency range of 0.1–100 Hz. Further also a infrasonic microphone of the type MK-224 with a resolution of 50 mV/Pa and a frequency range of 3–200 Hz is used. Since both sensors types are rather expensive a cheaper solution was found with an Electret microphone of the type KECG2742WBL-25-L (reduction of the system cost by a factor of ~ 10). This microphone with a sensitivity of -42 ± 3 dB has to be calibrated for the low frequency range which has been done by comparing the signals with the Chaperral sensor. As seismic sensor we used the Sercel SG-5 geophones with a sensitivity of 80 V/m/s and a natural frequency of 5 Hz and the Sensor NL SM-6 geophones with a sensitivity of 28 V/m/s and a natural frequency of 4.5 Hz. The sensor input signals are adapted to the ADC-range of the microcontroller by an amplifier circuit and are filtered by a band pass based on a RC-circuit with a cutoff frequency below 1 Hz to eliminate constant components and an upper cutoff frequency of around 150 Hz. The signals are recorded with a sample rate of 100 Hz where a 32× hardware oversampling is used to avoid aliasing effects.

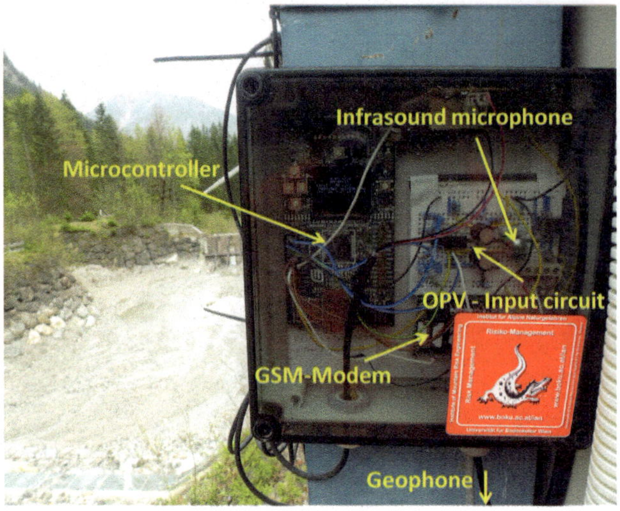

Fig. 1 System setup at the test site Dristenau with Electret microphone as infrasound sensor

For the signal processing and the execution of the detection algorithm a microcontroller of the type Stellaris Luminary Evaluation-board LM3S8962 with a 50 MHz ARM Cortex-M3 microprocessor is used. This microcontroller is also used as datalogger, which can store the sensor data on a micro-SD card for up to four month (16 GB card). It also offers an Ethernet connection for remote control and sending status massages and event alerts. The power consumption of this setup is below 1.5 W which makes this system very useful for standalone systems with solar power supply. Figure 1 shows a picture of a system setup installed at the tyroles test site Dristenau.

Detection Method

The developed detection algorithm is based on an analyses of the evolution in time of the frequency spectrum of the seismic and infrasound signals. Therefore the input signals are processed by fast Fourier transform [Bluestein FFT algorithm (Rabiner et al. 1969)] with 100 samples every second.

Afterwards different frequency bands of the infrasound and seismic signals are analyzed and have to fulfill different detection criteria for a specific time span (detection time) which is set to 20 s at the current version.

For the infrasound signal two different frequency bands are used for debris flows or debris floods since the frequency content depends on the viscosity of the process (Hübl et al. 2013; Kogelnig 2012). The average amplitude of the debris flow/debris flood frequency band has to be over a certain limit for the detection time whereby two thresholds are used to distinguish between different event sizes [Level 1 (L1): mostly small debris floods or higher discharge with sediment transport; Level 2 (L2): real debris flows and debris floods].

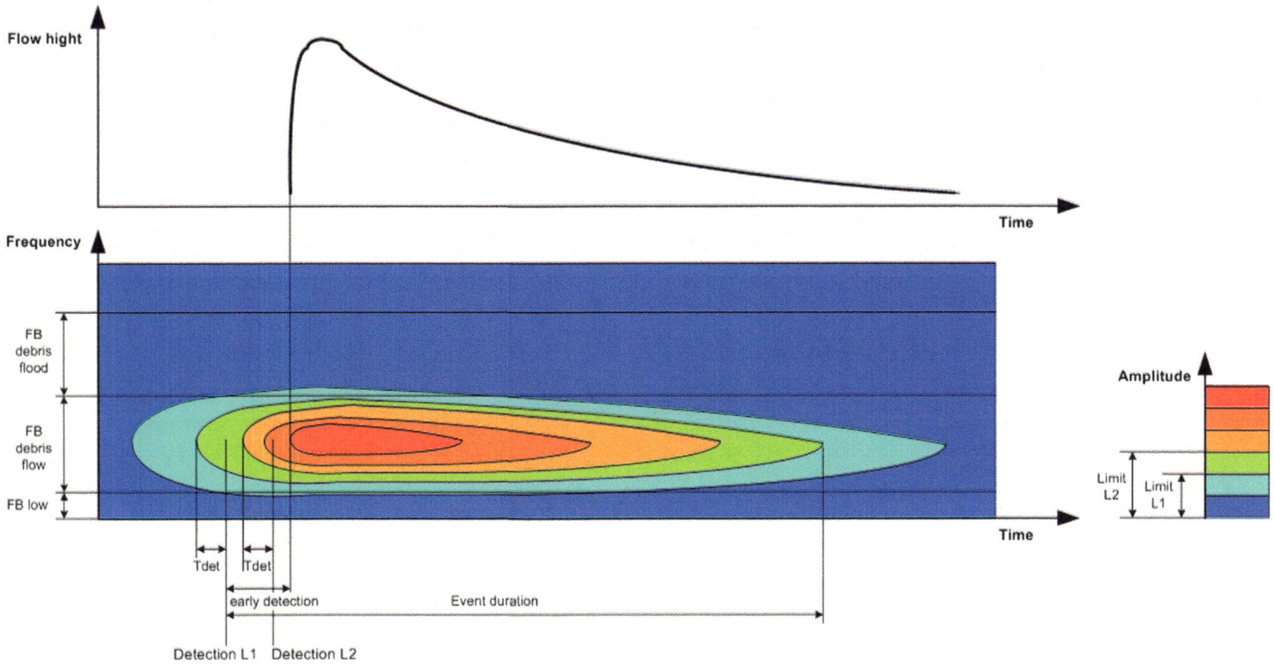

Fig. 2 Illustration of an event detection depicted in a running spectrum of a debris flow infrasound signal

To avoid false alarms due to wind a criteria is that the average infrasonic amplitudes of the debris flow/debris flood frequency band has to be above the average amplitude of the frequency band below, which is manly dominated by wind.

For the seismic signals only one frequency band is used for debris flows and debris floods, since the frequency-dependency of the amplitudes is not so significant than for the infrasound signal. As a further criteria the variance of the infrasound or seismic amplitudes has to be below a certain limit. Since mass movements produce broad banded seismic and infrasound signals compared to narrow banded signals from artificial signal sources, this criteria can be used to avoid false alarms. An illustration of this detection principle is shown in Fig. 2 in a running spectrum of a debris flow infrasound signal.

In a former version (Schimmel and Hübl 2015) also a fourth upper frequency band has been used for the infrasound signals but since debris flood signals with higher peak frequency have not been detected with this criteria, this frequency band has been omitted and a larger detection time was chosen.

Currently this detection method is tested on several test sites and a frequent fine tuning of the parameters is done to increase detection probability and reduce false alarms.

Event Detection Examples

This section shows an example of the seismic and infrasound signals of two different events and the application of the detection algorithm. The first event is a debris flow recorded on the 16.08.2015 at the test site Lattenbach (catchment area 5.3 km^2) and the second example shows a debris flood recorded on 08.07.2015 at the test site Dristenau (catchment area 9.9 km^2). Both test sites are located in Tyrol and operated by the Institute of Mountain Risk Engineering.

Debris Flow at Lattenbach

The debris flow at Lattenbach was recorded on 16.08.2015 with a peak discharge of 16 m^3/s, a total volume of 10,000 m^3 and the total duration of this event was approximately 1500 s. The max. infrasound amplitudes were up to 1.5 Pa and the max. seismic amplitudes were up to 200 µm/s. The event was detected by the detection algorithm at sec. 2859 for level 1 and at 2994 s for level 2. So the time between detection and passing of the of first surge (at 2960 s) at the sensor site was around 100 s. The infrasound and the seismic signals, as well as the flow height measured by a radar gauge is depicted in Fig. 3.

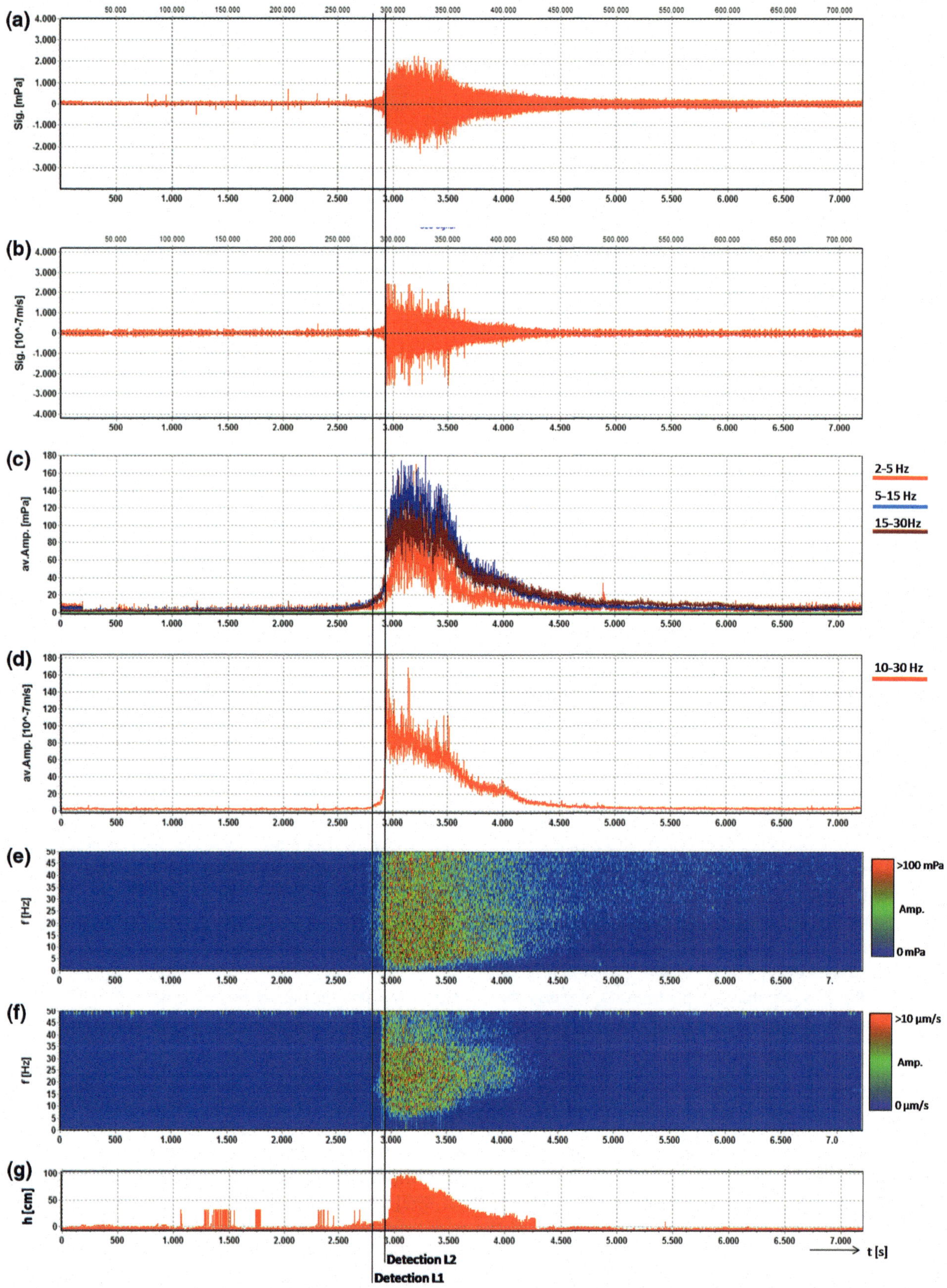

Fig. 3 Infrasound and seismic data of a debris flow monitored at the Lattenbach test site on 16.08.2015. Signals are represented with a common base of time. **a** Infrasound time series; **b** Seismogram; **c** Average amplitude of the three frequency bands of the infrasound signal; **d** Average amplitude of the frequency band of the seismic signal; **e** Running spectrum of the infrasound signal; **f** Running spectrum of the seismic signal; **g** Flow height; Lines: time of first detection based on infrasound and seismic data for level 1 and level 2

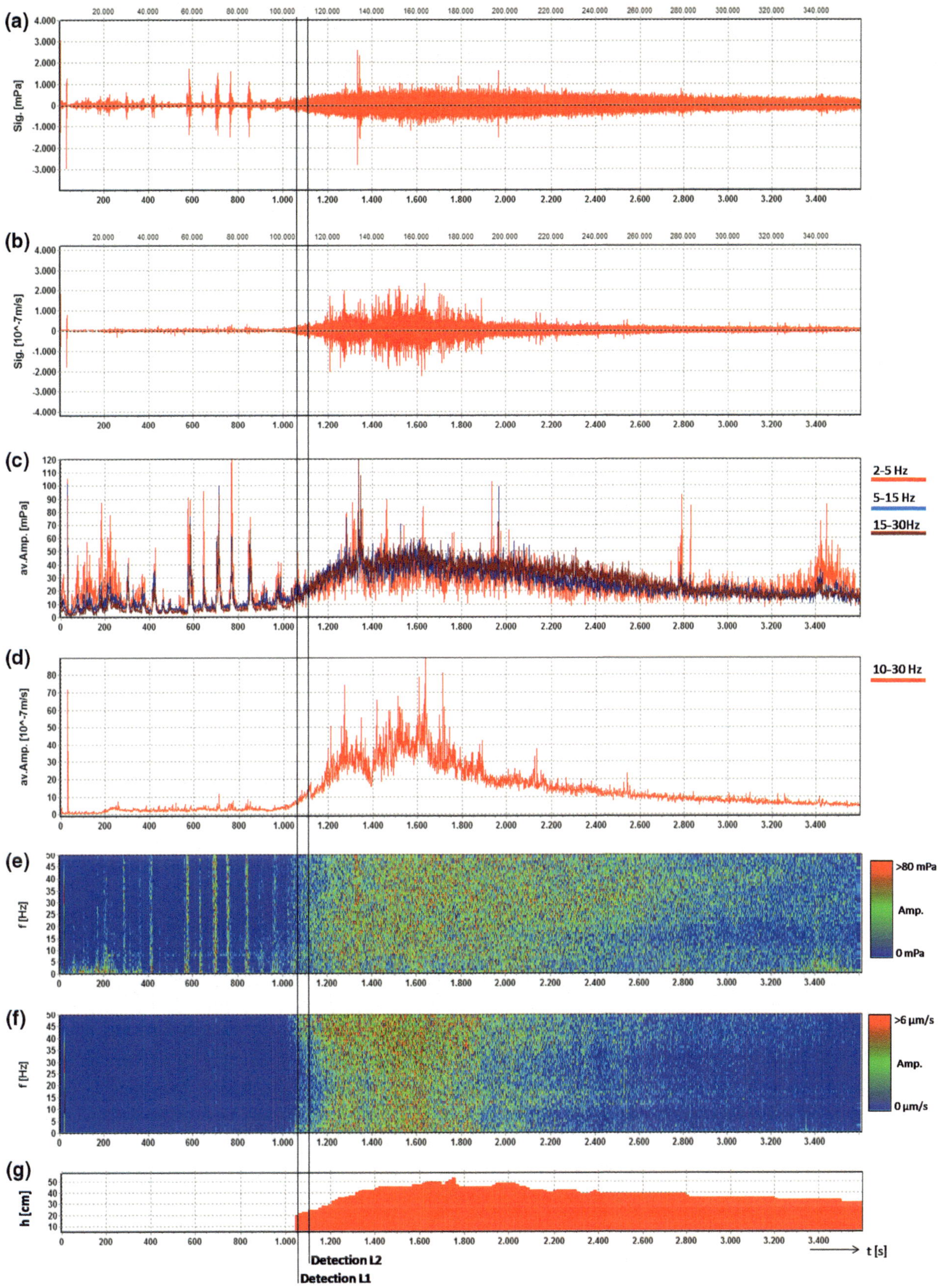

Fig. 4 Infrasound and seismic data of a debris flood monitored at the Dristenau test site on 08.07.2015. Signals are represented with a common base of time. **a** Infrasound time series; **b** Seismogram; **c** Average amplitude of the three frequency bands of the infrasound signal; **d** Average amplitude of the frequency band of the seismic signal; **e** Running spectrum of the infrasound signal; **f** Running spectrum of the seismic signal; **g** Flow height; Lines: time of first detection based on infrasound and seismic data for level 1 and level 2

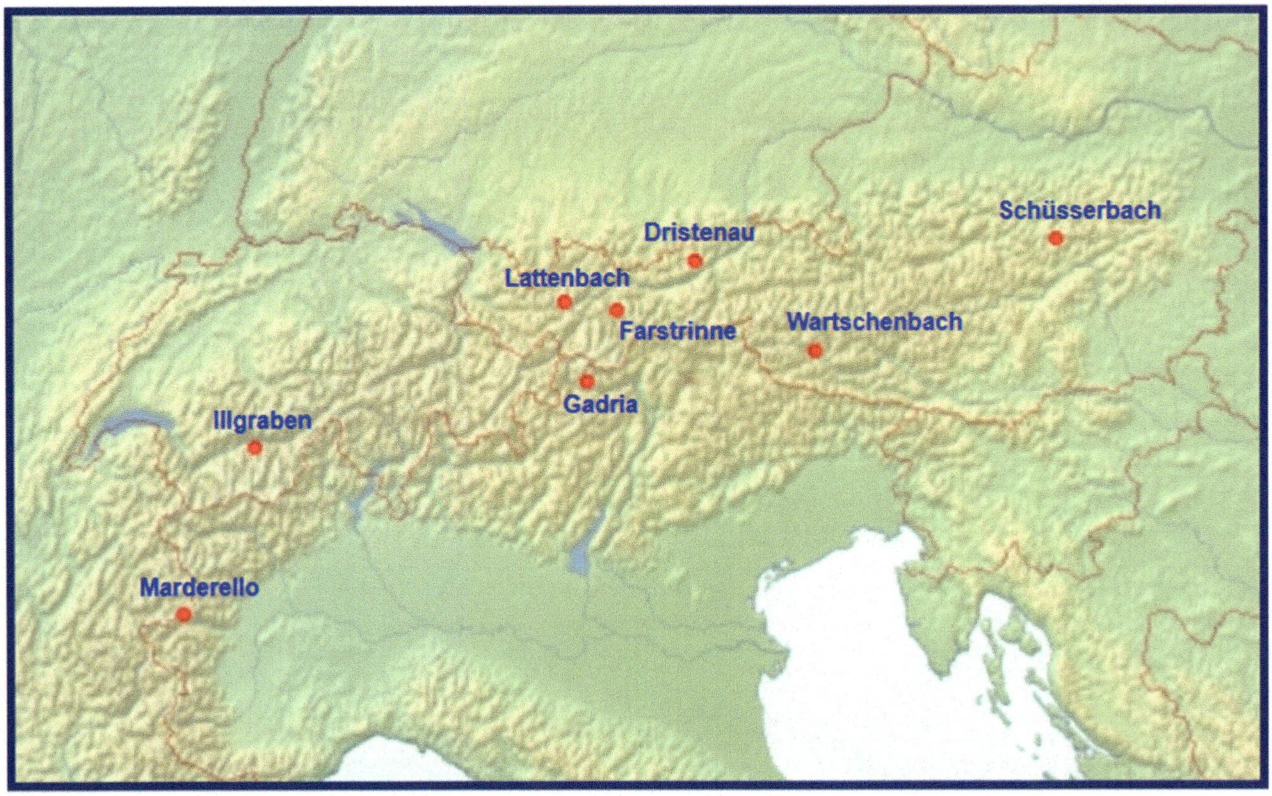

Fig. 5 Map of the test sites from 2013 to 2016

Debris Flood at Dristenau

The debris flood at Dristenau occurred on 08.07.2015 with a peak discharge of 3.5 m^3/s and a duration of approximately 2400 s. The max. infrasound amplitudes are up to 600 mPa and the max. seismic amplitudes are up to 100 µm/s. The event was detected by the algorithm at sec. 1149 for level 1 and at 1196 s for level 2 which means that this event was just detected as it passed by. Figure 4 shows the infrasound and seismic signals and the flow height (the flow height measurement started at level greater 20 cm)

The frequency distribution of the three infrasound frequency bands or the infrasound running spectrum reveals the difference between debris flow and debris flood signals. The infrasound signals of the debris flow at Lattenbach has its peak frequencies in the 5–15 Hz frequency band whereas the debris flood event at Dristenau has its peak frequencies occurring in the 15–30 Hz frequency band and above.

Results Event Detection

A long period of testing with high occurrence of different events is necessary for the development of the detection and identification method. Therefore this system is currently tested at five test sites in Austria, two in Italy and one in Switzerland. The map in Fig. 5 shows the location of the different test sites where this system has been installed in the last years.

Table 1 shows an overview of the number of events, detections and false alarms from 2013 to 2015. The results of the season 2016 were not available at the date of writing this paper. The event detections are split in smaller events (level 1 detections (L1), mostly small debris floods and higher discharge with sediment transport) and larger events [level 2 detections (L2), real debris flows and debris floods].

All larger events (L2) could be detected at every test site in the testing period and only four smaller events (level 1) could

Table 1 Detection results at the test sites since 2013

Test site	Year	Detected events	Detection L1	Detection L2	False alarms	Not classifiable detections	Not detected events
Lattenbach	2013	1	1	0	0	1	0
	2014	2	2	0	1	0	0
	2015	3	0	3	0	0	1
Dristenau	2013	18	14	4	0	1	0
	2014	7	7	0	0	0	0
	2015	12	9	3	0	0	1
Farstrinne	2013	0	0	0	1	0	0
	2014	2	0	2	0	0	0
	2015	1	0	1	0	0	0
Schüsserbach	2013	3	2	1	0	0	0
	2014	0	0	0	0	0	0
	2015	2	2	0	0	1	0
Wartschenbach	2013	5	5	0	0	9	0
	2015	1	1	0	0	0	0
Illgraben	2015	6	2	4	0	1	0
Gadria	2015	1	1	0	1	0	0
Marderello	2015	3	2	1	3	1	2
	Sum	67	48	19	6	14	4

not be detected. Only six false alarms occurred in the last years and 14 detections could not be clearly classified as event (nine of them at Wartschenbach due to technical problems).

Conclusion

This paper presents an approach for a detection system for debris flows and debris floods, which can be used as low cost and easy to install warning system for different kind of alpine mass movements.

The combination of infrasound and seismic sensors can increase the detection probability and reduce false alarms. So it was possible to detect all larger events in the period from 2013 to 2015 at eight different test sites, while only 6 false alarms were registered in that time. However the sensor equipment and the location of such a system has to be chosen carefully and the parameters of the detection algorithm has to be adapted on the application of the system and the background noise of the site.

First analyses of different event types and different magnitudes has shown a dependency of the peak frequency range on the viscosity and a relation of the maximum infrasound and seismic amplitudes to the event magnitude. But currently it is not possible to set up a common set of identification rules due to the wide variance of the produced signals. So a large databases of different well categorized events at different test sites will be necessary for a reliable event identification.

In summary this work confirmed that debris flows and debris floods produce seismic and infrasonic signals characteristics that are reproducible at very different experimental sites and under different environmental conditions and it shows promising results in the detection of alpine mass movements based on a combination of seismic and infrasound sensors.

Acknowledgements This work was financed in the first period by the Austrian Research Promotion Agency (FFG) Bridge-Project "Automatic detection of alpine mass movements" (No. 836474) and in the second period by the Austrian Academy of Science (ÖAW) Earth System Sciences (ESS) project "Identification of sediment-related disaster based on seismic and acoustic signals".

References

Arattano M (2003) Monitoring the presence of the debris-flow front and its velocity through ground vibrations detectors. In: Proceedings of the 3rd International Conference on Debris-Flow Hazards Mitigation: Mechanics, Prediction and Assessment. Millpress, Rotterdam, pp 731–743

Arattano M, Abancó C, Coviello V, Hürlimann M (2014) Processing the ground vibration signal produced by debris flows: the methods of amplitude and impulses compared. Comput Geosci 73:17–27

Burtin A, Hovius N, McArdell BW, Turowski JM, Vergne J (2014) Seismic constraints on dynamic links between geomorphic processes and routing of sediment in a steep mountain catchment. Earth Surf Dyn 2(1):21–33. doi:10.5194/esurf-2-21-2014

Burtin A, Hovius N, Turowski JM (2016) Seismic monitoring of torrential and fluvial processes. Earth Surf Dyn 4(2):285–307

Chou HT, Cheung YL, Zhang SC (2007) Calibration of infrasound monitoring systems and acoustic characteristics of debris-flow movements by field studies. Institute of Mountain Hazards and Environment, Chinese Academy of Science and Ministry of Water resources

Chou HT, Chang YL, Zhang SX (2010) Acoustic signals and geophone response of rainfall-induced debris flows. J Chin Inst Eng

Coviello V, Arattano M, Turconi L (2015) Detecting torrential processes from a distance with a seismic monitoring network. Nat Hazards 78(3):2055–2080

Huang C-J, Yin H-Y, Chen C-Y, Yeh C-H, Wang C-L (2007) Ground vibrations produced by rock motions and debris flows. J Geophys Res: Earth Surf 112:F02014. doi:10.1029/2005JF000437

Hübl J, Schimmel A, Kogelnig A, Suriñach E, Vilajosana I, McArdell BW (2013) A review on acoustic monitoring of debris flow. Int J Saf Secur Eng 3(2):105–115. ISSN 2041-9031

Kogelnig A (2012) Development of acoustic monitoring for alpine mass movements. Ph.D. Thesis, Institute of Mountain Risk Engineering, University of Natural Resources and Life Sciences (BOKU), Vienna

Kogelnig A, Hübl J, Suriñach E, Vilajosana I, Mcardell BW (2010) Infrasound produced by debris flow: propagation and frequency content evolution. Nat Hazards

Marchetti E, Ripepe M, Ulivieri G, Kogelnig A (2015) Infrasound array criteria for automatic detection and front velocity estimation of snow avalanches: towards a real-time early-warning system. Nat Hazards Earth Syst Sci 15:2545–2555

Rabiner LR, Schafer RW, Rader CM (1969) The chirp z-transform algorithm and its application. Bell Syst Tech J 48:1249–1292. doi:10.1002/j.1538-7305.1969.tb04268.x

Schimmel A, Hübl J (2014) Approach for an early warning system for debris flow based on acoustic signals. In: Lollino G et al (ed) Engineering geology for society and territory, vol 3. pp 55–58

Schimmel A, Hübl J (2015) Automatic detection of debris flows and debris floods based on a combination of infrasound and seismic signals. Landslides, online first. ISSN 1612-510X

Suriñach E, Kogelnig A, Vilajosana I, Hübl J, Hiller M, Dufour F (2009) Incoporación del la señal de infrasonido a la detección y estudio de aludes de nieve y flujostorrenciales, VII Simposio Nacinal sobre Taludes y LaderasInestables, Barcelona, Spain

Ulivieri G, Marchetti E, Ripepe M, Chiambretti I, Segor V (2012) Infrasonic monitoring of snow avalanches in the alps. In: Proceedings International Snow Science Workshop 2012, Anchorage, Alaska, pp 723–728

Zhang S, Hong Y, Yu B (2004) Detecting infrasound emission of debris flow for warning purpose, 10. Congress Interpraevement

Simulating the Formation Process of the Akatani Landslide Dam Induced by Rainfall in Kii Peninsula, Japan

Pham Van Tien, Kyoji Sassa, Kaoru Takara, Khang Dang, Le Hong Luong, and Nguyen Duc Ha

Abstract

The Akatani landslide triggered by heavy rainfall during Typhoon Talas on 4 September 2011 is one of 72 deep-seated catastrophic rock avalanches in Kii Peninsula, Japan. The landslide is about 900 m in length, 350 m in average width and 66.5 m of maximum depth of the sliding surface. A rapid movement of the landslide was downward the opposite valley and formed a natural reservoir that has a height of about 80 m and a volume of 10.2 million m^3. This paper presents preliminary results of the simulation of the formation process of the Akatani landslide dam by using ring shear apparatus incorporated with a computer simulation model LS-Rapid. Ring shear tests on sandstone-rich materials and mudstone-rich materials taken near the sliding surface indicated that a rapid landslide was triggered due to excess pore water pressure generation under shear displacement control tests and pore water pressure control tests. The pore water pressure ratio (r_u) due to rainfall was monitored from 0.33 to 0.37 in the ring shear tests on rainfall-induced landslides, approximately. Particularly, the formation process of the Akatani landslide dam and its rapid movement were well simulated by the computer model with physical soil parameters obtained from ring shear experiments. The actual ratio of pore water pressure triggering landslides was 0.35 in the computer simulation model. The results of the Akatani landslide simulation would be helpful to the understanding of failure process of deep-seated landslide induced by rainfall for future disaster mitigation and preparation in the area.

P. Van Tien (✉) · K. Takara · N.D. Ha
Graduate School of Engineering, Disaster Prevention Research Institute, Kyoto University, Gokasho, Uji, Kyoto, 611 0011, Japan
e-mail: phamtiengtvt@gmail.com

P. Van Tien · L.H. Luong
Ministry of Transport Institute of Transport Science and Technology, Hanoi, Vietnam
e-mail: lehongluong@gmail.com

K. Sassa · K. Dang
International Consortium on Landslides, 138-1, Tanaka Asukaicho, Sakyo-ku, 606-8226 Kyoto, Japan
e-mail: kyoji.sassa@gmail.com

K. Dang
VNU University of Science, Hanoi, Vietnam

M. Mikoš et al. (eds.), *Advancing Culture of Living with Landslides*,
DOI 10.1007/978-3-319-53483-1_59

Keywords
Landslide • Akatani • Rainfall • Mechanism • Ring shear apparatus • Computer simulation model

Introduction

Typhoon Talas in 2011, which produced heavy rainfall across the central of Japan, triggered more than 3000 landslides with a total volume of collapsed sediment approximately 100 million m^3 in Kii Peninsula, Japan (Hayashi et al. 2013; SABO 2013). The accumulative precipitation reached a record-breaking amount of 1812.5 mm at the Kamikitayama AMeDAS station in Nara Prefecture between August 31 and September 4 (JMA 2011). According to a report by the Fire and Disaster Management Agency on 28 September 2012, the disasters not only claimed 98 casualties but also led to 379 houses completely destroyed and 3159 houses partially destroyed. 17 out of 72 deep-seated catastrophic landslides were formed as natural landslide dams with the maximum height over than 100 m, which have put communities at potential risks from secondary hazards such as downstream flood, debris flow, overtopping flow, and erosion. The study area of Kii Peninsula and a google image of the Akatani landslides are shown in Fig. 1.

The study area in this research is the Akatani slope in the Totsukawa village, which was devastated by two large-scale deep-seated rock avalanches including the Akatani landslide and the Akatani East landslide. The Akatani landslide dam is located in upstream and about 1500 m far from the Akatani East in the north-east direction (Fig. 1). The dimensions of the Akatani landslide and the Akatani East landslide are about 900 and 800 m in length and 350 and 320 m in average width, respectively. Since the Akatani East landslide dam breached just after its formation, only the Akatani landslide has put potential risks of failure and overtopping to communities downstream. All mentions of the landslide dam in the Akatani area imply to the large-scale deep-seated Akatani landslide.

The Akatani landslide occurred on 4 September 2011 preceded by gravitationally deformation slope (Chigira et al. 2013) while the later took place on the previous day. The landslides dammed the river at the toe to create natural reservoirs. The maximum height and the volume of the Akatani calculated from DEM data before 2011 and DEM data after the sliding in September 2011 are about 80 m and 10.2 million m^3. The Akatani landslide dam has become the second largest landslide dam in Kii Peninsula. The impoundment capacity of Akatani dam is up to 5.5 million m^3 in a large catchment of 13.2 km^2 (SABO 2013). The maximum depth of the Akatani landslide is around 66 m (Fig. 2-right). After the 2011 catastrophic landslide, several re-slidings and debris flows with the maximum total depth of 30 m took place from October 2011 to August 2014 (Fig. 2-right). The calculated dimensions of the landslide dams above are based on DEM data prior to the 2011 event by Geospatial Information Authority of Japan (GSI) and DEM data generated from airborne laser scanning surveys by Kii Mountain District Sabo Office (SABO) in the years of 2011 and 2014.

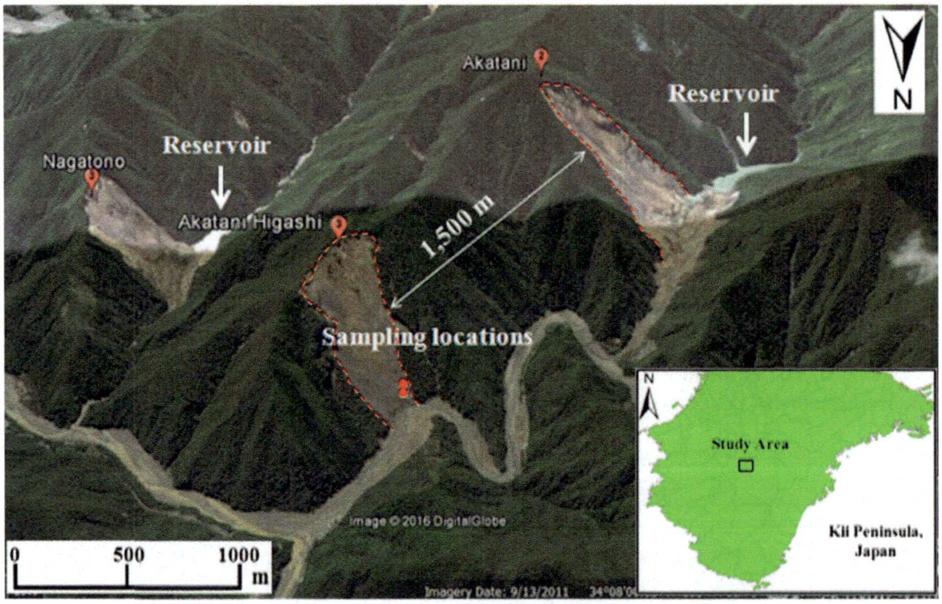

Fig. 1 A bird's-eye view of the Akatani landslides in Totsukawa village (by the Google Earth, 2011)

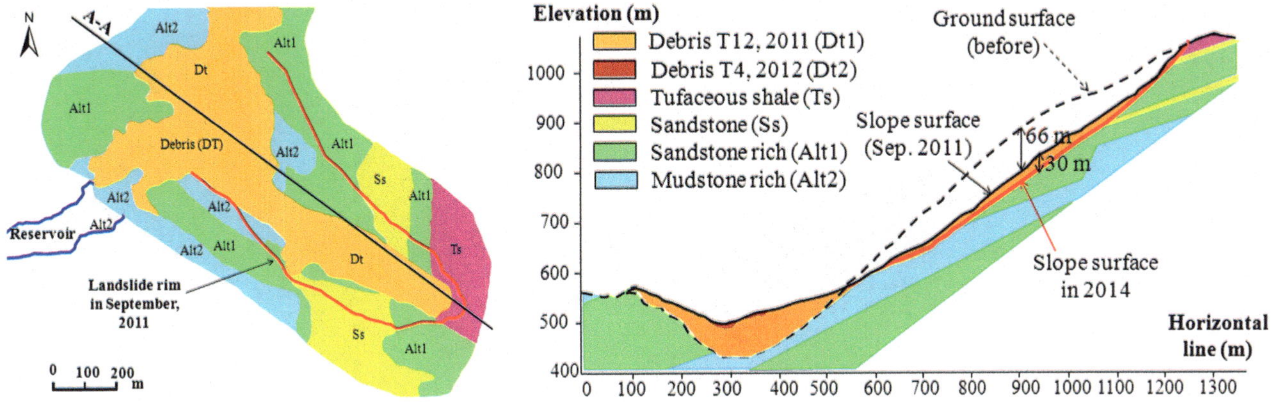

Fig. 2 Geological feature of the Akatani landslide (*left*) and a cross-section A-A (*right*) (re-drawn and modified from maps created by Kii Mountain District Sabo Office)

The geology of the Akatani slopes mainly composes materials of mudstone and sandstone, and a part of tuffaceous shale on the head scarp in Miyama Accretionary Complex, Hidaka-gawa Group of the Cretaceous Shimanto Terrane (Geological Survey of Japan 1998). The geological plan and a cross section of the Akatani landslide created by SABO in 2014 were re-drawn and modified as shown in Fig. 2.

This research was conducted to clarify the failure mechanism and simulate the formation process of the catastrophic Akatani landslide dam, which would be helpful for better understanding of the behavior of rainfall-induced deep-seated landslides with the similar geological conditions in the Kii Peninsula. Firstly, a site survey was carried out in detail to explore failure characteristics of slopes in the Akatani landslides area. Two soil samples of materials near the sliding surface were then taken for testing by using undrained high-stress dynamic-loading ring shear apparatus ICL-2. Finally, a computer model of the Akatani landslide with input parameters obtained by ring shear tests was conducted to simulate the formation and motion process of a rapid landslide induced by rainfall.

Site Investigation

Geological and Topographical Features

The Akatani slope has a deep topography with an inclination of 34° and elevations ranging from 370 to 1100 m. It is covered by dense and thick vegetation. A bird's-eye view of the surface of Akatani landslides is in a brown color because of oxidization process that taken place when the iron of subsurface materials was oxidized (as in Fig. 1). Site observations indicated that the Akatani area is characterized by broken formations and mixed rocks with wedge-shape discontinuities that were also reported by Chigira et al. (2013). The geology consists of interbedded sandstone and mudstone that are made of fine-grained materials (clay-sized and silt-sized grains) and coarse materials (boulders) in both (Figs. 3 and 4). The geological setting of the Akatani slope is likely the favourable condition for the build-up of pore water pressure (Chigira et al. 2013). It is because that allowed rainwater to percolate deeply into the slope resulting in an increase of groundwater level. In the Akatani landslide, there existed gullies in the head scarp on which groundwater and surface runoff flow down. The presence of a little water flowing out on the sliding surface implied the contribution of groundwater table to the landslide occurrence through build-up pore water pressure within the slope.

Sliding Surface

The Akatani landslides occurred along the northwest-dipping bedding planes (Chigira et al. 2013) (see in Figs. 1 and 2) and its sliding surface is undulatory and subparallel to the bedrock (Figs. 3c and 4a). Layers below the sliding plane contain the inter-bedding of fractured sandstone and mudstone. Materials in the shear zone were crushed and liquefied to be finer grained particles. Site survey indicated that the sliding surface was more likely to be formed along dip-slip faults or in thin weak zones of interbedded sandstone and mudstone (Fig. 4b).

Soil Sampling

The Akatani landslide was inaccessible due to its continuous falls of debris that reposed on the upper slope. However, the Akatani landslide and the Akatani East landslide occurred in the right flank of the Akatani valley where the slopes are located in the melange matrix of Cretaceous Accretionary Complex. A detailed site survey on the Akatani East landslide indicated the geological structure of the sliding body

Fig. 3 **a** A photograph of the Akatani landslide and its failure features; **b** A closed-up view of the head scarp with gullies; and **c** Broken-formation and the undulatory sliding surface

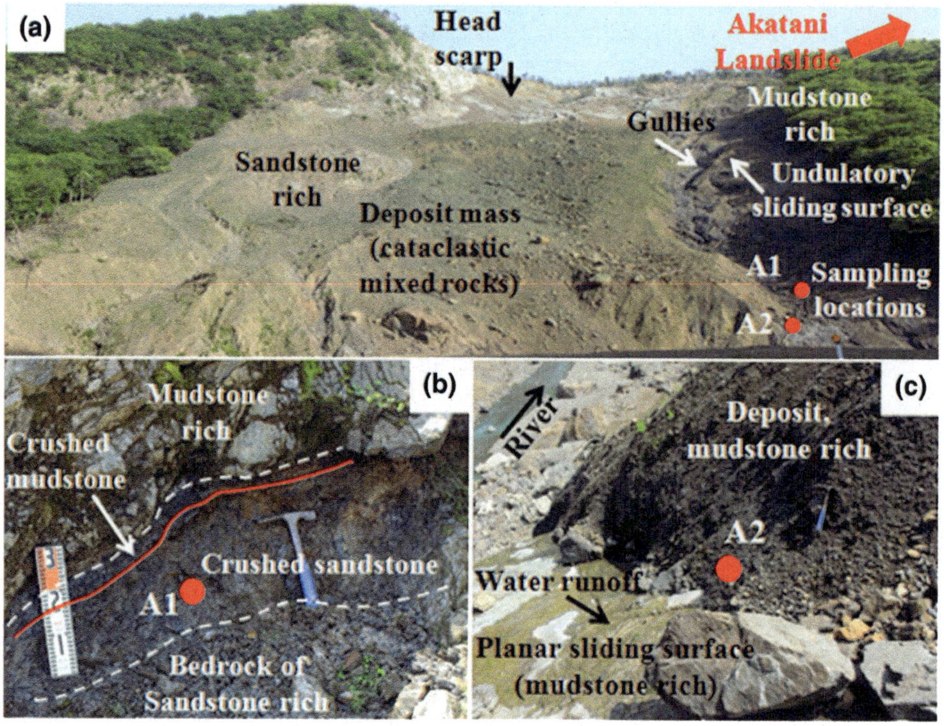

Fig. 4 **a** A photograph of the Akatani East landslide and its failure features; **b** Sampling location A1: Crushed layers of sandstone and mudstone in a fault; and **c** Sampling location A2: Deposit mass reposed on the sliding surface

also consists of fractured mixed rocks and broken formations of sandstone and mudstone. Under this circumstance, the geological features of the Akatani landslide are quite similar to the Akatani East landslide. Therefore, soil samples that have geological characteristics similar to materials of the sliding surface of Akatani landslide were collected at the sliding surface of the Akatani East landslide. As can be seen in the geological plan and cross section of the Akatani landslide (Fig. 2), the sliding occurred mainly in the layers of sandstone and mudstone, so it is needed to collect these two materials to perform laboratory tests for studying its failure mechanism.

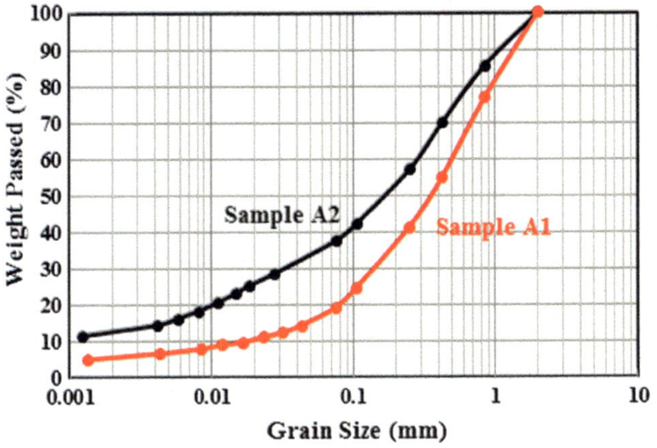

Fig. 5 Grain-size distribution of two samples

The first sampling location (A1) was on a fault gouge of interbedded sandstone-rich block (below) and mudstone-rich block (above) (Fig. 4b). As can be seen in Fig. 4b, there were two identified layers in the fault: the lower layer is a sandstone-rich material in light grey color and the upper one is a thinner layer of a finer-grained mudstone-rich material in a dark grey color. The upper layer and the lower layer may be materials produced from the large mudstone mass above and the bed rock of sandstone due to a less grain crushing during down sliding along the fault. The second sampling location (A2) was a dark grey mudstone-rich material staying between the strong bedrock and debris mass (Fig. 4c). By observing and doing a grain-size distribution analysis, we found that the thin layer of mudstone in the first sampling point is the same to the mudstone-rich material in the second location. Hence, we have only conducted laboratory testing for two samples, namely the sandstone-rich sample (hereinafter referred to as Sample A1) and the mudstone-rich sample (hereinafter referred to as Sample A2). The grey color of samples shows that soil materials were deoxidized completely when those were saturated by surface water flows (Fig. 4c). To perform ring shear tests, two soil samples passed through a 2 mm size sieve. The grain size distribution of two samples is denoted in the Fig. 5. As can be seen, both samples contain clay-size and silt-size materials and the Sample A2 is finer than the Sample A1. The dry unit weight and the saturated unit weight corresponding to the sample height under different consolidation stresses of two samples are shown in Fig. 6.

Test Results by Ring Shear Apparatus ICL-2

Ring Shear Device and Testing Procedure

The latest version of ring shear apparatus ICL-2 designed in 2013 was employed to investigate the initiation and motion mechanism of large-scale deep-seated landslides, which has a capacity until 3.0 Mpa of normal stress and pore water pressure in the undrained condition. This device allows observing quantitatively the entire process of formation, failure and motion at a high-velocity of given samples and monitoring well pore pressure changes and possible liquefaction for a large displacement at the steady state. An introduction to ICL-2 version and its fundamental concept for the landslide simulation are described by Sassa et al. (2014).

Testing procedure by using the ring shear device ICL-2 mainly consists of the following steps: saturating and setting samples; checking saturation degree by the measurement of the pore pressure parameter (B_D), and consolidating and shearing the sample. Firstly, soil samples were de-aired to be fully saturated in a vacuum tank in several days. Next, immediately after the shear box was completely prepared in order of doing the CO_2 and de-aired water circulation to let all bubbles of air come out, saturated samples were built into the shear box. Then, the saturation degree was checked indirectly using B_D parameter ($B_D = \Delta u/\Delta\sigma$), which calculated as the ratio of excess pore pressure increment and normal stress increment under undrained condition B_D

Fig. 6 Dry and saturated unit weights of two samples with its height under different consolidation stresses

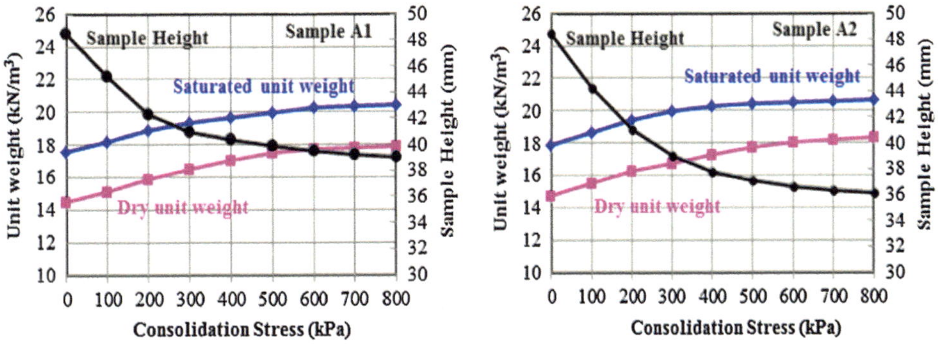

(Sassa et al. 2004). In this study, undrained tests were usually carried out with $B_D \geq 0.95$. Before starting tests, the sample was consolidated to the initial stress state at its sliding surface under drained condition. Finally, undrained ring shear tests on landslides were conducted in different modes of shearing including shear displacement control tests and pore water pressure control tests. For tests of rainfall-triggered landslides, the increment of pore pressure employs a back pressure control device with a servo-motor.

Test Results

(1) *Undrained Shear Displacement Control Tests*

Figure 7-left presents combined effective stress paths of undrained shear displacement control tests on Sample A1 at 400 and 800 kPa of normal stress that corresponds to 30–60 m in a depth of the sliding surface of the landslide. After consolidating the specimen under drained condition, the shear box was changed to the undrained condition and the sample was sheared under constant shear displacement of 0.01 cm/s in the undrained condition. During the testing, pore pressure change and shear strength reduction in the progress of shear displacement were monitored as illustrated in Fig. 7-right.

Test results indicated the stress path of samples reached the failure line at 39.5° during motion. A rapid motion of the landslide was initiated when a large excess pore water pressure generated and shear resistance dropped quickly. The sample was sheared until 1500 mm of shear displacement when the sample reached to the steady state. The steady-state shear resistances are 70 and 150 kPa for normal stresses of 400 and 800 kPa, respectively.

(2) *Rainfall-induced landslide simulation by pore water pressure control tests*

Pore water pressure control tests are the most important tests in this study for simulating rainfall-induced landslides. These tests were conducted on Sample A1 at 800 kPa of normal stress and 500 kPa of shear stress while those values for Sample A2 were 1000 and 620 kPa, respectively. The ratio of shear stress and normal stress is about 0.62 that corresponds to a natural slope angle of 34° approximately. The different normal stress shows a different loading on Sample A1 and Sample A2 from their depths at sliding surface. To do this simulation, both samples of sliding surface were saturated ($B_D = 0.96$) and consolidated to a pre-defined value of normal stress and shear stress in the drained condition. Then, pore water pressure acting on a sliding surface was produced by increasing gradually pore water pressure at the rate of 0.5 kPa/sec for Sample A1 and 1 kPa/s for Sample A2 until the failure for all tests. Test results of Sample A1 and Sample A2 are depicted in Figs. 8 and 9 with the stress path (left) and time series data (right).

Test results indicated that landslides occurred when the effective stress path reached the failure line at peak of 42.0° for Sample A1 and 41.1° for Sample A2. At the failure points, landslides were triggered at pore water pressure of 260 kPa for Sample A1 and about 370 kPa for Sample A2 that correspond to critical pore pressure ratios of $r_u = 0.33$ and $r_u = 0.37$ (r_u defines as a ratio of pore water pressure increment triggering failures and its normal stress). Rainfall-induced rapid landslides occurred due to excess pore water pressure generation and a sharp drop in shear resistance of both samples.

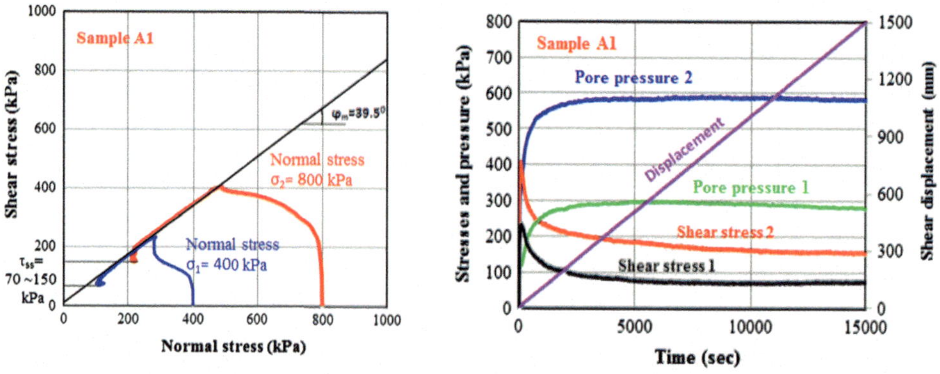

Fig. 7 Combined effective stress path (*left*) and shear strength reduction and pore water pressure generation in progress of shear displacement (*right*) of undrained shear displacement control tests (shearing velocity of 0.01 cm/s) on Sample A1 at normal stress σ_1 = 400 kPa with B_D = 0.95 and normal stress σ_2 = 800 kPa with B_D = 0.96

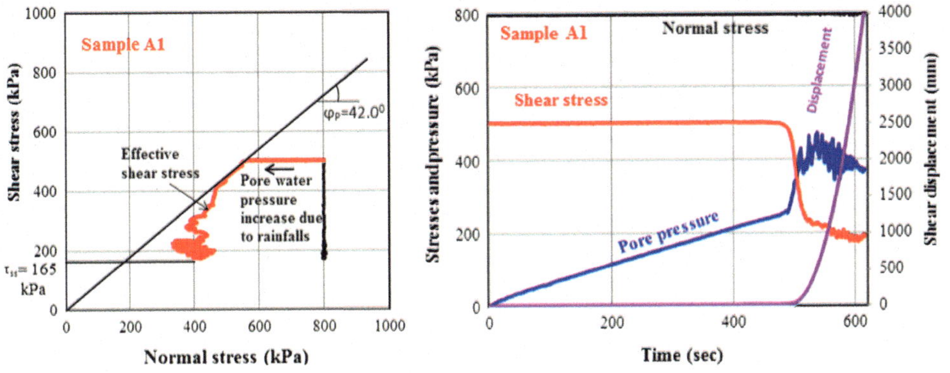

Fig. 8 Effective stress paths and time series of pore water pressure control test on Sample A1

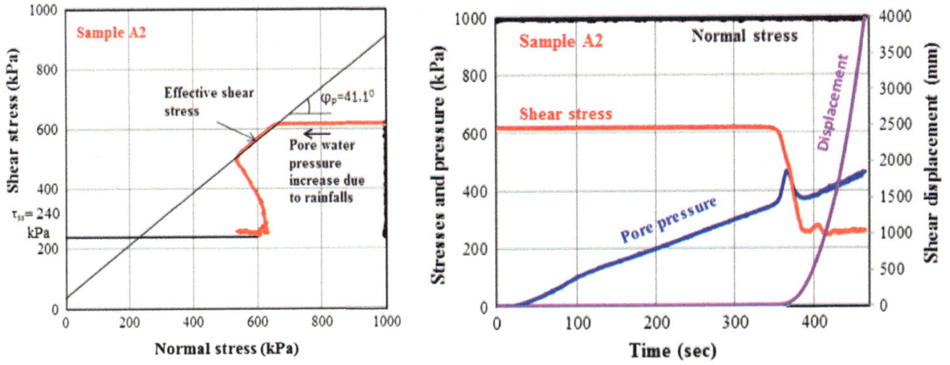

Fig. 9 Effective stress paths and time series of pore water pressure control test on Sample A2

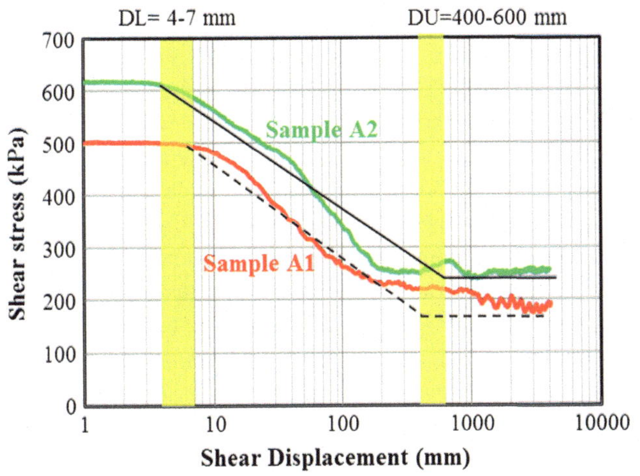

Fig. 10 Shear strength reduction in progress of shear displacement from pore water pressure tests

Shear strength reduction in the progress of shear displacement of ring shear tests on two samples are drawn in the Fig. 10. The results indicated that landslides were triggered at a critical shear displacement (DL) at peak strength from 4 to 7 mm and the failures reached the steady state at a shear displacement of 400–600 mm.

Simulation Result and Its Validation

Formation mechanism and post-failure motion process of the Akatani landslide were examined through a numerical modeling with LS-Rapid software that has been developed by Dr. Kyoji Sassa (International Consortium on Landslides). This computer model was integrated to simulate the whole sliding process of a certain slope from formation phase motion stage in post-failure. The basic concept of the integrated landslide simulation model (LS-RAPID) and its application to landslide simulations are presented by Sassa et al. (2010, 2014). In order to reproduce the rainfall-induced landslide, a triggering factor of pore water pressure generation was applied in this model. The slope surface mesh data of the Akatani landslides before sliding at 10 m resolution was employed, which is available to be downloaded from the Website of Geospatial Information Authority of Japan (GSI). The sliding surface of the landslide was built based on

Table 1 Soil parameters for a computer simulation model

Soil parameters	Value	Source
Shear strength parameters		
Total unit weight of the mass (γ_t)	20.5 kN/m^3	Test data
Unit weight of water (γ_w)	9.81 kN/m^3	Assumed
Lateral pressure ratio ($k = \sigma_h/\sigma_v$)	0.3–0.5	Estimated
Friction angle during motion (Φ_m)	39.5°	Test data
Peak friction angle at sliding surface (Φ_p)	41.1°	Test data
Steady state shear resistance (τ_{ss})	70–150 kPa	Test data
Peak cohesion at sliding surface (c_p)	250–300 kPa	Test data
Shear displacement at the start of strength reduction (D_L)	5 mm	Test data
Shear displacement at the end of strength reduction (D_U)	500 mm	Test data
Pore pressure ratio (r_u)	0.0–0.35	Test data
Pore pressure generation rate (B_{ss})	0.3–0.8	Estimated
Cohesion inside mass (c_i)	0.1	Assumed
Cohesion at sliding surface during motion (c_m)	0.1	Assumed
Parameters of the function for non-frictional energy consumption		
Coefficient for non-frictional energy consumption	1	See Sassa et al. (2010)
Threshold value of velocity (m/s)	70	Reference data
Threshold value of soil height (m)	100	Reference data

survey data for calculating sediment volume of sliding area that was created by Kii Mountain Office in 2012. Most parameters used in the model are obtained from ring shear tests while others were estimated from field investigation or were assumptions based on likely conditions of the landslide (Table 1). The explanation of important input parameters is presented as below.

Lateral pressure ratio (k): The ratio was estimated to be 0.3 in the upper slope and 0.5 in the lower part and the toe.

Steady-state shear resistance (τ_{ss}): We used a value of 150 kPa at a deep soil layer (Fig. 7) in the upper slope distributed mostly by sandstone-rich materials (Fig. 2a, b). Because, the right part and the left of the lower slope were dominated by sandstone-rich materials and mudstone-rich materials (Fig. 2a, b), respectively. Thus, 70 kPa obtained at the test of smaller normal stress (Fig. 7) and 120 kPa (as a half of 240 kPa in the Fig. 9) were selected at a shallow soil layer in the lower slope.

Excess pore pressure generation (Bss): The value Bss of each area on the slope depends on the degree of saturation resulting from groundwater level. Thus, we used the Bss of 0.3–0.4 for parts of the head scarp due to a less saturation and higher value of 0.5–0.7 for the middle part while a very high value of 0.8 in the toe and lower part of the slope.

Peak cohesion before motion (Cp): Due to a very deep sliding surface existed along bedding rocks of mudstones, we chose a value of 250–300 kPa for Cp.

Pore water pressure ratio (r_u): Obtained by ring shear simulation of rain-induced landslides, the average value $r_u = 0.35$ as a critical pore water pressure ratio was used with the time increment of 20 s and duration of 150 s.

The simulation results for the Akatani landslide are illustrated in Table 2. The blue ball zones present stable parts while red ball zones show unstable parts as any failure occurs.

In the Table 2a, the slope was still stable at pore pressure ratio (r_u) of 0.34. When pore pressure ratio rose to 0.35 (the critical value of pore water pressure), local failures occurred as denoted by a red point in the mesh (Table 2b). Soon after, failed points spread to surrounding area and a motion appeared from the middle of the slope. The upper part of the slope was drawn up by downward movement of its lower zone because of losing a support at the base. This process quickly continued and exacerbated the failure in a large area of the slope (Table 2c, d). Consequently, a mass movement of the slope accelerated a rapid landslide. The collapsed slope moved down to the opposite valley, blocked the river course and formed a dam at the toe (Table 2e). The validation of the model's result was done by comparing the

Table 2 Computer simulation's result of the Akatani landslide and its validation

(a) At 19.2 s, with pore water pressure ratio due to rainfall $r_u < 0.35$, the lope was still stable

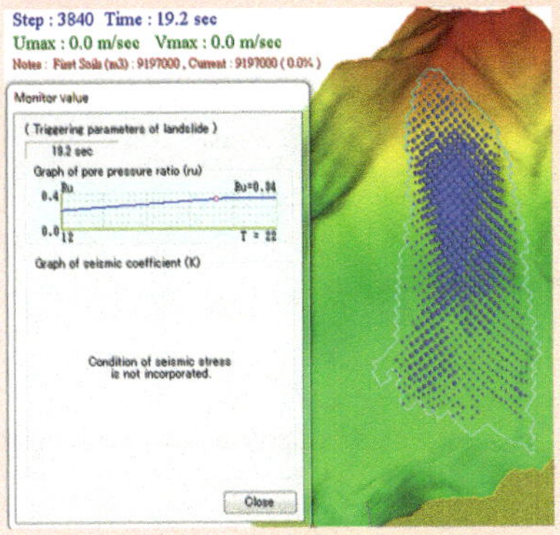

(b) At 21.8 s, when r_u reached to a value of 0.35, local failure occurred at the middle points of the slope

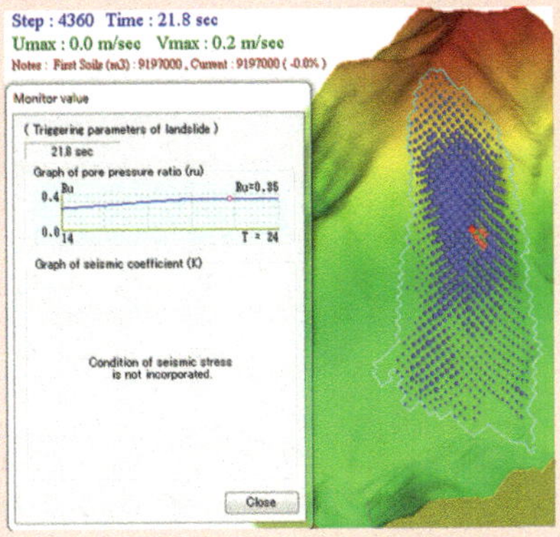

(c) At 34.3 s, progressive failures accelerated and were expaned to whole landslide body

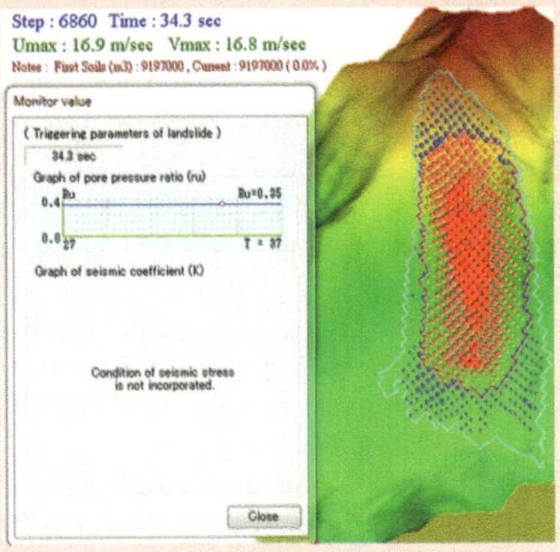

(d) At 63.1 s, soil mass started damming the river

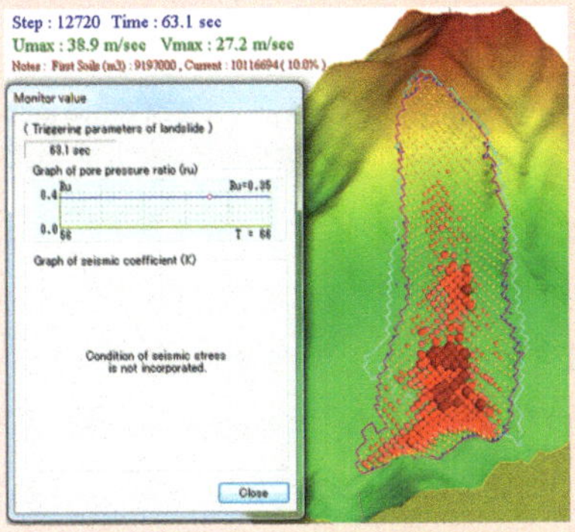

(e) At 119.2 s, the landslide ceased and formed a reservoir

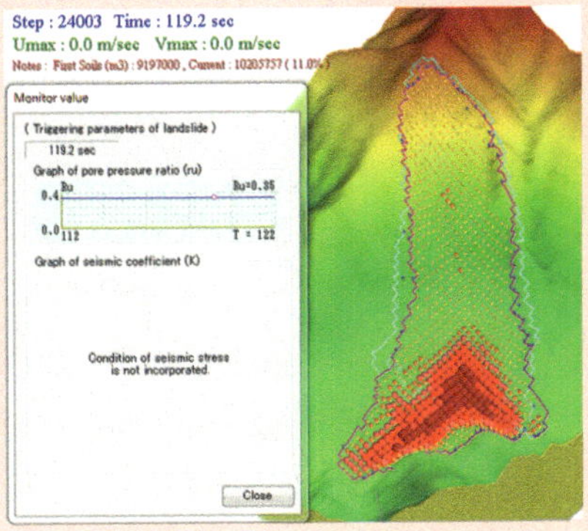

(f) An aerial photo of the landslide in September 2011 (SABO 2015)

Akatani landslide aerial photo (Table 2f) with the sliding area in the developed model.

Conclusions

The paper presents the interim results of the simulation of the formation progress for the deep-seated Akatani landslide dam. Some main geological and topographical characteristics of the landslides were also observed in a site investigation. We collected two samples near the sliding surface and conducted laboratory experiments by using ring shear apparatus ICL-2 to investigate the failure mechanism. Test results indicated that a rapid landslide was triggered due to excess pore water pressure generation and a significant loss of shear strength of Sample A1 and Sample A2 during shearing. Rainfall induced landslides at the pore water pressure ratio of 0.33–0.37 in ring shear tests approximately, but the specific value of the pore water pressure ratio triggering the landslide was 0.35 in the computer simulation model. The computer simulation model of the deep-seated rainfall-induced landslide due to an increase of ground water level brought by rainstorm was conducted well by using soil parameters obtained from ring shear tests on the samples of the sliding surface. The model produced the landslide at a rapid velocity and was examined as close as possible real sliding processes. This simulation is not only helpful for understanding the failure mechanism and motion processes of the deep-seated Akatani landslide, but it will be providing necessary technical knowledge for landslide disaster mitigation and preparedness for the Kii Peninsula area in the future.

Acknowledgements The author thanks Mr. HAYASHI, vice-director of Kii Mountain District Sabo Office for providing DEM data. Especially, I am grateful to Professor Hiroshi FUKUOKA (Niigata University), Mr. Tatsuya SHIBASAKI and Mr. OGAWAUCHI (senior engineers at JCE) and Mr. Hendy Setiawan (Kyoto University) for their cooperation and kindly supports during the field investigation. This research was financially supported by the research grant from Leading Graduate School Program on Global Survivability Studies in Kyoto University (GSS Program). We deeply acknowledge all these important supports during this study.

References

Chigira M, Tsou C, Matsushi Y, Hiraishi N, Matsuzawa M (2013) Topographic precursors and geological structures of deep-seated catastrophic landslides caused by Typhoon Talas. Geomorphology 201(2013):479–493

Geological Survey of Japan (1998) Geological map of Wakayama NI-53-15 at the scale 1:200,000. Available at https://gbank.gsj.jp/geonavi/geonavi.php#9,34.312,136.006

Geospatial Information Authority of Japan (GSI): Website for DEM data before the landslide. Available at http://www.gsi.go.jp/

Hayashi S, Uchida T, Okamoto A, Ishizuka T, Yamakoshi T, Morita K (2013) Countermeasures against landslide dams caused by Typhoon talas 2011. Tech Monitor, 20–26

Japan Meteorological Agency (JMA) (2011) Website for Rainfall data. Available at http://www.data.jma.go.jp/gmd/risk/obsdl/index.php

SABO (2013) A Pamphlet released in October, 2013 on Overview of the 2011 disaster induced by Typhoon No. 12, Implementation of an urgent investigation and countermeasures to disaster areas. Kii Mountain District SABO Office, Kinki Regional Development Bureau. The Ministry of Land, Infrastructure, Transport and Tourism (MLIT). http://www.kkr.mlit.go.jp/kiisanchi/outline/, 24 pp (in Japanese)

SABO (2014) A PDF document dated in July, 2014 on Deep-seated landslides occurred during Typhoon No. 12 in September 2011 in Akatani area. Kii Mountain District SABO Office, Kinki Regional Development Bureau. The Ministry of Land, Infrastructure, Transport and Tourism (MLIT), 15 pp (in Japanese)

SABO (2015) Documents (including reports, drawings, photographs and DEM data from airborne laser scanning surveys of the landslide) offered by Kii Mountain District SABO Office in May, 2015

Sassa K, Fukuoka H, Wang G, Ishikawa H (2004) Undrained dynamic-loading ring-shear apparatus and its application to landslide dynamics. Landslides 1(1):7–19

Sassa K, Nagai O, Solidum R, Yamazaki Y, Ohta H (2010) An integrated model simulating the initiation and motion of earthquake and rain induced rapid landslides and its application to the 2006 Leyte landslide. Landslides 7(3):219–236

Sassa K, Dang K, He B, Takara K, Inoue K, Nagai O (2014) A new high-stress undrained ring-shear apparatus and its application to the 1792 Unzen–Mayuyama megaslide in Japan. Landslides 11(5):827–842

Diversity of Materials in Landslide Bodies in the Vinodol Valley, Croatia

Sara Pajalić, Petra Đomlija, Vedran Jagodnik, and Željko Arbanas

Abstract

Numerous landslides of different types present common hazardous phenomena in the Vinodol Valley (64.57 km^2), situated near the City of Rijeka in coastal part of Croatia. During the previous and present geological investigations in the Vinodol Valley almost all landslide types were identified: falls, topples, slides and flows. The Vinodol Valley is characterized by irregular shape and a range of different landforms due to complex geological and geomorphological conditions. The inside of the valley is built of Paleogene siliciclastic (flysch) deposits, which is surrounded by the relatively steep carbonate borders composed of Upper and Cretaceous limestone. Along the most part of the NE border vertical rocky cliffs occur. The lower parts and the bottom of the valley (i.e., flysch deposits) are covered by heterogeneous Quaternary superficial deposits. In the Vinodol Valley more than 200 active and dormant landslides in soils were identified. Most of the active landslides are covered by dense forest vegetation. Original landslide topographies of dormant landslides are significantly modified by the anthropogenic agricultural activities. For this reason, the appropriate landslide identification and mapping method is the visual interpretation of the high-resolution LiDAR-derived imagery. Identified landslides are generally shallow to moderate shallow, with slip surface depth of several meters, and small to moderate small, with volumes in a range of 10^3–10^5 m^3. Due to the diversity of geological conditions in the Vinodol Valley and the large number of identified landslides, the index and classification properties of soils from the landslide-forming materials were investigated. Soil samples were taken and laboratory tested according to the rules of the European Soil Classification System. In this paper, results of preliminary investigations of the relationship between the different types of landslides and the landslide-forming materials are presented.

S. Pajalić
Faculty of Civil Engineering, University of Rijeka, 51000 Rijeka, Croatia
e-mail: sara.pajalic@gmail.com

P. Đomlija (✉) · V. Jagodnik · Ž. Arbanas
Department of Hydrotechnics and Geotechnics, Faculty of Civil Engineering, University of Rijeka, 51000 Rijeka, Croatia
e-mail: petra.domlija@uniri.hr

V. Jagodnik
e-mail: vedran.jagodnik@uniri.hr

Ž. Arbanas
e-mail: zeljko.arbanas@uniri.hr

M. Mikoš et al. (eds.), *Advancing Culture of Living with Landslides*,
DOI 10.1007/978-3-319-53483-1_60

Keywords
Landslide-forming materials • Soil classification • Landslide classification • Vinodol Valley

Introduction

The Vinodol Valley (area of 64.57 km^2) is situated in the coastal part of Croatia, in the Primorsko-Goranska County (Fig. 1). It has an elongated shape and stretches in the NW-SE direction nearly parallel to the eastern Adriatic coast. The area is mostly rural, with six major settlements and many small villages connected with relatively dense road network. It is populated by approximately 12,000 inhabitants and it is characterized by the long term tradition of agricultural activities (Mihalić and Arbanas 2013). The Vinodol Valley is characterized by complex geological conditions and specific morphological features. Hypsometrically lower parts and the bottom of the valley are built of Paleogene siliciclastic (flysch) deposits. The inside of the valley is surrounded by steep borders composed of Upper Cretaceous and Paleogene carbonate rocks.

Along the northeastern border in the central part and in the northwestern part of the valley high carstified limestone cliffs occur. Due to the complex geological conditions and intense weathering and geomorphological processes, the lower populated parts of the Vinodol Valley are mostly covered by the Quaternary superficial deposits.

Numerous landslides of different types and state of activity are common hazardous phenomena in the Vinodol Valley (Đomlija et al. 2014). Landslides occured individually, or interrelated to fluvial erosion processes (i.e., rill and gully erosion) and they often caused direct damages on infrastructures, public and private properties.

Research on the identification of hazardous phenomena in the Vinodol Valley has been undertaken with the aim to identify the types of landslides and erosion processes occurring in the siliciclastic and in the carbonate rock mass (Đomlija et al. 2014, 2016). During the previous and present geological and geomorphological investigations, almost all landslides types were identified: falls, topples, slides and flows. Only in soils, more than 200 landslides were identified (Đomlija et al. 2016). Geomorphological units of specific surface phenomena as results of the forming materials and hazardous processes were identified with the aim to depict relatively homogeneous portion of the land surface as the base of the further landslide inventory creation and analysis (Đomlija et al. 2016).

The Vinodol Valley is mostly covered by a dense forest vegetation. Most of the active landslides are located within deeply dissected erosional landscapes formed in soft flysch rock mass. Such steep denudational slopes situated in the central and in the SE part of the Vinodol Valley with prominent active erosion and sliding processes result in badland relief formation. Original landslide topography of dormant landslides and of landslides situated in urbanized areas are significantly modified by the erosion processes and anthropogenic activities. For these reasons, the appropriate landslide identification and mapping method is the visual

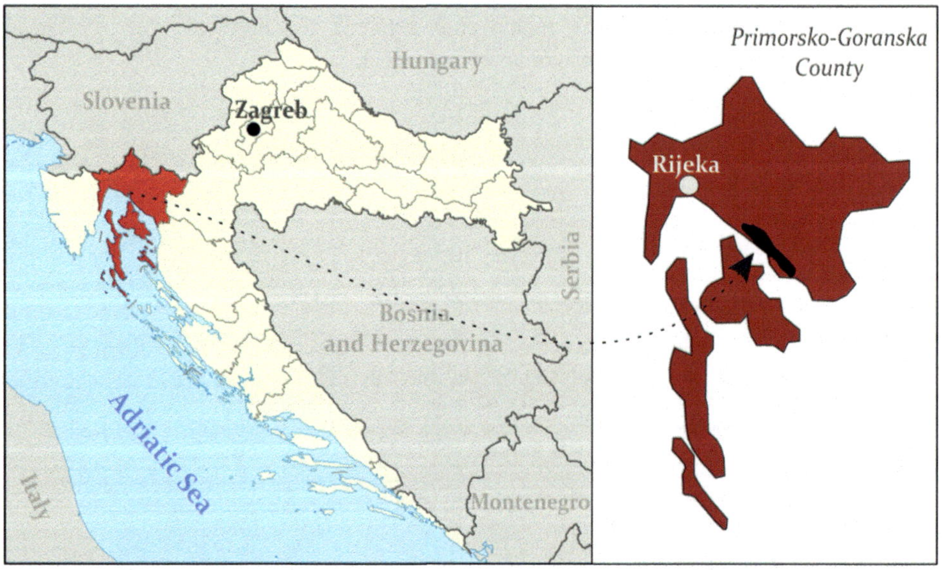

Fig. 1 Geographical position of the Vinodol Valley in the Primorsko-Goranska County, Croatia

interpretation of the high-resolution LiDAR-derived imagery.

Due to the diversity of geological conditions in the Vinodol Valley and the large number of identified landslides, the classification properties of soils from the superficial slope deposits, i.e., landslide forming-materials were investigated. Sampling of the soil material was performed from the individual active and dormant landslide bodies, and also from the superficial deposits at the locations without pronounced slope instabilities. Seventeen soil samples were laboratory tested according to the rules of the European Soil Classification System, (BS EN ISO 14688-1 2002). On the basis of the soil laboratory determination and the ESCS soil classification, soil materials were additionally classified according to the Varnes classification of landslide types modified by Hungr et al. (2014).

In this paper, results of preliminary investigations of the relationship between the different types of landslides and the landslide-forming materials are presented.

Study Area

Geological and Geomorphological Settings

The Vinodol Valley is characterized by complex geological and geomorphological conditions. The valley has irregular elongated shape and an asymmetrical cross section, with significantly longer northeastern slope. The length of the Vinodol Valley is approximately 22 km and the width is about 4 km. The prevailing elevations are in range between 100 and 200 m a.s.l., with an average elevation value of 283 m a.s.l.. Slope angles between 5° and 20° prevail, and the average slope angle is 16°. Geological settings are detail described in Blašković (1999), and later in Bernat et al. (2014) and Đomlija et al. (2016). General geomorphological subdivision of the Vinodol Valley and the preliminary geomorphological models of the northwestern, the central and the southeastern part of the Vinodol Valley are presented in Đomlija et al. (2016).

Figure 2 shows geomorphological units in the Vinodol Valley identified and mapped primarily on the basis of the visual interpretation of the high-resolution LiDAR imagery derived from the bare-earth DEM 1 m × 1 m. The criteria adopted for identification of specific geomorphological unit were morphography, morphometry, relief forming processes and the types of materials defined by the lithology of bedrock and superficial deposit (Đomlija et al. 2016). The SW and the NE border of the Vinodol Valley are built of Upper Cretaceous and Paleogene limestone and dolomites. The area of the SW border and the northern part of the NE border form the geomorphological unit of mountain slope (Fig. 2). Most of the area of the NE border of the Vinodol Valley forms geomorphological unit of rockwall characterized by subvertical carbonate slopes. The debris which has fallen form the uppermost parts of the mountain slope and which was transported downslope by debris and torrential flows formed in gullies in carbonate rock mass has accumulated beneath the unit of mountain slope in the NW part of the Vinodol Valley. Such debris accumulations form the geomorphological unit of proluvial sheet (Fig. 2). Scree deposits originating from rock falls and mechanical weathering of cliffs take forms of debris sheets or talus cones, and they represent the geomorphological unit of talus slope.

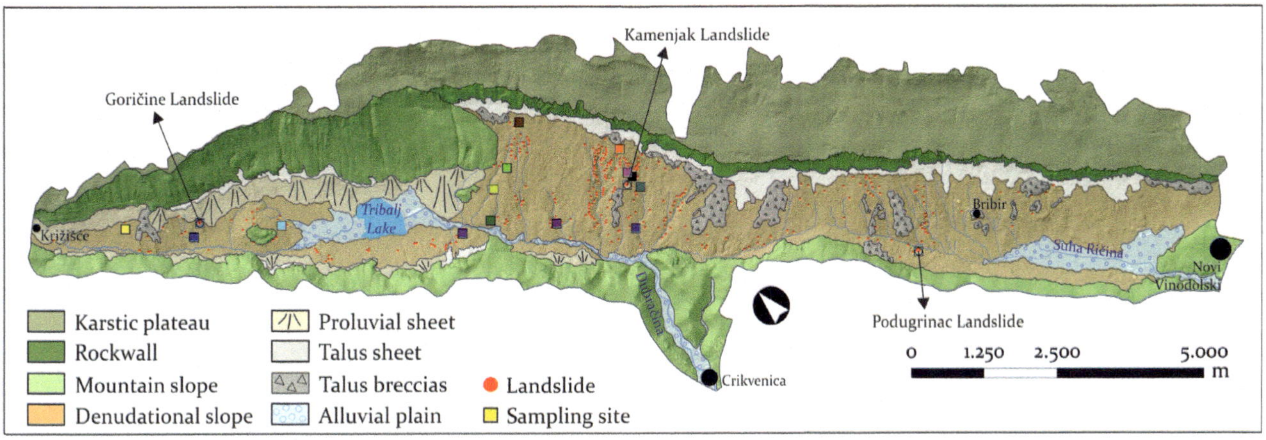

Fig. 2 Map of major geomorphological units identified the Vinodol Valley displayed on the hillshade map generated from the high-resolution DEM 1 m × 1 m. *Red dot labels* present locations of landslides identified in the geomorphological unit of denudational slope. *Squares* (in adjusted scale) present locations of soil sampling sites. List of sample labels attributed to the *square designations* are presented in the legend in the Figs. 5 and 6

The geomorphological unit of denudational slope (Fig. 2) is situated in the hypsometrical lower parts and in the bottom of the Vinodol Valley. The unit is composed of Paleogene flysch deposits (siltstones, marls and sandstones in alternation). The influence of fluvial erosion and mass wasting processes (i.e., denudation processes) on the predominantly soft siliciclastic rocks is significant in this geomorphological unit. As a result of these processes, the terrain has been dissected by several large and deep gullies formed in the central part and in the southeastern part of the Vinodol Valley. Flysch deposits crop out sporadically only in the steep lateral flanks of large gullies. In most of the area of the Vinodol Valley flysch rocks are covered by the Quaternary superficial deposits. Among superficial deposits lithified and semi-lithified talus breccia, residual soil and colluvial soil can differ.

Individual sedimentary bodies of the Quaternary talus breccia irregularly cover the surface of the denudational slope. In some parts of the Vinodol Valley breccia are situated between the gullies at the uppermost parts of their lateral flanks. Deposits of the weathered flysch rock mass lie on the boundary between the soil and the rock material. Residual soils of the flysch rock mass occur sporadically, and they are mostly cover by the colluvial soils of various thicknesses. Colluvial soils cover most parts of the denudational slope. Due to the gravitational transport and mixing of fragments and sediments originating from flysch and carbonate bedrock, superficial colluvial soils vary in their geotechnical properties and mechanical behavior, and are probably transitional between textural classes.

Landslides in Soil Materials

During the previous and present research of the hazardous processes in the Vinodol Valley more than 200 landslides in soils were identified, primarily on the basis of the visual inspection of the high-resolution LiDAR derived imagery. Preliminary landslide inventory of the area of the Dubračina River Basin encompassing the northwestern and the central part of the Vinodol Valley was presented by Đomlija et al. (2014). Landslides in superficial soil materials in the Vinodol Valley predominantelly occur in the area of the geomorphological unit of denudational slope (Fig. 2). Occurrence of landslides is significantly interrelated to zones of excessive gully erosion; e.g., numerous landslides are activated and located in the steep lateral flanks of deep gullies (Fig. 2).

Landslides in the Vinodol Valley can be generally divided into three main groups. Figure 3a shows the Kamenjak landslide in the central part of the Vinodol Valley, and several active or suspended landslides associated to zones of excessive gully erosion. This group of landslides encompasses numerous succesive or multiple, relativelly small and shallow planar slides. Landslide volumes are in range between 10^2 and 10^5 m^3, with estimated landslide depths of 1–3 m. Landslides occur on steep gully flanks covered by the veneer of colluvium or residual soil formed on the flysch bedrock. Landslide-forming materials can vary significantly due to local occurrence and instabilities of breccia's sedimentary bodies situated in the uppermost parts of some gully sub-basins (Fig. 3a). Many of these shallow slides become

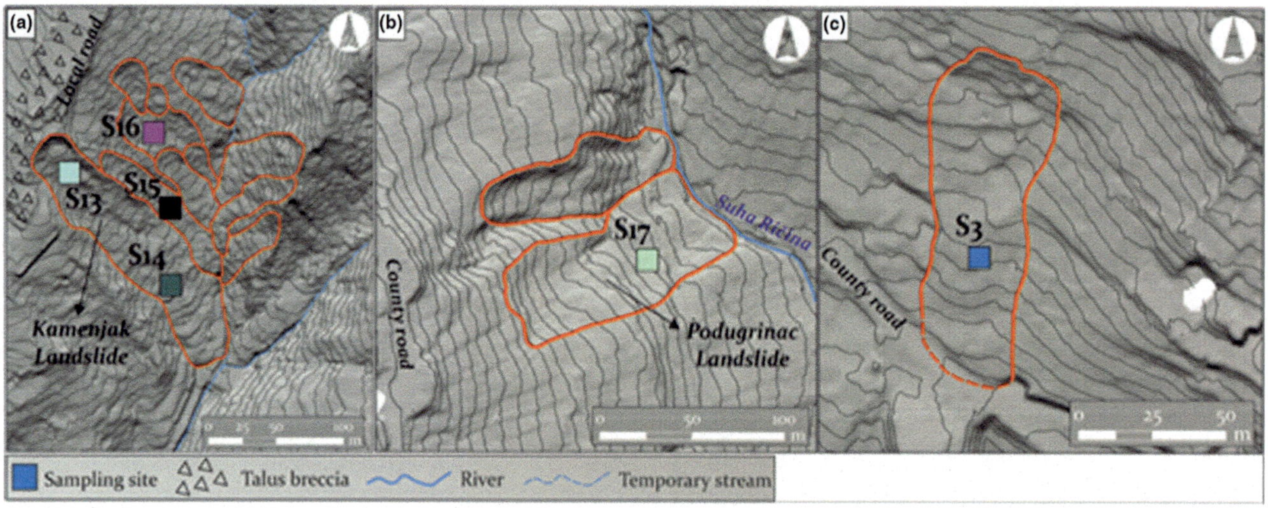

Fig. 3 Examples of different landslide features and topography of landslides identified in the superficial soil material in the Vinodol Valley, displayed on the hillshade map overlaid by the contour map generated with a 1 m contour spacing from the high-resolution DEM 1 m × 1 m: **a** Kamenjak landslide in the central part of the Vinodol Valley; **b** Podugrinac landslide in the SE part of the Vinodol Valley; **c** Goričine landslide in the NW part of the Vinodol Valley. Locations of the soil sampling sites from the individual landslide bodies are also presented

flow-like after the initial phase of planar sliding. Figure 3b shows two landslides situated near the Podugrinac settlement in the SE part of the Vinodol Valley. They present the group of suspended and dormant landslides situated on the river banks and initiated by erosional downcutting of slopes along Dubračina and Suha Ričina rivers. These landslides are mostly single or succesive planar slides. They are generally larger than the landslides situated in the flanks of gully sub-basins. Volumes are considered greater than 10^5 m^3, with estimated landslide depths of several meters. Figure 3c shows the topography of the dormant planar Goričine landslide situated in the urbanized area, in the vicinity of private houses and main county road in the NW part of the Vinodol Valley. The landslide has been probably activated by the influence of anthropogenic activities. Present anthropogenic activities have partially modified original landslide topography, and the outlined landslide contour is partially assumed. Estimated landslide depth is more than 5 m.

Data and Methods

Soil Sampling Techniques

Soil samples were collected from the superficial deposits, e.g., from the landslide bodies or from the stable slopes in the study area. Locations of total of 17 sampling sites are presented in Fig. 2. Locations of sampling sites of S3, S13, S14, S15, S16 and S17 are presented in Fig. 3. Sampling was performed in the area of the geomorphological unit of denudational slope, in which most of the landslides in Quaternary superficial deposits occur. Sampling was partially restricted by the dense forest vegetation, and by the complex topographic characteristics of the Vinodol Valley. Therefore, sampling locations were generally selected due to the best accessibility to the certain landslides identified by the visual interpretation of the high-resolution LiDAR-imagery, or to the nearest slope parts. The overview of general data on soil samples is presented in Table 1.

Four samples were collected in the NW part of the Vinodol Valley, 12 samples were collected in the central part, and one sample was collected in the SE part of the Vinodol Valley. 13 samples were collected from the landslide bodies, and four samples from the nearest accessible stable slopes (Table 1).

Samples were collected by applying one of the following techniques: (i) manual drilling technique, by using an auger and the sampler (Fig. 4a), and (ii) rectangular soil block sampling, by using the sampling box. Four samples which were collected from the stable slope parts (i.e., samples S5, S6, S9 and S12) were sampled by using the sampling box, and other 13 samples which were collected from the landslide bodies were sampled by using both the sampler and the sampling box. Prior to collecting samples by using the sampler, a small borehole was performed by using the manual auger.

Sampling depths vary due to the applied technique of the soil sampling (Table 1). Depths of the soil samples which were taken using the sampling box are in range between 0.50 and 1.0 m. Most of such samples were taken from the shallow depth of about 0.50 m (S1, S2, S5, S6, S9, S12, S17). Sampling depths of the soil samples which were taken

Table 1 Overview of general data on soil samples of the superficial deposits in the Vinodol Valley

Label	Sampling location	Sampling depth	Sampling technique
S1	Landslide body	0.50	Sampling box
S2	Landslide body	0.50	Sampling box
S3	Landslide body	1.50	Drill/sampler
S4	Landslide body	0.70	Drill/sampler
S5	Stable slope	0.50	Sampling box
S6	Stable slope	0.50	Sampling box
S7	Landslide body	1.20	Drill/sampler
S8	Landslide body	0.60	Sampling box
S9	Stable slope	0.50	Sampling box
S10	Landslide body	1.00	Sampling box
S11	Landslide body	0.60	Sampling box
S12	Stable slope	0.50	Sampling box
S13	Landslide body	0.80	Sampling box
S14	Landslide body	1.50	Drill/sampler
S15	Landslide body	0.80	Sampler box
S16	Landslide body	0.80	Sampler box
S17	Landslide body	0.50	Sampler box

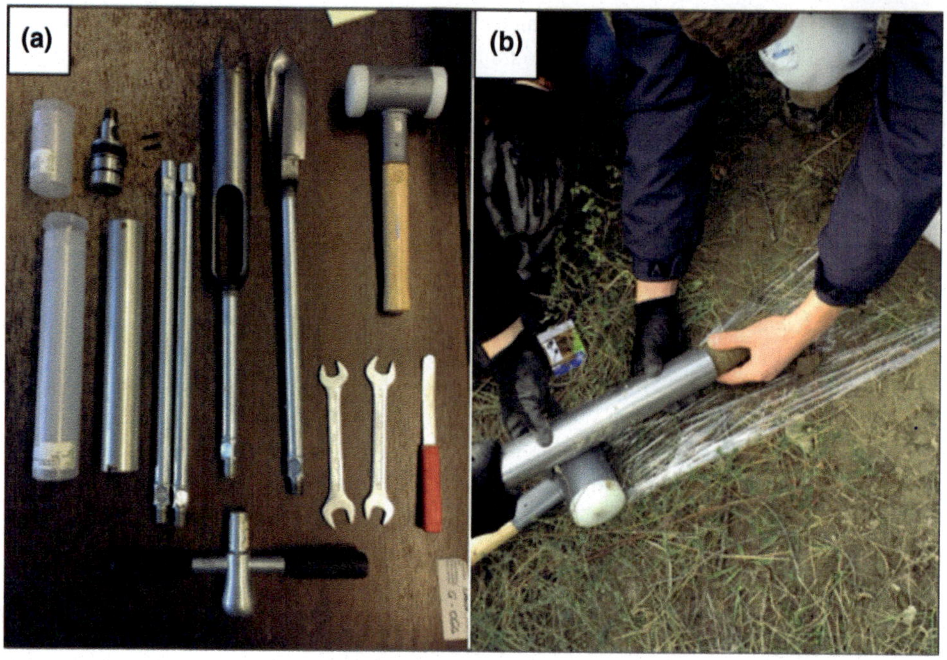

Fig. 4 Eijkelkamp sampling tool applied to the soil sampling in the Vinodol Valley: **a** manual drilling equipment; **b** field soil sampling

using the auger and sampler (i.e., S3, S4, S7, S14) are somewhat deeper and reach to 1.50 m.

Laboratory Testing Methods

Laboratory testing of the soil materials was performed in the Geotechnical laboratory of the Faculty of Civil Engineering, University of Rijeka. Two classification tests were performed: (1) sieve analysis and (2) Atterberg's limits. All classification tests were performed according to British Standard (BS 1377-2 2010).

13 of total 17 samples were sieved by using the wet sieving method, according to the rules of the BS 1377-2 (2010). Four samples (S13, S14, S15 and S16) were sieved according to the rules for the dry sieving method (Table 1). A series of sieves used in the both sieve analyses were in range between 20 mm and 0.063 mm. According to the BS 1377-2 (2010), preparation of the samples for the wet sieving analysis started 24 h prior the sieving. Samples were initially soaked in the solution of water and sodium hetamethaphosphate. The soil material was then shacked in a sieve shaker for the amount of time prescribed by the standard. After the shacking, sieves were left to drain, and samples were dried in the dryer on 105 °C for 24 h. The masses of dried sample retained on the sieves were weighted and recorded. Samples for the dry sieving analysis were initially oven dried and manually crushed by using the rubber mortar and the pestle. According to the standard (REF), samples were shacked, and soil masses retained on the sieves were weighted and recorded.

Liquid limit tests were performed by using the Casagrande's dish method. According to the standard (BS 1377-2 2010), each sample was prepared 24 h prior the test. After the liquid limit tests performance, plastic limit tests were performed and plasticity index for each sample was calculated. Due to the inaccuracy of the natural water content in the samples (e.g., the samples were collected either during dry weather or immediately after the rain), liquidity indexes were not calculated during research.

Proposed Classifications of the Soil Materials

Soil samples were laboratory tested according to the principles of the European Soil Classification System (ESCS), (BS EN ISO 14688-1 2002). The new ESCS classification for engineering purposes is developed on the soil principles for soil classification according to the European standard EN ISO 14688-2. Classification uses soil descriptions and soil symbols according to the European standard EN ISO 14688-1, and it generally compiles with guidelines defined by the USCS soil classification system in accordance with the US standard ASTM D 2487.

On the basis of the results of the laboratory soil determination and the applied ESCS soil classification, superficial soil materials were classified according to the Varnes classification of landslide types modified by Hungr et al. (2014). According to Hungr et al. (2014), the material division proposed by Varnes (1978), including the terms "rock, debris and earth", is compatible neither with geological terminology of materials distinguished by their origin, nor

Table 2 Landslide-forming material types according to Hungr et al. (2012, 2014)

Material name	Character descriptors	Corresponding USCS soil classes
Rock	Strong	–
	Weak	–
Clay	Stiff	GC, SC, CL, MH, CH, OL, and OH
	Soft	
	Sensitive	
Mud	Liquid	CL, CH, and CM
Silt, sand, gravel, boulders	Dry	ML
	Saturated	SW, SP, and SM
	Partly saturated	GW, GP, and GM
Debris	Dry	SW-GW
	Saturated	SM-GM
	Partly saturated	CL, CH, and CM
Peat	–	–
Ice	–	–

with geotechnical classification based on material properties. Hungr et al. (2014) proposed the modification of the definitions of landslide-forming materials to provide better compatibility with accepted geotechnical and geological terminology of rocks and soils. The new list of material types which replace the former threefold material classes used by Varnes (1978) is presented in Table 2.

Results

Grain Size Distribution and Atterberg's Limits

Results of the sieve analysis are presented in Fig. 5. 13 samples of superficial soil materials of the Vinodol Valley are displayed in the domain of the fine grained soils in the ternary plot. Among them, 10 samples from the landslide bodies and from the stable slopes are displayed in the domain of silts and clay (i.e., samples S1, S2, S4, S5, S6, S7, S8, S9, S10 and S11). Samples S3, S12 and S17 are displayed in a domain of granularly mixed materials, representing mixtures of coarse grains floating in a fine grained soil matrix. Samples S13, S14, S15 and S16, which locations of sampling sites are presented in the Fig. 3a, are displayed in a domain of the coarse grained soils. Content of the gravel size particles is significantly higher in the sample S13 than in the samples S14, S15 and S16.

Figure 6 shows results of the liquid and plastic limit tests. Three samples (i.e., S3, S12 and S17) are displayed under the A-line in the diagram. Nine samples (i.e., S1, S2, S4, S9, S11, S13, S14, S15 and S16) are displayed in a domain of medium plasticity clay, and five samples (i.e., S5, S6, S7, S8 and S10) are displayed in a domain of highly plasticity clay.

Classification of Landslide-Forming Materials

On the basis of the sieve and plasticity analyses, and of the procedure of the soil classifications described in Kovačević and Jurić-Kaćunić 2014, soil samples were classified both according to ESCS and USCS (Holtz et al. 2011) classification system. Results of the soil classifications are presented in Table 3.

Samples collected from the landslide bodies show significant diversity of soil materials. According to the ESCS classification, soils range from highly plasticity clays (ClH) to gravely clays (clGr). Location of samples situated in the vicinity to the breccia sedimentary bodies (i.e., samples S13, S14, S15 and S16) show significant amount of sandy particles and are classified as sandy clays (clSa) or poorly graded sandy clays (clSaP). Only sample S12 was classified as the silty soil of intermedium plasticity, which was collected from the stable part of the slope near the position of the relatively old dormant landslide.

Table 3 presents the diversity of classified soil material according to ESCS soil classification in comparison to the classification according to Hungr et al. (2014).

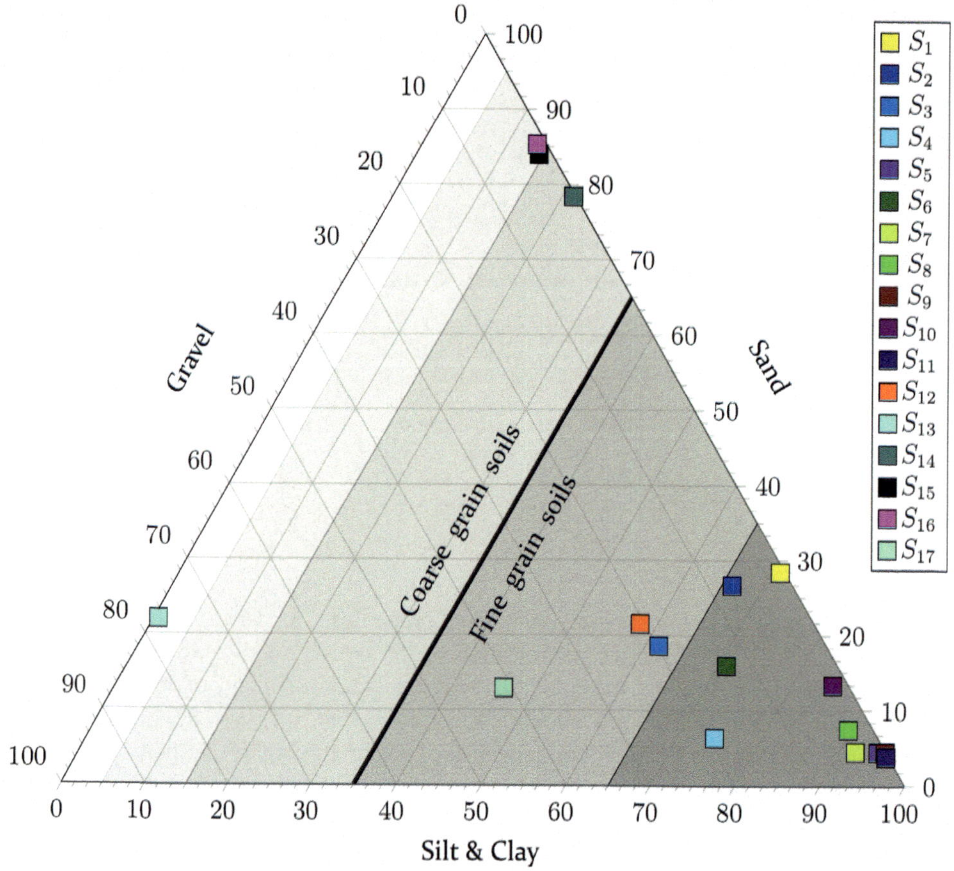

Fig. 5 Ternary plot of grain size distribution in the analysed samples from the superficial deposits in the Vinodol Valley

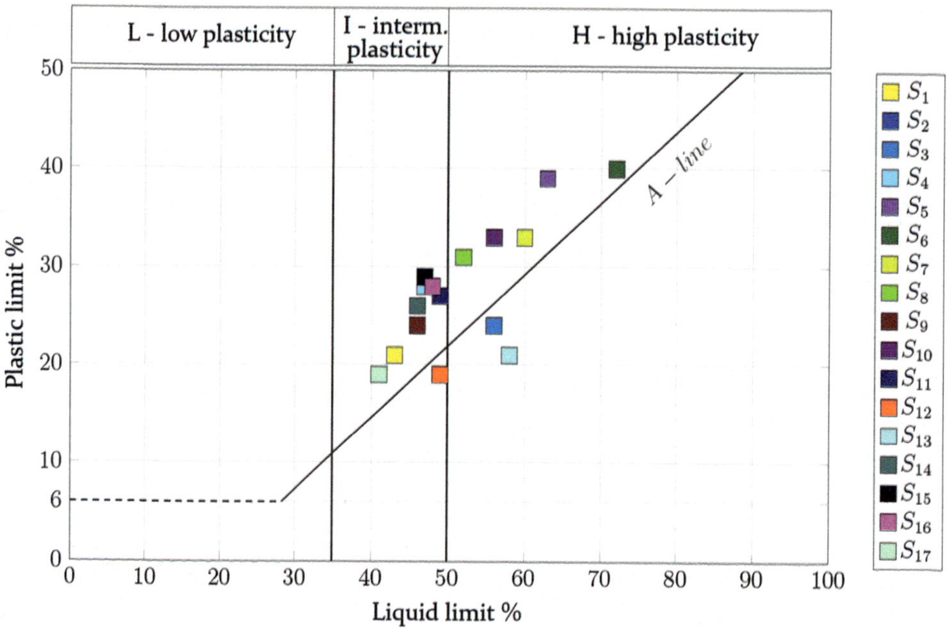

Fig. 6 Plasticity limits diagram of the analysed samples from the superficial deposits in the Vinodol Valley

Table 3 Soil materials from the superficial deposits of the Vinodol Valley classified according to the ESCS and USCS classification systems

Label	ESCS classification symbol	USCS classification symbol	Material type (Hungr et al. 2014)
S1	saClI	CL	Debris
S2	saClH	CH	Debris
S3	sagrClH	CH	Debris
S4	grClI	CL	Debris
S5	ClH	CH	Debris
S6	saClH	CH	Debris
S7	ClH	CH	Debris
S8	ClH	CH	Debris
S9	ClI	CL	Debris
S10	saClH	CH	Debris
S11	ClI	CL	Debris
S12	grSiI	ML	Debris
S13	saGrW	GP	Debris
S14	clSa	SC	Debris
S15	clSaP	SP	Debris
S16	clSaP	SP	Debris
S17	clGr	GC	Debris

Same materials determined as the landslide-forming material according to Hungr et al. (2014)

Discussion and Conclusions

During the previous and present researches of the hazardous processes in the Vinodol Valley, more than 200 landslides were identified on the basis of the visual interpretation of the high-resolution LiDAR-derived imagery. Most of the landslides occurred in the geomorphological unit of denudational slope, which is built of heterogeneous and soft flysch bedrock. Landslides are, in general, small to medium small with landslide volumes in a range between 10^3 and 10^5 m^3. Estimated landslide depths are in a range between one to several meters.

Due to the complex geological and morphological features of the Vinodol Valley, and to a significant influence of active geomorphological processes, heterogeneous superficial deposits of various thicknesses, grain size distribution and geotechnical properties are present at the surface of the slopes. In order to identify materials of landslide-forming materials and their properties influenced on slope stability, the initial investigation was oriented on determination of grain size distribution and Atterberg's limits of materials taken from landslide bodies and surrounded slope superficial deposits. In the first round of the investigation, seventeen soil samples were collected from the Quaternary superficial deposits of landslide-forming materials from different features of landslides along the whole Vinodol Valley. Due to the dense forests and complex morphological features of the Vinodol Valley, samples were collected according to the available accessibility of the landslide bodies or near of them. This approach was accepted to give a preliminary insight in materials in which landslides were occurred along the valley.

According to the ESCS classification, which was newly upgraded for the engineering purposes and which provides a wider range of soil types and sub-types, tested soils from the Vinodol Valley range from the highly plasticity clays (ClH) to the gravely clays (clGr) (Table 3). On the other hand, according to the commonly used USCS classification the most of tested samples are classified as highly plasticity clays (CH) or as poorly graded sands (SP). Only superficial deposits of the sample S12 were classified as intermediate plasticity silty gravels. Landslide-forming materials according to the updated Varnes classification of landslide types (Hungr et al. 2014) all correspond to the debris materials (Table 3). The word *debris* does not have clear equivalent in current geotechnical terminology (Hungr et al. 2012) but, together with *mud* has acquired important status in geology and must therefore be retained (Bates and Jackson 1984). Texturally, debris is a mixture of sand, gravel, cobbles and boulders, often with varying proportions of silt and a trace of clay (Hungr et al. 2012).

Varnes (1978) proposed criterion for debris as all materials containing more than 20% sizes coarser than sand. The last proposed Varnes landslide classification (Hungr et al. 2014) excludes the earth-type of the landslide-forming material, which may be more adequate to the material type of the tested samples, but it is not in accordance to current geotechnical classifications' soil groups (gravel, sand, silt and clay). According to original Varnes' classification of landslides (1978) and the obtained results of ESCS or USCS soil classification, most of the tested soils samples may be determined as the earth type of material.

Sieving procedure shows significant influence on the result based on the technique used. Samples S13, S14, S15 and S16, which are analysed according to dry sieving method, show significant amount of coarse grain material, unlike the rest of the samples tested using wet sieving method. The wet sieving method could have significant influence on the results of the samples S14, S15 and 16, transferring them to the fine grain part of ternary plot (Fig. 5). Sample S13, on the other hand, was collected nearly below the semi-lithified talus breccia situated at the upper most part of the steep gully flank in which the landslide was activated (Fig. 3a), and thus influenced by the coarse breccia material.

The identified diversity of landslide-forming materials based on classification and grain size distribution analysis conducted on only 17 taken samples in preliminary investigation indicates on likely occurrence of high diversity of strength properties of landslide-forming materials responsible for landslide initiations and occurrences. Based on these preliminary investigation results it would be possible to conclude that does not exist a unique simple pattern of conditions that lead to the initiation and appearance of landslides in the Vinodol Valley. The existing results indicate on a need for further field and laboratory investigation of landslide-forming materials on identified landslides in the Vinodol Valley. Simple and cheap classification and grain distribution testing of samples taken from numerous identified old and recent landslides would present a confidential base for adequate selection of more demanding and expensive soil testing program for better understanding of occurrence and development of landslides on flysch slopes of the Vinodol Valley.

Acknowledgements This work has been supported in part by Ministry of Science, Education and Sports of the Republic of Croatia under the project Research Infrastructure for Campus-based Laboratories at the University of Rijeka, number RC.2.2.06-0001. Project has been co-funded from the European Fund for Regional Development (ERDF).

References

Bates RL, Jackson JA (eds) (1984) Glossary of geology. American Geol. Inst., Falls Church, p 788

Bernat S, Đomlija P, Mihalić Arbanas S (2014) Slope movements and erosion phenomena in the Dubračina river basin: a geomorphological approach. In: Mihalić Arbanas S, Arbanas Ž (eds) Landslide and flood hazard assessment, Proceedings of the 1st Regional Symposium on Landslides in the Adriatic-Balkan Region. Zagreb, pp 79–84

Blašković I (1999) Tectonics of part of the Vinodol Valley within the model of the continental crust subduction. Geol Croat 52:153–189. doi:10.4154/GC.1999.13

BS 1377-2 (2010) Methods of test for soils for civil engineering purposes. Classification tests

BS EN ISO 14688-1 (2002) Geotechnical investigation and testing. Identification and classification of soil. Identification and description

Đomlija P, Bernat S, Arbanas Mihalić S, Benac Č (2014) Landslide Inventory in the Area of Dubračina River Basin (Croatia). In: Sassa K, Canuti P, Yin Y (eds) Landslide science for a safer geoenvironment, vol 2. Methods of landslide studies. Springer International Publishing, Cham, pp 837–842

Đomlija P, Bočić N, Mihalić Arbanas S (2016) Identification of geomorphological units and hazardous processes in the Vinodol Valley. In: Proceedings of the 2nd Regional Symposium on Landslides in the Adriatic-Balkan Region. Beograd

Holtz RD, Kovacs WD, Sheahan TC (2011) An introduction to geotechnical engineering. Pearson

Hungr O, Leroueil S, Picarelli L (2012) The Varnes classification of landslide types, an update. In: Eberhardt E, Froese C, Turner AK, Lerouieil S (eds) Proceeding of 11th International and 2nd North American Symposium on Landslides and Engineering Slopes: Protecting Society through Improved Understanding, Banff, Canada 3–8 June 2012. CRC Press, pp 47–58

Hungr O, Leroueil S, Picarelli L (2014) The Varnes classification of landslide types, an update. Landslides 11:167–194. doi:10.1007/s10346-013-0436-y

Kovačević MS, Jurić-Kaćunić, D (2014) European soil classification for engineering purposes. Građevinar 66(9):801–810. doi:10.14256/JCE.1077.2014 (In Croatian)

Mihalić S, Arbanas Ž (2013) The Croatian-Japanese joint research project on landslides: activities and public benefits. In: Sassa K, Rouhban B, Briceño S et al (eds) Landslides: global risk preparedness. Springer, Berlin, pp 333–349

Varnes DJ (1978) Slope movement types and processes. In: Schuster RL, Krizek RJ (eds) Special report 176: landslides: analysis and control. Transportation Research Board, Washington, pp 11–33

4thWLF2017 4th World Landslide Forum LJUBLJANA SLOVENIA EU

Small Flume Experiment on the Influence of Inflow Angle and Stream Gradient on Landslide-Triggered Debris Flow Sediment Movement

Hefryan Sukma Kharismalatri, Yoshiharu Ishikawa, Takashi Gomi, Katsushige Shiraki, and Taeko Wakahara

Abstract

Rainfall-induced landslide might transformed into more severe disaster, namely debris flow and natural dam which both holds serious threats on human life and material. The runout distance has crucial role for determining affected areas of a landslide. Our previous research found the correlation of inflow angle and stream gradient to transformation of landslide collapsed sediment either into natural dam or debris flow. This research intended to test our previous research result with a small flume experiment and aimed to analyze the influence of sediment inflow angle and stream gradient to the sediment deposition percentages as representative of runout distance and the possibility of natural dam formation. Soil samples were taken from landslide-triggered debris flow disaster initiation zone in Hiroshima (Hiroshima Pref.) and Izu Oshima (Tokyo Pref.), Japan which were induced by heavy rainfall. The small flume was 10 cm width and 15 cm height, the inflow segment angle was varied to 60° and 90°, and the stream segment gradient was varied to 10° and 15°. From the experiment results, stream gradient influence the sediment movement effectively rather than inflow angle, and it was sufficient to examine the possibility of collapsed sediment to form natural dam or debris flow. Soil samples from natural dam initiation zones and consideration of water content factor are essential for further experiment.

Keywords

Debris flow • Natural dam • Stream gradient • Small flume experiment

H.S. Kharismalatri (✉)
Department of Symbiotic Science of Environment and Natural Resources, Tokyo University of Agriculture and Technology, 3-5-8 Saiwaicho, Fuchu, Tokyo 183-8509, Japan
e-mail: kharismalatri@gmail.com

Y. Ishikawa · K. Shiraki
Department of Environment Conservation, Tokyo University of Agriculture and Technology, 3-5-8 Saiwaicho, Fuchu, Tokyo 183-8509, Japan
e-mail: y_ishi@cc.tuat.ac.jp

T. Gomi
Department of International Environmental and Agricultural Science, Tokyo University of Agriculture and Technology, 3-5-8 Saiwaicho, Fuchu, Tokyo 183-8509, Japan

T. Wakahara
Department of Ecoregion Science, Tokyo University of Agriculture and Technology, 3-5-8 Saiwaicho, Fuchu, Tokyo 183-8509, Japan

M. Mikoš et al. (eds.), *Advancing Culture of Living with Landslides*,
DOI 10.1007/978-3-319-53483-1_61

Introduction

Landslide which triggered by rainstorm might be categorized into two types; shallow landslide of about 1 m thick and deep-seated landslide of dozens of meters (Takahashi 2007). Despite the slip surface depth, distinction between two of them also depends on the scale, geology, and root-system resistance (e.g., Baron et al. 2005; Bruckl and Parotidis 2005; Uchida et al. 2011). Shallow landslide generally occurs within the strongest rainfall intensity, thus it contains plenty amount of water which driven the debris (collapsed sediment) to easily transformed into debris flow (Takahashi 2007) while sliding downwards along the slope, stream, or gully (Sassa and Wang 2005). Whilst deep-seated landslide tends to be larger than that of shallow landslides and, accordingly, the amount of collapsed sediment tends to be large as well. The collapsed sediment of deep-seated landslide might transform into either debris flow or natural dam (natural dam) (Ministry of Land, Infrastructure, Transport and Tourism (MLIT), n.d.).

Assessment of runout distance, i.e., the length travelled by landslide's collapsed sediment from the initiation zone until their lowest point of deposition zone, has crucial role for determining affected areas to develop sediment disaster preparedness, prevention and mitigation (D'Agostino et al. 2010; Strimbu 2011). Runout distance is controlled by properties of the sediment and characteristics of the movement path (Fannin and Bowman 2008). The approaches to estimate runout distance used one or a combination of the factors, namely sediment volume, mean gradient of the transportation zone, elevation difference from the starting point to the point where deposition begins, and travel angle (Bathrust et al. 1997).

Debris flow is known as the most powerful mechanism of transporting landslide sediment far downslope (Bathurst et al. 1997) and holds serious impact on human life and infrastructures since it moves rapidly, large in volume, destroy object without warning, and often occur without warning (Nishiguchi et al. 2012; Highland et al. 1997). Natural dam defined as natural blockage of river channels caused by landslide, having significant height and potentially causing inundation of water behind it (Canuti et al. 1998; MLIT 2006). Even though debris flow affected larger area with its long runout distance, natural dam holds further threats; back-flooding threaten upstream area upon the dam creation and when the dam breaks, which commonly due to overflowing of inundated water, large surges and debris flow threaten downstream area (e.g., Ermini and Casagli 2003; Inoue et al. 2012). Therefore, short runout distance of landslide collapsed sediment could not be neglected either.

On our previous analysis on landslide's topographical characteristics, we found out that landslides with inflow angle (the angle of collapsed sediment flow into the river/stream) less than 60° and stream gradient more than 10° were likely to transformed into debris flow, whilst landslides with inflow angle more than 60° and stream gradient smaller than 10° were likely to transformed into natural dam.

Methodologies that can predict of whether or not collapsed sediment of landslide have long runout distance, or in other words the sediment deposition percentages in downstream area, are increasingly needed. This research attempted to clarifies the influence of inflow angle and stream gradient to runout distance of sediments. Utilizing small flume experiments, this research aimed to analyze the influence of stream gradient and sediment inflow angle to the sediment deposition percentages as representative of runout distance.

Soil Samples

Landslide-triggered debris flow occurred in Hiroshima City (Hiroshima Pref.) in August 2014 and in Izu Oshima (Tokyo Pref.) in October 2013, both were caused by accumulative of heavy rainfall and led to large casualties and property damages. Izu Oshima is located about 120 km south of Tokyo and a part of a group of volcanic islands (Fig. 1). The geology of the island is dominated by pyroclastic fall deposits and lava flows (Izu Oshima Landslide Disaster Measure Exploratory Committee 2014), since Mount Mihara which located at the center of the island is frequently erupted. The average annual rainfall is 2800 mm, with maximum monthly rainfall of 600–800 mm (Izu Oshima Landslide Disaster Measure Exploratory Committee 2014).

Typhoon crosses Izu Oshima frequently, triggered sediment-related disaster, and caused large loss of life and materials, e.g., in 1948, 1958, 1972, 1981, and 1982 (Izu Oshima Landslide Disaster Measure Exploratory Committee 2014). On October 15–16, 2013, Typhoon Wipha brought maximum 24-hr rainfall of 824 mm to Izu Oshima (Yang et al. 2015) and triggered catastrophic sediment disaster. Soil samples were collected on Feb 1, 2015, 2 samples each at 4 failure zone locations in Motomachi District (Fig. 1). The initiation zone of the landslide was 500 m length, the flowing zone was 500 m, and the flooding and deposition zone was 1500 m length (Fig. 2). The debris flow brought woody debris to downstream and affected about a million m^2 area in Motomachi District. The disaster was classified into muddy-type debris flow, dominated by fine grain soil and sand of volcanic ash, the flow length approximately 2.5 km, and average stream gradient 10.7° (Oshima 2016). MLIT (2013) reported that in total, there were 36 deaths, 3 missing, 71 destroyed houses, and 132 severely damaged houses.

Hiroshima, which is located about 800 km southwest of Tokyo, is occasionally damaged by heavy rainfall-triggered

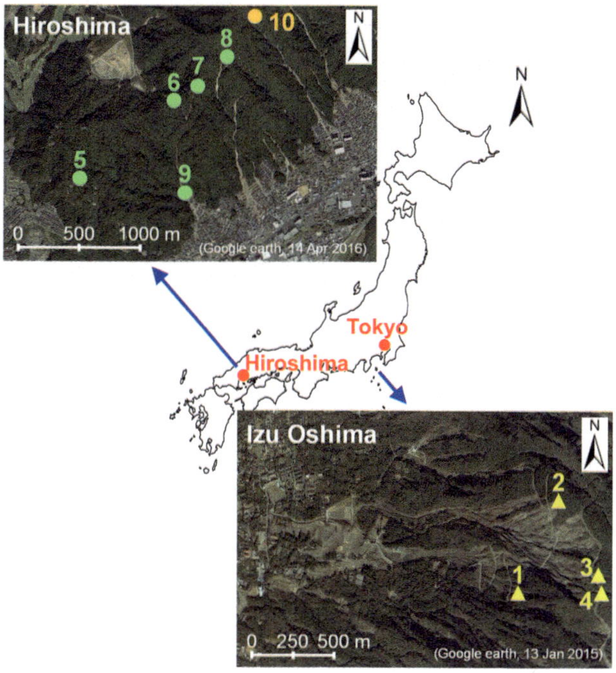

Fig. 1 Locations of soil sampling

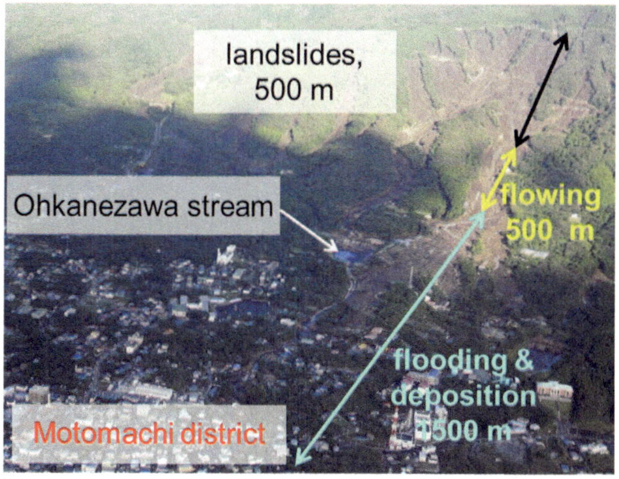

Fig. 2 Landslide-triggered debris flows in Motomachi District

sediment disaster. Following a heavy rainstorm on 29 June 1999, hundreds of landslides occurred in Kameyama, Asakita Ward, Hiroshima City (Wang et al. 2003). The debris flows were mostly triggered by small slides in source area which the volume enlarged while moving downslope (Kaibori et al. 1999; Ushiyama et al. 1999; Wang et al. 2003). The Asaminami Wards of Hiroshima City were severely damaged by rainstorm On August 20, 2014 and the sediment disasters induced by it. The rainstorm brought more than 200 mm of cumulative rainfall with maximum three hour rainfall is 236 mm from 2:00 to 5:00 a.m. and the recorded maximum amount of 287 mm (Wang et al. 2015).

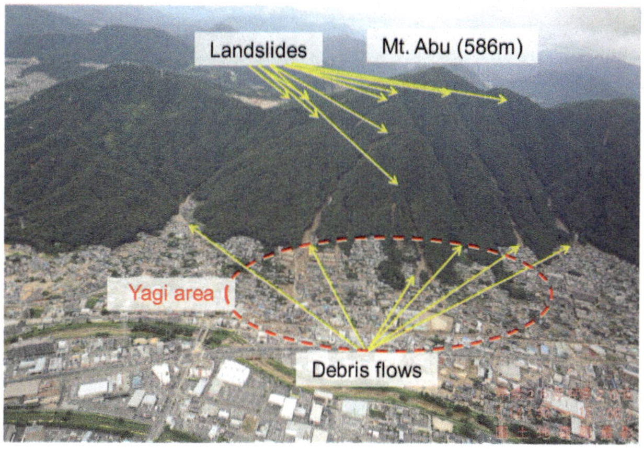

Fig. 3 Landslide-triggered debris flows in Asaminami Wards (photo taken by Geographical Survey Institute, Japan)

This rainstorm exceeded the 3 h rainfall with exceedance probability of over 500 years, which is 187 mm (Oshima 2016). In total, 166 cases of sediment-related disasters were occurred due to the rainstorm, including 107 debris flows and 59 slope failures (MLIT 2014a). The catastrophic disaster costs 73 deaths, 39 injuries, 123 destroyed houses, and 232 severely damaged houses in Hiroshima City (MLIT 2014b). Debris flows also destroyed prefectural apartments in Yagi area which was located about a kilometer from the landslide initiation zone (Fig. 3). In total, 12 soil samples were collected on April 6–7, 2015 from 6 failure zone locations in Asaminami Wards of Hiroshima City (Fig. 1). The soil was dominated by late cretaceous ganite (Hiroshima granite) on sampling locations number 5–9, and mélange matrix of middle to late Jurassic accretionary complex (weathered sedimentary rock) on sampling location number 10 (Oshima 2016; Wang et al. 2015).

Experimental Method

This research employed small flume to observe the influence of inflow angle and stream gradient to movement of landslide collapsed sediment. The flume was 10 cm width, 15 cm height, and consists of two segment; sediment inflow and stream segments (Fig. 4). The stream segment's gradient was varied to 10° and 15° (β), whilst the inflow segment's gradient (θ) was fixed on 45°. Furthermore, the angle between stream segment and inflow segment (inflow angle, α) was varied to 60° and 90°. The experiment condition is resumed in Table 1. In addition, the stream segment was 130 cm length and a bucket was placed in the end of it to catch the overflowed sediments.

The sediments were situated on its saturated condition to represent the disaster moment and positioned 30 cm upstream from the junction of the stream segment and inflow segment.

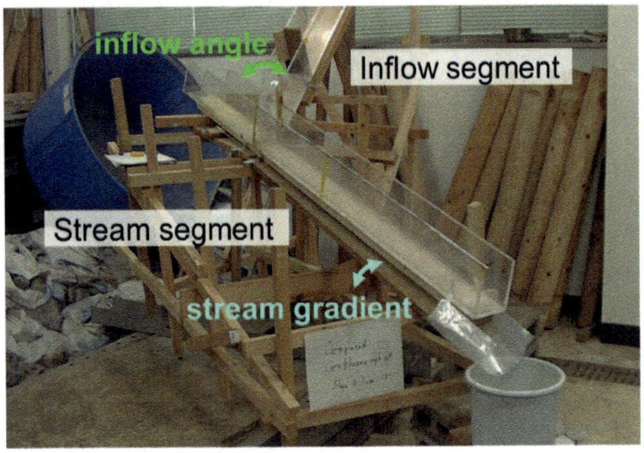

Fig. 4 Small flume instrument

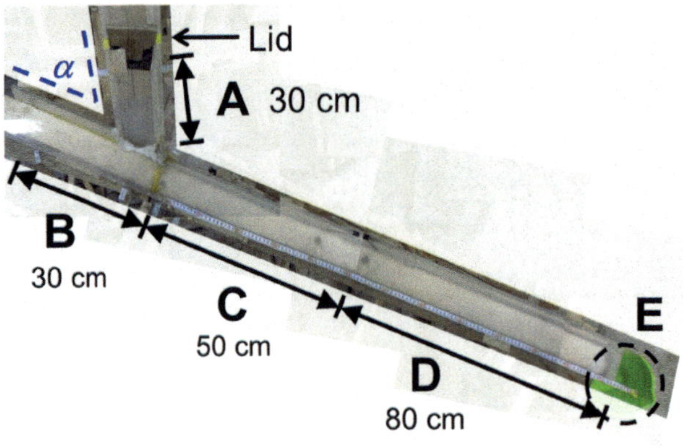

Fig. 5 Sections for sediment deposition calculation

Table 1 Conditions of the experiment

Soil samples	Izu Oshima: 8 samples Hiroshima granite: 9 samples Hiroshima sedimentary rock: 3 samples			
Water content	Saturation = 1.0			
θ	45°			
Inflow angle (α)	90°		60°	
Stream gradient (β)	15°	10°	15°	10°

The lid then opened and the sediment flew through the flume to the bucket. Deposition height and sediment collision height in junction section were measured. Afterwards, the soil in the bucket were collected, dried, and weighted. Sediment percentages remaining were then calculated.

Result and Discussion

Percentages of sediment deposition in five sections of the flume (Fig. 5) were calculated in order to understand the influence of inflow angle and stream gradient to runout distance of landslide collapsed sediment. The most sediment percentages were in section E or in the bucket for almost all experiment conditions (Fig. 6). Above all conditions, the inflow angle 90°—stream gradient 15° gave the most result of sediment percentage in the section E. For soil samples of Izu Oshima and Hiroshima sedimentary rock, stream gradient modification affected the sediment percentages effectively rather than inflow angle modification. While Hiroshima granite showed different result since in most of the experiments, soil samples were not reaching the section E, instead stopped at either section C or D and therefore the sediment percentage in section C or D was higher than section E.

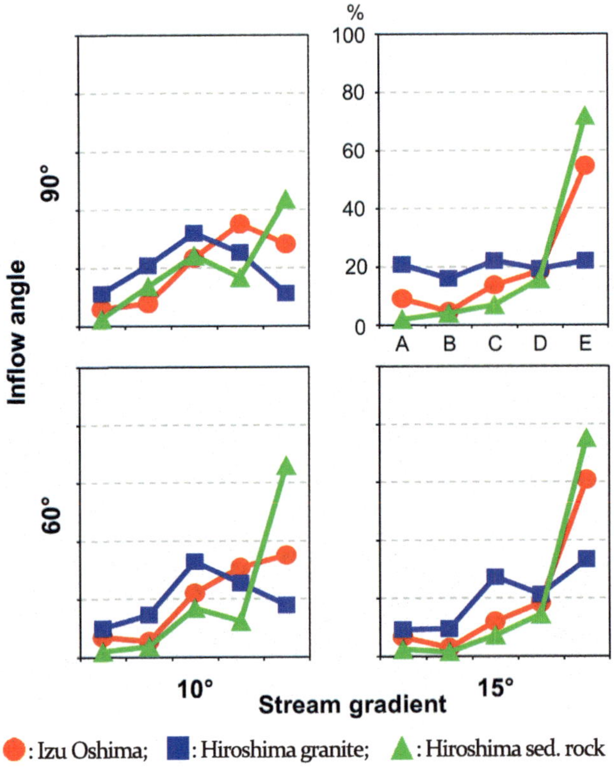

Fig. 6 Average sediment percentages in sections A–E

The result showed that stream gradient affect runout distance effectively rather than inflow angle for Izu Oshima soil samples (Fig. 7). For soil samples of Hiroshima granite and Hiroshima sedimentary rock, the modification did not give significant effect to sediment percentages in section E. For Izu Oshima soil samples, by reducing the inflow angle, it increased the sediment percentages in section E for about 7 and 6% when the stream gradient was 10° and 15° respectively. Whilst by reducing the stream gradient, it decreased the sediment percentages by 27 and 26% when the inflow

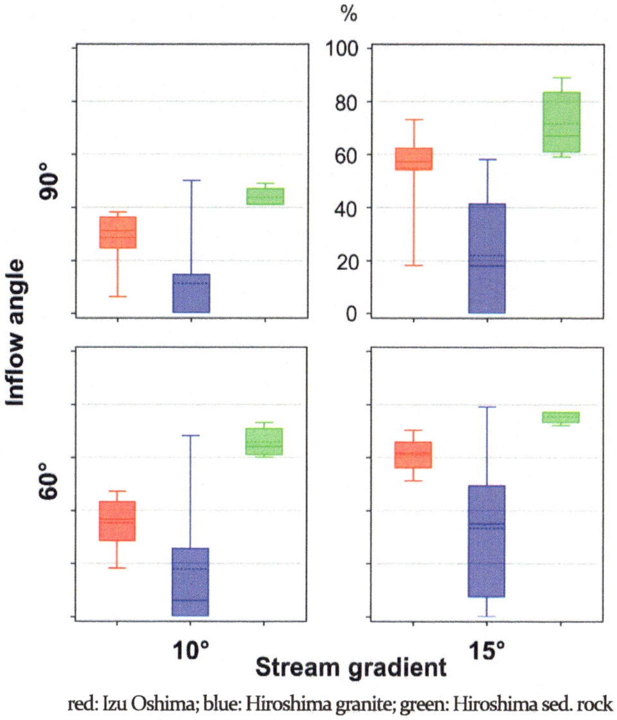

Fig. 7 Sediment deposition percentage in section E

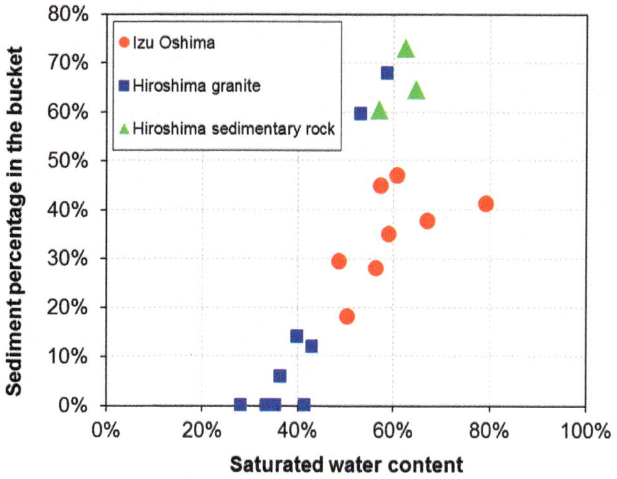

Fig. 8 Saturated water content and sediment percentage in the bucket at inflow angle 60°—stream gradient 10°

angle was 90° and 60° respectively. However, the dynamic of sediment percentage in section E of Hiroshima granite changed by 11%, for either inflow angle or stream gradient modification. Whilst the Hiroshima sedimentary rock has different behavior, the lowest sediment percentages found in inflow angle 90°—stream gradient 10°, and any modification on inflow angle or stream gradient from that increased the sediment percentages in section E by about 26%.

Sediment deposition percentage in the bucket was influenced by soil particle size. Most of the samples from Hiroshima granite could not reach the end of stream segment while samples from Hiroshima sedimentary rock occasionally formed blockage on the stream segment. Those two characteristics were never been found on any samples from Izu-Oshima. Moreover, water content also played a role in determining the sediment movement. As shown in Fig. 8, amount of sediment in the bucket tends to increase with the increasing of saturated water content. One of the main differences between debris flow and natural dam is water contents. Debris flows are generally contains abundant of water, and thus it much more fluid and reach longer distance than other landslide with the same volume (Iverson 1997).

The largest sediment percentage in the bucket was found in stream gradient 15°, regardless the inflow angle and soil samples. This implies that the steeper the stream gradient, the more soil to flow farther, the longer the sediment runout distance, and the larger the affected areas. Ikeya (1989) have stated that, in general, debris flow deposits on stream beds when the gradient falls to 3–10°. Further, Rickenmann (2005) resumed that observations from different catchment indicate that for many large debris flows deposition start once the stream gradient becomes smaller than 6–12°. Therefore, soil samples in the experiment own a large potential to flow down to flatter area.

Ministry of Construction (1987) examined 110 data of landslides in Japan, USA, Canada, Peru, New Zealand, and Pakistan, and found out that no natural dams occurred at stream with gradient more than 15°. The inflow angle influence was not analyzed in that research, but it assumed that stream gradient is sufficient to examine the possibility of collapsed sediment to form natural dam or debris flow. Furthermore, Rickenmann (2005) resumed that theoretical equations on predicting the runout distance, by considering gradient of runout stream, gradient of inflow stream/slope, inflow velocity, inflow depth, and friction of slope, without considering the inflow angle.

Takahashi (2007) classified debris flow as stony-type debris flow, turbulent-muddy-type debris flow, and viscous debris flow. The classification is a function of flow pattern (Reynolds number as benchmark), rock type (Bagnold number as benchmark), and geomorphic aspect of stream (flow depth/median diameter of solid particle as benchmark) (Takahashi 2009). The movement of soil from Izu Oshima was similar to muddy-type, whilst soil from Hiroshima was closer to stony-type debris flow. The Izu Oshima sediment movement was very turbulent, the collisions to the opposite of inflow segment were very rough and sometimes overflowed, and all soil samples were transported up to the bucket. Meanwhile, the Hiroshima sediment movement was not as turbulent as muddy-type, some soil samples could not reach the end of stream segment, and occasionally formed blockage on the stream segment.

Comparing the experimental result with the actual phenomenon, in experiment condition of stream gradient 10°, soil samples supposed to form natural dam. However, since the soil samples used in this research were originated from debris flow initiation zone, and natural dams were not generated from these types of soil, the sediment deposition on the stream segment could not be labeled as natural dam. Further, the actual phenomenon, both in Hiroshima and Izu Oshima, natural dam were not found.

To understand further regarding the influence of inflow angle and stream gradient to sediment movement, additional modification of inflow angle and stream gradient is needed, namely the modification of inflow angle to 30° and 0°, and stream gradient to 5°. Moreover, since the exact saturated water content of soil in the nature and water content at the disaster time could not be determined, variation of water content is needed in the experiment to understand the interaction between water content and modification of inflow angle—stream gradient to sediment movement.

Conclusion

Deposition of sediment in junction area was formed in several experiments. Stream gradient effectively influenced the percentage of sediment deposition in downstream rather than inflow angle, and it was sufficient to examine the possibility of collapsed sediment to form natural dam or debris flow. Sediment percentage in the downstream was increased with the increasing of stream gradient, but was not significantly affected by inflow angle modification. Muddy-type sediment movement was very turbulent, might flow longer and might have larger affected areas rather than stony-type sediment movements. For further research, the water content factor and soil samples from natural dam disaster needs to be considered in the experiment.

References

Baron I, Agliardi F, Ambrosi C, Crosta B (2005) Numerical analysis of deep-seated mass movements in the Magura Nappe; Flysch Belt of the Western Carpathians (Czech Republic). Nat Hazards Earth Sys Sci 5:367–374

Bathurst JC, Burto A, Ward TJ (1997) Debris flow run-out and landslide sediment delivery model test. J Hydraul Eng 123(5):410–419

Bruckl E, Parotidis M (2005) Prediction of slope instabilities due to deep-seated gravitational creep. Nat Hazards Earth Sys Sci 5:155–172

Canuti P, Casagli N, Ermini L, (1998) Inventory of natural dams in the Northern Apennine as a model for induced flood hazard forecasting. In: Stefanelli TC, Catani F, Casagli N, (2015) Geomorphological investigations on natural dams. Geoenviron Disasters 2(21)

D'Agostino V, Cesca M, Marchi L (2010) Field and laboratory investigations of runout distances of debris flows in the Dolomites (Eastern Italian Alps). Geomorphology 115:294–304

Ermini L, Casagli N (2003) Prediction of the behaviour of natural dams using a geomorphological dimensionless index. Earth Surf Proc Land 28:31–47

Fannin J, Bowman ET (2008) Debris flows—entrainment, deposition and travel distance. Geotech News 25(4):3–6

Highland LM, Ellen SD, Chistian SB, Brown WM (1997) Debris-flow hazards in United States. U.S. Geological Survey Fact Sheet 176–97. U.S. Geological Survey

Ikeya H (1989) Debris flow and its countermeasures in Japan. Bulletin of the International Association of Engineering Geology-Bulletin de l'Association Internationale de Géologie de l'Ingénieur 40(1):15–33

Inoue K, Mori T, Mizuyama T (2012) Three large historical natural dams and outburst disasters in the North Fossa Magna Area, Central Japan. Int J Erosion Control Eng 5(2):134–143

Iverson RM (1997) The physics of debris flow. Rev Geophys 35 (3):245–296

Izu-oshima Landslide Disaster Measure Exploratory Committee (2014) Report of Izu-oshima Landslide Disaster Measure Exploratory Committee. URL: http://www.kensetsu.metro.tokyo.jp/content/000006697.pdf. Last accessed 25 March 2016. In Japanese

Kaibori M, Ishikawa Y, Ushiyama M, Kubota T, Hiramatsu S, Fujita M, Miyoshi I, Yamashita Y (1999) Debris flow and slope failure disasters in Hiroshima Prefectures caused by heavy rainfall in June, 1999 (prompt report). J Erosion Control 52(3):34–43

Ministry of Construction, Chubu Regional Construction Bereau, River Section (1987) Collection of natural dams 1987

Ministry of Land, Infrastructure, Transport and Tourism, (n.d.) The features of deep-seated landslide. URL: http://www.mlit.go.jp/common/001019675.pdf. Last accessed 21 March 2016. In Japanese

Ministry of Land, Infrastructure, Transport and Tourism (2006) Explanatory of committee on crisis management of large-scale river blockage (natural dam). URL: http://www.mlit.go.jp/common/001024697.pdf. Last accessed 11 Dec 2015. In Japanese

Ministry of Land, Infrastructure, Transport and Tourism (2013) Sabo in the Kii Mountain District. Kii Mountain District Sabo Office, Kinki Regional Development Bureau, Nara

Ministry of Land, Infrastructure, Transport and Tourism (2014a) The correspondence situation to the landslide disaster occurred in Hiroshima-ken by a torrential downpour in August, 2014. URL: http://www.mlit.go.jp/river/sabo/H26_hiroshima/141031_hiroshimadosekiryu.pdf. Last accessed 24 March 2016. In Japanese

Ministry of Land, Infrastructure, Transport and Tourism (2014b) Report of countermeasure of slope land disasters in Hiroshima triggered by heavy rainfall in August, 2014. URL: http://www.mlit.go.jp/river/sabo/H26_hiroshima/141031_hiroshimadosekiryu.pdf. Last accessed 5 Jan 2015

Nishiguchi Y, Uchida T, Takezawa N, Ishizuka T, Mizuyama T (2012) Runout characteristics and grain size distribution of large-scale debris flows triggered by deep catastrophic landslides. Int J Japan Erosion Control Eng 5(1):16–26

Oshima H (2016) Differences in soil characteristics and flow behaviours of debris flows of 2013 at Izu Oshima and 2014 at Hiroshima. Bachelor thesis, Tokyo University of Agriculture and Technology

Rickenmann D (2005) Runout prediction methods. In: Jakob M, Hungr O (eds) Debris-flow hazards and related phenomena. Praxis, Springer, Berlin Heidelberg

Sassa K, Wang GH (2005) Mechanism of landslide-triggered debris flows: liquefaction phenomena due to the undrained loading of torrent deposits. In: Jakob M, Hungr O (eds) Debris-flow hazards and related phenomena. Praxis, Springer, Berlin Heidelberg

Strimbu B (2011) Modelling the travel distances of debris flwos and debris slides: quantifying hillside morphology. Ann For Res 54 (1):119–134

Takahashi T (2007) Debris flow: mechanics, prediction and countermeasures. Taylor & Francis Group, London

Takahashi T (2009) A review of Japanese debris flow research. Int J Erosion Control Eng 2:1

Uchida T, Yokoyama O, Suzuki R, Tamura K, Ishizuka T (2011) A New method for assessing deep Catastrophic landslide susceptibility. Int J Erosion Control Eng 4:2

Ushiyama M, Satohuka Y, Kaibori M (1999) Characteristics of heavy rainfall disasters in Hiroshima Prefecture on June 29, 1999. J Japan Soc Nat Disaster Sci 18:165–175

Wang G, Sassa K, Fukuoka H (2003) Downslope volume enlargement of a debris slide–debris flow in the 1999 Hiroshima, Japan, rainstorm. Eng Geol 69(3–4):309–330

Wang F, Wu YH, Yang H, Tanida Y, Kamei A (2015) Preliminary investigation of the 20 August 2014 debris flows triggered by a severe rainstorm in Hiroshima City, Japan. Geoenviron Disasters 2:17

Yang H, Wang F, Miyajima M (2015) Investigation of shallow landslides triggered by heavy rainfall during typhoon Wipha (2013), Izu Oshima Island, Japan. Geoenviron Disasters 2(15)

Relative Landslide Risk Assessment for the City of Valjevo

Katarina Andrejev, Jelka Krušić, Uroš Đurić, Miloš Marjanović, and Biljana Abolmasov

Abstract

This paper represents a relative landslide risk assessment of the City of Valjevo in Western Serbia. After the extreme rainfall during the May 2014, many new landslides were triggered, and Valjevo was one of the most affected areas in Serbia. The modeling was preceded by the data selection, and included ranging and preprocessing of the conditioning factors. The following eight factors were chosen as representative: stream distance, slope, lithology, elevation, distance from hydrogeological borders, land use, erodibility and aspect. Landslide susceptibility analysis was completed using the Analytical Hierarchy Process (AHP) multi-criteria method. Validation was performed by cross-referencing with an existing landslide inventory, which was made by field mapping and interpretation of satellite images. Finally, the relative risk was determined for the City of Valjevo by using a realistic population distribution model as a source for elements at risk. The results show the distribution of risk and suggest that 20% of the inhabited area falls into the high risk class, but this encompasses less than 5% of the total population.

Keywords

Landslides • Susceptibility map • Relative risk assessment • Validation

Introduction

During past decades landslides caused significant damage to population, material goods and the environment. Different natural phenomena and human disturbances trigger landslides. Natural triggers include meteorological changes, such as intense or prolonged rainfall or snowmelt, and rapid tectonic forcing, such as earthquakes or volcanic eruptions (Guzzetti et al. 2005). An intensive—and locally excessive—exploitation of the land, including development of new settlements, and construction of roads, railways, and other infrastructures have largely increased the impact of natural hazards both in industrialized and developing countries (Guzzetti et al. 2005). This variety of external stimulus causes a rapid increase in shear stress or decrease in shear strength of slope-forming materials (Dai et al. 2002). A contemporary approach to the assessment of hazards and risks of landslides includes a set of clearly defined methods that depend primarily on the purpose, the size of the area and the required accuracy of assessment (Cascini 2008; Corominas et al. 2014; Fell et al. 2008). Landslide susceptibility and hazard zoning can notably improve land-use planning, and thus can be considered an efficient way to reduce future

K. Andrejev (✉) · J. Krušić · M. Marjanović · B. Abolmasov
Faculty of Mining and Geology, University of Belgrade,
Djusina 7, 11000 Belgrade, Serbia
e-mail: katarina.andrejev@rgf.rs

J. Krušić
e-mail: jelka.krusic@rgf.bg.ac.rs

M. Marjanović
e-mail: milos.marjanovic@rgf.bg.ac.rs

B. Abolmasov
e-mail: biljana.abolmasov@rgf.bg.ac.rs

U. Đurić
Faculty of Civil Engineering, University of Belgrade,
Bul. Kralja Aleksandra 73, 11000 Belgrade, Serbia
e-mail: uros.djuric@rgf.rs

M. Mikoš et al. (eds.), *Advancing Culture of Living with Landslides*,
DOI 10.1007/978-3-319-53483-1_62

damage and loss of lives caused by landslides (Cascini 2008). Also, landslide susceptibility, hazard, and risk models could help mitigate or even avoid the unwanted consequences resulted from such hillslope mass movements (Komac 2006).

A wide variety of methods for assessing landslide susceptibility, hazard and risk are available and, to assist in risk management decisions. However, the methodologies implemented diverge significantly from country to country, and even within the same country (Corominas et al. 2014). It is also common that researchers stop at the level of landslides susceptibility model and rarely complete the entire risk modeling flow all the way. Herein, relatively simple approach for completing the relative landslide risk modeling process was demonstrated. It results in a relative risk assessment, since the data for more sophisticated landslide risk assessment were not available.

Methodology

Analytical Hierarchy Process (AHP)

The Analytical Hierarchy Process (AHP), (Satty 1980), is a theory of measurement concerned with deriving dominance priorities from paired comparisons of homogeneous elements with respect to a common criterion or attribute. This method allows quantification of expert assessment of the multi-criteria analysis (Komac 2006; Marjanovic et al. 2013). In the implementation of an AHP method for assessing the landslide susceptibility means you need to determine the impact of causal relationship parameters.

To simplify, AHP is a process where a few selected parameters (in this case: lithology, slope, aspect, land cover, elevation, erodibility, stream distance and distance from hydrogeological borders) define the final susceptibility model. Parameters that are used have an impact on a final model with their coefficients—weights (wi). Before creating a comparison of calculated models (Mi), the normalization of the results had to be done (we used a scale between 1 and 5). The sum of the weights of all the parameters is equal to 1, or 100% Eq. (1).

$$\sum_{i=1}^{n} w_i M_i = M_{AHP} \quad (1)$$

Parameter estimation in ideal conditions is provided by a team of experts who have experience independently to their criteria and then find final agreement. In this way, we get the averaged weights (points) for each parameter—the first matrix of AHP. Second step is to make new matrix and normalize weights of all parameters, i.e., find percent of each parameter's impact on landslide occurrence. Created matrix is inverse symmetric (aij•aji = 1) in respect to a main diagonal. For the final AHP model, the CR (Consistency Ratio) was calculated (Saaty and Vargas 2001). Comparisons made by this method are subjective and the AHP tolerates inconsistency through the amount of redundancy in the approach. If this consistency index fails to reach a required level, then answers to comparisons may be re-examined. The consistency index, CI, is calculated as Eq. (2):

$$\mathrm{CI} = (\lambda_{\max} - \mathrm{n})/\mathrm{n} - 1 \quad (2)$$

where $\lambda_{\max}$ is the Principal Eigen Value, value of the judgement matrix. This CI can be compared with that of a random matrix RI with equation Eq. (3):

$$\mathrm{CR} = \mathrm{CI}/\mathrm{RI} \quad (3)$$

The ratio derived, CI/RI, is termed the consistency ratio, CR. Saaty and Vargas (2001) suggest the value of CR should be less than 0.1 (Bhushan and Rai 2004).

Relative Risk Assessment

Landslide risk can be assessed qualitatively or quantitatively. Whether the qualitative or quantitative assessment is more suitable depends on both, the desired accuracy of the outcome and the nature of the problem, and should be compatible with the quality and quantity of available data. Generally, for a large area—moderate scale of 1:25,000 where the quality and quantity of available data are too meager for quantitative analysis, a qualitative risk assessment may be more applicable. In this case, landslide inventory had to be produced within a short period of time after heavy rainfall in May 2014. Therefore, so-called event-based landslide inventory (landslides caused by the same triggering event) was made for the territory of Valjevo (Corominas et al. 2014).

Too little information was available to carry out a sophisticated landslide risk assessment. In particular, there were not enough data on the probability of landslides of different magnitudes to make a quantitative risk assessment. Hence, relative (semi-quantitative) risk analysis was carried out through the combination of procedures, using the landslide susceptibility map and population distribution as basic elements at risk. Elements at risk considered in this work included population, and the quantifying parameter was the population density, i.e., the population is scaled according to the relative 0–1 range of population density. Ideally, landslide risk zoning maps are prepared using the hazard zoning maps, as well as the elements at risk, the spatial–temporal probability and vulnerability of elements at risk (Fell et al. 2008). Nevertheless, it is fairly complicated to obtain such maps at the regional scale. Therefore, the landslide risk analysis was performed within the GIS techniques, combining the following three layers: landslide susceptibility map, exposure to the higher susceptibility and registered landslides, and map of the exposed element population. The risk analysis that was carried out cannot be defined completely as qualitative nor quantitative. Obtained risk zones are defined and classified numerically, not in a descriptive manner. Still, the risk does not express a typical probability per certain return period. It was decided to determine this process of risk assessment as relative risk assessment regarding population density.

Case Study

The City of Valjevo occupies an area of 905 km^2 and it is located in western Serbia in the center of the Kolubara district (Fig. 1). It has a population of ~95000 inhabitants. Valjevo is situated along the valley the Kolubara River, formed by merging of two next biggest rivers in the study area, Jablanica and Gradac. The range of topographical elevation values varies between 122 m (northeastern part of the study area) and 1323 m (southwestern part), while the dominant topographical elevation range is between 240 and 430 m. Hence the morphology of terrain going from northeast to southwest, changes from flat-hilly to hilly-mountainous relief. Although the slope angle values range between 0° and 63°, the majority are between 5° and 12°. The climate of the study area can be characterized as moderate-continental climate, with elements of sub-humid climate. For an observation period of 1981–2010 the average annual amount of precipitation is 764.38 mm and, in extreme conditions (2014), 1211.27 mm.

In meteorological station Valjevo, maximum amount of rainfall per one day equaling 108.2 mm was recorded on 15th of May 2014. Due to severe event in 2014—heavy rainfall—many landslides were triggered causing significant damages. The field observations showed that there were 220 slide locations. Slides, combination of slides and flows, and flows are three main types of mass movements in the study area. Among the slide types, translational is dominant. Considering the stage of activity, approximately 98% of all landslides are active.

Geological setting of the study area is very complex. The oldest geological units are represented by the Jadar formation of Paleozoic age. Firstly, there are phyllites and quartz sandstones of Devonian age. Sediments of undivided Devonian and Carboniferous are meta-sandstones, shales, less argillite and phyllite. Carboniferous sediments and

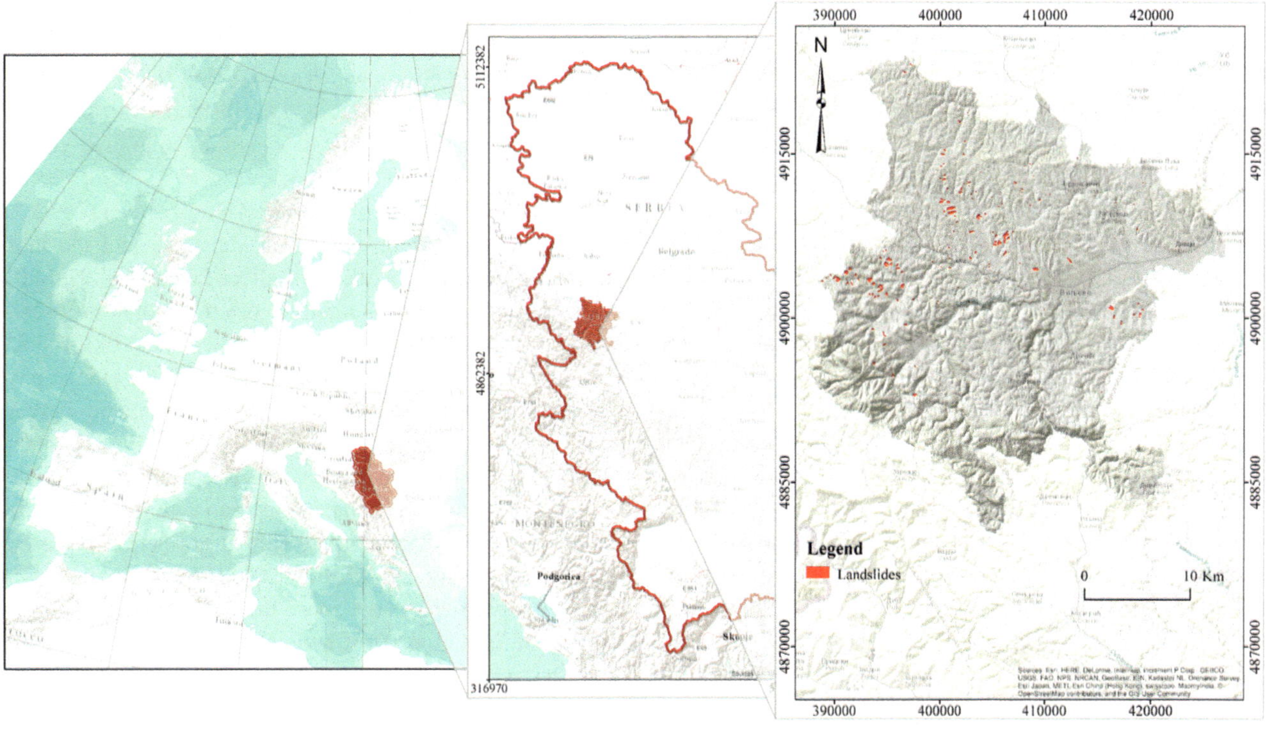

Fig. 1 Geographical position of the study area

metamorphic rocks are presented by limestone and argillite, as well as limestone, clay shale, rarely quartz sandstones. The sediments of Permian age are quartz sandstones, sandy—argillaceous schists and black bituminous limestone. These sediments directly pass into limestone, sandstone, argillaceous schist and dolomite of the Triassic age. Distribution of Jurassic formations is associated with the range of Podrinje—Valjevo mountains. The sedimentary-volcanogenic products of the diabase-chert formation are the most wide-spread. This formation is transgressively overlain by massive and thick-bedded limestone, marly limestone, marls and conglomerate of the Upper Cretaceous. Neogene formations consist of: conglomerate, marlstone, clay, sand and marly limestone. Quaternary sediments occupy a small area. They are made up of alluvial and terrace deposits in valleys of big rivers, talus, scree breccia and travertine.

Materials

The landslide inventory mapping is the most fundamental step in any landslide susceptibility and hazard assessment. The mapping of existing landslides is essential to study the relationship between the landslide distribution and the conditioning factors (Pourghasaemi et al. 2012).

Landslide database was made as a part of Project BEWARE, after the floods 2014. All landslides were identified by field mapping and interpretation of satellite images. Statistically, 42% of landslides are identified as active, almost the same percent, 41% as suspended, and 13% of landslides as reactivated. Landslide inventory is overlaying the AHP landslide susceptibility model, given in Fig. 4.

For landslides susceptibility assessment (LSA) eight conditioning factors were determined: slope, erodibility, lithology, distance from hydrogeological borders, stream distance, aspect, land use and elevation. The classification of each model (on scale 1–5) is statistically determined in regard to spatial distribution of landslides in certain areas in a relation to individual factor. The landslide inventory map was used for validation of the LSA model.

All models were prepared in raster format with the same 30 × 30 m resolution. The process of evaluation and creation of every individual conditioning factor model, which were used later on for the landslide susceptibility model, is briefly explained in the following text. Input models (Fig. 2) were classified, and description of each class is given in Table 1.

Slope (M1)—The main parameter of the slope stability analysis is the slope degree (Lee and Min 2001). The slope of the terrain model was derived from the digital elevation model (DEM). The areas with an inclination of 10–20° were classified as the most unfavorable, since statistically the largest number of occurrences of instability has been registered in this range. This map is divided into five slope categories (Fig. 2a).

Erodibility (M2) model was developed from (Gavrilovic 1971) methodology and with regard of database of registered flash floods. Five classes of potential erodibility were divided; from very low to very high potential to erodibility (Fig. 2b).

Lithology (M3)—Classification of geological factors is usually the most difficult and was based mainly on subjective judgement. For this assessment, the engineering-geological map of the study area at 1:300,000 scale was used. The landslides occurrence probability value is higher in crystalline metamorphic rocks, which typically have a thick crust of surface decomposition, and heterogeneous lacustrine sediments. The most of the active landslides are identified within this area. Smaller weighting factor is assigned to sandstone, flysch sediments, ultramafic rocks than lake sediments, while alluvial deposits had no influence (Fig. 2c).

Distance from hydrogeological borders (M4) is a parameter that was introduced to emphasize the rate of change of hydrogeological function in rock masses and its influence on the landsliding process. It is assumed that the zones closer to such border are more susceptible to landslide occurrence, as was confirmed during the field research. The map was also classified in five classes (Fig. 2d).

Stream distance (M5) model was made from a database with local watercourses and it is assumed that the value of the weighting factor decreases with distance from the local watercourse, and the model is classified into five classes (Fig. 2e).

Aspect (M6)—Some of the meteorological events such as the amount of rainfall, amount of sunshine, and the morphological structure of the area affect landslides. The hillsides receiving dense rainfall reach saturation faster; however, this is also related to filtering capacity of the slope controlled by various parameters such as slope topography, soil type, permeability, porosity, humidity, organic ingredients, land cover, and the climatic season (Pourghasaemi et al. 2012). Thus, there is different susceptibility to a variety of physical and mechanical decomposition, and water retention in the soil. It is assumed that northern and eastern slopes have the most unfavorable weights, while the southern and western have the most favorable, due to different degrees of insolation. This model is divided into five classes (Fig. 2f).

Land use (M7) model is derived from CORINE Land cover 2012 data, based on different types of vegetation

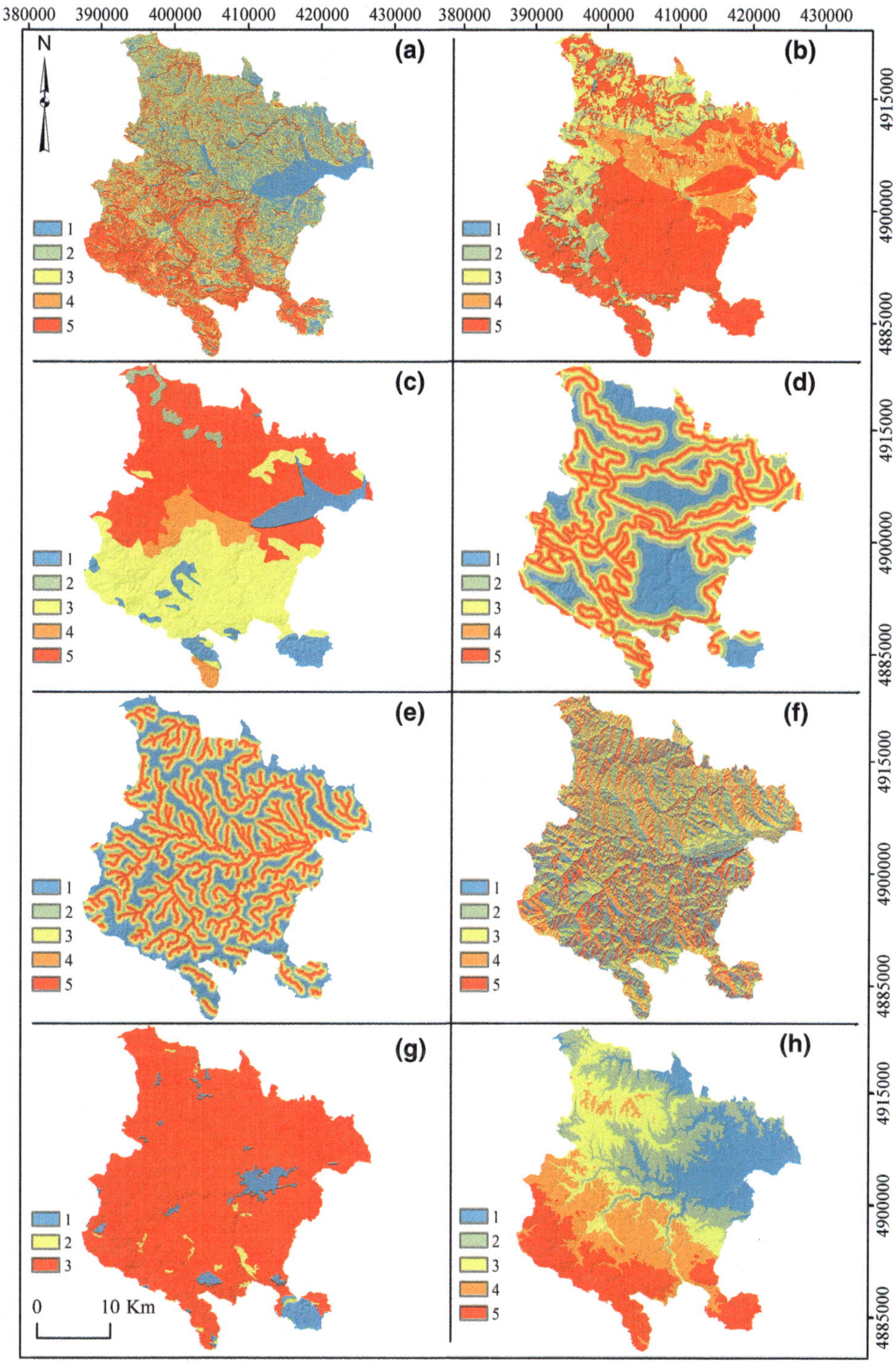

Fig. 2 Raster data sets **a** slope **b** erodibility **c** lithology **d** distance from hydrogeological borders **e** stream distance **f** aspect **g** land use **h** elevation

cover. It is assumed that different vegetation covers influence landslide susceptibility differently. This model is statistically divided into three classes (Fig. 2g).

Elevation (M8) model is reclassified DEM where classes are allocated on the basis of spatial distribution of the occurrence of landslides, and this map also is divided into five categories (Fig. 2h).

Results and Discussion

Landslide Susceptibility Assessment

The process includes several steps: (1) break down the problem into its component factors (according conditioning factors in this study); (2) arrange these factors in a hierarchic

Table 1 Description of classes within each conditioning factor model

Input models	Classes				
	1	2	3	4	5
Slope	0–5°	5–8°	8–11°	11–17°	17–63°
Erodibility	Very low	Low	Medium	High	Very high
Lithology	Proluvial-alluvial sediments, serpentinite, melafire, trachyte	Lake sediments	Limestones, flysch, ophiolitic melange, gabbroic rocks	Sandstone-carbonate rocks	Low crystalline metamorphic rocks, lacustrine sediments
Distance from hydrogeological borders	0–3.660 m	3.660–4.242 m	4.242–4.542 m	4.542–4.757 m	4.757–4.932 m
Stream distance	0–2.215 m	2.215–2.450 m	2.450–2.632 m	2.632–2.795 m	2.795–2.945 m
Aspect	−1–58.4°	584–1277°	1277–1999°	1999–2805°	2805–3597°
Land use	Artificial surfaces, coniferous and mixed forests	Natural grassland, transitional woodland	Non-irrigated arable land, pastures, complex cultivation patterns, land principally occupied by agriculture, broad-leaved culture		
Elevation	122–260 m	260–328 m	328–423 m	423–652 m	652–1338 m

order; (3) assign numerical values according to their subjective relevance to determine the relative importance of each factor; and (4) synthesize the rating to determine the priorities to be assigned to these factors (Saaty and Vargas 2001).

The modeling consisted in the correlation of mutual influence and giving the advantage to the factor which is more significant (Saaty and Vargas 2001). When the factor on the vertical axis is more important than the factor on the horizontal axis, this value varies between 1 and 9. Conversely, the value varies between the reciprocals 1/2 and 1/9. In this case, the most important factor is slope, and less important elevation (Table 2). The relationship in regard to the importance of each factor was constructed using the Eq. (4) which gave as a result Landslide susceptibility map (LSM) (Fig. 4).

$$\mathrm{LSM} = \mathrm{M1} \times \mathrm{W_{M1}} + \mathrm{M2} \times \mathrm{W_{M2}} + \mathrm{M3} \times \mathrm{W_{M3}} + \mathrm{M4} \times \mathrm{W_{M4}} + \mathrm{M5} \times \mathrm{W_{M5}} + \mathrm{M6} \times \mathrm{W_{M6}} + \mathrm{M7} \times \mathrm{W_{M7}} + \mathrm{M8} \times \mathrm{W_{M8}} \tag{4}$$

where: M1-slope; M2-erodibility; M3-litology; M4-distance from hydrogeological borders; M5-stream distance; M6-aspect; M7-land use and M8-elevation models and Wmn-percent weights for each model.

In the next phase, the matrix normalization was performed. There by, each value was divided by the total sum along the column. The result across rows is the impact on landslide susceptibility of each factor, given in percentage (Table 3).

Checking the consistency of obtained AHP model, it was established that the model is consistent, because λ_{max} = 8676, i.e., n = 8 and consistency index CR = 0.068 (<0.1) (Table 3).

Table 2 AHP comparison matrix

AHP	M1	M2	M3	M4	M5	M6	M7	M8
M1	1.0	2.00	2.00	2.00	3.00	4.00	5.00	9.00
M2	2.00	1.00	1.00	2.00	2.00	4.00	6.00	8.00
M3	0.50	1.00	1.00	2.00	2.00	3.00	6.00	8.00
M4	0.50	0.50	0.50	1.00	2.00	4.00	5.00	7.00
M5	0.33	0.50	0.50	0.50	1.00	2.00	4.00	6.00
M6	0.25	0.25	0.33	0.25	0.50	1.00	3.00	5.00
M7	0.20	0.17	0.17	0.20	0.25	0.33	1.00	4.00
M8	0.11	0.13	0.13	0.14	0.17	0.20	0.25	1.00
S	4.89	5.54	5.63	8.09	10.92	18.53	30.25	48.00

Table 3 Final (normalized) AHP matrix

	M1	M2	M3	M4	M5	M6	M7	M8	WMn(%)	CR
M1	0.20	0.36	0.36	0.25	0.27	0.22	0.17	0.19	0.25	1.230
M2	0.41	0.18	0.18	0.25	0.18	0.22	0.20	0.17	0.22	1.231
M3	0.10	0.18	0.18	0.25	0.18	0.16	0.20	0.17	0.18	0.997
M4	0.10	0.09	0.09	0.12	0.18	0.22	0.17	0.15	0.14	1.128
M5	0.07	0.09	0.09	0.06	0.09	0.11	0.13	0.13	0.10	1.045
M6	0.05	0.05	0.06	0.03	0.05	0.05	0.10	0.10	0.06	1.134
M7	0.04	0.03	0.03	0.02	0.02	0.02	0.03	0.08	0.04	1.068
M8	0.02	0.02	0.02	0.02	0.02	0.01	0.01	0.02	0.02	0.842
S	1.00	1.00	1.00	1.00	1.00	1.00	1.00	1.00	1.00	8.676

Table 4 Validation results of susceptibility model

Parameter	Equation	AHP
Sensitivity	TPR = TP/P = TP/(TP + FN)	0.609503
Specificity	SPC = TN/N = TN/(FP + TN)	0.770305
False Positive Rate (FPR)	FPR = FP/N = FP/(FP + TN)	0.229695
False Negative Rate (FNR)	FNR = FN/P = FN(TP + FN)	0.390497
Accuracy	ACC = (TP + TN)/(P + N)	0.769407

Validation of the results was made by Landslide inventory map collected by field mapping and analysis of satellite images of high resolution (Marjanović et al. 2016). Contingency Matrix, shows the different types of errors in the model (Table 4).

Where: True Positives are the instances where the model and the reference agree that landslide exists, True Negatives represent the instances where the model comply with the reference on non-landslide instances, while misclassifications are presented by False Positives (model claims landslides where they do not exist according to the reference) and False Negatives (vice versa) (Marjanovic et al. 2013).

The model is also estimated on the basis of ROC curve (Receiver Operating Characteristic), quantitatively based on the value of the parameter area under the curve AUC (Area Under the Curve), as well as, qualitatively based on the character and shape of the curve (Fawcett 2006). The AUC value of about 0.77 generally characterized model with good prediction (Fig. 3). Based on this result can be said that their usage is generally limited in high (regional) scale, because it is expected to have greater accuracy for prediction of landslides in detailed areas (Fig. 4).

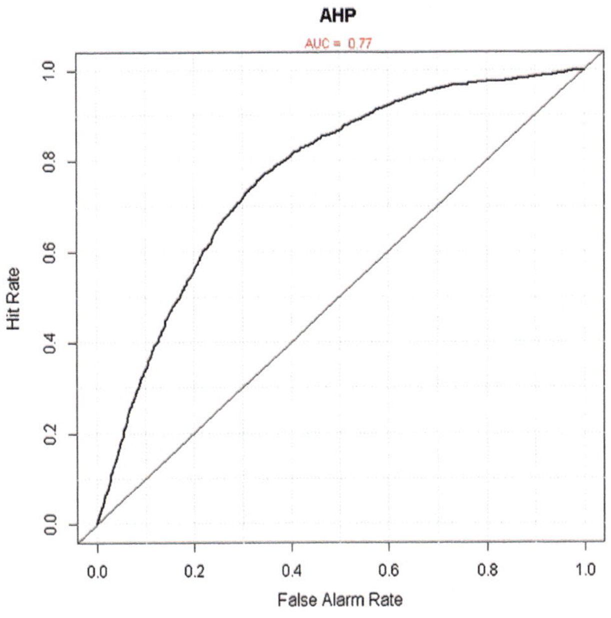

Fig. 3 Validation results given by ROC curve

Relative Risk Assessment

Due to the lack of a historic landslide inventory, the knowledge about: landslide causative factors, landslide mechanisms, runout behavior and temporal frequency, is still limited. Hence, considering available data, and all mentioned limitations, obtained results can be considered preliminary. Risk map is classified in five zones: very low (0–0.2), low (0.2–0.4), moderate (0.4–0.6), high (0.6–0.8) and very high risk (0.8–1.0). The biggest density of population is located in very low and low landslide susceptibility zones, and therefore this area have very low or low risk (Fig. 5). This applies for the most of the territory of the urban part of the city, which is the most densely populated part in the study region. Currently, the highest zones of risk are recognized within moderate and high susceptibility zones, in the northwestern and

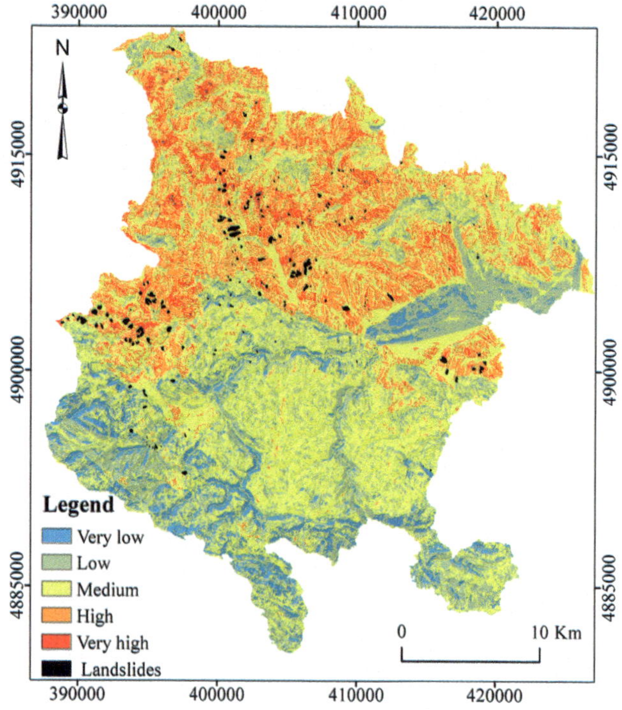

Fig. 4 Final AHP susceptibility model

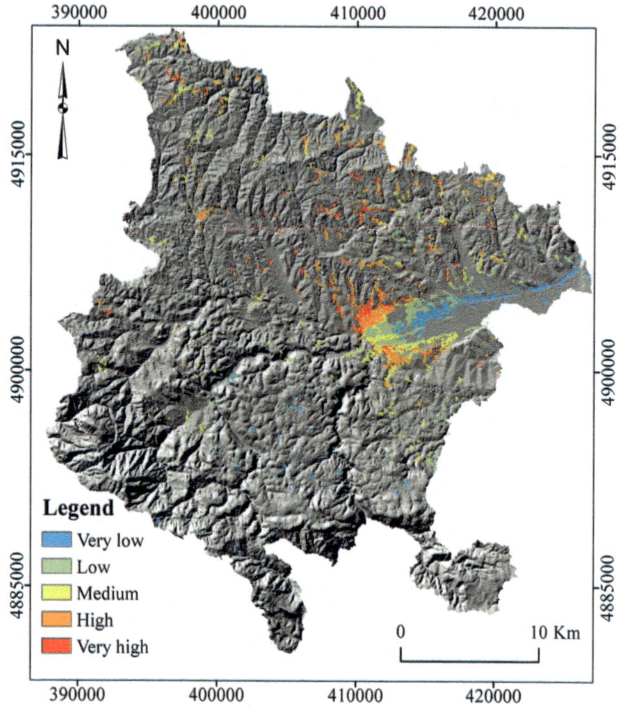

Fig. 5 Map of relative risk for the city of Valjevo

southern part of the study area. It should be emphasized, that the most of case study area has rural environments, with isolated less inhabited settlements, and relative risk is regarding density of population (relative risk assessment is carried out based on population density). Therefore, it is understandable why distribution of high risk zones is very diffuse and not always related to very high susceptibility class. To refine these results in the future more data should be collected.

Conclusion

In this paper, we propose preliminary landslide risk assessment methodology based on population density in medium scale. Landslide susceptibility was assessed by including 8 parameters: stream distance, slope, lithology, elevation, distance from hydrogeological borders, land use, erodibility and aspect. This model was validated, and results of the evaluation are presented by ROC curve with acceptable accuracy of the model (AUC = 0. 77). The risk map was obtained by overlaying landslide susceptibility map, exposure to the higher susceptibility and registered landslides, and map of the exposed element population. Applied relative risk assessment procedure does not express a probability of landslide occurring per certain return period, because it was based on susceptibility assessment. Hence, further activities will be to increase the completeness and quality of the available data, in order to improve reliability of risk mapping.

Acknowledgements This research was not possible without Project BEWARE (BEyond landslide aWAREness) funded by People of Japan and UNDP Office in Serbia (grant No 00094641). The project was implemented by the State Geological Survey of Serbia, and the University of Belgrade Faculty of Mining and Geology. All activities are supported by the Ministry of Energy and Mining, Government Office for Reconstruction and Flood Relief and Ministry for Education, Science and Technological Development of the Republic of Serbia Project No TR36009, too.

References

Bhushan N, Rai K (2004) Strategic decision making: applying the analytic hierarchy process. Springer, London (ISBN 978-1-85233-864-0), p 172

Cascini L (2008) Applicability of landslide susceptibility and hazard zoning at different scales. Eng Geol 102:164–177

Corominas J et al (2014) Recommendations for the quantitative analysis of landslide risk. Bull Eng Geol Env 73(2):209–263

Dai FC, Lee CF, Ngai YY (2002) Landslide risk assessment and management: an overview. Eng Geol 64:65–87

Fell R et al (2008) Guidelines for landslide susceptibility, hazard and risk zoning for land use planning. Eng Geol 102:85–98

Fawcett T (2006) An introduction to ROC analysis. Pattern Recogn Lett 27:861–874

Gavrilović S (1971) Inženjering o bujičnim tokovima i eroziji, Institut za eroziju, melioracije i vodoprivredu bujičnih tokova, Univerzitet u Beogradu, Šumarski fakultet i Časopis Izgradnja—specijalno izdanje

Guzzetti F, Reichenbach P, Cardinali M, Galli M, Ardizzone F (2005) Probabilistic landslide hazard assessment at the basin scale. Geomorphology 72:272–299

Komac M (2006) A landslide susceptibility model using the analytical hierarchy process method and multivariate statistics in perialpine Slovenia. Geomorphology 74:17–28

Lee S, Min K (2001) Statistical analysis of landslide susceptibility at Yongin, Korea. Environ Geol 40:1095–1113

Marjanović M, Abolmasov B, Đurić U, Zečević S (2013) Impact of geo-environmental factors on landslide susceptibility using an AHP method: a case study of Fruška Gora Mt. Serbia. Annales Geologiques de la Peninsule Balkanique 74:91–100

Marjanović et. al (2016) Coupling field and satellite data for an event-based landslide inventory. In: Proceedings of the 12th international symposium on landslides, 12–19 June 2016. Naples, Italy, pp 1361–1366

Pourghasaemi et al (2012) Application of fuzzy logic and analytical hierarchy process (AHP) to landslide susceptibility mapping at Haraz watershed. Iran Nat Hazards 63:965–996

Saaty TL, Vargas LG (2001) Models, methods, concepts and applications of the analytic hierarchy process. Kluwer, Springer, US, p 333

Varnes DJ (1984) Landslide hazard zonation: a review of principles and practice. Int Assoc Eng Geol, Paris, p 63

4thWLF2017
4th World Landslide Forum
LJUBLJANA SLOVENIA EU

Huge Slope Collapses Flashing the Andean Active Orogenic Front (Argentinean Precordillera 31–33°S)

Sebastián Junquera Torrado, Stella Maris Moreiras, and Sergio A. Sepúlveda

Abstract

The study area is located along the Andean active orogenic front comprising the most seismically active region of Argentina. Main Quaternary deformation is concentrated in this Western central part of the country associated with active faults linked to an intense shallow seismic activity (<35 km depth). During the last 150 years, the region has suffered at least six major earthquakes with a magnitude greater than Ms $\geq$ 7.0. The focus of this research is to analyse the landslide behaviour along this Andean active orogenic front. To that end, we carried out a landslide inventory along Precordillera (31°–33°S). We analysed type, size, activity grade and other morphological parameters of these landslides. We found huge collapses coincide with traces of active Quaternary faults in this region. However, landslides are clustered being denser splayed in the centre of study area. Furthermore, activity grade of such landslides is higher in this central zone decreasing gradually towards the north and the south. This central area is affected by the Juan Fernandez Ridge which is likely related to higher deformation rate.

Keywords

Quaternary • Neotectonics • Active fault • Earthquakes

Introduction

Distribution of landslides along active tectonic fronts is not a new finding among paleoseismologists and geomorphologists (Keefer 1987, 1994, 2000; Jibson and Keefer 1993; Jibson 1996; Jibson et al. 2006). This has been also achieved along the main ranges of the Andes (Fauqué et al. 2000; Moreiras 2006; Perucca and Angillieri 2008, 2009a, b; Moreiras and Sepúlveda 2009; Moreiras and Coronato 2010; Esper Angillieri 2011, 2012; Esper Angillieri and Perucca 2013; Esper Angillieri et al. 2014 among others). However, correlation between landslide behavior with characteristics of the active front has been rarely reached.

A seismic genesis is commonly suggested for main landslides in the Andes because of their relation with neotectonic activity, given the proximity of the failures to faults and statistical correlation of the landslide distribution with faults and shallow crustal seismic activity (Abele 1984; Antinao and Gosse 2009). Nonetheless, Quaternary activity of regional faults has been rarely proved by the lack of seismotectonic and palaeoseismological studies and the low frequency of large magnitude crustal earthquakes during the last century in the region (Moreiras and Sepúlveda 2015). Neotectonic studies focused on this dilemma are beginning in Argentina. Seismicity is extended

S. Junquera Torrado
CONICET—IANIGLA (CCT), Av. Dr. Ruiz Leal S/N, Parque Gral San Martín Mendoza, Argentina
e-mail: sjunquera@mendoza-conicet.gob.ar

S.M. Moreiras (✉)
Fac. Ciencias Agrarias, Geomorphology Group, IANIGLA–CONICET, Universidad Nacional de Cuyo, Av Ruiz Leal S/N Parque Gral. San Martin S/N. Ciudad, 5500 Mendoza, Argentina
e-mail: moreiras@mendoza-conicet.gob.ar

S.A. Sepúlveda
Departamento de Geología, Universidad de Chile, Plaza Ercilla 803, Santiago, Chile
e-mail: sesepulv@ing.uchile.cl

and
Instituto de Ciencias de la Ingeniería, Universidad de O′Higgins, Lib. Bernardo O′Higgins 611, Rancagua, Chile

M. Mikoš et al. (eds.), *Advancing Culture of Living with Landslides*,
DOI 10.1007/978-3-319-53483-1_63

to the Holocene in the Precordillera (Perucca and Moreiras 2008) based on preservation of seismites in palaeo-lakes dammed by two rock avalanches in the Acequión River (San Juan province, 32°SL). Rock avalanches clustered in the northern extreme of the Cordon del Plata (Cordillera Frontal) linked to evidences of Quaternary activity of the Carrera fault system prove the occurrence of M > 6 palaeo-earthquakes at least until the Late Pleistocene (Moreiras 2006; Moreiras et al. 2015).

In this research correlation between neotectonic activity and landslide spatial distribution is encouraged with the aim of understanding this relationship in the Andean Orogenic front.

Study Area

The study area corresponds to the Precordillera morphostructural unit being a north-south fold-and-thrust belt (Jordan et al. 1993) (Fig. 1). It is divided in four sectors: Eastern, Central, Western and Southern Precordillera (Cortés et al. 2005). The differences between these are based on their stratigraphic composition and structural nature. Stratigraphically, Eastern and Central Precordillera are composed of a thick sequence of Silurian, Devonian and Carboniferous rocks overlying Cambrian platform limestones. Western Precordillera is mainly composed of Cambrian and Ordovician ocean-floor sediments interbedded with basic and basaltic rocks. Southern Precordillera is composed of Lower Paleozoic rocks. West vergence characterized Eastern Precordillera; whereas Western, Central and Southern Precordillera show an east vergence in their structures.

Most of the Quaternary tectonic deformation in Argentina is gathered in the backarc region of the Andes. This fact is attributed to the flat-slab segment of the Nazca Plate produced due to the oblique subduction of the Juan Fernández ridge beneath the South American plate (Barazangi and Isacks 1976; Cahill and Isacks 1992; Anderson et al. 2007). This effect generates a shallow seismic activity (<35 km depth) (Smalley et al. 1993; Alvarado et al. 2009) that is evidently related to Quaternary faults. This highest seismic hazard region of the country recorded the most significant historical and instrumental seismicity. The largest Argentinean historical earthquake with an epicenter in the north of San Juan province (Ms: 7.5) occurred in this region. Quaternary deformation already known in Argentina is concentrated within this sector where several surface ruptures were reported during historical earthquake events such the case of the 1944 La Laja earthquake (Ms: 7.4) and the 1977 Caucete earthquake (Ms: 7.4).

Methodology

Initially, geological information was collected from Geological sheets published by SEGEMAR (Argentinean Mining Geological Service) at 1:100.000 and 1:250.000 scale. These geological sheets are Cerro Aconcagua 3369-I, Mendoza 3369-II, Potrerillos 3369-15 and San Juan 3169-IV. There is not published geological sheet of the Calingasta and Barreal region. Indeed, we included geological data collected by Quartino et al. (1971). This active tectonic segment of the Central Andes is strongly affected by Neotectonic activity. Data about Quaternary regional faults was collected to evaluate the relation between distribution of landslides and neotectonic activity in the study area (Costa et al. 2000, Casa et al. 2014).

A landslide inventory map was produced out using images from Google EarthTM for easy visualization. Manual digitalization of the identified landslides was released using QGIS software. We included in our inventory those landslides identified in previous works (Angillieri 2015; Pantano Zuñiga 2014). Landslide parameters were measured. Altitude data were obtained from ASTER Global Digital Elevation Model V002 (USGS 2015). The following parameters were estimated from the inventory:

- Area (At): total area that includes the ruptured zone and the area of the displaced mass (Ad).
- Perimeter (Pt): contour length of the lanslide.
- Perimeter of the displaced mass (Pd): contour length of the displaced material.
- Elevation (H): highest altitude is represented as maximum (Hmax). Lowest altitude is represented as minimum (Hmin).
- Height of the rupture zone (Hr): is calculated by the difference between Hmax and Hmin. It is the vertical distance between the crown and the foot.
- Total length (L): minimum distance from crown to landslide toe.
- Length of displaced mass (Ld): minimum distance from toe to top.
- Width of surface of rupture (Wr): maximum width between flanks of landslide perpendicular to the length of surface of rupture.
- Width of displaced material (Wd): maximum brendth of displaced mass perpendicular to length (Ld).
- Slope (S)
- Aspect (A): orientation with respect to geographic North.

Identified landslides were classified according to type of moment (Varnes 1978), depth of deformation (superficial or deep) and grade of activity (Crozier 1984) considering three categories: active, inactive, and dormant. An active landslide shows a main fresh scarp or reactivations are visible. Inactive

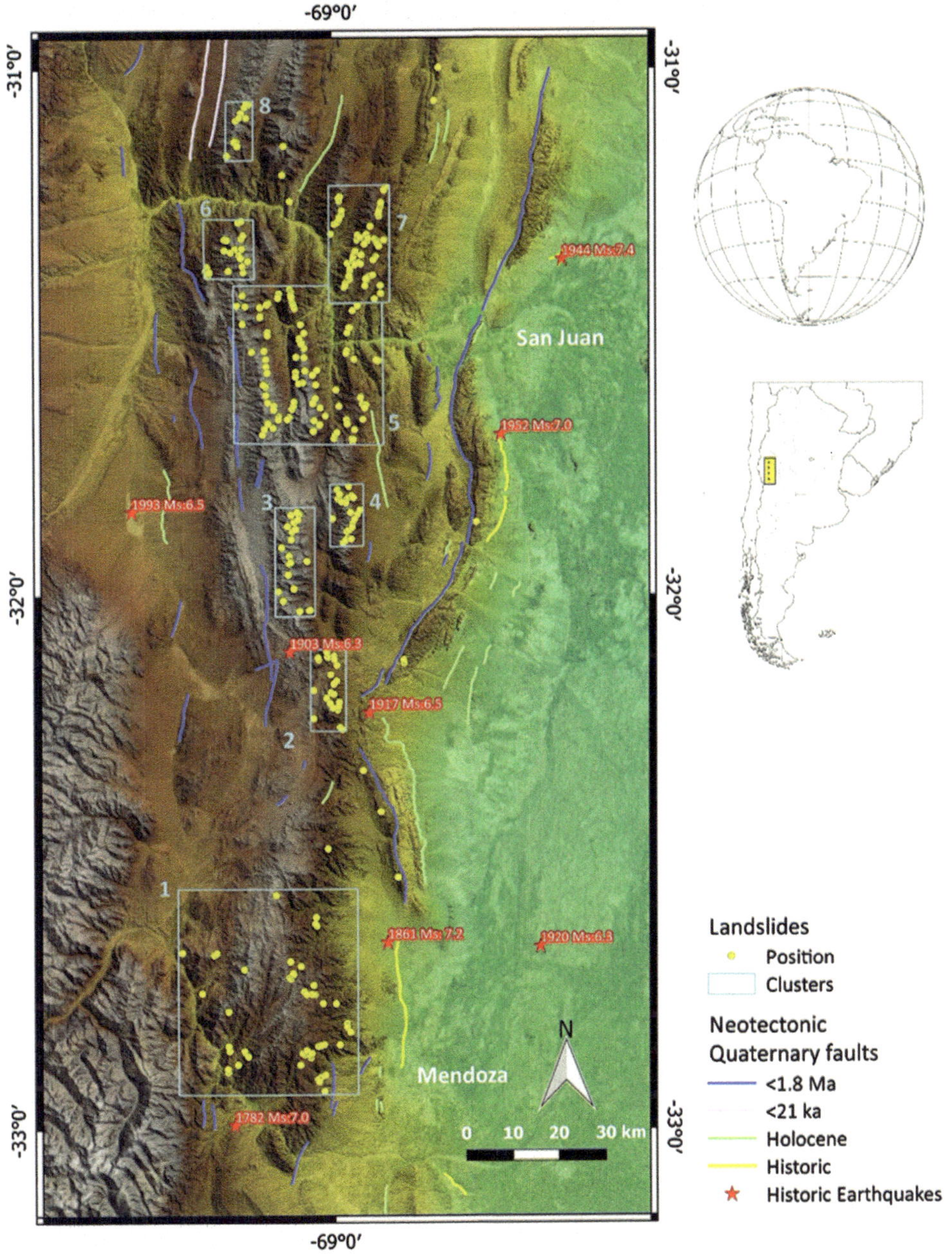

Fig. 1 Location map of the study area showing the landslide inventory and Quaternary faults between San Juan and Mendoza rivers in the Precordillera range

landslide does not show any evidence of activity. Whereas, landslides were classified as dormant when relicted reactivities are visible. The number of reactivations of each landslide was considered according to the grade of activity (Table 1).

Landslide Inventory

A total of 348 were identified in the study area including different types of movements. We distinguished 175 (=50%) rock slides, 78 (22%) debris slides, 72 (21%) debris-rock slides, and 23 (7%) complex events (Table 1). These landslides show very deep movements.

The primary lithology where more than 80% of landslides failured was detritic and sedimentary rocks such as sandstones, pelites, graywackes and shales and, to a lesser extent, metamorphic and volcanic rocks such as andesites, phyllites, schists, tuffs and volcanic breccias.

A total of 199 landslides are smaller than 0.25 km^2, which is a 57% of the total. A total of 114 landslides were identified with areas between 0.25 and 1 km^2. The 34 landslides remaining are greater than 1 km^2 reaching areas of 7.6 km^2. It is important to note that 25 of 34 landslides greater than 1 km^2 are actives.

The most common range of slope degree for 145 landslides is between 40° and 45° followed by 79 landslides

Table 1 Classification and parameters of each landslide clusters expressed in percentage

Cluster	1	2	3	4	5	6	7	8
N	55	26	23	27	103	21	70	11
Type								
Debris slide	27	23	22	11	19	38	17	55
D-Rock slid	18	23	39	7	20	24	23	18
Rock slide	44	50	35	81	53	38	54	0
Complex	11	4	4	0	7	0	6	27
Lithology								
(1) Fm CR	5							
(2) Fm LC	24							
(3) Fm Potre	7							
(4) Fm SMx	9							
(5) Fm VV	25							
(6) Gr Bon	4							
(7) Gr CHO	18							
(8) Gr USP	7	19						
(9) Cg SM		4						
(10) Fm CPA		15	74		29	76		100
(11) Fm Leon		4						
(12) Fm Mñ		12						
(13) Gr VV		46						
(14) Bl-O			26					
(15) Fm PN				100	57		57	
(16) Fm Alb					1			
(17) Fm LD					10		1	
(18) Fm SJ					3		30	
(19) Fm DP						24		
(20) Fm TC							10	
(21) Fm TMB							1	
Activity								
Active	2	35	57	74	37	33	21	18
Inactive	69	58	30	15	51	48	56	73
Dormant	29	8	13	11	12	19	23	9
Reactivities								
Yes	31	35	65	56	40	43	44	27
No	69	65	35	44	60	57	56	73
Depth								
Deep	53	38	61	56	51	33	37	18
Shallow	47	62	39	44	49	67	63	82

(1): Fm Cerro Redondo (andesites), (2): Fm Las Cabras (sandstones, pelites), (3): Fm Potrerillos (sandstones, conglomeratic sands), (4): Fm Santa Máxima (feldspathic sandstones, arkoses, pelites), (5): Fm Villavicencio (sandstones, pelites, shales), (6): Bonilla Group (carbonatic shales, phyllites, schists), (7): Choiyoi Group (lava, fenoandesitic tuffs, volcanic breccias), (8): Uspallata Group (conglomerates, sandstones, claystones, siltstones, tuffs), (9): Santa María Conglomerate (conglomerates), (10): Fm Cortaderas, Peñasco, Alojamiento (metasandstones, phyllites, marbles, sandstones, pelites), (11): Fm Leoncito (diamictites, siltstones, sandstones), (12): Fm Mariño (conglomerates, sandstones, clays, sandy tuffaceous clay), (13): Villavicencio Group (graywackes, shales, conglomerates), (14): Olistolitic blocks (olistolitic calcareous blocks), (15): Fm Punta Negra (graywackes, shales), (16): Fm Albarracín (sandstones, siltstones, conglomerates, pyroclastic flow deposits), (17): Fm La Dehesa (sandstones, siltstones, conglomerates), (18): Fm San Juan (limestones, marls), (19): Fm Don Polo (wackes, shales), (20): Fm Talacasto (sandstones, shales), (21): Fm Tambolar (sandstones, shales, marls, graywackes) between 45° and 50°. Only one slide collapsed on a slope greater than 60°.

Concerning topography the lowest and the highest altitude in the study area are 889 and 4380 m asl, respectively. The range of elevation where landslides failured is fairly wide. The most common altitude range was 2500–2750 m asl where 77 landslides collapsed. 54 landslides occurred in elevations between 2250–2500 m asl and 49 landslides between 2750–3000 m asl. Only 7 landslides collapsed at higher altitudes above 4000 m asl.

Slope aspect shows a remarkable failure orientation. The highest percentage (75%) of landslides show a NE-SE orientation. The rest are widely distributed at other orientations except the North that corresponds to warmest slope exposition.

Grade of Activity

Analysing the activity grade of identified landslides we found that 31% of them is active and 16% is dormant type. However, many of these landslides have been reactivated showing multiple movements, or at least they correspond to a successive style. Most of them show retrograde movement distribution. General aspect, preservation of primary geoforms and degree of degradation of landslide features suggest a Pleistocene age for initial collapses of these events. Otherwise, the lack of rock varnish development on the top of blocks, that in some cases should reach 6 m in size, could confirm this assumption. Furthermore, activity grade of landslides decreases gradually towards the north and the south from the centre of the study area.

Spatial Distribution of Landslides

Landslides are not randomly distributed in the study area. They are gathered in at least eight clusters in this segment of Precordillera (31–33°S). Only 12 landslides are located far away from any concentration falling out of any cluster (Fig. 1). We analysed the main features of each cluster (Table 1):

Cluster 1 is located in Precordillera of Mendoza. Apparently, landslides in this sector do not follow any pattern of distribution but seem to be more concentrated in the eastern part. This sector shows the lowest activity with only one active landslide representing 2%. The predominant type of landslide is rock slide and more than half of them were failed on sandstones and pelites.

The degree of activity increases by up to 35% in cluster 2. Landslides are shallower than sector 1. Conglomerates, shales, graywackes and sandstones are the main lithology where landslides failed. The predominant type is rock slide.

Cluster 3 shows a 57% activity rate. The main types of landslides are debris-rock slide and rock slide. Here, the percentage of deep slides and reactivities are the highest of the study area with 61% and 65%, respectively. The predominant lithologies of failures are sandstones, pelites and phyllites. This high sector has altitude of 4369 m asl., then 65% of the landslides were triggered higher than 3500 m asl.

Cluster 4 is located in Las Osamentas Range. The highest rate of activity is registered in this cluster with 74% of active landslides. Most of them are rock slides in graywackes and shales.

The greatest number of gathered landslides corresponds to cluster 5. The degree of activity decreases comparing with

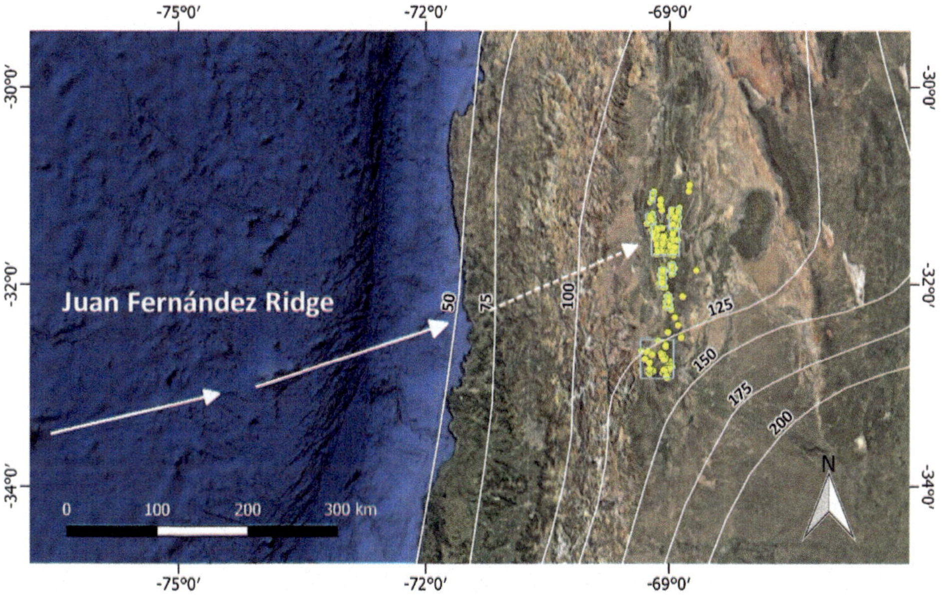

Fig. 2 Juan Fernandez Ridge overlapping more active landslide cluster

sector 4. The main type of landslide is rock slide; whereas there is the same percentage of shallow and deep slides. It is important to note that the biggest landslides are grouped in this sector with areas from 2 km^2 to 7 km^2.

In cluster 6 the grade of activity is a bit lower than sector 5 and the percentage of shallow landslides increases to 67%. The number of debris slides and rock slides is the same being the predominant type of landslide. 76% of the slides failed on sandstones, pelites and phyllites.

At the same latitude, we found the cluster 7 being the second sector with the most landslides. The grade of activity is low in this cluster with a 21% of active landslides. The main type of landslide is shallow rock slide. Failure associated lithologies are graywackes, shales, limestones and marls.

Cluster 8 is the second group with the lowest grade of activity after cluster 1. In turn, this sector contains just 11 landslides. The predominant type of landslide is debris slide and most of them are shallow. All of them failed on sandstones, pelites and phyllites in high altitudes.

Discussion and Conclusions

Despite that the spatial distribution of huge collapses coincides with traces of active Quaternary faults, our findings show that landslides are not randomly distributed along the study area. At least 8 landslide clusters were distinguished along the Precordillera region. However, landslide density is concentrated in the central region of study area coinciding with higher elevations of the Central Precordillera. Moreover, rate of activity is clearly evidenced in this central area where cluster 3 and 4 are identified. Landslide grade activity decreases gradually towards the north and south.

The most affected area in the Precordillera overlaps the alignment of Juan Fernandez Ridge (Fig. 2) associated with a greater regional deformation and likely linked to more active neotectonics. This fact could explain the concentration of huge collapses in this area. In particular, a seismic triggering is postulated for this landslides which seem to have been reactivated several times evidencing paleoseismic activity. Besides, this segment coincides with higher elevations of Central Precordillera justifying a greater acceleration as consequent of topographic effect.

Even though spatial distribution of landslides could be clearly conditioned by lithology, steeper slope and elevations; analyses of these parameters distribution is not enough to explain landslide clusters in the study area.

Acknowledgements This study is part of the Ph thesis of S. Junquera. Funds come from ANLAC program of National University of Cuyo leader by Prof. Moreiras and the project FONDECYT 1140317 leader by Prof. Sepúlveda.

References

Abele G (1984) Derrumbes de montaña y morenas en los Andes chilenos. Revista de Geografía Norte Grande 11:17–30

Alvarado P, Pardo M, Gilbert H, Miranda S, Anderson M, Saez M, Beck SL (2009) Flat-slab subduction and crustal models for the seismically active Sierras Pampeanas region of Argentina. In: Kay S, Ramos VA, Dickinson W (eds) MWR204: backbone of the Americas: shallow subduction, plateau uplift, and ridge and terrane collision. Geological Society of America, Boulder, Colorado, pp 261–278

Anderson M, Alvarado P, Zandt G, Beck S (2007) Geometry and brittle deformation of the subducting Nazca plate, central Chile and Argentina. Geophys J Int 171(1):419–434

Antinao JL, Gosse J (2009) Large rockslides in the Southern Central Andes of Chile (32–34.5 S): Tectonic control and significance for Quaternary landscape evolution. Geomorphology 104:117–133

Barazangi M, Isacks BL (1976) Spatial distribution of earthquakes and subduction of the Nazca plate beneath South America. Geology 4:686–692

Cahill T, Isacks BL (1992) Seismicity and shape of the subducted Nazca plate. J Geophys Res 97(B12):17503–17529

Casa A, Yamin M, Wright E, Costa C, Coppolecchia M, Cegarra M, Hongn F (eds) (2014) Deformaciones Cuaternarias de la República Argentina, Sistema de Información Geográfica. Instituto de Geología y Recursos Minerales, Servicio Geológico Minero Argentino, v2.0 en formato DVD. SIG SEGEMAR: http://sig.segemar.gov.ar. Accesed 30 Aug 2016

Cortés JM, Yamín M, Pasini M (2005) La Precordillera Sur, provincias de Mendoza y San Juan. 16° Congreso Geológico Argentino. Actas 1:395–402. La Plata.

Costa C, Machette M, Dart R, Bastías H, Paredes J, Perucca L, Tello G, Haller K (2000) Map and database of quaternary faults and folds in Argentina: U.S. Geological Survey Open-File Report (00-0108), p 75

Crozier MJ (1984) Field assessment of slope instability. In: Brunsden D, Prior DB (eds) Slope instability. London, Wiley, pp 103–42

Esper Angillieri MY (2011) Inventario de Procesos de Remoción en masa en un sector del Departamento Iglesia, San Juan, Argentina. Revista de la Asociación Geológica Argentina 68(2):225–232

Esper Angillieri MY (2012) Análisis de la vulnerabilidad por flujos en masa en la Provincia de San Juan (Oeste de Argentina). Revista de la Sociedad Geológica de España 25(3–4):145–156

Esper Angillieri MY, Perucca LP (2013) Mass movement in Cordón de las Osamentas, de La Flecha river basin, San Juan, Argentina. Quatern Int 301:150–157

Esper Angillieri MY, Perucca L, Rothis M, Tapia C, Vargas N (2014) Morphometric characterization and seismogenic sources relationships of a large scale rockslide. Quatern Int 352:92–99

Esper Angillieri MY (2015) Application of logistic regression and frequency ratio in the spatial distribution of debris-rockslides: Precordillera of San Juan, Argentina. Quatern Int 355:202–208

Fauqué L, Cortés JM, Folguera A, Etchverría M (2000) Avalanchas de rocas asociadas a neotectónica en el valle del río Mendoza, al sur de Uspallata. Revista Asociación Geológica Argentina 55(4):419–423

Jibson RW, Keefer DK (1993) Analysis of the seismic origin of landslides—examples from the New Madrid seismic zone. Geol Soc Am Bull 105:421–436

Jibson RW (1996) Use of landslides for paleoseismic analysis. Eng Geol 43(4):291–323

Jibson RW, Harp EL, Schulz W, Keefer DK (2006) Large rock avalanches triggered by the M 7.9 Denali Fault, Alaska, earthquake of 3 November 2002. Eng Geol 83(1–3):144–160

Jordan TE, Allmendinger RW, Damanti JF, Drake RE (1993) Chronology of mo-tion in a complete thrust belt: the Precordillera, 30–31°S, Andes Mountains. J Geol 101:135–156

Keefer DF (1987) Landslides as indicators of prehistoric earthquakes. Directions in paleoseismology. U.S. Geol Surv Open File Rep 87-673:178–180

Keefer DK (1994) The importance of earthquake-induced landslides to long term slope erosion and slope-failure hazards in seismically active regions. Geology 10:265–284

Keefer DK (2000) Statistical analysis of an earthquake-induced landslide distribution: the 1989 Loma Prieta, California event. Eng Geol Amsterdam 58:213–249

Moreiras SM (2006) Chronology of a Pleistocene rock avalanche probable linked to neotectonic, Cordón del Plata (Central Andes), Mendoza—Argentina. Quat Int 148(1):138–148

Moreiras SM, Sepúlveda SA (2009) Large paleolandslides in the Central Andes (32–33°S): new challenges. In: Proceedings XII Congreso Geológico Chileno, Santiago, paper S3_022

Moreiras SM, Coronato A (2010) Landslide processes in Argentina. Natural hazards and Human-Exacerbated Disasters in Latin-America. Geomorphol: Dev Earth Surf Process 301–331

Moreiras SM, Hermanns RL, Fauqué L (2015) Cosmogenic dating of rock avalanches constraining Quaternary stratigraphy and regional neotectonics in the Argentine Central Andes (32° S). Quat Sci Rev 112(15):45–58

Moreiras SM, Sepúlveda SA (2015) Megalandslides in the Andes of Central Chile and Argentina (32°–34°S) and potential hazards. Geodynamic Processes in the Andes of Central Chile and Argentina. Geol Soci London 399:329–344

Pantano Zuñiga AV (2014) Geomorfología, neotectónica y la peligrosidad geológica en la cuenca del Río Acequión, provincia de San Juan, Argentina. Tesis Doctoral. Facultad de Ciencias Naturales e I. M.L. Universidad Nacional de Tucumán

Perucca LP, Moreiras SM (2008) Indicative structures of paleo-seismicity in the Acequión region, San Juan province, Argentina. Geodinamica Acta 21(3):93–105

Perucca LP, Esper Angillieri MY (2008) La avalancha de rocas Las Majaditas. Caracterización geométrica y posible relación con eventos paleosísmicos. Precordillera de San Juan, Argentina. Revista Española de la Sociedad Geológica de España 21(1–2):35–47

Perucca LP, Esper Angillieri MY (2009a) El deslizamiento de rocas y detritos sobre el río Santa Cruz y el aluvión resultante por el colapso del dique natural, Andes Centrales de San Juan. Revista de la Asociación Geológica Argentina 65(3):571–585

Perucca LP, Esper Angillieri MY (2009b) Evolution of a debris-rock slide causing a natural dam: the flash flood of Río Santa Cruz, Province of San Juan. November 12, 2005. Nat Hazards 50(2):305–320

Quartino BJ, Zardini RA, Amos AJ (1971) Estudio y exploración geológica de la región Barreal—Calingasta (provincia de San Juan). Asociación Geológica Argentina, Monografía 1, Buenos Aires, p 184

Smalley RF Jr, Pujol J, Regnier M, Chiu JM, Chatelain JL, Isacks BL, Araujo M, Puebla N (1993) Basement seismicity beneath the Andean. Precordillera thin skinned thrust belt and implications for crustal and lithospheric behavior. Tectonics 12:63–76

USGS. ASTER global digital elevation Map V2. https://gdex.cr.usgs.gov/gdex. Accesed 30 Aug 2016

Varnes DJ (1978) Slope movement types and processes. In: Schuster RL, Krizek RJ (eds) Special Report 176: Landslides: Analysis and control. Transportation and Road Research Board, National Academy of Science, Washington DC., pp 11–33

Landslide Technology and Engineering in Support of Landslide Science

Kyoji Sassa
Executive Director of the International Consortium on Landslides

The World Landslide Forum (WLF) is the triennial conference of the International Consortium on Landslides (ICL) and the International Programme on Landslides (IPL). The IPL is a programme of the International Consortium on Landslides, managed by ICL and its supporting organizations: UNESCO, WMO, FAO, UNISDR, UNU, ICSU, WFEO and IUGS.

IPL and WLF contribute to the United Nations International Strategy for Disaster Reduction. The World Landslide Forum provides an information and academic-exchange platform for landslide researchers and practitioners. It creates a triennial opportunity to promote worldwide cooperation and share new theories, technologies and methods in the fields of landslide survey/investigation, monitoring, early warning, prevention, and emergency management. The forum's purpose is to present achievements of landslide-risk reduction in promoting the sustainable development of society.

Advancements in landslide science and disaster-risk reduction are supported by developments in landslide technology and engineering. Here we invited ICL supporters who support the publication of the international full-color journal "Landslides: Journal of the International Consortium on Landslides", the companies advertising in the seven volumes of "Landslide Science and Practice: Proceedings of the Second World Landslide Forum 2011" and the companies exhibiting at the Third World Landslide Forum 2014 to introduce their landslide technology and engineering. Eight companies applied to exhibit in this book, their names, addresses, contact information and a brief introduction are given below:

1. MARUI & Co. Ltd.
1-9-17 Goryo, Daito City, Osaka 574-0064, Japan
URL: http://marui-group.co.jp/en/index.html
Contact: hp-mail@marui-group.co.jp

MARUI & Co. Ltd is the leading manufacturing and sales company in Japan since 1920 of material testing machines for soil, rock, concrete, cement and asphalt. Marui engineers built and assisted in development of the series of stress and speed control ring-shear apparatus by DPRI and ICL to study landslides since 1982.

2. OSASI Technos, Inc.
65-3 Hongu-cho, Kochi City, Kochi 780-0945, Japan
URL: http://www.osasi.co.jp/en/
Contact: info-tokyo@osasi.co.jp

OSASI Technos, Inc. develops and markets the slope disaster monitoring system called OSASI Network System (OSNET). The monitoring devices use a built-in lithium battery and operate without external electricity supply in mountainous areas. The system enables a network of up to 64 units with up to 1 km distance between units. OSNET is suitable for quickly establishing monitoring systems on landslides in emergencies.

3. Okuyama Boring Co., Ltd.
10-39 Shimei-cho, Yokote City, Akita 013-0046, Japan
URL: http://www.okuyama.co.jp/
Contact: info@okuyama.co.jp

The Okuyama Boring Company Ltd specializes in landslide investigation, analysis of landslide mechanisms, and design of landslide remedial measures. The company uses its own software to analyze the initiation and motion of landslides, including the tsunami generated by landslides into reservoirs.

4. Japan Conservation Engineers & Co., Ltd.
3-18-5 Toranomon, Minato-ku, Tokyo 105-0001, Japan
URL: http://www.jce.co.jp
Contact: hasegawa@jce.co.jp

Japan Conservation Engineers & Co, Ltd develops landslide-simulation software and shear-testing apparatus,

M. Mikoš et al. (eds.), *Advancing Culture of Living with Landslides*,
DOI 10.1007/978-3-319-53483-1

including slip-surface direct-shear apparatus and ring-shear apparatus to measure the shear strength mobilized on the sliding surface of landslides. Japan Conservation Engineers is a consulting company for landslide investigation, reliable monitoring, data analysis and the design of landslide-risk reduction works.

5. KOKUSAI KOGYO Co., Ltd.
2 Rokuban-cho, Chiyoda-ku, Tokyo 102-0085, Japan
URL: http://www.kk-grp.jp/english/
Contact: overseas@kk-grp.jp

Kokusai Kogyo has undertaken aerial surveys, infrastructure development projects for road and harbor facilities, and landslide-disaster prevention and mitigation works since its foundation in 1947. The company has recently developed remote-sensing technology using the laser profiler, satellite synthetic aperture radar, and a new monitoring system called <Shamen-net> integrating GPS and other monitoring devices, all of which contribute to landslide-disaster prevention and mitigation.

6. GODAI KAIHATSU Corporation
1-35 Kuroda, KANAZAWA-City, ISHIKAWA Pref. 921-8051, Japan
URL: http://www.godai.co.jp
Contact: pp-sale@godai.co.jp

Development and sales of slope disaster prevention software and Internet services

7. OYO Corporation
Sumitomo Fudosan Kanda Building 9th Floor, 7 Kanda-Mitoshiro-cho, Chiyoda-ku, Tokyo 101-8486 Japan
URL: https://www.oyo.co.jp/english/
Contact: https://www.oyo.co.jp/english/contacts/

OYO Corporation was established on 2 May, 1957, located in Tokyo, Japan. The Corporation provides:

- Engineering works from soil investigation to design/construction management, which are accompanied with construction of civil engineering structures and building structures.
- Engineering works regarding investigation, analyses, projection, diagnoses, evaluation for landslides, slope failure, earthquake disasters, and storm and flood damage.
- Engineering works regarding investigation, analyses, projection, diagnoses, evaluation for environmental conservation/risks such as vibrations, noises and water quality, etc.
- Gathering, processing and selling information on the earth, such as ground surface, topography, environment, and disasters, etc.
- Development, manufacturing, sales, leases, and rental on various measuring equipment, software, systems.
- Development, manufacturing, and sales on security related equipment.

8. PROTEC ENGINEERING, INC.
5322-26, Oaza Hasugata, Seiro-machi, Kitakanbara-gun, Niigata 957-0106, Japan
URL: http://www.proteng.co.jp/en/
Contact: info@proteng.co.jp

PROTEC ENGINEERING have been developing several kind of products which protect human life and properties against natural disasters, mainly focusing on rock fall, slope failure and snow avalanche. The products include GEO ROCK WALL, ARC FENCE, SLOPE GUARD FENCE, and debris flow barriers.

Full-color presentations from these eight exhibitors focusing on their landslide technology are shown in the following pages. The progress of landslide science is supported by advances in landslide technology. The success of landslide risk-reduction measures needs effective landslide engineering. The International Consortium on Landslides seeks expressions of interest in contributing to "Landslide Technology and Engineering to Support Landslide Science" at the Fifth World Landslide Forum to be held on 2–6 November 2020, in Kyoto, Japan. We may call for presentations on landslide technology and engineering in the proceedings, as well as through exhibitions at the site. Those interested in this initiative are requested to contact the Secretariat of the International Consortium on Landslides <secretariat@iclhq.org>. We will send invitations to interested applicants when further details become available.

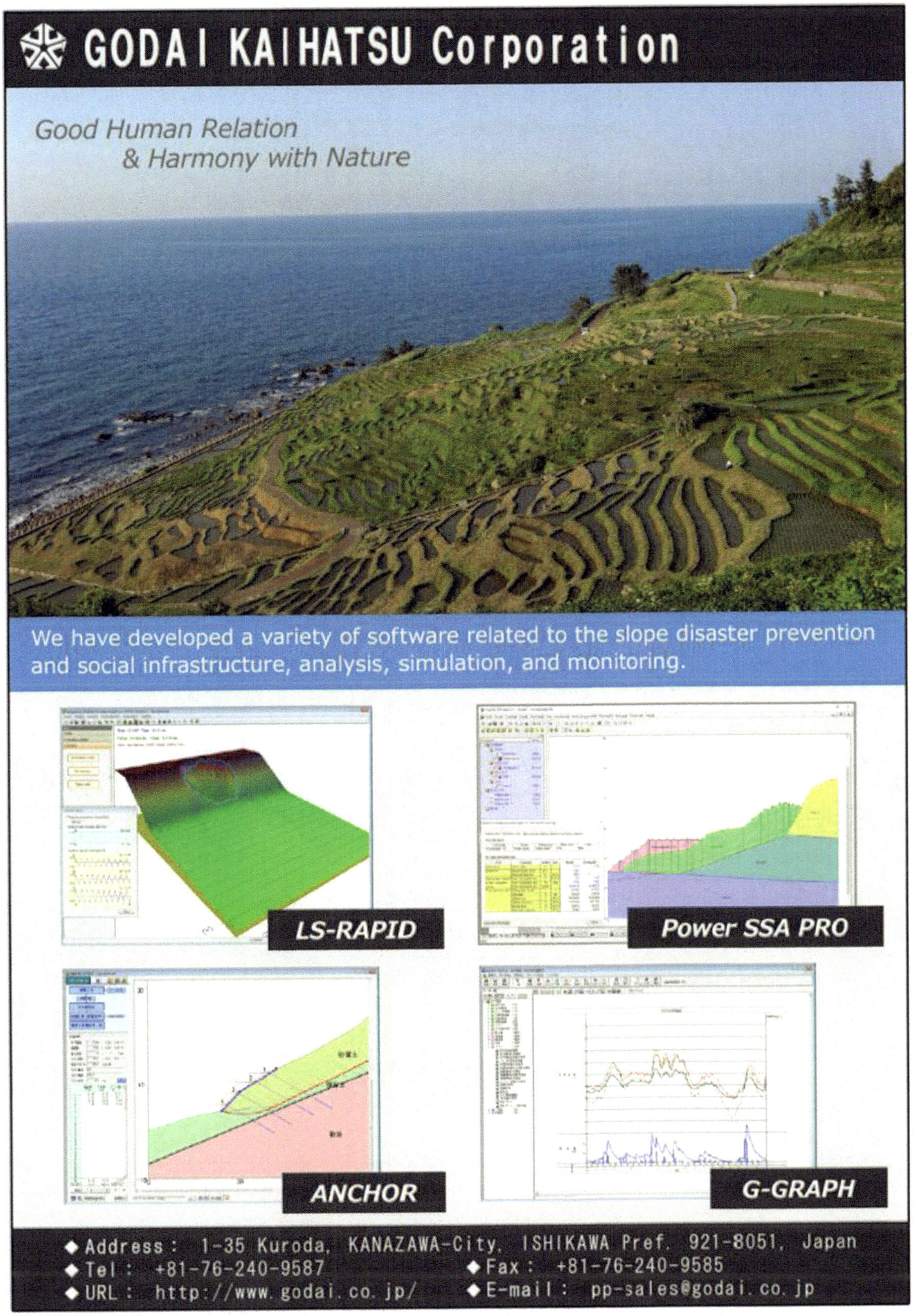
GODAI KAIHATSU Corporation
Good Human Relation
& Harmony with Nature
We have developed a variety of software related to the slope disaster prevention and social infrastructure, analysis, simulation, and monitoring.
LS-RAPID
Power SSA PRO
ANCHOR
G-GRAPH
◆Address : 1-35 Kuroda, KANAZAWA-City, ISHIKAWA Pref. 921-8051, Japan
◆Tel : +81-76-240-9587
◆Fax : +81-76-240-9585
◆URL : http://www.godai.co.jp/
◆E-mail : pp-sales@godai.co.jp

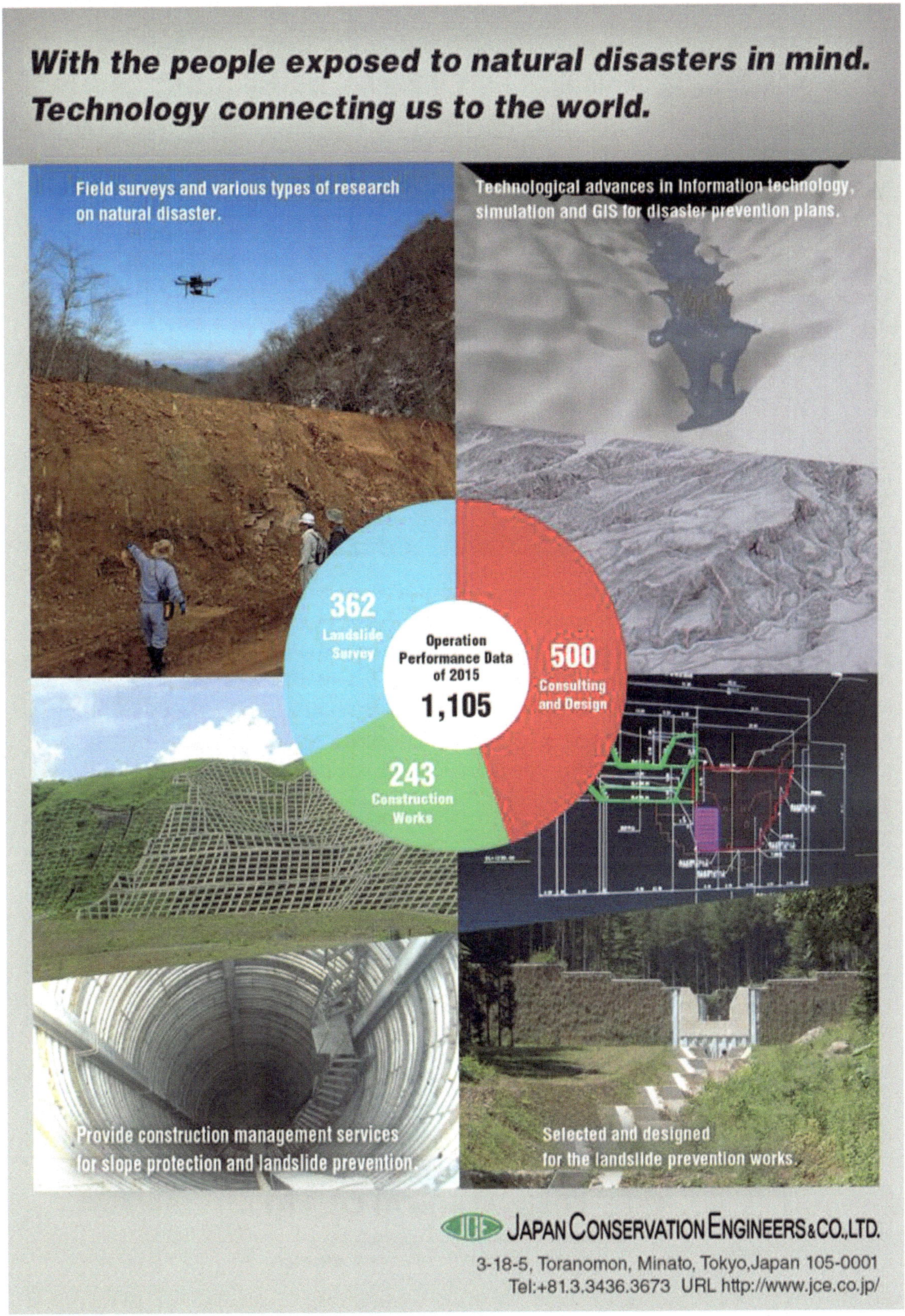

Green communities
~Towards a better future,
for people and the environment~
RADARSAT-2 Data and Products (c)MacDONALD, DETTWILER AND ASSOCIATES LTD. 2009, 2011 - All Rights Reserved
Kokusai Kogyo, as a leading company of geospatial information technologies, has been contributing to the improvement of public services with advanced measurement technologies and a wide range of consulting technologies. Kokusai Kogyo supports the development of "Green Communities" representing the new era public concerns on "environment and energy," "disaster prevention and mitigation" and "asset management" . Kokusai Kogyo offers advanced analysis of geospatial information consultancy for developing new government policies, maintaining and operating social infrastructures with safe and secure city planning, and building low-carbon cities.
Post-disaster image
Aerial oblique photograph of deep-seated landslide
Shamen-net
Shamen-net is a total monitoring system integrating GPS and other monitoring devices (Measurement precision: +/-1mm, on a real time basis)
Japan Asia Group
KOKUSAI KOGYO CO., LTD.
2 Rokubancho,Chiyoda-ku,Tokyo,Japan 102-0085
http://www.kkc.co.jp/english/ info_overseas@kk-grp.jp

OSASI
We pass on voices of the earth
Slope Disaster Monitoring System
OSNET
Low Power & Flexible Network System
OSASI
NETwork System
EXAMPLE SETUP 2
NetLG-001NE
Water Level Data Logger
SLG-10E
Extensometer Data Logger
OSNET is a network solution for disaster prevention monitoring developed by OSASI Technos. OSNET devices have an in-built lithium battery so that they can be reliably used in remote mountainous areas lacking power supply.
Networks of up to 64 devices with up to 1km
NetHB-1E
Six-port Hub
NetCT-1E
Controller
CF Card set
Portable PC
NetLG-401NE
4-Gauge Method Strain Gauge Data Logger
Twisted pair shielded cable
Connect Directly at the site or through modem and telephone
Portable PC
NetGW-1
RS232Converter
NetLG-301NE
Pipe Strain Meter/Water Level Data Logger
NetLG-201E
Precipitation Data Logger
NetAL-1E
Alarm Output Unit
AL-TypeBE
Alarm Unit
Revolving Light
Electronic Horne

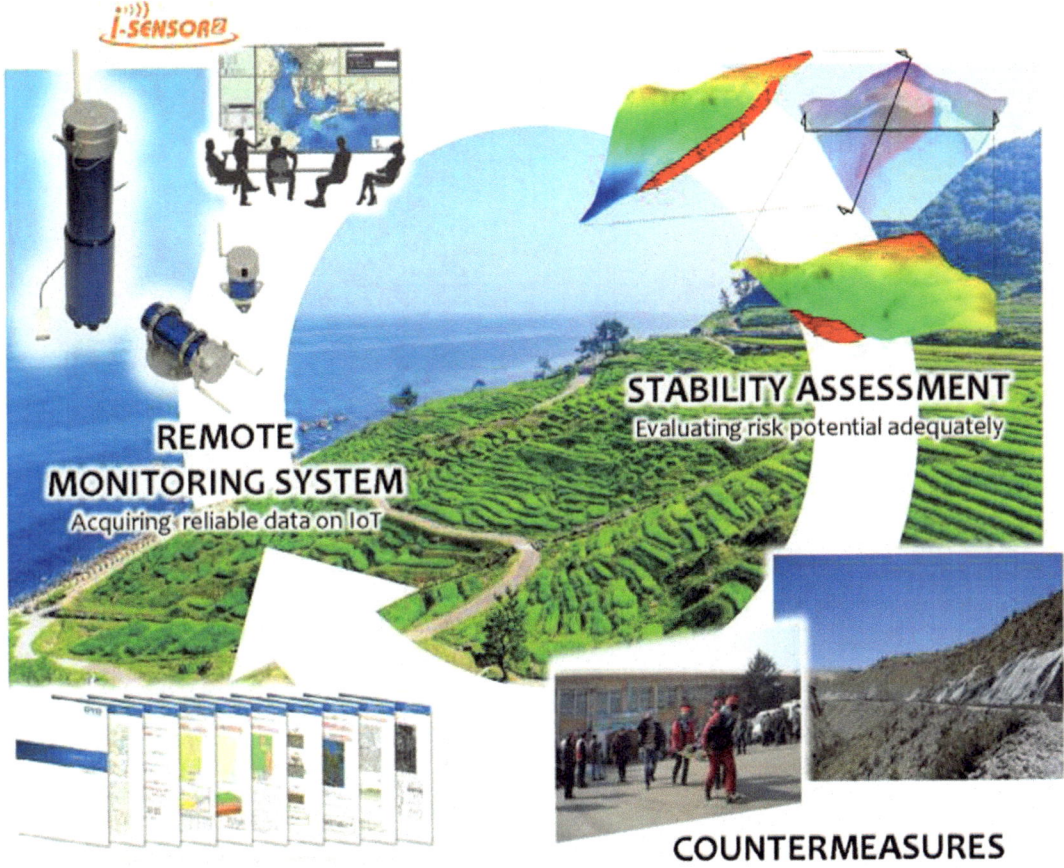

ICL Structure Table

International Consortium on Landslides

An international non-government and non-profit scientific organization promoting landslide research and capacity building for the benefit of society and the environment

President: Yueping Yin (China Geological Survey)
Vice Presidents: Irasema Alcantara-Ayara (UNAM), Mexico, Matjaz Mikos (University of Ljubljana), Slovenia
Dwikorita Karnawati (Gadjah Mada University, Indonesia)
Executive Director: Kyoji Sassa (Prof. Emeritus, Kyoto University, Japan), Treasurer: Kaoru Takara (Kyoto University, Japan)

ICL Supporting Organizations:

The United Nations Educational, Scientific and Cultural Organization (UNESCO) / The World Meteorological Organization (WMO) / The Food and Agriculture Organization of the United Nations (FAO) / The United Nations International Strategy for Disaster Reduction Secretariat (UNISDR) / The United Nations University (UNU) / International Council for Science (ICSU) / World Federation of Engineering Organizations (WFEO) / International Union of Geological Sciences (IUGS) / International Union of Geodesy and Geophysics (IUGG) / Government of Japan

ICL Members:

Albania Geological Survey / The Geotechnical Society of Bosnia and Herzegovina / Geological Survey of Canada / Chinese Academy of Sciences, Institute of Mountain Hazards and Environment / Northeast Forestry University, Institute of Cold Regions Science and Engineering, China / China Geological Survey / Nanjing Institute of Geography and Limnology, Chinese Academy of Sciences / Tongji University, College of Surveying and Geo-Informatics, China / Universidad Nacional de Columbia, Columbia / City of Zagreb, Emergency Management Office, Croatia / Croatian Landslide Group (Faculty of Civil Engineering, University of Rijeka and Faculty of Mining, Geology and Petroleum Engineering, University of Zagreb) / Charles University, Faculty of Science, Czech Republic / Institute of Rock Structure and Mechanics, Department of Engineering Geology, Czech Republic / Cairo University, Egypt / Joint Research Centre (JRC), European Commission / Technische Universitat Darmstadt, Institute and Laboratory of Geotechnics, Germany / National Environmental Agency, Department of Geology, Georgia / Universidad Nacional Autónoma de Honduras, UNAH, Honduras / Amrita Vishwa Vidyapeetham, Amrita University / National Institute of Disaster Management, India / University of Gadjah Mada, Indonesia / Parahyangan Catholic University, Indonesia / Research Center for Geotechnology, Indonesian Institute of Sciences, Indonesia / Building & Housing Research Center, Iran / University of Calabria, DIMES, CAMILAB, Italy / University of Firenze, Earth Sciences Department, Italy / Istituto de Ricerca per la Protezione Idrogeologica (IRPI), CNR, Italy / Italian Institute for Environmental Protection and Research (ISPRA) - Dept. Geological Survey, Italy / Forestry and Forest Product Research Institute, Japan / Japan Landslide Society / Kyoto University, Disaster Prevention Research Institute, Japan / Korea Forest Research Institute, Korea / Korea Infrastructure Safety & Technology Corporation, Korea / Korea Institute of Civil Engineering and Building Technology / Korea Institute of Geoscience and Mineral Resources (KIGAM) / Korean Society of Forest Engineering / Slope Engineering Branch, Public Works Department of Malaysia / Institute of Geography, National Autonomous University of Mexico (UNAM) / International Centre for Integrated Mountain Development (ICIMOD), Nepal / University of Nigeria, Department of Geology, Nigeria / International Centre for Geohazards (ICG) in Oslo, Norway /Grudec Ayar, Peru / Moscow State University, Department of Engineering and Ecological Geology, Russia / JSC "Hydroproject Institute", Russia / University of Belgrade, Faculty of Mining and Geology, Serbia / Comenius University, Faculty of Natural Sciences, Department of Engineering Geology, Slovakia / University of Ljubljana, Faculty of Civil and Geodetic Engineering (UL FGG), Slovenia / University of Ljubljana, Faculty of Natural Sciences and Engineering (UL NTF), Slovenia / Geological Survey of Slovenia / Central Engineering Consultancy Bureau (CECB), Sri Lanka / National Building Research National Organization, Sri Lanka /Taiwan University, Department of Civil Engineering, Chinese Taipei / Landslide group in National Central University from Graduate Institute of Applied Geology, Department of Civil Engineering, Center for Environmental Studies, Chinese Taipei/ Asian Disaster Preparedness Center, Thailand / Ministry of Agriculture and Cooperative, Land Development Department, Thailand / Institute of Telecommunication and Global Information Space, Ukraine / California State University, Fullerton & Tribhuvan University, Institute of Engineering, USA & Nepal / Institute of Transport Science and Technology, Vietnam / Vietnam Institute of Geosciences and Mineral Resources (VIGMR), Vietnam

ICL Supporters:

Marui & Co., Ltd., Osaka, Japan / Okuyama Boring Co., Ltd., Yokote, Japan / GODAI Development Corp., Kanazawa, Japan / Japan Conservation Engineers & Co., Ltd, Tokyo / Kokusai Kogyo Co., Ltd., Tokyo, Japan / Ohta Geo-Research Co., Ltd., Nishinomiya, Japan / OSASI Technos Inc., Kochi, Japan / OYO Corporation, Tokyo, Japan / Sabo Technical Center, Tokyo, Japan / Sakata Denki Co., Ltd., Tokyo, Japan

Contact:

International Consortium on Landslides, 138-1 Tanaka Asukai-cho, Sakyo-ku, Kyoto 606-8226, Japan
Web: http://icl.iplhq.org/, E-mail: secretariat@iclhq.org
Tel: +81-774-38-4834, +81-75-723-0640, Fax: +81-774-38-4019, +81-75-950-0910

M. Mikoš et al. (eds.), *Advancing Culture of Living with Landslides*,
DOI 10.1007/978-3-319-53483-1

Author Index

M. Mikoš et al. (eds.), *Advancing Culture of Living with Landslides*,
DOI 10.1007/978-3-319-53483-1

Printed by Printforce, the Netherlands